EVAPORITES
Their Evolution and Economics

JOHN WARREN

b

Blackwell
Science

© 1999 by
Blackwell Science Ltd
Editorial Offices:
Osney Mead, Oxford OX2 0EL
25 John Street, London WC1N 2BL
23 Ainslie Place, Edinburgh EH3 6AJ
350 Main Street, Malden
  MA 02148 5018, USA
54 University Street, Carlton
  Victoria 3053, Australia
10, rue Casimir Delavigne
  75006 Paris, France

Other Editorial Offices:
Blackwell Wissenschafts-Verlag GmbH
Kurfürstendamm 57
10707 Berlin, Germany

Blackwell Science KK
MG Kodenmacho Building
7–10 Kodenmacho Nihombashi
Chuo-ku, Tokyo 104, Japan

First published 1999

Set by the author using PageMaker v.6.5
Printed and bound in Great Britain
at the Alden Press, Oxford and
Northampton

The Blackwell Science logo is a
trade mark of Blackwell Science Ltd,
registered at the United Kingdom
Trade Marks Registry

For further information on
Blackwell Science, visit our website:
www.blackwell-science.com

DISTRIBUTORS

Marston Book Services Ltd
PO Box 269
Abingdon, Oxon OX14 4YN
(*Orders*: Tel: 01235 465500
         Fax: 01235 465555)

USA
Blackwell Science, Inc.
Commerce Place
350 Main Street
Malden, MA 02148 5018
(*Orders*: Tel: 800 759 6102
         781 388 8250
  Fax: 781 388 8255)

Canada
Login Brothers Book Company
324 Saulteaux Crescent
Winnipeg, Manitoba R3J 3T2
(*Orders*: Tel: 204 837-2987)

Australia
Blackwell Science Pty Ltd
54 University Street
Carlton, Victoria 3053
(*Orders*: Tel: 3 9347 0300
         Fax: 3 9347 5001)

A catalogue record for this title
is available from the British Library
and the Library of Congress

ISBN 0-632-05301-1 ✓

To Jennifer

"Where shall I begin, please your Majesty?" he asked.

"Begin at the beginning," the King said, gravely, "and go on till you come to the end: then stop."

White rabbit speaking to the King in Lewis Carroll's Alice in Wonderland.

Contents

Contents

CONTENTS

Contents

CONTENTS

Preface

My goal in writing this book is to give you a better understanding of what happens to evaporites in the subsurface. This is, after all, where most of the important things happen in evaporite basins. Diagenesis, metamorphism, brine and metal transport, hydrocarbon trapping, halokinesis, magmatic and hydrothermal interactions and sulphide precipitation are all subsurface processes. For more than 20 years I have attempted to understand what are the predictive fingerprints of economically significant types of evaporite associations in sedimentary, metasedimentary and igneous situations. For this book I did not want to pen an update of my first book (Warren, 1989), with its emphasis on depositional aspects and hydrocarbon plays. Any one wanting a book covering these topics can read my last effort or the more recent volume on evaporite sedimentology edited by Judy Melvin (1991). Judy's excellent book covers depositional aspects of evaporites and the hydrocarbon associations in much more detail than are mentioned in this book.

Only three chapters in this book deal with the depositional and diagenetic properties of actual salts - the sedimentology of evaporites both modern and ancient (Chapters 1, 3 and 7). The rest of the book brings together literature and ideas that, to my knowledge, have not yet been synthesised in any other book on evaporites. The hydrological stages of evaporite burial and uplift along with resulting hydrogeochemistries are covered in Chapter 2. Evaporite indicators, left after all the salts have dissolved, been replaced, or reprecipitated, are covered in Chapter 4. The characteristics and processes of evaporites that have flowed within extensional and compressional terrains, that is, salt tectonics, are discussed in Chapter 5. The properties and characteristics of meta-evaporites from greenschist to granulite grades are covered in Chapter 6. The last two chapters deal with particular evaporites and structures which are useful pointers for world-class ore deposits. The low temperature end of the economic spectrum associated with diagenesis and low grade metamorphism is covered in Chapter 8, while the much higher temperature end, associated with magmatism, high grade metamorphism,

hydrothermal salts and hydrothermal flushing is dealt with in Chapter 9. The significance of hydrocarbons and organic matter within an evaporite sequences gets a mention here and there throughout the book, especially in chapters 1, 4, 5, 8 and 9.

I would like to thank the following companies and individuals who pay me to do the things I enjoy most; geological research and training. Thanks to BHP Minerals, Pasminco, MIM, Woodside Offshore Petroleum, Homestake Gold, Earth Sciences at Flinders University, the National Centre of Petroleum Geology at Adelaide University, the Department of Petroleum Geosciences at the University of Brunei Darussalam and all the other groups and individuals who have allowed me to work with them in the past four years while I was writing this book. Thanks especially to Peter Muhling, Ron Smit and Oliver Warin of BHP for their interest and support. And thanks to the Earth Science Faculty at Flinders University, especially Alex Grady, for granting me an honorary professorial position. The spare hours spent at Flinders gave me access to some bright minds to test out some ideas for the book, as well as access to the excellent on-line library and database facilities of the Flinders University of South Australia.

I hope this is the first edition of what will become an evolving work that broadens the way explorationists and researchers will look at evaporites. Let me know what you, the reader, think of it and how its coverage can be improved.

John Warren, January, 1999
JK Resources Pty Ltd
Adelaide, Australia
<jwarren1@ozemail.com.au>

Current address:
Department of Petroleum Geosciences
University Brunei Darussalam
<jwarren@brunet.bn>

Chapter 1
Interpreting evaporite texture

My goal in writing this book with an emphasis on subsurface and metal-evaporite associations is to offer an improved understanding of ancient evaporites in ways that are not typically discussed in sedimentological texts. Models of evaporite depositional textures and facies, based on Holocene analogues, are already well documented with useful summaries to be found in Schreiber, 1988a; Warren, 1989, Melvin 1991; Kendall, 1992; Rouchy et al., 1994. All emphasise a close association with hydrocarbons. What I wanted to achieve in this book is an improved understanding of the metal-evaporite association and why geological models based solely on present day evaporite settings are no more than a partial key to the past. To do this I have tried to cover a much wider subsurface temperature spectrum than is relevant to the evaporite/hydrocarbon association. In addition to classic sedimentological approaches in the first half of the book, we will discuss meta-evaporites, igneous/hydrothermal associations, and evaporite/metal associations at temperature regimes ranging up to 1,000°C. I have also devoted a chapter to indicators of former evaporites ("vanished evaporites") in an attempt to familiarise readers with what ancient evaporites look like in outcrop and in much of the Precambrian. Another chapter attempts to draw together a very diverse literature base on the geology

of evaporites that serve as commodities. But, before we can look at any of these interesting associations and problems, we must first set up the conceptual framework by looking at textures and depositional settings of the more commonplace evaporites.

What is an evaporite?

I define an evaporite as a rock that was originally precipitated from a saturated surface or nearsurface brine by hydrologies driven by solar evaporation. This simple definition encompasses a wide range of chemically precipitated salts and includes alkali earth carbonates (Table 1.1). Some workers restrict the term evaporite to those salts formed directly via solar evaporation of hypersaline waters at the earth's surface. I would classify such evaporites as primary evaporites, that is, evaporites precipitated from a standing body of surface brine and retaining crystallographic evidence of the depositional process (e.g. bottom-nucleated and current-derived textures). My broad definition of the term evaporite means that, outside of a few Neogene examples, there are no completely "primary" evaporite beds (Figure 1.1).

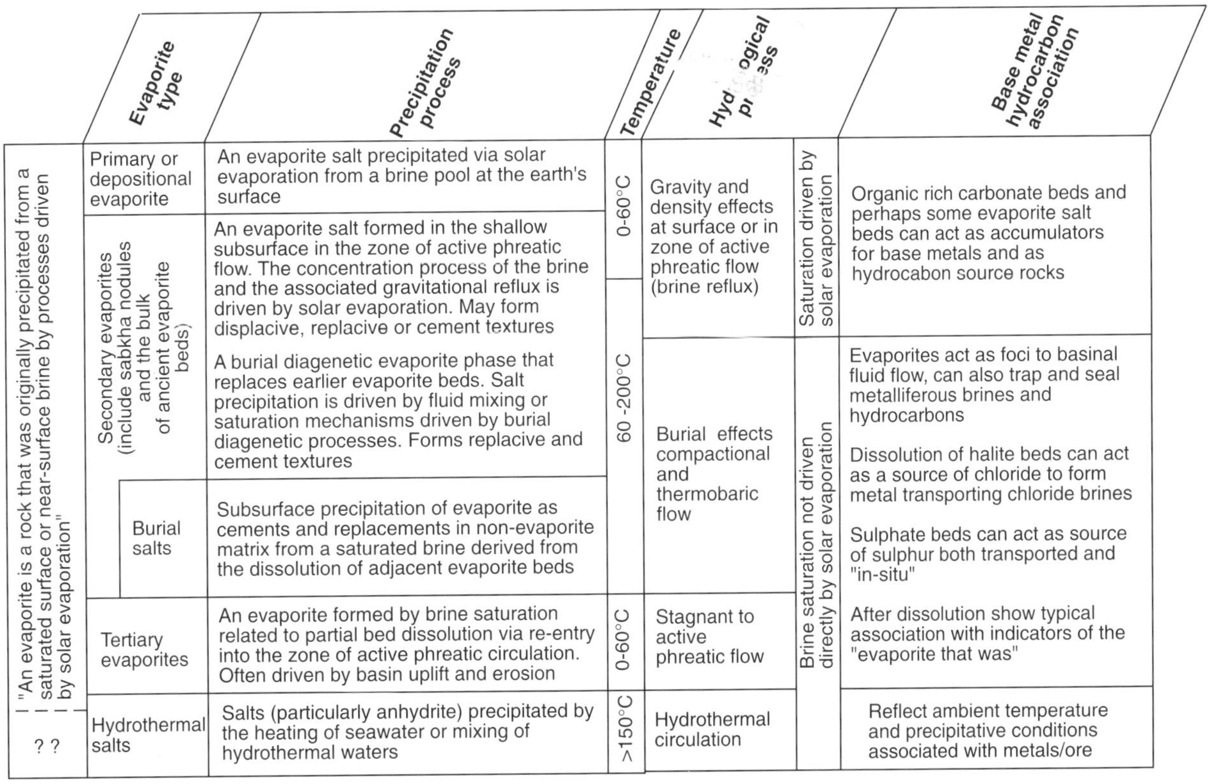

	Evaporite type	Precipitation process	Temperature	Hydrological process	Base metal hydrocarbon association
"An evaporite is a rock that was originally precipitated from a saturated surface or near-surface brine by solar evaporation" / processes driven by solar evaporation	Primary or depositional evaporite	An evaporite salt precipitated via solar evaporation from a brine pool at the earth's surface	0-60°C — Gravity and density effects at surface or in zone of active phreatic flow (brine reflux)	Saturation driven by solar evaporation	Organic rich carbonate beds and perhaps some evaporite salt beds can act as accumulators for base metals and as hydrocabon source rocks
	Secondary evaporites (include sabkha nodules and the bulk of ancient evaporite beds)	An evaporite salt formed in the shallow subsurface in the zone of active phreatic flow. The concentration process of the brine and the associated gravitational reflux is driven by solar evaporation. May form displacive, replacive or cement textures			
		A burial diagenetic evaporite phase that replaces earlier evaporite beds. Salt precipitation is driven by fluid mixing or saturation mechanisms driven by burial diagenetic processes. Forms replacive and cement textures	60 -200°C — Burial effects compactional and thermobaric flow	Brine saturation not driven directly by solar evaporation	Evaporites act as foci to basinal fluid flow, can also trap and seal metalliferous brines and hydrocarbons. Dissolution of halite beds can act as a source of chloride to form metal transporting chloride brines. Sulphate beds can act as source of sulphur both transported and "in-situ"
	Burial salts	Subsurface precipitation of evaporite as cements and replacements in non-evaporite matrix from a saturated brine derived from the dissolution of adjacent evaporite beds			
	Tertiary evaporites	An evaporite formed by brine saturation related to partial bed dissolution via re-entry into the zone of active phreatic circulation. Often driven by basin uplift and erosion	0-60°C — Stagnant to active phreatic flow		After dissolution show typical association with indicators of the "evaporite that was"
? ?	Hydrothermal salts	Salts (particularly anhydrite) precipitated by the heating of seawater or mixing of hydrothermal waters	>150°C — Hydrothermal circulation		Reflect ambient temperature and precipitative conditions associated with metals/ore

Figure 1.1. Classification of evaporite formation in the depositional-diagenetic-hydrothermal realms.

MINERAL	FORMULA	MINERAL	FORMULA
Anhydrite	$CaSO_4$	Leonhardtite	$MgSO_4.4H_2O$
Antarcticite	$CaCl_2.6H_2O$	Leonite	$MgSO_4.K_2SO_4.4H_2O$
Aphthitalite (glaserite)	$K_2SO_4.(Na,K)SO_4$	Loewite	$2MgSO_4.2Na_2SO_4.5H_2O$
Aragonite **	$CaCO_3$	Mg-calcite **	$(Mg_xCa_{1-x})CO_3$
Bassanite	$CaSO_4.1/2H_2O$	Magnesite**	$MgCO_3$
Bischofite	$MgCl_2.6H_2O$	Mirabilite	$Na_2SO_4.10H_2O$
Bloedite (astrakanite)	$Na_2SO_4.MgSO_4.4H_2O$	Nahcolite	$NaHCO_3$
Burkeite	$Na_2CO_3.2Na_2SO_4$	Natron	$Na_2CO_3.10H_2O$
Calcite**	$CaCO_3$	Pentahydrite	$MgSO_4.5H_2O$
Carnallite	$MgCl_2.KCl.6H_2O$	Pirssonite	$CaCO_3.Na_2CO_3.2H_2O$
Dolomite**	$Ca_{(1+x)}Mg_{(1-x)}(CO_3)_2$	Polyhalite	$2CaSO_4.MgSO_4.K_2SO_4.2H_2O$
Epsomite	$MgSO_4.7H_2O$	Rinneite	$FeCl_2.NaCl.3KCl$
Gaylussite	$CaCO_3.Na_2CO_3.5H_2O$	Sanderite	$MgSO_4.2H_2O$
Glauberite	$CaSO_4.Na_2SO_4$	Schoenite (picromerite)	$MgSO_4.K_2SO_4.6H_2O$
Gypsum	$CaSO_4.2H_2O$	Shortite	$2CaCO_3.Na_2CO_3$
Halite	$NaCl$	Sylvite	KCl
Hanksite	$9Na_2SO_4.2Na_2CO_3.KCl$	Syngenite	$CaSO_4.K_2SO_4.H_2O$
Hexahydrite	$MgSO_4.6H_2O$	Tachyhydrite	$CaCl_2.2MgCl_2.12H_2O$
Ikaite**	$CaCO_3.6H_2O$	Thernadite	Na_2SO_4
Kainite	$4MgSO_4.4KCl.11H_2O$	Thermonatrite	$NaCO_3.H_2O$
Kieserite	$MgSO_4.H_2O$	Trona	$NaHCO_3.Na_2CO_3$
Langbeinite	$2MgSO_4.K_2SO_4$	Van'thoffite	$MgSO_4.3Na_2SO_4$

Table 1.1. Major evaporite minerals. Less saline alkaline earth carbonates or evaporitic carbonates are indicated by **, the remainder are evaporite salts. Less common minerals, such as monohydrocalcite, hydromagnesite, the borates, nitrates and strontium salts are not listed here, but are listed and discussed in detail in Chapter 7 (see also Figure 4.33).

Almost every evaporite unit we see in the subsurface is diagenetically altered or secondary, although many still retain "ghosts" or relicts of a primary texture. Under this definition anhydrite nodules in a sabkha, including those in the modern Arabian (Persian) Gulf, are secondary peritidal or lacustrine cements and replacements. They grow in the capillary zone of the shallow subsurface as intrasediment nodules and crystals within a preexisting primary matrix of transported mudflat sediments.

Pre-Neogene evaporite beds are now almost all composed of secondary evaporites, often precipitated as crystals that replaced a primary evaporite. The replacements formed diagenetically, either as syndepositional or as burial diagenetic evaporite minerals that replaced, altered, or reprecipitated within primary evaporite beds. This overprint process took place once or numerous times in the days, weeks, months, or millennia of exposure, burial or uplift that followed primary precipitation. The high solubility of evaporite minerals means that, unlike quartzose and aluminosilicate sediments, evaporite beds are highly mobile in the subsurface. During burial, beds can flow through adjacent strata. The flowage mechanism can be: a) early diagenetic, as occurs during syndepositional fractionation, dissolution or reflux; b) later diagenetic, as occurs during complex burial-stage bed dissolution or reprecipitation, driven by subsurface fluid flow in the zone of free convection below the zone of overpressure; c) widespread and pervasive, as occurs during halokinesis (salt tectonics); and d) postdiagenetic and extending well into the metamorphic realm where daughter minerals, such as scapolite and tourmaline, can act as a source of volatiles and lubricants long after the precursor salts have gone.

Exhumed or uplifted evaporite beds also undergo alteration, dissolution and replacement as they re-enter the zone of active phreatic flow. Soluble components can then be reprecipitated in adjacent shales as alabastrine and satinspar gypsum or fibrous halite. Exhumed evaporites are classified as tertiary evaporites (Figure 1.1).

Earliest secondary evaporites are syndepositional precipitates, often forming cements and replacements even as the primary matrix accumulates around them. The process of pervasive ongoing rapid reactivity, with early formed salts continually backreacting and precipitating as their at-surface or pore brines evolve, is unique to evaporites (e.g. Table 2.4). Nonevaporitic sandstones, shales and limestones undergo diagenetic reactions when flushed by evolving pore fluids, but the diagenetic rock/fluid framework is slower to respond and requires years to millennia to overprint an original depositional texture. It is the ability of evaporites to flow easily, even as they backreact or dissolve, that distinguishes them from other types of sediment, and, as will be discussed in Chapters 8 and 9, is fundamental to the association of evaporites with many metal accumulations.

Evaporite mineralogy and texture may change, but all ancient secondary and tertiary evaporites occur within the volume of rock that was originally precipitated as the primary evaporite. In addition to these three categories there are two other classes of salts that are not "true evaporites". I term the two classes, burial salts and hydrothermal salts (Figure 1.1). Burial salts are made up of the mineral salts, such as anhydrite or halite, that do not necessarily occupy the same rock volume as the original evaporite. Their occurrence is related to subsurface fluid flow, hydrofracturing, brine mixing and brine cooling. As such, they do not necessarily precipitate via processes driven by solar evaporation. Burial salts are the commonplace authigenic cements and replacements that precipitate in a nonevaporite matrix from subsurface brines derived by dissolution of an adjacent evaporitic salt bed. Because of the proximity to a "true" evaporite bed, most authors would consider burial salts a form of "true" secondary evaporite.

In contrast, hydrothermal salts do not require a nearby dissolving evaporite to form. Hydrothermal anhydrite forms by heating of seawater or by the subsurface mixing of $CaSO_4$-saturated hydrothermal waters, either during ejection of hot hydrothermal water into a standing body of seawater, or during convective circulation (Figure 9.21). Hydrothermal salts are poorly studied, but often intimately intermixed with sulphides in areas of base-metal accumulations, such as the Kuroko ores in Japan or the exhalative brine deeps in the Red Sea. Hydrothermal anhydrite is a commonplace salt in many such active volcanogenic-hosted massive sulphide deposits.

Depositional significance of evaporites

The simplest subdivision of evaporite minerals is into evaporitic alkaline earth carbonates – aragonite, dolomite, low-Mg calcite and high-Mg calcite – and evaporite salts – gypsum, anhydrite, halite, trona, carnallite, etc. (Table 1.1). Primary evaporitic carbonates tend to form in the initial stages of brine concentration, whereas the other primary evaporite salts are precipitated in the more saline stages of concentration. Evaporitic carbonates can contain and preserve elevated levels of organic matter that subsequently generate hydrocarbons or act as reductants.

Evaporitic carbonates

Evaporitic alkaline earth carbonates are the first evaporite minerals to precipitate from a concentrating hypersaline surface water (see Chapter 2) and are usually composed of aragonite, high- and low-Mg-calcite, magnesite or even primary dolomite. The essential hydrological property of the depositional setting that forms evaporitic carbonates is that evaporative outflow exceeds inflow. This results in two characteristics of the depositional system, which hold also for the more saline evaporite salts. First, rapid changes in water level are possible, especially in the more marginward facies, leading to interlayering of strandline and subaqueous units. Under such a regime any subaqueously precipitated carbonate is liable to subaerial exposure and syndepositional subaerial diagenesis. Second, the solute content, especially the Mg/Ca ratio, of shallow hypersaline water fluctuates as the salinity fluctuates. For example, the Ca content of any brine is depleted by the early precipitation of calcite or low-Mg calcite. Subsequent carbonate precipitates from increasingly saline water will have a higher Mg/Ca ratio and tend to be dominated by high-Mg calcite, aragonite, magnesite, or even dolomite.

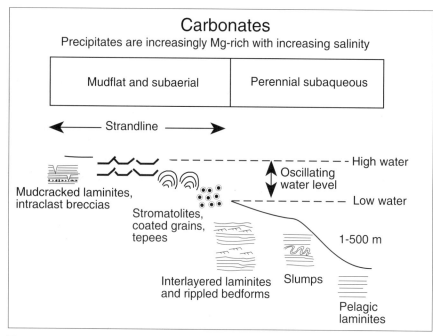

Carbonates
Precipitates are increasingly Mg-rich with increasing salinity

| Mudflat and subaerial | Perennial subaqueous |

Strandline

High water

Oscillating water level

Low water

Mudcracked laminites, intraclast breccias

Stromatolites, coated grains, tepees

1-500 m

Interlayered laminites and rippled bedforms

Slumps

Pelagic laminites

Figure 1.2. Depositional significance of evaporitic carbonate textures. Note the distinction between subaqueous and strandline (ephemeral water) indicators.

The ephemeral lakes of the Coorong clearly illustrate this salinity-related variability across adjacent brine lakes (von der Borch, 1965; Warren, 1990).

Evaporitic dolomite as a primary precipitate in a Holocene setting is rare and occurs only in a few documented areas, such as the Coorong ephemeral lakes in South Australia and Deep Spring Lake in California. More commonly, widespread evaporitic dolomite forms as a syndepositional replacement after an aragonite or Mg-calcite precursor (Land, 1985). Elevated levels of magnesium in the pore brines are required to replace $CaCO_3$ with dolomite and are often created by the precipitation of gypsum or anhydrite. A modern example of this type of "brine reflux" dolomite is seen in gypsiferous mudflat sediments of Lake Greenly playa, South Australia (Dutkiewicz and von der Borch, 1995). Ancient examples are discussed in Chapter 3.

Laminites

Mm-scale lamination is volumetrically the dominant sedimentary structure in modern and ancient evaporitic carbonates, but its origins are varied and complex (Figure 1.2). Beds dominated by such textures are called laminites. Many carbonate laminites form couplets or even triplets by the regular superposition of micrite with siliciclastic clay, organic matter or evaporite salts. These couplets and triplets are frequently referred to as varves, yet are not necessarily "true" varves composed of annual couplets.

As an example of a carbonate laminite, consider the deep bottom sediments of the Northern Basin in the Dead Sea, Israel. Evaporitic laminites, composed of alternating light and dark laminae, are accumulating beneath a density stratified lake brine that is more than 350 m deep. The whitish laminae are composed of stellate clusters of aragonite needles (5-10 μm diameter), which precipitated each summer at the air-brine interface and then sank. The darker laminae consist of clay minerals, quartz grains, detrital calcite and dolomite that washed in as suspended sediment from the surrounding highlands during occasional storm floods (Garber, 1980). Permanent stratification also characterises the waters of Lake Tanganyika, where the waters are anoxic below 250m depth and the deep bottom laminites (seasonal varves) contain between 7% and 11% total organic carbon (TOC; Cohen, 1989). Or consider the shallow lacustrine laminites of the more saline Holocene Coorong Lakes, Australia (Warren, 1988, 1990). Coorong laminites make up more than 80% of the sediment volume in the Coorong Lakes and are composed of alternating dark-grey organic-rich and light-grey organic-poor laminae. Mineralogies range from hydromagnesite to aragonite to Mg-calcite to dolomite. As in the Dead Sea, the Coorong laminites formed subaqueously on the floor of a chemically stratified lake. Unlike the 250+ metres depths of the Dead Sea and Lake Tanganyika, Coorong laminites accumulated as microbially entrained pelletal wackestone/mudstone beneath perennial waters less than 1-5 metres deep. Once on the bottom these pellets were easily eroded and redeposited as wave-driven bedload.

Laminites can also be preserved under conditions of ephemeral subaerial exposure. Algal-bound laminites (algal mats) crosscut by mudcracks characterise the algal channel and strandline facies in many modern and ancient evaporitic carbonates such as the sabkhas of the Arabian Gulf (Kendall and Skipwith, 1969). Modern and ancient laterally extensive carbonate laminites, without evidence of subaerial exposure, indicate subaqueous deposition under fluctuating surface water chemistries. They indicate water depths that may have ranged from a few decimetres to hundreds of metres and are not necessarily associated with deep water conditions (Warren, 1985; Kendall, 1992).

Strandlines: pisolites, tepees and stromatolites

The strandline in many evaporitic carbonate systems is characterised by an association of pisolites, ooids, crusts, tepees, domal stromatolites and mud-cracked cryptalgalaminites (Figure 1.2). It forms in areas of groundwater springs, mudflats and ponds, which define the fluctuating shorezone about a more permanent brine lake or seaway. This facies association is typically syndepositionally cemented into stacked layers of intraclast breccia, biolaminites, pisolites and tepees, often separated by salt units or their dissolution breccias.

Stromatolites and other microbialites

Microbial structures are commonplace features of the strandline of an evaporitic water body and textures range from thrombolites through travertines and tufas to cryptalgalaminites and stromatolites, with stromatolites being the most commonly recognised microbialite (Figure

1.2). A stromatolite is defined as a laminated organosedimentary structure with positive relief away from a point or limited surface of attachment. Stromatolite shapes range from columnar to domal to subspherical. Laminae define the successive positions of an agglutinating algal plexi (community), with each accreting layer produced by trapping and binding of sediment. In some settings this occurs in conjunction with direct carbonate precipitation onto or within trichomes and is driven by changes in ambient pCO_2 (Figure 1.3a). The dominant members of the community in a microbialite layer range from bacteria to cyanobacteria to algae with the proportions changing with depth in the sediment. The biotal components have also changed over the course of geological evolution (Figure 1.3b). Ancient flat-laminated mats where evidence of a biogenic origin is poor to nonexistent are sometimes called cryptalgalaminites.

A stromatolite can be considered to be a laminated agglutinated benthic microbialite that possesses topographic relief. It is part of a much wider group of poorly or crudely layered microbialites that can be found in a wide range of marine and nonmarine settings (Figure 1.3c; Riding, 1991). According to Riding, sediment trapping results in an agglutinated stromatolite characterised by a particulate microstructure that preserves few, if any, details

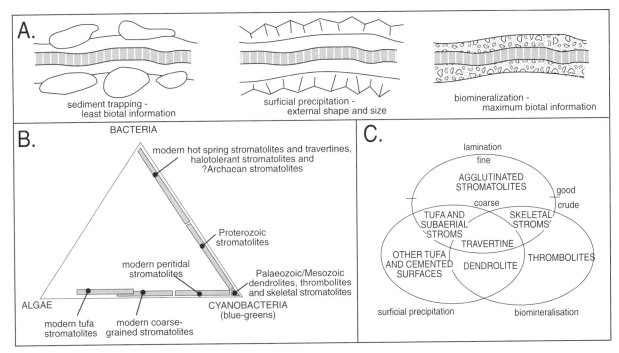

Figure 1.3. Microbialites. A) Styles of mineral accretion about a trichome. B) Biological components that construct various microbialites and the changes over geological setting and time. C) Classification of microbialites according to dominant constituents (after Riding, 1991).

of the original size, shape or orientation of the plexi responsible for capturing the sediment (Figure 1.3a). Biomineralized deposits, on the other hand, can retain considerable detail of the organisms, which are preserved as skeletal microfossils. In consequence, skeletal stromatolites, dendrolites and thrombolites reveal much more information about the microbes contributing to the structure. Mineralisation is especially important in nonmarine microbialites such as lacustrine tufas where inorganic $CaCO_3$ precipitation is often enhanced by CO_2 degassing of venting spring waters. Likewise travertines, which tend to form layered deposits of $CaCO_3$ near the outflow points of warm springs, can preserve details of the contributing microbes. Folk (1993) and many other workers in travertines have documented microtextures such as spheroidal dolomite, dumbbell shaped crystals, fine diurnal-layered dendrites and bushes that indicate a strong bacterial contribution to these deposits.

Precambrian stromatolites flourished in settings that ranged from deep open marine to supratidal to lacustrine, they were not confined to the more arid peritidal situations that characterise most Phanerozoic marine stromatolites. Since the early Phanerozoic the evolution of grazers into normal to mesohaline marine salinities means marine-associated stromatolites are best preserved under hypersaline conditions that are typically found in upper intertidal to supratidal zones. Such high stress environments tend to exclude many of the metazoan grazers that otherwise browse and destroy algal mats (Figure 1.3a). Modern areas of thick extensive marine stromatolites/cryptalgal laminites include the algal facies of the Arabian Gulf (Kendall and Skipwith, 1969) and the hypersaline intertidal flats of Shark Bay, Western Australia (Logan, 1987). There are still, however, some normal marine settings where stromatolites act as marine framework binders. For example, stromatolites take the place of coralline algae as binders within coral reefs as in Upper Miocene coral-stromatolite reefs in Spain (Riding et al., 1991). This unusual association seems to be related to normal marine conditions that existed immediately prior to the onset of basinwide evaporite deposition.

Unfortunately models of ancient stromatolites in the geological literature draw heavily on Holocene peritidal analogues. Many sedimentologists assume that well-preserved Phanerozoic stromatolites must have formed in intertidal to supratidal marine settings. As we shall see in Chapter 3, a marine-connected basin may not be the best way to model evaporitic carbonates associated with thick evaporites. An understanding of lacustrine microbial settings may be more useful in explaining the stromatolitic

and microbial diversity seen in ancient evaporitic carbonates.

In lacustrine settings the niches occupied by stromatolites and tufas cover a much wider range of salinities than in the marine realm. Microbialites constitute a substantial portion of the ephemeral-water and littoral settings that occupy the upper parts of carbonate-rich slope breaks about prograding lacustrine margins (Dean and Fouch, 1983). For example, marginal microbial bioherms reach thicknesses of 7 metres with diameters of 15 metres in the Oligocene Ries Basin (Riding, 1979; Arp, 1995). Fresh to brackish water stromatolites, along with oncoids and travertines occur in Plio-Pleistocene sediments of the East African Rift Valley lakes (Casanova, 1986), while freshwater to mesohaline stromatolites characterise Oligocene lake margins in France (Casanova and Nury, 1989). Mm-laminated manganiferous aragonitic stromatolites define the northwestern strandplain margin of the Northern Basin of the Dead Sea (Druckman, 1981) and similar features outline the carbonate margin of Marion Lake and other modern coastal salinas in South Australia (von der Borch et al., 1977; Warren, 1982a). Modern thrombolites occur in slightly deeper water (15-50m) about the edges of Lake Tanganyika (Cohen and Thouin, 1987). Evaporite-associated stromatolites and cryptalgalaminites have characterised the mudflat and littoral margins of saline and alkaline playa lakes throughout much of the Phanerozoic. Examples include the Tertiary of southern France (Truc, 1978), the Eocene Green River Formation of the USA (Surdam and Wolfbauer, 1974) and the Cambrian of South Australia (Southgate et al., 1989).

The organic plexi constructing a stromatolite in evaporitic setting is today made up of a halotolerant biota that, through the extrusion of a mucilaginous sheath, captures sediment of varying mineralogies. The most common biotal constituents are halobacteria and filamentous cyanobacteria (Figure 1.3b). This group can construct microbial forms ranging from stromatolites to tufas/travertines to thrombolites. Photosynthetic activity of the algae may also aid in the mineralisation. A few species of higher algae can also biomineralise carbonate within their cell walls (Figure 1.3c).

The mineralogy of a stromatolite is often dependent on what carbonate phase is precipitating in the ambient lake water. In many saline lakes the stromatolites and algal bioherms are aragonitic (e.g. Great Salt Lake, Eardley, 1966; Marion lake, South Australia, von der Borch et al., 1977) with carbonate precipitation perhaps facilitated by bacteria (Pedone and Folk, 1996). Calcitic and aragonitic algal tufa and mats often form at spring vents and can

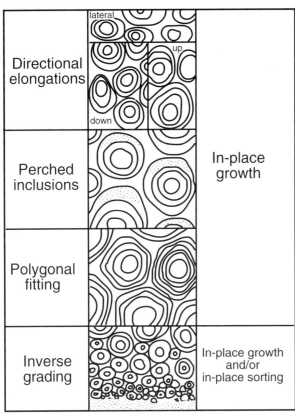

Figure 1.4. Fabrics of pisoids created by in-place processes. Although all these features have been interpreted as requiring vadose pore conditions, only downward directional elongation (micro-stalactite) require such an interstitial condition (after Esteban and Pray, 1983).

construct substantial spring mounds at outflow points in playa systems (Habermehl, 1988). An unusual but interesting variety of calcitic tufa is found in spring vents in Mono Lake and in Quaternary sediments of the Lohanton Basin, Nevada (Shearman et al., 1989; Council and Bennett, 1993). This thinolitic tufa was originally precipitated as the hexahydrate form of calcite (ikaite) about vents on the lake floor where degassing water temperatures were near freezing. Owing to ikaite's highly unstable nature at higher temperatures it was rapidly replaced by elongate lenticular and "hornlike" pseudomorphs, which are now composed of calcite (Shearman and Smith, 1985).

In the highly alkaline (pH>9) fresh to brackish waters of Salda Golu (Lake) in Turkey the subaqueous stromatolites of the lake margin are composed mostly of hydromagnesite along with entrapped diatom tests (Braithwaite and Zedef, 1994, 1996). Hydromagnesite-aragonite stromatolites dominate in the schizohaline ephemeral waters of North and South Stromatolite Lakes in the Coorong region of South Australia (Warren, 1988, 1990), while poorly

preserved hydromagnesite-magnesite stromatolites characterise much of the present-day playa surfaces of the Caribou Plateau in Canada (Renaut, 1993).

Vadose pisolites and other coated grains

Fenestral carbonates, sometimes intercalated with grainstones, are commonly interpreted as marine peritidal sediments (Shinn, 1983). Often the grainstones facies are dominated by coated grains, including oolites, and their wave-agitated "pelagic" mechanical original is clearly seen in well preserved crossbed sets. However, in some evaporitic settings these beds of coated grains have accreted to where the bed becomes a pisolitic rudstone that is often interlayered with evaporite solution breccias. These pisolitic beds show many textural features, such as bridged coats and polygonal fitting of pisoids, that imply they formed "in situ" under conditions of little or no wave-induced agitation.

Pisolite rudstones tend to be caught up in tepee structures and other layers and beds that also contain internal structures and cements indicative of alternating vadose and phreatic conditions. Such features include perched inclusions, inverse grading of pisolites, and pendulous cements (Figure 1.4; Dunham, 1972; Esteban and Pray, 1983; Handford et al., 1984). They indicate the depositional setting was part of a strandline, typified by the "to and fro" of the edge of a brine lake or seaway.

Pisolite crusts, with cements indicating they spent at least a part of their time above the water table, are sometimes termed vadose pisolites to distinguish them from permanently subaqueous "marine" pisolites or the pedogenically formed calcrete pisolites. The prefix marine is a little misleading when describing pisolite crusts in evaporitic settings. Modern examples of fitted-pisolith crusts are found atop peripheral seepage areas on the strandplain of evaporite-filled coastal lakes in western and southern Australia (Warren 1982a, 1983a; Handford et al., 1984) and in the spring-fed margins of large halite-filled continental salars in the Andean Altiplano (Risacher and Eugster, 1979; Jones and Renaut, 1994). These modern pisolite beds are part of a sedimentological association where pisolite crusts are forming mainly as abiogenic accretionary particles. They occur within confined saline groundwater seepage areas that are located on, or adjacent to, extensive carbonate mudflats about saline waterways. Such systems are not restricted to marine seepage settings.

Esteban and Pray (1983) argue vadose pisolites characterise the Permian island shoal and sandflat facies of west Texas. Such pisolitic crusts formed best within the central

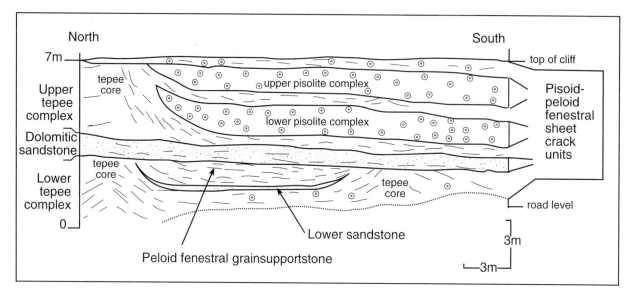

Figure 1.5. Cross section of a field locality of the Permian Tansill Formation at Hairpin Curve in Guadalupe National Park, west Texas. Note the prominent development of pisolites in intertepee depressions within the upper tepee complex and the erosional surfaces that truncate the tepee complexes (after Esteban and Pray, 1983).

depressions of decametre-diameter saucers in an extensive tepee-overprinted megapolygonal terrain. Warren (1983) argues the same vadose pisolites are part of a marine seepage facies about the edge of a widespread saltern/ mudflat complex that covered the platform interior.

Marine pisolites, with isopachous rim cements and less obvious vadose textures are much more commonplace in normal marine Proterozoic carbonates compared with their Phanerozoic counterparts (Swett and Knoll, 1989). The greater abundance of marine pisolites in the Proterozoic seas may reflect a different marine chemistry than today (see below).

Work by Gerdes et al. (1994) has shown that, in addition to pisolites, other carbonate grains such as ooids and peloids can precipitate "*in situ*" in evaporitic settings. They are found growing within modern microbial mats in hypersaline settings and require no wave-agitated or pelagic phase to precipitate concentric laminar grain coats. Laboratory experiments with unconsolidated muds, and comparisons with fenestral and grainstone textures in Guadalupian strata in west Texas by Mazzullo and Birdwell (1989), have demonstrated that intense "*in situ*" alteration of fenestral mudstones can create peloidal grainstones. They go on to argue that such diagenetic grainstones may be the precursors of some forms of pisolitic rudstone.

Tepees

Tepees are the buckled margins of saucer-like megapolygons in limestone or dolomite crusts that appear as an inverted "V" in vertical two dimensional exposures (Adams and Frenzel, 1950). Some authors have extended the use of the term tepee to pressure ridges in crusts of gypsum or halite, others have coined the term "petee" to describe the latter. I prefer to restrict usage of the term tepee to overthrust ridges in carbonate crusts. That is, I use the term nongenetically and to add a prefix to describe its genesis, e.g. groundwater/seepage tepee, seafloor tepee, sabkha tepee, caliche tepee (Kendall and Warren, 1987). Tepees are commonplace strandline features in both modern and ancient settings (Assereto and Kendall, 1977; Warren, 1982a, b, 1983a; Esteban and Pray, 1983; Logan, 1987; Kendall, 1989; Smoot and Lowenstein, 1991; Last, 1992; Jones and Renaut, 1994; Armstrong, 1995).

Tepees, along with vadose pisolites are characteristic features of cemented crusts in areas of the strandplain that experience wetting and drying cycles, groundwater seepage, fluctuating salinity conditions, and marked diurnal changes in surface and nearsurface temperatures. Successive episodes of tepee development often stack one atop the other and can be separated by erosion surfaces or carbonate-cemented sandstones. Pisoids appear to accumulate best within intertepee depressions or saucers (Figure 1.5;

Esteban and Pray, 1983). Subsequent expansion episodes, along with local buckling, may rework beds formed in these pisoid-rich depression into new tepees. When the pace of resurgence was strengthened during times of increased groundwater outflow, tepee crests were capped by domal stromatolites (Warren, 1982a; Kendall and Warren, 1987).

In ancient examples the tepee-influenced horizons are often interbedded with evaporite solution collapse breccia zones. This characteristic intercalation in ancient platform carbonates reflects the to and fro of the brine pool over the strandline carbonate facies at times when the platform was isolated from a surface connection and brine levels were fluctuating in response to the vagaries of evaporitic drawdown (Chapter 3). Some of the evaporite solution collapse may postdate the depositional hydrology of the cycle that formed the tepee, but, based on modern analogues such as Marion Lake, the schizohaline brine hydrology that forms the marginward seeps where tepee growth flourishes is also undersaturated with respect to halite and gypsum. Resurging waters in the seeps dissolve nearby evaporite beds as they form the tepee/pisolite beds (Warren 1982a). The end result of this ongoing dissolution is the collapse of the overlying tepee-affected strata. Any subsequent prograding of the seepage margin into the evaporite basin means the tepee zone is cannibalised from below and along its landward margin by its own groundwaters. All that may be preserved as evidence of this prograding hydrology within ancient platform cycles are terra rosa profiles mixed with remnant breccias composed of fragments of tepee crusts, stromatolite breccias and other residues of the strandplain (Assereto and Kendall, 1977; Bogoch et al., 1994).

The present and the past

An evaporitic carbonate depositional setting with its widespread $CaCO_3$ saturation and consequent early aragonite cement is distinctive and quite easily separated from Phanerozoic marine shelly carbonate shelf textures. The separation is much less distinctive in Mesoproterozoic strata or older. In fact, the best modern analogues for many Precambrian marine carbonate textures are probably to be found in Holocene evaporitic carbonates and not in modern normal marine carbonate settings. This dichotomy reflects two major events in earth history: 1) the evolution of an increasingly bicarbonate-depleted and sulphate-enriched ocean, and 2) the rise of a shelly macrofauna at the Cambrian-Precambrian boundary.

During the Precambrian $CaCO_3$ was extracted from seawater only by inorganic and microbial processes, the same two sets of processes that dominate in modern evaporitic carbonate settings. Archaean carbonate sedimentation entrained numerous and prolific giant botryoids of aragonite up to 1 metre in radius as well as widespread Mg-calcite precipitates (Grotzinger and Read, 1983, Grotzinger and Kasting, 1993; Grotzinger and Knoll, 1995). These beds sometimes formed cementstone sheets several metres thick with strike lengths of more than 100 km. By the Palaeoproterozoic, carbonate sediments entrained less spectacular masses of aragonite and calcite, although cementstone crusts and microdigitate stromatolites in tidal flats remained a characteristic style of carbonate platform aggradation. In contrast, Meso- through Neoproterozoic carbonate sedimentation saw a progressive decline in the precipitation of massive widespread carbonate cements and an increase in widespread precipitation of micritic whitings. Throughout the Precambrian, periodic high energy events ripped up and transported fragments of these cementstone crusts and whiting beds to redeposit them as intercalated intraclast conglomerates and breccias.

The dramatic rise of a skeleton-secreting and sediment ingesting macrofauna at the beginning of the Phanerozoic changed the nature of marine shelf sedimentation. The proportion of chemically precipitated mud and cement decreased and new sedimentary particles, such as abundant shells, faecal pellets and biologically derived muds, came to dominate the marine carbonate realm. The formation of widespread cementstone sheets on the seafloor ceased several hundred million years before the rise of the shelly macrofauna. It was only in the hypersaline environment that carbonate supersaturation persisted. There stromatolites and other halotolerant microbialites continued to flourish without the effects of bioturbation and gastropod grazing.

The secular decrease in the volume of seafloor precipitates reflects a change in the chemical state of the ocean. The much lower sulphate concentrations and much higher bicarbonate to calcium ratios reflecting elevated levels of CO_2 in the Archaean atmosphere (Chapter 2; Grotzinger and Kasting, 1993).

The organic-mesohaline association

Much evaporitic carbonate is micritic and devoid of recognisable fossil fragments. Historically, this lack of entrained fossils meant much evaporitic carbonate was considered to be a direct result of inorganic precipitation

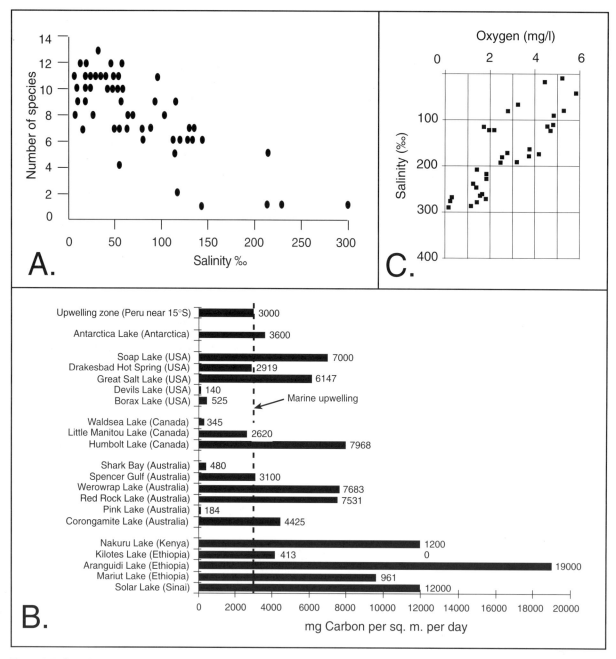

Figure 1.6. Organic matter in modern hypersaline settings. A) The effect of increasing salinity on metazoan species diversity in the saline lakes of the Coorong region, South Australia (after De Decker and Geddes, 1980). B) Organic productivity in various saline ecosystems. Typical marine upwelling zone (offshore Peru) is ≈3,000 mg C/m²day (dashed line). Compiled from sources listed in Warren (1986). C) Oxygen levels in present-day surface brines under increasing salinity. Modern seawater = 5.4 mg/l (after Warren, 1986).

("whitenings"). More recent research, published through the 1980s showed cyanobacteria, algae and halotolerant higher organisms, such as ostrocodes and brine shrimp, can thrive in surface waters at times of brine pool freshening and then aestivate or live on through salt-resistant egg cases (Figure 1.6a). Non-body-fossil evidence of this

activity in saline carbonate sediment is commonplace; e.g. pellets, amorphous algal or cyanobacterial residues, coccoid bacterial clusters, cryptalgalaminites and stromatolites.

When salinities are suitable in the surface waters, evaporite basins have some of the highest measured rates of organic productivity in the world (Figure 1.6b). This may reflect the episodic oscillations of nutrients and salinities that control the extremes of the "life and death" cycles in evaporite basins (Warren, 1986). The biota of euryhaline marine, or humid temperate and tropical continental lacustrine settings, is exposed to much less extreme living conditions. The same marked oscillations in salinity is also one of the main reasons that organic-rich sediments in evaporitic basins is laminated.

High productivity in the surface waters in an evaporite basin need not equate with high preservation levels of organic matter in the bottom sediments. Much of the improved preservation potential associated with "higher than marine" salinities reflects the effects of increasingly low levels of dissolved oxygen in the bottom and pore waters and its effect on the scavenging and grazing biota. Oxygen levels in natural brines tend to decrease with increasing salinity (Figure 1.6c). By the time the brine reaches halite saturation natural waters are dysaerobic to anaerobic (Warren, 1986). Thus bottom waters in any freestanding density-stratified brine lake tend to be anoxic brines that are largely isolated from the overlying less-dense more-oxygenated nutrient-rich nearsurface waters (e.g. Figure 2.9).

At lower brine salinities (vitasaline and penesaline) the dominant contributors to the modern evaporitic biomass are green algae and cyanobacteria. At higher salinities (>200‰) halophilic bacteria dominate the ecosystem (Figure 1.3b). In such settings the levels of preserved organic matter can be quite high in the bottom sediments. Modern brines also contain large amounts of dissolved organic matter (DOM), with levels that increase with increasing salinity (Hite and Anders, 1991). This material is often entrained within brine inclusions in halite and other salt crystals. The level of volatile fatty acid (VFA), especially acetic acid, is also high in these saline brines. High pore salinities also tend to exclude aerobic burrowing and grazing animals that would otherwise destroy the preserved microbial mats of hypersaline systems (Figure 1.6a).

Anoxic bottom brine is often populated by flourishing populations of sulphate-reducing bacteria. In bottoms where iron is present, abundant early framboidal pyrite is a commonplace byproduct of the abundant H_2S (Kribek, 1991; Sawlowicz, 1992). In an ancient drawdown basin this mesohaline brine stage is often the precedent of the main salt depositing phase. Thus organic-rich laminites, that may or may not entrain framboidal pyrite, tend to underlie thick salt beds.

Proximity of organic-rich carbonate laminites and bedded evaporites has lead some geologists to postulate that bedded salts are potentially also organic-rich sediments. This is not borne out by TOC analysis of evaporitic salts – total organic matter levels in ancient $CaSO_4$ and NaCl lithologies tend to be normal or depleted (Katz et al., 1987), although levels of DOM and VFA are high in the brine inclusions. The typically low organic matter content in bedded halite may reflect the higher depositional rates of evaporitic salts when compared with evaporitic carbonates. Under the same rate of supply of organics, the higher rates of halite accumulation dilutes the associated organic content compared to the carbonate sequences.

If one assumes the subaqueous lamination in evaporitic organic laminites is a varve (annual couplet), then knowing the organic content of this sediment allows a backcalculation of the preservation efficiency of the saline environment. The calculations of Hite and Anders (1991) gave preservation efficiencies of between 7 and 85%, values that are 2-10 times higher than typical maritime sediments.

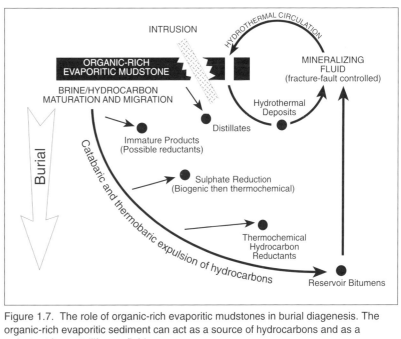

Figure 1.7. The role of organic-rich evaporitic mudstones in burial diagenesis. The organic-rich evaporitic sediment can act as a source of hydrocarbons and as a reductant for metalliferous fluids.

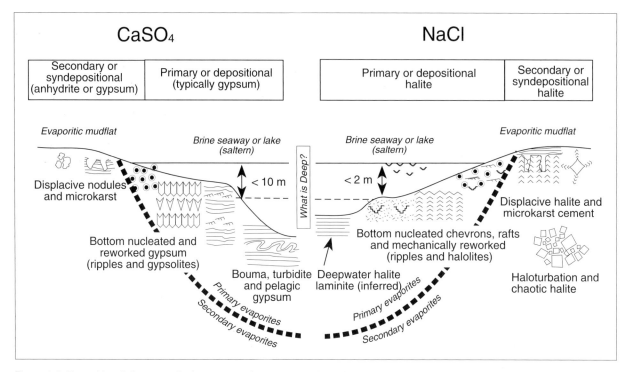

Figure 1.8. Depositional signature of primary evaporite textures and syndepositional secondary evaporites for CaSO$_4$ and NaCl salts. Secondary textures related to burial are shown in Figure 1.15.

Water depths above dense anoxic bottom brines, which preserve organic material, can be deep and below wave-base (Kirkland and Evans, 1981) or they can be shallow (Warren, 1986). During deposition these organic-rich carbonate mudstones can act as scavengers for metals and so preconcentrate both metals and sulphur. Once organic-rich carbonate laminites are deeply buried, they can be flushed by convecting thermobaric and hydrothermal fluids. Then they can act as a source of hydrocarbons, metals and organometallic complexes (Figure 1.7). Other organic-rich mudstones, that are located in less deeply buried more marginward basin positions, can act as reductant zones to hot metalliferous chloride brines moving up along fault conduits from deeper levels. These less deeply buried basin-margin regions are typically located near growth faults or fractures that define basin edge, and are important sites of exhalative and shallow "inhalative" ores. Economic aspects of organic maturation, sulphate reduction and diagenetic metal-evaporite associations are discussed in Chapter 8.

Primary evaporite salts

Evaporite salts are deposited in a brine pan or seaway after the brine has precipitated the alkaline earth carbonates.

The salts, dominantly CaSO$_4$ or halite, show a range of crystal textures that indicate stability, depth and permanence of the precipitating brine (Figure 1.8).

Simple monomineralic carbonate, sulphate or halite layers and laminae, as well as beds composed of complex interlayered mixtures, dominate the subaqueous deposits of many saline lakes and seaways. Chemical "settling" laminae are a commonplace subaqueous stratification style, some laminae outline underlying crystal morphologies, others are flat-planar and can be correlated for long distances across the basin. In addition, biolaminae including microbialite bedding and accretionary tufa and travertine deposits, can accumulate during less saline intervals. Due to the dominance of chemical sedimentary processes operating in these brine bodies, and the early interlocking and cementation of growth aligned crystals, mechanically reworked laminae are less common in subaqueous evaporitic settings than in their carbonate or siliciclastic counterparts.

Widespread layers or crusts composed of cm-dm sized aligned bottom-nucleated crystals are one of the most widely recognised shoal-water textures. Depending on the salinity of the precipitating brine this bottom-nucleated unit is typically composed of halite chevrons or aligned gypsum crystals. Beneath deeper waters, silt-sized crystals can accumulate on the bottom as a "rain from heaven" or pelagic deposit. A range of slope deposits, dominated by reworked shoal water crystals, accumulates in the region separating the shallow from the basinal (Chapter 3).

Growth-aligned gypsum beds

A few thousand years ago subaqueous crystals of gypsum, up to 1 metre tall, grew in stable bottom waters of shallow (<10m deep) Holocene brine pans of southern Yorke Peninsula, South Australia (Warren, 1982b). The crystals coalesced into crusts and beds dominated by swallowtail and palmate gypsum crystals. Similar giant growth-aligned subaqueous gypsum with single crystals up to 10 metres tall have been documented in the Lower Miocene evaporites of Poland (Babel, 1990) and the various Late Miocene Messinian sub-basins of the Mediterranean (Rouchy et al., 1994).

Large gypsum crystals that make up the bottom-grown beds typically entrain curved growth faces interspersed among parallel-sided prisms. Very few natural gypsum beds are made up of growth-aligned crystals formed as aggregates of upward-oriented simple twins, that is, they are not conjoins of the simple prisms shown in Figure 1.9a. In fact, the crystal forms illustrated in Figure 1.9a, and given as typical gypsum morphologies in most mineralogy texts, are the least common forms of gypsum in natural subaqueous systems.

Where the curved faces encapsulate the crystal, a single lensoidal crystal results (Figure 1.9b). The term "bird-beak" gypsum is sometimes used to describe its distinctive lenticular shape. Its outline can be used with pseudomorphs to separate gypsum from the truncated lenticular "axe-head and keg barrel" forms of diagenetic anhydrite. Individual lenticular gypsum crystals range from sand-size to boulder-size. Lenticular sand-sized

forms make up the uppermost lake sediments and the gypsum lunettes about the salt flat edges of coastal and continental salt lake in Australia (Warren, 1982b). Holocene lenticular gypsum crystals are intimately associated with the algal sediments of Ras el Shetan mudflats in the Gulf of Aqaba, Egypt (Aref, 1998a). The reason why the curved face comes to dominate the crystal shape of gypsum is not well understood. On the basis of a number of laboratory-based growth experiments using brine columns separated by gel growth media, Cody (1979, 1991) argued the lenticular shape reflects the presence of particular organic

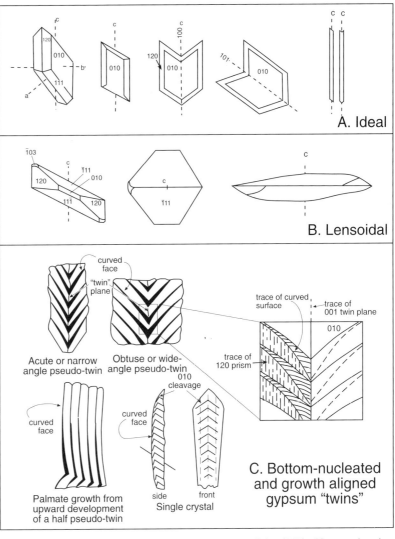

Figure 1.9. Gypsum. A) Prismatic forms of gypsum parallel to (010) with arrow head twins along (100) or (101) and extreme acicular forms of prism and twin on the right. B) Lenticular form of gypsum. Note how the lenticular form (on right) is dominated by the curved faces that are much less important on a typical prism as seen in the side and top views of a crystal (on left). C) Various forms of subaqueous "zig-zag" gypsum growth as described in text (after Shearman, 1978; Orti-Cabo and Shearman, 1977; Rouchy et al., 1994).

compounds in the parent brine. In contrast, Warren (1982b) found that masses of sand-sized lenticular crystals occur in seasonally vadose settings in the numerous coastal salinas of southern Australia. He argued the lenticular shape tends to form best in the upper seasonally vadose parts of the brine pan fill where pristine parallel-sided gypsarenite prisms were subject to periodic short-term dissolution and regeneration.

In addition to this lenticular form, the more central and perennially subaqueous areas in a gypsum system are made up of masses of large parallel-sided megacrystals that all show varying levels of expression of this curved face (Figure 1.9c). Individual crystals range in size from dm to metres in length with some Miocene crystals in Poland more than 10 metres long (Babel, 1990). Varying degrees of curvature lead to the growth of distinctive gypsum twins with acute (narrow) to obtuse (wide) angles along the twin plane. Doug Shearman has described this form of overlapping curvature along the twin plane as "arms of Siva" gypsum. Strong vertical development of the upper half of some twins creates beds of upward aligned gypsum crystals showing strong curvature of the intercrystalline faces. Such forms are sometimes called palmate gypsum. Some purists would argue that most of these natural forms of growth aligned gypsum are not true twins (Babel, 1990). The varying degrees of curvature entrained in the crystal generates changes in the apparent twin angle, even in a single crystal. This breaks the rule of constancy of twin angle, leading some to refer to them as pseudo-twins (e.g. Schreiber, 1988b). Whatever the crystallographic semantics, these large aligned gypsum crystals only grow beneath shallow subaqueous brines.

Growth-aligned coarse-grained gypsum beds form beneath perennial subaqueous brine pans that are density stratified for more than 6-8 months of the annual depositional cycle in Holocene coastal lakes in South Australia (Warren, 1982b). In the early stages of infill of the salina, the perennial bottom brine was always saturated with respect to gypsum. Salinity conditions were stable and allowed the upper euhedral surface of the aggrading crystal bed to preserve the upper outlines of crystals in each growth stage. If the gypsum precipitating today in Lake Inneston, South Australia, is indicative of the subaqueous growth style, then throughout much of their growth, any large upward facing gypsum crystals are covered by a microbial mat or film that captures aragonite pellets. Rapid bottom growth of gypsum probably only takes place when stratification of the brine column is lost via seasonal concentration of the fresher surface water body to the same salinity as the dense gypsum-saturated bottom waters.

Once this happens the whole water column is supersaturated with respect to gypsum and gypsum growth begins anew on a metres-deep brine pool floor (see Figure 2.9 for quantification of analogous hydrology). As the aggrading euhedral gypsum bed surface encloses $CaCO_3$ impurities, it creates zig-zag laminae within the bed.

The $CaCO_3$ pellets that form the laminae are deposited during the seasonal evaporation of a surface lens of freshened water that each winter and spring forms atop the bottom brine. The pellets are created by a pelagic and grazing biota that flourishes during the freshened stage of the surface water body (spring-early summer). In the Lake Inneston and Deep Lake salinas the pelleting is a result of the feeding activities of ubiquitous ostrocodes and the Southern Hemisphere brine shrimp (*Parartemia zietziana*). When salinities in the surface waters are suitable this biota spreads out over the whole brine lake. When as salinities of the surface waters increase to levels that can no longer support a metazoan population, mass mortality kills them. At the same time the whole brine column equilibrates and mixes so that gypsum precipitation begins again across the pool floor. During such supersaline times small living populations of brine shrimp and ostrocodes still remain and flourish about seawater-fed seeps in the tepee and mound-spring dominated salina margins.

Over millennia the depositional surface of the gypsum bed aggrades as gypsum infills the brine lake (Warren, 1982b). As the top of the gypsum bed aggrades it reaches a level where even the dense bottom brines are seasonally freshened. Then the perennial bottom waters, as well as surface waters, are undersaturated with respect to gypsum for a number of months each year. At that time the upper gypsum surface dissolves to form a flat planation surface. Now the seasonal growth of the aggrading bed surface encloses $CaCO_3$ to form a flat mm-laminated gypsum bed.

Growth-aligned halite beds

Holocene halite crusts and beds composed of bottom-nucleated aligned chevrons form subaqueously in water depths of less than two metres in Lake McLeod, Western Australia (Logan, 1987) as well as in many other modern brine pans worldwide (Handford, 1991; Smoot and Lowenstein, 1991). Chevron halite crusts deposited in shallow to ephemeral brine conditions are also preserved in ancient thick halite beds, such as the Ordovician-Silurian Mallowa Salt of the Canning Basin (Cathro et al., 1992) and the Permian San Andres Formation of west Texas (Hovorka, 1987). The chevron structure is the preserved outline of successive crystal growth edges

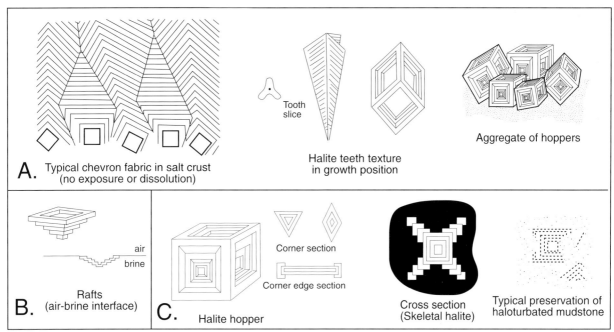

A. Typical chevron fabric in salt crust (no exposure or dissolution)

Tooth slice

Halite teeth texture in growth position

Aggregate of hoppers

B. Rafts (air-brine interface)

air

brine

C. Halite hopper

Corner section

Corner edge section

Cross section (Skeletal halite)

Typical preservation of haloturbated mudstone

Figure 1.10. Halite. A) Growth aligned forms of subaqueous halite and typical chevron growth fabric outlined by inclusions in the crystals. Those crystallite cubes with upward-aligned edges, rather than faces, come to dominate the growth fabric as a series of merged chevron crystals. Other bottom growth forms include upward-growing halite teeth and aggregates of hoppers on the bottom that were created by overgrowths on sunken rafts. B) Halite raft growth, where crystals hang from the air-brine interface. C) Various forms of skeletal or intrasediment halite hoppers. Notice how the outline of the halite shape (or its pseudomorph can vary from rectangular to triangular depending on the plane of intersection of the viewing plane with the crystal hopper (In part after Shearman 1970; Rouchy et al., 1994).

outlined by brine inclusions. These inclusions impart a cloudy or milky appearance to the crystals in core, a feature that is often used to distinguish primary halite from clear or spar-like crystals of secondary or diagenetic halite.

Crusts precipitate today on the floors of shallow brine pools as upward-growing chevrons, cubes and cornet-shaped crystals (Figure 1.10a). Crystals can grow as new aligned crystals or can form as syntaxial overgrowths on founder rafts and other cumulate crystals that have settled to the pan floor. As for all isotropic crystals growing in crowded conditions, the crystal edges of halite cubes that point upward into the brine grow more rapidly than crystals with upward-pointing crystal faces. The upward pointing "V's" of the edges quickly dominate to form the chevron texture that characterises many subaqueous halite crusts (Figure 1.10a). Less common are thin crust intervals composed of upward-facing cornets and plates. There the crystals grew by the upward aggradation of the horizontally aligned faces, and perhaps indicate less crowded conditions of bottom growth (Handford, 1991).

Over time, decimetre-to-metre thick halite crusts and beds can stack one atop the other to form evaporite units that can be hundreds of metres thick. When conditions are suitable,

huge thicknesses of evaporite can accumulate in very short time frames. For example, a more than one-km-thick sequence of halite has accumulated in the Afar Depression, Ethiopia, in less than 10,000 years. No other type of sediment, carbonate or siliciclastic can aggrade as rapidly as an evaporite salt bed (Warren, 1991).

Cumulates, rafts, coated grains and mechanical reworking

When surface water salinities are suitable, crystals can grow via solar concentration at the air-brine interface. Crystals expand to form rafts that are held at the interface by surface tension (Figure 1.10b). They float and grow until the crystal mass becomes so heavy that it exceeds the holding ability of the forces of surface tension and sinks to the bottom. Modern rafts of halite can be centimetres to decimetres across before they sink to the bottom. Crystals settling to the bottom of the brine pool form beds of cumulate crystals (e.g. "salt and pepper" halite and massive gypsisiltites).

Periodic influxes of freshened surface waters, followed by evaporative concentration, can produce laminites composed of bottom cumulates, which also contain increasing

15

fractions of the more saline minerals. Such "rain-from-heaven" laminites are composed of alternate carbonate-sulphate and sulphate-halite couplets and triplets. Rafts and other cumulate crystals in very shallow to ephemeral brine pans may be blown or carried by bottom currents to the strandline to accumulate as subaerially exposed masses of crystals.

Alternatively, increasing wind and wave action can break up the rafts while they are out in the brine lake and still well away from the strandline. These remnants then sink to mix with the smaller crystals making up the cumulate beds of the brine lake floor. Later, when the lake shallows and further desiccates, these cumulate beds are covered by crusts and beds of bottom-nucleated chevron halite or swallowtail gypsum. In ephemeral brine settings this upward transition from cumulates to crusts is a commonplace textural sequence.

Ancient cumulate crystal beds not covered by bottom nucleated crusts were mechanically reworked, via storm and wave-induced bottom currents, into crossbeds and ripples. This tended to happen on the bottoms of ancient perennial brine seaways and lakes that were above wavebase and covered by brines bodies that were less than 5-10 m deep. In even shallower nearshore positions that were subject to higher wave-energies, fine-grained cumulate crystals formed the nuclei to concentric ooid coatings composed of gypsum and halite (gypsolite and halolite respectively). Today, such coated grain textures are precipitating immediately basinward of the strandline of Lake Asal in the African Rift Valley and along the southern end in the Dead Sea (Castanier et al., 1992).

Floating crystal rafts precipitating from dense brines, such as from halite or carnallite saturated brines, can continue to grow and coalesce as crystal aggregates at the air-brine interface. Rather than sinking as cumulates when they reach silt to sand size, crystals can aggregate into rafts that are decimetres across. Rafts can then be blown to the edge of the brine body to accumulate as strandline cumulates, such as are found today in carnallite beds about the brine pan edge of Lake Qaidam in China. Whenever the lake water body freshens and expands, the dissolution of these strandline beds can generate dense brine plumes that alter underlying bedded salts (Casas et al., 1992; Hovorka, 1992; Schubel and Lowenstein, 1997).

Alternatively, some of the low-solubility evaporite salts can accumulate as wind-reworked vadose deposits around the margins of the brine lake. Today the undisturbed salinas of coastal southern Australia are in the penultimate stages of brine pan fill. These are the same pans that had in the earlier stages of their Holocene infill precipitated large bottom-aligned gypsum crystals. But now, with the Holocene aggradation of the gypsum fill to near sealevel elevations, the lake surface is at hydrological equilibrium and can aggrade no further. There is now a much smaller volume of surface water to evaporate each year (Warren, 1982b). Surface brines come to saturation much more quickly and surface waters completely dry up so that rapid, multiple nucleation of gypsum crystallites takes place across the floor of the brine pool. Beds of sand-sized gypsum (gypsarenite) now accumulate on the floor of ephemeral brine lakes, only to be subject to later desiccation and ongoing meteoric dissolution.

Each year, as the lakes dry and the water table falls, the upper part of this sand-sized sediment column leaves the capillary zone as it is blown into dunes that line the margins of the lake (lunettes). This capillary zone deflation is the same hydrological process that forms the characteristic truncation or "Stokes surfaces" described in Chapter 2. As the gypsum in these dunes is elevated above the lake water table it is subject to the same set of pedogenic processes that form soil features atop and within the nearby calcareous dune systems (Warren, 1983b). Gypsum rhizoliths and bioturbation structures, as well as gypsum soils (gypsites) and aeolian crossbeds typify the dune sediment (Warren 1982b). Calcareous body fossil fragments, with the exception of wind-reworked algal oogonia and other calcispheres, are rare to absent. Today such "bioturbated" and aeolian gypsum deposits accumulate in Holocene salina settings where at least a portion of the annual hydrological cycle is characterised by vadose conditions.

Occurrences of modern vadose evaporites are largely composed of gypsum. Other salts, with their inherently higher solubilities, tend to dissolve too rapidly in vadose situations and so do not form thick preservable accumulations. Exceptions to this generalisation of only the less saline pedogenic salts being preserved can be found in the hyperarid Atacama Desert, South America. There pedogenic accumulations of nitrates and iodates, along with halite and gypsum, are accumulating in the desert regolith; and have been doing so since the mid Miocene, when uplift of the Andes first created desert conditions (see nitrates).

Pedogenic gypsum horizons can sometimes stack into substantial thicknesses. For example, there is a 25 metre thick gypcrete (the sulphate equivalent of a calcrete) hosted in Miocene alluvial fan sediments in the Calama Basin of northern Chile. The gypcrete is a single horizon in the alluvial bajada and it surrounds an ancient endoheic basin. It formed via stacking of gypsum-cemented soil horizons (Hartley and May, 1998). It was probably preserved, despite its high susceptibility to meteoric recycling, by being covered by a 9.5 Ma carapace of ignimbrite. This relatively impervious volcaniclastic has isolated it from any downward-percolating meteoric waters.

Although body fossils other than algal structures are rare, many subaerial and subaqueous gypsum units preserve abundant trace fossils that, in addition to track and trails, also include rhizoliths and other pedotubules. Rhizoliths form at times of freshening, they indicate periodic subaerial exposure and colonisation by terrestrial plants. Rodriguezaranda and Calvo (1998) identify six associations of trace fossils in the evaporitic sediments of the continental Tertiary basins of Spain: (1) networks of small rhizoliths; (2) large rhizoliths; (3) tangle-patterned small burrows; (4) isolated large burrows; (5) L-shaped traces; and (6) vertebrate tracks. The rhizoliths occur both in the marginal areas of hypersaline lakes and across lakes characterised by ephemeral but moderately high salinity waters. In either setting, pedoturbation indicates colonization by grasses and bushes and so indicates vadose conditions at the sediment surface that were tied to a drying of the lake and a lowering of the water table.

According to Rodriguezaranda and Calvo (1998) burrowing invertebrates were especially active in those Spanish evaporite lakes where gypsum was the main mineral phase. They even define a separate facies in these lakes called the "bioturbated gypsum deposit". It is typified by tangle-patterned small burrows and minor isolated large burrows; they are interpreted as burrows of insect larvae, probably chironomids, coleopterans and annelids. Similar organisms typify modern lake environments with water concentrations averaging 100-150 g/l, they flourish in lake areas subject to frequent drying and terrestrial encroachment.

At the other end of the mechanically-reworked evaporite spectrum are the continuously subaqueous "rain-from-heaven" accumulations, which are composed of laminar couplets of sand- and silt-sized evaporite crystals. Crystals originally precipitated at the air-brine interface or within the upper metre or two of the perennial brine body. This uppermost part of a deep brine body is most subject to freshening or concentration. The resulting finely layered and laminated bottom cumulates are thought to typify many ancient "deepwater" evaporite-laminites. Such deposits are often found in basinal positions located well away from the evaporite platform margin. In some ancient basinal settings, such as in the Permian Castile Formation of west Texas, the water depths on the evaporite seaway bottom were sub-wave base and sediments lay beneath anoxic gypsum-saturated bottom waters. Such laminites were not reworked by a bottom dwelling benthos, so individual $CaSO_4$ lamina in the Castile Formation can be correlated laterally for distances of more than 100 km (Dean and Anderson, 1982).

Laminar pelagic cumulates on the deep bottom of the brine seaway could be reworked by currents and gravity. Slumped and current-reworked deepwater anhydrites have been recognised in slope and rise areas near many ancient evaporite platform margins, e.g. Zechstein basinal evaporites in NW Europe (Schlager and Bolz, 1977), the Triassic lacustrine slope deposits of Northern Chile (Bell, 1997) and the Middle Miocene (Belayim Formation) evaporites of the Red Sea Rift (Rouchy et al., 1995). These deepwater reworked sequences can be recognised by the entrained turbidite and Bouma textures. The source for this reworked anhydrite was either clastic $CaSO_4$ crystals eroded from the adjacent, but much shallower, platform or gravitationally unstable "pelagic" salts. These deepwater sequences are discussed in "depth" in Chapter 3.

Can crusts that show evidence of pulsed and aligned growth precipitate in deep water?

With both gypsum and halite crusts, the high degree of supersaturation required for primary growth at the floor of the brine pool means that brine depths are typically shallow during crust growth. Maximum water depths are measured in metres for modern gypsum crusts and decimetres to centimetres for halite crusts. A halite chevron or a gypsum crystal in a crust is outlined by successive crystal growth pulses. Each pulse is characterised either by thin layers of less saline salts or by inclusions of brine that were poikilitically enclosed during crystal precipitation. This style of microlaminated crystal layering reflects changes in crystal precipitation rates induced by short-term changes (daily to monthly) in salinity on the subaqueous floor of the brine pan (Warren 1982b; Handford, 1991).

Bottom brines in deep water are too dense and too stable to change with the rapidity required to form the mm-cm scale discontinuities that characterise the internal structure of growth-aligned crust fabrics. Hence laminated bottom-nucleated textures do not form as widespread precipitates in deep water. Deep waters in brine seaways create a well buffered chemical system (see various authors in Melvin (1991) for full discussion). If this is not a truism then one must answer, what mechanism acted over the floor of ancient brine seaways, which were up to hundreds of metres deep, to rapidly displace or cool dense halite-saturated bottom waters with short-lived pulses of less-dense freshened or cooler water that slowed the rate of crystal growth? Freshened water will always float atop a dense brine. The volume of less dense water required to displace and dilute a brine column hundreds of metres deep, or the time required for complete mixing to the bottom, excludes a deepwater origin for most finely layered bottom-nucleated fabrics.

If the crystal crust is growing near a deepwater hydrothermal spring, there is a possibility of pulsing crystal growth in the bottom waters due to temperature fluctuations or changes in the rate and saturation of the spring outflow. This scenario is not volumetrically important in terms of ancient growth-aligned evaporite textures, but is economically interesting and is discussed in later chapters dealing with the evaporite/base-metal association.

Secondary evaporites

Most ancient evaporites show strong evidence of secondary or diagenetic textural overprints. Providing a brine is saturated with respect to a particular evaporite mineral it has the potential to precipitate that mineral. This potential exists both at the surface and in the subsurface. So far, we have discussed the primary processes of evaporite precipitation in the context of single-phase primary evaporite crystals deposited on the floor, or about the margins, of brine pans, lakes, salterns or saline seaways. However, evaporites also precipitate as intrasediment crystal growths and replacements, or as cements in preexisting evaporite and nonevaporite hosts.

Processes of secondary evaporite formation are reflected in crystals and textures that indicate displacive growth, recrystallisation, back reactions and replacement (see Chapter 2). Such events indicate a superimposed brine chemistry that differs in terms of chemical composition, or salinity, or temperature from the brine chemistry that formed the primary or preexisting evaporitic sediments.

Sabkha (intrasediment) salts

The well-studied nodular anhydrites and gypsums of the Abu Dhabi sabkha in the Arabian Gulf are probably the most widely recognised examples of syndepositional secondary evaporites (see Warren (1991) for a summary of

literature). The nodules form displacively and replacively from concentrated pore fluids in the capillary and upper phreatic zones beneath the sabkha surface (Figure 1.11). These intrasediment crystal growths precipitate in a matrix of upper intertidal and supratidal sediments. With continued $CaSO_4$ saturation in the pores, the gypsum and anhydrite nodules grow and coalesce to form the enterolithic fold and chickenwire textures so typical of the supratidal portion of a sabkha sequence.

Anhydrite nodules in a sabkha grow "from the inside out." They grow as new laths of anhydrite form between older laths (Shearman and Fuller, 1969). As successive laths crystallise in a nodule, they displace, rotate or break the older laths and the growing nodule pushes aside the adjacent hot matrix. This process is illustrated schematically in Figure 1.11a, where the position of each successive new lath is indicated by a double-ended arrow. The end result of this "inside-out" growth is a near pure anhydrite nodule, with laths about the edge of the nodule rotated into a subparallel alignment with the nodule margin. Many nodules are constructed of a number of subsets of this subparallel arrangement, implying large nodules grow by accretion of smaller nodules.

Skeletal or displacement halite crystals (also termed pagoda halite) have characteristic entrained matrix inclusions that outline successive growth steps. Crystals show an incomplete development of the cube face, with local intense development of the cube edges into dendritic or pagoda-like morphologies. Rapid crystal growth often entrains some of the mud matrix within the halite, usually parallel to the cube faces. Such rapid growth is thought to indicate a rapidly attained and high degree of supersaturation with respect to halite (Arthurton, 1973; Southgate, 1982). As such, dendritic or pagoda forms cannot grow in deeper burial environments where rate of salinity fluctuation and the kinetics of the reaction are much slower and the surrounding matrix is more likely to be indurated. Displacive halite is thought to grow by capillary evaporation of supersaturated pore waters in soft brine-soaked mudflat or sabkha sediments (Figures 1.8 and 1.10c). Such hoppers are found growing in carbonate muds along the western shore of the Dead Sea (Gornitz and Schreiber, 1981). Similar halite crystals are found down to metres below the surface of present day mudflats in Bristol Dry Lake in California (Handford, 1991). Growing about the edges of the same lake are unusual capillary pavements of growth-aligned seepage gypsum (Rosen and Warren, 1990).

The amount and variety of intrasediment salt is highly variable. Some mudstones contain scattered crystals and

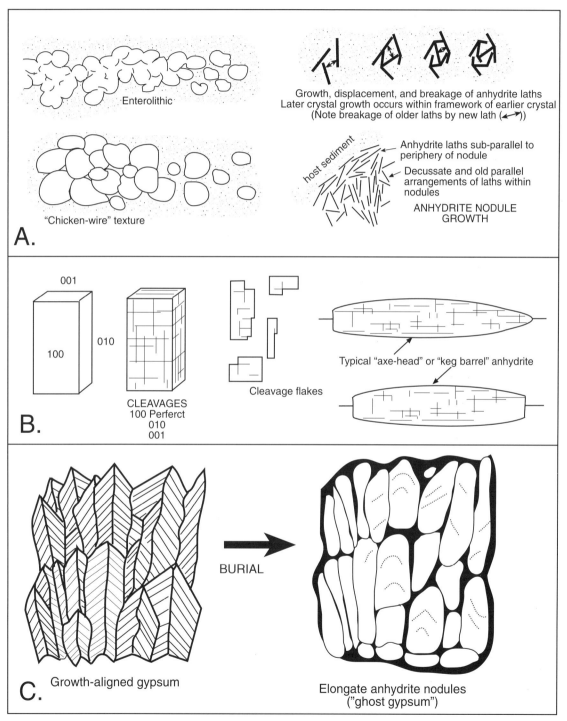

Figure 1.11. Anhydrite characteristics. A) Modes of anhydrite nodule growth as chicken wire and enterolithic anhydrite. Also shows mode of anhydrite growth as successive new anhydrite laths break and rotate earlier formed laths. The end result is a nodule with a margin composed of laths aligned subparallel to the edge of a nodule. Many nodules preserve clusters of subparallel laths showing large nodules grow by accretion of smaller nodules. B) Crystallographic characteristics of anhydrite laths note well developed orthorhombic cleavage. Large isolated anhydrite spar crystals in mudstones sometimes have a lenticular form. It or its pseudomorphs can be distinguished from lenticular gypsum by a typical "axe-head" and "beer-keg morphology" created by its well developed cleavages at one or both tends of the lens. C) Preservation of gypsum ghosts as elongate anhydrite nodules created during burial. (After Shearman and Fuller, 1969; Warren and Kendall, 1985).

nodules that make up less than 1% of the rock volume. Others are made up of approximately equal volumes of host matrix and evaporite salts, while yet others are characterised by layers composed almost entirely of intrasediment salts. The thickness of the layer over which intrasediment salts can precipitate is controlled by the style of brine supply. If the salts are precipitating by capillary evaporation beneath a subaerially exposed mudflat then the thickness is controlled by the pore throat size of the host sediment. Capillary zones can be metres thick in fine clays, while they are only centimetres thick in sands. If the intrasediment salts are precipitating by the downward percolation of dense brines from a subaqueous water body then the thickness is dependent on the rate of diffusion or reflux into the underlying clay. Handford (1991) estimates intrasediment salts can precipitate in soft muds down to depths of 1-2 metres below the sediment-brine interface.

Syndepositional karst

Crusts and beds dominated by primary subaqueous textures can be partially dissolved by more dilute surface brines. Freshening can be generated by a rise in the brine-pool level associated with rainfall, by seawater flooding, or by morning dew dissolution associated with the complete drying of the brine pan. The resultant dilution creates a truncating dissolution surface that crosscuts crystal faces in both subaerial and freshened subaqueous settings (Figure 1.12). Vertical to subhorizontal karst cavities typically form beneath the truncation surface under subaerial conditions. Cavities are then filled by clear evaporite cement sometimes atop a cavity-lining rind of detrital mud and clay. In chevron halite beds these cavities are often filled with a clear inclusion-poor halite spar cement. Void infill is often related to the depositional event that precipitated an immediately overlying chevron halite or swallowtail gypsum layer (Handford, 1991; Warren, 1982a). Using bromine signatures, Cathro et al. (1992) found

the Devonian brines that precipitated the cloudy primary halite in the Mallowa Salt of the Canning Basin were indistinguishable from the signatures in the secondary spar-like halite that filled voids in the primary halite, implying an early origin from the same depositional/syndepositional brine.

Intensity and frequency of dissolution overprints are largely determined by brine depth, input of dilute waters and the permanency of the brine body. Most modern and ancient halite salt pan crusts preserve evidence of syndepositional dissolution in the form of dissolution surfaces and halite-spar-filled cavities. Lowenstein and Hardie (1985)

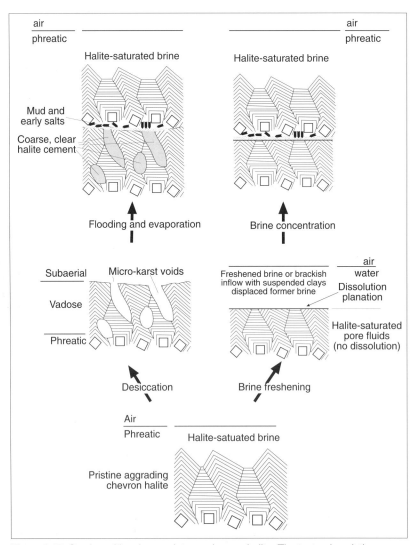

Figure 1.12. Syndepositional overprints on chevron halite. The textural evolution illustrated on the left shows the formation of vadose microkarst, while that on the right illustrates the effect of bottom freshening on the floor of a perennial brine pool and the formation of a planation surface with no vadose karst development.

emphasised the presence of such features in a halite bed is the unequivocal signature of the ephemeral surface waters that typify salt pan deposition. The texture reflects the dissolution events that accompany subaerial exposure of the salt crust and the periodic flooding of the halite pan by fresh to brackish meteoric waters. Dissolution cavities are less likely to occur in bottom nucleated beds deposited in evaporite basins covered by deeper perennial brine bodies where any freshened inflow will tend to float atop the denser halite-saturated brine body until its density equalises that of the underlying brine (see Chapter 2). If brine freshening does occur in a perennial brine situation then a planation surface without vadose cavities is the likely result (Figure 1.12; see also Chapter 2)

Shallow textural re-equilibration

Dense saturated bottom brines percolating down through a shallowly buried primary evaporite unit can create diagenetic havoc with the original texture. For example, saturated bottom brines percolating through a buried cumulate halite deposit can convert it into a tightly cemented coarsely crystalline halite mosaic retaining little evidence of the primary texture (Smoot and Lowenstein, 1991; Cathro et al., 1992). Dense halite-saturated brines flushing down through shallowly buried subaqueous gypsum beds that are dominated by bottom-nucleated gypsum crystals can set up diagenetic conditions where the gypsum is replaced by halite/anhydrite pseudomorphs of the original aligned gypsum. In Permian evaporites of west Texas the replacement mimics the crystal shapes in beds composed of bottom-nucleated gypsum precursor, as well replacing internal layering within crystals. This happened early during reflux of halite saturated brines into gypsum beds, where the finer grained gypsum matrix had already converted to nodular anhydrite. The high salinity brines drove a process of gypsum dissolution and halite replacement at depths of less than two metres beneath the brine-covered sediment surface (Figures 1.13 and 2.8; Hovorka, 1992).

In a similar fashion, changes in the near surface hydrology within the modern saline mudflats of Salt Flat playa in west Texas, have "sabkha-ized" a subaqueous laminated lacustrine gypsum bed (Hussain and Warren, 1989). The process was driven by the superposition of a Holocene capillary hydrology onto subaqueous gypsiferous laminites originally laid down in a Pleistocene salt lake. This capillary zone overprint precipitated nodular displacive gypsum within a matrix of laminated gypsum/dolomite. The change

in hydrology indicates a change in climate from more humid Late Pleistocene to more arid Holocene conditions. It also set up shallow subsurface pore chemistries, whereby lenticular gypsum crystals were pseudomorphed and replaced by halite spar.

Percolation of waters periodically saturated and undersaturated with a particular evaporite salt can also create textural havoc in a nonsalt matrix. For example, displacement halite grows via capillary evaporation during the supersaturated phase of pore flushing of the saline mudflat facies surrounding many brine pans. The growth of this halite disturbs and rotates primary mechanical structures in the original sediment matrix (laminae, ripples, mudcracks, etc.). With each freshening of the pore fluid,

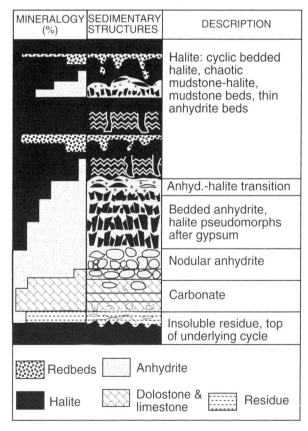

Figure 1.13. Typical depositional cycle in the Permian evaporites of the Texas Panhandle where gypsum (rather than anhydrite) was the dominant sulphate precipitated, even at salinities near halite saturation. Large gypsum crystals grew vertically on floors of brine pools, surrounded by a matrix of autochthonous transported gypsum sand and silt. Most or all of the primary gypsum was altered to anhydrite or leached and replaced by halite in the very shallow (<2m) burial environment under the influence of halite-saturated diagenetic brines (after Hovorka, 1992).

perhaps by seepage following storm runoff, the pore waters become undersaturated with respect to halite and displacement halite dissolves. Sediment matrix then collapses into the voids thus formed, further disrupting primary structures. As pore fluids are again concentrated by capillary evaporation, displacement (intrasediment) halite again starts to form, with similar results. This process is known as haloturbation. If the sediment is preserved with halite still present, it is termed chaotic halite. If the halite is completely dissolved, all that may be left is a haloturbated massive mudstone (Figure 1.10c).

Owing to the mixing of waters of two salinities, reflux dolomitisation beneath a platform evaporite can takes place concomitantly with CaCO$_3$ dissolution (Chapter 2). Relative rate of limestone solution compared to dolomite growth is a likely control on porosity levels during the later stages of the reflux dolomitisation process (Sun, 1995). If the rate of dolomite growth keeps pace with the rate of CaCO$_3$ dissolution, then the mimetic replacement of micrite matrix and micritised grains occurs with little change in porosity levels, as in many ancient evaporitic peritidal dolomites. If the rate of reflux dissolution exceeds the rate of dolomite replacement, as in open marine platform carbonates beneath the seaward edge of a platform evaporite, then there is large scale skeletal aragonite dissolution and an increase in effective porosity in the dolomitised interval compared to precursor limestone. During shallow subsurface brine reflux in the Permian in the Palo Duro Basin, Texas, the flow of such refluxing waters dissolved marine carbonate to form anhydrite-filled moulds (Bein and Land, 1983), while in the Levelland-Slaughter interval it enhanced reservoir porosity (Elliott and Warren, 1989).

Porosity loss

Syndepositional alteration, cementation and porosity loss, in a halite bed continues into the shallow subsurface until the bed can no longer support fluid flushing. This occurs once all effective porosity is occluded by compaction and

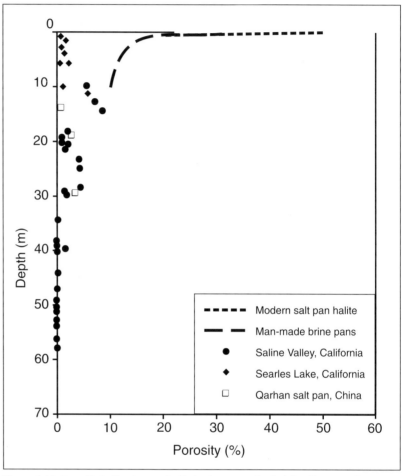

Figure 1.14. Porosity in Quaternary halite beds versus depth (after Casas and Lowenstein, 1989). All effective porosity is lost by 70 m burial.

secondary evaporite cementation. In halite-dominated sequences this loss of effective porosity is early and shallow. Casas and Lowenstein (1989) in a study of Quaternary halite deposition showed that Quaternary halite layers at about 10 m below the surface have porosities of <10% and that layers at depths below 45m are tightly cemented without visible porosity (Figure 1.14). By 100m of burial, halite units are tight and impervious. Diagenetic porosity occlusion by surface-driven hydrology in these halite beds is essentially over. No further porosity change can occur until beds start to dissolve or the dihedral intercrystalline angle of halite begins to change as the beds approach metamorphic conditions (Chapter 6).

From a few hundred metres burial until it reaches the zone of metamorphism, a halite unit maintains permeabilities that are measured in no more than microdarcies or nanodarcies. Thus halite beds and allochthonous sheets provide a highly effective seal to sediments beneath. To move closer to the surface, trapped basinal fluids must either flow laterally around such a salt barrier or dissolve it (Chapter 2). If the trapped pore fluids are hydrocarbons then no dissolution can occur as halite is virtually insoluble in gas and oil. Retention of pore fluids below a salt barrier means porosity in the host sediment is maintained to much greater depths and that sediments below a salt barrier in an actively subsiding basin are typically overpressured.

The pervasive early loss of porosity from more than 50% near the surface to zero by 100m burial also has important implications for textural and geochemical studies of ancient bedded halite. Casas and Lowenstein (1989) showed that this porosity loss was entirely due to early post depositional diagenetic cementation by clear halite. Other mechanisms of porosity loss by chemical or mechanical compaction were dismissed for lack of evidence. There were no strained or broken crystals, pressure solution boundaries nor stylolites, and patches of chevrons preserved in among the mosaic halite spar were all undeformed.

Many Phanerozoic halite beds that have not undergone halokinetic deformation also contain patches of relict depositional and early diagenetic textures, which have remained unchanged for hundreds of millions of years. For example, the Ordovician-Silurian Mallowa Salt (formerly Carribuddy Formation) in the Canning Basin, West Australia, retains aligned halite chevrons crosscut by syndepositional karst and overprint of coarse mosaic recrystallised halite (Cathro et al., 1992). Such textures are thought to be syndepositional, yet are not dominated by large volumes of pristine chevron halite. Rather they preserve evidence of early porosity loss brought on by numerous syndepositional cycles of dissolution and recementation. Lowenstein and Hardie (1985) showed that patchy remnants of chevrons and cornets in a dominantly coarse-grained mosaic of halite spar is not necessarily a burial fabric. All "mature" salt pan halites have this fabric within tens of metres of the depositional surface.

The mechanisms that allow a shallow brine to become supersaturated with respect to halite and to precipitate intercrystalline cements/replacements are still not well understood. One possible mechanism is brine mixing. When compositionally distinct brines mix in the subsurface a new halite-saturated brine may result. This is the case when $MgCl_2$ brines are mixed with $CaCl_2$-rich brines (Lerman, 1970; Raup, 1970). Casas and Lowenstein (1989) have questioned whether this is the dominant mechanism that forms pervasive halite cements in most shallowly buried playa sediments. Rather, they suggest a simple

temperature drop is the dominant control. Surface brines in halite-forming surface waters can be heated to temperatures in excess of 60 - 70°C, while groundwater brines typically have temperatures of 15 - 30°C. They suggest that the cooling associated with the downward reflux of a surface water is all that is required to saturate a brine and so precipitate widespread halite cements.

Deeper burial - mosaic halite?

The formation of pervasive mosaic halite early in the burial history of a salt bed creates a problem in recognising those parts of a halite unit that have been subjected to later burial recrystallisation. A deeper burial environment is also thought to precipitate mosaic halite spar. Diagenetic textures in most ancient bedded halite units are dominated by sutured mosaic textures, where grain boundaries suture in the same fashion as neomorphic pseudosparite in calcite (Spencer and Lowenstein, 1990). Individual crystals display curved boundaries that meet at triple junctions with angles approaching 120°, and interstitial impurities concentrate along the crystal boundaries (Figure 1.15).

Mosaic texture is comparable to the polygonal equigranular mosaic texture that forms in annealing metals, where cooling grains tend to optimise their size, shape and orientation in order to minimise energy. The same mosaic texture can also be found in a soap foam as bubble boundaries with 120° triple junctions. The exact mechanism that controls the formation of mosaic textures in halite is still not well understood. It forms as a pervasive recrystallisation overprint in flowing halite that is undergoing tectonic compression or halokinesis. In flowing salt that has approached the upper part of a growing salt structure the crystals tend to flatten in the direction of maximum stress and this flattening can be used to distinguish it from other forms of mosaic halite (see Chapter 5).

Limpid dolomite in mosaic halite

The formation of late, ordered stoichiometric limpid dolomite along the contact between bedded halites and interbedded mudstones is an unusual outcome of burial diagenesis in halite-mudstone successions (Gao et al, 1990). In the Permian example documented by these authors, the limpid dolomite is characterised by a low Sr content. This they interpret as indicating that meteoric or structural waters were a source of ions for the dolomite precipitate. Their suggested mechanism was dolomite crystallisation under favourable sulphate-reduction

23

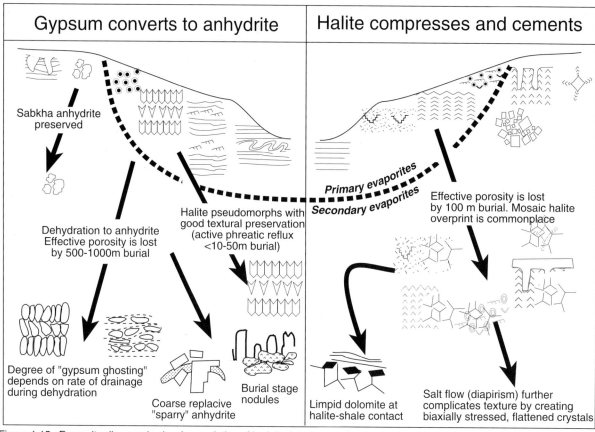

Figure 1.15. Evaporite diagenesis showing evolution of burial-related secondary evaporite textures (includes burial salts).

conditions during burial compaction. Compaction mixed the waters, released from the dewatering mudstones, with brines derived by the dissolution or recrystallisation of adjacent impervious halite beds. Limpid dolomite spar precipitated along the halite-shale contact as the escaping shale waters caused the halite at the contact to dissolve and recrystallise as mosaic spar.

Burial of sulphate evaporites

Gypsum and other bedded evaporite units containing salts with entrained structural water, which is lost during burial, show an inherently more complicated textural response to burial than the simple mosaic textural overprint of halite beds (Figure 1.15; Warren, 1991, p. 167 ff.). Gypsum releases structural water as it converts with burial to nodular mosaic anhydrite (see Figure 2.12). Carnallite ($KCl.MgCl_2.6H_2O$) releases Mg and water as it converts to sylvite (KCl) by incongruent dissolution associated with burial diagenesis or phreatic flushing by meteoric waters. Loss of structural water during burial dehydration means

syndepositional textures in the converting bed are modified as they are flushed by this escaping water.

Nodular anhydrite and water loss

Gypsum rather than anhydrite is the most common form of sedimentary $CaSO_4$ at earth surface temperatures. Intense solar heating in very arid regions can dehydrate gypsum to form anhydrite at the surface. Gypsum also converts to anhydrite in the capillary zones of sabkhas and evaporitic mudflats where the gypsum is bathed in highly saline NaCl brines. But in ancient saline giants both of these processes were relatively insignificant in forming the widespread beds of massive nodular anhydrite that extend across tens of thousands of square kilometres in an ancient evaporite basin. There the thick beds of widespread nodular anhydrite were formed in early burial as thick beds of primary gypsum were compacted and converted to anhydrite (Warren, 1991).

When gypsum is buried and ambient temperature rises above 50-60°C, it converts to nodular anhydrite. The depth of transformation is between a few metres to more than a kilometre. The exact depth in any region will depend on lithostatic pressure, local geothermal gradient and pore brine salinity (Figure 1.15). For example, in the Salton Sea region with its saline pore brines and high geothermal gradients, the conversion from gypsum to anhydrite occurs at 80-105°C and at depths of 200-250m (Figure 6.9; Osborn, 1989). Similar anhydrite conversion was noted in cores from the Pleistocene evaporites in nearby Bristol Dry Lake, California; but here the isotopic evidence suggests sulphate-reducing bacteria also play a role in the conversion (Rosen and Warren, 1990).

If the nearsurface pore-fluid salinity approaches halite saturation, the conversion from gypsum to anhydrite can occur at shallower depths and much lower temperatures ≈35 - 45°C. Such temperatures occur as shallow as 1-2 metres below the depositional surface of a halite-saturated brine pool and lie in the zone of active phreatic flow or brine reflux (Figure 2.8). At such depths the buried gypsum bed retains well-developed intercrystalline porosity and inherently high permeabilities. This facilitates the leaching of the gypsum crystals and their infill by clear halite spar, which pseudomorphs the various forms of the precursor gypsum (e.g. Figure 1.13; Hovorka, 1992, Warren, 1991).

With less saline brines in the pores of a buried $CaSO_4$ bed the conversion of gypsum to anhydrite may not occur until burial depths of hundreds of metres. By such depths, the natural compaction has greatly depleted intercrystalline porosity in the gypsum bed. Release of water from an already compressed and cemented gypsum will create 40% water-filled porosity in an otherwise impervious anhydrite unit. Over the long term this water is not retained but escapes into the subjacent sediments. A likely effect within the converting $CaSO_4$ bed is a decrease in its inherent strength; and in compressive situations, an increased ability to act as a lubricating horizon or décollement (Chapter 2). If the water cannot drain freely from the dewatering gypsum, the bed may convert to a quicksand-like consistency. This explains some of the intense deformation and enterolithic textures with flow orientations that are commonplace even in flat-lying and otherwise undeformed ancient bedded anhydrites.

As it converts, the more deeply buried and compressed $CaSO_4$ unit loses almost all of its original texture. Rare indistinct elongate, and partially flow-aligned, nodular "ghosts" may be all that remain of the primary bottom-nucleated textures in many ancient anhydrite beds (Figure

1.11c). Loss of water of crystallisation also affects units subjacent to the dehydrating gypsum/anhydrite. The escaping water is saturated with respect to $CaSO_4$, but undersaturated with respect to other evaporite minerals, including any nearby carbonates, as well as halite and the bittern salts (Chapter 2).

Even though gypsum and other hydrated saline minerals show more complicated burial diagenetic histories than halite units, and a greater propensity for destruction of their primary textures during burial dehydration, porosity loss in these units is essential complete by burial depths of 500m. After this, these anhydrous units, like halite at even shallower depths, are essentially impervious and act as aquitards and pressure seals to any crossflow of pore water, hydrocarbons or metals. Evaporite salt beds below 500-1000 metres typically respond to further burial by flowage and focused dissolution.

Burial stage anhydrite nodules

Subsurface sulphate-rich waters derived from deeply buried dissolving evaporites can precipitate anhydrite nodules in adjacent nonevaporitic strata. Such nodules are a form of burial salt fed by saline compactional and thermobaric waters. Superficially these nodules can resemble those of a sabkha anhydrite, but there are a number of distinguishing features (Machel, 1993; Machel and Burton, 1991; Warren, 1991). Most obvious is that the burial nodules precipitate in porosity zones that are not depositionally defined. They can occur within or adjacent to faults, or within open marine shelf carbonates, which are not intertidal/supratidal units. They often occur within, or juxtaposed to, dark, bituminous argillaceous seams and pockets, or within zones of anastomose bituminous veinlets and carbonaceous haloes, all of which are the result of pressure solution and stylolitisation. They typically replace the host rock and crosscut stylolite boundaries. Thus, they enclose material generated during burial diagenesis, such as authigenic pyrite and saddle dolomite, and show little evidence of mechanical compaction. They are often associated with coarsely crystalline, late-stage poikilotopic and pore-filling anhydrite spar cements.

In Chapter 4, the processes that control burial-stage evaporite nodule and authigenic cement formation are discussed in detail. Owing to its importance to the petroleum and minerals industries, this topic of burial salts will become one of the rapidly expanding areas of evaporite research in the next decade.

25

Tertiary evaporites: uplift and conversion

At the other extreme of the burial cycle is the conversion of exhumed and uplifted evaporite beds. Most common is the conversion of anhydrite beds into diagenetically regenerated gypsum (Figure 1.16). Uplifted halite beds often first undergo differential dissolution to leave behind layers of residual nodular anhydrite, which are then dissolved by ongoing meteoric flushing (Chapter 4). Two commonplace gypsum fabrics are the result of exhumation: coarse porphyroblastic gypsum and fine-grained alabastrine gypsum (Holliday, 1970; Warren et al., 1990). Porphyroblastic gypsum often retains dispersed relics of the precursor anhydrite. Crystal forms are coarse, up to two cm across, and can be euhedral or anhedral, thin and acicular or thick and stubby. Porphyroblasts often aggregate into cm-scale rosettes or blebs with acicular gypsum rinds. This creates a texture that is sometimes described as "daisywheel gypsum". Because of the destructive nature of porphyroblastic overprinting, interpreting original depositional environment from nearsurface CaSO$_4$ units, composed entirely of diagenetically regenerated gypsum, is next to impossible.

The other form of bedded gypsum, generated by anhydrite rewatering to gypsum in the zone of active phreatic flow, is fine-grained alabastrine gypsum. Individual gypsum grains are typically less than 50 μm, with grain boundaries that range from poorly defined to equidimensional granuloblastic. The change from porphyroblastic to alabastrine may be depth-related (Warren et al., 1990). Porphyroblastic gypsum defines the re-emergence of nodular anhydrite from the stagnant phreatic into the deep portion of the zone of active phreatic flow. Alabastrine gypsum forms in the zone of more active phreatic flow. This re-emergence is also associated with the formation of gypsum karst and the formation of evaporite-dissolution breccias (e.g. Figure 4.15).

Excess amounts of trace elements, especially strontium and boron, are released from some bedded anhydrites as they reconvert to gypsum. The released elements may precipitate in the regenerating alabastrine gypsum as celestite or boron-bearing minerals, such as proberite, ulexite, tyreskite and priceite.

Fibrous gypsum and halite (satin spars)

Veins and fractures that are filled with fibrous satin spar CaSO$_4$ (typically gypsum) or halite are widespread in the mudstones and shales adjacent to bedded evaporite units that show evidence of widespread dissolution. Many fractures are subhorizontal and lie roughly parallel to the contact with the bedded evaporite unit. Other, less common fractures, range up to subvertical. The fracture filling is usually zoned by two or more parallel layers of either fibrous CaSO$_4$ or fibrous halite. Fracture-fill crystals are oriented with their long axes perpendicular to the fracture walls. Coarse calcite crystals occasionally fill the centre of the fracture. Most fracture fills are monomineralic and gypsum is the dominant mineral in most nearsurface fracture systems. Internal fracture zonation is pronounced and reflects episodic opening and filling of the fractures. Crystal textures, including the sigmoidal deformation of some fracture filling crystals, indicate that the many satin spar fractures were infilled as they opened (El Tabakh et al., 1998a).

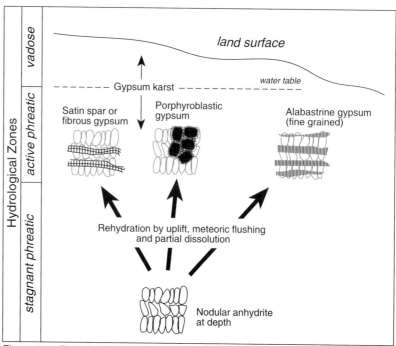

Figure 1.16. Rehydration fabrics in calcium sulphate from uplift, meteoric flushing and dissolution. These are the tertiary evaporites of Figure 1.1.

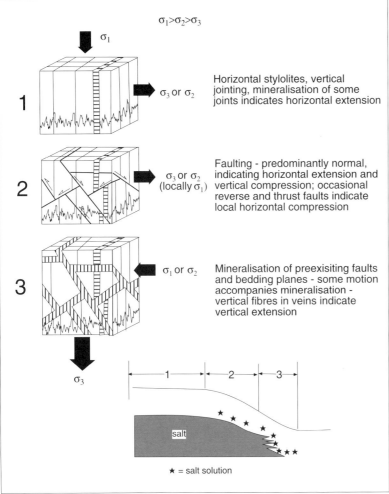

$\sigma_1 > \sigma_2 > \sigma_3$

1 — Horizontal stylolites, vertical jointing, mineralisation of some joints indicates horizontal extension — σ_3 or σ_2

2 — Faulting - predominantly normal, indicating horizontal extension and vertical compression; occasional reverse and thrust faults indicate local horizontal compression — σ_3 or σ_2 (locally σ_1)

3 — Mineralisation of preexisiting faults and bedding planes - some motion accompanies mineralisation - vertical fibres in veins indicate vertical extension — σ_1 or σ_2

★ = salt solution

Figure 1.17. Conceptual model of deformation above salt dissolution zone showing structural sequence and proposed location relative to dissolution front. Stage 1 occurs prior to salt dissolution and is characterised by jointing resulting from normal burial. Late stage 1 and stage 2 occur as salt dissolution begins. Early stage 2 is characterised by a few vertical gypsum veins with horizontal fibres. Later stage 2 is characterised by normal faults and uncommon reverse faults. In stage 2 layer parallel extension results from the onset of dissolution and subsidence. Stage 3 is characterised by gypsum veining of bedding planes and faults. In stage 3 vertical extension results from widespread dissolution and collapse of underlying evaporites (after Gustavson et al., 1994).

feature of hydration, which may have been due to hydrofracturing by overpressured confined groundwater as exhumed anhydrite reconverted to gypsum. Progressive sediment hydrofracturing may take place because of continuous unroofing. Unloading and release of overburden weight will cause the formation of near surface fractures that are parallel to the earth's surface. Thus fibrous halite and satin spar $CaSO_4$ are commonly associated with, but not exclusive to, zones of exhuming evaporites (Figure 1.17). It is not yet known if satinspar filled fractures form at greater depths (1000s of metres) associated with overpressuring and compactional or thermobaric waters.

Summary

Evaporitic settings can accumulate evaporitic carbonates as well as more saline salts dominated by gypsum and halite beds. Evaporitic carbonates accumulate across the whole range of brine depths from ephemeral to shallow perennial to deep. Carbonate laminites tend to dominate in the deeper brine areas, while a whole range of textures and features, including laminites, form in the shallow and strandline settings. Strandline regions are often characterised by pisolites and algal structures of various types, ranging from cryptalgalaminites to stromatolites and tufas. The ephemeral water and groundwater seepage systems that characterise strandlines facilitate the cementation and ongoing growth of carbonate crusts. Over time, these marginward carbonate crusts can grow and deform into overthrust V-shaped carbonate ridges called tepees.

Evaporite textures of the more saline primary evaporites, such as gypsum and halite, reflect the depth and stability of the brine that precipitates them. Small evaporite prisms tend to accumulate in water characterised by rapidly fluctuating bottom brine conditions. The prisms are often mechanically reworked into rippled and ooid-like accumulations or captured in algal mats. Such deposits can

Fibrous halite and satin spar formation are linked to mechanisms of fracture formation, which are in turn tied to chemically remobilised evaporites and an active hydrological system. Fractures filled with satin spar have been linked to the generation of open fractures induced by subsidence associated with salt dissolution (Figure 1.17; Gustavson et al., 1994). The opening of the fractures may be due to: evaporite dissolution, sediment unloading, or hydraulic overpressure. Superimposed on these processes may be the force of crystallisation of gypsum. Shearman et al. (1972) suggested that displacive satin spar veining is a be deposited in brine depths ranging down to hundreds of metres, but most accumulate in shallow brines less than a metre or two deep. It is the rapid salinity fluctuations associated with very shallow waters that favour rapid crystallite formation and multiple nucleation.

The typical evaporite accumulating on the floor of a stable brine pool is composed of upward or growth-aligned crystals. Depending on the stability, salinity, and permanence of the overlying bottom brine, various mineralogical and dissolution surfaces can be superimposed on the upward growing prisms. When the overlying brine column is not subject to marked salinity fluctuations, then pristine crystal outlines are made up of stacked halite chevrons or zig-zag gypsums, with varying degrees of face curvature. When the overlying perennial brine body mixes and freshens, subaqueous crystal truncation surfaces can form. Similar surfaces can form where the brine body dries up. Desiccation-related dissolution surfaces are distinguished from subaqueous truncation surfaces by an inherent abundance of vadose, karst and geopetal features. These three are all features that form during complete drying and falling of the water table into the evaporite bed.

But evaporites do not just form on the bottom of a brine lake or seaway. They can also form where porewaters attain supersaturation in a preexisting nonevaporite matrix (secondary evaporites). Probably the best documented textures of this type are sabkha nodules composed of anhydrite and gypsum in the supratidal sediments of the Arabian Gulf. Large crystals of displacive halite form in a similar fashion beneath the mudflats of the Arabian Gulf, where the capillary pore fluids are at halite supersaturation. Dense brines, created during the formation of primary and early secondary salts, can sink into the underlying sediments where they can facilitate backreaction and alteration processes in the now buried evaporite beds

At greater depths of burial, evaporite salts can reprecipitate as pore-filling cements and burial salts. Many primary and early secondary textures are overprinted at this stage by dewatering and recrystallisation processes associated with compactional dewatering and thermobaric alteration. Brines created by the subsurface dissolution of evaporites may also reprecipitate as burial salts in nearby nonevaporite beds.

If evaporites beds survive the rigours of deeper burial, they can be uplifted and re-enter the active phreatic zone where they are once again subject to dissolution and recrystallisation. Satin spar fractures and alabastrine textures are common indicators of this style of alteration. Such evaporites can be classified as tertiary evaporites.

All of the textural and re-equilibration processes outlined in this chapter reflect the ongoing evolution of the brines that bathe the beds throughout their formation, burial and dissolution. Brine hydrology and its evolution is the fundamental control on evaporites. In the next chapter we look at how brine evolution controls evaporites from the time they precipitate until the time they are completely dissolved.

Chapter 2
Brine evolution and mineralogy

Brine types and classification

An evaporite is defined in Chapter 1 as "a rock that was originally precipitated from a saturated surface or nearsurface brine by hydrologies driven by solar evaporation". All evaporites first precipitate at or near the earth's surface and within the zone of active phreatic flow. With burial, the salts are dissolved or altered by the throughflushing of both active-phreatic and basin-derived brines. The definition makes no assumptions as to the origin of the parent brine; it may be marine, nonmarine or a hybrid. In the past many geologists assumed that all large ancient salt bodies were created by the evaporation of marine waters. Today, there is a growing body of evidence that, in terms of their water source, some large ancient evaporites, previously considered as marine deposits, are actually hybrids fed by a combination of marine and nonmarine waters or are derived from evaporation of totally nonmarine waters. Subsurface dissolution of older evaporites and hydrothermal flushing of buried evaporite beds are increasingly seen as important brine sources (Hardie, 1984, 1991).

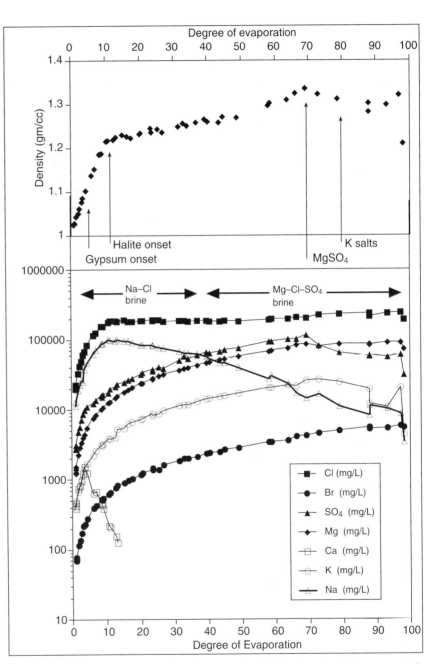

Figure 2.1. Evaporation pathway of modern seawater showing how density increases and the proportions of various ions in solution change as the brine concentration increases. Recompiled from data in appendices in McCaffrey et al., 1987.

Marine brines

Today the chemical make-up and the proportions of the major ions in seawater are near constant in all the world's oceans. Modern seawater is dominated by Na^+ and Cl^-, with lesser amounts of SO_4^{2-}, Mg^{2+}, Ca^{2+}, K^+, CO_3^{2-} and HCO_3^- (Figure 2.1). Using the classification of Eugster and Hardie (1978), modern seawater is a Na–(Mg)–Cl–(SO₄) water, with a density of 1.03 gm/cc and a salinity of 35 ± 2‰. When seawater evaporates, primary evaporite salts precipitate from increasingly concentrated hypersaline waters (Figure

2.1; Table 2.1). The first mineral to precipitate is $CaCO_3$, usually as aragonite; this begins in mesohaline waters as the brine reaches twice the concentration of seawater (40 to 60‰) and achieves a density ≈1.10 gm/cc. As the brine continues to concentrate and approaches about five times the concentration of seawater, that is 130 to 160‰, gypsum ($CaSO_4 \cdot 2H_2O$) precipitates from penesaline waters with densities ≈1.13 gm/cc. At 10 to 12 times the original seawater concentration (340 to 360‰) and densities ≈1.22 gm/cc, halite (NaCl) drops out of the supersaline waters. After halite, the bittern salts (potassium or magnesium sulphates/chlorides) precipitate from supersaline waters at concentrations more than 70-90 times that of the original seawater. By this stage of concentration the brine density is in excess of 1.30 gm/cc and the brines approach the viscosity of liquid honey.

Evaporation of seawater produces two types of brine: Na-Cl brines and $Mg-Cl-SO_4$ brines. In the salinity range 35-330‰ Na-Cl brines dominate, while at higher salinities the brines become progressively dominated by $Mg-Cl-SO_4$, a result of the preferential removal of Na by halite precipitation. While Mg becomes the dominant cation at higher salinities, followed by K, chloride remains the dominant anion throughout the evaporation sequence (Figure 2.1).

Ideally, supersaline seawater evaporated to complete dryness should deposit a suite of Mg and K minerals that include: epsomite/astrakanite, kainite, carnallite, kieserite and bischofite (Usiglio, 1849; Borchert and Muir, 1964). The type of bittern salt and the salinity at which it precipitates should vary according to temperature and levels of impurities in the residual brines. Harvie and Weare (1980) calculate the order of onset for the marine bittern salts as: glauberite, polyhalite, epsomite, hexahydrite, kieserite, carnallite, bischofite. In reality, mixing with freshened recharge about the basin's margin and the influence of humid air layers lingering above the brine body mean a standing body of seawater-derived brine rarely reaches the bittern stage.

Much of our understanding of the marine evaporation series comes from studies of closed system conditions, either in the laboratory, or in artificial salt pans (Usiglio, 1849; McCaffrey et al., 1987). Most published marine brine evolution chemistries come from measurements taken in man-made solar crystallisers, which are designed for the progressive crystallisation or processing of suitably fractionated brines. In such salt works the pan basement is typically an artificially compressed clay bed, sometimes covered by black plastic liners to reduce natural brine loss via seepage. Concentrating brine liquors are pumped from one evaporation pan to the next with monocrystalline impurity-free precipitates or liquors as the typical goal. Hydrologies in ancient evaporite seaways rarely approached such ideal conditions. Natural systems were associated with leaky basements, brine reflux, back reactions, and brine renewal or freshening at any time in the evaporation cycle.

Sanford and Wood (1991) calculated the changes in fluid composition and in the sequence of salts that would precipitate under an open system where bottom brine slowly discharged through underlying sediments. They determined that the leakage ratio (ratio of water outflow to inflow) has a profound effect on both the evaporite mineral suites precipitated and the thickness (or volume) of a deposit. In a leaky system there was a significant reduction in the maximum salinities of the surface brines and it was difficult to precipitate and preserve potash or bittern salts at the surface. The geological reality is that ancient potash salts are typically not preserved as bottom-nucleated primary precipitates. Textures in most natural bittern salts, such as carnallite and sylvite, indicate nearsurface brine cooling and backreactions drove their crystallisation. Modern primary potash salts are best preserved as secondary cements in highly restricted continental hybrid or nonmarine settings (Chapter 7).

Brine Stage (Hypersaline)	Mineral Precipitate	Salinity (‰)	Degree of Evaporation	Water Loss (%)	Density (gm/cc)
Normal marine	Seawater	35	1x	0	1.04
Mesohaline or vitahaline	Alkaline earth carbonates	35-140	1-4x	0-75	1.04 -1.10
Penesaline	$CaSO_4$ (gyp./anhyd.)	140-250	4-7x	75-85	1.10-1.126
Penesaline	$CaSO_4$ ± Halite	250-350	7-11x	85-90	1.126-1.214
Supersaline	Halite (NaCl)	>350	>11x	>90	>1.214
Supersaline	Bittern salts K-Mg Salts	variable	>60x	≈99	>1.29

Table 2.1. Mineral paragenesis and classification of concentrating seawater.

Flux Ratio	Na	K	Ca	Mg	Cl	SO$_4$	HCO$_3$	Order of appearance of saline minerals on evaporative concentration
0.96	490	3.7	4.8	111	548	59.0	2.7	CaCO$_3$-gypsum-anhydrite-glauberite-halite-polyhalite-epsomite-hexahydrite-kieserite-kainite-carnallite-bischofite
1.00	470	10.2	20.3	106	548	53.9	0.6	CaCO$_3$-gypsum-anhydrite-glauberite-halite-polyhalite-epsomite-hexahydrite-kieserite-carnallite-bischofite
1.05	448	17.6	36.2	101	548	53.8	0.6	CaCO$_3$-gypsum-anhydrite-halite-polyhalite-sylvite-carnallite-kieserite-bischofite
1.10	427	24.3	51.8	97	548	51.4	0.4	CaCO$_3$-gypsum-anhydrite-halite-sylvite-carnallite-bischofite-tachyhydrite
1.25	376	41.2	91.4	85	548	45.3	0.3	CaCO$_3$-gypsum-anhydrite-halite-sylvite-carnallite-antarcticite-tachyhydrite

Table 2.2. Predicted composition of seawater (meq l^{-1}) as a function of variations in the ratio of mid-ocean ridge hydrothermal brine flux compared to river water flux into the ocean, and the saline mineral paragenesis of the resulting seawater (after Spencer and Hardie, 1990). Only the first appearance of each mineral is listed (see Figure 2.4 for backreaction effects).

Surface saturations in modern marine-margin brines, such as in the Abu Dhabi sabkhat or the coastal salinas of western and southern Australia, are further complicated by high humidities associated with onshore winds. Elevated above-brine humidity creates a natural buffer to evaporation above the lake or sabkha. Precipitated bittern salts do not accumulate and much of the halite itself is recycled to the seaward flushing groundwaters (Kinsman, 1976).

Given the formation of Mg-Cl-SO$_4$ brines above 330‰, ancient salt beds characterised by abundant MgSO$_4$ components are taken to indicate a marine brine source. But were the ionic proportions of Phanerozoic seawater constant? The ionic proportions, and so the precipitation order of ancient seawater bitterns, may have depended on variations in the flux rates of river inflow relative to flux rates through mid-ocean ridges (Spencer and Hardie, 1990; Hardie, 1996). Simple mixing models show that relatively small changes in the flux of mid-oceanic hydrothermal brines can result in significant changes in the Mg/Ca, Na/K and SO$_4$/Cl ratios of highly concentrated seawater (Table 2.2). If changes in hydrothermal flux did occur, they were probably related to secular changes in the rate of seafloor spreading. Changes of molal ratios in seawater would have generated significant changes in the type and order of potash minerals precipitated. For example, Spencer and Hardie's model predicts that an increase of only 10% in the flux of mid-ocean ridge hydrothermal brine over today's value would precipitate sylvite and calcium-chloride salts instead of the Mg-sulphate minerals expected during bittern evaporation of modern sea water (Table 2.2). And yet many workers, including earlier work by Hardie himself, argue that bittern salts lacking MgSO$_4$ components indicate a nonmarine source. The possibility of changes in Phanerozoic seawater chemistry confuses a mineralogical distinction between marine and nonmarine brine sources (see the potash problem).

Nonmarine brines

A more diverse, even less predictable, suite of evaporite minerals precipitates during evaporation of continental waters. Unlike marine-derived brines, evaporating continental waters do not draw on the near isochemical reservoir of the ocean. Rivers and groundwaters are the source of most of the ions that are ultimately deposited as evaporite salts in nonmarine settings. In a closed hydrological system the composition of nonmarine brines depends on the lithologies that are leached in the drainage basin surrounding a salt lake (Eugster and Hardie, 1978; Eugster 1980). Flow through limestone aquifers produces inflow waters rich in Ca and HCO$_3$; dolomite dissolution generates Mg; igneous and metamorphic matrices yield silica-rich Ca–Na–HCO$_3$ waters. Pyritic shales and other sulphide-rich sediments will contribute sulphate ions, whereas basic and ultrabasic rocks tend to produce alkaline Mg–HCO$_3$ waters (Table 2.3).

In all, Eugster and Hardie (1978) distinguished five major water types in closed continental evaporite basins:

1. Ca–Mg–Na–(K)–Cl

2. Na–(Ca)–SO$_4$–Cl

3. Mg–Na–(Ca)–SO$_4$–Cl

4. Na–CO$_3$–Cl

5. Na–CO$_3$–SO$_4$–Cl waters

As any one of these waters concentrates within a particular evaporite basin, it deposits a characteristic suite of evaporite minerals (Figure 2.2; Table 2.3). First precipitates are the alkaline earth carbonates: low-magnesian calcite, high-magnesian calcite, aragonite, and dolomite. The mineralogy of this initial precipitate depends on the Mg/Ca ratio of the parent brine. The subsequent evaporation pathway of a basin brine is determined by the proportions of calcium, magnesium, and bicarbonate ions in the brackish inflow waters.

If the lake waters are enriched in HCO$_3$ compared to Mg and Ca (i.e. HCO$_3$ >> Ca+Mg), then the brine follows path I (Figure 2.2). Ca and Mg are depleted during the initial precipitation of alkaline earth carbonates leaving excess HCO$_3$. As HCO$_3$ is the next most abundant ion in these waters it combines with Na in the next stage of concentration. Sodium carbonate minerals, such as trona and nahcolite, precipitate. Little or no gypsum can form from pathway-I brines as Ca is completely used up during the preceding alkaline earth carbonate stage. In contrast, during the evaporation of modern seawater, all the HCO$_3$ is depleted in the initial alkaline earth precipitates (mostly aragonite and Mg-calcite). The excess Ca then combines with SO$_4$ to form gypsum. It is chemically impossible for sodium carbonate to form by evaporation of modern seawater (Table 2.3). The assumption that the proportions of calcium to bicarbonate in seawater have not changed much in the last 600-800 m.y. implies that trona salts or

their pseudomorphs indicate nonmarine settings throughout the Phanerozoic and much of the Neoproterozoic.

If initial inflow waters have (Ca + Mg) >> HCO$_3$ then, after the initial evaporitic carbonate precipitates, the brines become enriched in the alkaline earths but depleted in HCO$_3$ and CO$_3$. If the relative volume of HCO$_3$ is low, little carbonate can precipitate with further concentration. Brine evolution follows path II, where excess alkaline earths (Ca, Mg), left after depletion of carbonate, combines with the sulphate ions to precipitate large volumes of sulphates (gypsum and/or epsomite). Path II precipitates a suite of salts and bitterns similar to that derived from seawater (Table 2.3). If the ratio of (Ca+Mg)/HCO$_3$ is near unity (path III) carbonate precipitation can be extensive and voluminous. As Ca is progressively removed there is a progressive increase in the Mg/Ca ratio of the residual brine until large volumes of high-Mg calcites, dolomites and even magnesites precipitate.

Such evolved path III continental brines in combination with varying degrees of basement leakage explain the diversity of evaporitic carbonate mineralogies in the lakes of the Coorong, South Australia (Warren, 1990). Individual lakes, less that 1 km apart and filled with up to 6 m of Holocene carbonates, show mineralogies that range from low Mg-calcite to high Mg-calcite to calcian dolomite to magnesian dolomite±magnesite to aragonite±hydro-magnesite to aragonite+gypsum. Calcitic and calcian dolomites tend to dominate lakes of lower average annual salinity, while the magnesian carbonate phases, along with aragonite/gypsum, tend to dominate in lakes of higher average salinities. During desiccated stages of the late

Brine Composition and Source	Major Saline Minerals	Key Minerals
Na-K-CO$_3$-Cl-SO$_4$ Non-marine waters (mainly meteoric)	alkaline earth carbonates, mirabilite, thenardite, trona, nahcolite, natron, thermonatrite, shortite, halite.	Na$_2$CO$_3$ minerals
Na-K-Mg-SO$_4$-Cl Seawater, non-marine water or hybrids	alkaline earth carbonates, gypsum, anhydrite, mirabilite, thenardite, glauberite, polyhalite, epsomite, bloedite, kainite, halite, carnallite, sylvite, bischofite.	MgSO$_4$ and Na$_2$SO$_4$ minerals
Na-K-Mg-Ca-Cl Non-marine waters (hydrothermal and basinal brines)	alkaline earth carbonates, gypsum, anhydrite, sylvite, carnallite, bischofite, tachyhydrite, antarcticite.	KCl±CaCl$_2$ minerals in the absence of Na$_2$SO$_4$ and MgSO$_4$ minerals

Table 2.3. Mineral occurrences and parent brines (after Hardie, 1991).

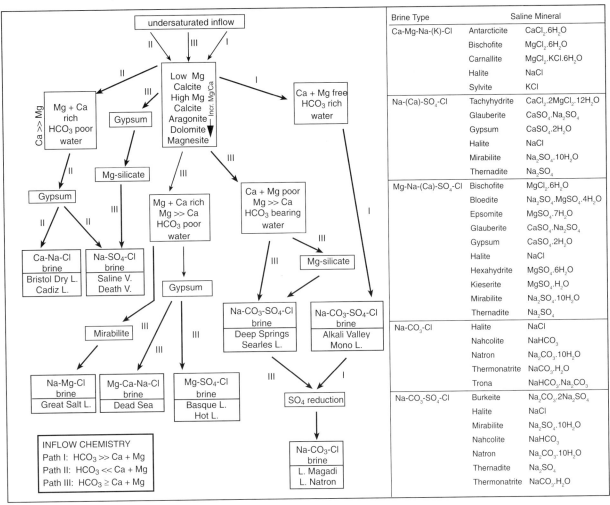

Figure 2.2. Hydrologic classification and brine evolution pathways of concentrating nonmarine waters and a listing of major evaporite minerals associated with the different brine types (after Eugster and Hardie, 1978). L = Lake, V = Valley.

summer, halite-MgSO$_4$ efflorescences form atop carbonate mudflats in many Coorong lakes, but are flushed from the system by winter rains. Salinities in all lakes are schizohaline; they are brackish in the winter and hypersaline in the summer and show order-of-magnitude salinity fluctuations within their annual summer/winter cycles. Average residence times of the waters in the different lakes ultimately control mineralogic assemblage. Those lakes with leakier basements allow a more rapid throughflush of waters and so show lower long term salinities and so accumulate less saline precipitates.

Mineral variation in the Coorong lakes illustrates the importance of relative rates of brine loss (residence time or leakage) in controlling mineralogy in a hydrologically open system. It is illustrative of a general observation, also underlined by Sandford and Wood (1991) and Rosen

(1994), namely that there are very few truly "closed" hydrological systems in evaporitic basins. All natural groundwater systems in salt-accumulating areas leak to varying degrees. Geological evidence for leakage includes the formation of reflux dolomite, the loss of halite porosity by 30 metres burial, the near surface dissolution of salt beds, and the diverse suite of early burial alterations and backreactions shown in evaporitic sediment.

Mixed marine-nonmarine brines

Mixing of marine and nonmarine groundwater sources in many ancient evaporite basins was encouraged by the high degree of isolation or "continentality" required for the accumulation of thick evaporite sequences. Syndepositional fluctuations in groundwater base level and groundwater head, perhaps driven by sealevel changes and tectonism, meant many ancient evaporite systems were influenced by both marine and nonmarine brines during the time that they were accumulating evaporite salts. That is, a marine basin may have been supplied by marine seepage, overflow, or even occasional surface breakthrough along a narrow ephemeral surface inlet, but, if the basin was to accumulate thick salt beds, it was most likely a regional low surrounded by highly continental sediments and hydrologies with its margins defined by active faults. Water sources in such basins were typically a combination of seawater seepage, continental inflow, and hydrothermal spring outflows.

Recent work in Australia has shown that marine aerosols can supply substantial quantities of ions to the precipitative system in continental salt lake depressions located hundreds of kilometres from the continental margin. The Murray Basin, in southeast Australia, contains a 500m thick sequence of Palaeocene to Recent sediments deposited in fluvial to marine settings (Jones et al., 1993). The upper 100m in the central part of the basin now entrains large volumes of subsurface brines with salinities in excess of 100,000 mg/l. These type II brines have major solute compositions and proportions very much like that of evaporated marine waters. Their isotopic composition, however, is that of evaporated meteoric water. This led Jones et al. (1993) to the conclusion that the brines have been derived by the longterm accumulation of marine aerosols, mixed with continental waters. The marine salts are blown into the basin either as dry salt or meteoric precipitation. Evapotranspiration in this semi-arid basin greatly exceeds rainfall and so solutes accumulate as brackish nearsurface waters. These shallow waters are further concentrated via capillary evaporation and outflow into saline lakes. Regionally, the shallow groundwater flow regime is sluggish and not capable of dispelling these brines, so they infiltrate and accumulate above a regional aquitard.

Many of our ideas on the nature of continental evaporite brines and mineralogy have been derived, from excellent work in the high topography, "closed-basin" rain shadow lacustrine systems of the Basin and Range province of the southwest USA. There the bedrock lithology in the inflow area is a major contributing factor to variations in ionic proportions of the final brine (Figure 2.2). In contrast, in the semi-arid, low topography, more hydrologically open systems that feed into large inland playas of Australia, the more saline groundwaters often have widespread seawater-like Na-Cl dominated ionic proportions in their shallow brines. The brine chemistry varies little, even across regions of highly variable bedrock lithologies. In contrast to the closed basins of the high-relief land surface of the USA, many Australian playas are hydrologically open; the outcropping parts of local or regional discharge systems within regionally extensive aquifer systems. Aridity and deeply weathered soils have characterised these areas since the later Tertiary. Playas in such systems are locally called "boinkas" (Macumber, 1991).

Jankowski and Jacobson (1989, 1990) and Jacobson and Jankowski (1989) showed that in such systems the amount of rainfall on a given part of a catchment area is an important determinant of the chemical composition of the playa brine. In the Lake Amadeus region of central Australia, higher rainfall regions generate playa inflow waters that are bicarbonate-dominated, while lighter rainfall regions generate chloride-dominated waters. The differences in brine chemistry can be related to the rate of cycling of carbonate within the catchment profile. More carbonate is dissolved in the near surface where rainfall is high, and so it contributes a greater proportion to the brine composition. Higher rainfall regions also flush greater volumes of water through the same profile in a given time and so are characterised by shorter periods per given water volume for rock/fluid interaction. In low rainfall regions chloride tends to remain mobile in the catchment profile, while carbonate is more likely to remain fixed as carbonate minerals in the soil profile. The buildup of the more soluble salts over time in regionally extensive hydrological discharge systems can be impressive. Salt-rich soil profiles near Lake Eyre, South Australia sit atop a widespread flat-lying and deeply weathered pyritic Cretaceous shale. Much of the profile contains a pedogenic unit of epsomite, which has a thickness measured in metres and an area measured in thousands of square kilometres (Lock, 1986).

Difficulties in distinguishing a marine or nonmarine origin are further complicated by ongoing dissolution of older exhumed evaporites, which can act as chemical feeds into purely continental closed basins. This can create a marine-like signature in the ionic proportions (Table 2.3; Na-K-Mg-SO$_4$-Cl brines). For example, carnallite is a modern lake precipitate in Qaidam Basin, northern Qinghai-Xizang (Tibet) Plateau, China (Figure 7.2). This basin lies within a large, tectonically active region, totally isolated from the ocean and with basin-fill composed of clastic and evaporite sediments. Its modern at-surface discharge playas are subject to intense evaporation and are characterised by hypersaline brines as well as potash and borate evaporites (Vengosh et al., 1995). The chemical composition of the dissolved solutes in the modern brines and waters reveals three main sources. (1) Inflow of hot springs enriched in sodium, sulphate and boron. Evaporation of these waters leads to a high Na/Cl ratio (>1), a Na-Cl-SO$_4$ brine and an evaporite mineral assemblage of halite-mirabilite-borate (Lakes Daqaidam and Xiaoqaidam). (2) Inflow surface river waters, which are modified by preferential dissolution of halite and potassium and magnesium salts. Evaporation of such waters creates Na-(Mg)-Cl brines with low Na/Cl (<1), Br/Cl, Li/Cl and B/Cl ratios. (3) Ca-Cl subsurface brines with chemistries controlled by both salt dissolution and dolomitization processes. Evaporation and salt crystallisation of mixtures of brines (2) and (3) leads to a "marine-like" brine (Na-Mg-Cl brine with Na/Cl ratio <<1). Although more than 1,000 km from the nearest ocean the precipitated sequence of salts is similar to that predicted for progressive evaporation of seawater (Lake Dabuxum and Qarhan playa: halite-sylvite-carnallite-bischofite).

One way to distinguish the nonmarine origin of these marine-like brines and their precipitates is by boron isotope analysis (Vengosh et al., 1995). The δ^{11}B values of the input waters to the Qaidam Basin range from -0.7 to +10.9‰, while brines in salt lakes range from +0.5 to +15.0‰. Both groups overlap and are similar to those of associated granitic rocks (δ^{11}B = -2.4 to +3.7‰) and hence indicate the nonmarine origin of these fluids. The highest δ^{11}B values in the Qaidam Basin are associated with fluids with low B/Li ratios, indicating selective removal of elemental boron and ^{10}B by adsorption onto clay minerals. The magnitude of ^{11}B enrichment due to adsorption differs from 15-20‰, and thus nonmarine brines are easily distinguished from marine-derived brines. Seawater-derived brines have much higher δ^{11}B signatures in the range +39 to +59‰ (Figure 6.8).

Back reactions

Not just hybrid waters complicate the interpretation of brine source in ancient evaporite salts. Salts can backreact with the parent brine as they crystallise both at the surface and in the subsurface. Hence, bittern chemistry and associated brine evolution is only poorly modelled by studying brine evolution in man-made seawater ponds. In such salt works the evolved brines are separated from earlier precipitated mineral species by sluice gates and pumps. The aim is to create harvestable monomineralic or bimineralic salts and brines in the various fractionating

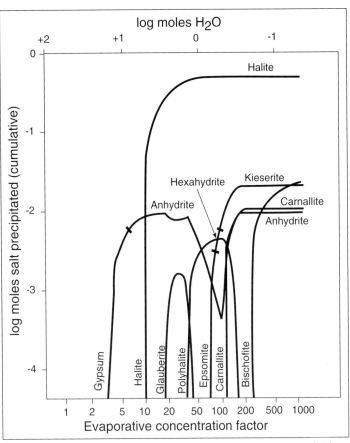

Figure 2.3. Evaporative concentration of modern seawater simulated using the computer program of Harvie and Weare (1980) showing cumulative masses of minerals precipitated and resorbed by backreaction as evaporation proceeds. Initial seawater has a concentration factor of 1, with an initial mass of 1 kg of H$_2$O (after Hardie, 1991).

pans. Backreaction either does not occur or is kept to a minimum.

In natural systems the secondary salts, such as sylvite, glauberite and polyhalite, form syndepositionally via interaction between highly evolved nearsurface brines and earlier formed minerals (Figure 2.3; Table 2.4; Harvie and Weare, 1980). For example, during the equilibrium evaporation of seawater at 25°C, glauberite and polyhalite are predicted to form by backreaction of the brine with earlier formed anhydrite and gypsum. Using a computer model, Harvie et al. (1980) explained much of the variability present in the bittern salts of the Permian Zechstein deposits in Germany via a combination of equilibrium batch evaporation interspersed with times of fractionation and backreaction. As seawater concentrated to slightly below halite saturation, anhydrite replaced gypsum. As evaporation continued, anhydrite was partially replaced by glauberite. Glauberite in turn was replaced by polyhalite

Secondary mineral phase	Syndepositional origin (<30°C)
Anhydrite $CaSO_4$	Dehydration of early formed gypsum as a_{H2O} of evaporating brine decreases.
Bloedite $Na_2SO_4.MgSO_4.4H_2O$	Back-reaction between previously precipitated epsomite and concentrating brine.
Gaylussite $CaCO_3.Na_2CO_3.5H_2O$	Back-reaction of evaporating alkaline brine with early formed aragonite or calcite.
Glauberite $CaSO_4.Na_2SO_4$	Back-reaction of evaporating Na_2SO_4-brine with early formed gypsum or halite.
Kieserite $MgSO_4.H_2O$	Dehydration of early formed hexahydrite as a_{H2O} of evaporating brine decreases.
Polyhalite $2CaSO_4.MgSO_4.K_2SO_4.2H_2O$	Back-reaction of evaporating K-Mg-SO_4 brine with early formed gypsum, anhydrite or glauberite.
Sylvite KCl	Incongruent dissolution of carnallite in undersaturated waters.

Table 2.4. Examples of diagenetic (secondary) evaporites that form easily below 30°C (after Hardie, 1990; Sanchezmoral et al., 1998).

and levels of anhydrite were further reduced. These replacements changed the residual brine composition so that kainite did not supersaturate, a problem that had plagued earlier geochemical models of marine-derived potash precipitation in the Zechstein Series. The Harvie et al. model not only explains the absence of primary kainite in the Zechstein deposits, but also explains the pseudomorphic replacement of gypsum and anhydrite by glauberite or polyhalite and the replacement of glauberite by polyhalite. Previous authors had attributed these replacements to burial-stage thermal and solution metamorphism (Borchert and Muir, 1964).

Likewise, beds of polyhalite in the Lower Werra Anhydrite (Zechstein - Upper Permian) of northern Poland, are a very early replacement of anhydrite. They formed within platform sulphates by diagenetic backreactions with highly evolved marine brines (Peryt et al., 1998). The formation of polyhalite was preceded by the syndepositional anhydritisation of the original gypsum deposit, which in places still preserves its primary textures. This anhydritisation on the platform and its slopes was a reaction of the precipitated gypsum with descending brines, fed by later drawdown basins that were adjacent to the sulphate

platform. These brines, which were potassium and magnesium rich and nearly saturated with respect to halite, reacted with anhydrite to precipitate polyhalite along the slopes of the Zdrada Platform. The oxygen and sulphur isotopic compositions of sulphate evaporites indicate that marine solutions were the only source of sulphate ions supplied to the Zechstein Basin. Anhydrite was transformed to polyhalite by reaction with marine brines, which were more concentrated than those precipitating the precursor calcium sulphates.

Thus, backreactions occur both at the sediment-brine interface and within shallow subsurface brine plumes that displace and mix with less dense underlying pore brines. Pseudomorphs and reaction rims are common indicators of this process, as are relict mineral cores encased in the mineral that replaced it. Documented examples include pseudomorphs of halite after subaqueous gypsum (Figure 1.13), reaction rims of polyhalite and glauberite around earlier gypsum or anhydrite, relicts of carnallite in secondary sylvite, and pseudomorphs of sylvite plus kieserite after earlier langbeinite (Table 2.4).

Hypersaline groundwaters can also react with adjacent nonevaporite minerals to form indicator clays and authigenic phases. For example, in reactive volcanogenic margins associated with many rift-valley evaporites there are suites of authigenic zeolites and clays. Many formed by the diagenetic reaction of nearsurface saline pore waters with metastable clays derived by weathering and alteration of labile volcanics. The best studied authigenic phases are the sodium silicates, such as magadiite; the borosilicates, such as searlesite; and the zeolites of the alkaline saline lakes, such as clinoptilolite, erionite and analcime (Table 7.10). Sepiolite (meerschaum) clays form beds and nodules in saline lake deposits in the Miocene Eskisehir Lake of Turkey. These Mg-rich clays were deposited by direct precipitation out of silica-saturated saline and alkaline lake waters (Ece and Coban, 1994). The rather unusual inflow water was derived by the weathering of a Cretaceous ophiolite complex that contained abundant serpentinite and magnesite. Many of the saline lakes in this region of

Turkey also contain widespread evaporitic micrites dominated by primary magnesite.

The potash dilemma: marine versus nonmarine

Potash evaporites, traditionally interpreted as marine salts, fall into two categories: 1) potash deposits with $MgSO_4$ salts, such as polyhalite, kieserite and kainite; and 2) potash deposits with halite, sylvite and carnallite, and entirely free or very poor in the magnesium-sulphate salts. This latter group makes up more than 60% of ancient potash deposits (Hardie, 1990, 1991). The former group may well be marine-derived as they contain the bittern suite predicted by marine evaporation and backreaction (Figures 2.1, 2.3), but the latter group must have precipitated from Na-Ca-Mg-K-Cl brines with ionic proportions quite different from that of concentrated modern seawater.

The $MgSO_4$-depleted potash evaporites have been explained by some as modified marine evaporites thought to result from backreactions during burial diagenesis of normal marine evaporites (Borchert, 1977; Dean, 1978; Wilson and Long, 1993). If so then they were derived from concentrated seawater that was altered via; a) dolomitisation, b) sulphate-reducing bacterial action, c) mixing of brines with calcium bicarbonate-rich river waters, or d) rock fluid interactions during deep burial diagenesis (see later).

Alternatively, Hardie (1990) suggested that they formed from sulphate-depleted inflow waters that seeped into the evaporite basin via springs and faults. The springs were fed either by $CaCl_2$-rich hot hydrothermal brines or by the cooling of resurging deep basinal brines. Such fault-fed deeply-circulating evaporite brines occur today in the Dead Sea, Qaidam Basin, Salton Sea and the Danakil Depression. The upwelling of the brines in these regions is driven either by thermally induced density instabilities related to magma emplacement, or by the creation of topographic gradients that force basinal brines to the surface. Ayora et al. (1994) demonstrated that such a continental Ca–Cl system operated during deposition of sylvite and carnallite in the upper Eocene basin of Navarra, southern Pyrenees, Spain.

Thirdly, there is the distinct possibility that seawater chemistry has evolved throughout the Phanerozoic and that this evolution defined whether or not $MgSO_4$ salts were typical marine evaporites (Table 2.2). Recent work by Kovalevich et al., (1998) on inclusions in primary-bedded halite from many evaporite formations of northern Pangaea shows that during the Phanerozoic the chemical composition of marine brines was oscillating significantly between the Na-K-Mg-Ca-Cl type and the Na-K-Mg-Cl-SO_4 type. The former does not precipitate $MgSO_4$ salts when concentrated, the latter does. They interpreted these changes as corresponding to the chemical evolution of the Phanerozoic ocean. They argue these oceanic oscillations also correlate in time with earlier suggested secular changes in the mineralogies of marine nonskeletal limestones and potash evaporites. In the past the concentration of the Ca-ion did not exceed its present concentration in marine water by a factor of three. The timing of the increase was synchronous with a decrease in the SO_4-ion concentration which was as much as three times lower when compared to the present concentration of that ion in seawater.

The significance of the potash association with $MgSO_4$ salts is not yet resolved.

Fluid and brine movement in sedimentary basins

After primary evaporite precipitation, ongoing pore water flow drives nearsurface and burial diagenesis, both regions where evaporites dissolve and reprecipitate as their pore brine chemistries evolve. Fluid flow also transports heat, causes dissolution and precipitation of authigenic nonevaporite minerals (including feldspars, quartz overgrowths, carbonates, clays and metal sulphides), or influences petroleum migration. Sediments filling sedimentary basins are typically inhomogeneous at scales from the microscopic to the macroscopic. Quantitative modelling of subsurface fluid flow needs adequate three dimensional description of rock properties, which should include porosity, permeability, compressibility and thermal conductivity maps for the whole sedimentary column and perhaps portions of the basement sequence (Bethke, 1989). Our present subsurface technologies and rudimentary paradigms for fluid flow modelling cannot adequately deal with this complexity. Our current basin-scale fluid flow models are qualitative or at best semi-quantitative. Nonetheless the regimes show hydrological end-members characterised by distinct relative positions, hydrogeo-chemistries, and flow dynamics.

The hydrologic framework of a large depositionally active evaporite basin consists of subsurface waters and brines moving within several hydrologic regimes: 1) the active phreatic-depositional, 2) the compactional, 3) the

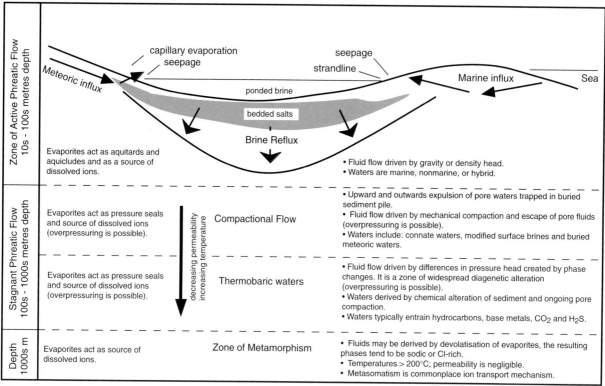

Figure 2.4. Hydrological regimes and processes in evaporitic systems.

thermobaric, and 4) the active-phreatic-exhumation/uplift regime (Figure 2.4). Boundaries are typically indistinct and transitional.

Active phreatic regime

All of the primary and syndepositional textures found in extensive ancient evaporite beds first precipitated from brines outcropping at the top of the active phreatic zone. It is the only hydrological setting that can supply the density or gravity head necessary to drive, pond and concentrate the large volumes of fluids required. The active phreatic flow regime encompasses the zones of meteoric flow, seepage, brine ponding and brine reflux (Figure 2.4). In most large evaporite seaways or lakes the surface of the brine body is the outcropping expression of the regional water table. Areas of permanent surface and nearsurface brine typically indicate a discharge zone. In arid continental settings such outcropping groundwater outflow zones create discharge playas.

Shallow subsurface brines and meteoric waters in the zone of active phreatic flow move under the influence of gravity, density/salinity or temperature gradients. Where areas of

strong topographic relief define the margin of a perennial pan or seaway, as in a continental rift valley or in a transtensional basin-and-range setting, unconfined and confined meteoric water typically flows and seeps into the edges of a brine saturated depression. Further out, in the more central and topographically lower brine-saturated parts of the basin, density-driven brine reflux, and not meteoric throughput, is the dominant process driving brine flow (Figure 2.4). In a large evaporite drawdown basin, such as in the Mediterranean in the Late Miocene, the level of the discharge zone and the surface of the adjacent brine seaway was at times located thousands of metres below sea level. Gravity seepage then drove the inflow of huge volumes of marine/meteoric waters into the depression (Figure 3.6b).

Water table levels control aggradation

Unconfined meteoric water flowing into the edges of an evaporitic depression has a potentiometric head defined by the position of the regional water table. The unconfined meteoric water table surrounding a coastal marine-seepage lake is typically a few centimetres higher than sea level. In a continental saline lake or in an ancient drawdown saline giant, any meteoric water table about the basin margin was higher than the brine level within the evaporite-accumulating depression (Figure 2.5a). Once meteoric/marine groundwater seeps into outflow zones it comes into contact with more saline and denser pore brines of the salt-accumulating central depression. On average, the outflowing waters of the mudflat or bajada margin are less dense than hypersaline waters in the brine-saturated centre. About the strandline, they typically form a shallow fresher groundwater layer floating atop denser hypersaline pore water fed in part by lateral seepage from the brine lake centre (Figure 2.5a).

Without ongoing meteoric recharge, freshened groundwater outflow dissipates via capillary evaporation in the mudflats, and alkaline earth carbonates precipitate as surface crusts or as nearsurface pore fills (Figure 2.5a). Capillary evaporation continues to lower the mudflat water table during a prolonged dry spell until flow directions may be reversed as brines seep in from the basin centre to replace the lost and drawndown pore fluids. At the same time any salts, crystallising as surface efflorescences or in the shallow subsurface of this exposed mudflat, evolve into more saline precipitates.

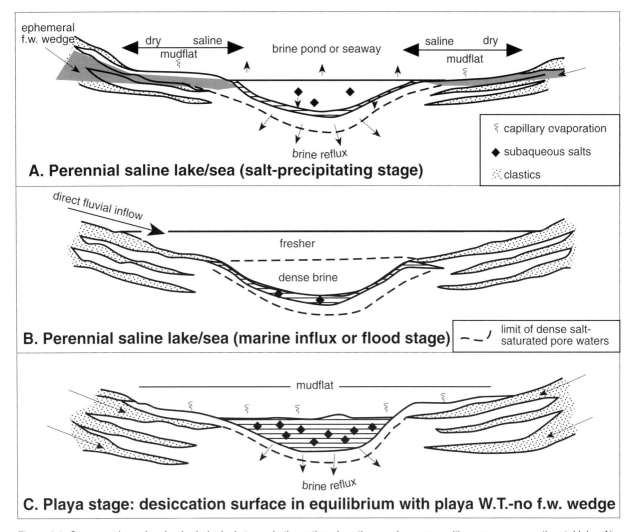

Figure 2.5. Cross sections showing hydrological stages in the active phreatic zone in an evaporitic seaway or a continental lake. A) Perennial saline lake accumulating widespread subaqueous evaporites from a nonstratified brine body. Capillary evaporation of fresher water lenses about the lake margin. B) Perennial saline lake or water full stage with subaqueous brine lens isolated from solar evaporation processes by a layer of surface water. This shuts down widespread crystallisation in lower brine body. C) Playa stage where ongoing evaporite fill, or lowering of regional water level, means the sediment surface is mostly subaerial and in capillary equilibrium with saline lake water table (W.T.). This is a classic evaporitic mudflat or sabkha hydrology. No fresh water (f.w.) lens exists until after the next flood event.

With the next inflush of meteoric recharge the water table rises in this marginward zone and pore fluids refreshen. At the same time the more saline capillary salts dissolve, along with any of the more soluble surface efflorescences (Figure 2.5b). In the central depression a wedge of freshened water floats atop a stable body of dense brine. This coverage by a blanket of freshened water atop the brine lake or pan stops further subaqueous salt accumulating on the brine bottom until the freshened wedge is destroyed by ongoing evaporation. Once the upper brine/water mass concentrates to where its density equalises that of the lower brine mass, the two bodies can mix and bottom precipitation of salt begins anew.

Thus marginward subaerial zones of most evaporite basins are schizohaline, while salinities at the brine-sediment interface within perennial brine-covered depressions tend to be more stable. Sediment textures in schizohaline zones are dominated by re-equilibration reactions, backreactions, dissolution-reprecipitation and haloturbation textures (Warren, 1991). As the whole evaporite basin aggrades to its hydrological equilibrium level (Figure 2.5c) the more central brine seaway or lake converts into an ephemeral capillary mudflat. This induces schizohaline textures in the uppermost parts of each shallowing hydrological cycle across the whole discharge zone (Figure 2.5c).

Groundwater outflow zones

Groundwater outflow or seepage zones are areas characterised by mixtures of strandline and spring deposits. They are found about the edges of large saline basins or about intrabasinal artesian outflow seeps. The outflow hydrology supplying the seep may be unconfined, artesian or hydrothermal and the feed may be marine, nonmarine or hybrid. Not only do salinities fluctuate in these marginward zones, but the water table level and the associated thickness of the capillary zone will also fluctuate. Oscillating spring-fed systems today dominate the margins of many continental and marine-margin evaporite basins. These include: Lake McLeod in Western Australia, the Marion Lake complex in South Australia, the alluvial fan and fluvial edges of lacustrine sediments in the Basin and Range of the southwest USA, and the continental side of the Abu Dhabi sabkhat of the Arabian Gulf (Lowenstein and Hardie, 1985; Rosen and Warren, 1990; Warren, 1991).

Recent work using strontium isotopes as tracers in the seepage-fed playa mudflats of Lake Tyrell, Australia, has shown that the three-dimensional interaction between the reflux process and unconfined meteoric margins is more complicated than simple downward percolation after mixing (Lyons et al., 1995). Rather, the pattern of mixing of meteoric and playa brines is a complex series of interfingers with mixing boundaries created via a number of density-driven convective plumes.

As well as open or unconfined aquifer outflow about the seepage edge of an evaporite basin, meteoric outflow, fed by confined or artesian aquifers, may also seep into the more central parts of the brine basin to form intrasalt bed outflow regions. This is especially obvious in ancient saline drawdown basins (see Chapter 3). As less dense confined waters rise through the brine body, typically along faults or inactive reefs, they create crosscutting diagenetic zones characterised by freshened pore waters and haloes of less saline minerals. This occurs where: 1) a groundwater aquifer approaches the sediment surface-as in areas of breached anticlinal folds below the lake or seaway; 2) where faults or fractures breach the confined aquifer and the overlying sediment seal; 3) where porous aquifers abut less porous basinal evaporitic sediment, as occurs during deposition of saline giants where former pinnacle reefs become seawater seeps on the subaerial floors of the desiccated drawdown seaways (Kendall, 1989).

Spring mounds, fed by artesian outflow waters in Lake Eyre, Australia, form Quaternary carbonate seepage structures that rise up to 45 metres above the surrounding hypersaline playa floor. Mounds are composed of micritic marls, invertebrate remains, algal tufa, pisolites and micrite-cemented aeolian sands and are surrounded by gypsiferous mudflats (Habermehl, 1988). The position of many of the larger spring mounds within Lake Frome, South Australia, is controlled by a north-south trending fault cutting across the eastern side of the lake (Draper and Jensen, 1976). On a larger scale, as in the ancient saline giants that characterised the Elk Point Basin in the Middle Devonian of western Canada, ancient artesian seawater outflows deposited suites of saline strandline haloes about subaerially exposed pinnacle reef crests (Kendall, 1989).

Brine ponding and reflux

Concentrated surface brines sink into the sediments by a process known as brine reflux. Reflux occurs when ponded concentrating brines atop the floor of an evaporitic seaway or lake become dense enough to displace underlying pore fluids and so percolate into the underlying succession (Figures 2.4, 2.5). Descending brines can have chemistries that are supersaturated with different mineral phases to the matrix through which they are flowing and so drive various replacement and backreaction processes such as dolomitisation and salt replacement (pseudomorphs).

The first use of evaporite-generated brine reflux was by Adams and Rhodes (1960) to explain extensive lagoonal and reefal dolomites associated with evaporites of Guadalupian age in the Permian Basin of west Texas (Figure 2.6). Reflux dolomites form when, "hypersaline brines eventually become heavy enough to displace the connate waters and seep slowly downward through the slightly permeable carbonates at the lagoon floor". When $CaCO_3$ and gypsum precipitated in hypersaline areas on such platforms, the solution density was as much as 1.2 gm/cc. The Mg/Ca ratios of the remaining brines had risen, and these dense Mg-enriched solutions sank through the underlying platform sediments (Figures 2.1, 2.6). As dense brines seeped basinward through underlying platform limestones, they displaced subsurface pore fluids that were originally marine-derived waters with initial densities around 1.03 gm/cc. In this way large volumes of Mg-rich brine passed through previously deposited shelf limestones and precipitated dolomite in the former shelf limestone both as a pore fill and as a replacement.

Their model explained the dolomitization of the Permian reef complex of west Texas; where backreef, shelf and lagoonal carbonates are intensely dolomitised, while the shelf edge carbonates, such as the Capitan Reef, are not. They envisioned dolomitization took place via brine reflux through the porous sediments to depths of several hundred metres. The feasibility of the model has been mathematically confirmed by hydrological modelling of Kaufman (1994) and Shields and Brady (1995). Garber et al. (1990) also used a model of reflux of brines, derived from the very saline Salado Formation, to explain magnesite, not dolomite, in the same basin. They argued that when a brine has a Mg/Ca ratio > 40 it will precipitate a magnesite replacement rather than the more commonly observed dolomite. Similar reflux processes, driven by the deposition of Zechstein evaporites, have precipitated authigenic magnesite cements in siliciclastic dune sands of Rotliegende reservoirs in the North Sea (Purvis, 1989).

Most brine reflux models published to date utilise a platform evaporite setting. Examples include: the Cambrian Ouldburra Formation in the Officer Basin, Australia (Kamali et al., 1995); the Devonian Birdbear (Nisku) Formation of Canada (Whittaker and Mountjoy, 1996); Permian San Andres Formation, USA (Leary and Vogt, 1986; Ruppel and Cander 1988; Elliott and Warren, 1989); the Permian Tansill and Yates Formations of the Central Basin Platform, USA (Andreason, 1992; Garber et al., 1990); the Jurassic Smackover and Haynesville Formations of the Gulf of Mexico (Moore et al., 1988); the Lower Cretaceous Edwards Formation of Texas (Fisher and Rodda, 1969) and the Middle Palaeocene Beda Formation, of the Sirt Basin, Libya (Garea and Braithwaite, 1996). Hydrological geometries of reflux dolomite associated with drawdown basins are different and are discussed in Chapter 3 after the various styles of saline giant are presented.

Reflux dolomitization in a platform setting can often be distinctive, even in core, as the intensity of total dolomitization decreases the further down core or laterally one moves from the evaporite-carbonate contact (Figure 2.7). If not completely re-equilibrated during later burial dolomitization, the oxygen isotope signatures in platform dolomites formed by reflux typically follow a lightening trend away from the evaporite unit; that is, the further one moves from the contact, the less was the sediment's isotopic signature influenced by the heavier reflux-

landward   basinward

Reflux of dense brines
dolomitises adjacent carbonates

platform evaporites   platform carbonates

100 m

100s - 1,000s of km

Figure 2.6. Schematic illustrating the process of brine reflux creating secondary dolomite in a carbonate platform (after Warren, 1991).

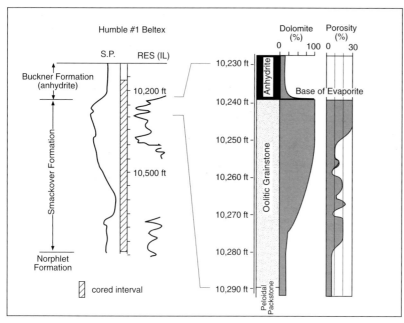

Figure 2.7. Wireline signature and the interpreted core description of the Humble #1 Beltex Well, Bowie County, Texas. Shows a close relationship between Buckner anhydrite and dolomitisation in the underlying Smackover Formation. Note changes in proportion of dolomite and porosity in terms of proximity to evaporite contact (after Moore et al., 1988).

derived brines. If the sediments are not extensively recrystallised by later burial dolomitization, trace elements, along with stratigraphic position, can give another clue to a reflux association. In a study of dolomite from Neogene to Permian age, Sass and Bein (1988) found that evaporative dolomites associated with gypsum/anhydrite tend to show 50-57% $CaCO_3$ and the highest levels of sodium (up to 2700 ppm). Marine nonevaporitic subtidal dolomites have a similar range of compositions, but sodium concentrations of only 150-350 ppm.

A more recently recognised effect of reflux in a saltern system is to drive backreactions and replacement processes in salt beds beneath an evolving brine body. Hovorka (1992) recognised reflux of halite saturated brines through gypsum beds as the dominant control in the formation of halite pseudomorphs of subaqueous gypsum (Figure 2.8). The whole process is driven by the fact that gypsum deposition is favoured by low temperatures and low salinities, while anhydrite precipitation is favoured by higher temperatures and higher salinities (Figure 2.8). At the warm temperatures that typify modern low-latitude brine pools (25 to 50°C), the gypsum-to-anhydrite transition is predicted to occur somewhere between the initial precipitation of gypsum and initial precipitation of halite. In contrast, brines at halite saturation are undersaturated

with respect to gypsum, while saturated with respect to halite. Textural relationships in such settings where bottom-aligned gypsum is replaced by halite are shown to be controlled by nearsurface diagenesis and related to brine pool evolution stages 1, 2 and 3 (Figure 1.13; Hovorka, 1992). During Stage 1 the perennial brine body is four to six times the concentration of seawater and is at gypsum saturation (Figure 2.8). In stage 2 brines are at the upper end of the calcium sulphate field (6-10 x seawater) and are in the anhydrite stability field. Metastable gypsum continues to precipitate on the brine pool floor, but reflux of this brine converts gypsum to nodular anhydrite. By stage 3 the surface brines lie in the halite saturation field (>11x seawater) so that halite is precipitating on the floor of the brine pool. The refluxing brine is now undersaturated with respect to gypsum so that it is dissolved in the shallow subsurface and the resulting voids are filled by halite cements. This replacement process takes place in the first few metres of burial. As thick halite beds are tightly cemented by 100m of burial, surface-driven brine reflux and associated backreactions in ancient salt beds are most intense in the first few tens of metres of burial.

Permanency of the reflux system

Dense refluxing brine plumes and stagnant brine reservoirs beneath areas accumulating bedded evaporites are not readily displaced by a lateral influx of less dense meteoric groundwater (Figure 2.5a). Rather any fresher water inflow tends to float atop more dense brines until it is dissipated by evaporation or concentration and brine mixing. The volume of rainfall/freshwater required to drive displacive meteoric-dominated shallow unconfined-phreatic wedges into these saline depressions, versus the longterm aridity required to accumulate and preserve widespread surface and thick nearsurface salt beds, means the two hydrologic subregimes cannot easily coexist within the salt-accumulating areas of the basin.

The occasional surface input of freshened waters into evaporitic seaways (salterns) or perennial saline lakes tends to create stratified surface water bodies with less-

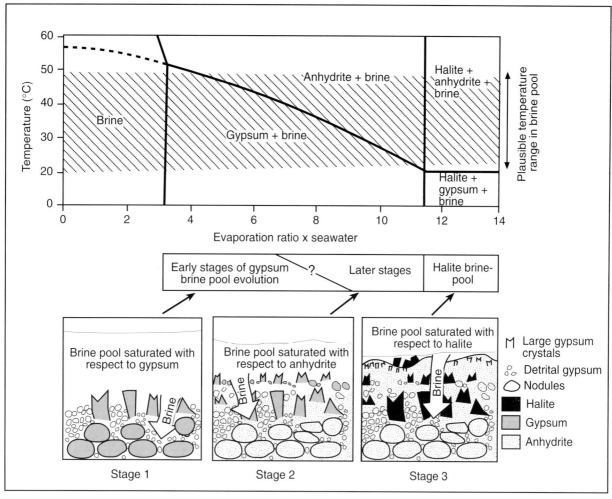

Figure 2.8. Gypsum replacement by halite during brine reflux. Upper part shows phase relationships between gypsum and anhydrite in NaCl brine. Lower part shows how the alteration and replacement is dependent on the evolution of the overlying brine supplying waters that are refluxing through the gypsum bed. Stage 1: Large gypsum crystals grow vertically from the floor of the brine pool, and gypsarenite accumulates between the crystals. Gypsum is stable in diagenetic fluids concentrated three to approximately six times seawater concentration: alteration to form nodules occurs in shallow subsurface environments. Stage 2: As salinity increases from six to ten times seawater concentration, gypsum continues to precipitate metastably, but is dehydrated to thermodynamically stable anhydrite in the shallow diagenetic environment. Stage 3: When the brine pool evolves to concentrations about 10 times seawater concentration, halite begins to precipitate. Halite-saturated brines introduced into the shallow burial environment dissolve any remaining gypsum and replace it with halite (after Hovorka, 1992).

dense lower-salinity waters floating atop more-dense saline bottom brines (Figure 2.5b). If there are large salinity contrasts between the two surface water masses, they tend not to mix until the densities of the two water masses equilibrate. Such seasonal brine stratification is observed today in many coastal salinas in Australia, in perennial saline lakes in the African Rift Valley and in the Dead Sea, Israel.

Lake Hayward, a coastal salina in Western Australia, typifies such a stratified brine system and has ionic proportions in the bottom brines similar to that of seawater. Carbonate muds are accumulating on the microbially bound floor of its small perennial brine pool (2-3 m deep) under a limnology that is density and thermally stratified for much of the year (Figure 2.9; Rosen et al., 1995). Meteoric inflow creates a well-defined longterm mixolimnion; this is the upper, less dense and cooler water mass in the lake, with salinities ranging from 50,000 to 210,000 ppm. It exists from late autumn to early summer (May to February) as it floats atop a lower denser and warmer water mass (monolimnion) with salinities in the range 150,000 to 210,000 ppm (Figure 2.9).

Stratification disappears for a few months from midsummer to mid-to-late-autumn. During the summer the waters of the upper water mass evaporate and concentrate as their bicarbonate content steadily increases. From the onset of stratification in late autumn, across mid winter and on into to early summer the temperature trends of the two water masses are parallel. They form a heliothermic system where the lower water mass is some 15-20°C hotter than the upper water mass. By mid summer (e.g. January 1992) lower water mass begins to cool, while the temperature of the upper water mass continues to rise. Once the temperatures (and densities) of the two water masses equalise they mix as the lake overturns.

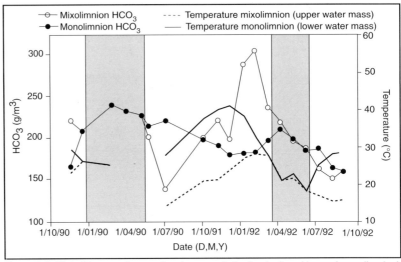

Figure 2.9. Plot of temperature regime in upper (mixolimnion) and lower (monolimnion) water masses and bicarbonate content in Lake Hayward, Western Australia. As the lake mixes, the thermal and density stratification disappears. Immediately prior to mixing the bicarbonate concentration decreases suggesting precipitation of calcium carbonate (aragonite) (after Rosen et al., 1995).

Waters of the lower water mass (the monolimnion) now come into contact with the atmosphere once more. While the water masses are stratified, the chemocline and the thermocline are sharply defined across a 10cm interface with a salinity contrast that may be as much as 135,000 - 140,000 ppm and a temperature difference of up to 19 °C.

The time of mixing is immediately preceded by a sharp fall in the level of bicarbonate in the mixolimnion, suggesting the precipitation of calcium carbonate (mostly aragonite) occurs in the late summer to autumn. After mixing the stratification begins to reform with the next influx of meteoric waters onto the lake surface. In the early stages of setup leading into the longterm stratification it appears stratification is not stable and the water bodies may remix before stable stratification sets in by mid winter (e.g. May-June 1992; Figure 2.9).

Although its daily temperature/stratification record is probably the best documented example of the hydrology of a naturally stratified brine pool, Lake Hayward has a number of limitations when attempting to use it as an analogue for the hydrology of ancient stratified brine seaways. Unfortunately Lake Hayward is not saline enough to accumulate substantial thicknesses of salts more saline than aragonite, which it accumulates as benthic microbially bound micrites (Coshell and Rosen, 1994). It is not a hydrological analogue for bottom-nucleated salts, such as gypsum or halite, that dominate the central portions of ancient evaporite seaways and lakes. Hayward also suffers

from the scale limitations that all modern marine-associated evaporite analogues suffer from (see Chapter 3). Its small aerial extent (≈ 0.6 km^2) prohibits a direct hydrological comparison with bedded salt accumulations in ancient saline giants (areas $> 10^6$ km^2). Its centripetal meteoric inflow hydrology, driven by a high annual rainfall, means it cannot be directly compared to ancient brine pools where strandlines moved up to hundreds of kilometres in a wet-dry cycle and where unfractionated meteoric waters were much less significant in the freshening process.

In ancient evaporitic seaways and lakes the style of hydrology is best interpreted from the preserved depositional textures (Lowenstein and Hardie, 1985; Warren and Kendall, 1985). If stratified brines set up atop a growing crystallisation surface in an ancient seaway, solar driven crystal growth of gypsum or halite in the bottom waters slowed or stopped. Once the water body was stratified, no solar mechanism was available to concentrate the bottom waters further (as in Lake Haywood today). Subaqueous crystallisation could only continue in zones of brine mixing near seeps. Once freshened inflow ceased, ongoing evaporation concentrated the upper brine body until its density matched that of the lower brine body. The two water bodies then mixed and crystal growth on the bottom could begin anew. This type of bottom crystal precipitation, where the crystallisation surface was never in contact with an undersaturated bottom water, deposited euhedral growth aligned crystals on the sediment floor (e.g. chevron halite or aligned swallowtail gypsum).

If, on the other hand, the freshened water body did come into contact with the aggrading crystallisation surface of the salt bed then horizontal planation surfaces developed at the aggrading salt bed surface. This occurred both in saline basins subject to complete desiccation and in those parts of evaporite seaways covered by perennial brine sheets (tens of cm deep) that were subject to periodic freshening (see Figure 1.12). A desiccation stage, followed by a freshened water flood, creates a near horizontal planation surface with intercrystalline karst, which marks a time when the water table was lowered into the salt bed immediately prior to a subsequent freshening event (trona in Lake Magadi and chevron halite crusts on the Baja peninsula; Shearman, 1970). In contrast, freshening of perennial bottom brines creates a horizontal planation surface atop the salt bed with little or no intercrystalline karsting (laminated coarse-grained gypsum in Marion Lake; Warren, 1982b).

In both cases the freshened surface water sheet only dissolves the uppermost portion of the salt bed, it does not displace the much denser pore brines that saturated the bulk of the underlying salt bed (Figure 2.5c). A planation surface defines the top of saturated pore brines. Vadose karst in a desiccated salt pan can only extend down to the top of this dense pore brine layer. Below it the pore brines are saturated, above it the waters are undersaturated or nonexistent.

These flat, laterally extensive dissolution surfaces typically truncate the tops of previously growing crystals, be they trona, halite, gypsum or any other bottom nucleating salt (Figure 1.12). As the freshened water body begins to concentrate it may precipitate a less saline mineral phase, which settles onto the planation surface. It may also be the time when clays, formerly suspended in the flood waters, begin to flocculate and sink through increasingly saline surface waters. As the freshened water body continues to concentrate into salinities that precipitate the dominant salt phase the aggrading crystals that underlie the planation surface can now poikilitically enclose the less saline precipitates (e.g. laminae of aragonite pellets in laminated coarse-grained gypsum; Warren, 1982b). In the potash-depositing playas of Qaidam Basin the oxygen isotope signature of inclusions in chevron halite record multiple episodes of freshening and concentration during the accumulation of each salt bed (Yang et al., 1995).

The likelihood of more invasive nearsurface meteoric flushing, leaching and interaction with bedded salts increases where an advancing alluvial wedge progrades over a shallowly buried evaporite bed. Meteoric waters prograde with the wedge and can dissolve, alter and backreact with underlying and adjacent bedded salts. The tight impervious nature of the shallowly buried salt means most of the reactions occur near the periphery of the evaporite bed. This leads to karstification and backreaction mostly within a few metres of the dissolving salt edge. Syndepositional diagenesis associated with prograding alluvial wedges is most intense beneath the strandline adjacent to the alluvial-wedge.

In the potash evaporites of NE Thailand this set of processes has altered the salts at the contact between redbed clastics and the underlying halite/carnallite units. It, and the associated karstification, locally converted carnallite to sylvite, with nodular borates as one of the byproducts. This prograding wedge of alluvial fines also loaded the bedded evaporites causing them to flow into salt pillows. Pillow growth, in turn, created topographic highs on the lake floor, which then influenced the local groundwater flow and potash brine evolution and ponding (Figure 7.11b).

Groundwater controlled truncation surfaces

Because evaporites form from discharging evaporating brines and groundwaters, any longterm water table fluctuations in an evaporite basin can influence sedimentation style and mineralogy on scales coarser than intrabed truncation layers. This is really a longer term reflection of the same dissolution/aggradation surfaces described in the preceding section and in Chapter 1. Longterm aggrading water tables (or regional subsidence) enable the capture of formerly vadose sediment by the damp, sticky to wet conditions that characterise the capillary and phreatic zones and so facilitate the retention of formerly vadose and nonevaporite sediment in the depositional system - sediment that would otherwise be blown away. Falling water tables enable the entry of vadose environments into formerly phreatic sediments and the dissolution of salt beds or aeolian deflation of sediments that come to lie above the water table.

Where such changes in the base levels of erosion and deflation have been recognised in aeolian sands they are called "Stokes surfaces" (Fryberger et al., 1988). Where they are recognised in mudflat or sabkha sequences they create a well defined surface that defines the top of the capillary zone of each sabkha depositional cycle (Warren and Kendall, 1985; Warren, 1991). If packages of sediments aggrade and stack as the longterm water table rises (or the basin floor subsides), then a succession of such large scale truncation surfaces can be preserved atop a number of depositional cycles (e.g. truncation surfaces in Figure 1.5).

Locality	Geological setting of Stokes surfaces	References
Arabian Gulf sabkhat United Arab Emirates	Coastal saltflats prograding into Arabian Gulf	Patterson and Kinsman, 1981 Warren, 1991
White Sands dune field, New Mexico, USA	Continental saltflat and gypsum dunes deflated from Lake Lucero	Loope, 1984 Fryberger et al., 1988
Jafurah Sand Sea, Saudi Arabia	Interdunal saltflats in sand sea (erg) migrating into the Arabian Gulf	Freyberger et al., 1988
Guerrerro Negro dune field, Mexico	Coastal barchan dunes migrating landward over lagoonal sediments	Freyberger et al., 1988
Great Salt Lake desert, Utah, USA	Continental dunes and lacustrine saltflats	Stokes, 1968

Table 2.5. Some examples of modern capillary deflation surfaces or Stokes surfaces.

Truncation surfaces are easily recognised in aeolian systems by the sharp planation of aeolian crossbeds with only a thin veneer of overlying sediment (Fryberger et al., 1988). In sabkha sequences the preserved erosion surfaces are an integral part of the capillary hydrology that controls deposition and will often truncate anhydrite nodules or enterolithic folds in the upper parts of the supratidal package (Warren and Kendall, 1985). An aggrading water table in sand-rich aeolian environments promotes precipitation of evaporite cements in the capillary zone and the upper part of the phreatic zone. Such salt-cemented units are sometimes called salcretes. Fryberger et al. (1988) record massive phreatic gypsum cementation zones in the White Sands dune field, New Mexico, that act as a limit to aeolian deflation and so create extensive near-horizontal planation surfaces.

Recognition of such surfaces has broader scale implications in terms of regional hydrological evolution over time frames measured in thousands of years. Most areas of formation of modern Stokes surfaces are either a coastal groundwater response to the last rise in sea level some 6,000 years ago, or a hydrological response in deserts related to marked climatic changes from the Late Pleistocene to the present, usually toward increasing aridity (Table 2.5). The former indicate an aggrading phreatic hydrology while the later indicate a degrading hydrology.

One of the most detailed hydrological studies of the hydrology associated with increasing aridity and a degrading hydrology was conducted on the inland playas of Central Australia in and around Lake Amadeus (Jacobson et al., 1989). There a longterm reduction in groundwater levels associated with increasing aridity led to a change in the status of some of the playas from discharge playas to recharge playas via a process known as playa capture.

It operates when a fall in the level of regional groundwater, perhaps driven by the deepening of one playa floor leads to the removal of one or more nearby playas from the discharge level of regional groundwater outflow. The concept is based upon studies of discharge playas near Curtin Springs in central Australia (Figure 2.10; Jacobson and Jankowski, 1989). Groundwater head is known to have decayed over several thousand years and has changed the nature of sedimentation. A few thousand years ago Samphire Lake was a discharge playa and it still retains thick deposits of groundwater-derived lacustrine gypsum. Today the discharge system is no longer active in the lake and the regional water table lies metres below the playa floor. Samphire Lake has become a vegetation-covered recharge playa where salts are no longer accumulating. Today with lowering of the water table the regional discharge is focused into Spring Lake and Glauberite Lake. This rendered the Samphire Lake surface sediment susceptible to aeolian deflation and vegetation coverage. Over time Glauberite Lake will become the primary focus of groundwater activity in this region and will ultimately form a regional groundwater sump.

Longterm changes in groundwater head can result in a redistribution of salt accumulating sinks within a sedimentary basin undergoing a lowering of the hydrological base level. Such changes in hydrological base level also have implications for the distribution of marginal versus basin-centre evaporite accumulations in saline giants undergoing regional drawdown. They also create regional mega-surfaces or bounding surfaces that are recognisable on seismic sections through continental and marginal marine sequences. If they are incorrectly interpreted as eustatically induced contacts, rather than watertable or climatic responses, they can greatly complicated any attempt at age fitting a sealevel curve to the sequence stratigraphy. Both these topics are discussed further in Chapter 3.

Economic implications of the active phreatic zone

The subsurface contact between the reflux hydrology of the sabkha system and its more marginward meteoric system has been proposed as a possible control on copper precipitation (Kyle, 1991). Renfro (1974) in a study of the ore-forming mechanisms in the Zambian Copper Belt pointed out that Precambrian sabkha sediments host the ore. Modern coastal sabkhas are nourished by surface and subsurface flow of landward migrating, low-Eh high-pH seawater and by seaward migrating, high-Eh low-pH meteoric or continental water. Commonly sabkhas are bordered on the seaward side by intertidal mudflats and lagoons that are covered by sediment-binding cyanobacteria. As the Precambrian sabkha migrated basinward, continental groundwater from the hinterland moved laterally and upward through buried intertidal microbial mats containing strongly reducing pore fluids. The inflowing continental water initially would have been acid and oxidised and thus able to transport iron and base metals as chloro complexes. Sulphide minerals precipitated as interstitial phases once this metalliferous fluid passed through the hydrogen-sulphide-charged mats.

Renfro (1974) noted that mineralisation in the Zambian Copper Belt was dependent on the quantity of available reductant, duration of sabkha process, and chemistry of the metal-bearing meteoric water. Garlick (1981) published a modification to the diagenetic model of Renfro, in which he proposed that metalliferous groundwaters flowed laterally across the sabkha through the intertidal zone to discharge offshore where sulphidic microbial mats fixed metals in the lower intertidal and subtidal zones.

Both Renfro's and Garlick's syndepositional ore genesis models have been questioned. Significant quantities of metals need to be supplied within the short times available for individual depositional cycles of up to 10,000 to 200,000 years per cycle (Eugster, 1985). To overcome this time/flux problem Ferguson and Skyring (1995) postulated that some ancient evaporite-associated active phreatic ore deposits involved a preconcentration stage where metalliferous redbed groundwaters evaporated in a discharge playa or marine continental sabkha prior to downgradient flow into the coastal sabkha outflow zone. Haynes and Bloom (1987) used simulation models to demonstrate that nearsurface mineralizing waters of sabkhas were probably end-members in a spectrum of ore-bearing subsurface fluids capable of forming stratiform copper deposits. The fO_2 in such waters was calculated to be in the range 10^{-35} to 10^{-55} bars ($\approx 10^{-30}$ to 10^{-50} Pa) and capable of transporting 100s to 1000s ppm Cu, Co, Fe.

The ability of active phreatic waters to transport metals can be seen in hydrological studies centred around Lake Tyrell in the Murray Basin (Fegan et al., 1992; Lyons et al., 1992). There the highest metal concentrations occur in the low-pH, high-Eh regional groundwaters that feed the oxic, metal-rich, low-pH groundwater seeps about the margin of

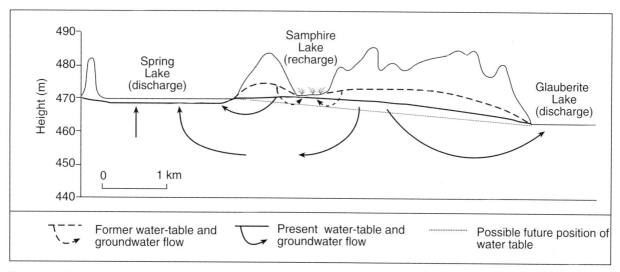

Figure 2.10. Schematic of playa capture or abandonment via decay of groundwater head. Samphire Lake was once a discharge playa (dashed line indicates former water table) that was accumulating gypsum, but it was converted to a recharge playa as Glauberite Lake progressively captured the regional groundwater flow system (after Jacobson and Jankowski, 1989).

this discharge playa. Much lower metal values typify reflux brines beneath the lake salt pan. The source of these metals in the regional groundwaters is thought to be the dissolution of a suite of heavy minerals within the regional aquifer, the Pliocene Parilla Sand. As acid groundwaters flow toward the lake centre, some metals are removed near the lake edge. An analysis of leachates from the sequential extractions within the top 70 cm of the lake floor sediments indicates that most of the Cu, Pb, Zn, Fe and Mn are adsorbed onto Fe-oxides. Copper, and to a lesser extent Mn, are bound to organics or sulphides in discrete, thin layers, while Pb is deposited on exchangeable sites in upper layers, possibly as sulphates. Overall, total metal concentrations are not enriched in the lake sediments, which may be due to low pHs and/or limited sulphate reduction in the system. The metal concentrations in the spring outflow region indicate an extremely dynamic system, where, due to changes in the hydrologic regime, metals can be either removed or solubilized at different times of the year. According to the authors, the Lake Tyrell system may represent an ore-forming system immediately prior to extensive precipitation and accumulation of metal sulphides.

Deep brine reflux beneath evaporite-forming areas was proposed by Lange and Murray (1977) as a mechanism for displacing the underlying deep warm basinal brines that would then flow upward and outward to precipitate MVT ore about the basin margin. However, when explaining stratiform base metal accumulations, such localised unidirectional syndepositional flow models of metal remobilisation in the active phreatic zone are not as popular as they were in the 1970s. Many current genetic models give lesser importance to the active phreatic regime in terms of the precipitation of large volumes of base metals. Instead they stress deeper regional or basinal fluid flow systems that are associated with the compactional and thermobaric regimes, especially when coupled with convective flushing and deep evaporite dissolution (Chapter 8).

Uraniferous caliche ores in Western Australia can, however, be fully explained by mobilisation and evaporative precipitation in the zone of active phreatic flow associated with saline pans. The driving mechanism is capillary evaporation. Uraniferous calcretes in Western Australia are typically found in evaporative discharge zones. Groundwater carries oxidised metal complexes, including uranium, as it weathers adjacent lateritic terrains that were derived from radiogenic or "hot" granitic basement. Subsequent partial evaporation of this slow moving groundwater concentrates the uranyl ions by decomplexing the soluble uranyl-carbonate complexes, which in turn promotes the formation of carnotite-rich capillary calcrete (Mann and Deutscher, 1978). Because they are evaporatively derived, uraniferous calcrete deposits tend be found in capillary zones in seepage regions where the water table approaches the land surface. Seeps are often located in stream channels about the margins of saline seeps and lakes. Such areas pass downdip into stratiform subsurface capillary zones where soil gypsum, and sometimes displacive halite, are accumulating.

Compactional regime

The zone of compactional water lies beneath the zone of active phreatic circulation in an active sedimentary basin (Figures 2.4, 2.11a). The compactional regime is characterised by the upward and outward expulsion of pore waters formerly trapped within the compacting sediment pile. Compactional waters include connate waters, modified brines that were buried with the compacting evaporitic sediment pile, and meteoric waters buried within shales, sand or carbonate aquifers.

Compactional fluid flow is an early-burial fluid migration process in many young subsiding basins where it dominates sections shallower than 1,500-2,000m with ages of up to a few million years. Differences in pressure head between adjacent areas are the main driving force of compactional flow. During the earliest stages of sediment compaction, most compactional fluids tend to move vertically upward through still porous overlying strata (Magara, 1986). Then, in the later stages of compaction, as interbedded clays become relatively impermeable, the principal direction of fluid migration may become stratigraphically controlled by interbedded permeable beds composed of less compactable cemented sandstones or carbonates.

In contrast, a basin entraining an evaporite bed contains a highly effective aquitard/aquiclude almost as soon the salt is deposited. Evaporites are impermeable once buried to depths of a few hundred metres. If water-saturated sediments are compacting beneath a thick unit of bedded evaporites, it deflects the otherwise upward flow of compacting waters. Salt beds focus fluid flow laterally along their undersides (Figure 2.11a). Much escaping compactional water is initially undersaturated with respect to the evaporite salt. The underside quickly becomes a zone of salt dissolution and a site of insoluble residues. Escaping compactional water is forced to flow obliquely toward the basin margin along the underbelly of the dissolving evaporite bed. At the same time ongoing salt dissolution increases the salinity of the laterally flowing waters (Figure 2.11a). This has two effects: 1) the formation of stratiform diagenetic basal beds composed of insoluble residues (anhydrites, clays and carbonates) and 2) the set up of early sub-salt fluid convection.

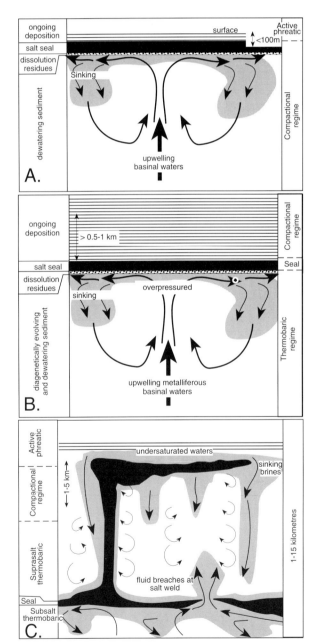

A.

B.

C.

Figure 2.11. Brine flow regimes. A) Compactional regime. B) Thermobaric regime. C) Flow regimes in diapiric province, also shows potential region for escape of hot possibly metalliferous fluids at breach zones in flowing salt bed (salt welds, growth faults, fault welds, etc.).

Lateral flow of escaping basinal waters can continue beneath the salt bed until the salt bed is breached or the laterally flowing waters reaches the edge of the salt unit. Flow may then once again become strongly vertical. As a general rule of thumb, thick evaporite units in a subsiding basin will focus and pond upward-escaping compactional waters and restrict regions of brine escape to areas atop fractures, dissolution-induced breaches, or dissolution edges of the evaporite unit. Halokinesis through its inherent thickening and thinning of bedded salt may help focus fluid escape to breached salt withdrawal basins between growing salt anticlines and diapirs (Figure 2.11c).

Thermobaric regime

Thermobaric fluid flow occurs in the deeper part of a sedimentary basin where the temperatures and pressures are greatest. In its upper portion, the thermobaric zone is intimately related to compaction flow (Figure 2.11b, c). It is the zone of diagenetic alteration where significant volumes of "new" thermobaric water may be released by dehydration reactions of clay minerals or by dehydration of hydrated evaporite minerals such as gypsum or carnallite. Fluids in the thermobaric regime move in response to pressure gradients generated by phase changes, such as the generation of hydrocarbons, or the release of mineral-bound water, or ongoing lithostatic loading. Density contrasts created by salinity or temperature gradients also drive convective flow of pore water. Chemical changes in the sediment matrix under this regime mean thermobaric waters frequently contain methane, carbon dioxide and hydrogen sulphide derived from the thermal alteration of organic matter, as well as metals leached from basinal shales. The low permeabilities of compacted sediment lying between the impervious evaporite beds in this regime mean typical rates of fluid flow are slow (mm-dm/year).

Evaporite brines as lubricants and catalysts

There is an interesting effect as a thick gypsum bed is buried and passes into the thermobaric zone. As the temperature rises above 60°C gypsum is transformed to nodular anhydrite. In a hydrologically open compacting system the process is complete by a depth of 1000 m, corresponding to a total pressure of ≈300 atm (30 MPa). Transformation may occur at much shallower depths if the pore fluids are saline, or may not be completed until much greater depths if the unit is overpressured.

As the gypsum is transformed to anhydrite, it releases water and so creates a water-filled porosity equivalent to

49

38% of the original gypsum volume. Such water ultimately seeps into adjacent strata and escapes from the unit. This escaping water of crystallisation can lubricate nearby growth faults and other relatively shallow slippage features, including nappes. If the water of hydration cannot drain freely from the dewatering gypsum, the bed converts to a quicksand-like consistency (Warren, 1991). Based on the common presence of anhydrite in décollement sutures in the European Alps and elsewhere (e.g. Jordan and Nuesch, 1989; Jordan et al., 1990) it seems a common effect of sulphate dehydration is an initial decrease in the intrinsic strength of the $CaSO_4$ unit and an enhanced ability to act as a lubricant.

Laboratory testing of mechanical strength during the dehydration of gypsum ($CaSO_4.1/2H_2O$) under closed drainage supports these observations (Figure 2.12). There is a large decrease in differential stress under undrained conditions at temperatures $\approx 110°C$, indicating high pore pressures induced by gypsum dehydration (Heard and Rubey, 1966; Murrel and Ismail, 1976). Thus a non-draining dehydrating gypsum bed can act as a zone of weakness or décollement. However, in the laboratory the reaction product is bassanite not anhydrite; a mineral that, like dolomite, is very difficult to manufacture under the kinetic restraints of the laboratory, but is abundant in the real world.

Building on earlier work but now using varying degrees of pore fluid drainage, the experiments of Olgaard et al. (1995) and Ko et al. (1995) clearly show that under well drained conditions the weakening of the gypsum tends to occur only in the early stages of the dehydration reaction. Their experiments were run at varying temperatures (23 - 150°C), confining pressures that mimic free to limited drainage (0.1-200 MPa), and strain rates ($7x10^{-7}$ to $6x10^{-5}$ s^{-1}). As more porosity was created by ongoing dehydration there followed a rapid decrease in pore pressure. The experimental peak in pore pressure corresponded to an observed weakening and embrittlement in the gypsum. Under drained conditions this weakening occurred in the first 1% of the reaction. Ultimate

strength was recovered within 3% of the reaction and ultimately exceed that of pure gypsum. They concluded that, unless the surrounding rocks had very low permeabilities, the pore pressures during real world dehydration could only equal or exceed lithostatic pressures during the earliest stage of the gypsum dehydration reaction.

Microstructural observations of Ko et al. (1995, 1997) showed that the initial stage of loss of water of crystallisation creates isolated areas of dehydrated gypsum surrounded by new fluid-filled porosity. Excess pore pressures develop as long as this porosity is not interconnected or fluid is produced faster than it can escape through any newly forming interconnected porosity. Under slow fluid leakage conditions an excess pore pressure first develops, which leads to weakening; but this is followed by strengthening (shown by time arrow in Figure 2.12). This timing of the development of minimum strength and maximum pore pressure is controlled by the drainage evolution of the various mosaics or networks of dehydrated gypsum within

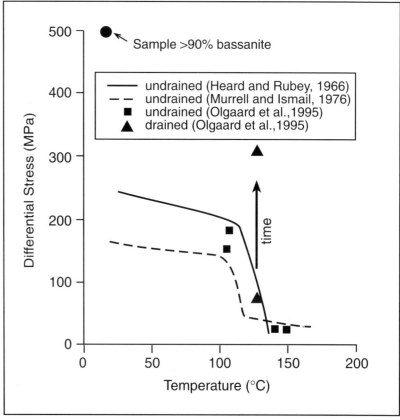

Figure 2.12. Plot showing differential strength of dehydrating gypsum decreases above 110°C in undrained conditions where the water of crystallisation cannot freely escape. Under drained conditions there is an initial decrease in differential strength followed by an increase. This is shown by samples at either end of time arrow (after Ko et al., 1995).

the converting gypsum unit. While the dewatered regions are isolated, a pore pressure can build up, which, as it becomes more pervasive, weakens the gypsum. But, as the locally dehydrated intervals start to fuse with each other, pore fluids can now flow from one area to another and ultimately escape into the adjacent non-calcium sulphate surrounds. This strengthening of the dewatering gypsum over time marks an internal transition from undrained to drained behaviour.

As well as acting as a lubricant, escaping water of crystallisation can act as a catalyst in dissolution-reprecipitation reactions within units subjacent to the dewatering evaporites. As buried gypsum dewaters to anhydrite the escaping water is saturated with respect to $CaSO_4$ but undersaturated with respect to other evaporite minerals, including any nearby carbonates (Sass and Ben-Yaakov, 1977) as well as any halite and the bittern salts. As such water seeps through the surrounding rocks it can leach minerals to create secondary porosity or bring about mineralogical transformations. Throughflow of such waters dissolved marine carbonate grains to form anhydrite-filled moulds during Permian brine reflux in the shallow subsurface of the Palo Duro Basin, Texas (Bein and Land, 1982).

With complete dewatering, a gypsum bed that is 100 m thick will convert to a 62 m anhydrite bed (Borchert and Muir, 1964; p. 133). This releases 48.6 m^3 of water per square centimetre of cross-sectional area. At 30°C this volume of $CaSO_4$-saturated water can dissolve 8.1 m^3 of halite or 54 m^3 of carnallite or convert 81 m^3 of carnallite to sylvite:

$$KMgCl_3.6H_2O + 4H_2O \longrightarrow KCl + Mg^{2+} + 2Cl^- + 10H_2O$$

Similar fluid rock interactions in the subsurface obviously influence many of the minerals and burial cements we now see in evaporites and adjacent beds. Basinal fluids derived from dissolving evaporites may also exert a substantial influence on the diagenetic and metallogenic evolution of the sediment pile (Chapters 8 and 9).

Burial dissolution and brine convection

As water-bearing sediments compact in a sedimentary basin, their connate or subsurface waters flow toward the sediment surface. The maximum solubility of halite in water is around 250-300‰ so that salt beds flushed by these escaping waters or by seawater-derived pore fluids are liable to dissolution. Ongoing dissolution of the underside of the evaporite bed in both the compactional and the thermobaric regimes generates various dissolution

residues. It also generates convective brine flushing within underlying or adjacent strata (Figures 2.11 and 2.13).

Undersides of thick evaporite salt beds tend to be dissolved by undersaturated basinal waters that pond or flow beneath it. Over time the ongoing dissolution forms a dense subsurface brine halo or curtain directly beneath the seal. The brine becomes so saline and dense that, even as it starts to move laterally beneath the underbelly of the salt, its increasing density causes it to sink back into the underlying sediments. As the brine sinks it reheats, flows laterally and then rises to return to the underside of the evaporite bed to complete the convection cell (Figure 2.11a,b). Such density-driven convection is one mechanism of depositing and remobilising secondary salts (burial salts) in beds that were originally nonevaporitic. Ongoing dissolution of the underbelly of an evaporite bed can even establish convective circulation beneath an evaporite-induced pressure seal. A natural pressure cooker!

The modelling of plumes beneath salt beds and sub-horizontal allochthonous salt sheets, such as occur in the Gulf of Mexico, shows free convection has the potential to be a significant mechanism in both salt dissolution and mass transport, even if the sediments beneath the salt have permeabilities as low as 0.01 md (Figure 2.13; Sarkar et al., 1995). For such sediment it will take some 10 My for the sub-salt convection to develop to full strength. Once active, an average of 3-5m of salt is dissolved off the underbelly every million years, with a maximum darcy flux of 3 mm yr^{-1}. In the Gulf of Mexico the matrix permeabilities of overpressured shales beneath allochthonous salt in the Eugene Island region are much higher than 0.01 md, and so the flow and dissolution rates are even higher (Sarkar et al., 1995).

As salinity builds, these chloride-rich density-derived convective flow cells, which are an inherent part of the hydrology beneath the evaporite seal, can recycle the same pore water over and over through the underlying sediment column. This is a pervasive cycling mechanism that moves chloride-rich brines through base-metal-containing shales or redbed sands that lie beneath buried salt beds and sheets. When such waters break out, via major growth faults or salt welds, from beneath the salt beds and sheets into suitably prepared ground, the associated cooling and brine mixing can precipitate metal sulphides. It also drives other diagenetic reactions such as secondary porosity formation and the precipitation of authigenic cements (Figure 2.11c).

Dissolution occurs along the flank of a salt stock or diapir, the resulting dense pore fluids move into adjacent upturned sedimentary beds (Figure 2.11c). Any cavity created by

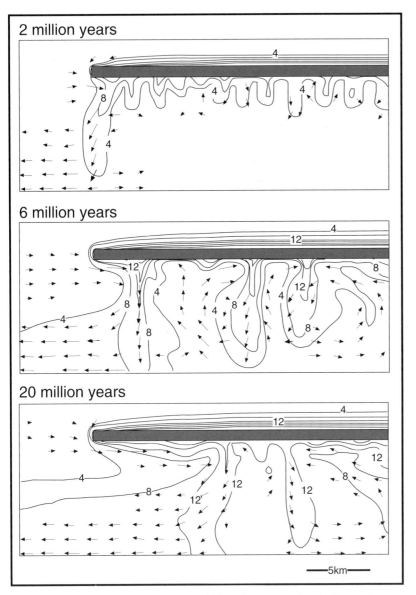

Figure 2.13. Modelled temporal evolution of brine plumes associated with ongoing dissolution of a salt sheet in a shale matrix at times of 2, 6 and 20 My. Salt sheet is grey (after Sarkar et al., 1995). Contours of plumes are in salinity difference: calculated as pore water salinity minus seawater salinity (3.5%). Numbers are in wt %. The leftmost plume is mostly fed by ongoing dissolution of the salt edge. Vertical exaggeration is 2.

salt loss associated with the undersides of subhorizontal salt geometries, steeply dipping beds about a salt stem create a plumbing system with a tendency to self drain and so facilitate further dissolution of the stem.

Hydrocarbons are much less soluble in a brine than in fresher water. This means hydrocarbons carried in basinal waters tend to "salt out", or exsolve, in the vicinity of a salt stock or along the underside of a salt bed or allochthon. "Salting out" is a direct result of the increase in the salinity of the carrier fluid as the salt body is approached. The accumulation of oil and gas along the underside of a salt seal, or adjacent to a salt stock, drastically slows the rate of solution. Halite is near insoluble when in direct contact with exsolved hydrocarbons rather than pore waters. Such conditions help explain why hydrocarbon legs adjacent to a salt stem have a propensity to stack one atop the other with the stem acting as a highly efficient seal. These salt-stem sealed, steeply-inclined reservoir beds are the most highly productive regions per unit area of all the various evaporite traps (Warren, 1989).

Preservation of bed contacts

Edge-associated dissolution and brine remobilisation means that the basal and the upper contacts of a thick evaporite bed dissolve from the edges, like a melting block of ice, to leave behind thin dissolution residues (see Chapter 4 for a full discussion of these deposits). The escaping brines seep through adjacent nonevaporite sediments, such as shales or carbonates, where they precipitate secondary salts and drive recrystallisation processes, including incongruent salt dissolution.

Buried evaporite beds in contact with adjacent aquifers will be sites of phreatic flushing as soon as the active phreatic brine reflux hydrology starts to dissipate. Dissolution from the edges occurs as the evaporite bed enters the zone of compactional water flow, and continues

dissolution of the salt stem is filled, either by salt as it rises in the stem, or, if dissolution is rapid enough, by the collapse of sedimentary beds against the stem to form "downthrown" blocks. As dissolution of the salt stem proceeds, the high salinity (high density) pore brines create a sinking brine halo about the salt stem. Such high salinity waters tend to displace less saline basinal pore waters and flow down dip within the more porous portions of the steeply dipping beds. Unlike the self-limiting rate of as long as the adjacent nonevaporite sediments retain some permeability. This explains why the upper and lower contacts of many thick ancient evaporite beds are not depositional contacts. Rather they are zones characterised by dissolution residues, recrystallisation and pressure solution textures. For the same reason, well preserved primary and syndepositional textures in bedded evaporites are rarely found at the contacts with adjacent nonevaporite lithologies. Depositional fabrics are best preserved internally within an evaporite bed, away from the edge zones where more aggressive fluids are eating into and altering the original evaporite textures. Diagenetic waters continue to dissolve or "melt" evaporite bed contacts until the evaporite bed dissolves or all effective permeability is lost in the adjacent beds (see Chapter 4).

Potash beds at the top of the Jurassic Louann Salt in the Gulf of Mexico are excellent examples of the diagenetic effects of escaping thermobaric waters on the contacts of subsurface evaporite beds (Figure 2.14; Land et al., 1995; Land, 1995a, b). There sylvite occurs atop a thick halite succession that lacks MgSO$_4$ salts, giving them a mineralogy that, according to Hardie (1990), makes them a prime candidate for a nonmarine, possibly hydrothermal, brine

source. Land et al. disagree and argue for a diagenetic origin of the remaining potash phases. They conclude that the huge volume of halite (Louann Salt) in the Gulf of Mexico ($\approx 1.55 \times 10^6$ km^3; Figure 3.3), along with a time of precipitation that is much less than 20 million years, requires a huge influx of chloride that could only have been derived from a world ocean. They find that the marine-like bromine concentrations, and the ^{87}Sr/^{86}Sr ratios in whole rock samples, indicate that nearly all the solid phases in the halite zone were derived from a marine source (Figure 2.14). A hydrothermal source should have produced ^{87}Sr/^{86}Sr ratios much less like those of Jurassic seawater.

As an alternative to a nonmarine brine source, they argue that the present bittern zone is a diagenetically modified "insoluble residue" after dissolution of the soluble marine bittern phases that were once present at the top of the Louann Formation. Sylvite is notably less soluble than the other bittern salts with which it may once have been deposited (kieserite, epsomite, carnallite, etc.). In other words, the highly soluble MgSO$_4$ salts that were once present at the contact have been dissolved and flushed from the system by fluid flow focused along the evaporite edge. Diagenetic modification of the bittern suite may

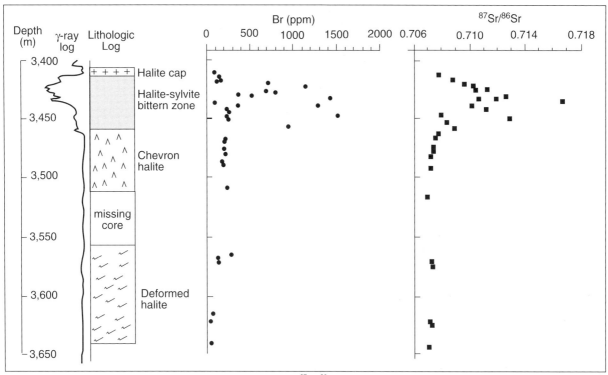

Figure 2.14. Gamma-ray log, lithologic log, Br concentration and ^{87}Sr/^{86}Sr ratio in whole rock samples from the cored interval through the Louann Salt in the Champion-Klepac #1 well (after Land et al., 1995). The well was drilled to a total depth of 4,800 m. Based on the monotonous gamma log, pure halite, at least 1,050 m thick, underlies the cored interval.

have begun early, as it does in other potash deposits (e.g. Figure 7.11b). However, the Rb/Sr analysis and work on inclusions in the Louann halite argue for a much longer term subsurface evolution involving formation waters that had experienced considerable rock-water interaction.

Salts and fluid flow

As long as sediments adjacent to a subsiding/flowing salt bed can maintain some degree of permeability, much of the more interesting subsurface hydrology in an active sedimentary basin centres about thick buried sequences of slowly dissolving salt. The ongoing dissolution and alteration of these beds creates chemical and pressure gradients that exert significant influence of the style and patterns of subsurface fluid flow. This in turn influences location and nature of hydrocarbon/metal accumulations.

Evaporites as pressure seals

Evaporite beds, due to their inherent lack of permeability (Figure 1.14), tend to retard the free vertical escape of compactional water and thermobaric waters. If further sediment is deposited atop the evaporite bed, little fluid escapes through the evaporite bed and the pore fluid pressure rises in sediments beneath the evaporite seal (Burrus and Audebert, 1990; Lerche and Petersen, 1995). Once pore pressure exceeds that of a free standing column of water of the same height, the sediment lenses within or below the evaporite bed become overpressured (geopressured). Hence, any actively subsiding and rapidly filling sedimentary basin containing widespread thick bedded evaporites is prone to early overpressuring, with the base of the evaporite bed likely to become the top of a shallow zone of geopressure. Bedded evaporites in this situation can be thought of as pressure seals. Fluid may leak from sediments beneath an evaporite bed, but leakage is typically focused into zones atop salt welds, fractures, or above growth faults that thin or cut through the evaporite bed (Figure 2.11c). If pore pressure near the base of a salt bed continues to rise to where it equalises the lithostatic pressure then hydrofracturing occurs. Growth-pulsed fibrous cements of halite and gypsum may then precipitate in new fracture nets centred in the lower portions of the evaporite.

Evaporites can maintain the ability to form pressure seals well into the burial history of a sedimentary basin. The highest measured bottom-hole pressures (formation pressures) in the world, and some of the highest wellhead flowing pressures (measured while wells are in production),

are found in the Gulf of Mexico in Jurassic carbonates sealed by Jurassic evaporites. Unlike many overpressured sandstones in compacting Tertiary shale basins, these high pore pressures occur in evaporite-sealed carbonates that are not particularly young. They occur in the Jurassic Smackover Formation reservoirs in gas and condensate fields of the Central Mississippi salt basin in the Gulf of Mexico region (North, 1985). There salt rollers in the Jurassic Louann Salt are associated with deeply buried growth faults that cut the overlying Smackover carbonate reservoirs. The faults are cut off above the Smackover at the overlying anhydrite seal (Haynesville and equivalents) and fail to penetrate overlying strata. Down to depths of 3,600m the fluid pressure gradients approximate hydrostatic (10.5 kPa m^{-1} or 0.465 psi/ft). By 6,000m a powerfully overpressured zone has been entered and the gradient has increased to 20.35 kPa m^{-1} or 1.0 psi/ft. This approximates the lithostatic (overburden) stress for that depth. The gas accumulations in the Smackover clearly illustrate the ability of bedded evaporites to maintain seal integrity over time and into pressure fields approaching those that induce hydrofracturing.

An evaporite's ability to act as a pressure seal obviously begins in the zone of compactional water, but can continue well into the thermobaric regime (Figure 2.11). The pressure seal will be destroyed once the evaporite bed dissolves or its lateral continuity is breached. Interstitial water is expelled from deeper zones subject to the highest pressures and migrates toward zones with a weaker potential associated with breaches in the evaporite seal. Zones of breaching may then act as a focus for escaping pressured and metal-rich basinal fluids (Figure 2.11c; Chapter 8).

Salt-related temperature anomalies and brine flow

Halite is thermally conductive when compared to most rock types. At room temperature the thermal conductivity of salt is typically three times that of adjacent shales (Figure 2.15a). At 150°C its thermal conductivity is about two thirds of its 25°C value and still more than twice that of an adjacent now compacted shale. Thus a salt bed or salt structure, with its much higher conductivity, exerts a strong influence on the thermal profile of an evaporite-entraining basin.

When the salt remains as a bed, the effect of the conductive layer is to raise the temperature gradient in sediments atop the salt bed, to lower the temperature gradient in the salt layer and to make beds below the salt layer cooler than they would be if the salt was not present (Mello et al., 1995). For example, with a 1 km thick salt bed encased in a shale sequence, the sediments below the salt are 20°C cooler than they would be if the salt were not present (Figure 2.16a). According to Mello et al. (1995) this cooling effect is proportional to the thickness of the salt layer and all subsalt sediments are subject to non-transient cooling effects that last as long as the salt remains. Such a temperature difference obviously effects the organic maturation levels of sediments below the salt layer.

If the salt bed has flowed and a salt diapir is present, but with its crest buried well below the sediment surface, the conductive spine of salt sets up a thermal dipole with a positive anomaly atop the spine and a negative anomaly is created at its base (Figure 2.16b). Sediments atop the spine are hotter than they would be without the salt, while sediments below the structure are cooler. This obviously effects rates of thermal maturation of organic matter in sediments atop and below the structure. Effects of the upper part of the thermal dipole on vitrinite reflectance above buried salt structures have been recognised in the oil industry for many years. Vitrinite reflectance is typically elevated atop buried salt structures. However, the cooling effects on beds below a salt bed, diapir or sheet were largely ignored until the work of Mello et al. (1995).

If the diapir crest is at the sedimentation surface, as it is during the passive salt diapir stage (Chapter 5), the structure acts as a thermal monopole. The salt spine acts as a sink that channels heat to the surface (Figure 2.16c). In this situation temperatures around and beneath the salt spine and its source layer are less than would exist in a shale basin without salt. The amplitude of the negative anomaly locally reaches -85°C and its cooling effect extends out for more than three times the diameter of the salt spine (Mello et al., 1995)

The contrast in thermal regimes where a thermal dipole is created when the salt crest is deeply buried versus a monopole where the crest is at the surface raises the question, how close to the surface need a diapir crest be before it starts to act as a heat sink rather than a heat source? Modelling by Mello et al. (1995) suggests the following rule of thumb; the maximum negative anomaly amplitude is reduced 10-15°C for every 250 m of increase in the depth to the top of the salt spine (assumes a shale host). The amount of heat drained from a basin is thus sensitive to the depth to the top of a salt structure and the length of time for which the salt remains at or near the surface before the structure is buried. Basins characterised by syndepositional salt flow and the maintenance of salt spines near the surface will produce the coldest situation with respect to

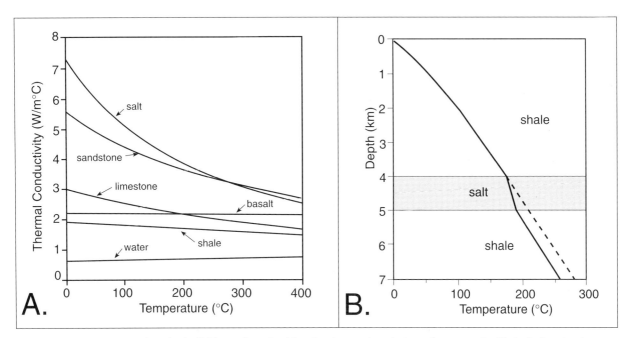

Figure 2.15. Thermal properties of salt. A) Thermal conductivity of various rock grain types (zero-porosity lithologies) and water as a function of temperature. B) Vertical steady-state temperature gradient for a shale section with salt (solid line) and without salt (dashed line). The temperature plot (A) was used to construct the steady state profiles of Figures 2.16 and 2.17 (after Mello et al., 1995).

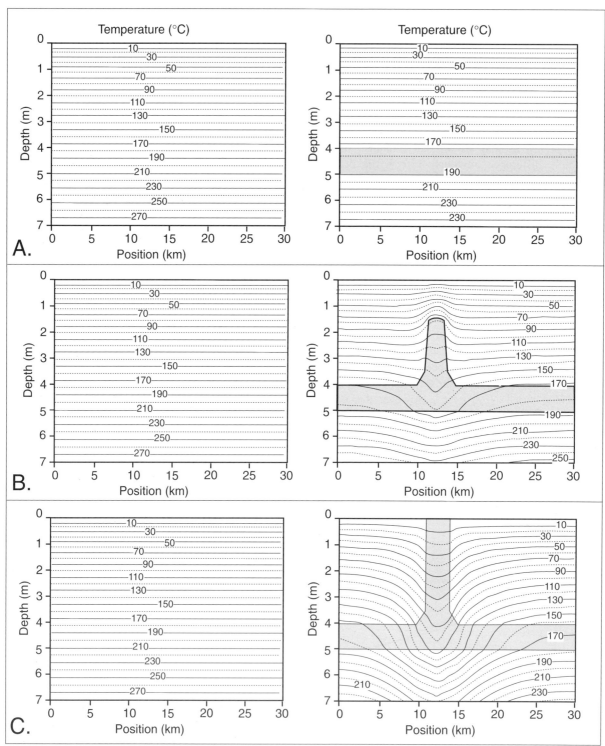

Figure 2.16. Model of steady-state temperature distribution for a shale section containing salt with varying geometries (shown on right hand side) versus the same shale section with no salt (left hand side). All calculations use shale and salt thermal properties and gradients given in Figure 2.15. A) Salt layer, note that sediments below the salt layer are 20°C cooler than in a shale section with no salt. B) Salt diapir that is connected to its source layer and is not in contact with the surface. Notice the elevated temperatures atop the salt spine and the reduced temperatures beneath the spine - a thermal dipole. C) Salt diapir that is connected to its source layer and is also at the sediment surface. The salt spine acts as a highly efficient heat drain that collects and channels heat from around and beneath the spine and delivers it to the surface. In this situation there is no heating associated with the spine, only cooling - a thermal monopole (after Mello et al., 1995).

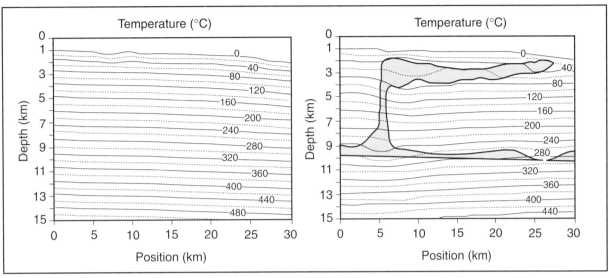

Figure 2.17. Model of steady state temperature distribution created by the emplacement of an allochthonous salt sheet (right hand side) versus profile in a shale section with no salt (left side) Three major negative anomalies are associated with this geometry: a zone of cooling beneath the sheet; a negative or cooling anomaly associated with the dome, which feeds the sheet; and a zone of cooling beneath the sheet (after Mello et al., 1995).

subsalt sediments. This mechanism may help explain why many oil-prolific salt basins still produce large amounts of hydrocarbons today, when thermal calculations that do not account for salt effects suggest that most of the known source rocks in the basin should be overmature (Mello et al., 1995).

Of course the simple spines, modelled in Figure 2.16b and c, are usually part of a much more voluminous allochthonous salt complex, where salt sheets form layers or tiered canopies at a number of different stratigraphic levels. Figure 2.17 presents a thermal model for such a system based on structures in the Gulf of Mexico (e.g. Figure 5.1). The salt sheet is modelled as a steady-state regime in a shale host, it lies within 500 m of the sediment-water interface and still connected with the mother salt bed. It should generate three negative thermal anomalies (Mello et al., 1995):

• a negative thermal anomaly beneath the salt source layer (mother bed);
• a thermal anomaly associated with the salt dome, which feeds the allochthonous salt sheet;
• a thermal anomaly below the salt tongue itself.

There are also some positive anomalies above the salt dome and its tongues, but these are small (<15°C) compared with the magnitude of the negative thermal anomalies. The minimum anomaly with the greatest magnitude is ≈-82°C located directly beneath the salt dome (Figure 2.17; position

5 km, depth 10km). The salt source layer or mother bed contributes only -10°C of the thermal anomaly in the deepest parts of the section. Constructive interference between the salt sheet and the feeding salt spine creates a local thermal anomaly of -55°C at the inside inflection point of the two structures (Figure 2.17; position 7.5 km and depth 4.5 km).

Temperature models are an indication of fluid flow patterns in the subsurface; as much subsurface heat in a sedimentary basin is transported by fluid flow. Thermal anomalies that existed around salt structures will obviously drive brine flow in the subsalt and suprasalt sediments. But fluid flow rates around salt structures can also be enhanced by density flow created by the dissolution of salt.

Like dissolution from the underbelly of a salt layer (see Figure 2.13), the dissolution atop and adjacent to a salt spine can create convecting density plumes (Figure 2.11c). Undersaturated basinal waters coming into contact with the edges of the salt spine dissolve the salt to generate a gravity or density fluid drive. In this case the thermal drive created by the thermal anomalies in the suprasalt sediments interact with the density drive to speed up flow rates in fluids convecting adjacent to the salt spine. If the crest of a salt diapir has risen to where it lies once again in the active phreatic flow regime, intense dissolution-generated plumes can set up at even shallower depths. Dissolution of the salt in both scenarios generates dense plumes of descending chloride-rich brine.

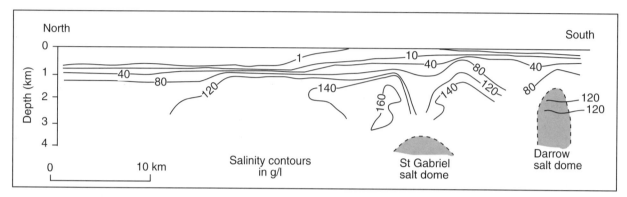

Figure 2.18. Cross section of part of the Louisiana Gulf Coast showing the shallow salinity plumes (g/l) generated by subsurface brine dispersion away from the dissolving crests of two salt domes (after Hanor, 1994).

In the Gulf of Mexico the density inversion created by the nearsurface dissolution of a diapir crest acting in conjunction with a thermal drive is sufficient to drive large scale convective cells with flow rates of metres per year through beds adjacent to the diapir (Figure 2.18; Hanor, 1994). This is two to three orders of magnitude faster than the mm/yr rates that are otherwise typical of the thermobaric zone.

The diagenetic importance of these suprasalt flow cells is clearly demonstrated in a study of sediment alteration on the west flank of the West Hackberry Dome, Louisiana Gulf Coast (McManus and Hanor, 1993). Drilling has documented the existence of 5×10^{10} kg of authigenic calcite-pyrite cement in Miocene sands at depths of 1.4 to 2.1 km in a 1.5 x 1.5 km area adjacent to the salt spine. The Sr, C and S isotopic compositions of the cements show Ca and S were derived from dissolution of a salt-dome anhydrite cap and that carbonate was derived both by thermochemical oxidation of methane and by sulphate reduction, possibly at temperatures as low as 70°C. Constraints on the maximum aqueous concentrations of Ca, which can be produced by dissolving diapiric salt, require that more than 5×10^9 m^3 of aqueous fluid, equivalent to a fluid volume/pore volume ratio of >250:1, were involved in destroying salt, transporting Ca and SO$_4$, and precipitating these cements. Thus the presence of such cements about a salt structure requires a dynamic subsurface mass-transport regime involving either large volumes of throughflushing fluid or fluids that are extensively recirculated.

Flow in hydrologically mature basins

Compactional and thermobaric regimes are active as long as sediments are accumulating in the sedimentary basin. With the cessation of sedimentation/subsidence, or with tectonic uplift, compaction driven flow decreases and

ultimately ceases. Overpressure disappears. Strata in such hydrologically mature basins are increasingly flushed by meteoric waters fed by recharge along the uplifted margin of the basin. This type of gravity driven circulation is sometimes termed centripetal flow, as it encompasses flow into the basin centre from the surrounding highland areas. It is opposite to centrifugal flow, which characterises the compactional and thermobaric regimes

When a sedimentologically mature basin undergoes collision and overthrusting, basin inversion will occur. At the same time the growing overthrust belt can act as a huge press or "squeegee" driving thermobaric waters out of the deeper parts of the basin into the shallower basin margins. Some authors have explained MVT sulphides as forming in the basin margin zones, where these escaping metalliferous "squeegee" waters mix with topographically driven meteoric waters (Oliver, 1986). This process is thought to precipitate sulphide accumulations hundreds of kilometres in front of the active thrust.

Inversion-related faulting, which juxtaposes potential aquifers against pressured or dissolving evaporite beds, can also mobilise thermobaric brines into a deep aquifer system to precipitate late-stage authigenic evaporite cements. This is clearly seen in the Rotliegende Formation in Leman Field in the Southern Permian Basin of the North Sea (e.g. Figure 4.28; Gaupp et al., 1993; Sullivan et al., 1994). Coarse-grained burial-stage poikilotopic anhydrite cements increase in abundance in the vicinity of faults cutting the Rotliegende sands. Isotopic data show this late stage anhydrite precipitated at maximum burial of 3.5 - 4 km during the Late Jurassic- Early Cretaceous, that is, at the start of basin inversion. The data indicate an unusual, short-lived catastrophic event of cross-formational fluid flow along fractures and reactivated faults induced by basin inversion. Late stage magnesite and dolomite cements may also be related to influx of the same subsurface brines.

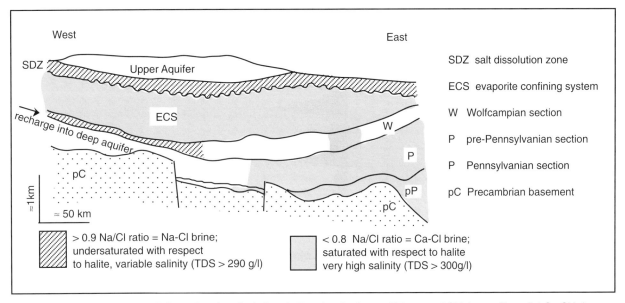

Figure 2.19. Na/Cl molal ratios define subsurface hydrology in Permian Basin, west Texas and Oklahoma. The relict Ca-Cl brines are held in place by the low permeabilities of the evaporite confining zone, while Na-Cl brines derived by halite dissolution are to be found in the salt dissolution zone (SDZ) and the Wolfcampian section (W) of the deep basin aquifer (after Bein and Dutton, 1993).

After inversion or uplift, the evaporite beds can re-enter the zone of active phreatic flow or deep meteoric circulation (artesian flow). Salt dissolution in such areas can feed brine into adjacent deeply circulating confined aquifers located beneath thick well preserved evaporite intervals. In the Permian Basin of west Texas and New Mexico the deeply circulating Na–Cl brines in arkosic sandstone aquifers acquired their high salinity through the dissolution of halite by meteoric waters as young as 5-10 Ma (Figure 2.19; Bein and Dutton, 1993). This initially fresh, meteoric water moved into the western half of the deep-basin brine aquifer through nearsurface salt dissolution zones and then downward as Na–Cl brines seeped along the deep aquifer to depths of thousands of metres (SDZ zones in Figure 2.19). Above this deep aquifer are relict Ca–Cl basinal brines that are Permian mixtures of descending evaporitic brines and connate Permian and Pennsylvanian seawaters. Little or no meteoric Na–Cl brine has moved into this Ca–Cl brine zone hosted by the evaporite confining system (ECS). The lack of penetration reflects the extremely low permeabilities of the deeply buried halite, shale and anhydrite- or halite-cemented strata that dominate the ECS (a "block of ice" analogue). Ages of the Ca–Cl brine in the ECS, based on ^{4}He and ^{40}Ar analyses of carbonates in the central part of the deep aquifer range from 100 to 300 Ma (Zaikowski et al., 1987), making it a true "fossil" groundwater.

Deep meteorically driven dissolution of gypsum/anhydrite beds can provide the sulphate-ion source for the bacterial production of reduced sulphur. This reduced sulphur may then be oxidised to form economic accumulations of natural sulphur (Figure 7.36b).

Precambrian oceanic chemistry using evaporites

Given that the mineralogical evolution of a concentrating seawater brine reflects the major ionic proportions present in seawater (Figure 2.1), questions then arise. Can we use our understanding of chemical controls on evaporite precipitates to map the evolution of major ion proportions in seawater since the early Archaean? How constant was the chemistry of the world's oceans since the early Archaean? Do the major evaporite precipitates reflect the possibility of a primordial reducing atmosphere?

Some authors postulate that there have been no significant changes in the major ion proportions in seawater for the past 3.5 Ga (Holland, 1984). Others postulate the Archaean was dominantly a time of little or no atmospheric oxygen and that ocean waters were reducing fluids (Krupp et al., 1994). Yet others postulate that the bicarbonate to calcium ratio was so high in Archaean and Palaeoproterozoic seawater that all the calcium was used up in widespread abiotic marine aragonite and Mg-calcite precipitates (Grotzinger and Kasting, 1993). Still others researchers

59

have postulated major changes in oceanic chemistry throughout the Phanerozoic and Precambrian related to changes in style and rate of sea floor spreading rates. Evaporites and their pseudomorphs when plotted against time should give some indirect clues to help solve this dilemma.

Given a paucity of preserved evaporite salts prior to the Neoproterozoic, our discussion must centre around mineralogical evidence left as evaporite pseudomorphs. Our scale of resolution is far less than when modelling ancient brine chemistries based on preserved salts. Possible changes in Phanerozoic seawater chemistries, as indicated by preserved suites of evaporites, were discussed earlier in this chapter (e.g. Table 2.2).

Pseudomorphs, especially of halite hoppers, occur in marine rocks as old as early Archaean, but are far more common, as are the actual salts, in Proterozoic and Phanerozoic strata (Figure 2.20). Halite (the mineral) or its pseudomorphs characterise areas of widespread chemical sedimentation from the Archaean to the present. $CaSO_4$ distribution is more enigmatic. Possibly the oldest documented $CaSO_4$ pseudomorphs are cm-sized growth-aligned barytes that are interpreted to replace bottom-nucleated gypsum in 3.45 billion year old metasediments in the Pilbara/North Pole region of Western Australia (Barley et al., 1979; Lowe, 1983; Buick and Dunlop, 1990). Similar growth aligned baryte crystals, perhaps pseudomorphing swallowtail gypsum, formed in the Nondweni greenstone belts of South Africa some 3.4 billion years old (Wilson and Versfeld, 1994).

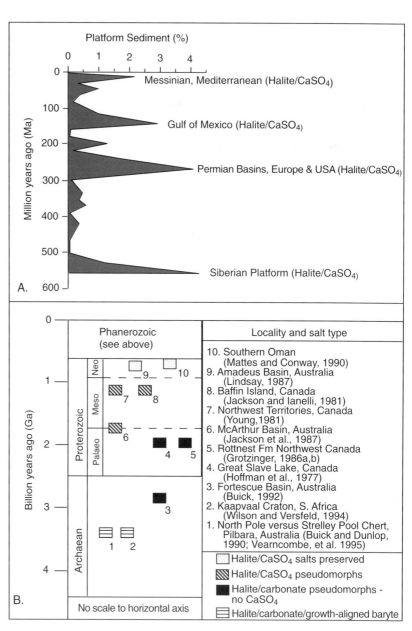

Figure 2.20. Evaporite occurrence over time. A) Phanerozoic evaporites as a percentage of total platform sediment, lists dominant deposits contributing to the major peaks (compiled from data in Ronov et al., 1980; Warren, 1989; Railsback, 1992). B) Precambrian evaporite occurrences and mineralogies compiled from listed sources.

Sequences in both regions are now completely silicified or barytised. At the time they were first documented, the recognition of shallow-water Archaean sulphate evaporite pseudomorphs at North Pole, Pilbara Craton, caused a re-evaluation of models of a totally reducing Archaean atmosphere (Dimroth and Kimberley, 1975; Clemmey and

Badham, 1982). The presence of free sulphate in surface water brines of the Archaean world implied oxygenated hydrosphere conditions, at least locally.

However, in both the Pilbara and the South African sequences there are no actual sulphate evaporites preserved, only growth-aligned crystal textures now preserved as baryte. Textures in baryte ore from Frasnian sediments in

Chaudfontaine, Belgium, are near identical to those observed at North Pole, Australia. The Belgian crystals have been shown by Dejonghe (1990) to be primary bottom-nucleated baryte with no precursor mineral phase.

Some workers in the Pilbara feel that the growth-aligned baryte in the region is a primary precipitate and formed in the vicinity of hydrothermal vents (Vearncombe et al., 1995; Nijman et al., 1998). As such, it is not secondary after gypsum. A similar hydrothermal discharge model has been developed for the Barberton Greenstone belts (de Ronde et al., 1994, 1996). Even if the aligned baryte crystals are primary, their formation requires sulphate to be present and so some oxygen must have been present on the seafloor, at least in the vicinity of the depositional site. Even so, the level of Archaean sulphate in the world ocean was probably no more than a few percent of the current levels and remained so until the evolution of an oxygen-producing biota in the Proterozoic (Habicht and Canfield, 1996).

According to Nijman et al. (1998) the occurrence of the North Pole baryte in sedimentary mounds atop growth faults argues that sulphate was derived by the boiling of escaping vent waters. That is, the escaping vent fluids were enriched with Ba, Si and sulphide. As these hydrothermal waters vented onto the Archaean seafloor in waters some 50 metres deep, they boiled or violently degassed about the vents. Consequent mixing with a normally stratified seawater, caused instantaneous oxidization into sulphate that then with cooling combined with the barium to form growth aligned baryte crystals on the seafloor (see Chapters 8 and 9 for discussions of seawater vent systems - smokers).

The origin of Archaean barytes in South Africa and Australia is still equivocal but tending more and more toward a hydrothermal consensus. At this stage of our understanding, these barytes cannot be used to support an evaporite paragenesis of gypsum in the Archaean. Widespread beds composed of calcium sulphate minerals or their pseudomorphs are scarce to absent in rocks older than 1.8 Ga (Grotzinger and Kasting, 1993). That is, with possible exception of the ambiguous Archaean baryte pseudomorphs described above, pseudomorphs after bedded or massive gypsum/anhydrite formed in evaporitic environments are absent from the Archaean and much of the Palaeoproterozoic record. In contrast, halite pseudomorphs are to be found throughout this interval (e.g. Boulter and Glover, 1986).

Then, some 1.6-1.7 Ga, widespread examples of pseudomorphs after calcium sulphate appear in the rock record (Figure 2.14). Pseudomorphs of widespread calcium sulphate beds first appeared some 1.7 Ga in the Late Palaeoproterozoic to Mesoproterozoic sediments of the McArthur Basin, Northern Territory, Australia, and in rocks of Great Slave Lake in northern Canada (Figure 2.20). Other widespread sulphate pseudomorphs occur in the 1.2 Ga Borden Basin, Baffin Island, and the 1.2 Ga Amundsen Basin in the Canadian Arctic Archipelago. In all these basins the former sulphate evaporites formed deposits up to tens of metres thick, with lateral extents measured in hundreds of square kilometres. These sulphate-entraining sediments were laid down in shallow marine, coastal, and alluvial environments under an oxygenated Meso- to Neoproterozoic atmosphere (Jackson et al., 1987; Walker et al., 1977).

Neoproterozoic sulphate evaporites that include preserved anhydrites, along with widespread bedded halite, occur in the Bitter Springs Formation of the Amadeus basin, central Australia, its equivalents in the Officer Basin, the Callana beds of the Flinders Ranges and the Infracambrian salt basins of the Arabian (Persian) Gulf (Wells, 1980; Cooper, 1991; Mattes and Conway, 1990; Edgell, 1991).

Prior to 1.9 Ga the paucity of preserved marine salts other than halite pseudomorphs, and the paucity of $CaSO_4$ indicators means that the chemical composition of the world's surface waters was not necessarily identical to that of the Phanerozoic ocean (Figure 2.20). The evolution of the evaporite pseudomorphs supports Grotzinger and Kasting's (1993) argument that HCO_3/Ca ratios were much higher in the Archaean and the Palaeoproterozoic oceans. During progressive concentration of these seawaters, gypsum/anhydrite precipitation was rare or impossible as there was no calcium left in the brine. The absence of gypsum in Archaean marine precipitates does not hinge on the absence of sulphate in marine waters and hence has little to do with Archaean oxygen levels.

Since 2 Ga the evolution of marine evaporites begins to reflect the same predictable succession of precipitated salts (at least up to the bittern stage) as is found in modern seawater. Marine waters appear to have maintained similar proportions of major ions throughout the Phanerozoic and the Proterozoic back to 1.8-1.9 Ga. Prior to this, the Archaean ocean was a Na–Cl-HCO_3 ocean and not the Na–Cl ocean of today (Kempe and Degens, 1985; Maisonneuve, 1982). A corollary of this argument, is that trona, along with halite, constituted the two most volumetrically important marine evaporite minerals in the Archaean. According to Grotzinger and Kasting (1993) all seawater calcium was deposited as marine cementstones and other alkaline earth precipitates well before the levels of bicarbonate were depleted. Thus, carrying their argument into Archaean evaporitic settings, means that the excess bicarbonate present in any seawater fed to an evaporite forming area must have combined with sodium to form trona and other sodium bicarbonate precipitates as well as widespread halite. Archaean marine evaporites had a chemistry similar those found in modern volcanogenic rift lakes, such as Lake Magadi and Lake Natron (pathway I brines). This rather different marine and brine chemistry may also explain the preponderance of widespread stratiform albitites as early diagenetic replacement of labile volcaniclastics/zeolites in Proterozoic and Archaean volcanogenic/greenstone terranes (see Chapters 6 and 9). In modern seawater-fed evaporite terranes the two most widespread evaporites are gypsum and halite. Thus, a pseudomorph-based model of Archaean seawater evolution is in marked contrast to brine evolution of modern seawater. Today bicarbonate is depleted early in the evaporation series and the presence of trona in an evaporite basin can be used as an unequivocal indicator of nonmarine waters.

The relative scarcity of actual Pre-Phanerozoic salts, not pseudomorphs, especially in the Archaean has been used by some to argue that conditions were less favourable for widespread evaporite deposition (Cloud, 1972). Others, including I, feel that the relative scarcity of preserved evaporites in older sequences reflects the greater likelihood of fluid flushing, evaporite dissolution and metasomatism in progressively older rocks. It may well be that sulphate evaporites were less common in the Archaean, and that sodium carbonates mixed with halite were dominant evaporite salts in seawater-fed saline giants, but widespread evaporite deposition from chloride-dominated brines did occur throughout the Archaean. A paucity of preserved bedded evaporite salts in the Precambrian reflects an increased probability of partial or complete evaporite dissolution, remobilisation and metasomatism with increasing geological age.

It also means that we have metallogenic provinces worldwide that were created by evaporite-associated metalliferous chloride brines. Once meta-evaporites and "the evaporite that was" are included in our understanding of metallogenesis, evaporites and their associated brines become significant contributors to the metallogeny of many of world's premier mineralised provinces. World-class Precambrian ore deposits, such as occur in the Mt Isa Block, the Broken Hill Block and the McArthur Basin, are there because of dissolution and metal leaching associated with salt flow and the ultimate disappearance of former evaporite beds.

Chapter 3
Evaporite basins and their stratigraphic evolution

Is the present the key to the past in evaporite studies? This chapter looks first at the distribution of modern evaporite-depositing areas and then at the much more extensive evaporite basins of the past, many of which have no modern analogue in size or depositional diversity. Our inability to compare all aspects of evaporites past and present, reflects a lack of longterm shallow seas or perennial lakes in modern arid geological settings. The implications of this lack of strict uniformitarianism are discussed, along with problems inherent in many of the correlation assumptions used in sequence strati-graphic interpretations of evaporite basins. This analysis underlines the importance of worldwide sealevel differences in icehouse and greenhouse earth. Geometries of ancient depositional systems expressed in terms of sealevel oscillations do not work as well in interpreting ancient saline giants as they do in interpreting evaporites laid down on marine platforms. It is much more useful to consider the internal tectonics and hydrologies of the large evaporite basins as the fundamental controls of basin-scale evaporite geometries, mineralogies and bounding surfaces.

Distribution of modern evaporites

Modern bedded evaporite deposits typically accumulate in saline lake and mudflat environments within ground-water discharge regions in the arid and semi-arid deserts of the world (Figure 3.1a; Table 3.1). Coastal deposits of evaporites also occur in the same desert belts in areas fed by marine seepage

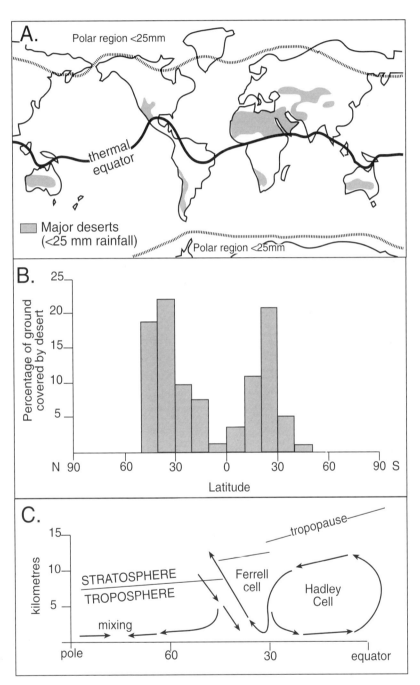

Figure 3.1. A) World distribution of modern deserts as determined by plotting areas with less than 25 cm annual precipitation. Also shows polar regions with less than 25 cm precipitation. B) Latitudinal distribution of the world's modern deserts. C) Longitudinal cross section through the earth's atmosphere showing major circulation cells. Belts of cool dry descending air at 30° N and S of the equator create the main arid zones of the world (after Warren, 1989).

into isolated coastal depressions or mudflats. Continental playa deposits, such as Lake Eyre and Salar de Uyuni, typically contain much larger areas of salts than coastal deposits, but still do not approach the aerial extent or thicknesses of their ancient counterparts (Figures 3.2, 3.3). Evaporite salts also form as lakes precipitates and efflorescences in the cold polar deserts of Antarctica, but the amount of salt in these regions pales to insignificance when compared with the volumes forming in settings closer to the equator. Brine mixing and freezing, rather than direct solar concentration, plays a much more important role during crystallisation of the polar deposits (Carlson et al., 1990).

Salt-accumulating desert playas form as discharge areas within endoheic (internal drainage) basins; where more water is leaving the basin via evaporation than is entering it as rainfall, surface, or subsurface inflow (Table 3.1). They are found in: 1) tectonic basins that include fault-defined intermontane basins and intracratonic structural sags, 2) interdune or inter-draa depressions about the edges of sand seas or ergs, 3) wind deflation hollows, 4) abandoned fluvial valleys 5) bolide-impact or volcanic craters, or 6) in interior drainage depressions formed by a combination of the preceding. The greatest present-day volumes of salts are to be found accumulating within continental tectonic basins located in the world's desert belts.

Modern deserts cover large portions of the landsurface in two belts between 15 and 45° north and south of the equator (Figure 3.1b). This region lies beneath the cold dry descending air of Hadley Cells, which set up the high pressure belts of the subtropical horse latitudes. At the broadest scale, deserts are the result of large scale atmospheric circulation driven by varying intensities of solar irradiation at the earth's surface (Figure 3.1c). Solar radiation is most intense directly above the equator and lessens toward the poles. The equatorial belt experiences greater insolation than the adjacent latitudes, so the air warms and rises creating equatorial belts of low pressure. As it rises it cools, losing much of its water vapour as rain to the tropical jungles. While it moves further away from the equator, this dry air further cools and compresses, finally, it sinks back to the earth's surface at around 30° latitude north and south of the equator. There a belt of cool descending dry air is reheated as it returns to earth so garnering a much enhanced potential to absorb any free water from the land surface: and so the major desert belts of the world are formed (Figure 3.1c).

Of course, the earth-scale distribution of these Holocene climatic belts reflects current atmospheric dynamics. When the polar icecaps expanded, as they did in the last glacial maximum, some 18-25 Ka, then the circulation belts were pushed and compressed toward the equator. This increased the intensity of circulation and altered the latitudinal distribution of the climate belts. It had a marked effect on the style of deposition within deserts, so that almost all modern continental playas have experienced numerous water-full versus dry stages in the last few hundred thousand years.

Continent	Region	Location and setting	Notable features
Africa	East Africa	Continental Danakil Depression, Lake Magadi, Lake Natron, Lake Kivu, Lake Tanganyika, Lake Turkana, Lake Urima	Continental evaporites. In addition to large volumes of halite and $CaSO_4$, potash salts and trona are important salts in particular lakes with mineral evolution dependent on inflow chemistry
	North Africa	Continental Lake Chad, chotts of northern Africa. Coastal Coastal sabkhas and salinas of Algeria, Tunisia, Libya, Morocco, the Nile delta and the Red Sea	Sodium carbonate is the dominant primary mineral in Lake Chad, while sodium sulphate and sodium chloride occur in subordinate amounts. Potassium and magnesium salts occur in near negligible amounts Coastal sabkhas and salinas dominated by gypsum and anhydrite with small volumes of potash in modern chotts and sabkhas of Tunisia
	South Africa	Continental Pretoria, Otjiwalundo and other salt pans, interdunal sabkhas of Namibia	Thousands of small wind-aligned salt pans, often with downwind lunettes, dot the arid landsurface. They form preferentially on a clayey substrate that is made up of the Ecca and Dwyka Shales in South Africa, the Karoo in Namibia and the Kalahari Beds in Botswana
	Middle East	Continental Dead Sea Coastal Coastal sabkhas and salinas of the Arabian Gulf	Dead Sea is accumulating mostly crust halite in the shallow southern basin and deepwater carbonate laminites in the northern basin Nodular calcium sulphate with minor halite in classic coastal sabkha successions of Abu Dhabi
Antarctica		Continental Lake Bonney, Taylor Valley; Lake Vanda, Wright Valley	Carbonate minerals predominate in the Taylor Valley, while sulfate minerals predominate in the Wright Valley. Halite and thenardite are widespread in both areas. Trona and thermonatrite are found exclusively in the East Taylor Valley, and soda nitre, bloedite, and darapskite in the Wright Valley
Asia	Central Asia	Continental Qiadam Basin, Gulf of Kara Boguz, Lake Inder	Potash salts (especially carnallite) are accumulating as karstic fills in bedded halite in perennial lakes and playas of the Qaidam Basin. Borates occur in Lake Inder.
	Southern Asia	Continental Iranian playas; Van Golu, Salda Golu & Toz Golu, Turkey; Didwana Lake and other playas of Thar desert, India. Coastal Coastal sabkhas of Ranns of Kutch, India	Iranian playas show complex facies interactions with rising salt diapirs. Turkish lakes are accumulating large volumes of magnesium carbonates and dolomite

Table 3.1. Continued over

Continent	Region	Location and setting	Notable features
Australia	Eastern	**Continental** Lake Tyrell, Lake George, Maar Lakes, Lake Victoria	Continental salt lakes (playas), some with gypsum/clay lunettes that rim brine pans; others, especially those in volcanic maars, are characterised by Mg-rich carbonate sediments & crusts
	Central	**Continental** Lake Amadeus (Karinga Creek lakes), Lake Eyre, Lake Frome, Lake Torrens	Continental salt lakes (playas) typically with gypsum lunettes that rim deposit's brine pan. Lunettes were much more active in the Late Pleistocene. Salt crusts are now largely recycling through the lakes, with few massive accumulations. There is a fresher artesian aquifer beneath many continental lakes. Mostly halite and gypsum accumulate in the lakes, minor thernadite/mirabilite accumulating in the Karinga Creek area.
		Coastal Marion Lake complex, Lake Macdonnell, Streaky Bay Lakes, Fisherman Point sabkhas	Coastal lakes or salinas are classic Holocene areas for coarse-grained gypsum beds up to 10-15m thick, pisolite seepage pavements, modern tepees, mm-laminated domal stromatolites.
	Western	**Continental** Lake Lefroy, Lake Cowan, Lake Dundas, Lake Ballard, Lake Carey, Lake Raeside	Continental playa lakes within elongate and sinuous Tertiary-age palaeodrainage channels on the Yilgarn Craton. Evaporitic fill in lake is up to 10 m thick and mostly gypsum with lesser halite. Wind and wave-driven currents rework sediments that are then deflated by the wind to be redeposited in marginal gypsum dunes (lunettes).
		Coastal Lake Mcleod, Hutt Lagoon, Leeman Lagoon	Coastal lakes or salinas with gypsum and halite in beds up to 10-15m thick in Lake McLeod
Europe	Spain	**Continental** Playas of Los Monegros and Bajo Aragon, Laguna de Tirez in La Mancha region	Glauberite and other sodium sulphates accumulating in modern playas.
North Americas	Central Canada	**Continental** Ceylon Lake, Freefight Lake, Deadmoose Lake, Little Manitou Lake and many other saline lakes and playas on the Great Plain of central Canada	Salt pan and perennial subaqueous evaporites containing mirabilite, thenardite, and bloedite. Some lakes, such as Freefight and Deadmoose, are accumulating salts in perennial "deep" waters.
	Northern United States	**Continental** Saline lakes and interdunal sabkhas, Nebraska and North Dakota	Salt pans containing sodium sulphate salts, as in Central Canada.
	Southern United States	**Continental** Salt Flat playa, west Texas, playas in the Rio Grande Rift **Coastal** Laguna Madre	Pleistocene continental lacustrine accumulations of mostly gypsum±halite, now deflated and overprinted by Holocene sabkha signatures. Coastal mudflats with displacive gypsum
	Western Unites States	**Continental (Basin and Range)** Death Valley, Saline Valley, Deep Springs lake, Searles Lake, Bristol Dry Lake, Great Salt Lake, Owens Lake, Salton Sea	Quaternary continental lakes with diverse range of salts including thick halites, lesser anhydrites, trona, glauberite and other salts depending on chemistry of inflow waters. Holocene is a time of saline mudflats, at other times in the Pleistocene the same depressions were perennial saline lakes.
South Americas		**Continental** Salar de Atacama, Salar Uyini, Salar Coipasa, Pastos Grandes, Laguna Salinas, Salar de Pintados	Large volumes of continental evaporites accumulated within the central Andes during Neogene uplift of the Altiplano-Puna plateau with associated development of the Andean volcanic arc. Halite and gypsum are dominant minerals, along with local and economically important borates, nitrates and potash salts.

Table 3.1. Quaternary examples of evaporite accumulations.

Many modern midlatitude discharge playas show evidence of longterm drier arid conditions today compared with earlier times. For example, the strontium-isotopic composition of preserved gypsum in Lake Frome, a large discharge playa in South Australia, records periods of high rainfall ("wet") at 3-6, 12-15, and >17 Ka, with "dry" periods at ≈10 and ≈17 Ka (Ullman and Collerson, 1994). In the last interglacial (early in oxygen isotope stage 5), an enlarged Lake Eyre maintained a perennial water body up to 25 m deep (Magee et al., 1995). Subsequently, as the climate deteriorated, there were a number of dry periods separating successively less effective wet phases, culminating in the deposition of a substantial halite salt crust around the time of the last glacial maximum. Lake Eyre attained its present saline mudflat/ephemeral playa status only some 3-4 Ka. The drying trend over the last hundred thousand years corresponds to a longterm lessening of monsoon intensity in the northern part of Australia, the main water source area for waters flowing south into the Lake Eyre Basin.

In Death Valley, California, there were dry period mudflats from 0 to 10 Ka and 60 to 100 Ka characterised by abundant glauberite and gypsum with minor calcite (Li et al., 1997). In contrast, the wet period from 10 to 60 Ka was typified by halite and mud layers with relatively abundant calcite. This episode includes sediments deposited some 10-35 Ka, when Death Valley contained a perennial water body. The changes in evaporite mineralogy are thought to indicate a predominately Na–Ca spring-fed inflow during dry climatic periods and an increased volume of bicarbonate-rich river waters in the wet periods.

Even under the drier climates that characterise modern playas the volumes of water entering and leaving the basin can be substantial. For example, in a single flood in 1974 more than 39.3 km^3 of surface runoff flowed into Lake Eyre, a discharge playa in the Lake Eyre Basin, South Australia. In the following two years more than 39.5 km^3 of water evaporated from the same lake (Kotwicki, 1986). Historically, the annual inflow of the Jordan River into the Dead Sea is some 1.25 km^3, a water volume that is about 80% of the total inflow. Each year some 1.58 km^3 is evaporated from the Dead Sea (Hardie et al., 1978).

In addition to longterm Quaternary climate change under time frames of tens of thousands of years, local topographically induced variation in climate can enhance

65

conditions suitable for desert and playa formation. For example, the playas and midlatitude deserts of central Asia and central Australia are in part a product of their isolation from the nearest ocean. Continentality is the term that describes the geographic isolation of a land area by large distances from the oceans. It explains why the highly continental Takla Makan desert of central China is the only hyperarid (<25mm rainfall) desert outside the world's subtropical high pressure belts.

Deserts also form in the rain shadows of high mountain ranges; for example, the deserts of Patagonia lie on the lee side of the Andes, while the deserts of Nevada and Utah are in the rain shadow of the Sierra Nevada. Rain shadows form where air masses move up and over mountain ranges. As the air rises it cools, moisture develops into clouds and falls as rain on the side of the mountain range that faces into the prevailing wind. On the other side of the range, the subsiding air is dry and its moisture-bearing capacity increases further as it sinks and warms.

Cold upwelling ocean currents help account for the deserts of Baja California, as well as southwest Africa (Benguela Current). The Peru Current, as it upwells along the west coast of South America, further dries the Atacama Desert. Areas of cold upwelling ocean currents create cool onshore winds with a lowered moisture bearing capacity. As such winds pass overland, they warm and so have their moisture-bearing capacity increased. The winds take up moisture from the land surface and create nearshore deserts. Near the coast these hyperarid strips are frequently engulfed in a sea fog, created where the cold seawater sits beneath a blanket of warm air. In such areas fog, rather than rain, is the main supplier of water to the desert. Fogs and sea breezes moderate air temperatures in this type of desert.

At finer climatic scales, variations in effective evaporation rates influence mineralogy and preservation potential. Aerodynamically, the evaporation rate of a surface brine is a product of wind velocity and the difference in water vapour pressure between the main air mass and the air above the surface of the evaporating brine (Myers and Bonython, 1958). Water vapour pressure over a brine decreases with increasing salinity. At very high salinities the pressure difference between the brine and the air immediately above the brine can become so small that evaporation effectively stops; and so the higher salinities necessary to precipitate the bittern salts may not be reached and so less saline salts accumulate within the basin.

In nature, a complete humidity induced shutdown happens only rarely; the air above an evaporating brine body is usually in continual motion, driven by winds blowing across the basin. In reality, the limited lateral extents of most Quaternary brine bodies mean the vapour pressure and air humidity over a brine body, and hence the salinity of a surface or near surface capillary brine, are determined by the humidity of air blown in from the surrounding region. In present-day coastal areas beneath humid marine air, such as the sabkhas of the Arabian (Persian) Gulf and the coastal salinas of southern Australia, a tendency exists for the lower salinity evaporites, gypsum and halite, to be the sole precipitates and to preserve mainly gypsum/anhydrite as the sabkha evaporites (Kinsman, 1976; Warren, 1989). Most modern coastal evaporite areas rarely accumulate salts more saline than halite.

Intrabasin-scale variations in temperature and effective evaporation can be influenced by the nature of the deposited sediment. When a salt crust covers the surface of a mud flat that is undergoing capillary evaporation there is a decrease in evaporation efficiency. The decrease is not related to humidity change but to the increased albedo created by the surface of a white salt crust. Effective evaporation rates from salt-free mudflats in Lake Frome range from 90 to 230 mm/year and average 170 mm/year (Allison and Barnes, 1985). In areas covered by salt crust in the same lake the effective evaporation rate falls to 5-30 mm/year (Ullman, 1995).

All these icecap and locally induced climatic effects emphasize that the present day settings are not direct analogues for many environments where larger and thicker evaporites accumulated in the past. For example, to produce and accumulate primary bittern salts, such as polyhalite, epsomite, carnallite and kieserite, in a modern system requires a degree of aridity so extreme that it is difficult to envisage any modern settings reaching the bittern stage for long enough periods to accumulate substantial thicknesses of bittern salts. The only modern area of substantial potash salt deposition occurs in the Qaidam Basin of western China, some 1500 km from the nearest ocean (Kezao and Bowler, 1985). Even there, the potash comes mostly from the dissolution of earlier subcropping Plio-Miocene potash originally deposited as bedded lake evaporites in continental transform depressions (see Chapter 7). To understand how substantial thicknesses of evaporite accumulate as widespread potash and salt beds with lateral extents measured in hundreds or thousands of kilometres, one must study pre-Quaternary deposits.

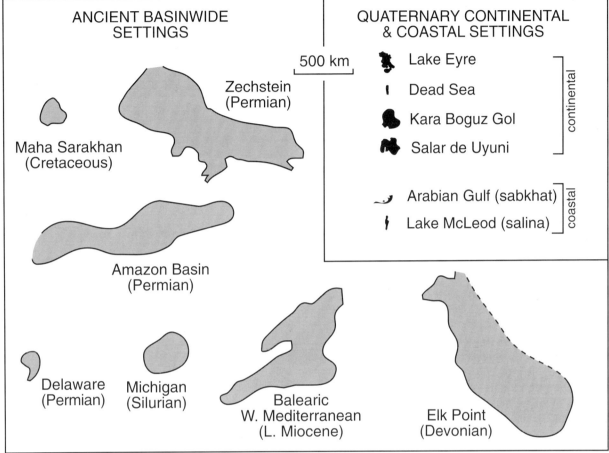

Figure 3.2. Comparison of aerial extent of Quaternary and ancient evaporite deposits (in part after James and Kendall, 1992).

Distribution of ancient evaporites

There is no Holocene proof of a desiccated ocean basin, yet in the Late Miocene (Messinian) salinity crisis the deep depressions on the Mediterranean floor were filled with 2-km-thick sequences of halite and gypsum in less than 300,000 years (Balearic Basin; Figure 3.2). These Messinian evaporites were laid down on the floors of a chain of shallow saline seas, some of which were 2,000 metres below ambient sea level, and at times a seafloor desert extended from Spain to Israel and from northern Italy to Libya (Cita, 1983; Hsu et al., 1973). Similarly there is no Holocene proof of evaporitic brine-reflux dolomitisation, simply because there is no Holocene example of a whole platform being covered by an evaporite deposit. Yet in ancient evaporite basins, reflux dolomitization was a widespread and pervasive mechanism that yielded substantial intercrystalline porosity in platform carbonates (Moore, 1989; Warren, 1991; Sun, 1995).

Compared with evaporite deposits of today, the wider extent, greater depositional and diagenetic diversity, and greater thickness of evaporites deposited on ancient platforms and across whole basins was due to three main factors:

• *Warmer worldwide climate* created wider latitudinal belts for evaporite deposition and preservation. For example, there is a clear diminution in evaporite-depositing latitudes from the Jurassic to the Cretaceous, coinciding with a change from an arid to a more humid worldwide climate (Gordon, 1975). Yet, even in the Cretaceous, with its transition from evaporites to coal at around 50° latitude, the extent of evaporite deposition was still considerably wider than today (Habicht, 1979).

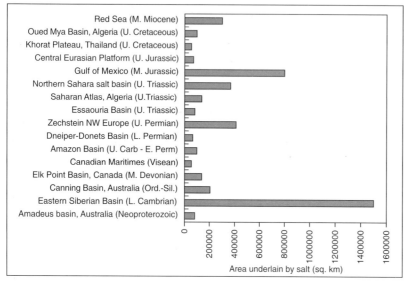

Figure 3.3. Plot of the areas underlain by salt in various significant ancient evaporite deposits, which are also called saline giants (data derived from Hardie, 1990; Zharkov, 1981).

evaporite-filled depression (Hsu et al., 1973; Kashfi, 1985).

Building blocks of ancient salt beds

Most ancient evaporite deposits have thicknesses and horizontal extents that are two to three orders of magnitude greater than those of Quaternary evaporites (Figures 3.2, 3.3). A study of ancient evaporite deposits underscores the limitations of a strict application of the Law of Uniformitarianism. Processes and textures of deposition and early diagenesis can be studied in modern settings (as in Chapter 1), but models based solely on Quaternary analogues are not adequate explanations.

Whatever the basin's regional or tectonic setting, the depositional signature of ancient evaporites can be broken down into three main building blocks: mudflats, salterns and deeper water slope/basin deposits.

Evaporitic mudflat sediments form laterally extensive units (10 - 100m thick) composed of stacked, shoaling-upward, matrix-dominated cycles (1 - 5 m thick). Individual salt horizons in the unit are not easily correlated due to the limited continuity of individual salt pans in the depositional setting (Figure 3.4a; Warren, 1991). Salts were deposited as mosaics of dry and saline mudflats (sabkhas), which separated local brine-filled depressions (salinas) in both continental and marine settings. The evaporite-hosting matrix was typically deposited by storm-induced, not tidal, currents that fed sediments and surface waters onto the mudflat both from the land and the basin. As the stormwaters began to dry up, windblown cm-thick evaporating brine sheets oozed and trickled back and forth atop the flat surface of the mudflat. The vast extent of the mudflat and resulting high humidity in the overlying air greatly slowed the rate of evaporation.

Thin sheets of brine probably covered extensive areas of the mudflats for weeks or months after a storm, and deposited thin layers and stringers a few centimetres thick of sub-aqueous gypsum or halite atop the freshly deposited silts and sands. Later, the brine sheet dried up, or was blown to another locality, leaving the freshly deposited subaqueous crust to the vagaries of air, sun, wind and

• *Huge shallow epeiric seas* often formed large inland seaways in arid areas, so creating areas of extensive platform and intracratonic evaporites. This style of evaporite deposit was commonplace during times of longterm tectonic quiescence and during low amplitude sealevel fluctuations of greenhouse earth. It was often associated with deposition of extensive carbonate platforms (Hay et al., 1988). Shoal-water strata of the backreef of the Permian Basin of west Texas and New Mexico were deposited in this fashion, as was the Cretaceous Ferry Lake Anhydrite of east Texas, the Red River Formation of the Williston Basin, Wyoming, and the Red Heart Dolomite of the Georgina Basin, Australia.

• *Set up of the correct tectonic and climatic conditions* for the formation of extensive evaporite deposits extending across whole depositional basins (e.g. saline giants such as Castile/Salado formations of west Texas and the Zechstein strata of Northwest Europe). These basinwide domains formed during tectonically active times, when a combination of climate and tectonism restricted seawater inflow to vast ocean-margin and lacustrine basins that had formed by incipient rifting, continental collision, and/or transtensional faulting. The last such saline giant formed 5.5 million years ago (late Miocene), when the collision of Africa and Eurasia converted the Mediterranean-Tethyan Sea into a chain of shoal-water evaporite lakes and the Zagros Basin into an diagenesis. Renewed exposure of the mudflat surface led to the growth of additional intrasediment evaporites, carbonate crusts, the alteration of the gypsum to anhydrite, and the partial or complete dissolution of the surficial brine sheet deposit. Throughout this time the evaporite unit was subject also to periodic influxes of aeolian silt blown in from adjacent dry mudflats.

The strandplain brines were ephemeral, or ponded as shallow surface waters within local salina depressions atop the mudflat. Hence, bedding planes in mudflat units often contain exposure features, such as intraclast breccias, mudcracks, erosional and karstic surfaces. Displacive evaporites, nodular and enterolithic $CaSO_4$, discoidal gypsum or anhydrite, and skeletal halite, grew chaotically through the saline mudflat units. In deeper surface depressions on the mudflats there were small but long-lasting brine pools (salinas - a few to tens of kilometres across and a few dm to metres deep) where brine tended to pond for months or years. Areas of longer-lived brines deposited lenses of salina evaporites up to a few metres thick. On the "landward side" the evaporitic mudflat beds were often interlayered with aeolian or fluvial braidplain/sheetwash sediments, which episodically prograded out over the mudflat.

The term tidal-flat or peritidal evaporite is inappropriate for these gigantic mudflats, as there is little or no tidal exchange about the edges of huge inland salt lakes or the more landward portions of epeiric seaways (Hallam, 1981; Warren, 1991). Strandplain evaporite is apropos describing the hydrodynamics of these vast expanses of evaporitic mudflats.

Examples of evaporitic mudflat-dominated deposits include; the Minnelusa Formation in Wyoming (Achauer,

1982); the seal facies in the Permian Basal Seven Rivers Formation in Yates field, west Texas (Warren, 1991); the northern regions of the Palo Duro Basin during deposition of the Upper San Andres Formation, Texas Panhandle (Hovorka, 1987); the more landward parts of the Lower Clear Fork Formation of Texas (Handford, 1981); the Cambrian Red Heart Dolomite of the Georgina Basin, Australia (Nicolaides, 1995); the Archaean Wittenoom Dolomite of the Hamersley Group, Western Australia (Kargel et al., 1996).

Salterns were areas where widespread subaqueous shoal-water evaporite deposition took place in both marine-fed and continental settings (Figure 3.4b). Saltern describes extensive shallow-subaqueous evaporite beds that form continuous depositional units across hundreds of kilometres in the hypersaline portions of ancient evaporite lakes or seaways (Warren, 1991). There are no modern counterparts in terms of scale or location, although shallow-water evaporite textures can be observed in many salinas along the coasts of Australia and the Mediterranean. Saltern sediments were typically cyclic and made up of shoaling units 2 to 50 m thick, sometimes capped by mudflat sequences. Cycles tended to be widespread and individual salt beds correlatable over distances of tens to hundreds of kilometres.

Individual beds were dominated by subaqueous, often laminated or layered evaporites. Crystals were precipitated from perennial brine sheets that were a few tens of centimetres or more deep and were deposited as single crystals, rafts, clusters and crusts (see Chapter 1). The end result was one or more superimposed shoaling-upward beds composed of one or two dominant minerals, usually halite or gypsum (now anhydrite). Horizontal laminae in the beds were often crosscut or interrupted by coarse-

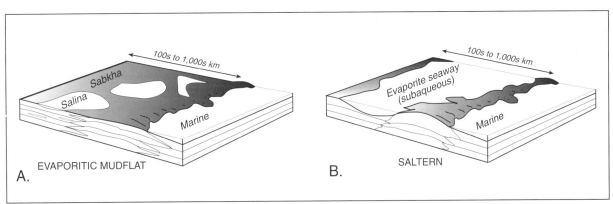

Figure 3.4. Depositional geometries of: A) evaporitic mudflat and B) saltern (After Warren, 1989,1991).

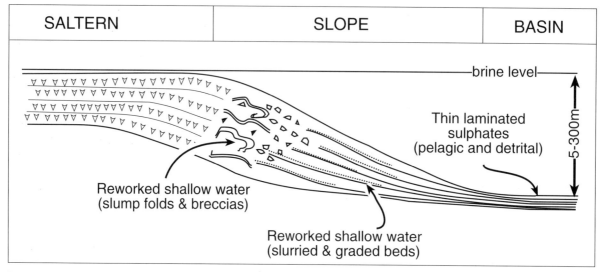

Figure 3.5. Depositional setting and characteristics of deep water evaporites (in part after Schlager and Bolz, 1977; Warren 1989).

grained to giant, vertically aligned crystals of calcium sulphate or halite. Many of the smaller crystals in the beds were reworked by wind and storms into subaqueous traction deposits characterized by graded and reverse-graded beds, rippled and crossbedded layers, and algal structures. The graded beds and the rippled layers probably recorded waning flood conditions (Hardie and Eugster, 1971). Reverse-graded beds recorded either episodes of storm surging (increasing energy levels) across the saltern, or episodes of brine dilution that had induced gypsum recrystallisation and slower crystal nucleation in the upper part of the laminae. Occasional freshening of the saltern was demonstrated by the occurrence of thin carbonate layers (calcite or dolomite) within the evaporite beds. Such interbeds often entrained cryptalgalaminites, micritic domal stromatolites, pellets from fossil brine shrimp and ostrocodes and rare vertebrate tracks.

Ancient examples of saltern deposits include the Cretaceous Ferry Lake Anhydrite in the northern Gulf of Mexico (Pittman, 1985), the P_1 and P_2 evaporites of the San Andres Formation on the northwest shelf of the Permian Basin, the Lower San Andres Formation of the Texas Panhandle (Elliott and Warren, 1989), the upper Visean anhydrites of southern Belgium (De Putter et al., 1994), and the Hith Anhydrite of the Middle East (Alsharhan and Kendall, 1994). In the subsurface and in wire line logs, an ability to correlate clean evaporites over large areas, and to define the mineralogical purity as more than 80% evaporite (on wireline crossplots), tends to separate saltern evaporites from mudflat evaporites (Warren, 1991).

Deepwater evaporites are usually only found in the deeper brine portions of basinwide evaporite successions and encompass slope and basin deposits (Figure 3.5). Fine-grained laminites accumulated in the central deeper parts of the basin, separated and intercalated with less saline salts as well as carbonates or organic matter. The fine layering reflects changes in saturation states of the surface brines or variations in rate of sediment supply and grain size from the more marginward slope deposit feeds. These "deepwater" laminites accumulated in two main ways. Either as "rain from heaven" with crystallites first forming in the upper supersaturated metre or two of the brine column and then settling to the deep brine seafloor. Or they are the distal puffs of evaporite turbidites and debris flows that characterise the basin slope. Slope deposits are mechanically reworked from saltern or mudflat sediments and are characterised by textures typical of deeper water, such as turbidite fans, Bouma cycles, slumps and debris flows.

Studies of basin-centre and slope evaporites in the Permian of Delaware and Zechstein basins, the Visean evaporites of the Canadian Atlantic, and the Miocene of Red Sea rift in the Yemen Republic indicate these ancient deepwater sequences can be interpreted in much the same way that deepwater deposits are interpreted in carbonate and siliciclastic depositional systems (Anderson et al., 1972; Clark, 1980; Richter-Bernburg, 1986; Schlager and Bolz, 1977; Schenk et al., 1994; El Anbaawy et al., 1992; Rouchy et al., 1995). The main difference with other deepwater clastics relates to water depth and the relative importance of "rain from heaven" versus resedimented shelf deposits in depositing the fine laminites of the basin centre.

A lack of a benthonic microfauna, and the need for varying degrees of drawdown to precipitate thick evaporite sequences, means water depth estimates are difficult in ancient slope and basin evaporites. Richter-Bernburg (1957) using a deep "water full" basin model for the Zechstein evaporites estimates water depths for anhydrite deposition of more than 1000 metres in the basin centre. More recent work by Schreiber (1988b) postulates a minimum relief of 100-200m for the shelf to base-of-slope transition during anhydrite deposition. Warren (1989, 1991) argues that determinations of deep versus shallow in a drawdown basin are difficult, but favours water depths of no more than 100-200m. Kendall (1992) argues deep water evaporites typically form in water depths greater than 20-40 metres; but, like Warren, notes that deep-water-appearing evaporites can form in brines only a few metres deep.

The mechanism driving resedimentation is also contentious; Schlager and Bolz (1977) favour turbidity currents, while Clark (1980) favours a control by storm activity. Unlike the deepwater anhydrites in these preceding examples, resedimented ancient halites are not readily recognised and their interpretation stems from their entrained nonevaporite clasts rather than from the textures in the finely layered halites (Czapowski, 1987). There are no widespread deepwater sulphates or halites forming today. The best modern analogues for deep water evaporites are in the much smaller deepwater (≈30m) sodium-sulphate lakes of Canada (Last, 1994) and the finely-laminated bottom carbonates of the Northern basin of the Dead Sea (≈300m; Garber, 1980; Ben-Avraham et al., 1993). Deeper bottom waters in many ancient evaporite basins were density-stratified and anoxic, or at best dysaerobic (Kirkland and Evans, 1981; Hite and Anders, 1991). Bottom sediments often entrained significant volumes of pelagic organic matter and, with suitable burial, yielded hydrocarbons or ultimately were converted to graphitic schists. The bottom sediments in the deepwater gypsum of the Gibellina Basin contain up to 7% TOC. At present the organics entrained in the evaporitic carbonates of the Sicilian basins are still thermally immature (Warren, 1986).

Evaporites: broad scale models

Saltern and mudflat evaporites, along with slope and basin deposits, construct three interrelated regional depositional settings in either continental or marine-fed basins:

Continental playa/lacustrine deposits These are constructed of stratiform salt units, often with the greater volume of salts forming in halite-dominated evaporitic mudflats and saltpans. Deposits range in thickness from metres to hundreds of metres, with lateral extents measured in tens to hundreds of kilometres. Such salt beds are separated and often surrounded by deposits of lacustrine muds, alluvial fans, ephemeral streams, sheet floods, aeolian sands and redbeds. Typically the salts accumulate in hydrologically closed or highly restricted discharge basins, with a shallow water table in the saline mudflat area and a more-or-less centrally located depression of ephemeral ponded brine that is characterised by a mosaic of salt pans. Evaporite mineralogy is dependent on composition of inflowing waters; suitable settings can accumulate thick sequences of trona, glauberite and thernadite as well as more typical sequences of halite and anhydrite.

Platform evaporites Stratiform beds (<50 m thick, often < 5 to 10 m thick) mostly of halite and anhydrite/gypsum deposited as mixed evaporitic mudflat and saltern evaporites, sometimes with lesser volumes of the bittern salts. Typically deposited across large marine platform areas that pass basinward into deeper water sediments (Figure 3.6a). In marine-margin epeiric settings these platform evaporites are typically intercalated with shoal-water marine-influenced carbonate shelf/ramp sediments, which in turn pass basinward across a subaerial sill into deepwater marine carbonates and landward into arid zone continental siliciclastics and mudflats.

Basinwide evaporites Thick, basinwide units (often > 50 m thick) of deepwater/shallow water evaporite deposits containing textural evidence of many different depositional settings, including mudflat, saltern, slope and basin (Figure 3.6b). When evaporite deposition took place the whole basin was evaporitic, often with a distinctive platform-slope-basin profile (e.g. Werra Anhydrite of northern Europe; Richter-Bernburg, 1986). Dominant mineralogies are anhydrite and halite, sometimes with bittern salts accumulating in the most saline parts of the basin. Let us now look in a little more detail at textural characteristics in these regional settings that allow a determination of depositional setting using core or outcrop.

Continental evaporites

Continental evaporite depositional settings have excellent Quaternary analogues that are discussed in detail along with case histories and representative textures and vertical sequences in Schreiber (1988a), Warren (1989), Melvin

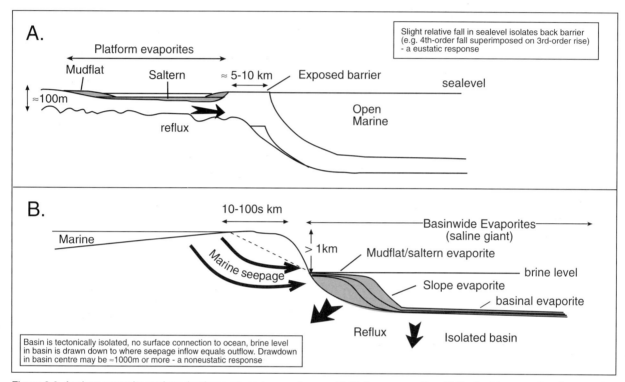

Figure 3.6. Ancient evaporite settings that have no modern analogues. A) Platform evaporites. B) Basinwide evaporites (see text for explanation).

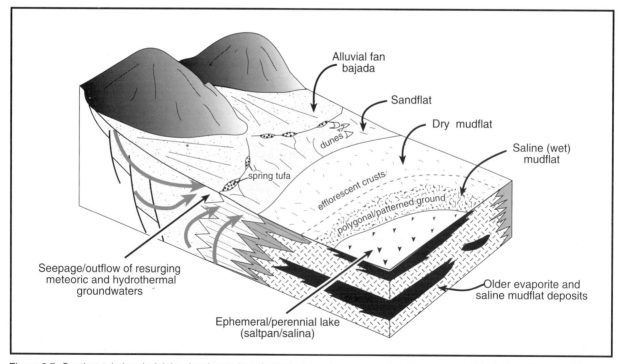

Figure 3.7. Continental playa/salt lake showing various facies belts and processes of solute supply that include surface runoff, meteoric and hydrothermal groundwaters (in part after Kendall, 1992; Eugster and Hardie, 1978).

(1991) and Kendall (1992). Three arid-zone depositional complexes entrain continental evaporites (Hardie et al., 1978):

- Alluvial fan–ephemeral saline lakes.
- Ephemeral stream floodplain - dune field– ephemeral saline lakes.
- Perennial stream floodplain–perennial saline lakes.

The *alluvial fan–ephemeral saline lake* depositional complex occurs in fault-bound, orogenic basins in tectonically active areas. Inflow into this setting is predominantly from groundwater, hydrothermal waters and occasional storm runoff. Due to the importance of seepage supplied from the adjacent ranges and bounding faults, this setting is one that often contains extensive continental mudflats that occur within a well defined sedimentary transition from the basin margin to the basin centre (Figure 3.7):

alluvial fan/ephemeral stream/floodplain —> sand flat —> dry mudflat—> saline mudflat —> salt pan

The dry mudflat and the saline mudflat are sometimes grouped and referred to as the ephemeral saline lake facies belt.

This depositional complex is probably the best documented type of Quaternary continental evaporite system. Examples include intermontane playas in the Basin and Range of North America and central British Columbia, some East African rift valleys and playas on the Tibetan Plateau and Mongolia. Ancient counterparts of this style of deposit include Wilkins Peak Member of the Eocene Green River Formation, Utah, the Tertiary borate deposits of Turkey and Cambrian Lake Parakeelya of the Officer Basin, South Australia (Fischer and Roberts, 1991; Helvaci, 1995; Southgate et al., 1989).

The *ephemeral stream floodplain - dune field–ephemeral saline lake* complex is found in tectonically stable areas associated with arid-zone subtropical high-pressure belts. Continental mudflats occur in this lacustrine complex, but to a lesser extent than in the alluvial fan– ephemeral saline lake complex. Most of the sediments in this setting are composed of evaporite-free sand flats or mud flats. The salt that does occur in the systems tends to be recycled as 1-3m thick crusts that dissolve and reprecipitate with each flood event. Haloturbated muds and shales, rather than preserved thick salt beds, indicate the more saline parts of this depositional setting. Inflow is via artesian seepage, storm runoff or long distance ephemeral floods, with intense storm and flood events often spaced years apart. In Lake Eyre, a modern example of this setting, the major floods are fed by long distance southward transport of flood waters from intense monsoonal outbursts in northern Australia. The dune fields in such complexes are derived by aeolian reworking of the floodplain and lake sediments during the dry periods between storms. Quaternary examples include Lake Amadeus, Lake Callabonna, Lake Eyre, Lake Frome, Lake Gairdner and Lake Torrens in inland Australia. Ancient counterparts include portions of the Lower Permian Rotliegendes of northern Europe, the Late Triassic of northwest Europe (Figure 3.8) and the Palaeoproterozoic Whitworth Formation of the Mt Isa Block, Australia (Glennie, 1989a,b; Talbot et al., 1994; Simpson and Eriksson, 1993).

Perennial stream floodplain–perennial saline lake depositional complexes occur in several tectonic settings, with deposits typically lacking extensive mudflats. Modern

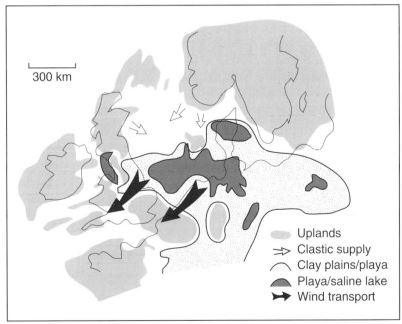

300 km

Uplands
Clastic supply
Clay plains/playa
Playa/saline lake
Wind transport

Figure 3.8. Late Triassic palaeogeography of northwest Europe showing how deposition was predominantly in an endoheic basin and wind was blowing dust from mudflats about large discharge playas (after Talbot et al., 1994).

perennial saline lakes are characterized by steep slopes surrounding the lakes, this allows saline water to pond in an arid climate. Any sabkhas that are present are localized narrow muddy embayments about the lake shoreline. A lack of extensive mudflats is the result of steep basin sides and the associated lack of an extensive capillary zone. Quaternary examples include: the Dead Sea, Israel; Great Salt Lake, USA; Salda Lake, Turkey; Lake Tanganyika, East Africa.

Water inflow is more permanent than the other two types listed above and is localized to rivers and a few arroyos as well as fault-fed springs. It tends to be from point sources, and not diffuse about the lake periphery. Strandlines tend not fluctuate as rapidly as in ephemeral lakes, although longterm cycles of more humid and drier climatic intervals are often reflected in the mineralogies and diagenetic alteration of the lake sediments. In modern situations, the sedimentary facies fringing the lake are commonly old lake-bottom deposits and are not the depositionally active sand/mud flats that characterise the preceding two settings. Such exposed lacustrine sediments located peripheral to the perennial lake can be "sabkharized" by longterm falls in the lake water level (Salt Flat Graben; Hussain and Warren, 1989). Ancient counterparts include Tertiary-age magnesite deposits of Lake Kunwarara in Queensland, Australia (Schmid, 1987), the Visean Loch Macumber of Atlantic Canada (Schenk et al., 1994) and possibly the Jurassic Todilto Formation of New Mexico (Kirkland et al., 1995).

Platform evaporites

Saltern and mudflat evaporites on ancient platforms were laid down in one of two settings, either on a *ramp* or on a *shelf* (Figure 3.9). Platform evaporites were typically deposited as part of a series of stacked shoaling cycles, the evaporite stage was repeatedly interbedded with open-marine shoal-water carbonates or continental siliciclastics. The only exception to this association was a platform that was drowned soon after

evaporite deposition, there the platform evaporites were covered by dark low-energy wackestones and mudstones deposited in a deepwater sediment-starved basin.

Evaporitic ramps

Evaporitic ramps were characterized by very low slopes into open-marine waters with no substantial break in slope on the marine portion of the platform. Slopes down the ramp were as low as 1 in 50,000 compared with modern shelf gradients of between 1 in 500 to 1 in 2500 (Ahr, 1973; Read, 1985; Burchette and Wright, 1992). Modern geomorphic counterparts are the carbonate ramps of the southern Arabian Gulf and the Yucatan Platform, but there are no modern geological counterparts to ancient evaporite-covered ramps. Extensive shallow epicontinental or epeiric seas were typical features on platform ramps and were first

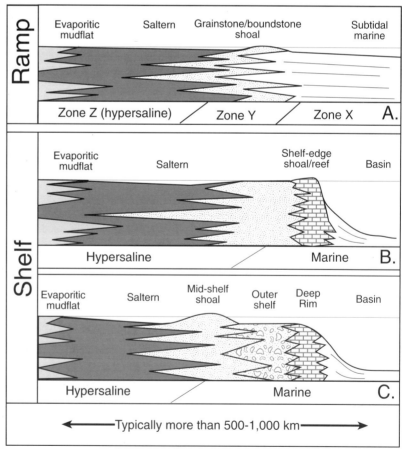

Figure 3.9. Evaporite depositional styles - ramp and shelf. A) Gently deepening ramp with restriction created by exposed zone Y shoal. B) Rimmed shelf with restriction supplied by exposed reef/shoal that defines the platform margin. C) Rimmed shelf with deep rim and restriction supplied by exposed mid-shelf shoal (in Warren, 1991).

modelled by Shaw (1964) and Irwin (1965). Irwin's epicontinental-sea model proposes there were three zones of varying hydraulic energy on ancient platform ramps (Figure 3.9a):

• *Zone X* was a wide low-energy region situated below wavebase on the open-marine side of the platform so that sediments were mainly bioturbated marine mudstones and wackestones. It was often hundreds of kilometres wide and locally characterized by limited water circulation with occasional stagnation. Now and then this led to the deposition of organic-rich laminated sediments, especially in the deeper parts of intrashelf basins (Ayres et al., 1982).

• *Zone Y* was a narrow, intermediate to high-energy belt beginning where waves first bottomed on the seafloor, it extended landward to the limit of any tidal action. Sediments in this zone included sand and grainstone shoals, reef mounds and patch reefs.

• *Zone Z* was an extremely shallow, restricted low-energy region on the landward side of zone Y. In arid settings, where zone Z was isolated from zone X, extensive saltern and mudflat evaporites were deposited. At times when zone Z and zone X had a surface connection through zone Y, the basinward portion of zone Z was covered by restricted marine sediments.

Much of the marine wave energy dissipated some distance basinward from zone Z as it spent itself against the shoals, bars, and reefs of zone Y. Shallow wave-agitated facies of zone Y passed down the slope into deeper-water subtidal open-marine facies of zone X without a rapid change in slope angle. Behind the zone Y shoal, the zone Z slopes were even less than on the rest of the ramp and had typical bottom angles of only a few seconds of arc. Thus, huge areas of zone Z were covered by extremely shallow water, much of it only tens of centimetres deep. During evaporite deposition in zone Z there was no effective surface connection to the ocean and so tidal effects were negligible. Zone Z water levels fluctuated most rapidly during intense storms or hurricanes as sheets of sediment-laden seawater were washed landward over zone Y into zone Z, or during rare episodes of flash-flooding that were generated by torrential rainstorms in the continental hinterland. Both marine and meteoric influx freshened the brines of the mudflat and the saltern as sheets of water migrated tens of kilometres across the strandplain. At the same time, a new

pulse of marine or continental clastic sediment was delivered to zone Z.

Evaporite-forming areas in zone Z were often hundreds of kilometres wide, far wider than any evaporitic area on the modern land surface. Today, the evaporite-depositing area in the Arabian Gulf (a modern-day ramp counterpart) is only 10 to 15 km wide (Figure 3.2). During the Permian, in times of widespread evaporite deposition on the Palo Duro Basin and the northwestern shelf of west Texas, one could have travelled 50 to 150 km south and seaward from the strandplain of the brine sheet and never crossed through water that was much more than a metre or two deep, nor less saline than gypsum-saturated (Handford, 1981; Hovorka, 1987; Elliott and Warren, 1989).

On most ramps there was some degree of topographic relief about zone Y, and some restriction of water flow across it (Figure 3.9a). Vertical sequences of evaporites deposited in zone Z were very much dependent on the relative physiography of zone Y. If zone Y formed a well developed shoreline with extensive washover flats behind, then zone Z was an evaporitic mudflat; a mosaic of sabkhas and salinas (Figure 3.4b). If on the other hand, zone Y was a shoal backed by an evaporite lagoon, then the succession in zone Z would consist of saltern evaporites with an increasing proportion of subaerial mudflat indicators (desiccation cracks, karst, erosion features, displacement textures) toward the top of the unit and toward the strandplain (Figure 3.4a).

Units P₁ and P₂ of the San Andres Formation in the northwestern platform of the Permian Basin and the Palo Duro Basin illustrate the saltern-ramp end member (Bein and Land, 1982; Elliott and Warren, 1989). The northern portions of the Lower Clear Fork Formation of the Palo Duro Basin illustrate the mudflat-ramp end member; it was an evaporitic mudflat passing northward into a desert complex of wadis and dunes (Handford, 1981, 1991). Southward the Clear Fork ramp passed into saltern deposits and then into open-marine carbonates of zones Y and X (Ramage, 1987). Relationships across the Clear Fork ramp are indicative of the facies relationships of many ancient siliciclastic/evaporite ramps, where extensive areas of saltern evaporites were typically backed by evaporitic mudflats or sand-sheets of arid-zone siliciclastics and evaporitic carbonates. Little if any open-marine carbonate matrix was ever transported this far back onto the ramp, even during intense hurricane events, and water conditions were too saline for most carbonate-secreting organisms with the possible exception of cyanobacterial and halobacterial mats.

75

Evaporitic rimmed-shelves

Ancient rimmed-shelves were shallow-water carbonate platforms with an outer wave-agitated edge marked by a pronounced increase in slope (commonly a few degrees to more than 60°) into deep water (Read, 1985). During marine carbonate growth the rim was continuous to semicontinuous, characterized by high-energy reefs, reef mounds, and the intervening grainstone shoals and flats. Rimmed shelves require biogenic binding and cementation to maintain an elevated rim, they cannot form on siliciclastic platforms nor can they flourish if evaporites are depositing in the hinterland. Both these scenarios set up inimical seaward-flushing waters. During evaporite deposition, the platform rim was subaerially exposed and the area behind the rim was often composed of seepage-fed evaporitic carbonate flats that passed landward into salterns and then into mudflat evaporites or arid zone siliciclastics (Figures 3.6b and 3.9b,c).

The degree of restriction in the evaporite depositing area was controlled by the continuity of the rim. To form an evaporitic unit in the lagoonal area behind the rim/shoal required the crest of the rim to be at or above sea level, with any channels cutting across the rim and its sandflat to be choked with sediment and so prevent the free exchange of marine and back-rim waters. Otherwise, like the carbonate platforms of today, conditions behind the shelf rim or shoal would have never become saline enough to deposit evaporite beds and, at best, restricted marine carbonates would have been deposited.

Two evaporite end members were deposited landward of the rim, either saltern sequences or mudflat-dominated successions (Figure 3.4). The Cretaceous Ferry Lake Anhydrite in East Texas is an excellent example of the saltern end member (Loucks and Longman, 1982). It is composed of stacked laterally extensive units of subaqueous anhydrite after gypsum. The evaporites are intercalated with normal-marine-lagoonal bioclastic/oolitic wackestone and the whole package lies behind a platform rim composed of rudistid grainstones and reefs.

Basinwide evaporites

Basinwide evaporites are thick basin-filling evaporite units that extend across large areas of a sedimentary basin and contain textural evidence of a variety of depositional settings ranging from deepwater to saltern to mudflat (Figures 3.6b, 3.10). They have no modern analogue in scale or diversity and are often described as "saline giants". Basin-central evaporites is another term to describe this grouping of deposits (Kendall, 1992). I prefer the term basinwide as the evaporite sequence can fill the basin centre and then climb out over the old basin margin (e.g. Delaware Basin). Basinwide evaporites constitute the sedimentary fill in many ancient evaporite basins, such as the Delaware Basin of west Texas, the Zechstein Basin of Europe, the Louann Salt Basins in the Gulf of Mexico, the Hormuz Salt Basins in the Arabian Gulf and the Late Miocene Messinian Subbasins in the Mediterranean (Figure 3.11).

The total evaporite fill in such basins was often hundreds to thousands of metres thick and composed of stacked "brining upward" evaporite cycles tens to hundreds of metres thick. Saltern evaporites were common in the "platform" areas of the basin, or across the whole depositional area of the basin in the latter stages of infill or at times of basinwide evaporative drawdown (Figures 3.6a, 3.10). Compared with saltern units in marine-margin platform evaporites, the saltern units in basinwide evaporite basins tend to stack into much thicker units (>100m) that are not restricted to the platform but extend at times across the whole basin. In the basin-centre position they are sometimes intercalated with deeper-water slope/basin evaporites. The latter group were typically deposited during the early stages of each episode of basinwide evaporite infill and pass rapidly up into saltern and mudflat deposits. Stacking of saltern cycles in a basinwide sequence indicates the basins did not fill in a single depositional episode, but

Figure 3.10. Venn plot showing the diversity of environmental settings possible in basinwide evaporites.

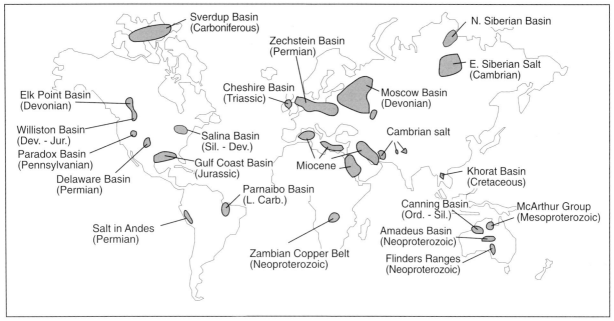

Figure 3.11. Map showing location and age of some of major basinwide evaporite deposits. All of the Proterozoic examples, except the Amadeus Basin, are characterised by dissolution breccias and pseudomorphs, not the actual salts. Owing to the small size of the map, basins are drawn for visual clarity not to exact scale (see Figures 3.2 and 3.3).

as a number of brining-upward and/or shoaling-upward depositional cycles.

Basinwide evaporites were deposited in three main settings (Figure 3.12; Warren, 1989; Kendall, 1992):

- Deep water–deep basin
- Shallow water–shallow basin
- Shallow water–deep basin

Deep water–deep basin evaporites have basin centres dominated by "deepwater" evaporites composed mostly of finely laminated salts, where individual laminar grouping can be correlated over wide areas (Figure 3.12a). Basin slopes are dominated by reworked saltern evaporites deposited as slumps and turbidites. Seawater flows into the basin as a perennial seepage; if the basin remains isolated, the evaporite unit can fill the basin to a water level just below that of the supplying ocean. Consequently, the evaporite unit "in toto" is a shallowing upward succession often composed of smaller shallowing-upward cycles. The basin fill starts off dominated by deepwater laminated and resedimented salts that pass up section into shallow-water salterns, mudflats, or continental playas. The Castile Formation and the transition into the overlying Salado Formation in the Delaware Basin in west Texas is a good example of this passage from predominantly deep to predominantly shallow. The basin-centre started off

relatively deep with the deposition of the varved sediments of the Castile Formation, and with time filled and shallowed into the shoal-water/playa deposits of the Salado Formation. Any sediments that immediately postdate the evaporites of this overall shallowing sequence will also be shoal water or mudflat deposits. This is the case with the shoal-water/mudflat deposits of Rustler Dolomite where it overlies the Salado Formation in the Permian Basin, west Texas. Other good examples of deeper water deposition occur in the basin-centre deposits of the north and south Zechstein basins of NW Europe.

The *shallow water–shallow basin* is formed under conditions of ongoing subsidence where the rate of evaporite aggradation can keep pace with tectonic subsidence (Figure 3.12b). The basin starts off shallow and remains shallow throughout the deposition of the evaporite sequence. Sedimentary signatures are dominated by interfingering saltern and mudflat successions often crosscut by karstic erosional surfaces. The whole package is composed of a number of shoaling-upward evaporite cycles. Shallow water–shallow basin evaporites are often found as a rift valley or aulocogen fill. The basal evaporite succession in the rift is continental and passes up section into marine evaporites as the rift basin opens to marine seepage across a series of restrictive transform and transfer faults (Warren, 1989).

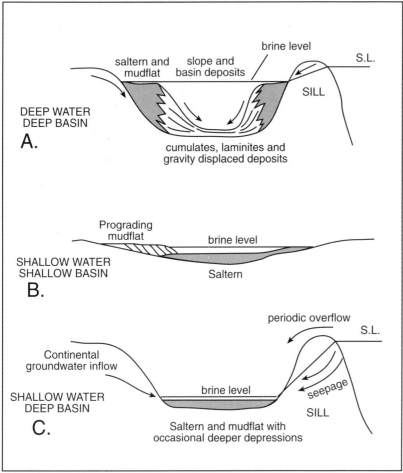

saltern and mudflat

slope and basin deposits

brine level

S.L.

SILL

DEEP WATER
DEEP BASIN

A.

cumulates, laminites and
gravity displaced deposits

Prograding
mudflat

brine level

SHALLOW WATER
SHALLOW BASIN

B.

Saltern

periodic overflow

S.L.

Continental
groundwater inflow

brine level

seepage

SHALLOW WATER
DEEP BASIN

C.

SILL

Saltern and mudflat with
occasional deeper depressions

Figure 3.12. Basinwide evaporite settings (modified from Warren, 1989; Kendall, 1992).

supplied by evaporative drawdown, pulling seawater into the basin as a series of marginward seeps perhaps interrupted by the occasional waterfall (Figure 3.12c). The evaporite unit was often underlain by deepwater marine sediments deposited before the basin depression was isolated from a continuous surface supply of seawater. Likewise the evaporite succession is often overlain by deepwater sediments, which are deposited once open-marine waters gain entrance to the basin at the end of evaporite deposition. Messinian evaporites in the various sub-basins that underlie the present deep floor of the Mediterranean Sea are probably the best documented example of this type of basinwide evaporite.

In reality, the interpretation of basin-filling evaporites is never as simple nor as predictable as the three-end-member classification suggests; it should only be used as a guide, not a gospel. Changes in brine level, tectonics or climate, play havoc with simplistic models of evaporite basin type. Relative water level in an evaporite basin is never constant, it rises and falls in response to intensity of tectonism, climate, sea level change, playa capture and changes in the degree of seepage and surface runoff in the inflow zones. Large evaporite provinces such as the Mediterranean Basin during the Late Miocene, or the Permian Basins of Texas and northern Europe in the Late Permian, are actually composed of a series of sub-basins. Each sub-basin in an evaporitic province responds in its own fashion, depending on the topography, tectonics, and climate. During the Messinian in the Mediterranean some evaporite basins were filling with predominantly shallow-water gypsum, while at the same time subjacent basins were filling with deepwater successions and yet others were filling with potash salts.

High resolution chronological analysis has shown that not all the Mediterranean sub-basins are time synchronous (Clauzon et al., 1996; Butler et al., 1995). Two time-separate styles of evaporite deposition characterise the Messinian interval: 1) marginward areas around the periphery of the Mediterranean, versus 2) more basinal or

This is the general sequence beneath Atlantic continental margins (Evans, 1978) and the sequence that is currently forming in the Ethiopian sector of the African Rift. It also characterises most intracratonic evaporite/epeiric basins, such as in the Michigan and Williston Basins in the US and the Canning Basin in Australia. Unlike the other basinwide evaporite deposits these intracratonic salts are commonly intercalated with shallow-water open-marine carbonates and siliciclastics. Using data from one or two wells it can be very difficult to distinguish this type of basinwide evaporite from a marine-margin platform evaporite. The distinction requires a thorough knowledge of the basin's tectonic history and is often best resolved using regional seismics.

The *shallow water–deep basin* sequence is dominated by shallow water evaporites deposited as stacked saltern and mudflat cycles in a basin with a base level hundreds to thousands of metres below sea level. This type of basin is central regions (Figure 3.13). Their diachroneity was deduced from the stratigraphic relationships linking these evaporites to a major Messinian erosional surface (M-reflector). During the first phase (from 5.75 to 5.60 Ma), the deposition of evaporites took place in Spain and Sicily and other marginward areas in response to a modest sealevel fall; in the second interval (from 5.60 to 5.32 Ma), more central parts of the Mediterranean basin became isolated. During this latter period, the deposition of basinwide evaporites across the deeper part of the Mediterranean took place along with the cutting of the Mediterranean canyons and the formation of a widespread erosional surface. The situation in Sicily was further complicated by mineralogy evolving in a series of uplifting foreland basins during active thrusting. In other

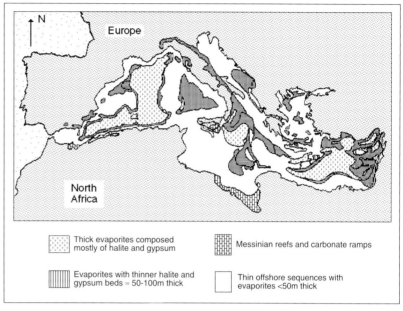

Figure 3.13. Distribution of onshore and offshore Messinian evaporites (after Warren, 1989).

marginal basins, such as in Cyprus, the changes in the level of drawdown were related to ongoing transcurrent faulting and rifting (see later in this chapter). Much of the tectonic drive was a result of the collision of Africa with Eurasia, which ultimately caused the whole Late Miocene Mediterranean basin to become isolated and evaporitic.

Thus, complexity of evaporite style, changing with time, intrabasin location and tectonics, does not lend itself to a simple three part classification. Nor can we come to the modern for counterparts of all the depositional styles. Basinwide evaporites are simply not found in the present epoch, as current interactions between tectonics, hydrology and climate are not capable of forming a saline giant.

The importance of hydrology

Continental evaporites accumulate in endoheic or closed basins. Marine-fed evaporites require a high degree of isolation from the ocean to produce thick sequences of both basinwide and platform evaporites. Lucia (1972) studied the interrelationship between maximum salinity in an evaporite basin and the basin's water balance (water loss versus gain). He showed that for halite to precipitate any surface inlet from the ocean must be at least one hundred million times smaller in cross-sectional area than the surface area of the basin; for gypsum it must be at least

a million times smaller (Figure 3.14). That is, any basin accumulating thick beds of evaporites effectively lacks a continuous surface connection to the ocean or other standing body of brackish water.

Maiklem (1971) in a study of evaporative drawdown in the Elk Point Basin of Canada showed that even with a high permeability of 73 darcies in the Pesqu'ile barrier it did not have sufficient transmissivity to prevent evaporative drawdown. Kendall (1989) recalculated his figures using more realistic values of 0.9-2.7 darcies for permeability in the barrier and concluded that evaporative losses were 200-800 times greater than the rate of supply. Any saline giant precipitating bedded salts is subject to evaporative drawdown and seepage influx of marine-fed phreatic waters (Figure 3.12a,c). A corollary is that the time of thick evaporite accumulation correlates to a time of unconformity or hiatus formation in the adjacent nonevaporite lithologies making up the surrounds of the depression.

An evaporite basin is always located in a large isolated depression where more water is being lost by evaporation than entering as seepage inflow and runoff. A potentiometric gradient (elevation head) exists between the floor of the basin and its surrounds. Such large deep holes in the landsurface attract not just marine seepage but also ground and formation seepage from the zones adjacent to the hole. In a drawdown saline-giant basin with a predominantly

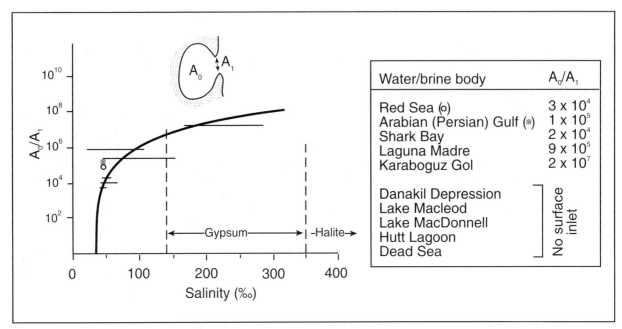

Figure 3.14. Relationship between the ratios of the cross-sectional area of a surface inlet and the surface area of the associated hypersaline basin plotted against salinity in the basin, with examples from several evaporitic basins. Note that Quaternary examples that are accumulating substantial evaporite beds have no surface connection and are fed either by continental inflow or by marine seepage (modified from Handford, 1991; Lucia, 1972).

marine source, the seawater enters mostly as resurging brines through springs in the basin floor. In a saline giant fed by nonmarine water, the water enters by a combination of occasional storm runoff and rainfall, as well as by groundwater and hydrothermal springs.

Once water ponds in an evaporitic depression it must remain for some time to concentrate at the surface or just below the land surface. Otherwise the depression cannot accumulate thick sequences of salt. Water would seep and evaporate away before it attains salinities high enough to precipitate salts. The precipitation of an initial salt layer, usually as capillary cements in the basin floor sediments, creates an aquitard that slows the rate of downward brine loss via density-driven reflux. The lowering of permeability in turn creates a hydrological curtain, where a slowly leaking dense brine plume builds up in the basin floor and bathes the body of the accumulating evaporites in saturated and supersaturated waters. I call this the zone of cement-driven slowing of brine flow the brine curtain (Figure 3.15). Today this process isolates the lowermost regions of Holocene salinas of western and southern Australia (Warren, 1991). Rates of brine escape or brine residence time within the brine curtain control many of the backreactions discussed in Chapter 2 (e.g. halite pseudomorphs after subaqueous gypsum - Figures 1.13 and 2.8, Hovorka, 1992; carnallite in Lake Qaidam -

Figure 7.3c, Casas et al., 1992; and the mineralogy of the Holocene Coorong Lakes - Warren, 1990).

Once the brine curtain hydrology is established and actively cycling hypersaline groundwaters, it is not easily displaced by less-dense marine or meteoric groundwaters. The curtain creates an autochthonous hydrology of stable hypersaline pore waters in which the evaporites can continue to accumulate and secondary cements and alteration haloes can form (Figure 3.15). Once formed, it also establishes a basin-centre reflux plume; with wedge-like margins that force most of the less dense reflux and meteoric groundwaters to enter the basin as a series of seeps and springs about the edge of the curtain. Upon burial the evaporite cements associated with the base of the curtain are typically the first part of the evaporite sequence to be dissolved by resurging basinal and compactional waters.

Locally, the brine curtain may be breached well out into the salt-saturated parts of the basin. This occurs atop areas of exceptionally strong upwelling or artesian outflow. In some ancient drawdown basins this localised outflow occurs in zone of pinnacle reefs and can be associated with the formation of local haloes of anomalous less saline mineralogies. For example, an apron of secondary anhydrite fringes carbonate mound springs in Lake Frome and is in turn surrounded by a laterally extensive halite plain/crust.

Unlike the hydrologies of most ancient basinwide evaporites, the brine curtain in many Quaternary continental lakes in central Australia ends not far below the surface of the lake sediments. In lakes such as Lake Eyre and Lake Amadeus, the brine curtain is destroyed at depths of a few to tens of metres by dilution from resurging brackish artesian waters that underlie most of the lake bottoms (Figure 2.10; Jankowski and Jacobson, 1990; Schmid, 1988). This brackish upwelling regional groundwater flushes salts from nearsurface lake sediments. It prevents stacking of thick bedded salt within the playa depression. Any buried salt beds are subject to dissolution along their underside by ongoing brackish water inflow. This open hydrology maintains most of the salt budget in these lakes in a liquid or dissolved ionic form. The only solid salt to accumulate in such lakes is a continually recycling salt crust that is never more a few metres thick. During times of complete desiccation it can cover the lowest parts of the lake depression. Even this is partially or completely redissolved during occasional lake floods. Where resurging artesian waters break though the brine curtain to the lake surface, as along the fault line that cuts across Lake Frome, they feed prominent lines of spring mounds on the lake floor.

Logan (1987) and Kendall (1992) use the term "hydroseal" to describe the lowermost salt layer generated by the brine curtain in an evaporite sequence. They interpret it, and not the formative reflux hydrology, as the fundamental cause of evaporite accumulation and the varying mineralogies in the basin. The notion of a "hydroseal" is not strictly correct, as secondary or capillary salts precipitating in any zone of refluxing brine must be by definition forming in an aquitard and not an aquiclude. It slowly leaks, to varying degrees, across the whole basin floor. Even when the salt bed becomes totally cemented it still dissolves from its underside to supply brine to a deeply buried brine curtain (see Chapter 2). As the brine slowly seeps down through the brine curtain it displaces and mixes with underlying compactional and basinal waters, as well as any more marginward meteoric or marine inflows. The term "seal" is not appropriate and the notion of the base of an evaporite bed as a total seal, which is not subject to dissolution, is one of the reasons why the significance of basal anhydrites and other diagenetic deposits, such as some of the Kupferschiefer ores, were not attributed to an inherent evaporite dissolution hydrology (Chapters 4, 8).

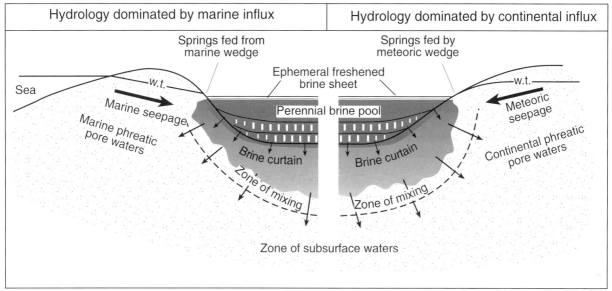

Figure 3.15. Schematic of a brine curtain in a marine-fed and a meteoric-fed evaporitic setting (V:H ≈ 10:1). Once established the brine curtain is actively maintained by ongoing evaporation and reflux, as well as by the impervious nature of the salt bed (highly effective aquitard). It is difficult to displace the curtain via the marginal influx of the less dense marine of meteoric waters. These tend to float atop the denser waters of the curtain and move into the salt-filled depression via groundwater wedges that feed spring seeps about its edges. Short-term high-volume influxes of fresher waters tend to form a floating brine sheet across the brine body, which disappears once it is concentrated to where its salinity and density equals that of the underlying brine body. A longer term fluctuation in the volume of freshened inflow can cause changes in the mineral equilibrium that is depositing salts on the perennial brine covered floor. It can lead either to a freshening or a salinity increase in the waters of the brine curtain. Such longterm changes in flow volume favour dissolution, alteration or backreaction of preexisting mineral phases, as can complete drying of the brine body, which may follow a longterm regional lowering of the regional water table (see Chapter 2).

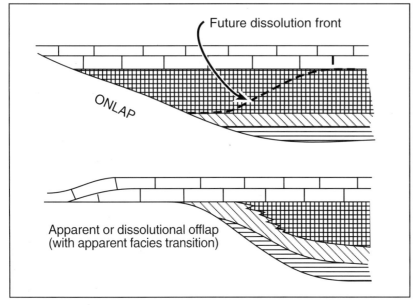

Figure 3.16. Dissolution of evaporites about the margin of a basin can create an apparent offlapping stratigraphy from an onlap situation (after Kendall, 1988).

Problems in correlation

The preceding hydrological analysis shows there can be problems when attempting intra- and inter-evaporite correlations across large distances. Shaw (1977) in a landmark paper argues that a classic silled-basin model, with evaporites accumulating in the basin centre and a restricted surface connection to the ocean defined by growing reefs, is hydrostatically unsound. He goes on to demonstrate that in such a scenario deepwater evaporites cannot accumulate. Kendall (1988) goes on to illustrate that only when a basin is totally isolated while depositing salts, are deepwater evaporites possible (Figure 3.12a). He also points out that for deepwater salts to accumulate the whole basin must be simultaneously at saturation with a single mineral phase. Otherwise the inclination of the brine halocline, required to deposit gypsum in one area of the brine lake and halite in another area of the same brine body, is inherently unstable. Such a fluid system cannot exist over the longterm in a single brine body, as a body of liquid cannot support shear. The inclined halocline required by this model would spontaneously collapse to regain its gravitational equilibrium. Once it had restratified to a horizontal halocline the precipitation of salts in the deeper brine body ceases until the salinities of the upper and lower brine bodies equalise (Chapter 2).

In terms of stratigraphy, this means a thick subaqueous bedded halite intersection in one well is an unlikely time correlative of a thick subaqueous anhydrite/gypsum bed in

an adjacent well a few kilometres away. The only exception is if they were deposited in nonconnected water bodies or in two separate sub-basins. Monominer-alogic subaqueous evaporite beds are perhaps one of the few examples of true "layer cake" stratigraphy. Many of the supposed correlations and basin reconstructions in the literature, which propose that deep subaqueous halite in the basin-centre is time-equivalent to a shallower subaqueous basin marginward anhydrite and to a restrictive but growing reef carbonate in the seawater inflow area, are erroneous. They indicate a lack of understanding of the physical requirements of drawdown hydrology in hypersaline systems. By implication, the geological inter-pretations and correlations in the same regions will be marked by the nonrecognition of the significance of erosion surfaces, hiati or dissolution horizons in adjacent wells or in the carbonates or other marine strata that occupy the basin margin.

Kendall (1988) lists some further factors that should be addressed when correlating evaporite units or applying the principles of sequence stratigraphy to evaporite basins:

1. A lack of stratigraphic control; evaporite depositing areas tend to be populated by halotolerant organisms that have low biostratigraphic potential. This means that biostratigraphic faunal analysis can be used to bracket the evaporite group (using fossils in nonevaporite strata above and below the unit of interest); but intra-evaporite subdivision is near impossible.

2. The rapidity of evaporite deposition; evaporites accumulate at some of the most rapid sedimentation rates known. Even if fossils are present in cyclic units that separate a number of evaporite beds, the very short time required to deposit thick evaporite sequences means the biota are incapable of evolving fast enough to be used for biostratigraphic control of the evaporites.

3. Complexity of stratigraphic relationships; stratigraphic relation-ships between the surrounding nonevaporite lithologies and the evaporites are typically dynamic and problematic. Their hydrology and rapid rate of deposition mean evaporites accumulate so rapidly that lateral

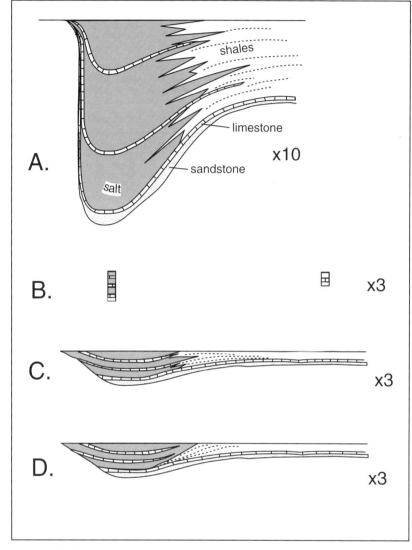

Figure 3.17. Effect of vertical exaggeration on correlation between thick evaporite sequences and marginal clastics or carbonates (after Busson, 1980; Kendall, 1988). A) Conventional correlation with significant vertical exaggeration showing lateral equivalency of evaporites and marginward clastics or carbonates. B) Original data drawn with less vertical exaggeration. C) Conventional correlation with less vertical exaggeration (hydrologically unlikely in an evaporite basin accumulating thick salt beds). D) Alternative correlation, which assumes drawdown is require and so there is no equivalency between salt beds and marginward clastics or carbonates (hydrologically likely in drawdown basin). Alternative D) is reasonable to most geologists when not suffering from a psychological need to draw near-horizontal correlation lines. It implies times of thick salt accumulation were distinct from times of carbonate deposition.

evidence of thinning, especially about the edges of a thick evaporite succession where flushing by meteoric and basinal waters is most likely to occur. If the evaporite edge dissolves completely, its remnants (thin dissolution breccias) may be missed, and a whole episode of basin history and structural evolution ignored. Even when evaporites remain in the basin, the greater degree of marginward dissolution means the current distribution of mineralogies may bear no relation to their original distribution in a basin (Figure 3.16). Removal of the marginal portion of a halite bed that overstepped an earlier anhydrite unit can convert an onlapping situation to an apparent offlapping situation. Likewise many "bull's-eye" basins may be a result of marginward dissolution of more soluble evaporite layers.

5. Use of vertically exaggerated cross section; their training makes most geologists reluctant to draw steeply inclined bed-to-bed correlations when more horizontal possibilities are available (Figure 3.17; Kendall, 1988; Busson, 1980). A propensity for large vertical exaggeration in basin cross sections leads to a tendency to draw correlation lines that show transitional facies between the evaporite beds and the surrounding nonevaporites, and to impose age equivalence of evaporites and nonevaporites. But the inherent hydrological constraints required to form a platform or basinwide evaporite fill means salt beds tend to accumulate rapidly in isolated depressions on the land's surface. Times of evaporite deposition correlate to times of isolation and so to times of nondeposition or gaps (unconformities) in adjacent nonevaporite strata. In such situations we should be quite happy to draw inclined correlation lines showing that at the time of evaporite deposition there was little or no deposition taking place in the adjacent nonevaporite strata (Figure 3.17d).

correlatives to the evaporite deposit are at best thin, or more typically, are surfaces of erosion or nondeposition.

4. A high likelihood of partial to complete subsurface dissolution, almost from the time the salts accumulate. This means most ancient evaporite deposits show clear

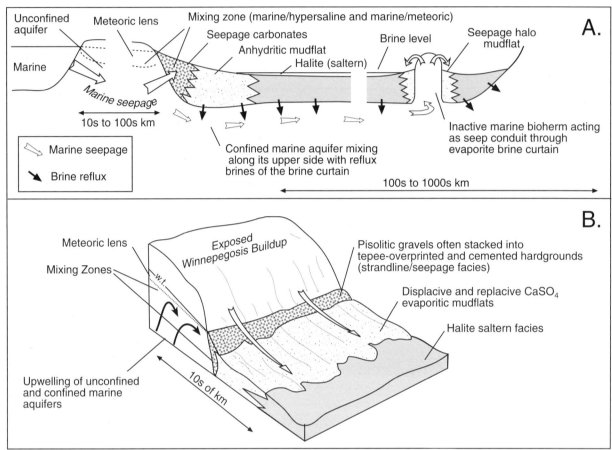

Figure 3.18. Evaporative drawdown dolomitization. A) Hydrology of basin showing seepage facies and halite in basin centre. Note the potential for the mixing of seawater seepage with refluxing brines both about the margin of the drawdown basin and beneath the reflux curtain across much of the basin floor. Diagram is not drawn to any consistent lateral scale. B) Close-up of facies pattern associated with seepage outflow zones and the potential positions of marine/meteoric and marine/hypersaline mixing zones (in part after Kendall, 1989).

Dolomite in a drawdown basin

Models of basin isolation have also implications for evaporite-associated dolomitization models. In contrast to the well documented reflux dolomites of platform evaporite associations (Chapter 2), models of dolomitising hydrologies in saline giants, are a little more complicated and are still being developed. Brine and seawater seepage flow was driven by evaporative drawdown of the order of hundreds to thousands of metres (Figure 3.6b). One can only infer, not observe, the hydrological and diagenetic character of such basinwide systems by studying the more accessible examples of such platform and basinwide evaporites as in the Messinian evaporite/dolomite associations of the circum-Mediterranean (Sun and Esteban, 1994; Meyers et al., 1997), the Permian Zechsteinkalk dolomites associated with basinwide evaporites of NW Europe (Clark, 1980) and the Devonian basinwide

evaporites and associated dolomites of the Keg River-Pine Point Presqu'ile of Canada (Maiklem, 1971; Davies and Ludlum, 1973; Bebout and Maiklem, 1973; Skall, 1975; Kendall, 1989).

Maiklem (1971) and Kendall (1989) pointed out the potential for widespread dolomitisation via seawater seepage beneath subaerially exposed slopes about a drawdown basin margin (Figure 3.18). Drawdown basins have the potential to flush huge volumes of seawater through the basin edge. The groundwater heads in such systems are not measured in metres but hundreds of metres, and if one accepts that seawater flushing has the potential to dolomitise marine limestone then the potential for dolomitisation is huge within aquifers about the margin of an evaporite drawdown basin. Maiklem (1971) also noted the potential for mixing zone dolomitisation in the regions of drawdown about the basin margin. This diagenetic zone develops where a zone of meteoric water floats atop the seawater conduit. The schematic, Figure 3.18a, is drawn assuming an unconfined aquifer in the interval separating the marine from the seepage zone. The reality is much more likely to be a combination of confined and unconfined aquifers feeding predominantly seawater into the basin margin. This zone of mixing will obviously move to and fro over the exposed basin edge as the brine level in the basin responds to changes in tectonics and climate.

Outflow points of these predominantly seawater brines develop a characteristic set of pisolitic/tepee cementstone lithologies - the strandline/seepage facies described in Chapter 1. Kendall (1992) shows how this carbonate seepage facies can be used to define seawater and mixed-brine outflow points in the Devonian Keg River Basin. These spring-aprons were mostly located about exposed former shelf slopes and bioherms (Figure 3.18b). When part of an inactive pinnacle reef or other aerially exposed bioherm, these permeable carbonates could act as confined aquifer outflow zones that were capable of breaching the active reflux curtain on the basin floor (Figure 3.18a). Slightly down dip of the seepage facies were evaporitic mudflats dominated by anhydrite and in the lowermost sections of the basin floor were widespread areas of perennial surface brine (the halite seaways in the Keg River Basin). Away from seawater and meteoric discharge seeps, the downward reflux associated with the brine curtain moved Mg-rich waters into underlying sediments. This water ultimately mixed with deeply circulating seawater seepage fluids and other subsurface waters (Figure 3.18a). Such a basinwide hydrology creates both reflux and mixing zone chemistries, which have the potential to dolomitise any limestone host in contact with the outer parts of the brine curtain.

The detail of such drawdown mechanisms can be seen in dolomite distribution within Neogene evaporitic sub-basins of the Mediterranean. For example, Meyers et al. (1997) and Lu and Meyers (1998) have documented pervasive dolomites that cut across intra-Miocene sequence boundaries in the Nijar Platform and also occur in the youngest Messinian reef/ oolitic units. The dolomites do not transect the Miocene-Pliocene unconformity, although they do occur as clasts in overlying basal Pliocene lime grainstones. This constrains the dolomitization as being latest Miocene and pre-Pliocene in age, suggesting this widespread dolomitization occurred within 600,000 years and possibly in less than 200,000 years.

Carbon and oxygen isotopes of Nijar dolomites show marked positive covariation, with most values ranging from +5.4 to -1.2‰ $\delta^{18}O_{PDB}$ and +2.5 to -4.3‰ $\delta^{13}C_{PDB}$ (Meyers et al., 1997). The highest $\delta^{18}O$ values are heavier than reasonable estimates of Messinian normal-salinity seawater dolomites, suggesting the involvement of evaporative brines in the dolomitisation process. Calculations of covariations of O and C isotopes during water-rock interaction and fluid-fluid mixing argue against recrystallisation of Nijar dolomites, an interpretation also supported by their Ca richness (49-57 mol %). Mixing calculations show that four-fifths of Nijar data are consistent with dolomitization via mixtures of freshwater and a few tens of per cent of marine-derived evaporative brine, while one-fifth are consistent with mixtures of freshwater and normal-salinity seawater. Modelling of covariations of $^{87}Sr/^{86}Sr$ and Sr/Ca ratios that developed during mixing, further supports involvement of evaporative brine, and suggests that any freshwaters involved in the dolomitization had low Sr/Ca ratios and relatively high Sr and Ca concentrations. Na, Cl, and SO_4 concentrations in Nijar dolomites (200-1700 ppm, 300-600 ppm, and 600-6500 ppm, respectively) also argue for involvement of evaporative brines, and for mixtures of brine with freshwater and/or seawater.

Meyers et al. (1997) and Lu and Meyers (1998) proposed that most dolomitization occurred after Messinian reef formation during multiple relative sealevel/brine level changes associated with evaporative drawdown. Marine carbonate deposition of the Nijar shelf occurred during high sealevels, and evaporite deposition during lowered basin water levels associated with evaporative drawdown. During drawdown, basin brines with 5x- to 6x-seawater concentrations (gypsum precipitating) developed on the floor of the Nijar Basin. Brines mixed with freshwater and seawater mainly in a groundwater setting about the basin margin as fluctuating base levels of the brine curtain drove before it a zone of freshened/meteoric water seepages.

In the Zechstein of NW Europe the dolomitized algal-bryozoal reefs, oolites and intertidal to shallow subtidal skeletal grainstones to mudstones occur adjacent to the margin of the Zechstein Basin. They are interlayered with the platform portion of basinwide evaporites composed of anhydrite and halite. The latter were deposited during the complete isolation and drawdown of the Zechstein Basin. It was at this time of drawdown that reflux dolomites replaced the underlying Zechstein limestones (Clark, 1980). At the same time the regression of facies associated with the drawdown of the brine base level caused the encroachment and displacement of formerly marine to hypersaline porewaters by meteoric waters in the dolomites of the Magnesian Limestone (Kaldi and Gidman, 1982).

Classic sequence stratigraphy in an evaporite basin?

Sequence stratigraphy is the study of rock relationships within a chronostratigraphic framework of repetitive, genetically related strata bound by surfaces of erosion or nondeposition, or their correlative conformities (Posamentier and James, 1993). Sequences are thus defined as conformable successions of genetically related strata, bound at the top and bottom by unconformities and their correlative conformities (Van Wagoner et al., 1988, 1990; Tucker, 1993). Unconformities are surfaces of erosion or nondeposition that represent a significant time gap.

An important application of sequence stratigraphy involves the construction of age models for given stratigraphic successions based on the correlation of local stratigraphy with the global cycle chart (Posamentier and James, 1993). Different orders of sea level change are recognised based on the time interval encapsulated in the sea level cycle (Figure 3.19; Table 3.2). There is still much debate as to controls of the magnitude and frequency of the various sea level changes effecting sedimentation packages. Some authors (e.g. Miall, 1997) question the reality and the scientific basis for the construction of these eustatic age curves. Another more generally accepted application of sequence stratigraphy is lithology prediction based on interpretation of cyclicity in the rock record.

The use of sequence stratigraphic principles in the construction of age models is based on the assumption that stratigraphic successions preserved in a basin are a function primarily of eustatic rather than local tectonic fluctuations. Thus in classic sequence stratigraphy the major control on deposition is assumed to be relative sea level change, determined by rates of eustatic sea level variation and tectonic subsidence. Particular depositional tracts are developed during specific phases of the sea level curve: lowstand systems tract (LST), transgressive systems tract (TST), highstand systems tract (HST).

Depositional sequences on shallow carbonate/evaporite platforms range from 1 to 10 my in duration and reflect longterm changes in accommodation linked to third order global eustatic cycles on cratons and passive margins (Read and Horbury, 1993). They may also reflect extensional and compressive phases in tectonically active regions. Sequences are bounded updip by regional unconformities and can be internally divided into LSTs, TSTs and HSTs. The internal architecture of these systems tracts differs substantially between ramp and platform systems (Loucks and Sarg, 1993). Larger packages of sequences make up supersequences, which are seismically resolvable, while the sequences themselves are made up of sub-seismic-resolution parasequences (metre- to decametre-scale depositional cycles), which may be erosionally bound by more local scale unconformities.

Is sea level important in interpreting platform and basinwide evaporites?

When attempting to interpret eustatic effects (age models), a requirement of "no surface connection to the ocean to allow thick sequences of basinwide and platform evaporites to accumulate" plays havoc with the fundamental eustatic paradigms of sequence stratigraphy. Brine levels in evaporite-depositing areas fluctuate widely and there are no simple relationships between sea level changes going on outside the basin and the geometry of the evaporite body (Kendall, 1992).

There are numerous volumes and papers now published that deal with high resolution sequence stratigraphy of arid zone carbonate shelves and ramps and how they differ from siliciclastic sequences (Read et al., 1995; Budd et al., 1995; Loucks and Sarg, 1993; Tucker, 1993; Posamentier et al., 1993; James and Kendall, 1992). The details will not

Order	Years	Tectonoeustatic	Rifting and thermal subsidence	Global eustatic	In-plane stress	Glacioeustatic, tectonic, sedimentary
1st	10^8					
2nd	10^7	↕	↑	↕		
3rd	10^6		↓	↓		
4th	10^5				↕	↑
5th	10^4					↕

Table 3.2. Orders of sealevel change and likely driving mechanisms (after Tucker, 1993).

be repeated here and the interested reader should refer to these volumes. There are only a few published papers that deal specifically with the sequence stratigraphy of evaporite systems; key papers include, Tucker, 1991; Goodall et al., 1992; Strohmenger et al., 1996a,b. The results are contentious and Strohmenger's papers deal mostly with the carbonate (?marine-connected) stage of the evaporite basin. This is understandable as it is in these units that potential hydrocarbon reservoir facies are to be found.

All evaporite basins with a marine seepage feed have a brine level that is below sea level. It is, after all, drawdown that drives brine influx. The relative difference in water level across the isolating barrier may be no more than a few to tens of metres, as it was in the platform evaporites of Guadalupian backshoal strata in west Texas, or it may be thousands of metres as it was in the late Miocene salinity crisis in the Mediterranean and the early stages of the Permian salinity crisis in the Zechstein Basin of NW Europe. Brine salinities and the rate of salinity change in a brine-filled depression, as well as the absolute brine level in an isolated basin, are controlled by the climate, the rate of brine leakage through the underlying brine curtain and the rate of seepage replenishment of the inflow waters and brines about the depression's margin.

If one wishes to apply the concepts of sequence stratigraphy to determining position on a global sea level curve using age-modelling then marine-fed platform evaporites must be distinguished from basinwide evaporites. At our present level of understanding we can probably fit marine-margin platform evaporites into a global sea level context. Their formation is largely dependent on marine seepage through a carbonate barrier. It requires platform isolation

brought about by a slight fall in sea level (Figure 3.6a). The evaporites can fill the depression behind the barrier; but cannot, as they are seepage fed, rise any higher than sealevel. The isolating barriers in turn have geometries and

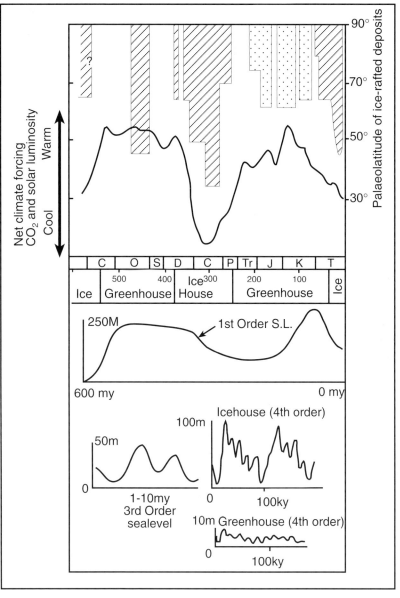

Figure 3.19. Relationship between icehouse and greenhouse conditions, eustasy and global CO_2 (after Read and Horbury, 1993). Upper half of figure shows palaeolatitudinal extent of continental ice-rafted deposits (crosshatched) and marine ice-rafted debris (dots). The curve beneath plots net forcing of climate due to variations in CO_2 and solar luminosity. Times of general icehouse and greenhouse earth are shown on the horizontal axis. Lower half of diagram shows first order Vail sealevel curve in upper part. Bottom diagram shows typical third order sealevel curve with 1-10 my period. Schematic 4th order with superimposed 5th order sealevel curves are drawn for icehouse and greenhouse conditions.

distributions across a platform that are controlled by sea level changes (Read et al., 1995; Loucks and Sarg, 1993).

In contrast, basinwide evaporite units, with a predominantly seepage feed, are more problematic in terms of the effects of sea level fluctuations on evaporite geometries and mineralogies. The problem will remain until the hydrologies of large scale drawdown systems are better understood. Formation of basinwide evaporite sequences are largely tied to a suitable juxtaposition of tectonics and climate. In such a scenario the decametre sealevel fluctuations on the marine side of the isolating barrier are largely irrelevant to evaporite depositional styles within the lowermost portions of an isolated basin experiencing a drawdown that is up to 1,500 - 2,000 metres below sea level. Likewise continental evaporites are not directly tied to the world sea level curve. In both basinwide and continental evaporite scenarios the application of sequence stratigraphic modelling is more useful in a genetic rather than an age (world sealevel) context.

Icehouse and greenhouse sealevels

Sequences in seismic stratigraphy are unconformity-bound packages generated by high amplitude sealevel changes with the bounding surfaces produced during relative sea level falls (Tucker, 1993). High amplitude sea level changes producing regional sequence boundaries can occur on several time scales depending on the presence of absence of polar ice caps. At times of no polar ice cap (greenhouse earth) the high amplitude changes that produce sea level oscillations of several tens of metres are a response to second and third order processes, while fourth-/fifth order processes produce only metre-scale relative sea level changes and associated parasequences (Figure 3.19). In contrast, when there are polar ice caps (icehouse earth) the high amplitude sea level changes (typically greater than 100 metres) are brought about by fourth-/fifth order processes. The driving force for these high amplitude/high frequency oscillations is change in polar ice volume driven by Milankovitch rhythms. During icehouse times the same longer term second-/third-order fluctuations in sea level that dominate the greenhouse style still occur, but they now have a lower amplitude and longer period than the glacially driven changes. Thus the glacially driven cycles swamp these longer term effects, so that in ice house times fourth-order sequences are produced and are arranged into sequence sets (Tucker, 1993). Greenhouse periods produce third order sequences constructed by fourth-/fifth order parasequences.

Thus the Neogene sedimentary marine record is an icehouse record where fourth-order sequences dominate the continental margin as unconformity-bound, relatively thin (5-30m thick), stratal packages. Other older icehouse intervals include the late Precambrian and the Permo-Carboniferous (Figure 3.19). In contrast in greenhouse times (Late Cambrian to Middle Ordovician, Silurian to Early Devonian, Late Permian to Oligocene), the stratal packages consist of much thicker unconformity-bound sequences (50m to hundreds of metres thick) deposited as third order cycles made up of fourth-fifth order parasequences that are 1 to 10 metres thick.

Marine-margin platform evaporites

Marine carbonate platform systems are now understood to show a different set of responses to sea level change compared with siliciclastics (Figure 3.20). This is because almost all normal-marine carbonate sediment has a biogenic origin. Unlike siliciclastic shelves, carbonate shelves most actively aggrade and then prograde both seaward and landward during times when the continental margin is covered by shallow seas. That is, those most rapidly growing parts of the platform, usually the reef crest and associated grainstone belts, aggrade most rapidly when part of a transgressive systems tract (Figure 3.20e). The total platform progrades most rapidly when part of a highstand systems tract (Figure 3.20f). When part of a lowstand systems tract little or no growth can take place in either deep or shallow water (Figure 3.20d).

Unlike siliciclastics, carbonate sedimentation rates cannot be assumed to be near-constant across a platform. The variation reflects the strong environmental controls imposed on biogenic sedimentation rates. Changes in geometric stacking patterns in carbonates do not necessarily directly reflect changes in the rates of relative sea level rise as environmental changes are not always directly linked to changes in sea level (James and Kendall, 1992; Hunt and Tucker, 1992).

In a marine carbonate system the TST is the time when the maximum growth potential of the carbonate platform is realised and the parts of the open marine platform with the greatest growth potential typically aggrade to sea level and then begin if prograde both seaward and into the lagoon (Figure 3.20e). In contrast, evaporite deposition occurs on a platform when an evaporite-depositing depression is isolated from a surface connection with the sea (Figure 3.20h). On a ramp, the barrier is usually a buildup of some sort of continuous zone Y shoal, which separates the open marine of zone X from the evaporite lagoon/mudflat of

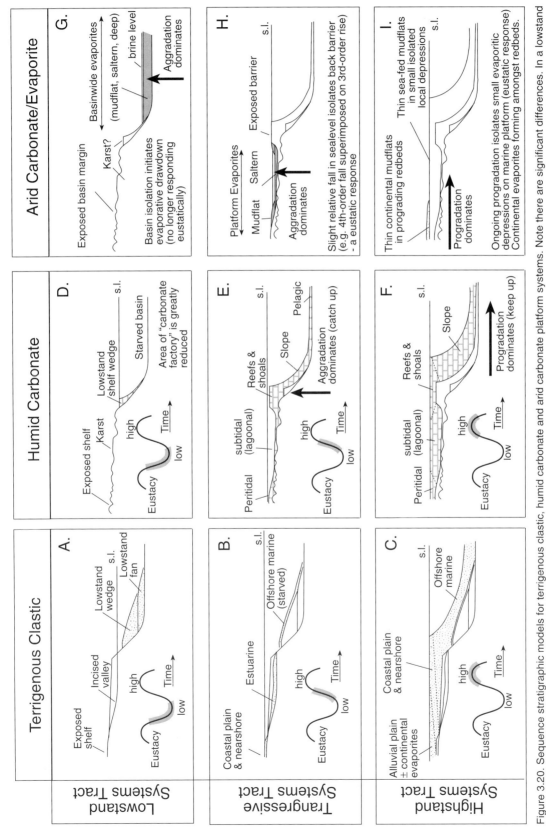

Figure 3.20. Sequence stratigraphic models for terrigenous clastic, humid carbonate and arid carbonate platform systems. Note there are significant differences. In a lowstand systems tract there are substantial volumes of terrigenous sediment deposited at the foot of slope, but relatively minor carbonate deposition if sealevel is below the shelf break (lowstand shelf wedge). In an isolated evaporite basin large volumes of basinwide evaporites can accumulate as a lowstand systems tract. In a transgressive systems tract carbonate sedimentation in a humid setting reaches its highest growth potential. If a transgressive systems tract is also subject to a higher order fall that isolates a seepage depression on the shelf, then thick sequences of platform evaporites can form. In a highstand systems tract only thin mudflat evaporite units can form, while thick sequences of platform clastics and carbonates can accumulate. See text for full discussion of this figure.

zone Z. On a rimmed shelf it is typically the reef belt or the associated grainstone/pisolite flat that have built up to sea level.

Isolation of the barrier crest and the initiation of thick platform evaporite precipitation are usually initiated by a slight fall in sea level following a time of vertical aggradation (catch up) in the more rapidly growing parts of the carbonate platform; e.g. a fourth order fall superimposed on a third order rise. Thick platform evaporites are thus most likely to form as part of a transgressive systems tract (third order) in a greenhouse earth. Any slight fall in sea level will be marked by an unconformity across the topographic high (rim/shoal), which isolates the marine from the evaporite-depositing areas (Figure 3.20h). Thinner evaporative mudflat (sabkha/salina) units may form as part of a highstand systems tract where ongoing progradation of the shelf (keep up) isolates small depressions further landward (Figure 3.20i).

When platform evaporites were deposited, the subaerial isolation either of the rim/grainstone buildups on a rimmed shelf or of the zone Y shoal facies on a ramp were responsible for the restriction of the lagoon to the point where evaporites could accumulate. Any occasional channels cutting across the rim or shoal had to be sufficiently restricted to prevent the free surface exchange of marine and evaporite waters. Otherwise, like the carbonate shoals and reefs on platforms of today, subaqueous conditions behind the shelf rim or shoal would have never become saline enough to deposit subaqueous evaporites and only thin capillary zone sabkha caps (< 1m thick) could prograde across the landward feather edge of marine lagoon sediments. Prior to isolation, a suite of normal marine to restricted marine sediments had been accumulating in the lagoon, so that in a vertical profile normal marine bioturbated shelf wackestones and mudstones typically underlie the platform evaporites.

Once an evaporite lagoon is isolated, two end members can be deposited behind the rim, either saltern sequences or mudflat (sabkha/salina)-dominated successions (Figure 3.4). The dominance of one or the other reflects the response of evaporite drawdown hydrology to changes in relative sealevel on the platform. An evaporitic mudflat mosaic forms when there is little difference in elevation between the barrier crest and the evaporitic depression (former lagoon floor). In core from such a prograding setting the sediments of the rim /shoal (oolite/grainstone/pisolite/boundstone) pass directly upward into thin intercalated sabkha/salina units or stacked sabkha cycles and back to shoal facies, which make up the base of the next cycle, with no intervening saltern sequences.

If the rim/shoal aggrades rapidly and the backshoal platform does not keep pace with a rise in sea level, the early shoal-water carbonates deposited on the lagoonal side of the rim are not all that different from those on the seaward side of the rim. This is especially true where the rim is incipiently drowned. While the rim was growing and catching up to a new, higher sea level the waters in the lagoon could freely exchange with the open ocean and so a diverse open-marine biota flourishes within the lagoon. When the rim catches up to the new sea level and coalesces into an isolating barrier, conditions in the carbonate-accumulating lagoon become more restricted. As the exchange with open ocean waters is reduced; sediments are now deposited as restricted-marine carbonates. With complete restriction across the rim or shoal, the lagoonal waters become sufficiently saline to deposit evaporites and fill the drawdown lagoon with a shoaling-upward saltern cycle perhaps capped by a mudflat sequence. The saltern evaporite unit is as thick as the lagoon was deep at the inception of evaporite deposition (Figure 3.20). Such subaqueous saltern evaporites are continuous across the evaporite lagoon and so are easily correlated from well to well as metre- to decametre-thick clean evaporites. The ten or so subaqueous cycles of the Cretaceous Ferry Lake Anhydrite in the northern Gulf of Mexico region (Pittman, 1985) and the Permian San Andres cycles on the northern margin of the Delaware Basin (Elliott and Warren, 1989) are examples of stacked saltern beds, while the base of the Late Permian Seven Rivers Formation atop Yates Field on the Central Basin Platform in west Texas is an example of the stacked mudflat mosaic (Warren, 1991).

Platform evaporites favoured by a greenhouse earth

The high amplitude rapid to-and-fro of the icehouse sea across Neogene platforms has discouraged the deposition of extensive shallow water platform evaporites (Warren, 1989, 1991). Platform evaporites tend to form under greenhouse earth conditions characterised by lower amplitude sea level fluctuations. The high-amplitude high-frequency sealevel changes of icehouse earth encourage local high-relief "catch-up" reefs and banks due to ongoing rapid flooding and exposure of the platform, along with substantial unconformities and pervasive meteoric diagenesis deep in the reefs. Hence icehouse carbonate facies show considerable lateral thickness variations (Read and Horbury, 1993).

Under an icehouse scenario the platform reefs and shoals typically do not develop into laterally continuous strike belts. Large isolated back-platform depressions with hydrologies suitable for platform evaporites do not easily form. Any highstand evaporite facies are likely to be little more than narrow sea-marginal bands, as they are today. Sea level frequently falls off the platform (>100m fall) in icehouse earth and so any local areas of sea-marginal peritidal evaporite deposits that may have formed are subject to widespread dissolution via vadose flushing (as is seen in the worldwide lack of well preserved Pleistocene marine sabkhat and salinas - Warren, 1991).

In contrast, the low amplitude low frequency sea level oscillations that characterise epeiric seaways and shelves in greenhouse earth intervals favour the formation of large highly restricted areas on the inner shelf portions of many platforms. Carbonate platform rims could more than keep pace with the low amplitude sea level fluctuations, and so wide and flat carbonate platforms with huge back-platform lagoons were commonplace (epeiric seas). A slight fall in sea level meant seepage waters easily ponded in large shallow shelf depressions. In other areas where epeiric seaways covered large inland portions of the continent, the formation of a tectonic or shelf rim barrier followed by a fall in sea level meant that the entire intracratonic basin could become evaporitic.

Thus, stacked platform evaporites tend to form in greenhouse times as rapid, but thick, saline infills atop marine lagoon sediments. They are part of an overall TST, but only precipitate during short intervals characterised by higher-order low amplitude falls in sea level. At such times the platform crest is subaerial. Today the isolating effects of slight sealevel falls, as well as any possible formation of an ongoing laterally continuous carbonate platform rim, are swamped by the overriding influence of rapid, high-amplitude, icesheet-induced sealevel changes across relatively narrow continental shelves. Conditions suitable for the formation of platform evaporites simply do not exist under today's icehouse regime.

Platform evaporites, sea level curves and diagenesis

Platform evaporites can exert a substantial influence on carbonate mineralogy, porosity preservation and seal capacity of each platform cycle. Brine reflux in the brine curtain beneath an accumulating platform evaporite bed can dolomitise substantial portions of the marine carbonates of each cycle and so generate intercrystalline and vuggy porosity. At the same time the updip evaporite bed creates an excellent seal to the porous high energy grainstones of the restrictive shoal or reef belt (Moore, 1989; Elliott and Warren, 1989). Some of the largest oil fields in the world have formed because of the sealing effects of platform evaporites. Ghawar in Saudi Arabia, the world's largest oil field, produces from Upper Jurassic Arab cycles; each of the four producing intervals has a platform evaporite seal and the whole structure is sealed beneath the Hith Anhydrite; likewise the giant Jurassic fields of Abu Dhabi. Similarly most of the large fields in the Upper Permian San Andres-Grayburg Formations (e.g. Yates Field, Texas) owe their occurrence to platform evaporite seals, as do Early Silurian oil reservoirs in the Interlake Formation of the Williston Basin. Platform evaporites also seal the Triassic cycles of Syria and Iraq, and the Jurassic reservoirs of North Africa and the Gulf Coast.

In studying platform evaporite sequences worldwide the effects of later subsurface dissolution can greatly complicate any high resolution stratigraphic interpretation of the stratigraphy. Often this dissolution occurs as the platform evaporite passes from an active phreatic hypersaline hydrology into a compactional or thermobaric regime. It creates thin dissolution breccias that extend over large portions of the platform. Such breccias can be confused with subaerial exposure surfaces and humid/temperate phreatic karst. Unless the significance of evaporite dissolution breccias and the timing of dissolution in relation to burial are recognised, whole episodes of platform isolation may be miscorrelated.

The detailed subsurface facies work by Andreason (1992) clearly shows the importance of recognising evaporite breccias when undertaking detailed correlation of sequence boundaries across carbonate-evaporite platforms. Without it, understanding the position of hydrocarbon reservoirs in the Upper Permian Yates Formation would be difficult. Many published outcrop-based studies of high resolution stratigraphy in carbonate-evaporite platforms are problematic because dissolution breccias, which tend to be thin and recessive, are largely unrecognised or if recognised misinterpreted as a subaerial exposure surface. And yet, a one-metre thick breccia may mark the position of a former 10-50m thick salt bed. This is especially true in Precambrian platforms where biostratigraphic checks on correlation are not possible. Criteria that define evaporite dissolution breccias are discussed in Chapter 4.

Basinwide evaporites

Thick basin-filling evaporites (more than 500 - 1,000m thick and hundreds of kilometres across) accumulated in both greenhouse and icehouse periods. They did so, not because of fluctuations in sea level, but as a result of tectonic isolation of large parts of the earth's surface from the direct surface influx of the world's ocean waters (Kendall, 1989, 1992; Warren, 1989). Their fills tend to be autocyclic and tectonically induced, although subsequent fluctuations in sealevel may have aided erosional breaching and refilling of the depression by seawater episodes. Such areas exemplify the thick salt successions in the Miocene sediments of the Red Sea and the Gulf of Suez, the Jurassic salts in the Gulf of Mexico, the Permian of Delaware Basin west Texas and the Zechstein basins, NW Europe, the Devonian of western Canada, the Fars-Gascharan sequences of the Middle East and the Late Miocene Messinian successions in the Mediterranean. These saline giants accumulated in basin-scale depressions in arid climate belts and were marked by no surface connection with the ocean. They formed under tectonic settings of incipient continental rifting, or continent-continent collision, or continental transtension. Supply of brine to these saline giants was a combination of varying proportions of marine seepage-drawdown, continental runoff and volcanogenic/ hydrothermal formation waters (e.g. Hardie, 1990).

Once isolated, relative sea level may play a role in influencing hydrological base level and rate of brine supply in a marine-fed drawdown basin. A relatively high sea level will create a steeper potentiometric gradient and hence encourages more rapid replenishment of brine and a greater volume of brine seeping through the permeable barriers that separate the brine depression from the open ocean (e.g. Kendall, 1992). When relative sea level is lower, this decreases the potentiometric gradient and slows the rate of brine flow. A fall in sea level also causes the sea coast to retreat from the isolated depression and so further decreases the rate of sea water supply.

Autocyclic mechanisms, along with the formation of brine curtains or hydroseals of varying efficiencies, and the vagaries of climate and tectonics in influencing the supply of nonmarine waters, all interact to swamp the influence of sealevel oscillations on the other side of the barrier. For example, the aggradational infill of a tectonic depression by salt beds that create their own underlying aquitard or brine curtain has the same effect as the change in sealevel on potentiometric gradient of the inflow; as the sequence aggrades it decreases the potentiometric gradient and slows the rate of marine recharge.

Assuming no changes in the volume of nonmarine and hydrothermal inflows, and a stable sea level, the effect of salt aggradation is to fill in the depression and so decrease the potentiometric gradient over time. The effect was noted by Warren (1982b) and used to explain the changes in the infill cycle textures in Holocene gypsum-filled salinas in coastal southern Australia. Logan (1987) discusses similar hydrological transitions in controlling vertical sequences and the transition from halite to gypsum with aggradation in Lake McLeod, Western Australia. Kendall (1989) used such Holocene analogues to interpret the upward transition from anhydrites to carbonates in the Middle Devonian of northern Alberta (Muskeg and Sulphur Point Formations). Previously the transition was interpreted as reflecting a sea level rise, but Kendall argued it would more likely reflect a transition related to sea level fall and a decreasing volume of marine inflow into the isolated basin. His explanation better explains the terminal nonmarine bed that makes up the overlying Watt Mountain unit.

In summary, unless a rise in sea level breaches the tectonic sill and converts the basin once more to a restricted seaway, decametre-scale sea level fluctuations exert a relatively minor influence on hydrology and hence the detailed stratigraphy of a basinwide evaporite unit. While the basin is isolated much of the fill is autocyclic, it reflects changes in the hydrological base level that have more to do with fluctuations in tectonics and climate as they effect inflow/seepage hydrology. Changes in relative sea level that are occurring outside the evaporite basin and probably across a barrier that is hundreds or thousands of kilometres from the drawdown strandline are largely irrelevant.

Basinwide sequence stratigraphy?

Given this rudimentary understanding of basin hydrology and isolation, let us now look at some of the implications in terms of high resolution stratigraphic analysis in basinwide evaporite deposits. Documented examples are few and far between, and tend to concentrate on patterns in the carbonate facies that separate the major salt beds. High resolution interpretation of basins that contain, or once contained, thick salt beds is more challenging because:

• outcrop studies of geometries of evaporite units are near useless as most of the evaporite is dissolved well before it approaches the land surface - differential collapse and flow during dissolution means detailed internal relationships and depositional geometries are lost not just at outcrop but also in the shallow subsurface,

• seismic analysis tied to well data in bedded saline giants is rarely undertaken because very few oil companies wish to characterise the internal stratigraphy of what is an undeniable seal facies,

• deeply buried thick salt sequences are subject to flow into diapiric and other structures - this largely destroys the stratigraphic integrity of internal reflectors in halokinetic parts of the basin (see Chapter 5).

Tucker's (1991) work on the Zechstein is to date the only paper to address the principles of sequence stratigraphy in terms of processes active in basinwide evaporite precipitates. He derives two end member models to interpret the evaporite carbonate relationships in saline giants: Model 1, incomplete drawdown with marginal gypsum wedges and basinal laminated gypsum; and Model 2, complete drawdown with halite basin-fills. However, his hydrological constraints are not clearly stated and his assumption that marginal gypsum wedges typically form in relation to a lowered sealevel (his Figure 2) is probably not correct (see Goodall et al., 1992).

Figure 3.21 assimilates some of the concepts of Tucker (1991) as well as the work of Kendall (1988) and my own work. It is a preliminary attempt to place basinwide evaporite geometries and overall basin fills into a sequence stratigraphic context. Unlike Tucker (1991) it is based on the notion that in an isolated basin the drawdown hydrology is largely independent of decametre scale sealevel fluctuations that are going on outside the basin. Most of the internal decametre-scale evaporite geometries in a saline giant are autocyclic, while larger scale geometries reflect ongoing interactions between tectonics and climate. Only when the basin is connected at the surface to the world's oceans do the geometries of the carbonate units reflect a eustatic control.

In the time prior to its isolation the basin has a surface connection to the world's ocean (Figure 3.21a). The depositional response at this time is a classic marine carbonate response and the only widespread evaporites that precipitate are typical marine-margin saltern/mudflat evaporites that are intercalated with normal marine platform sediments (see Figures 3.6a and 3.20 for interpretations of this style). Pinnacle reefs and mud mounds may grow on the shelf foreslope during the marine TST (e.g. right-hand part of Figure 3.21a). In the basin centre, thin (condensed) pelagic deepwater carbonates and hemipelagic mudrocks will be deposited, mostly as thin finely-laminated units that pass laterally into slope deposits about the basin edge.

Once sealevel falls below the sill height —either through a eustatic change in sea level or, more likely, through regional tectonic movement that cuts off the basin from a surface connection with the world's oceans— evaporative drawdown will quickly lower the brine level in the evaporite basin to where it is well below the old carbonate shelf break (Figure 3.21b). Under an arid climate, water within the now isolated basin will rapidly become hypersaline. Seawater, along with formation, continental, and hydrothermal waters, now seeps into the depression. It exits through permeable, now subaerially exposed, formerly deepwater carbonate slopes and former pinnacle reefs. These outflow zones generate spring and tufa deposits with characteristic textures such as pisolites, tepees, tufas and stromatolites. In the still relatively deep subaqueous basin centre, organic-rich mesohaline carbonate laminites typically mark the onset of basin isolation. They continue to form until a dynamic equilibrium is reached, where as much water exits the basin as enters it. At what stage this equilibrium is reached in the drawdown of the basin waters will vary from basin to basin. In some it may be reached while substantial brine depths are maintained in the deeper parts of the basin and gypsum may be the main crystallising phase. In other basins the drawdown may not reach hydrological equilibrium until the brines are at halite saturation and water depths are only a few metres.

Gypsum forms across the whole basin when the drawndown brine body reaches saturation with respect to $CaSO_4$. Gypsum precipitates best in the "chemical factory" of the shallow basin edge, as a new saltern/mudflat wedge begins to build into the basin; while gypsum laminites aggrade in the deeper basin centre (Figure 3.21b). If the basin does not subside greatly as it becomes isolated, the level of this new gypsum wedge (platform) will first form atop old slope carbonates and below the old carbonate shelf break. With respect to sealevel outside the isolated basin this is a lowstand deposit, but with respect to the subsequent drawdown hydrology of the basin it is a highstand system tract.

Resedimentation of the saltern/mudflat gypsum into deeper water by storms, slope failures, debris flows and turbidity currents will deposit graded beds, slumps and breccias as gypsum slope deposits and fine-grained "rain-from-heaven" gypsum laminites in the basin centre. If this situation is hydrologically stable, that is hydrological inflow balances outflow at gypsum-precipitating salinities, and the situation is maintained for time frames $\approx 10^5$ years or more, the whole basin evaporite fill may be composed of gypsum. Such was the case during precipitation of large portions of the Permian Castile Formation of Texas (e.g. Kendall and

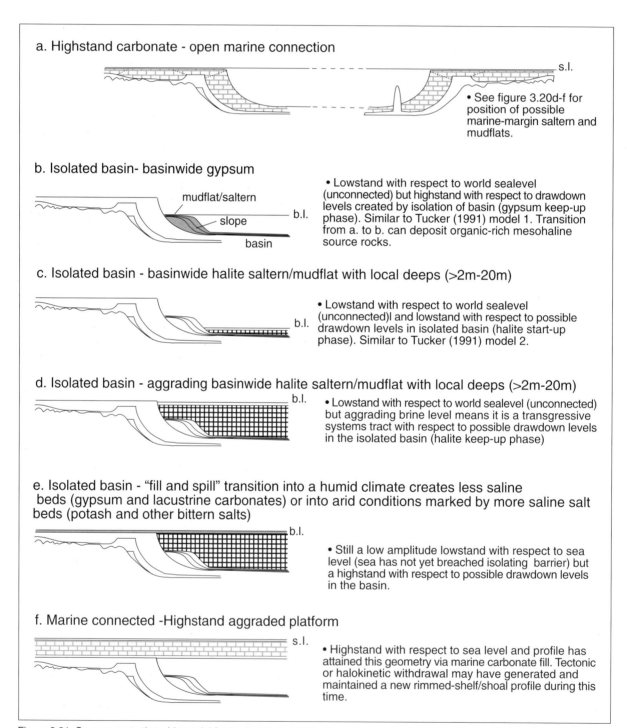

a. Highstand carbonate - open marine connection

s.l.

• See figure 3.20d-f for position of possible marine-margin saltern and mudflats.

b. Isolated basin- basinwide gypsum

mudflat/saltern
slope
basin
b.l.

• Lowstand with respect to world sealevel (unconnected) but highstand with respect to drawdown levels created by isolation of basin (gypsum keep-up phase). Similar to Tucker (1991) model 1. Transition from a. to b. can deposit organic-rich mesohaline source rocks.

c. Isolated basin - basinwide halite saltern/mudflat with local deeps (>2m-20m)

b.l.

• Lowstand with respect to world sealevel (unconnected)l and lowstand with respect to possible drawdown levels in isolated basin (halite start-up phase). Similar to Tucker (1991) model 2.

d. Isolated basin - aggrading basinwide halite saltern/mudflat with local deeps (>2m-20m)

b.l.

• Lowstand with respect to world sealevel (unconnected) but aggrading brine level means it is a transgressive systems tract with respect to possible drawdown levels in the isolated basin (halite keep-up phase)

e. Isolated basin - "fill and spill" transition into a humid climate creates less saline beds (gypsum and lacustrine carbonates) or into arid conditions marked by more saline salt beds (potash and other bittern salts)

b.l.

• Still a low amplitude lowstand with respect to sea level (sea has not yet breached isolating barrier) but a highstand with respect to possible drawdown levels in the basin.

f. Marine connected -Highstand aggraded platform

s.l.

• Highstand with respect to sea level and profile has attained this geometry via marine carbonate fill. Tectonic or halokinetic withdrawal may have generated and maintained a new rimmed-shelf/shoal profile during this time.

Figure 3.21. Sequence stratigraphic model for basinwide evaporite deposition in a basin undergoing complete isolation from a surface connection to the world's oceans and subsequent drawdown.

Harwood, 1989). Metre-scale sedimentary cycles (parasequences) will be developed within the prograding gypsum wedge reflecting higher-frequency low-amplitude (5th-6th-order, i.e. $10^4/10^3$ year) autocyclic brine-level changes within the basin.

If hydrological equilibrium is still not maintained in the basin, and brine losses continue to exceed inflows, the brine level will continue to be drawn down and the salinity of the residual brine will increase. Saltern gypsum will begin to precipitate in the basin centre if the brine depth drops below 10-30 m. If a hydrological balance is still not reached and the brine level continues to fall, the brine body next reaches halite saturation (Figure 3.21c). Now water depths are less than a few tens of metres to metres as saltern halite is deposited across the whole basin floor (Figure 3.21c). Locally there may be deeper-water depressions on the basin floor that fill with fine-grained laminated "salt-and-pepper-textured" deepwater halite (this only requires brine depths >2m, if Lake McLeod is used as an analogue; Warren, 1991). This stage of the sedimentation is still a lowstand with respect to sealevel outside the basin, and is now a lowstand systems tract with respect to the basin hydrology.

Once halite units begin to precipitate across the basin floor, the degree of permeability loss in salts below the precipitation surface increases drastically, reflecting a near complete loss of porosity in metre-decametre thick halite beds (Figure 1.10). The brine curtain and its cement aquitard can now support a permanent halite-saturated brine body, which can maintain a near permanent position atop the primary crystallisation surface with only occasional episodes of complete desiccation or rainfall-induced surface freshening. Thus the halite body begins to aggrade as a series of subaqueous metre-thick autocyclic parasequences, with each cycle deposited in time frames of 10^2-10^3 years. The maintenance of this aggrading hydrology allows the basin to fill with predominantly saltern halite (Figure 3.21d). With respect to sealevel outside the isolated basin it is a lowstand system, but in terms of the intrabasin geometries it is a transgressive systems tract. But it is a transgressive systems tract that has no parallel in the marine realm. It is an autocyclic transgression that aggrades as it maintains and grows its own impervious base from slowly refluxing saturated brines. There is no suitable eustatic term to describe this process, as there is no concept of autocyclic transgression in the classic sequence stratigraphic literature. Tucker (1991) terms this stage the lowstand basin-fill-halite; I am more comfortable calling it an intrabasinal transgressive systems tract.

As the halite fill continues to aggrade it climbs over the old gypsum wedge and, if the basin is subsiding but still isolated, it can also climb over the old carbonate shelf margin (Figure 3.21e). As the fill tops an older platform margin it reaches a vast flat plain. Before being retained the permanent basin waters must now spread over a much wider area of an older platform top. I call this the "fill and spill" stage of the basin's sedimentation and hydrology. It is still a lowstand systems tract with respect to external sea level, but a transgressive systems tract with respect to

intrabasin geometries. It is typically a mudflat system, although salterns may still occur. The fill and spill point horizon is often marked by a change in mineralogy, which indicates a change in depositional hydrology.

In a semi-arid system this spill point is often defined by precipitation of beds of less saline evaporites mineralogy such as gypsum or lacustrine carbonates. In a semi-arid setting, rainfall and meteoric influx are an increasingly important part of the ponded surface waters as they spread outside the former less aerially extensive depression. If fill and spill occurs in a very arid setting the effect of meteoric waters is minimal and the system now becomes subject to much higher degrees of desiccation. In such a system surface waters and refluxing brines may pass into bittern saturation. Accumulations of syndepositional bittern or potash salts are likely under such hyper-arid conditions. A hydrological transition to "fill and spill" under a hyper-arid scenario explains why the most economic potash cycles in the Salado Formation of the Delaware Basin are largely confined to regions immediately landward of the old carbonate platform margin (see Chapter 7). Fill and spill transitions, defined by changes in mineralogy, can also occur as an aggrading halite body breaches the top of an earlier gypsum shelf-wedge

Once again the transition is autocyclic and dependent on climate within the basin. The system is still a lowstand with respect to sealevel outside the basin. In terms of the intrabasin geometry it is a transgressive systems tract either atop the older HST gypsum wedge or atop the former HST of the older carbonate margin. Tucker (1991) also recognises this TST (his early TST gypsum or TST retrogradational sabkha evaporite lagoons), but he only allows for a $CaSO_4$ mineralogy and he relates it to a surface reconnection with the world's oceans. Unlike Tucker (1991), I feel that this widespread thin unit is deposited subsealevel in a still isolated basin where surface reconnection to the ocean has yet to occur. I also emphasise that this TST is not always gypsum/anhydrite, it can be composed of a multiplicity of mineralogies depending on the degree of climatic restriction in the basin. Kendall (1992; his Figure 40) also recognises this stage, but he relates the mineralogical transition to less saline minerals to a transgression beyond the hydroseal or basal aquitard, this allows greater volumes of brine to be lost from the system, and so average salinities in the surface brines are lowered. This is a valid concept, but it is not complete as it does not explain why some basins are actually precipitating more hypersaline minerals and bittern salts in the upper parts of the halite basinfill.

Once the basin is reconnected to the open ocean, the formerly exposed carbonate platforms are reflooded and a carbonate TST re-established. In this scenario any subsequent platform evaporite deposits can be considered equivalent to the marine marginal platform evaporites discussed in the preceding section and their internal geometries behind the isolating platform rim or shoal may well once again reflect eustatic controls (Figure 3.20f). Because the aggraded evaporite surface is a flat plain at the time of marine reflooding, the onset of marine carbonate deposition is characterised by ramp deposition. If the carbonate biota also contain appropriate binders and reef formers the profile quickly evolves into a rimmed carbonate shelf. The transition to a rimmed shelf in many ancient basins undergoing regional extension is focused by halokinetically induced shallowing of the seafloor (see salt tectonics chapter).

Some examples

Many of the carbonate intervals that separate thick salt units in saline giants may indicate times of non-isolation, when seawater breached the isolating barrier and the water level in the basin was sealevel. Such carbonates can be used to age model global sealevels. The basinwide evaporites are more problematic. Two of the best documented areas where high resolution sequence analysis has been attempted are the gypsum evaporites of Messinian age in the marginal basins of the Mediterranean and the basinwide anhydrite/halite evaporites of Permian Zechstein in NW Europe. Both these saline giants contain bittern intervals that are rich in magnesian sulphate salts such as kainite, polyhalite and kieserite (Kuehn and Hsu, 1978; Decima and Wezel, 1973; Garcia-Veigas et al., 1995; Harvie et al., 1980) and so, following the brine chemistry arguments in Chapter 2, the basins are likely to have seawater seepage as the main brine source.

Messinian evaporites of the

Mediterranean

In the Messinian we have excellent onshore exposures and core from the more marginward basins in Sicily and Spain (Figure 3.13). There is little sampling or facies interpretation of evaporites in the deeper parts of the Mediterranean Sea, where thick sequences were deposited during maximum drawdown. These deeper more basin-central sub-basins in the Mediterranean Sea now lie below more than 2,000m of water. In the Zechstein Evaporite Series, the entrained

carbonates are onshore gas plays in Germany and the Netherlands, and so we have much better core sampling and seismic analysis of both basin centre and basin margin evaporites. But in this sequence the stratigraphic interpretation is largely in the prospective carbonate intervals and does not deal in any detail with the internal stratigraphy of the evaporites themselves.

In the icehouse world of Messinian evaporites in the Mediterranean, there were two styles of evaporite deposition, one was characterised by high level, marginward circum-Mediterranean deposits (Figure 3.22). The second was characterised by much larger, thicker, basin-centre deposits that were tectonically instigated by the complete isolation of the Mediterranean from adjacent oceans.

According to Clauzon et al. (1996) the first evaporite depositing episode (5.75-5.60 Ma) took place during a cooling period associated with a weak global sea level fall that isolated numerous tectonically induced circum-Mediterranean marginal basins in Spain, Libya, Morocco, the Apennines, Sicily, and Cyprus. Isolation of these marginal basins occurred while the adjacent Mediterranean Sea still retained a direct surface connection with the Atlantic (Time 2; Figure 3.22).

The next period of evaporite deposition (5.60-5.32 Ma) occurred as basinwide evaporites were deposited in a series of giant evaporite sub-basins (hundreds to thousands of kilometres across) and are now located beneath the present deep Mediterranean floor (Time 3; Figure 3.22). Away from halokinetic areas, none of these areas outcrop and only their uppermost portions have been sampled by a few wells in the DSDP (Deep Sea Drilling Project) programme (Figure 3.13). At the time they formed there was a huge drop of some 1500m in water levels, and widespread brine lakes of the Mediterranean were completely isolated from any surface connection with adjacent Atlantic or Tethyan remnants. Deposition of thick saltern and deeper-water evaporites was coeval with cutting of canyons and formation of the Messinian unconformity about the margin of the Mediterranean (Clauzon et al., 1996; Riding et al., 1998). The isolation was driven by the ongoing collision tectonics as Africa collided with Eurasia. At the time of the isolation there was a worldwide warming trend and a likely global sea level high, but tectonic isolation completely negated the effects of eustacy within the Mediterranean (Figure 3.22).

The last decade of stratigraphic work clearly shows a time distinct separation of the two evaporite styles (Riding et al., 1998). Onset of evaporite deposition in marginal

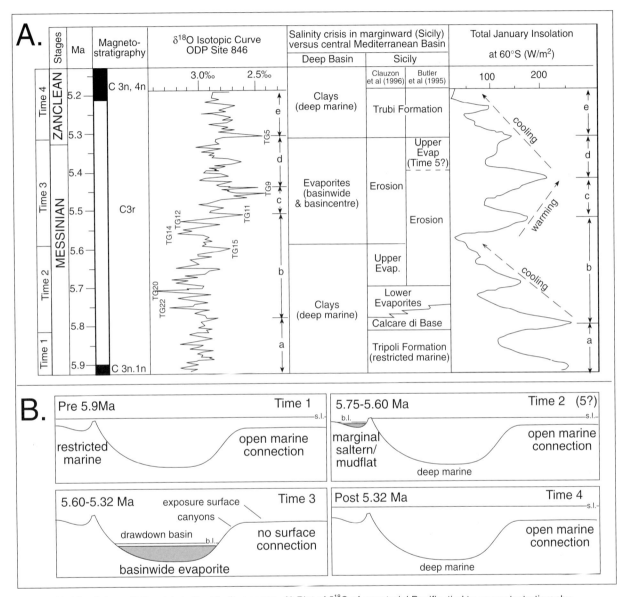

Figure 3.22. Messinian salinity crisis in the Mediterranean. A) Plot of δ¹⁸O of equatorial Pacific tied to magnetostratigraphy, Messinian stratigraphy and modelled January insolation for that time frame (modified after Clauzon et al. (1996) and Butler et al. (1995)). Note the difference in interpretation of the upper evaporite in Sicily. B) Schematic times sequence showing mode of formation of marginal and basinwide evaporites in the Mediterranean. Note that the interpretation of Butler et al. (1995) requires a time 5.

basins was eustatically driven and often preceded by cyclic Late Miocene carbonate ramp or reef cycles; which are readily interpreted using the principles of high-resolution carbonate sequence stratigraphy (e.g. in Spain and Italy-Martin et al., 1996; Esteban, 1996: in Morocco- Benson et al., 1997; Cunningham et al., 1997). Once the marginward basins were isolated, ongoing tectonics and autocyclic mechanisms controlled the chemistry, cyclicity and textural styles in the various evaporite basins.

For example, a complex system of transcurrent faults, with both transpressive and transtensional characteristics, controlled depositional geometry in the marginal basins of the Betic Cordillera of SE Spain (Michalzik, 1996). There is no evidence for a cataract-like connection of these basins with a strong unidirectional flow regime before and during the formation of Messinian evaporites as first suggested by Hsu et al. (1973). The most frequently occurring gypsum lithofacies are related to shallow-water depositional environments and include: crystalline selenitic gypsum ('grasslike gypsum'), laminated gypsum and crossbedded gypsarenite. Only minor sabkha-type chickenwire alabastrine gypsum were identified by Michalzik (op. cit.). Evenly laminated gypsum units, together with graded gypsarenites (turbidites), gypsrudites (debrites) and slumps, are restricted to basins or parts of basins with considerable palaeoslopes. Evidence for diagenetic overprinting includes the formation of giant gypsum twin crystals, 'super cones' and alabastrine nodular gypsum. The accumulation of the total package of Messinian evaporites in SE Spain was instigated by a superimposed third-order sealevel lowstand. Smaller sedimentary units, which make up the parasequences at the scale of shoaling decametre-cycles or less, fall well into the range of fourth- to sixth-order cyclic and rhythmic successions. They are clearly related to basin dynamics or to autogenetic processes of the local facies zone (facies dynamics). None of the smaller-scale cycles (fourth and higher order) are related to high-frequency eustasy (Michalzik, 1996).

The depocentres of the marginal basins in Sicily are in synclines related to underlying thrust structures of the frontal part of the Maghreian chain (Figure 3.23; Butler et al., 1995). Prior to isolation from a surface connection to the ocean, these basins were hydrodynamically linked through the thrust foredeeps to world sealevel. Thus Late Miocene reef and ramp precursors in the various Sicilian basins were linked eustatically. Initial isolation and drawdown of the Mediterranean base level is marked across the thrust belt by first cycle carbonates and evaporites on the structural highs and lows, respectively. Vast accumulations of halite and potassium salts (up to 1200 m) and restricted to growing thrust synclines were deposited in this first marginward stage of evaporite deposition. At that time the different evaporite signatures were related to different water conditions, reflecting various continental and marine seepages and circulations across various sub-basins (Butler et al., 1995). Internal cyclicity (parasequences) reflect high-frequency autocyclic variations in brine level, and in my opinion are not eustatic. Thrusting provided accommodation space for evaporites and also controlled the water pathways into the desiccating basins. In a similar fashion Messinian sediments of the northern Marchean Apennines form part of a syn-orogenic foreland basin succession. They underwent shortening by thrust-related folding soon after deposition (Mazzoli, 1994).

Clauzon et al. (1996) group the deposition of the Upper and Lower Evaporites of Sicily into a single period from 5.75 to 5.60 Ma. The initial eustatically induced marginward isolation was then followed by complete isolation and drawdown of the whole Mediterranean (Figure 3.22a). In contrast, and although they use only relative time, Butler et al. recognise a major sequence boundary between the Upper and Lower Evaporites of Sicily (also called the First and Second cycles). They interpret this sequence boundary as corresponding to the major drawdown of the whole Mediterranean and the formation of a major regional unconformity, which now surrounds all these basinwide deposits (M reflector). They go on to interpret the Upper Evaporite (Second Cycle - mostly saltern gypsum) as being deposited in the isolated marginal seepage basins of Sicily once the Mediterranean was again flooded by seawater entering via a newly re-established surface connection (Figure 3.22b). The contact with the upper evaporites and the overlying deepwater marine Trubi Formation is a maximum flooding surface (mfs) in both scenarios.

In the evaporite basins of Cyprus the salts were deposited in a series of semi-isolated small marginward basins not far below ambient sealevel (Robertson et al., 1995) For example, evaporites in south Cyprus (Maroni Basin) formed in elongate basins between compressional lineaments instigated by Early Miocene thrusting in a tectonic situation similar to that just described for Sicily. In the sub-basins of west, southwest and south Cyprus, large scale slumping on saltern gypsum into deeper water formed megarudite debris flows, which are thought to have been triggered by one or several phases of extensional faulting.

This level of understanding of the inherent, and still not fully resolved, complexities in the Messinian has taken more than three decades of detailed scientific work, mostly in exposed marginward stratigraphies, which are still largely preserved as primary gypsum. The work and the resolution is ongoing. With the exception of the initial third order fall in worldwide sealevel, all these deposits were isolated via active tectonism. Their finer stratigraphic geometries are not reflective of eustacy but of drawdown and autocyclicity. Much larger volumes of basinwide evaporites are preserved beneath the present deepwater centres of the various Mediterranean basins. Bedded sections currently lie beneath hundreds of metres of Pliocene and Pleistocene sediment; only halokinetic intervals lie within a few tens of metres of the modern deep seafloor. With the exception of a few DSDP holes that intersected the uppermost few metres of bedded salts, the internal stratigraphies of these saline giants are largely unsampled and uninterpreted. And yet we know deposition of 1-2 km-thick basinwide salts took place in less than 300,000 years. By definition, the total 1+ km package of the Messinian

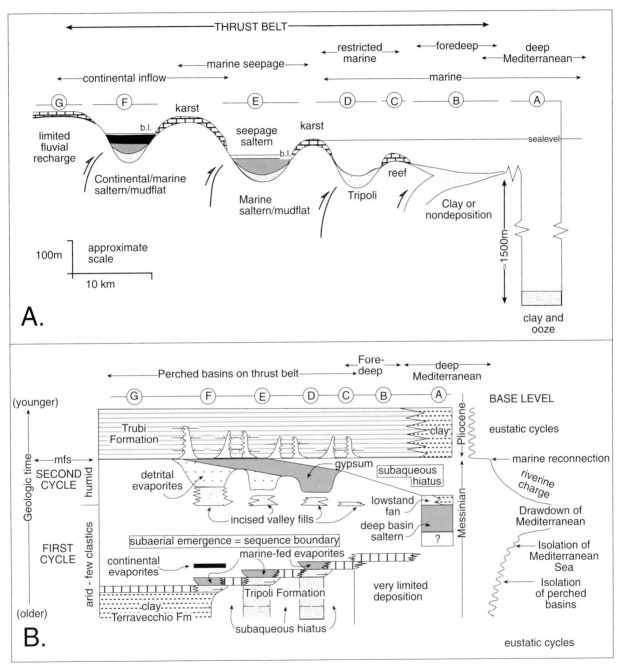

Figure 3.23. Marginal thrust top evaporites in Sicily (modified after Butler et al., 1995). A) Regional model for stratigraphy across different thrust top basins during deposition of the Lower Evaporites (see Figure 3.22). Note the evaporite forming basins are isolated from a surface connection with the Mediterranean, but have a seepage connection. Restricted surface marine circulation facilitates deposition of the Tripoli Formation. Progressive thrusting leads to progressive reduction of import of marine inflow and finally to deposition of potash evaporites under conditions of hyperaridity and predominantly continental recharge. B) A relative chronostratigraphic model for Messinian and Lower Pliocene deposits of Sicily and their correlation with basinwide deposits of the central Mediterranean sub-basins. Note that deposition of basinwide evaporites on the floor of the Mediterranean corresponds to an unconformity (sequence boundary) that separates the Lower and Upper Evaporites. This interpretation contrasts with that of Clauzon et al. (1996).

sequence must be a fourth order sequence (Table 3.2). It implies that internal reflectors and cycles seen in seismic sections through thick basinwide evaporites, must reflect fifth and sixth order time frames.

Any eustatic curves, based on seismic that encapsulates the very high accumulation rates of basinwide evaporites, will outline very high frequency responses. Even if we ignore the fact that ancient basinwide evaporites are depositional systems with internal responses not related to sea level changes, comparison of curves derived from internal geometries of basinwide evaporites with eustatic curves derived from the worldwide study of various marine sequences is simply not possible. The internal reflectors and responses that we can resolve seismically within a basinwide evaporite unit are responding to water level changes that are far too short-term to be compared with the eustatic information captured in classic Exxon-style sealevel curves. We cannot correlate the various peaks and troughs outlined by the depositional tracts in either system, they are responding to different time-based water-level frequencies. The highest seismically resolvable frequencies in a marine sequence are third, and perhaps the occasional fourth, order response. In basinwide evaporite sequence we are looking at seismically resolvable bounding surfaces that are responses to fifth or sixth order cycles.

Errors must arise if intrabasinal correlations attempt to tie thick basinwide evaporite deposits to what are mistakenly considered time-equivalent eustatic curves based on studies of seismic sections through marine carbonates either within or outside the basin. What it most clearly shows is that, using the most recent example we have of a saline giant, the deposition of the one km-plus evaporite unit in the various basin centres corresponds to a two dimensional bounding surface or centripetal unconformity immediately outside of the evaporite depositing region. This should be the starting point for any seismic-based correlation of the evaporites (or subsequent dissolution breccias) with adjacent normal marine sediments.

So, is it also a questionable approach to attempt a sequence stratigraphic age comparison between the marine sediments and deposits sandwiched between the various salt beds of a seawater-fed saline giant? Unless we clearly define all sequence boundaries and realise that thick beds of evaporites equate with regional unconformities in surrounding sediments, we may be comparing chalk and cheese. And yet some authors have been courageous enough to attempt internal sequence stratigraphy of ancient saline giants based on various eustatic scales. We shall now look at such a study in the Zechstein evaporites.

Sequence stratigraphy in the Zechstein

Zechstein sedimentation commenced when the sub sealevel Rotliegendes Basin was flooded through a combination of rifting and eustatic sea level rise (Glennie and Buller, 1983; Taylor, 1990). In the classic terminology, four main evaporite cycles and rudimentary fifth and sixth cycles (Z1-Z6) then precipitated (Figure 3.24). An ideal classic Zechstein cycle starts with a transgressive nonevaporitic "shale", followed by carbonates and culminates in thick evaporites (mostly halite/anhydrite). The most complete classic cycles are found in the Z1 and Z2 cycles where lithologies have been related to higher-order glacio-eustatic sealevel fluctuations by Strohmeyer et al. (1996a,b).

Tucker (1991) published an alternative to the classic Zechstein stratigraphy based on what he interpreted as third order sequences (ZS1-ZS7). In it he correlates evaporites with sealevel lowstands (LSTs) and carbonates with sealevel highstands (TSTs and HSTs). Tucker's approach, in tying evaporites to lowstands, is opposite to that of the classic approach, but more realistic in terms of evaporite hydrology. His classification was criticised firstly by Goodall et al. (1992) on the basis he assumed sealevel exerted a one-for-one control on evaporite mineralogy and position, and then by Strohmeyer et al. (1996a) as his classification did not respect exposure surfaces and stratal geometries. Strohmeyer et al. (1996a,b) went on to refine a sequence stratigraphic approach for the basal Zechstein (Z1 and Z2) using facies analysis geometric relationships and stacking patterns as seen in seismic from the subsurface of Germany (Figure 3.25).

However, Strohmeyer et al. found that due to the poor quality of internal reflectors in the Zechstein sequence they could not use the classic approach of interpreting the major depositional units from seismic, then establishing a seismic stratigraphic model and then tying well data to seismic. Rather they had to first establish a geological model based on detailed stratigraphic analysis of core and outcrop. Then they developed a sequence stratigraphic framework and then interpreted the intra-Zechstein reflectors. Their analysis (Figure 3.25) shows that reworked sandstones of the Weissliegendes and/or the Zechstein conglomerate (Z1C) record the initial transgression of the Zechstein sea and overlie the Zechstein sequence boundary (ZSB1). Transgressive systems tract (TST) deposits of the first Zechstein sequence (ZS1) are also represented by deeper marine carbonates of the Mutterflöz (T1Ca). The overlying Kupferschiefer is interpreted as a condensed section (CS) indicating maximum flooding (mfs) of the first Zechstein sequence. The bulk of the shallow water

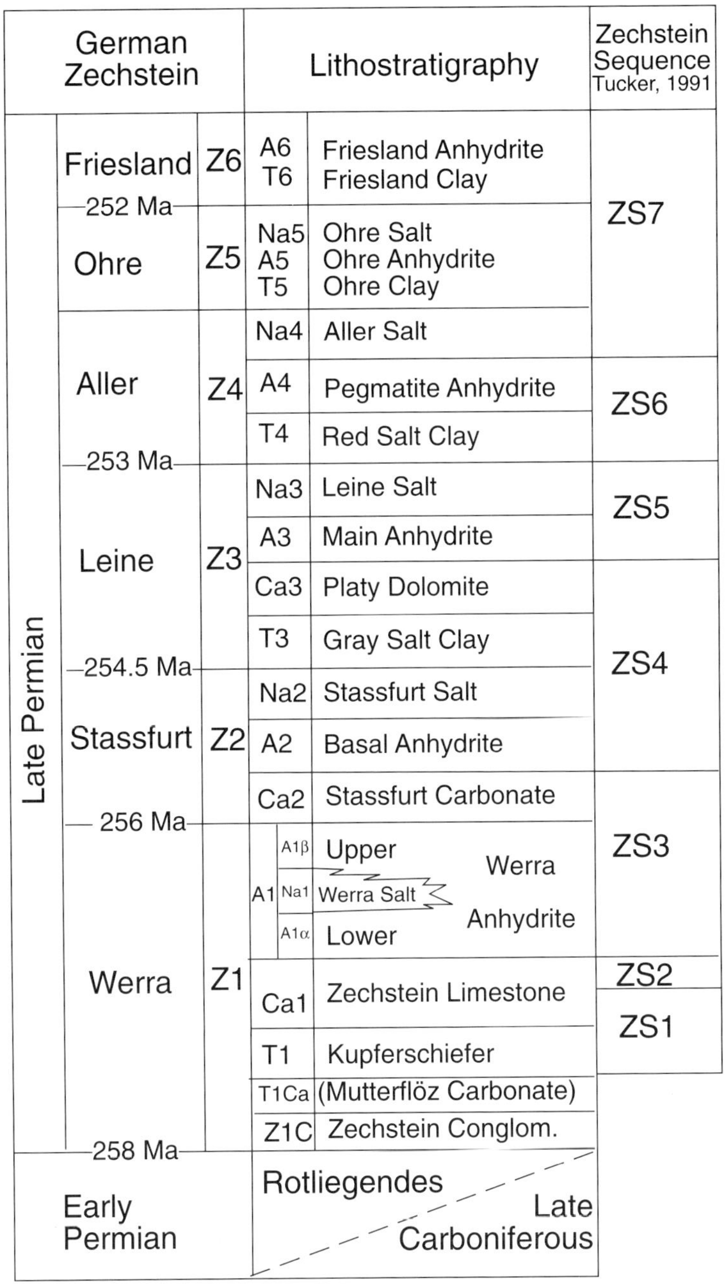

German Zechstein			Lithostratigraphy		Zechstein Sequence Tucker, 1991
Friesland	Z6	A6 T6	Friesland Anhydrite Friesland Clay		ZS7
—252 Ma—					
Ohre	Z5	Na5 A5 T5	Ohre Salt Ohre Anhydrite Ohre Clay		
Aller	Z4	Na4	Aller Salt		
		A4	Pegmatite Anhydrite		ZS6
		T4	Red Salt Clay		
—253 Ma—					
Leine	Z3	Na3	Leine Salt		ZS5
		A3	Main Anhydrite		
		Ca3	Platy Dolomite		ZS4
		T3	Gray Salt Clay		
—254.5 Ma—					
		Na2	Stassfurt Salt		
Stassfurt	Z2	A2	Basal Anhydrite		
—256 Ma—		Ca2	Stassfurt Carbonate		
		A1β	Upper	Werra	ZS3
		A1 Na1	Werra Salt	Anhydrite	
		A1α	Lower		
Werra	Z1	Ca1	Zechstein Limestone		ZS2
					ZS1
		T1	Kupferschiefer		
		T1Ca	(Mutterflöz Carbonate)		
		Z1C	Zechstein Conglom.		
—258 Ma—					
Early Permian			Rotliegendes	Late Carboniferous	

(left margin: Late Permian)

Figure 3.24. Lithostratigraphy of the Zechstein Series in Germany (after Strohmenger et al. 1996a, b, and sources therein). Shown are the classic Zechstein cycles (Z1-Z6) and the Zechstein sequences (ZS1-ZS7) as proposed by Tucker (1991).

Zechstein limestone (Ca1) is interpreted as a highstand systems tract (HST). It is separated from the thin uppermost part of the Zechstein Limestone by a karst horizon corresponding to an erosive sequence boundary (ZSB2). Correlatable shallow water carbonates (oncolites and stromatolites) indicate lowstand systems tracts (LST) of the second Zechstein sequence (ZS2). The uppermost part of the Zechstein Limestone represents the transgressive systems tract of the second Zechstein sequence (ZS2). According to Strohmeyer et al. (1996a,b) there is no indication of a sequence boundary at the top of the Ca-1 platform carbonates (contrasts with Tucker 1991). They argue Ca1 carbonates, in basinal position, show a deepening upward tendency toward the superimposed deposits of the A1 sulphates. They place the maximum flooding surface (mfs) of the second Zechstein sequence at the top of the Ca1 platform carbonates. The overlying anhydrites of the A-1 sulphate platform are interpreted as predominantly highstand systems tract deposits displaying an erosive sequence boundary (ZSB3) at its top. This is in direct contradiction with Tucker (1991) and according to Strohmeyer et al. (1996a,b) it is only in the slope and basinal environments that the A1 also represents lowstand wedge deposits of the third Zechstein sequence (ZS3).

This contradiction underlies the difficulties in sequence stratigraphic analysis of evaporite deposits using models derived from geometries of marine systems. The terms highstand systems tract and lowstand systems tract are largely defined by whether the shelf of the basin that is under consideration is covered by water. Many interpretations also assume a surface connection to the ocean and so direct eustatic responses in brine base level in the basin. We have already discussed the likely errors inherent in this assumption. In the marine realm we are talking about eustatic sea level amplitudes of no more than a hundred or so metres in icehouse periods and less than decametres in greenhouse periods. There is an inherent assumption that in greenhouse and icehouse times the prime locus of sedimentation is near a shoreline either on the shelf or just basinward of the shelf break. In an evaporite basin that is undergoing drawdown we are discussing isolation rapidly followed by a water level fall of more than 500-1,000 metres prior to widespread evaporite formation. Any strandline in the system will likely be well out on the old continental slope. During complete drawdown the whole highstand shelf and slope as well as part of the basin floor is exposed and water levels

101

even in the deepest parts of the basin are mostly less than a few tens of metres and no more than one or two hundred metres.

Sequence boundaries such as the ZSB2 are more problematic in such a drawdown scenario (Figure 3.24). Strohmeyer argues that the Ca1 carbonates, in basinal position, show a deepening upward tendency toward the superimposed deposits of the A1 sulphates. What he is observing is a gradational transition. The term deepening is probably too categorical in this scenario. If the basin is isolated so that thick gypsum/anhydrite can accumulate, then the brine level must be lowered with respect to the old shelf. The A1 profile above the ZSB2 sequence boundary is most likely subaqueous throughout, and is a "brining upward" profile. The sequence does not deepen upward, it is continuously subaqueous but brines upward as the regional water level is lowered. That is, the average annual salinity of the brine body is increasing up section and ongoing lowering of the brine level does not stop until an effective brine curtain/inflow feedback is established. Stacking patterns laid down in such scenarios are not well understood but should not be forced to fit the classic paradigms of marine sequence stratigraphy. We are looking at a system that probably was at any time deeper or shallower depending on the vagaries of climate and tectonics and largely independent of eustacy. All we can say in terms of the A1-sulphates is that at no time during the deposition of the sulphate phase was the water deeper than a few hundred metres and that it was mostly less than tens of metres deep (Warren, 1989; Kendall, 1992). During the preceding carbonate platform phase the water body was obviously less saline and probably deeper on average than during sulphate deposition.

The Zechstein 2 Carbonate (Ca2) overlies the A1-Sulphate and according to Strohmeyer et al (1996a,b) encompasses transgressive systems tract as well as highstand systems tract deposits of Zechstein sequence ZS3. The maximum flooding of the Ca2 is correlated with the flooding of the A1-sulphate platform. Therefore, Ca2-platform carbonates (Main Dolomite) are dominated by shallow-water highstand deposits of the third Zechstein sequence (ZS3). Deposits of the Ca2 transgressive systems tract are coated-grain and composite-grain packstones/grainstones that contain reworked anhydrite clasts of the underlying A1.

The transition of shallow-water Ca2 carbonates into the overlying anhydrites of the A2 is often gradational (Strohmeyer et al., 1996a, b). Carbonates, showing typical Ca2 facies (e.g. algal-laminated coated-grain and composite-grain wackestones/ packstones and algal bindstones), are often intercalated with the anhydrites of the A2. Furthermore, according to Strohmenger, the anhydrites of the A2 often display ghost structures of Ca2 facies. The lower part of the A2 is interpreted by them as sabkha deposits (A2-sabkha) indicating the emergence of the Ca2 platform at the end of Ca2 time. The stratigraphic limit between the Ca2 and A2 is highly arbitrary and conventionally drawn in wells by the last up section occurrence of anhydrite. According to Strohmeyer et al (1996a,b) the fourth sequence boundary of the Basal Zechstein (ZSB4) occurs not on top of the Ca2 shallow-water carbonates, but somewhere within the A2-platform deposits (once again in contrast to the model of Tucker, 1991). The lower part of the A2 (anhydritised Ca2 facies) is interpreted as the time equivalent of shallow-water Ca2 carbonates that overlie deeper-marine upper slope carbonates in a basinward position, corresponding to the lowstand systems tract (lowstand wedge) of the fourth Zechstein sequence (ZS4). The uppermost part of the A2 (pure salina-type anhydrites on top of anhydrites showing ghost structures of Ca2-platform facies) is thought to represent the transgressive systems tract of the fourth Zechstein sequence (ZS4). The relatively thick anhydrite accumulations of the A2 (approx. 100 m), which occur basinward of the Ca2 slope, are thought to represent both lowstand systems tract as well as transgressive systems tract deposits of the fourth Zechstein sequence (ZS4).

According to Strohmeyer et al. (1996a,b) the basinward progradation of the third-order Basal Zechstein depositional sequences is forced by the tectono-eustatic, first- and second-order sealevel lowstand in the latest Permian. The cyclicities observed within the different systems tracts are thought to be caused by higher order sealevel fluctuations with durations of 0.08-0.5 Ma (fourth-order cycles), 0.03-0.08 Ma (fifth-order cycles) or 0.01-0.03 Ma (sixth-order cycles and related to orbitally controlled Milankovitch cycles. Strohmeyer et al. (1996a,b) go on to argue that the 0.1 and/or 0.4 Ma Milankovitch cycles indicate to high-amplitude sealevel fluctuations throughout the Late Permian. They draw analogies to small-scale sequences as described from the Yates Formation (United States, Permian basin) by Borer and Harris (1991) and the Wegener Halvø Formation (Jameson Land basin, East Greenland) by Scholle et al. (1993).

In my opinion there is a problem in comparing eustatic sealevel models, derived from platform evaporites of the Permian of Greenland and west Texas, to the basinwide drawndown evaporites of the Zechstein. The work by Strohmeyer et al., (1996a,b) is an excellent facies breakdown of the stratigraphy. Where I have a problem is

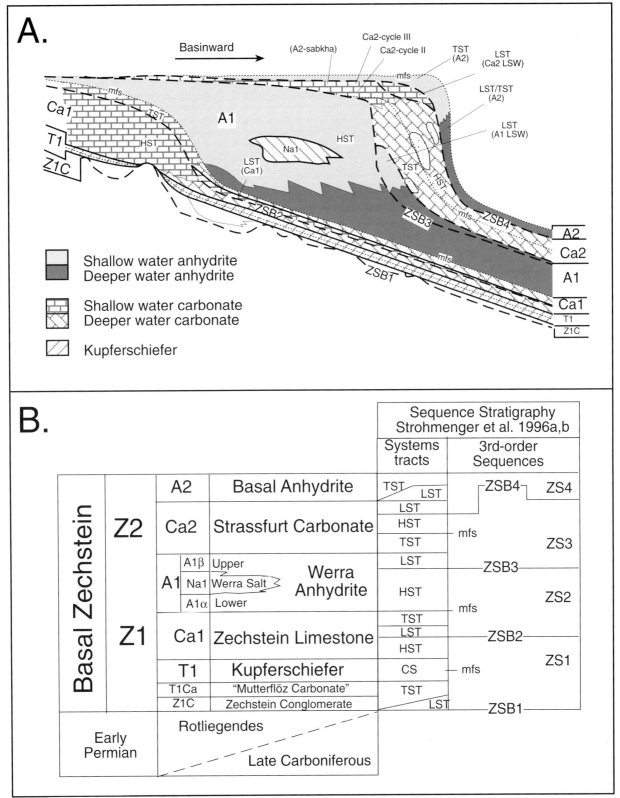

Figure 3.25. Zechstein evaporites of northwestern Germany (after Strohmenger et al., 1996a,b and sources therein). A) Schematic of basal Zechstein relationships on the southern margin of the Zechstein Basin. B) Lithostratigraphy of basal Zechstein showing classic subdivision (Z1-Z2) and Strohmenger et al. (1996a) subdivision (ZS1-ZS4). ZSB = Zechstein sequence boundary.

accepting the leap of faith in applying marine-derived Vail-curve age models to the brine level fluctuations that controlled the shelf to basin geometries of the Basal Zechstein. This approach ignores the need for complete lack of a surface connection with the ocean to deposit thick evaporite sequences. Furthermore, if we look to analogies in the marine-fed saline giants sub-basins of the Messinian, then sedimentation rates used by Strohmeyer et al. for evaporite packages, such as the A1 sulphates, are much too slow. Thick basin-centre evaporites require particular conditions of tectonic isolation to form, when this does occur the evaporites are laid down very rapidly as they were in the Mediterranean.

In summary, work by evaporite sedimentologists in modern and ancient successions shown that kilometre-thick sequences will be deposited in less than 300,000 years and that infill only happens at times of complete basin isolation when there is no surface connection to the world's oceans.

Quo vadis high-resolution sequence stratigraphic analysis in saline giants?

Chapter 4
Indicators and effects of dissolution: "the evaporite that was"

There are more evaporite dissolution breccias preserved in the rock record than there are preserved salt beds. Complete dissolution means salt no longer remains, partial or preferential dissolution means the vertical or lateral extent of the remaining salt sequence is much reduced and the more soluble salts have gone. Bed dissolution, as well as nodule and crystal replacement, can be syndepositional and driven by active phreatic flow. Or it can occur later, driven by compactional and thermobaric flushing. Or it can occur during uplift associated with collision, inversion and over-thrusting as buried salts beds are flushed by deeply circulating meteoric waters. Dissolution also occurs in the more enigmatic meta-evaporitic and igneous realms, where it can be accompanied by mineral transformations such as scapolitisation, albitisation and tourmalinisation (Chapters 6 and 9).

Brines generated at a dissolution front can move into adjacent more permeable intervals where they can precipitate new authigenic evaporite cements or be carried up along flow escape conduits, such as faults and fractures, into higher cooler portions of the sediment pile. Although dissolutional hydrology is transitory, the evidence of former evaporite beds –the evaporite that was– is permanent and ranges from dissolution breccias and residues, to replaced evaporite crystals and nodules, to authigenic evaporite cements near flow conduits such as faults and fractures (Table 4.1).

Evaporite solution breccias and dissolution residues

Evaporite solution breccias and their residuals are important keys in understanding the diagenetic history of a sedimentary basin and may be the only evidence of a former evaporite unit (McWhae, 1953; Middleton, 1961; Stanton, 1966; Swennen et al., 1990). Evaporite solution breccias (syn. evaporite collapse breccias) form when the subsurface removal of evaporite salts allows the intercalated or overlying rock to settle and become fragmented. Dissolution residues are typically found within the breccias and are fine-grained accumulations of detrital and insoluble materials that were once entrained in the dissolving salt unit. They are usually made up of quartzofeldspathic silts and muds or of dolomite and calcite rhombs. In situations where a multimineralogic salt unit has undergone partial dissolution the residue may be an accumulation of less

soluble salts, e.g. a layer of anhydrite nodules, which were once suspended, in a now dissolved halite bed.

The impervious nature of buried evaporites means that a salt bed dissolves like a melting block of ice. That is, it shrinks via a series of retreating fronts, all moving from the edge inward. Retreat is most rapid along fronts in contact with large volumes of undersaturated pore waters, as occurs near active faults or about the feather edge of an evaporite bed. Salt solution can create cavities into which fall clasts of the adjacent lithology — an evaporite solution breccia. As dissolution eats inward it also creates free flowing residues of sand- to mud-sized insolubles, which in their turn infiltrate, and sometimes fill, the spaces between the clasts. Interclast detritus is sometimes the only preserved evidence of materials that were originally entrained within the evaporite bed.

Breccia characteristics

The typical product of the dissolution of an evaporite bed is a stratiform evaporite solution breccia; a horizon usually dominated by cm-metre angular clasts in a mud-sand matrix. Until now, commonly used sandstone and carbonate breccia classifications emphasise energy levels during deposition. They cannot adequately address the character of solution breccias, many of which are diagenetic rather than depositional. As we are specifically interested in breccias associated with the dissolution of evaporites, let us look first at a nongenetic breccia classification, then at how evaporite solution breccias relate to other types of sedimentary breccia.

Breccias are a rock type that ranges across many sedimentary and tectonic regimes. Some sedimentary breccias can be considered "proto-conglomerates" that is, deposits with clasts that are not transported sufficiently to become rounded (Figure 4.1c). Breccias of this kind are commonplace among limestones, sandstones and shales. They can be created in the depositional setting by glacial and alluvial outwash, by waves acting on a reef or sea cliff, and by slope failure. As a group, this type of breccia tends to reflect rapid changes in transport energy, with clasts reflecting the lithologic diversity of the source area. Many chert breccias are the residuals of the weathering of a cherty carbonate terrain. Breccias with a lower clast diversity and clasts broken form nearby lithologies can be created by diagenetic, plutonic, volcanic and tectonic processes, or by the impact of extraterrestrial objects (bolides).

INDICATOR	EXAMPLE	REFERENCE
Solution collapse breccias	Lower Carboniferous, Vesder, Belgium	Swennen et al. (1990)
	Devonian Carbonates, western Canada	Stanton (1966)
	Cretaceous Kirschberg breccia, Texas	Warren et al. (1990)
	Devonian, Cadjebut area, Lennard Shelf	Warren and Kempton (1997)
Preferential dissolution residues	Miocene gypsum breccias in halite, Poland	Babel (1991)
	Basal Anhydrite, Khorat Plateau, Thailand	El Tabakh et al. (1998); Warren (1997)
	Anhydrite residues via solution of Maha Sarakham Salts, Khorat Plateau, Thailand	Sessler (1990); Utha-aroon et al. (1995)
Rauhwacke	Quaternary-weathered brecciated detachment horizons in nappes of European Alps	Schaad (1995); Warrak (1974)
Silicified evaporites	Late Palaeozoic carbonates, Bear Island, Svalbard	Folk and Siedlecka (1974)
	Mississippian strata, Kentucky and Tennessee	Milliken (1979)
	Permian Park City Fm., Wyoming	Ulmer-Scholle and Scholle (1994)
Calcitised evaporites	Permian Zechstein EZ-1 Carbonates, Europe	Harwood (1980)
	Permian Capitan Fm., Texas and New Mexico	Scholle et al. (1992)
Calcitised dolomites (dedolomite)	Permian Raisby Fm. (Zechstein), NE England	Lee and Harwood (1989); Lee (1994)
	Permian Tansill Fm, New Mexico	Lucia (1961)
	Mesozoic limestones, French Jura	Shearman et al. (1961)
Celestite nodules	Salina Dolostones, Ohio	Carlson (1987)
	Purbeckian of Dorset, UK	Salter and West (1966)
Fluorite cements	Eocene mudstones, Florida	Cook et al., (1985)
	Visean evaporites, France	Rouchy (1986)
Baryte nodules and cements	Arid soils, central Australia	Sullivan and Koppi (1993)
	Magnet Cove baryte, Nova Scotia	Mossman & Brown (1986)
Authigenic anhydrite/halite cements	Cretaceous Pearsall and Glen Rose Fm., Texas	Woronick and Land (1985)
	Permian Rotliegende Sands, North Sea, Europe	Sullivan et al. (1994)
	Jurassic Smackover Fm., USA	Heydari and Moore (1989)
	Devonian Keg River Fm, Alberta	Aulstead and Spencer (1985)
	Ordovician Glenwood Fm, Michigan Basin	Simo et al. (1994)
	Ordovician Trenton Group, Ontario	Middleton et al. (1993)

Table 4.1. Some documented examples of evaporite indicators.

Breccias are composed of angular clasts, typically rock fragments of granule or pebble size or larger. When classifying breccia one should ask the following (Morrow, 1982):

- Are clasts of a single lithic type (monomict/oligomict) or of widely variable types (polymict); are lithic types in clasts similar or different from the rocks bounding the breccia body; are clasts composed of durable rock fragments or of a mixture of hard and soft clasts?
- Are clasts closely fitted, loosely fitted, or spatially independent, i.e., crackle - mosaic - packbreccia - floatbreccia (Figure 4.1a)?
- Do clast shapes and clast-to-clast contacts indicate either plasticity of clasts or brittle behaviour at the time of sedimentation? What is the nature of the clast edge (Figure 4.1b)?
- Are spaces between clasts filled with cement or with detrital matrix; i.e. clast-support or matrix-support?
- Does a weathering rind or coat extend all around a clast, or is it developed only on one side, or is there no indication of weathering?
- What is the nature and style of clast size and size grading; evidence for bedding; imbrication or other spacial alignment of clasts?
- What is the overall size and shape of the breccia deposit; is there a sharp lower contact and/or sharp upper contact; is the breccia body discordant (trans-

"THE EVAPORITE THAT WAS:" INDICATORS AND EFFECTS

stratal) or concordant (interstratal) in relation to bounding rocks?

Packbreccias are clast-supported, while floatbreccia clasts float in a finer matrix of granules, sands or mud. The distinction between crackle, mosaic and rubble breccia depends on whether the clasts are fitted, mildly disoriented or totally independent (Figure 4.1a). Crackle denotes a fabric in which there is little relative displacement of fragments. Mosaic implies fragments are largely but not wholly displaced. Rubble breccias are completely disordered and no fragment boundaries match. Some packbreccias have particulate matrices, others are cemented by spar or silica indicating clast support. Elements other

than fabric, such as the size and shape of the breccia body, its stratigraphic setting, the depositional setting of adjacent sediments and lateral equivalents to the breccia provide evidence of origin. Intrastratal breccias indicate former bedded salts, while trans-stratal breccias characterise breccia chimneys and halokinetic dissolution breccias.

Evaporite solution or collapse breccias tend to form diagenetically as a buried evaporite bed is flushed by undersaturated subsurface waters. Clasts in the resulting breccia can be derived from a lithology that once overlay the dissolving salt bed or from thin stringers and beds that were entrained in the dissolving salt. In many dissolution breccias the distance travelled by the clasts before being

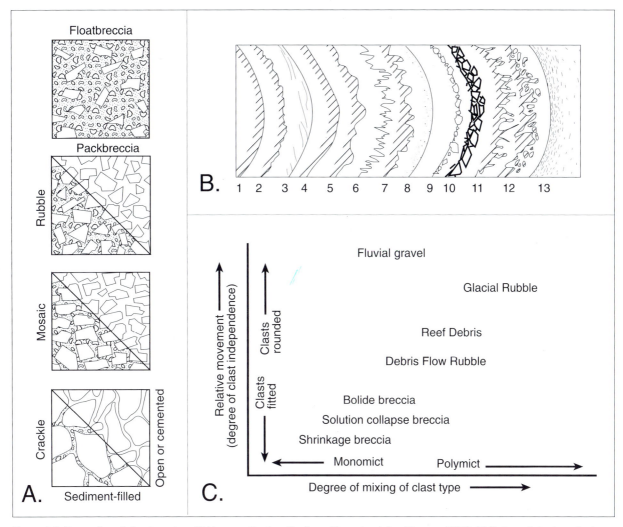

Figure 4.1. Evaporite solution breccias. A) Nongenetic classification of breccias (after Morrow, 1982). B) Nature of clast contacts; 1. smooth, 2. slightly pitted, 3. striated, polished, 4. faceted, 5. solution rounded and pitted, 6. stylolithic, 7. ragged, 8. weathering rim or alteration rind, 9. microfragment armouring, 10. fragments in process of breaking out, 11. jagged, 12. crumbled, 13. diffuse, gradational (after Laznicka, 1988). C) Breccia classification(after von Engelhardt et al., 1977).

107

redeposited is negligible, often only centimetres to a few metres. Timing of clast movement may be episodic or a single widespread instantaneous event. Clasts in an evaporite solution breccia may have moved in concert, like a school of fish into a cavity, or they may have gently foundered with little breakage. The distance of movement is dependent on the size and lateral extent of the solution cavity into which the clasts move and whether the clasts originally resided in thin interbeds within the dissolving evaporite or if they were part of a thick strong lithology that once overlay the dissolving salt bed. If the cavity is only a few millimetres to centimetres high, the overburden founders gradually and gently and so clast orientation and fitting is little disturbed by the process of dissolution. Crackle and mosaic breccias will be commonplace. If large local cavities develop, the collapse is more likely to be catastrophic and the result to be dominated by disturbed rubble or packbreccias.

Evaporite solution collapse breccias

Stanton (1966) in his benchmark paper on formation of evaporite dissolution breccias demonstrated solution, subsidence, and evaporite flowage acted simultaneously during subsurface brecciation. The resulting evaporite solution breccias in the Upper Devonian Leduc Formation, Alberta Basin, Canada, are made up of cm-dm angular dolomite clasts in a matrix of dolomite mud or anhydrite. He concluded that anhydrite-supported breccias are a transitional stage in the dissolution of interbedded halite-anhydrite-dolomite. Angular fragments of the dolomite overburden and interburden in an anhydrite-supported breccia (floatbreccia) are loosely supported by solution-thinned, but still abundant, nodular anhydrite with textures indicating viscous flowage during dissolution. As the breccia forms and flows, dolomite fragments are continuously broken and rebroken while the supporting anhydrite matrix carries a high proportion of loose, whole or broken dolomite rhombs (silt-sand size crystals). Ultimately, a packbreccia forms by the complete solution of the anhydrite, it consists of tightly packed dolomite fragments and clasts within a matrix of sand-sized detrital rhombs and dolomite microspar cement.

Evaporite solution breccia beds typically have a planar to slightly irregular, continuous lower contact, which is often coated with insoluble residues (Figure 4.2). Sometimes rocks in the footwall beneath the breccia bed are locally veined by infiltrations of satin spar and poikilotopic gypsum or anhydrite. Smith (1972) argues the fractures hosting such footwall stockworks were created by the pressure of

crystallisation of the evaporite itself, others envisage a more passive process of cement fill within preexisting joints or fractures. The footwall surface is usually covered by a few centimetres of mud- to sand-sized insoluble residues. The thickness of a pure evaporite that dissolves to produce a thin layer of insoluble residues may be substantial. For example, the Hartlepool Anhydrite contains only 0.5% - 0.8% insolubles (Smith, 1972). In this situation complete solution of a bed 100m thick would produce a layer of insolubles a mere 50-80 cm thick. In many breccia beds the insoluble residues now make up the matrix between clasts that were derived by the breakup of thin beds of carbonates or clastics. These are the fragmented remains of thin intrasalt beds that were once supported and interlayered by the now dissolved salt.

Insoluble residues typically consist of detrital quartz, authigenic chert, silt, clay and dolomite rhombs, which are later cemented by authigenic phases (e.g. calcite, dolomite or silica cements). Many evaporite beds originally entrained dispersed syndepositional silt-sized dolomite rhombs. When the encasing salts dissolve, a dissolution residue or matrix is left behind; it is dominated typically by solution-disaggregated dolomite crystal silts and sands and is sometimes called "chalky dolomite". Its high degree of initial intercrystalline porosity makes it an excellent potential aquifer for later mineralising solutions or hydrocarbons. Siliciclastic and carbonate sands bound by early evaporite cements (e.g. halite cements in aeolian, wadi, or oolite sands) can also be leached during burial diagenesis. In extreme cases the result is a disaggregated freeflowing subsurface sand with depositional level permeabilities, which remain until it is recemented by subsurface processes.

The basal zone of insoluble residues passes up into the main breccia body, an interval dominated by packbreccia or floatbreccia. Breccia fragments and clasts in the main breccia can be angular or solution-rounded (Figure 4.1b). Almost all the rubble texture is generated by the collapse of overlying or interbedded lithologies into cavities created by dissolving salt beds. Where the original sequence was made up of interbedded salts and other lithologies, the resulting clast rubble typically collapses in sequential order, so that the rubble pile retains the crude stratigraphic order of the nonevaporite interbeds.

Clasts are made up of varying amounts of sandstone, dolomite, limestone, mudstone and chert fragments. The main breccia may be polymict or monomict (syn. oligomict), depending on the nature of the interbeds and the overburden. Breccias can be clast- or matrix-supported (packbreccia

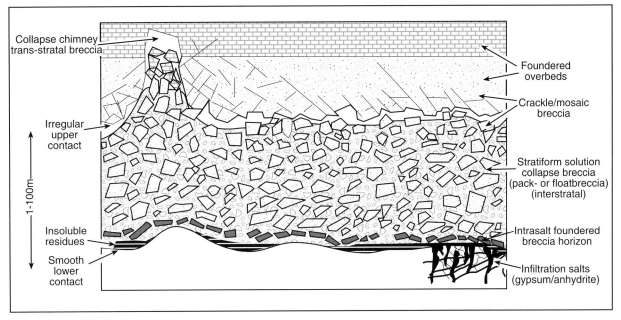

Figure 4.2. Lithological distribution in an evaporite dissolution triggered collapse breccia. See text for explanation of facies. Left half of figure illustrates relationships in a trans-stratal chimney, right half in a stratiform solution collapse breccia.

versus floatbreccia) depending on the amount of residue entrained in the evaporitic interval prior to dissolution. Pure evaporites, interbedded with indurated dolomites and other nonevaporite strata, tend to produce clast-supported packbreccia fabrics. Evaporite sequences interbedded with, or underlying, nonconsolidated sediments, or evaporites with a high content of dispersed impurities, tend to produce matrix-supported floatbreccia fabrics.

Unlike its lower contact, the upper contact of the main breccia interval is typically gradational, progressing from rubble breccia in the main body into mosaic, crackle then ultimately into unaltered overburden. The contrast between a sharp lower contact and a gradational upper contact is one of the most reliable characteristics of a solution breccia. It can be used both in outcrop and in the subsurface to separate a solution breccia from a depositional breccia (with depositional drape features along its upper contact) and a tectonic breccia (with cataclastic and typically symmetrical contacts).

Where complete removal of less soluble evaporites has not yet occurred, evaporite salts, such as flow-textured anhydrite nodules, may be part of the matrix within the main breccia interval. This is the situation described by Stanton (1966; see above). Local reprecipitation of evaporites into the breccia matrix may also generate large poikilotopic crystals, which encase the dissolutional residues. Such cements are often coarse-grained anhydrites

with a somewhat anhedral form (see later). Pseudomorphs of now dissolved evaporite minerals and nodules, composed of calcite, silica or celestite, may also remain either in the matrix or within clasts.

After the breccia has formed, both matrix and clasts are further cemented or altered during ongoing subsurface fluid flushing. The cement may be a fine-grained crystalline cement of equidimensional calcite crystals, as described by Swennen et al. (1990) in the Belle Roche breccias in Belgium, or a sparry coarsely crystalline dolomite. Coarse sparry or saddle dolomite is often associated with Mississippi-Valley-type deposits (Olson, 1984; Warren and Kempton, 1997). Dolomite burial cements in evaporite solution breccias are often complex and multistage. For example, there are six burial-related stages of dolomite formation and two episodes of breccia formation preserved in the evaporite solution breccias of the Tripolis Unit, Mainalon Mountain, Greece (Papioanou and Carotsieris, 1993; Karakitsios and Pomonipapaioannou, 1998) .

Flow of water through cavities during breccia collapse can rework fine-grained matrix material. Localised cm-dm cavities can fill with waterlain laminae that are typically overprinted by water escape features, slumps and, less commonly, with ripples and crossbeds. If the cavity lies in the vadose zone, or if a deeper phreatic cavity is filled by gas, such as methane or H_2S, then the cavity may be ornamented with microstalactites and other speleothems.

Calcitic vadose speleothems ornament cavities in dissolution breccias in the Kirschberg Evaporite breccia of central Texas (Warren et al., 1990), while pendulous sphalerite stalactites, precipitated in H$_2$S/methane-filled bathyphreatic cavities, typify the evaporite-replacing breccia ores in the Cadjebut Pb-Zn mine, Australia (Warren and Kempton,1997).

As a thick widespread evaporite bed dissolves, it does so more rapidly and pervasively in zones of greatest fluid access. This can be along a fault-bound edge, along a thinned feather edge, along its upper or lower contact, or along joints and fracture planes that cut the bed. As it dissolves, carbonate cements can precipitate as cement rinds within stratiform cavities, or along cross-cutting fracture planes. Alternating episodes of dissolving, fracturing and cementing will form veined fracture infills and fragments within the disappearing bed. With further collapse, these cements may be rebrecciated and cemented numerous times into internally complex clasts and fragments. Ongoing repetition of this process can create clasts made up of nothing but multiple episodes of vein and cavity-fill cements (Figure 4.3). Some fracture and block-defining cements can remain, and, with the disappearance of the supporting evaporite, evolve to form an open lacework or cellular grid of cemented materials and residues called a boxwork limestone. Such boxworks are especially obvious in nearsurface dissolution settings, as in rauhwacke terrains or in zones of growing tepee structures. Once a rock is more deeply buried and compressed, the boxwork either collapses or is only preserved within massive cements. For example, even after burial, local relics of boxwork texture still remain beneath the V's of the Permian tepees of west Texas and the Triassic tepees of Italy (Kendall and Warren, 1987).

The lower Visean Belle Roche breccia (east Belgium) is a classic multistage

evaporite dissolution collapse breccia (Figure 4.3; Swennen et al., 1990). Its origin as a dissolution breccia is indicated by: a) the sharp lower contact of the breccia and a gradual transition into the overlying strata, b) the presence of semicontinuous beds within the breccia giving it a crude 'stratification', and c) the existence of several types of evaporite pseudomorphs that are now composed of calcite, dolomite and silica. The majority of the sedimentary carbonates making up the breccia fragments show primary

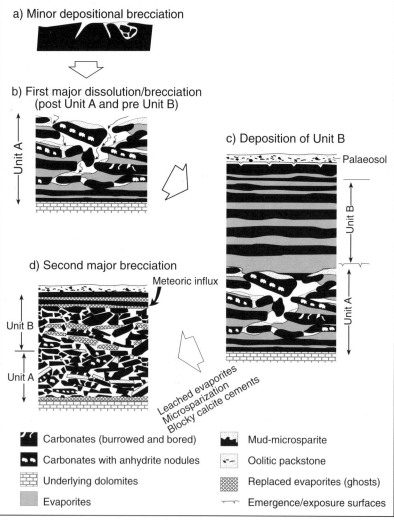

Figure 4.3. Sequential brecciation episodes in the lower Visean Belle Roche breccia (after Swennen et al., 1990). A) Minor syndepositional dissolution and fracture brecciation mainly related to shrinkage. B) This occurred shortly after deposition of bedded evaporites of unit A and involved incompletely lithified clasts, infiltration of freeflowing mud with ooids. C) Resumption of evaporitic sedimentation to deposit unit B with paleosol cap. D) Second major brecciation event driven by meteoric influx. It overprinted both unit B and unit A. The brecciation process was dominantly by foundering rather than cavern collapse, and so stratigraphic order was retained in the breccia stack. Foundering means the overlying beds parallel the general stratification.

features indicating they were originally deposited under hypersaline subaqueous conditions and were probably interbedded with the now dissolved evaporite beds.

Most Belle Roche breccia fragments have an interlocking packbreccia fabric showing a subparallel clast alignment that indicates foundering by gradual subsidence rather than catastrophic collapse into a large cavity (Figure 4.3b,c; Swennen et al., 1990). Multiple brecciation episodes are obvious in the lower breccia (Unit A), where brecciated and veined fragments occur within a microsparite (neomorphosed mud) matrix. There were at least two major brecc-iation episodes (Units A and B) driven by the successive dissolution of inter-layered evaporites. The second brecciation event related to infil-tration of meteoric water and to the complete dissolution of any remaining evaporites (Figure 4.3d). Meteoric infiltration was probably triggered by uplift related to an orogenic event at the end of the Visean (Sudetic orogenic phase).

Some of the breccias in this region were also dolomitized in active-phreatic mixing zones by processes facilitated by bacterial metabolism (Nielsen et al., 1997). High degrees of dolomite supersaturation were produced in the anaerobic to dysaerobic zones of bacterial sulphate reduction. The sulphate came from the dissolving evaporites and the organics from the adjacent interbeds. The resulting dolomite precipitates show characteristic bacterial morphologies, such as spheroidal and dumb-bell shaped crystals. The whole breccia was finally cemented by a blocky calcite under deeper burial conditions.

Breccia geometries

Evaporite solution breccia beds can be syndepositional, or form much later after millions of years and hundreds to thousands of metres of burial, halokinesis, and uplift. Breccia beds tend to be widespread and stratabound, with geometries that reflect the distribution of the precursor evaporite (Figure 4.4). If the precursor layer was part of a stacked cyclic sequence of platform evaporites interlayered with carbonates or siliciclastics, the breccias tend to stack into interstratal cyclic horizons. This is the most easily recognised style of evaporite solution breccia. If the precursor evaporite was part of a halokinetic or remobilised sequence then the solution breccia is trans-stratal and follows the distribution of former salt stems, salt welds, or thrust planes.

Intensity of brecciation and the general geometry of the foundered strata in the breccia bed can vary widely from place to place, even within a single unit (e.g. Smith, 1972). The geometry of collapse is related to variations in:

a) Inflow geometry of the solution front: water may enter along a fracture, fault, blanket aquifer, reef, diapir margin, etc. The timing of development of the solution front may in turn be related to episodes of seismic or tectonic instability (e.g. seismic pumping driving episodic basin dewatering, or periodic uplift or sea level lowering creating a gravity head to drive meteoric flow).

b) Rate of evaporite solution: this is dependent on the rate of formation water crossflow and the saturation state of the crossflow. This is in turn controlled by climate, tectonics, regional uplift, development of major unconformities, etc.

c) Confining pressure: related to thickness of overburden at the time of collapse. This is in turn related to the age of collapse (burial/uplift timing).

Stratiform breccias

The simplest and most general styles of interstratal collapse are typified by blanket aquifers and broad areas of gentle sagging where diffuse undersaturated water flow over large areas has slowly removed evaporite salts (Table 4.2; Figures 2.11a,b and 4.4; Hauptdolomit). The blanket aquifer can lie above, below, or even within the dissolving evaporite bed. Overburden brecciation in such gently foundering regions is minor so that fossils and sedimentary structures are mostly preserved in blocks with near upright orientations. Larger interbed and overburden clasts tend to be tabular, and only slightly separated or disorientated, i.e., dominated by mosaic to crackle breccias.

In areas of more focused dissolution where large caverns and solution planes have form within the dissolving evaporite beds, the whole of an overlying sequence can be reduced to a rubble breccia — a mass made up of completely disarticulated and rotated (commonly rectangular) blocks and fragments ranging from silts to blocks many metres across. In still other areas the foundered strata can be bound by sharp folds or dislocations that are in effect normal faults. Such areas of intense brecciation are often the end product of localised collapse and can grade laterally and vertically into almost unbrecciated equivalents. These dislocations can be sharp and clean, or associated with areas of intense diagenetic alteration such as calcitisation and precipitation of ferran carbonates. This latter group of intensely altered and replaced breccia intervals within a carbonate or sandstone host is part of a larger group of

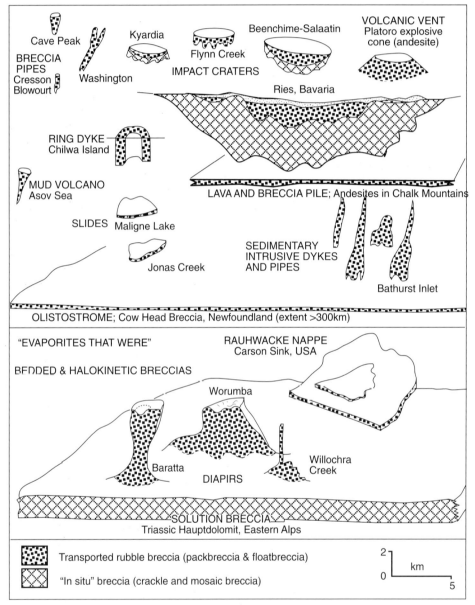

burial cycle, even while immediately overlying strata are accumulating, tend form a ridge and furrow dissolution geometry that influences the thickness and depositional style of overlying sediments. Overlying clastic beds are no more than only partially lithified when the underlying evaporite dissolves and so are highly disturbed by ongoing collapse of their substrate. This forms characteristic intrabed slumps, water escape textures and other soft sediment deformation features in the beds immediately above the dissolution horizon. Breccia clasts in such scenarios typically show rounded and crumbly edges.

In contrast, bed collapse after induration tends to create breccias dominated by structures indicative of more brittle collapse failure. This group includes trans-stratal joint and chimney stopes that may have tens to hundreds of metres of relief, as well as pervasive interstratal angular breccias derived from overlying strata where the clasts show little or nor evidence of internal deformation.

Figure 4.4. Size and shape of various breccia bodies that occur in sedimentary successions (modified in part after Laznicka, 1988). Evaporite-related ("evaporites that were") geometries are illustrated in the lower part of the figure.

fault-focused, often metalliferous, deposits with hydrologies driven by evaporite solution. They are considered further in Chapters 8 and 9.

Hydrological controls

Salt beds that dissolve early tend to create characteristic breccia textures and soft sediment slumps in the overburden that can be used to distinguish a syndepositional from later modes of formation. Evaporites that dissolve early in their

Early dissolution of a nearsurface evaporite tends to be a response to facies-controlled permeability corridors in the blanketing uncon-fined aquifers. That is, areas of preferential ingress into the evaporite bed are controlled by depositional permeabilities within the overburden (e.g. ingress beneath porous and permeable sand-rich fluvial channels versus relatively impervious mud-rich floodplain

Features of an interstratal dissolution breccia
• Sharp lower contact and irregular upper contact to breccia bed.
• Clasts derived from overlying lithologies.
• Breccia body shows a crude "stratification", with clasts of differing lithologies often stacked in original stratigraphic order.
• Depositional textures in some breccia fragments indicate hypersalinity during deposition.
• Zone of insoluble residues (silt, chalky dolomite, etc.) lies atop lower contact, and sometimes has infiltrated into fractures beneath.
• Evaporite cements in fractures beneath lower contact.
• Several types of evaporite pseudomorphs (nodules, radial textures, length slow chalcedony, etc.) are present in clasts or matrix and interclast cements.

Table 4.2. Features of an interstratal breccia created by foundering (most common evaporite solution breccia)

sediments). In contrast, the hydrologies of later dissolutional systems tend to be more directional and focused by secondary permeability features. By then the overlying beds are indurated and relatively impervious so that major zones of undersaturated infiltration or crossflow are controlled by faults or fractures as well as regional tilting

or sag. Water arriving at the site of dissolution can be either from meteoric or basinal sources.

The feedback between early shallow salt dissolution and partially consolidated overlying sediment styles can be traced for over 35 km of the outcrop length in the Blomiden Formation (Late Triassic) of the Mesozoic Fundy rift basin of Nova Scotia (Ackermann et al., 1995). This unit exemplifies syn-dissolutional salt loss that was induced by shallow groundwater freshening tied to fluvial sedimentation. Fluvial sands of variable thickness (unit C) lie atop a highly chaotic and slumped mudstone with evaporite nodules in its lower parts (unit B; Figure 4.5). This chaotic mudstone unit preserves much of the evidence of syndepositional collapse of the evaporite substrate. It is characterized by highly disrupted bedding that is commonly cut by small (<0.5 m) domino-style synsedimentary normal faults. It also preserves evidence of downward movement of fines, of geopetal structures, of variable thicknesses related to local hollows after evaporite solution, and of irregular, partially faulted contacts with the overlying unit.

It is locally overlain by a fluvial sandstone (unit C) of varying thickness, that is in turn overlain conformably by flat bedded mudstones, which still preserve a syndepositional sand patch fabric. Unit B is underlain by a normally faulted massive mudstone (unit A). Although the thickness of the fluvial sandstone (unit C) is highly

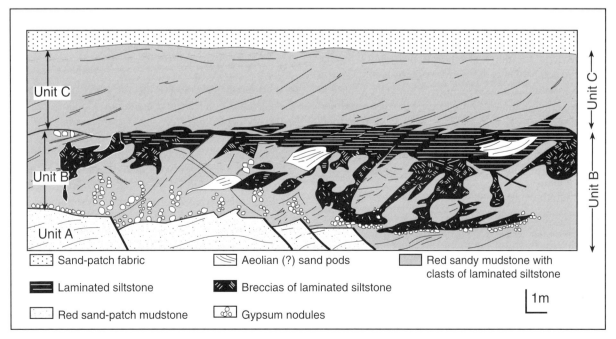

Figure 4.5. Syndepositional collapse features related to dissolution of an underlying evaporite bed in the Lower Blomiden Formation (Late Triassic) near Blomiden, Nova Scotia. See text for discussion of units A-C (after Ackermann et al., 1995).

113

variable, the overlying flat-bedded mudstone exhibits only gentle regional dip, hence the bulk of the evaporite layer must have dissolved prior to the deposition of the flat bedded mudstones. Like unit B, the sands of unit C exhibit numerous soft sediment deformation features including; dewatering structures, convoluted bedding, kink bands, and convergent small-scale fault fans. The frequency and intensity of these features increases dramatically above low points in its the base, showing unit C sands were laid down within local solution depressions that were still growing as sediment was deposited. Unit B (the chaotic unit) is characterised by blocks of laminated claystone, which become increasingly disrupted and brecciated downward toward its base where locally abundant gypsum nodules are still preserved. Lenses and fissure fills of mudstone to coarse sandstone within the chaotic unit often entrain clasts of laminated mudstone originally derived from immediately overlying beds.

Both the chaotic unit and the overlying fluvial sands were deposited as an underlying evaporite layer underwent syndepositional subsidence and collapse. Groundwaters driving the dissolution were probably tied to the same river system that was depositing units B and C. The former presence of evaporites in this unit is now indicated solely by residual gypsum nodules within the lower part of unit B (Figure 4.5).

In contrast, breccias that form after induration of overlying strata tend to be combinations of interstratal and trans-stratal geometries with breccia clasts that are made up of well-indurated angular blocks. Post induration interstratal breccias tend to follow the lateral distribution of the evaporite bed from which they formed, while trans-stratal (trans-formational) breccias tend to cross cut the strata and form by a combination of chimney growth and overburden stoping (Figure 4.3). Soft sediment deformation features are less pervasive in post-induration breccias and are typically restricted to intervals where insoluble residues accumulated in joint-defined fluid-filled cavities.

Post-induration interstratal breccia beds tend to have sharp flat bases and uneven, wavy to transitional blocky upper surfaces. If the dissolved evaporite was pure with few intrasalt beds the resulting rubble breccia unit is thin, but widespread. For example, a clean 100 m thick bed of salt may, when dissolved, leave behind a residue bed and rubble breccia that is no more than 0.5 - 2 m thick, but that residue may form a continuous marker horizon with a lateral extent of hundreds of kilometres. Core recovery from such intervals is often difficult. Even with good core recovery, recognising beds composed of little more than

solution residues is difficult, they are easily misinterpreted as unconformities, tectonites, or syndepositional intraclast breccias. Preserved evaporite indicators, such as "cabbage-head" chert or calcite concretions, are one way to define the evaporite association of such breccias (see later in this chapter)

Spectacular breccia-filled trans-stratal solution chimneys, created by evaporite dissolution and joint-controlled stoping, dominate Silurian outcrops in the Mackinac Straits area of Michigan and western Lake Erie (Landes, 1945; Carlson 1992). Collapse features in the Mackinac Straits first began to form in the Silurian-Devonian as Silurian Salina salt dissolved and the overlying indurated strata collapsed into the voids. There are three styles of breccia in the region: 1) megabreccia beds in which blocks up to several hundred feet in size moved intact except for some minor tilting; 2) relatively thin intraformational or intrastratal breccia developed along the bedding planes both regionally and within layers in the megabreccia blocks; 3) thicker trans-stratal breccias extending vertically from a few feet to several hundred metres and abutted by undisturbed bedded strata. The thickest trans-stratal breccias extend vertically up to 460 m and at one site the individual blocks have fallen 290 m. Most of the intense evaporite karstification and celestite replacement took place in the Late Silurian. Some caverns have recently been reactivated on a much smaller scale (Carlson, 1992). Modern breccia chimneys do not necessarily break through to the surface, but may stop growing when volume expansion of the brecciated material is sufficient to fill the salt cavity. Such present day "blind" salt karst may only be located using geophysical techniques.

Evaporite collapse does not just form breccias composed of nonevaporitic clasts and residuum. Collapse breccias made up of fragmented less-soluble salts can also accumulate. For example, angular blocks of bedded gypsum can form cavity fills and breccia beds as supporting halite or bittern beds are dissolved. Such breccia beds can show characteristic smooth bottoms and more irregular tops. Sulphate collapse breccias typify the outcropping margin of the Permian Castile Formation of the Delaware Basin in west Texas and New Mexico, and the Middle Miocene (Badenian) gypsum laminites of southern Poland (Babel, 1991 - see gypsum karst).

The early, and the post-induration, dissolution events discussed so far indicate a subsurface superposition of a less saline hydrology at some time after the cessation of the active phreatic hydrology that deposited the salt bed. However, dissolution collapse can also occur much earlier,

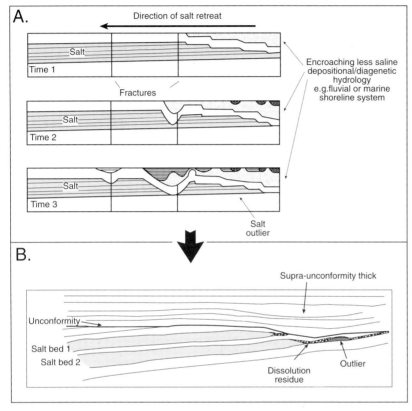

A.

Direction of salt retreat

Salt
Time 1

Fractures

Salt
Time 2

Encroaching less saline
depositional/diagenetic
hydrology
e.g.fluvial or marine
shoreline system

Salt
Time 3

Salt
outlier

B.

Supra-unconformity thick

Unconformity

Salt bed 1
Salt bed 2

Dissolution
residue

Outlier

Figure 4.6. Subcrop dissolution creates characteristic sag geometries in overlying strata (modified in part from Anderson et al., 1994; Warren, 1997).

even during the accumulation of the primary salt unit. This style of syndepositional karsting and the resultant textures were described in Chapter 1. There is an even earlier style of dissolution where more soluble layers are syndepositionally removed by seasonal freshening of the same brine body that is depositing the salt. For example, in one of the first uses of the term '"the salt that was" Schreiber and Schreiber (1977) described Miocene gypsum beds in Sicily where early halite layers were dissolved, even as the gypsum bed was accreting.

Evolution of a breccia bed

Subsurface salt dissolution spreads outwards into a salt bed from various entry points to create a number of advancing collapse or foundering fronts (Anderson and Knapp, 1993). Evaporite beds undergoing subsurface solution in the active phreatic zone will dissolve most rapidly along those boundaries that are most subject to crossflow by large volumes of undersaturated waters. Unconfined active phreatic aquifers tend to eat away at the upper surface of a more deeply buried salt bed. Confined aquifers, as well as compactional waters or thermobaric

convection cells, tend to eat into the evaporite beds from below (see Chapter 2). Fractures and planes of weakness, or areas of more soluble minerals that cut across the regional extent of an evaporite bed, act as preferred subsurface entry points of under-saturated waters into and through the bed (Figure 4.6a). Point or line-sourced groundwater entry into an evaporite bed creates dissolution chimneys in the foundering unit above the dissolving evaporite, while blanketing aquifers or upwelling zones tend to create more widespread zones of gentler foundering..

An established dissolution front adjacent to a subcropping and perhaps gently dipping salt bed tends to migrate laterally into the basin and to carry with it a depression in the land surface. This migrating depression tends to focus fluvial activity and so enhance further dissolution along the updip edge of the evaporite bed. Lateral migration continues until the subsurface salt bed is deeper than the undersaturated base of the zone of active phreatic flow. Thus, edge-inward dissolution in regions of subcropping thick evaporite beds creates dissolution moats or depressions about the edges of a dipping and dissolving evaporite beds (Figure 4.6a). In a continental setting this style of solution creates a characteristic topography with stacked fluvial depression adjacent to a salt dissolution edge. A topographic high typically forms above the as yet undissolved salt. When preserved, such geometries of sediment thicks adjacent to a salt feather edge are of sufficiently large scale to be recognisable in seismic lines (Figure 4.6b).

The ground surface immediately above and about the high atop a dissolving salt bed is characterised by hummocky and blind valley karst. This geomorphology constitutes the present land surfaces about the edges of the Hutchinson Salt Member in Kansas and the Maha Sarakham Formation in NE Thailand (Anderson et al., 1994; Supajanya and Friederich, 1992). In Kansas, since the late Tertiary, the exhumed shallow-dipping salt beds of the Permian Hutchinson Salt Member has been dissolved more-or-less continuously along its active eastern margin. Dissolution is driven by sustained contact with unconfined,

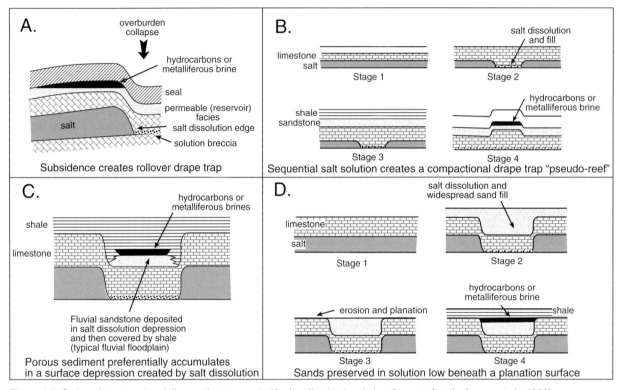

Figure 4.7. Styles of structural and diagenetic trap created by localised salt solution (in part after Anderson et al., 1988).

undersaturated groundwater. The associated westward migration of the eastern margin has resulted in surface subsidence and a contemporaneous valley-fill of Quaternary alluvium. In places, these alluvium deposits extend more than 25 km to the east of the present-day edge of the main body of contiguous rock salt. The margin has receded this distance over the past several million years. Rock salt is leaching vertically from the top down, and horizontally along the uppermost remnant bedded soluble layer(s). As a result, the eastern margin of the salt thickens gradually (up to 90 m) in a stepwise manner from east to west for distances on the order 5-15 km. The steps are formed by the less soluble intrasalt beds. Locally the salt member is preferentially leached along NNE-trending palaeoshear zones to leave behind thick stepped outliers.

Salt bed leaching typically underscores a set of self-perpetuating processes. Fractures created by the collapse of overlying strata into accommodation space generated by salt cavities also provide sinks or additional conduits for the percolation or upwelling of undersaturated waters. This in turn instigate yet more salt solution in the vicinity of the sink.

As a salt bed disappears it can create depositional, diagenetic or structural (drape) traps for hydrocarbons and

metalliferous fluids (Figure 4.6; Anderson et al., 1988). Traps can form early within the depositional setting, or much later during catabaric salt removal. Compactional drapes can create a rollover anticline that is closed over the edge of a residual salt body (Figure 4.7a). Local solution or collapse can form a sediment-filled sink that transforms to an inverted thick with the ongoing solution of the surrounding salt (a "pseudo-reef" - Figure 4.7b). Or it can form as a focused sand infill within a solution-induced depression on the land surface (Figure 4.7c): or it can form as a remnant of a formerly widespread sand unit that was ultimately preserved only in sediment-filled depressions within the salt bed (Figure 4.7d). The Hummingbird reef structure in Saskatchewan is a well documented example of a "pseudo-reef" (Smith and Pullen, 1967). The Berry Field in Lower Cretaceous fluvial sandstones of Alberta is an example of a reservoir created by sand deposition at the edge of a dissolving salt unit (Hopkins, 1987). Compared with most stratigraphic traps in Cretaceous sands in the region, the Berry Field is unusual in its occurrence on the flank of a local structural high within a larger regional structural depression. This unusual synclinal setting for the reservoir is a direct indication of its origin as a fluvial system within a landscape position controlled by salt solution. For more information on salt solution traps the interested reader is referred to the excellent summary paper by Anderson and Knapp (1993).

Diapiric solution breccias

Salt diapir (halokinetic) breccias ultimately contain no salt but outline a former diapir geometry. The only evidence of the original evaporites may be a few pseudomorphs, after halite or calcium sulphate, either in the clasts or in the interclast matrix. The breccia itself is an allochthonous assortment of angular clasts surrounded by a matrix made up of insoluble residues ("rock flour") typically dominated by carbonate, clays, sericite and chlorite. Diapiric solution breccias are currently forming in and around subcropping salt diapirs in the Arabian Gulf. Classic examples, where no salt remains, outcrop extensively in the Neoproterozoic strata of the Flinders Ranges of South Australia, the Macdonnell Ranges of the Amadeus Basin of Central Australia, and the Damara Orogen of southern Africa (Mount, 1975; Lemon, 1985; Kennedy, 1993; Behr et al., 1983).

Halokinetic breccias can also form along former thrust or glide planes after the dissolution of the halite or calcium sulphate horizons that lubricated the thrust. Such breccias are termed rauhwacke and are discussed later in this chapter. Solution breccias in the Damara Orogen encompass processes of fluidisation and cataclasis within mobile evaporites under greenschist and higher metamorphic grades. A full discussion of this sequence, along with other metamorphosed evaporite indicators, is to be found in Chapter 6, while the processes of salt diapirism are presented in Chapter 5.

Clasts in halokinetic breccias are always trans-stratal and tend to be dominated by polymict associations of allochthonous clasts. Such clasts are combinations of; a) material that once formed intrasalt beds, b) of material that was plucked or was enclosed by flowing salt, and c) of material that was derived from the collapse of overburden as the active diapir deflated and dissolved (Figure 4.8). Internally, owing to the complex internal folding that accompanies salt flow, the breccia bodies outlining the former diapir are much less likely to preserve original stratigraphic layering of any intrasalt beds.

Breakage of siltstones and carbonate intrabeds within a halokinetic salt bed begins once the encasing salt begins to move (Figure 4.8). Some workers also argue flowing salt can pluck large blocks from strata adjacent to a salt wall or atop the flowing mother salt bed. This is not typical of modern outcropping salt flows in the Arabian Gulf, where most of the intrasalt clasts come either from within or below the mother salt bed (see below). Once caught up in the flowing salt, clasts are then carried along within the salt flow, to accumulate as breccia masses in stratigraphically higher regions where the encasing salt dissolves in "at-surface" or near surface positions. Thus the clasts in diapiric breccias (as in the breccia trains of the former salt allochthons of Algeria) often include minerals, such as magnesian-talc, sodian-scapolite and dravite-uvite tourmaline, which indicate the clasts were carried by the salt flow from depths where the metamorphic/diagenetic grade was higher than the current stratigraphic level of the breccia

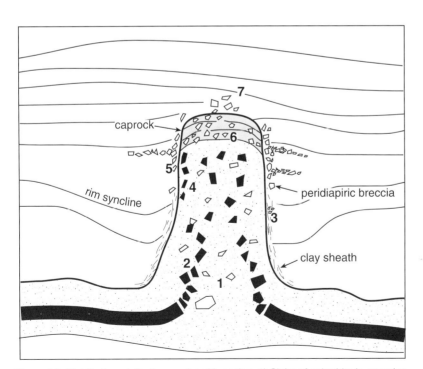

Figure 4.8. Distribution of diapir associated breccias. 1) Slabs of anhydrite in granular halite; 2) internal flowage breccia formed by dismemberment of competent internal beds; 3) breccia pockets in "clay sheath"; 4) transported flowage breccia of wallrock fragments incorporated into outer zones of stock; 5) peridiapiric breccia, extent is greater in zones of former salt glaciers; 6) caprock breccia; 7) false caprock breccia, a dissolution breccia generated by salt collapse (after Laznicka, 1988).

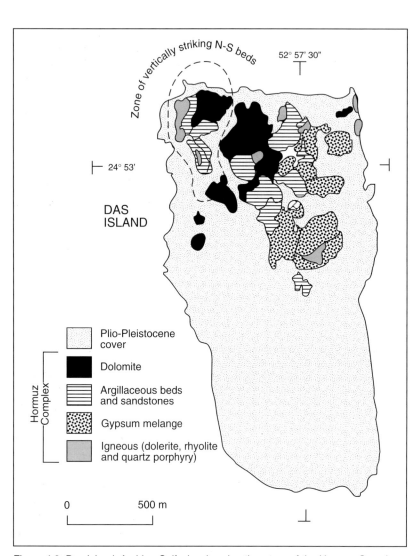

Figure 4.9. Das Island, Arabian Gulf, showing chaotic nature of the Hormuz Complex; it includes blocks of ferruginised and rafted material that are hundreds of metres across (after Kent, 1987). See Figure 5.29 for location.

originally deposited as a basinwide evaporite at a time near the Precambrian/Cambrian boundary and has since been mobilised numerous times into salt structures (Kent, 1987).

The range of clasts carried in that part of the Hormuz Complex that outcrops as island cores within the Arabian Gulf is diverse, but uniform over much of its outcropping extent. In no case are there any clasts derived from the cubic kilometres of indurated Palaeozoic and later sediment through which it flowed on it passage to the surface (Kent, 1987). One of the most characteristic raft lithologies is a black foetid finely laminated and bedded dolomite that is often coarsely recrystallised and shows evidence of deep burial diagenesis. It may well have been deposited in hypersaline settings that were also responsible for the deposition of the thick salt beds that flowed to become the Hormuz Complex. Blocks composed of sandstones and siltstones are also widespread, they are mostly reddish, grey or greenish grey and variably indurated. With a breccia body both the dolomites and the siliciclastics can outcrop as rafted blocks up to several hundred metres across (Figure 4.9).

Igneous rocks are another important component of the rafted blocks in the breccia. In the Gulf islands the igneous rafts tend to be dominated by acid tuffs (trachytes and rhyolites) along with acid intrusives and dolerites. On the Iranian mainland the igneous rafts are referred to as "greenstone" blocks, as they are dominated more by basalts and dolerites. Throughout the Gulf and on the Iranian mainland there are large volumes of red and specular haematite present in the outcropping Hormuz Complex. Some outcropping rafts of silicified shale are also crosscut by malachite stained joints. On Hormuz Island the outermost ring of the salt plug itself is made up of haematitic tuffs, locally so rich in iron oxide that it is mined as a source of pigment (Kent, 1987). The significance of an Fe-rich phase in terms of basin metal transport, along with dissolution-derived chloride brines, is discussed further in Chapter 9.

(see Chapter 6). This observation is useful when core has intersected a breccia body and an attempt is made to distinguish interstratal breccias and trans-stratal chimneys from halokinetic breccias and welds.

Diapiric dissolution breccias define many outcropping salt plugs in the Arabian Gulf, in what is one of the world's largest salt provinces and also one of its largest hydrocarbon provinces. In fact, most of the oil and gas fields in the region formed under the influence of salt flow (see Chapter 5). Diapirs outcrop/subcrop to form what is locally called the Hormuz Complex — a polygenetic assemblage of diverse lithologies that includes both the nonevaporite blocks carried along by salt flow, as well as the supporting salt and gypsum/anhydrite matrix. The Hormuz was

Most of the diapir-cored Gulf Islands have experienced little salt movement since the mid Miocene, and the Hormuz Complex is now eroded and largely covered by Neogene carbonates. In contrast, halokinetic breccias are forming today atop and peripheral to namakiers (salt glaciers) of the Zagros chain of Iran (Talbot and Jarvis, 1984). Where an active salt tongue subcrops or lies beneath a dissolution-derived carapace, the clasts accumulate by breakup of the soft overburden as well as from stacking of fragments and rafts formerly within the now dissolved salt flow. The process is akin to that of ablation tills in an ice glacier, except the dissolving phase is salt rather than ice. The resulting namakier (salt glacier) breccia is composed of a coarse, unsorted, heterolithic (polymict) rubble supported by a fine-grained calcareous matrix.

In the namakier breccias of the coastal parts of the Zagros chain this jumble of exotic material is a combination of Hormuz Complex materials brought to the surface by upwelling salt and remnants of Neogene marine carbonates that once covered now-dissolved salt tongues. Thus, ongoing dissolution of namakier salt leaves behind a rubble moraine at the level of the former salt allochthon that is a combination of exotic salt-rafted blocks from lower in the stratigraphy and of sediments that were deposited at the time the salt was flowing at the surface. Differentiation of these two populations of clasts is useful when attempting to time episodes of at-surface salt in Phanerozoic successions adjacent to former salt stems.

Belous et al. (1984) described redeposited breccia trains composed of fragments of Devonian basalt and limestone at several stratigraphic levels in sediments surrounding the Late Palaeozoic Adamovsk salt stock in the southern Ukraine. They attributed the breccia train to lateral redeposition of salt-lifted blocks during periods of stock unroofing, possibly via a salt glacier mechanism. Drilling in 1991 in the Tapley Hill Formation on the flank of the Blinman Diapir (Flinders Ranges, South Australia) intersected an anhydrite-cemented breccia with most clasts not derived from the enclosing shale (Cooper, 1991). The breccia is interpreted as a the residue of a salt glacier flowing on the seafloor under relatively deep water.

Neoproterozoic diapiric breccias crop out throughout the Flinders Ranges of South Australia (Figure 4.10a; Dalgarno and Johnson, 1968). Many breccias occur in anticlinal cores but are still located at or near the level of their mother salt beds (Callanna Beds) and so are possible former salt pillows or glide planes rather then true diapirs. Others, such as the Oratunga and Wirrealpa Diapirs show subcircular patterns of outcrop at stratigraphic positions

well above the level of the original mother salt bed. Almost all the diapirs line up along major regional shears and faults, indicating that the mother salt bed was flowing while the basin was under extension, a condition that instigates salt flow in almost all of the world's diapiric provinces (see Chapter 5). Lemon (1985, 1988) clearly illustrates synkinematic controls on sedimentary facies adjacent to many of these breccias (Figure 4.10b,c).

Typical breccias are made up of heterogenous rock types set in a matrix of strongly folded and disrupted beds that are dominated by weathered pale-grey siltstones and fine sandstones. The clasts are composed of a polymict mixture of dolomites with gypsum and anhydrite pseudomorphs, of dolomitic sandstones, of siltstones and sandstones with displacive halite casts along bedding planes, as well as diabase and basic volcanics, including amygdaloidal basalts (Lemon, 1985). The margins of many former diapirs in the Flinders Ranges are also characterised by numerous copper shows (see Chapter 8).

Within the breccia mass these diapiric breccias show a banded aligned fabric with shaping, disruption, mixing, and alignment of clasts that must be attributed to flow. Rafts of coherent sediment up to 500 metres across are found within the breccia masses (Figure 4.10b, c). Bedding within rafts near the edge of the breccia mass is commonly near vertical and subparallel to the diapir margin, while rafts nearer the centre of the breccia mass are more randomly distributed, a feature that indicates frictional alignment of large clasts near the diapir edge. There is a surprising similarity between the size and lithology of rafts found atop Neogene diapir crests in the Arabian Gulf and those in breccia masses in the Flinders Ranges. The steeply dipping bedding in rafts located near the diapir margin is also a feature of both areas (Figure 4.9 c.f. Figure 4.10b).

Matrix minerals in the Flinders Range's breccias include: carbonate, chlorite, feldspars (both detrital and authigenic), haematite, magnesio-riebeckite, quartz, stilpnomelane, and talc. Mount (1975) interpreted them as zeolite facies minerals formed during breccia emplacement via reactions involving hypersaline fluids. He argued conditions during breccia formation were aqueous, low pressure, and hypersaline. Mineral forming reactions occurred in an oxidative low-grade metamorphic environment replete with CO_2, with open system throughput and low temperatures (150° - 300°C). He concluded this palaeohydrochemical system had strong affinities to natural hydrothermal and geothermal systems. It indicates a later burial stage compared to the current salt-filled diagenetic systems of the Arabian Gulf. The minerals formed at a time

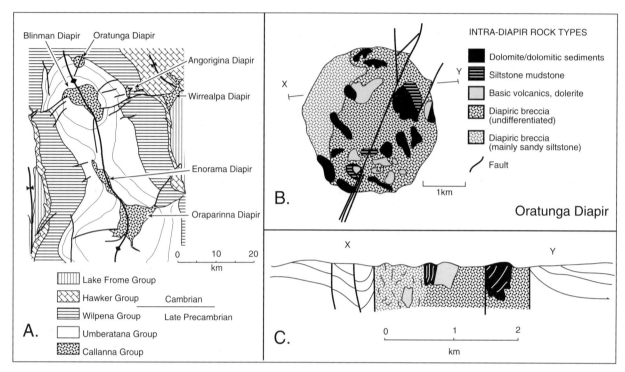

Figure 4.10 Halokinetic breccias, Flinders Ranges, Australia. A) Regional distribution of halokinetic breccia in the Callanna Formation. B) Distribution of large lithoclasts or rafts in outcropping breccia matrix that define the Oratunga Diapir. C) East-west cross section of Oratunga Diapir (modified from Lemon, 1985).

when most of the salt had already dissolved from the system and only the breccia body remained to outline the former salt geometry (see also Damara Orogen, Chapter 6).

The late stage replacement features shown by the halokinetic breccias of the Flinders Ranges also means that where the crestal, rather than the stem, portions of the diapir breccia crop out they tend to show classic "falling diapir" geometries (pers. obs.). Such examples include the Wirrealpa and Oraparinna Diapirs. The falling diapir stage is characterised by crestal collapse and the formation of fault-defined graben thicks in the overlying sediments (Chapter 5). As the diapir was collapsing, the underlying salt was thinning and dissolving. This facilitated the growth-fault focused escape of formerly confined basinal fluids on to the seafloor. This mechanism explains the formation of large fault-fed baryte masses atop many former diapir crests, along with the various small copper shows found in association with the diapir breccia margins.

Mount (1975) noted breccia/host rock contacts were invariably abrupt and coincided with planes of weakness in the host strata. The contacts, despite synformational sculpting and quarrying by invading diapir material, match across some diapir stems. Host strata were rarely brecciated

or upturned against a diapir, illustrating again the passive nature of the intrusion. In my opinion his observations explain why material from strata located above the salt bed are not broken into rafts and transported to the surface. Salt emplacement is predominantly a passive process into preexisting planes of entry. Where contact deformation has occurred, it usually predates diapirism and was due to glide plane faulting that later controlled breccia geometry. Mount (op. cit.) termed this style of contact 'permitted intrusion'. He argued it occurred under extension of the cover or overburden, perhaps induced by regional extension. Alteration of the host rock adjacent to contacts was absent but for minor dolomitisation and sideritisation in certain areas.

Unlike prevailing views of forcible injection for diapir breccias in the Flinders Ranges, Mount in the 1970s stressed the passive flow of their mobile breccia precursor (salt) into tensional structures. He concluded 'diapir', in terms of the then popular notion of active piercement, was not the most appropriate term for Flinders Range's breccias. His understanding of the dominantly passive nature of the diapiric process preceded by almost two decades our current models of brittle overburden atop weak flowing salt (see Chapter 5). In 1975 he stated: "The overburden was relatively brittle, its weight the prime driving force to the intrusions. Mobility of source material, rather than factors such as density, was paramount to diapirism. The mobility is explained by the former presence of saline evaporites in the interstices of the breccias."

Thrust/detachment breccias and rauhwacke

Rauhwacke horizons (also known locally as cornieule or cargneule) are the weathered and leached remnants of what were once 'weak' salt-entraining tectonised layers. Under stress, their evaporitic precursors acted as rheologic inhomogeneities that provided surfaces of décollement. That is, rauhwacke units are weathered outcrop expressions of evaporite-lubricated thrust planes, glide planes and detachment terranes. They are commonplace in the Jura Mountains in Switzerland, the Atlas Mountains in Morocco and the Salt Range in Pakistan. Rauhwacke forms once the more soluble evaporite components are leached from a tectonic breccia, or as the sheared evaporite and its surrounds are metasomatically converted to less soluble salts, such as calcite spar or dolomite. Rauhwacke breccias in outcrop are dominated by carbonate and clastic fragments in a cataclastic matrix. Their subsurface equivalents are interbedded thin evaporitic units, tectonic breccias, and partially altered residues. Boxwork and cellular textures are commonplace in outcrop.

In older studies, the rauhwacke itself was considered to have facilitated décollement during Alpine tectonics. Now it is known that evaporite-lubricated protoliths of the rauhwacke acted as detachment horizons and incompetent layers during thrusting and folding. As well as the tectonite specific usage of the term rauhwacke that I follow in this book, the term is sometimes used by other authors in a more general way to describe any evaporite solution breccia with angular cavities or fragment outlines (e.g. Schreiber, 1988; Karakitsios and Pomonipapaioannou, 1998).

Often the development of a salt-lubricated tectonite breccia is intimately related to regional detachments, which are also forming diapirs (see Chapter 5). The distinction between rauhwacke and diapiric breccias is mostly geometric. Rauhwacke and their tectonite precursors tend to be dominantly interstratal stratiform surfaces of tectonic brecciation and solution collapse. They often separate a brittle-faulted hanging wall from a less altered footwall succession (thin-skinned tectonics), and only occasionally

cut up-section. In contrast, diapiric solution breccias tend to be trans-stratal with geometries outlining the shape of the precursor salt diapir. When diapirs, allochthons and detachment surfaces are part of the same salt-lubricated system, the two breccia styles merge and become near impossible to separate.

Halite, or sometimes anhydrite beds, are the main lubricants during the formation of stratiform tectonite layers. The inherent lack of strength of an evaporite, when compared to other lithologies, focuses the greater part of the detachment or thrust plane into the salt-entraining interval. Jordan and Nuesch, 1989 and Jordan et al., 1990 describe deformed anhydrite textures preserved in the main Alpine thrust plane. These are the deeper less-leached equivalents of nearer-surface zones where meteoric leaching has created rauhwacke from their evaporitic tectonite precursor.

In the Austrian Alps, Spötl (1989b) describes tectonised evaporitic breccias that created, and are preserved in, faults and thrusts of the Permian Haselgebirge Formation. All these tectonic breccias flowed and fragmented during the Alpine Orogeny (Middle Cretaceous - Late Eocene) and are part a giant regional tectonic melange. The Haselgebirge breccias are composed of polymict clasts (from cm to several tens of metres in diameter) that in the subsurface are often embedded in a halite matrix. The breccia matrix contains varying amounts of clay and silt impurities (normally in the range of 15-55 %) plus minor quantities of anhydrite. In addition to clasts derived from the tectonic breakup of the initial intrasalt beds (shales, silts, sandstones, anhydrites, scarce pyroclastic rocks), nappe transport drove huge blocks of faulted Mesozoic carbonates, marls and shales into the underlying visco-plastic Haselgebirge. Like their brecciated surrounds, these partially or completely isolated overburden blocks show evidence of severe strain that includes; internal fragmentation, stretching, translation and rotation. Overburden pressure seems to have controlled the degree of brecciation and slickensides within these detachments.

In the subsurface the tectonised "Haselgebirge" breccias can sit within intervals of recrystallised halite units still preserving what at first sight appear to be depositional lamination. Closer inspection of the layered halites defines preserved isoclinal noses and overfolds, clearly illustrating the tectonic origin of the laminae (Spötl, 1989b). Pervasive salt flow and recrystallisation means any relics of primary structures in the halite beds are almost all destroyed. Preserved remnants of sedimentary features in the Haselgebirge Formation only remain within blocks and fragments composed of carbonate and anhydrite.

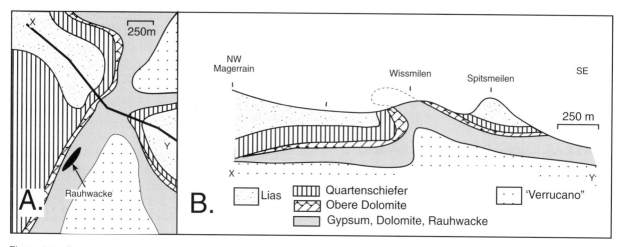

Figure 4.11. Rauhwacke developed atop subcropping gypsum (after anhydrite) in the European Alps (from Schaad, 1995). A) Plan view; B) Cross section X-Y.

These fragments were once more competent intrasalt beds. They were progressively fragmented by tectonism and flowage, and now make up the majority of clasts in the Haselgebirge breccias. The breccias preserve evidence of their progressive tectonic evolution ranging from: limited ductile deformation —> brittle fracturing —> boudinage —> individual rotation of newly developed clasts —> development of localised matrix-supported breccias (not laterally persistent) —> laterally persistent matrix breccias with increasingly homogenous grain sizes (Spötl, 1989b). In the early stages of intrasalt bed breakup the clasts were angular. With ongoing clast deformation and rotation, the clasts become rounded and develop slickensided boundaries. The brecciation of the intrasalt beds reflects the different responses of anhydrite or carbonate to strain, when compared with halite. In response to shear stresses of much less than 1 MPa at geologically significant stain rates, halite typically flows and recrystallises via pressure solution (Carter and Hansen, 1983; also Chapter 5). Due to their higher strength, the shale, carbonate and anhydrite intrabeds respond brittlely at the same strain rates.

Surface-related karstification of different tectonised protoliths leads to the formation of structurally and geometrically distinct rauhwacke. Schaad (1995) identified two end member evaporitic protoliths associated with rauhwacke development in the Swiss Alps (Figure 4.11): a dolomite-bearing gypsum, and a gypsum-bearing dolomite. Dolomite-bearing gypsum rauhwacke develops as karst-fill sediment within the thrust plane, and are typified by unstructured, often polymictic units, with geometries that reflect the shape of the precursor karst cavity. Gypsum-bearing dolomite rauhwacke makes up stratiform geometries that are dominated by dolomite

breccias with crackle to mosaic textures. This form of rauhwacke is a solution-collapse breccia developed after the solution of widespread tectonised evaporite beds. In certain cases, the karstification of the thrust plane evaporites was aided by the synformational fluvial or fluvioglacial processes of edge retreat, much in the same way as gently-dipping evaporite beds dissolve in the near subsurface (see earlier).

Rauhwacke sheets (carbonate thrust breccias) also define outcrop of tectonic slices in nappe piles now dominated by pelitic rocks of Early Mesozoic age in the Carson Sink region of Nevada (Muttlebury and Lovelock Formations; Speed, 1975; Speed and Clayton, 1975). There the most extensive carbonate breccia nappe, the Muttlebury Nappe, extends over an area of 100 km^2 (Figure 4.4). Outcropping rauhwacke sheets are interlayered and interlensed with non-breccia marble slabs. The coarse grain size and granoblastic texture of the marbles suggests an origin by metamorphism of at least moderate grade. But the degree of apparent recrystallisation in the marbles contrasts with a lack of recrystallisation and metamorphic alteration in the intercalated limestones and rauhwacke breccias. That is, the apparent metamorphic grade of these marbles is seemingly incompatible with that of the adjacent unmetamorphosed sedimentary carbonates.

The origin of the marble beds interlayered with rauhwacke is enigmatic and best resolved using stable isotopic signatures (Speed and Clayton, 1975). Carbon and oxygen isotope ratios of calcites from various carbonate rocks in the Carson Sink, including the marbles, fall into two distinct trends. One has nearly constant δ^{13}C and variable δ^{18}O; the other has nearly constant δ^{18}O, and a δ^{13}C signature that varies between -7 and -25 ‰ (PDB). The first trend includes all the fine-grained carbonate rocks and calcarenites of the region, whereas the second encompasses all analyses of the problematic marbles. Speed and Clayton's (1975) interpretation of the isotopic trend of the marble is that the precursor to the marble was a coarse-grained gypsum or anhydrite-lubricated décollement that was associated with active thrust and glide planes, much like the anhydrite lubricated Alpine thrust described by Jordan et al. (1990). The calcium sulphates were subsequently calcitised either by biogenic carbon derived from sources outside of the sulphate beds, probably from nearby euxinic limestones, or perhaps from hydrocarbons sourced from nearby pelitic terranes. Biogenic carbon could have been transported as hydrocarbons in an aqueous medium or as dissolved carbon species oxidised in the source beds. Speed and Clayton argue the process of marble formation was similar to that now forming the calcitised and sulphurous cap rocks of diapirs in the Gulf of Mexico (see Chapter 7).

Thus the rauhwacke, interlayered with these enigmatic marbles in the Muttlebury Formation, are intraformational tectonic breccias. They were created by meteoric dissolution of those sulphate-lubricated thrust intervals that remained after calcitisation (biogenic marble formation) of some calcium sulphate intervals was complete. Calcite and quartz sand are commonplace matrix components in the rauhwacke breccia; they are interpreted as accumulations of insoluble residues and stringers that were originally disseminated through the sulphate beds.

Preferential dissolution: residues of less-soluble salts

Subsurface flushing of the more soluble salts from an evaporite bed can leave behind layers of less soluble salts. Gypsum/ anhydrite caprocks atop diapirs are widely recognised examples of preferential dissolution. A less well recognised form of preferential dissolution occurs where undersaturated groundwaters flush the bed contacts of stratiform salt units. Such a set of processes effects halite beds in Maha Sarakham evaporites beneath the Khorat Plateau of NE Thailand (Sessler, 1990; Utha-aroon et al., 1995; Warren, 1997; and Chapter 7 in this volume).

The Maha Sarakham sediments of NE Thailand are of upper-middle to late Cretaceous age and unconformably overlie continental deposits of the Khorat Group. Maha Sarakham sediments were deposited as a thick sequence

(up to 1000 m) of basinwide continental evaporites (Utha-aroon, 1992; El Tabakh et al., 1995). The salt tends to be stratiform about the basin edge and halokinetic in the thicker, deeper beds of the present centre (Figure 4.12a). Three thick evaporite units (informally called the Lower, Middle and Upper Salt), or their residual equivalents, are present throughout the basin (Figure 4.12b). Evaporite minerals in the various beds include: sulphates, rock salt and potash minerals. Potash is found mostly in the upper part of the Lower Salt unit, but saltern halite is by far the most common salt and many beds preserve pristine chevron textures. The salterns occasionally freshened so that widespread gypsum (now anhydrite) nodules and intrasalt beds were deposited.

Thick clastic redbeds of mudstone separate each of the three main salt beds. These are informally called the Upper, Middle and Lower Clastics. In core the clastic units are homogeneous clay beds, with locally laminated and stromatolitic horizons, infrequent nodular caliche horizons, occasional anhydrite nodules, root structures and mudcracks. Near their contact with adjacent salt beds there are fractures and fissures filled with satinspar anhydrite and halite, indicating dissolution of a portion of the adjacent salt bed (Chapter1).

The three main salt units that make up the Maha Sarakham Formation are laterally continuous and are traceable basinwide. Nevertheless, due to an ongoing interplay between salt dissolution and flowage almost from the time the salt beds were deposited until today, a lithostratigraphic correlation from one place to another is not a straightforward exercise (Figure 4.12b).

The earliest evidence for salt dissolution is cm-dm karst-pits filled by coarse recrystallised halite cement. This is a syndepositional dissolution texture identical to that in the modern Qaidam Basin (see Chapters 1, 7). Salt dissolution next occurred in the shallow subsurface at bed boundaries where clastic units were in contact with bedded salts. Waters undersaturated with respect to halite, moved through redbed aquifers, initially as the clastics underwent deep weathering then as the buried clastics acted as conduits for compactional and basinal waters. This style of clastic-groundwater-driven dissolution left behind accreting residue textures at each major salt/clastic contact. Where the dissolving salt bed was composed mostly of clean halite all that now remains is a thin laminar residue of accreted insoluble clay residues. Due to the highly saline and anoxic nature of the brine, the iron in these zones is reduced and where the salt bed is still actively dissolving the colour of the adjacent clastic has been altered from red to green.

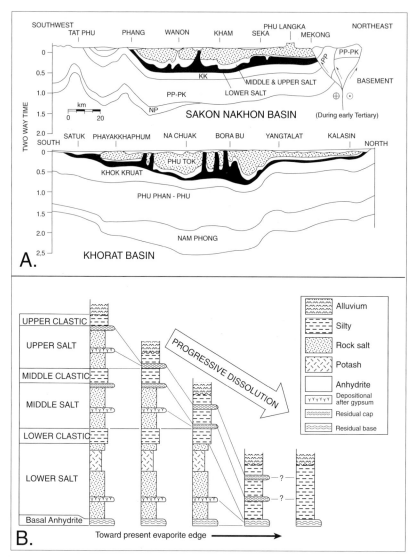

Figure 4.12. A) Regional subsurface structure of Maha Sarakham Formation showing preferential development of diapirs in lower salt toward the basin centre and dissolution toward structural edges of basin. B) Progressive dissolution of halite and accretion of residual anhydrite beds toward the present basin edge (after Utha-aroon et al., 1995). See Figure 7.9a for locality and detailed stratigraphy.

and dewatering Khorat Group sediments (see Basal anhydrite).

Each CaSO$_4$ residual bed is made up of thinly layered to laminated light grey to off-white anhydrite/gypsum nodules and micronodules. Individual layers may be as thick as 2 cm. Lamina/layer orientation ranges from horizontal to vertical, with clays and organic matter defining the laminae and the nodule edges. Where the anhydrite is undergoing further dissolution the beds are broken and brecciated with interclast matrix derived from the adjacent clastic member. Brecciated textures are most common in residues of the Upper Salt member as this unit is most subject to modern nearsurface meteoric leaching. If the Upper Salt member is still present in the stratigraphy it protects the Middle Salt member from surface driven dissolution.

Within the Middle Salt member is a basinwide dm-m-thick anhydrite marker bed. Complete dissolution of the halite encasing the marker bed of the Middle Salt member, and associated gentle foundering of the overlying middle clastic member, stacks the marker bed along with other anhydrite nodules and stringers in their original stratigraphic order (Utha-aroon et al., 1995). This style of oriented solution stacking of the intrasalt sulphate beds even preserve the subaqueous gypsum pseudomorphs of the anhydrite marker bed in their original growth position. Recognition of preferential dissolution occurring in a basin's stratigraphy allows a correlation across areas where halite beds have disappeared and only the anhydrite residuals remain (Figure 4.12b). The final stage of salt dissolution occurs when the anhydrite residuals completely dissolve and brecciate to leave the three clastic members stacked one atop the other.

Where the Upper and Middle Salt members have completely dissolved the present-day landsurface is defined by a topographic depression adjacent to a blind valley karst system atop a suprasalt high. Fluvial systems flow into this

In other regions, where the dissolving halite contained numerous anhydrite nodules and stringers, the resulting dissolution residue is composed of stacked anhydrites that can be partially rehydrated to gypsum. These residual beds can occur along both the upper and lower dissolving bed contacts of all three salt units. At the contacts of the Upper and Middle Salt members the CaSO$_4$ residuals tend to be dm-m thick on the upper bed contact, but only cm thick on the underside. On the underside of the Lower Salt member, the CaSO$_4$ residual is much thicker and probably reflects a much greater focused throughflushing from the underlying depression and feed subsurface waters to the dissolution front, which in turn focuses and stacks Quaternary fluvial channel sands (Figure 4.6a). Some geomorphologists working in the salt solution terrains of NE Thailand have not recognised the effects of salt solution on the stacking of fluvial systems. Instead, they have interpreted the thickened sections of Quaternary fluvial sediments as indicating a deepening of the river thalweg and related it to Quaternary uplift. The effects of preferential dissolution further emphasizes the pitfalls of using simplistic notions of lateral correlation between thin marginward anhydrite and thick basin-centre halite (Chapter 3).

Basal anhydrites: subsurface solution residues?

Preferential dissolution in NE Thailand doesn't just create anhydrite residues in the active phreatic zone, it also occurs where the underside of this saline giant was bathed in compactional and thermobaric waters (Figure 4.13). Such a basal bed is characterised by a near constant basinwide thickness and consistent stratigraphic textures across the underside of large portions of an evaporitic basin. In many evaporite basins of the world, similar constant thickness beds are often described as basal anhydrites and assigned a primary depositional origin.

The underside of the thick Lower Salt of the Maha Sarakham, NE Thailand is the type area for thin, distinctive basinwide basal anhydrite bed composed of nodular anhydrite with no obvious sedimentary precursor (El Tabakh et al., 1998b; Warren, 1997). This basal anhydrite passes downward, via a thin cm-thick organic-rich residue layer, into underlying nonevaporite sediment made up of pedogenically cemented fluvial sands and shales of the Khok Kruat Group. Previous workers have assigned a depositional origin to this widespread, near-constant thickness basal anhydrite. In contrast, recent work interprets it as a diagenetic dissolution facies. It is a pressure-welded nodular anhydrite unit that accreted in the subsurface as a buildup of insoluble residues on the underside of a basinwide dissolving halite sequence (Figure 4.13a). The process

was driven by the escape of undersaturated compactional and thermobaric waters. It is, in effect, an upside-down caprock (Figure 4.13b).

Unlike depositional anhydrites which typically thicken into the low stand wedge, the basal anhydrite has a near constant thickness across the entire basin (Figure 4.14). There are a few wells where the anhydrite is thicker than 1-2 metres, but there the anhydrite shows evidence of flow thickening and the wells are typically located at or near the dissolution edge of the Lower Salt. The near constant thickness of the anhydrite reflects the intrinsic self-limiting permeability of a nodular anhydrite bed accreting beneath a salt overburden. That is, the overlying Lower Salt unit is dissolving along its underside leaving behind a nodular anhydrite residue. The nodular residue continues to accrete along its top by gathering anhydrite nodules via preferential solution of the salt host. These nodules were originally suspended within the now dissolving bedded halite. At the same time as new nodules are accreting along the upper side of the basal anhydrite, the anhydrite may be dissolving along its underside.

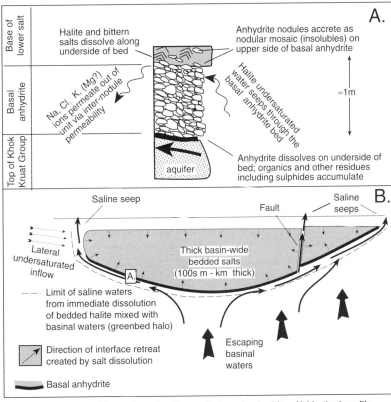

Figure 4.13. Mode of formation of a diagenetic basal anhydrite. A) Vertical profile through a dissolution-generated basal anhydrite. B) Schematic cross section with box showing relative position of vertical profile (modified from Warren, 1997).

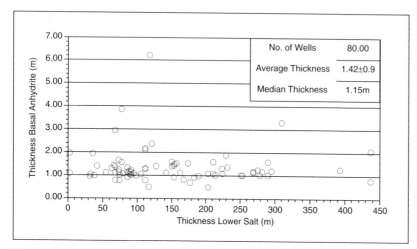

Figure 4.14. Crossplot of thickness of the basal anhydrite unit versus thickness of the overlying "Lower Salt" in the Maha Sarakham Formation, Thailand.

The overall dissolution process is driven by escaping compactional waters flushing along the underbelly of the Lower Salt unit (Figure 4.13b). If the basal anhydrite accretes to where it is more than a metre or two thick it starts to act as an aquitard for the throughflushing waters that are seeping through its internodular porosity. Its thickness reaches a point where undersaturated compactional waters can no longer flow through it, via its internodule permeability, to reach the halite-anhydrite dissolution front. Hence, halite undersaturated waters can no longer reach the zone of halite dissolution. Once undersaturated waters no longer access the halite dissolution front, the anhydrite accretion slows and even shuts down along the upper side of the unit. Even so, compacting undersaturated waters continue to flow along the underside of the basal anhydrite and so thinning of the anhydrite from its underside continues. Once sufficient anhydrite dissolves from its underside, undersaturated waters can again reach the halite and dissolve it anew. Thus the basinwide near constant thickness of the basal anhydrite can be thought of as reflecting a buffer between the rate of anhydrite nodule accretion on its upper surface and the rate of anhydrite loss on its lower surface. In effect a basal anhydrite has a "half-life". The actual thickness is determined by the inherent transmissivity of the material between the nodules in the basal anhydrite.

The microlaminar cm-thick organic residue on the underside of the basal anhydrite is composed of the insolubles that remain once the anhydrite is dissolved. It also contains micronodules and fissure fills of metal sulphides such as pyrite and bornite. Coarse burial-stage calcian dolomite also forms in the internodular zones of the basal anhydrite. This model, explaining some basal anhydrites as a form of

basinwide dissolution residue, may well have implications for other evaporite basins that possess a basal anhydrite unit with near constant thickness and show little or no preserved gypsum ghosts (e.g. portions of the Zechstein/Kupfer-schiefer; Chapter 8).

Gypsum karst

Babel (1991) describes preferential intra-evaporite dissolution processes in gypsum of the Middle Miocene (Badenian), southern Poland. There gypsum-clast breccias form by the early diagenetic dissolution of halite beds within laminated gypsum. Sometimes the effected interval is a single cm-thick bed where halite dissolution is interpreted to have occurred under a freshening interval that only effected sediments just below the sediment-brine interface. Other times the dissolution formed crackle and rubble zones that are tens of metres thick and must have occurred after substantial thicknesses of indurated gypsum and halite had accumulated. Dissolution in some areas took place while portions of the gypsum beds were still water-saturated and soft, leading to complicated soft-sediment deformation features. Dissolution in other areas formed after the gypsum was indurated to form classic collapse breccia intervals with smooth bases and irregular upper surfaces. Residuals within the gypsum are defined by very fine-grained gypsum and clay horizons that typically also define the base of a collapse interval. The breccias pass gradually into faint- or unlaminated gypsum, known as alabaster or compact gypsum.

Similar alabastrine and compact gypsums are forming by modern meteoric flushing in a gypsum karst within a 10-metre (30 feet) thick $CaSO_4$ bed that is part of the Cretaceous Kirschberg Evaporite Member exposed in Fredericksberg Gypsum Quarry (Warren et al., 1990). Alabaster is precipitating in perched phreatic portions of vadose caverns within gypsum beds that lie beneath a substantial collapse crackle breccia. Breccia clasts are derived from the overlying Edwards Formation dolomite and residual beds of silt to sand sized dolomite rhombs ("chalky dolomite"). The associated gypsum speleothems within the dissolving gypsum bed are a form of tertiary evaporite (as defined in Chapter 1).

Gypsum fill in the phreatic portion of the gypsum caves is made up of silt-sand sized euhedral prisms accumulating with a delicate "pack-of-card" texture, which indicates little or no transport after settling out of the cavern brine. Gypsum is also reprecipitating in gypsum karst cavities with speleothem textures that include stalactites and stalagmites. In other gypsum caves in the same unit, with more active water flows under predominantly active phreatic conditions, the cave walls tend to be smooth and unornamented.

The following three-stage sequence of gypsum karstification during exhumation was documented by Warren et al. (1990; Figure 4.15):

• *Formation of nodular "daisy-head" gypsum* As uplift raised nodular anhydrite into the zone of rehydration, replacement by gypsum was at first slow but pervasive, with near stagnant meteoric phreatic fluids oozing through the calcium sulphate bed. The main permeability pathways were provided by fractures and the intercrystalline porosity in dolomite films surrounding the calcium sulphate nodules and along thin dolomite stringers. Mm-cm-sized, aligned gypsum crystals nucleated off these flow paths and grew by replacement of the anhydrite. With further uplift there was a change to more active hydrologic conditions, which modified the rate of secondary gypsum growth. The remaining gypsum grew rapidly as a microcrystalline alabaster. The adjustment in tertiary gypsum type as the unit was exhumed also reflects more pronounced temperature and saturation changes tied to increased groundwater flow.

• *Karst structures and calcite/ gypsum speleothems* As the gypsum bed passed into the vadose zone, the process of gravity-driven karstification began in earnest. The gypsum now was subject to widespread dissolution that was focused into vertical pipes or caves with in the gypsum bed. Gypsum and calcite speleothems, mainly in the form of popcorn and flowstone, lined the pipes and caves.

• *Brecciation and complete loss of gypsum* Complete dissolution of the gypsum bed produced local zones of collapse breccia, with circumclast radial and botryoidal calcite cement. Collapse occurred gradually. Locally the overlying beds were disrupted and gently folded, regionally the overlying beds were little disrupted by widespread foundering associated with the complete flushing of gypsum. A residual zone of pulverulent dolomite rhombs ("chalky dolomite") now defines the cap to the dissolving gypsum bed. It is composed mostly of the dolomite rhombs that were formerly suspended within now dissolved gypsum. Some of this unconsolidated

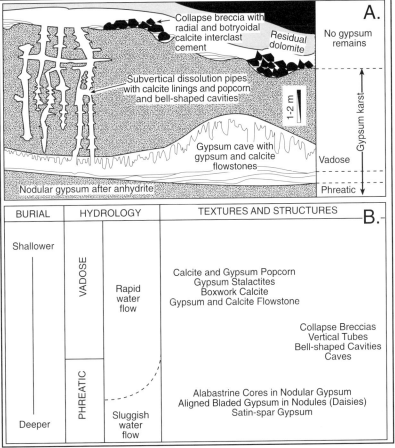

Figure 4.15. Model karst and related calcite textures developed though sequential complete dissolution and retexturing of an uplifted gypsum bed - based on textures in the subcrop of the Fredericksberg Gypsum in central Texas. A) Schematic cross section through the dissolving gypsum bed. B) Textural evolution related to groundwater regime (modified from Warren et al., 1990).

dolomite residuum is washed into kartsholes and caverns that crosscut across the gypsum to accumulate with classic "vadose silt" textures.

Interestingly, in this region there is abundant calcite precipitation as a void fill in the rubble breccia cavities and pipe linings, but there is no discernable conversion of the chalky dolomite to calcite. The dolomite remains well ordered and near stoichiometric. The lack of dedolomitisation raises the question: Is it only a commonplace process in evaporite beds on the way down in the burial cycle, when newly formed dolomite is more likely to be a less ordered highly metastable form of dolomite (see calcitisation discussion later in this chapter)?

The end result of unroofing an ancient evaporite bed in this humid region, even though the bed never makes it to the surface, is a stratiform chalky calcitic/dolomitic residuum layer that underlies a gently folded and collapsed packbreccia bed. This marks the former position of a regional decametre-thick $CaSO_4$ bed. Now its former

extent is defined by a much thinner interstratal bed characterised by brecciated and rebrecciated fragments of calcite popcorn, boxwork and botryoidal calcite, all intermixed fragments of the overlying strata and the dolomite residuum. Little if any of the original evaporite salt remains in the final breccia.

Economic solution breccias

As they form, or after they form, evaporite solution breccias can act as foci for base metal and hydrocarbon accumulations. Where the dissolving bed contains sulphate evaporites, it can act as a source of sulphur in base metal sulphide precipitates and as a locus for ore deposition. This is clearly seen in the ores of the Cadjebut Mine on the Devonian Lennard Shelf, West Australia (Warren and Kempton, 1997). The Cadjebut zinc mine is situated at the southern margin of the Emanuel Range and is hosted by platform carbonates of probable Givetian age near the base

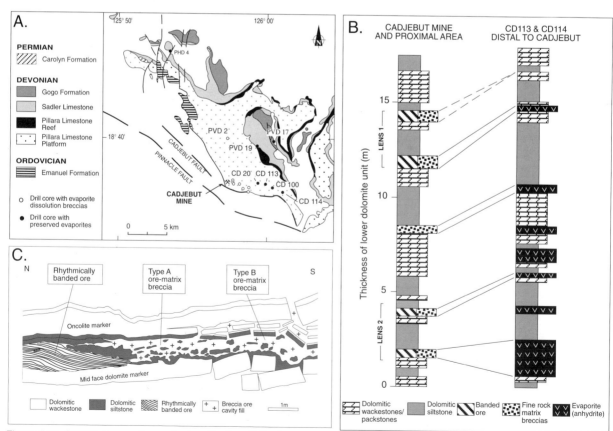

Figure 4.16. Cadjebut evaporites and mineralisation. A) Regional mine geology showing cored wells with and without evaporites. B) Correlation of ore lenses to solution breccias to evaporites in more distal positions from mine. C) Mapped ore face showing the different styles of ore texture in the mine. This ore horizon is laterally equivalent to an evaporite dissolution breccia, which in turn passes into equivalent anhydrite beds (after Warren and Kempton, 1997).

of the Pillara Limestone (Figure 4.16a). The orebody lies approximately 500 m to the north of the Cadjebut Fault and trends subparallel to it over a strike length of 3 km. Surface expression of the orebody is the "No Way Gossan," which crops out 100 m to the west of the mine portal. The orebody is roughly linear in shape, trending toward the northwest and dipping with the host dolomite at 5°-10° to the northeast.

Cadjebut Mine had pre-mining reserves of 3.8 million tonnes of 17% combined zinc and lead. Ore grade mineralization (Zn-Pb) occurs as two stratabound lenses, referred to as "lens 1" and "lens 2" (Figure 4.16b), ranging in thickness from 4 to 6 m. The width of the orebody varies from 50 to 150 m, as defined by a cut off grade of 7% lead-zinc, and is flanked to the north and south by hypogene alteration haloes of marcasite, calcite and baryte (Tompkins et al., 1994). The contact between ore grade mineralization (>7% Zn+Pb) and the alteration halo is usually abrupt, over distances of a few metres. In section, the ore-bearing lenses are stacked vertically above one another and separated by about 5 to 6 m of barren or weakly mineralized dolostone (Figure 4.16b). In plan view, both ore lenses have a broad sinusoidal shape and display the same geometrical congruence.

Two dominant ore types are present (Figure 4.16c): (a) stratiform, rhythmically banded zinc-rich ore; and (b) stratiform to stratabound, lead-rich breccia-hosted cavity-fill ore. The rhythmically banded ore occurs as four stratiform horizons, two throughout lens 2 and two in the northern part of lens 1. Rhythmically banded baryte and/or rhythmically banded marcasite form a halo around the orebody associated with the rhythmically banded zinc-rich ore.

There is a one-to-one correlation of sulphide-mineralised horizons in the Cadjebut Mine with evaporite dissolution breccia horizons in the immediate area around the mine, and with remnant bedded anhydrite units well away from the mine (Figure 4.16b). Evaporite beds were the precursor lithology to the sulphide ore lenses at Cadjebut. The stratabound linear nature of the Cadjebut deposit reflects a time when upwelling metalliferous basinal fluids interacted with bedded stratiform nodular anhydrite beds and trapped hydrocarbons to precipitate lead and zinc sulphides. Current structure in the mine region still shows a gentle anticline, which Warren and Kempton (1997) interpret as a partial artifact of the mineralisation structure, enhanced by the sequential loss of the surrounding

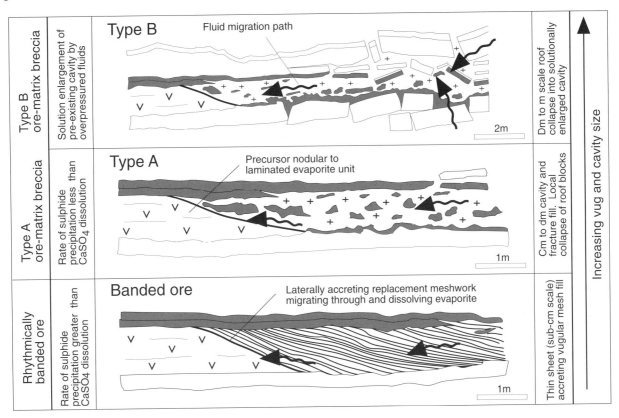

Figure 4.17. Cadjebut MVT ore, NW Australia. Schematic showing control on ore texture related to rate of supply of metalliferous fluid versus rate of evaporite dissolution (after Warren and Kempton, 1997).

evaporites, after ore had locally replaced a bedded anhydrite precursor. The linear nature of the anticlinal structure implies some form focusing of ore fluids and hydrocarbons, either atop an active underlying fault, or within a preexisting anticlinal trap, perhaps created as a rollover drape on the edge of a evaporite dissolving front (e.g. Figure 4.7a). Hydrocarbon residues (bitumens) are still present in the ore zones.

The rate of evaporite dissolution versus the rate of supply of mineralising fluid controlled the large scale textures within the Pb-Zn ores. Lead-zinc mineralisation in the Cadjebut Mine occurs as three distinct ore types (Figure 4.16c):

• stratiform rhythmically banded zinc-rich ore,
• stratiform Type A ore-matrix breccia ore, and
• crosscutting Type B ore matrix breccia ore.

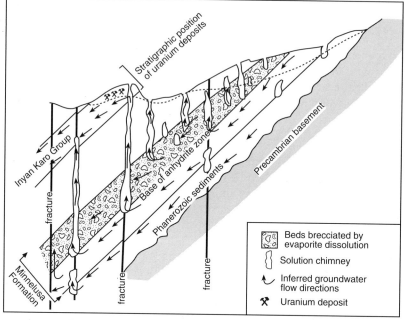

Figure 4.18. Infiltrated uranium mineralisation in the Cretaceous Inyan Kara Group sandstones, southern Black Hills (South Dakota). Solution collapse chimneys and breccias in the Late Palaeozoic Minnelusa Formation transport uraniferous groundwaters to precipitation sites (modified from Gott et al., 1974).

The rhythmically banded ore formed under conditions where the potential for sulphide precipitation exceeded the rate of evaporite dissolution. This occurred in cm-sized laminar dissolution meshworks that ate their way into the anhydrite beds (Figure 4.17). Millimetre-laminated colloform to pendulous microstalactitic sphalerite were the dominant precipitates. Type A ore matrix breccia formed in irregular cm-dm stratabound cavities. These formed at times when the rate of sulphide precipitation was slower than the rate of sulphate dissolution. Such cavities were not sufficiently large to instigate catastrophic roof collapse. Type B ore-matrix breccia ores formed in dm-metre sized crosscutting cavities, fractures and veins. Void space was sufficiently large to allow the single-event collapse and rotation of metre-sized roof blocks into the voids. Type B ore-matrix breccias are lead-rich compared to the banded ore and are interpreted by Tompkins et al. (1994) to post date the formation of banded ore. All the Cadjebut ore textures were created from sulphate supplied by the dissolving evaporites across mm-dm-sized replacement fronts or meshworks under deep bathyphreatic

and typically overpressured conditions (Warren and Kempton, 1997).

Evaporite solution breccias can also focus the deep meteoric flow of radiogenic groundwaters and so control the distribution of economic uranium deposits. For example, in the uranium district of southern Black Hills, South Dakota, Gott et al. (1974) showed how an extensive blanket of collapse breccia and trans-stratal breccia chimney fills were produced by the dissolution of gypsum and anhydrite beds in the upper part of the Late Palaeozoic Minnelusa Formation. In some places up to 83 m of evaporites were removed and numerous breccia chimneys were created that extend up to 430 m vertically into Cretaceous continental sandstones of the Inyan Kara Group (Figure 4.18). The pipe fill consists of downward displaced sandstone blocks that were later cemented by calcite. Gott et al. argue that the breccia chimneys created a plumbing system through which artesian oxidised waters leached and transported uranium from deeply buried uraniferous sands and granites up into the Cretaceous section. There they encountered organic-entraining Cretaceous sands and so precipitated carnotite-dominated infiltrations along redox interfaces.

Other evaporite indicators

Aside from dissolution breccias, stratiform residues, and some styles of sulphide mineralisation a number of other diagenetic textures indicate the former presence of evaporites (Table 4.2). Replacement of evaporite nodules by silica or calcite are perhaps the most widely recognised indicators of the former presence of evaporites and are sometimes described as "vanished evaporites". Fabrics are often spectacular and include both nodules and splays of silicified and calcitised evaporite minerals, calcitised evaporitic dolomites (dedolomites), and dissolution and solution enlargement fabrics.

Sabkha and mudflat evaporites typically form as isolated crystals or nodules in beds dominated by a more permeable nonevaporite matrix (mudstones, siltstones). Unlike breccias formed by the boundary dissolution of massive impervious evaporite beds, such nodules are hand-specimen scale features encased in a potential aquifer. During burial they are not isolated from the effects of near surface and basinal waters. As undersaturated pore fluids flush through the surrounding nonevaporite matrix it dissolves or replaces the salts. Pseudomorphic cavities are then filled or cemented by less saline minerals such as silica, various carbonates, baryte or celestite. Characteristic nodular shapes and crystal outlines of the evaporite precursor are retained by the replacing or cementing mineral phase at scales that can be recognised in both core and outcrop.

Silicified evaporites

The diagenetic replacement of ancient evaporites, especially anhydrite nodules, by silica has been widely reported (West, 1964; Tucker, 1976a,b; Radke and Mathis, 1980; Arbey, 1980; Friedman and Shukla, 1980). Quartzine[1] often infills chert nodules, with overall outlines near identical to their anhydrite/gypsum precursor (West, 1964). In addition, length slow chalcedony (lutecite[2]) is often found in association with evaporite deposits (Folk and Pittman, 1971; Siedlecka, 1972; Chowns and Elkins, 1974). Until recently, there were no documented examples of the process of evaporite replacement by quartz in Quaternary sediments. Now autochthonous, doubly terminated,

euhedral megaquartz crystals have been observed infilling voids in a dolomite matrix, as well as forming overgrowths on detrital quartz grains, in a gypsum- and anhydrite-bearing Pleistocene sabkha dolomite sequence in the Arabian Gulf (Chafetz and Zhang, 1998). These siliceous sabkha precipitates, whose silica source is probably recycled biogenic material, are forming within metres of the present sediment surface. Individual quartz crystals commonly attain lengths of 1 mm. Many quartz crystals faces preserve shallow dolomite mould impressions or they partly, or completely, engulf dolomite rhombohedra. However, this process of replacement is early, to see the full suite of replacement textures and the variations in timing of the replacement means one must study ancient evaporite sequences.

Milliken (1979) summarised the typical petrographic features of silica that replaces $CaSO_4$ nodules in Mississippian sediments of southern Kentucky and northern Tennessee (Figure 4.19). The nodule typically has a knobbly irregular surface. Diagnostic internal textures include: 1) length-slow chalcedony after lathlike evaporites especially anhydrite; 2) quartzine; and 3) small amounts of lutecite associated either with megaquartz that shows strong undulose extinction, or with euhedral megaquartz (Chowns and Elkins, 1974). The megaquartz often encloses small blebs of residual anhydrite.

Many sulphate nodules are silicified in a multistage process that involves both replacement and void filling (West, 1964; Chowns and Elkins, 1974). The process commences about the margins of a nodule (stage 1) with a volume for volume replacement of anhydrite by microcrystalline quartz. It normally ends with the growth of euhedral drusy quartz crystals into a central vug (stage 2 and 3). This mode of replacement exemplifies textures seen in texture style A in Figure 4.19.

Stage 1 chalcedony mimics or pseudomorphs the felted lath textures of the precursor anhydrite in the outer portion of the nodules. The pseudomorphs occur as radiating or decussate aggregates with a distinctive flow pattern. Identical decussate and flow textures occur in laths that make up sabkha anhydrite nodules and define their explosive mode of growth (Figure 1.11a). As well as the lath microtexture, outlines of larger crystals that predated anhydritisation and silicification may also be preserved by the nodule margin, these crystal outlines vary from prismatic to bladed. Many silicified nodules still retain the knobbly cauliflower outline of its precursor anhydrite. Other nodule edges preserve the interfacial outlines of gypsum or trona or borate precursors.

---

1) Quartzine is a form of chalcedony composed of fibres having a positive crystallographic elongation parallel to the c axis.

2) Lutecite is a form of chalcedony characterised by fibres that are seemingly elongated about 30°C to the c axis.

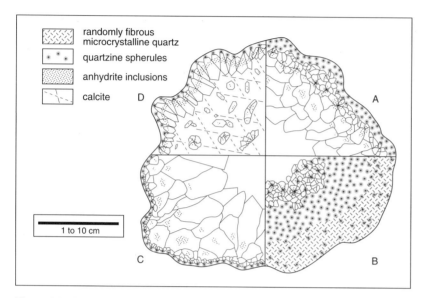

Figure 4.19. Texture in silicified anhydrite nodule. A) Most common style of nodule with thin rim of spherulitic quartzine followed inward by megaquartz characterised by increasing crystal size and with frequent anhydrite inclusions. B) Nodule with a large proportion of fibrous quartz. Thick rim of spherulitic quartz passes inward to strongly undulose megaquartz. Some of the megaquartz has anhydritic nuclei. C) Nodule with a thin rim of spherulitic quartzine passing inward to a large proportion of megaquartz. D) Nodule with megaquartz rim and an interior filled with calcite. Calcite may have filled a cavity or replaced earlier silica. Cavity fill is often defined by rim of authigenic baryte or sulphides (modified from Milliken, 1979).

Stage 2 microquartz and quartz fill can assume euhedral faces as they grow into voids created by the dissolution of the nodule. At the same time the quartz may continue to engulf and pseudomorph small areas of residual anhydrite or other less common evaporite salts. Quartz crystals precipitated at this time are commonly zoned, with more anhydrite inclusions within the inner zone of the pseudomorph. Some quartz crystals are doubly terminated and probably grew via the support of a dissolving meshwork of anhydrite. With the final dissolution of the supporting mesh these quartz crystals sometimes dropped to the floor of the void to create a geopetal indicator.

Stage 3, the final stage of the void fill is typified by the precipitation of coarse drusy euhedral quartz with no included anhydrite. This coarse quartz resembles coarse vein quartz.

Sometimes the processes of void fill may be arrested to leave a hollow core in the silica-lined geode. The void may be filled by a different burial stage cement such as baryte, sparry carbonates (e.g. ferroan dolomite and calcite) or even metal sulphides. This is the case with the large (up to 1 m diameter) silicified anhydrite nodules of Proterozoic Mallapunyah Formation of the McArthur Basin in Northern

Australia where baryte, then metal sulphides and then sparry calcite typify the latter stages of void fill (pers. obs.). Such geodes are excellent indicators of burial cement stratigraphy in a mudstone matrix that otherwise preserves few indicators of the evolving pore fluid chemistry.

Replaced nodules can also be filled by various styles of the coarser-grained megaquartz, yet still retain the outline of the precursor evaporite nodule (Figure 4.20). Work on diagenetic timing of silicified $CaSO_4$ nodules (e.g. Milliken, 1979; Geeslin and Chafetz, 1982; Gao and Land, 1991; Ulmer-Scholle and Scholle, 1994) shows that most silica replacement begins with shallow burial, either in the zone of active phreatic flow or in the upper portion of the zone of compactional flow (probably at depths of less than 500-1000 m). Early silica replacement in the zone of active phreatic flow is indicated by a lack of compressional flattening of the nodule, by the preservation of delicate surface ornamentation, such as "cauliflower chert" surfaces, and the preservation of compactional drapes around subspherical replaced nodules. If replacement of an anhydrite nodule occurs later in the burial cycle, many nodules have by then become flattened or sluggy. Milliken's (1979) isotopic evidence implies much silica replacement in the nodules she studied was relatively early in the burial cycle. The silica was supplied by throughflushing pore fluids with compositions ranging from seawater to mixed meteoric-seawater. Of course, nodule replacement by silica or calcite does not have to happen on the way down in the burial cycle, it may also happen during uplift back into the telogenetic realm where the strata have once again entered the zone of active phreatic flow.

Reactions that control the precipitation of silica as it replaces sulphate in early diagenesis are not well understood. Birnbaum and Wireman (1985) argue that microbial degradation of organic matter is important in forming silica precipitates in an evaporite deposit. They demonstrated, through experiment, the influence of bacterial sulphate reduction on silica solubility. The ability of sulphate-reducing bacteria to remove silica from solution is related to local changes in pH and to hydrogen bonding

"THE EVAPORITE THAT WAS:" INDICATORS AND EFFECTS

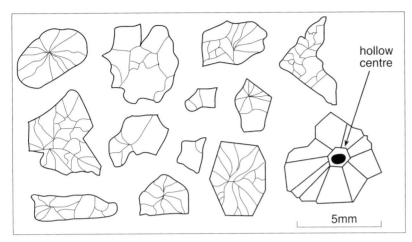

hollow centre

5mm

Figure 4.20. Sketches showing various shapes of the quartz nodules and internal arrangements of megaquartz crystals from replaced anhydrite nodules in the Mesoproterozoic Wollogorang Formation of northern Australia and the Neoproterozoic Gaissa Sandstone/Porsanger Dolomite of Finnmark, Arctic Norway (in part modified from Tucker, 1976b).

within amorphous silica followed by polymerization to higher weight molecules. During silica replacement of sulphate evaporites, the pore fluid becomes depleted in dissolved sulphate as it is reduced to H_2S by the action of anaerobic sulphate-reducing bacteria metabolising sulphate from an anhydrite or gypsum substrate. Where this selective dissolution of the sulphate occurs in the presence of amorphous silica the reaction is accompanied by the precipitation of silica. Hence the microscale mimicry of the lath outline in the outer parts of many nodules. According to Birnbaum and Wireman it reflects bacterially mediated silica replacement of the nodule in relatively shallow burial settings where bacteria can flourish.

Abiological processes, including thermochemical sulphate reduction, are important at greater depths where bacteria no longer survive. Providing matrix permeability is retained through burial, silica replacement can be ongoing into the thermobaric stage and can even continue into uplift stages associated with hydrological maturity. Replacement under a thermobaric regime is frequently indicated by the preservation of hydrocarbon inclusions in the infilling silica cement.

Overall, the texture of silica infill or replacement in a nodule is dependent on the rate of sulphate dissolution, the timing of silica precipitation and the rate of silica supply. Some nodules are dominated by the early *lit par lit* replacement textures (styles A and C in Figure 4.19), others have textures indicating silica cements growing into an open phreatic void left after the complete dissolution of the $CaSO_4$. Such nodules may still retain a hollow centre

where the anhydrite once resided (Figure 4.20). When a silica-filled geode did not start to accumulate silica until after all the $CaSO_4$ dissolved, the main evidence for an evaporite precursor comes from the shape of the replaced nodule and its stratigraphic position within the evaporitic depositional sequence (e.g. in supratidal to upper intertidal facies in a sabkha peritidal cycle).

Not all the anhydrite nodules, now replaced by silica, were syndepositional evaporites. Maliva (1987) showed that the anhydrite, now replaced by quartz geodes in the Sanders Group of Indiana, actually precipitated in the subsurface while its surrounding matrix of normal marine Sanders Group sediment was still unlithified (Figure 4.21). The anhydrite nodules formed in the subsurface during early burial as hypersaline brines sank into the normal marine limestones of the Ramp Creek and Harrodsburg Formations. Silica subsequently replaced the anhydrite. These geodes are almost invariably associated with the development of reflux dolomite.

Nor are all silicified nodules in evaporitic lacustrine sequences after $CaSO_4$ nodules. In the saline alkali lakes of the African rift valley, magadiite ($NaSi_7O_{13}(OH)_3.3H_2O$) is a common sodium silicate associated with volcanogenic sediments (Eugster, 1969). It forms beds and nodules in sediments outcropping about the margins of many of the trona-containing lakes in the East African Rift, including Lake Magadi. These magadiite nodules weather into cherts typically described as Magadi-style cherts. Evaporite minerals likely to be associated with Magadi-type cherts are the sodium carbonates (trona, gaylussite or pirssonite); gypsum is not usually present under the strongly alkaline conditions required for Magadi-type cherts to form (Hardie and Eugster, 1970). Other sodium silicates such as kenyaite and Na-Al silicate gels and zeolites are also present locally. Conversion of magadiite to bedded and nodular chert takes place close to the sediment surface and is related either to the leaching of sodium by dilute waters in shallow environments or to spontaneous conversion in deeper brine-saturated zones. The released sodium may contribute to the precipitation of trona, but not all magadiite beds need be associated with trona.

Ancient Magadi-type chert nodules crop out in lacustrine rocks of the Middle Devonian (Middle Old Red Sandstone)

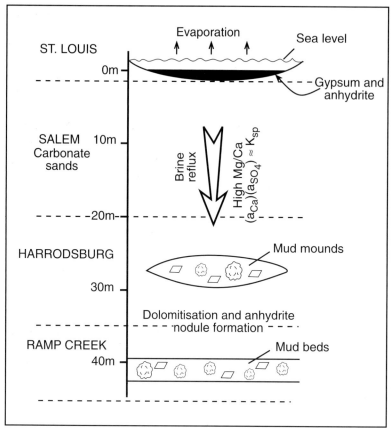

Figure 4.21. Precipitation of anhydrite and gypsum during St. Louis Limestone deposition formed a dense magnesium- and sulphate-rich brine that percolated downward and seaward through the then unlithified sands of the Salem Limestone. Dolomitisation of carbonate mud mounds and beds in the Harrodsburg and Ramp Creek Formations released calcium causing pore waters to become supersaturated with respect to anhydrite, resulting in the precipitation of anhydrite nodules that were later silicified (after Maliva, 1987).

extrusion forms; 2) contraction features, such as reticulation cracks and polygonal ridges, due to the loss of volume in the transition to chert. However, it has yet to be shown that these textures are unique to magadiite-chert transitions; similar surface textures are to be found in anhydrite nodules. In ancient strata an ability to separate gypsum/anhydrite pseudomorphs from trona/natron features in sediments hosting the cherts could be useful. Both gypsum and trona have a monoclinic bladed form, and both form either as nodules or as grasslike beds of growth-aligned prisms. Once the actual salt has gone, their distinction in hand specimen using pseudomorphs can be very difficult. In thin section, Schubel and Simonson (1990) showed that chert with a magadiite precursor has a characteristic rectilinear pattern of quartz crystal orientations visible in thin section, which they attributed to pseudomorphing after the original spherulitic magadiite structure.

As mentioned earlier, Folk and Pittman (1971) and Folk and Siedlecka (1974) concluded that length-slow chalcedony (lutecite) in chert nodules is a useful indicator of the former presence of evaporites (anhydrite or magadiite). But its occasional presence in nonevaporitic regimes, such as turbidites and basalts, has shed some suspicion on its absolute reliability. Heaney (1995) further refined the significance of lutecite by showing it is largely composed of a form of silica called moganite. Moganite is abundant in, but not restricted to, cherts formed in evaporitic environments. Nonevaporitic silica typically contains between 5 and 15 wt% moganite, whereas undeniably evaporitic specimens contain between 20 and 75 wt% moganite. No chert from nonevaporitic settings has been found to contain more than 25 wt% moganite. Consequently, enhanced moganite concentration (> 20 wt%) in microcrystalline silica is a valuable indicator for vanished evaporites. Heaney noted a frequent association of moganite with "Magadi-type" chert and suggested that moganite is a diagenetic alteration product of the hydrous Na-silicate magadiite. Moganite is a metastable phase and all samples analysed by Heaney that are older than Cretaceous contain little or no moganite. The transformation of magadiite to moganite must occur over hundreds to

Orcadian Basin, Scotland (Parnell, 1986). There chert nodules are replacive within organic rich laminates, the deposits of a stratified lake. Many exhibit soft sediment deformation features, including sediment injection along shrinkage fractures within the nodules, and preserved stellate quartz pseudomorphs after an evaporite mineral, possibly trona. Nodules were nucleated about vertical cracks in the host sediment and show polygonal or parallel aligned patterns. Thus the precipitation of the magadiite precursor must have occurred from nearsurface alkaline groundwaters flushing through partially consolidated sediments, rather than within the lake water column.

Surdam et al. (1972) listed the following textures as indicators of chert replacing magadiite: 1) preservation of the soft-sediment deformation features of the putty-like magadiite, such as enterolithic folding, lobate nodular protrusions, casts of mudcracks and trona crystals, and thousands of years in alkaline lake systems, whereas the inversion of metastable moganite to quartz requires tens of millions of years. When dealing with older samples, lutecite is only a reliable evaporite indicator when it occurs in suitable host lithologies and in association with other indicators of "the evaporite that was".

Calcitisation and dedolomitisation

Flushing by surface-derived, low-sulphate meteoric waters brings on evaporite rehydration, dissolution and replacement. In such settings, calcite cements typify nearsurface replacements, while anhydrite hydration and dissolution characterise the deeper portions of the same meteoric flow system. But, like silicification, the calcitisation of evaporite crystals and nodules can occur at any time in the burial cycle, whenever the throughflushing waters are undersaturated with respect to the evaporite salts and supersaturated with respect to $CaCO_3$.

In South Australia, hollow aragonite pseudomorphs of Holocene gypsum grow in the zone of active phreatic flushing about edges of coastal salinas where bedded lacustrine gypsum of the Marion Lake complex is flushed by less-saline meteoric/marine pore waters (von der Borch et al., 1977; Warren, 1982a). Similar calcite pseudomorphs after bottom-nucleated gypsum have been recognised in the Lower Carboniferous of the Vesder region, Belgium (Figure 4.3; Swennen et al., 1981). Later in the burial history of an evaporite bed, compaction- or thermobaric-derived waters can leach $CaSO_4$ or other salts from the system to create characteristic voids that are then filled by burial-stage calcite.

Carbonate replacement of calcium sulphate can be driven by various hydrological processes and at various depths in the subsurface (Pierre and Rouchy, 1988):

- Sulphate dissolution during flushing by bicarbonate-bearing waters with bicarbonate derived from:
  - dissolved atmospheric CO_2,
  - oxidation of organic compounds (marine or terrestrial),
  - thermal decarboxylation of organic matter.
- Bacterial sulphate reduction in organic rich sediments. The process of sulphate reduction may also be abiotic and thermochemically induced.

Bicarbonate ions can also be generated by the decarboxylation of organic matter. The late stage calcite replacement of evaporite nodules postdates silicification in Tosi Chert, Wyoming, and was associated with hydrocarbon migration (Ulmer-Scholle and Scholle, 1994). Pseudomorphic replacement by calcite or dolomite of euhedral trona and shortite crystals in the saline facies of the Cambrian Lake Parakeelya Member, South Australia, also took place during burial diagenesis as basinal brines flushed through the system at temperatures of up to 110°C. The catabaric dissolution and removal of the sodium carbonate salts took place either coincident with or following the generation of hydrocarbons (Southgate et al., 1989).

Calcitisation of evaporite nodules is most widely documented as occurring when meteoric waters of the active phreatic zone react with an evaporite bed (Harwood, 1980; Lee and Harwood, 1989; Warren et al., 1990; Scholle et al., 1992). Meteoric flushing can occur soon after an evaporite unit is deposited or it can occur after deeply buried evaporite units are uplifted back into the zone of active meteoric throughflow. In hydrologically mature sedimentary basins, such uplift-related meteoric flow is forced by head-driven circulation from high-elevation recharge areas can extend to depths of thousands of metres.

One of the best documented examples of calcitised evaporites comes from the onshore Permian Raisby Formation of the UK, where alteration processes can be studied both in outcrop and in core. There, the outcropping dolomites contain nodular cavities, pores and pseudomorph voids, which are lined or filled by calcite spar. Deep borehole cores in the Raisby Formation still retain nodules and euhedral crystals of gypsum and anhydrite that are near identical in size and shape to the cavities and pseudomorphs now seen in outcrop. This indicates that calcium sulphates were once abundant in the outcropping Raisby Formation, but almost all the original salts have since been removed by meteoric groundwaters.

Calcitisation of evaporite nodules is still active in appropriate zones of groundwater inflow (Figure 4.22; Lee and Harwood, 1989; Lee, 1994, 1995). Early in the diagenetic history of the Raisby Formation, primary gypsum was replaced by anhydrite nodules. Later, coarser replacive burial anhydrite formed, following both pressure-solution and further dehydration of any residual primary gypsum. During further burial, some of the anhydrite in the nodules was replaced by catabaric baryte, dolomite and sphalerite. Upon uplift of the Raisby Formation during the Tertiary, any remaining nodular anhydrite rehydrated to porphyroblastic gypsum. With further exhumation the

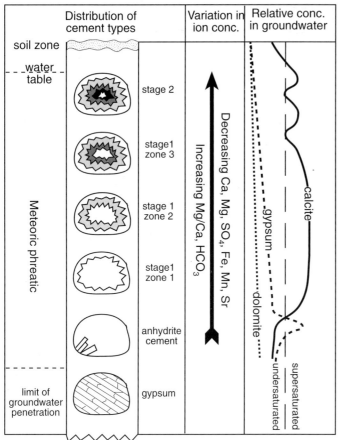

Distribution of cement types		Variation in ion conc.	Relative conc. in groundwater

soil zone

water table

Meteoric phreatic

stage 2

stage1 zone 3

stage 1 zone 2

stage1 zone 1

anhydrite cement

gypsum

limit of groundwater penetration

Decreasing Ca, Mg, SO₄, Fe, Mn, Sr

Increasing Mg/Ca, HCO₃

calcite

gypsum

dolomite

undersaturated

supersaturated

Figure 4.22. Nodule evolution in the Raisby Formation, UK. Vertical axis indicates distance from the land surface, but may also indicate evolution through time of any one cavity. The lowermost cavity is filled with gypsum derived by rehydration of buried anhydrite. Contact with meteoric water causes dissolution of gypsum followed by local anhydrite pore cement that is then replaced by calcite. Succeeding calcite fill of cavities indicates increasing oxygenation and activity of groundwater system (after Lee and Harwood, 1989).

gypsum was then dissolved by meteoric groundwaters and multiple stages of meteoric calcite cement precipitated within the voids (Figure 4.22). All meteoric phreatic waters in this part of the UK are currently undersaturated with respect to dolomite and saturated with respect to calcite. Thus the calcite void fills in the Raisby Formation are made up of clear, inclusion-free, non-ferran, low-magnesian calcite.

Prior to the calcite infill of the voids and after the rehydration of all buried anhydrite to gypsum there was an enigmatic episode of what is a nearsurface anhydrite cement. This early pore-fill is now replaced by calcite and encased in zone 1 calcite. Its blocky rectilinear morphology precludes either a gypsum or a baryte precursor. The pore water chemistry, which favoured anhydrite over gypsum in what

must have been a shallow unroofing setting, is not well understood. It may be related to a short term increase in salinity of nearsurface pore waters brought on by widespread meteoric dissolution of large volumes of nearby halite.

The succeeding calcite infill occurred in two stages; Stage 1 crystals are circumvoid equant, while stage 2 crystals are coarser, circumvoid and columnar (Figure 4.22). Under cathodoluminescence the stage 2 columnar crystals are nonluminescent and alternate with internal sediment layers, while stage 1 crystals show three geochemically distinct cathodoluminescence zones related to increasing oxygenation of the pore fluids (Figure 4.23). This successive zonation reflects ongoing exposure and increasing flow activity in the phreatic zone. Stage 1 – zone 1 (bright orange) cement has high Mn and low Fe and Mg. Stage 1 – zone 2 (dull orange-luminescent) has relatively high Fe and Mn and low Mg. While Stage 1 – zone 3 (nonluminescent) has negligible Mn and little Fe. Stage 2 cements have much greater Mg than stage 1, while the orange luminescing cements (zones 1 and 2) contain much higher levels of Sr than zone 3 or stage 2 cements (Lee and Harwood, 1989). This geochemical and cathodoluminescent zonation is typical of increasing oxygenation, associated with the passage into an increasingly oxygenated and active groundwater system.

Often associated with the process of calcitisation of former evaporite nodules is the process of dedolomitisation of the host matrix. Dedolomitisation describes the conversion of a dolomite back to a limestone by the growth of calcite rhombs within the dolomite matrix. This process can continue until the dolomite is converted to a limestone or it may be arrested when the calcitisation of the matrix is no more than partial. Dedolomitisation has long been interpreted as a product of evaporite diagenesis and is typically associated with calcium-rich groundwaters derived from a nearby dissolving CaSO₄ unit (Lucia, 1961; Warrak, 1974; Lee, 1994). This is the emphasis of the dedolomitisation discussion in this chapter; but it should be recognised that dedolomitisation can also occur under the influence of calcium-rich basinal waters, where the elevated levels of calcium are independent of evaporite dissolution. Land and Prezbindowski (1981) documented dedolomitisation from hot, calcium-rich brines moving up into a dolomite reservoir in the Edwards Group of the Gulf

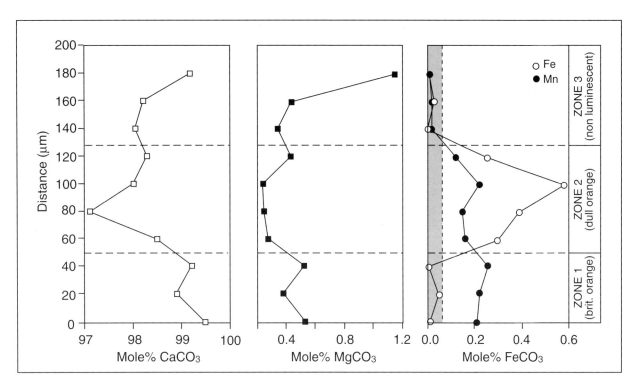

Figure 4.23. Microprobe traverse through a Stage 1 cement crystal growing inward from the nodule wall showing distinct geochemical differences between Zone 1 (bright orange), Zone 2 (dull orange) and zone 3 (nonluminescent). Shaded areas indicate values are below reliable detection limits (after Lee and Harwood, 1989).

Coast. They attributed the high calcium content of these brines to the albitisation of plagioclase in downdip sandstones.

Evaporite-associated dedolomites are often associated with evaporite dissolution breccias that indicate the former position of the now dissolved bed (Lee, 1994). Dedolomite under this scenario forms via the reaction of calcium sulphate-rich solutions with a preexisting dolomite to produce calcite with magnesium sulphate as a possible byproduct. The latter is rarely preserved, as it is highly soluble, and either remains as dissolved ions in the escaping waters or is quickly redissolved and flushed by throughflowing groundwaters (Shearman et al., 1961). The $CaSO_4$ dissolution process is often driven by meteoric flushing of nearsurface oxidising waters. In such settings former ferroan dolomites are preferentially replaced and the resulting calcitised dolomites are outlined by intervals stained red with iron oxides and hydroxides.

Dedolomites formed by meteoric flushing typically occur beneath subaerial exposure surfaces or unconformities. A well exposed and studied example of early dedolomite, associated with karstification of a shallowly buried sulphate system, occurs in the Miocene lacustrine sequences of the

Madrid Basin in central Spain (Cañaveras et al., 1996). There, a variety of calcite fabrics, initiated by dedolomitisation, formed in association with a well defined solution collapse breccia (Figure 4.24). Dominant dedolomite textures are made up of sutured calcites and radial-fibrous calcites, the latter consisting of pseudospherulite mosaics and fibrous crusts. Other subordinate dedolomite fabrics consist of micro- to mesocrystalline mosaics of rhombic, occasionally zoned calcites, as well as reworked pseudospherulite crystals.

The diagenetic carbonate zone (DCZ) that hosts these textures overlies and grades laterally into lacustrine dolomite (lower dolomite unit - LDU) and evaporites (evaporitic unit - EU), and in turn is capped by a palaeokarst or exposure surface (Figure 4.24a; Cañaveras et al., 1996). The boundary between the DCZ and the LDU is transitional and is defined by a grading in the %$MgCO_3$ profile (Figure 4.24b). Mg values range from 0.1 to 1.2% in the dedolomites of the DCZ and micrites of upper limestone unit (ULU) and from 8 to 14% in the underlying dolomites of the LDU (Figure 4.24b).

Geochemical evidence indicates dedolomitisation occurred in the shallow subsurface (< 40 m depth) via the

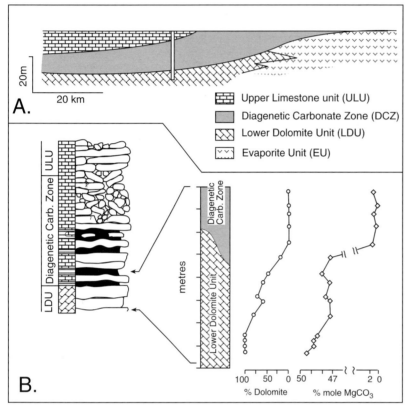

Figure 4.24. Distribution of the main subunits in Middle to Upper Miocene lacustrine units of the Madrid Basin, central Spain. Shows the position of the solution collapse breccia in controlling the dedolomites and the nature of the contact between the lower dolomite unit (LDU) and the dedolomites of the diagenetic carbonate zone (modified from Cañaveras et al., 1996).

throughflushing of oxidizing meteoric-derived ground waters (Cañaveras et al., 1996). $\delta^{18}O$ values of the dedolomites in the DCZ average -6.7‰ and range from -6.1 to -7‰, while $\delta^{13}C$ averages -8.2‰ and ranges from -7.4 to -9.2‰. The $\delta^{18}O$ values in the LDU average -0.3‰ while $\delta^{13}C$ averages -6.1‰. Isotope values in the ULU range from -6.4 to -6.8‰ in $\delta^{18}O$ and from -8.2 to -8.9‰ in $\delta^{13}C$ (Figure 4.25a). The values of the dedolomites are consistent with precipitation from meteoric waters and give precipitation temperatures of 8 to 12°C. Vertical trends in $\delta^{18}O$ and $\delta^{13}C$ can be used to define the average position of the water table in the DCZ at the time the dedolomites were forming (Figure 4.25b). This too supports a palaeokarst model characterized by an irregular shallow water table and a narrow vadose zone (Cañaveras et al., 1996). The diagenetic system behaved as an open system for nearly all trace elements analysed; but in the shallower zone (vadose zone) the system was partially closed with respect to strontium. The dedolomitisation was driven by two combined hydro-graphical patterns: authigenic recharge through limestones and allogenic recharge.

Dedolomites can also be formed via evaporite dissolution in the deeper subsurface where dissolution-influenced basinal brines are escaping into the basin margin. Budai et al. (1984) suggested a subsurface setting for the burial dedolomites of the Mississippian Madison Group of Wyoming and Utah. Two stages of fracture-related dedolomite are present, one that precedes and one that postdates regional dolomitisation. The early dedolomite occurred as active groundwaters flushed evaporitic sequences during shallow burial. The later post-stylolitisation fracture-associated dedolomite records chemistries and isotopic signatures of deep burial and tectonic-vein mineralisation. Associated with this late dedolomite are calcitised anhydrite nodules and stylolite-bitumen-related dedolomites that also indicate burial-related episodes of dedolomitisation. This episode of dedolomitisation took place during hydrocarbon maturation as basinal brines were moving through the Madison reservoir. Then hot Ca-enriched brines were able to dissolve anhydrite and dolomite while at the same time precipitating calcite along stylolitised seams and within anhydrite nodules.

Celestite as an indicator?

Celestite ($SrSO_4$) occurs as crystals, nodules and cements in many sedimentary rocks, particularly evaporitic dolomites, but also in dolomitic limestones and marls. $SrSO_4$ forms either as a primary precipitate from aqueous solutions or, more usually, by the interaction of gypsum or anhydrite with Sr-rich waters (Kushnir, 1986). In this case the dissolution of a calcium sulphate bed supplies the SO_4 that then combines with the Sr to precipitate as celestite. Prior to dissolution, Sr is believed to occupy the calcium sites in the gypsum lattice where it can be enriched ten times over the Sr content in enclosing dolostones (Carlson, 1983).

Not all waters capable of forming celestite are evaporite-derived. Documented sources of strontium-rich waters that are capable of precipitating celestite include:

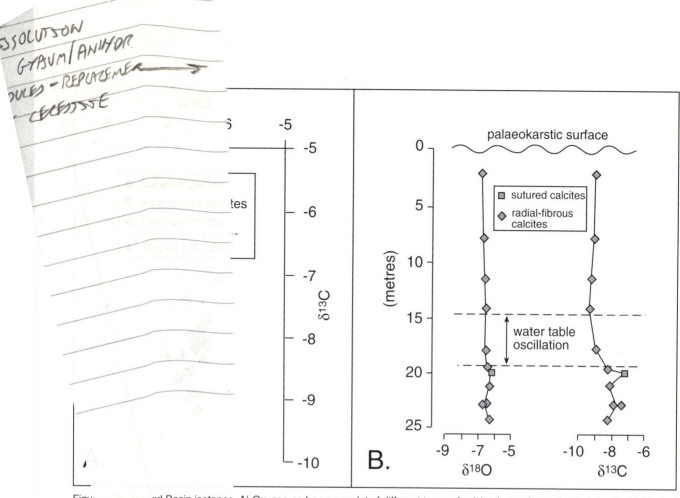

Handwritten notes (partially visible, upper left):
DISSOLUTION
GYPSUM/ANHYDR.
NODULES - REPLACEMENT →
- CELESTITE

Figure ... ...drid Basin isotopes. A) Oxygen-carbon crossplot of different types of calcite that make up the dedolomite. B) Vertical trends in δ^{18}O and δ^{13}C in section of the Diagenetic Carbonate Zone (DCZ). The zone of carbon depletion between 15 and 20 metres below the palaeokarst surface corresponds to a trend of more constant δ^{18}O values above this interval and more variable values below. These observations are consistent with the preservation of a palaeo-water table within the DCZ (after Cañaveras et al., 1996).

- marine-derived brines (Frazier, 1975; Baker and Bloomer, 1988),
- waters formed during the conversion of aragonite to calcite (Nickless et al., 1975),
- waters formed during the dolomitisation of limestones (Wood and Shaw, 1976),
- waters formed during the conversion of gypsum to anhydrite (Kushnir, 1982),
- waters formed during the dissolution of subaerially exposed gypsum (Carlson, 1987).
- resurgent basinal waters that have leached Sr from feldspars and clays in arkosic redbeds (Scholle et al., 1990).

Internal textures preserved within a celestite nodule or aggregate depend on the ratio of rock to fluid interaction and the timing of flushing by Sr-rich waters. Kushnir (1986) argues that finely divided celestite crystals can only precipitate from highly supersaturated brines with Sr levels measured in hundreds mg/l. Coarsely crystalline celestite can precipitate from less saline waters, such are found in meteoric environments adjacent to leaching CaSO$_4$ beds.

Evaporite-associated celestite may replace or pseudomorph individual evaporite crystals, it may be isopachous and passively fill vugs and nodular cavities created by the dissolution of an evaporite precursor, or it may form interclast cements in dissolution breccias. West (1973) discusses the evaporite-celestite association in detail and the interested reader is referred to this benchmark paper. For a more recent and general summary of celestite occurrence in sediments worldwide, Scholle et al. (1990) is recommended.

Carlson (1987) documented replacement of evaporites by celestite in northwestern Ohio in association with evaporite dissolution and karstification. The celestite replacements are more durable than the original evaporite salts that have long since vanished. They are found along the western margin of the Ohio (Cayugan) basin within the Greenfield Dolomite and undifferentiated Salina dolostones. Replacement styles include: lenticular and prismatic crystals after gypsum or nodules of anhydrite or laminar beds replacing laminar evaporites. The prismatic replacements are exceptionally well preserved, with

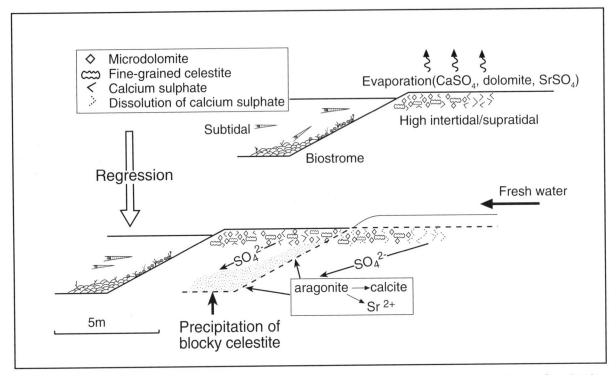

Figure 4.26. Formation of early celestite replacement in the Brattstad Member of the Steinsfjord Formation, Norway. Strontium is derived by freshwater leaching of anhydrite and aragonite (after Olaussen, 1981).

euhedral, deeply embayed outlines and internally zoned growth bands that entrain numerous inclusions of dolostone and anhydrite. Crystal-wide consistency of the optical data for the anhydrite in the inclusions of the prisms indicates that at the time of celestite replacement they replaced single gypsum crystals and not felted anhydrite mats. In contrast the nodular and laminar pseudomorphs display features, such as chickenwire and enterolithic structures, that indicate an anhydrite precursor. Replacement of sabkha anhydrite by celestite was documented by Salter and West (1966) in the Purbeckian of Dorset, along with perfect pseudomorphs after lenticular gypsum (West, 1964).

Early, almost syndepositional, replacement by celestite occurred in both the vadose and the active phreatic zone of the evaporitic Silurian Steinsfjord Formation of the Oslo Region, Norway (Olaussen, 1981). Marl and carbonate host sediments were deposited in supratidal, intertidal, and restricted subtidal environments. Celestite occurs in two distinct textural associations within these peritidal sediments: 1) as aggregates of tiny celestite crystals in nodular dolomicrites of the high intertidal facies, or 2) as anhedral or euhedral prismatic crystals infilling intrapores and interpores in the low intertidal to subtidal facies. Celestite is thought to have precipitated from SO_4-enriched porewaters via the solution of calcium sulphate. Solution

occurred early during evaporitic mudflat progradation as it was flushed by land derived meteoric waters (Figure 4.26). Released sulphate then reacted with strontium generated by an aragonite to dolomicrite transformation within the high intertidal and an aragonite to calcite transformation in the low intertidal-subtidal.

An unusual late-stage celestite is intimately associated with hydrothermal $CaSO_4$ (now gypsum) in the Asaka gypsum mine (Koriyama City, Japan). The origin of this Kuroko-style hydrothermal $CaSO_4$ is discussed in detail in Chapter 9. Celestite occurs on the sides of fibrous (satinspar) gypsum in the mine area. It is interpreted by Matsubara et al. (1992) as a hydration product of strontian anhydrite (SrO up to 0.94 wt %) that also occurs in the same area. The hydration is meteorically driven and involves the preferential conversion of the strontian anhydrite to gypsum, where the $SrSO_4$ molecule remains intact during the conversion and eventually crystallises as celestite.

Baryte or fluorite as indicators?

Fluorite is not uncommon as a minor phase in mineralised sedimentary carbonate sequences. It reflects a carbonate's high susceptibility to replacement when flushed by low temperature, somewhat acidic, hydrothermal and basinal fluids. For example, fluorite is a minor late stage diagenetic cement in Devonian carbonates in the Keg River Formation, Canada (Aulstead and Spencer, 1985). The fluorite was carried to the site of precipitation by rising basinal brines that had flushed deeply buried evaporites. Similarly, the isotopic signatures of authigenic fluorites in Triassic carbonates of Europe show the supplying basinal brines were sourced from underlying or tectonically juxtaposed Permian Haselgebirge evaporites (Götzinger and Grum, 1992).

But fluorite is not solely sourced in evaporites, nor deposited only in a carbonate host. Fluorite can be found in fissure veins and stockworks in granitic rocks, greisens, pegmatitic carbonatites, metamorphics, and igneous rocks that include volcanic sublimates (Hodge, 1986). These are obviously not tied to an evaporite source. Nor are the vein and stratified (manto) fluorite deposits that are associated with areas of rifting and faulting where fluoride-rich hydrothermal fluids migrate up from the lower crust or mantle to replace carbonate hosts (Van Alstine, 1976).

Even though it is not evaporite specific, fluorite can form as a widespread, but volumetrically minor, early authigenic mineral in some evaporitic matrices. Its precipitation may be related to solar concentration of a surface brine and not to the presence of hydrothermal pore waters. For example, fluorite occurs as sparse isotropic euhedral cubes (50-200 μm diameter) within primary gypsum nodules at a depth of 235 - 240 m below the present land surface in Eocene carbonate mudstones of Florida. Celestite also occurs in the same sediment. Cook et al. (1985) argue the fluorite was syndepositional with the precipitation of gypsum in the capillary zone of a sabkha mudflat. If so, it was ultimately derived by the evaporative concentration of normal seawater. Additional fluoride in the parent pore waters may have been derived by fluvial transport of salts weathered from Appalachian volcanics. Cook et al.'s conclusion is supported by the work of Kazakov and Sokolova (1950) who documented sedimentary fluorite in various marine carbonates/evaporites in Russia in hosts that range in age from Cambrian to Cretaceous. In their experiments they concluded that fluorite can precipitate from seawater brines concentrated three to four times by volume, especially if an additional (fluvial of volcanic) source of dissolved fluoride was available in the depositional setting.

Crocker (1979) argues that granular blockspar ores in the Proterozoic Transvaal Supergroup of South Africa indicate primary synsedimentary endogenic fluorite mineralisation and are a special precipitive stage associated with the dolomite evaporite facies. Rouchy (1986) documented an association between fluorite, native sulphur and Visean evaporites. In an earlier paper, Amieux (1980) showed fluorite was a characteristic diagenetic stage in the black shale to evaporite transitional facies within the Ludian (Lower Oligocene) sediments of the Mormoiron Basin, Vaucluse, southeastern France. There the major fluorite precipitation phase postdated major dolomitisation and gypsification but preceded celestite and silicification.

Baryte, like fluorite, is found in a wide range of lithologies and is not specific to evaporitic sequences. At 25°C and 1 bar, the value of the equilibrium constant, K, for the replacement reaction:

$$CaSO_4 + Ba^{2+} <=> BaSO_4 + Ca^{2+}$$

as calculated from Gibb's Free Energy values, and assuming pure solids is:

$$K = 10^{5.5} = a_{Ca^{2+}} / a_{Ba^{2+}}$$

Thus, it is thermodynamically feasible for baryte to replace calcium sulphate at earth surface temperatures whenever the activity ratio of $(a_{Ca^{2+}}/a_{Ba^{2+}})$ is less than $10^{5.5}$, a condition met in many subsurface waters (Mossman and Brown, 1986). At more elevated temperatures the limiting ratio is even less due to the retrograde solubility of anhydrite.

Its principal documented occurrence is as a gangue mineral in hydrothermal veins, which commonly also entrain metalliferous ores (see Chapter 9). Baryte also forms veins, layers and cavity fills in black smokers or as disseminated crystals, layers and nodules in sedimentary rocks. It is this last mode of occurrence that is of interest in terms of an evaporite indicator mineral.

Baryte can directly replace gypsum very early in the diagenetic history. Loosely arranged platy crystals of baryte form pseudomorphs after lenticular gypsum in Quaternary soils from central Australia (Sullivan and Koppi, 1993). In older more economically significant sedimentary deposits, evaporitic baryte typically replaces nodular forms with irregular and coalesced outlines reminiscent of anhydrite or gypsum. In several such cases, isotopic analysis indicates the sulphate was remobilised from a calcium sulphate precursor such as, caprock sulphate (von Gehlen et al., 1962), sedimentary gypsum (Kesler et al., 1988; Parafinuik, 1989), or anhydrite-rich evaporites (Mossman and Brown, 1986; Frimmel and Papesch, 1990).

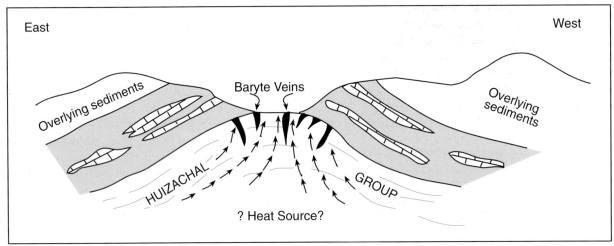

Figure 4.27. Schematic E-W cross section through the Galeana district showing localisation of baryte veins in the Huizachal red beds beneath the Olvido gypsum (grey). Baryte formed as ascending Ba-rich waters ponded beneath the gypsiferous seal to the anticlinal crest (modified from Kesler et al., 1988).

Baryte in the Magnet Cove baryte mine region, Nova Scotia, formed with nodular textures identical to those of sabkha calcium sulphate (Mossman and Brown, 1986; Chapter 8). Light coloured baryte nodules, hosted in a matrix of ferroan dolomite, show complete gradations from chickenwire to enterolithic. Hosting carbonates were originally deposited in intertidal and supratidal settings. The baryte is interpreted as a direct replacement of precursor sabkha anhydrite.

De Brodtkorb et al. (1982) described economic celestite-baryte deposits of the Early Cretaceous from the extra-Andean region of Neuquen, west central Argentina. The ores are located in the sub-Andean Mesozoic belt and show a conspicuous stratabound character directly related to the presence of evaporitic facies. Isotopic $^{87}Sr/^{86}Sr$ data confirm an evaporitic origin. Host sediments formed during a transition of epicontinental shelf environments from intertidal to penesaline supratidal settings. The celestite always occurs in saline and hypersaline facies. Syngenetic nodular beds occur, as do epigenetic stratabound ores of *in situ* mobilized celestite with stalactite growth and cavity infilling textures along with epigenetic vein-type ores formed by Ba-enriched solutions.

In the case of the Galeana Baryte District of Mexico the gypsum beds of the Olvido Formation formed an anticlinal trap to the underlying Huizachal Group that hosts the baryte in veins in a limestone host (Kesler et al, 1988). The baryte formed when ascending barium-rich basinal brines ponded beneath, and partially dissolved, the impervious gypsiferous Olvido Formation. After formation, the

anticlinal crest was erosionally breached to expose baryte at the surface (Figure 4.27).

Concretions of baryte in sandstones can create rosettes known as "sand rosettes" or "desert roses," where the baryte acts as a poikilitic cement that encases sand grains. Such baryte is thought to have been precipitated by reactions between soluble sulphate (usually from nearby dissolved gypsum or anhydrite) and basinal waters carrying $BaCl_2$. Baryte is a commonplace burial cement in reservoir sandstones where two basinal brines mix. The hydrodynamics of such deep mixing is often driven by tectonics, for example, the inversion tectonics of the North Sea (see later in this chapter). Oxidised pyrite can also act as a source of sulphate in subsurface waters, especially in the vicinity of organic-rich shales.

Baryte is also a widespread authigenic phase in sandstones of the Salt Wash Member in the Late Jurassic Morrison Formation on the Colorado Plateau (Breit et al., 1990). There it forms as intergranular cements, poikilotopic cements, euhedral crystals and replacements of detrital grains. The area that contains abundant baryte in the sandstone host coincides with the extent of evaporites in the underlying Pennsylvanian Hermosa Formation. A genetic link between the baryte and the dissolving evaporites is reinforced by similar ranges in sulphur isotope data in both lithologies ($\delta^{34}S$ range from +8‰ to +14‰). Sulphate, derived from the dissolving evaporites, is thought to have risen along faults into the Morrison Formation; a process perhaps aided by halokinetic focusing. Evidence of a fault conduit for the Ba-rich waters is further supported by $^{87}Sr/^{86}Sr$ ratios in the baryte cements, which tend to decrease with increasing distance from the supplying fault (from 0.7103 to 0.7084; Breit et al., 1990). This trend reflects the mixing of radiogenic strontium from waters ascending the faults ($^{87}Sr/^{86}Sr = 0.7100$), with strontium in waters derived by dissolution of early diagenetic carbonate cements in the sands ($^{87}Sr/^{86}Sr = 0.7080$). As the $^{87}Sr/^{86}Sr$ values along the faults exceed those in the underlying Hermosa anhydrite it is thought that radiogenic strontium was also added to the ascending waters by reactions between the rising brine and the arkosic horizons that intersect the fault conduit. These same arkosic sands are also a source of the copper in ores in the nearby Lisbon Valley (Figure 8.20)

Baryte can also be associated with meta-evaporite sequences. Stratabound baryte-hyalophane sulphidic orebodies, that were recently discovered at Rozna (in the Moldanubian zone of the Bohemian Massif, Czech Republic), are hosted within the high-grade metamorphosed complex (Kribek et al., 1996). The host entrains a meta-evaporitic succession and is composed of migmatitic biotite or sillimanite-biotite gneiss and amphibolite, with rare intercalations of marble, calcsilicate gneiss and lenses of coarsely crystalline anhydrite. Similar sequences are discussed in Chapter 6, which deals with meta-evaporites.

Authigenic burial salts

Highly saline basinal brines, generated by the dissolution of evaporites, can precipitate new salts in strata adjacent to the dissolving evaporite or along escape conduits for the brine such as faults and fractures. These authigenic burial salts are typically composed of coarsely crystalline anhydrite or halite and even magnesite (Table 4.1). Burial salts typically form coarse-grained cements that poikilitically enclose grains or fill tectonically induced fractures. Near-identical coarse poikilotopic evaporite cements can also form during syndepositional brine reflux within brine-saturated pores beneath an accumulating evaporite (Purvis, 1989). These syndepositional reflux salts, according to the discussions of Chapter 1, are true evaporites in that they formed in a hydrology driven by solar evaporation. In contrast, burial salts that form in brines generated by the subsurface dissolution of salt beds are not true evaporites. They were not directly precipitated by processes driven by solar evaporation, nor do they replace the volume occupied by primary and secondary evaporite salts. They are, in reality, evaporite transformation salts or burial salts. Distinguishing early reflux evaporite cements from later burial salts often requires detailed laboratory-based petrographic and isotopic study.

In addition to burial salts, other nonevaporite calcium sulphates can precipitate in the subsurface. For example, intergranular anhydrite cements can form if unmodified connate seawater is simply heated above 150°C (see Chapter 9). Or, aqueous sulphides can be oxidised to sulphates which then precipitate as an authigenic $CaSO_4$ phase. This process is relatively uncommon in the deep subsurface where suitable reducing conditions tend to dominate (Dworkin and Land, 1994). It is a much more commonplace process during weathering of subcropping pyritic shales and mine waste piles. We will concentrate in this section on the dissolution-related burial salts within both siliciclastic and carbonate hosts.

Anhydrite is a late stage burial salt in quartzose Rotliegende reservoir sands in Leman Field, North Sea (Figure 4.28a; Sullivan et al., 1994). Poikilotopic anhydrite is best developed along coarser-grained sand laminae suggesting their inherent higher permeabilities influenced sites of precipitation. The same anhydrite cement also encloses late authigenic ankerite and pressure-solved cemented grains, showing its deep burial origin. The anhydrite in the fractures cements zones up to 0.5cm wide and 0.5m long in the core. Cement crystals are made up of bladed coarsely crystalline anhydrite that has grown subparallel to the fracture walls. Cemented coarse-grained laminae and anhydrite-filled fractures frequently intersect in the core suggesting both styles represent the same generation of cement.

An interpretation of timing is best resolved using Sr and O isotopic analysis of the texturally late poikilotopic intergranular and fracture-filling authigenic anhydrite cements within the Rotliegende. Oxygen isotope composition of the anhydrite cements indicates precipitation temperatures between 120°C and 140°C from waters with a $\delta^{18}O$ composition similar to the present day formation waters (approximately 0‰ SMOW; standard mean ocean water). These are the maximum inferred burial temperatures for the Rotliegende Sandstone in the Leman Field. Such conditions would have occurred during the Late Jurassic-Early Cretaceous at the start of basin inversion at maximum burial depths of 3.5-4 km (Figure 4.28b). When corrected for their 150 Ma precipitation age, the Sr isotope ratios in the various anhydrite burial cements, and their Sr contents, are highly variable. $(^{87}Sr/^{86}Sr)_{150}$ ratios range from 0.70797 to 0.71129 and Sr contents from 351 to 2952 ppm (Figure 4.28c). Fracture-filling anhydrite also traverses this entire range, implying an origin from the same fluid. There appears to be a general trend of increasing Sr ratio with increasing Sr content of the anhydrite burial salts

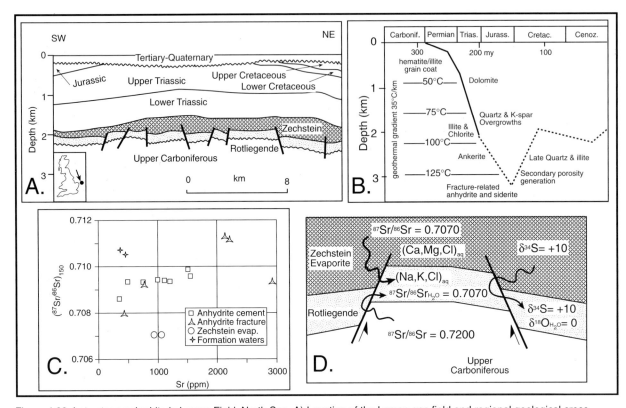

Figure 4.28. Late stage anhydrite in Leman Field, North Sea. A) Location of the Leman gas field and regional geological cross section. B) Subsidence history of Rotliegende reservoir sandstone showing time and temperature of anhydrite precipitation postdates other diagenetic events and precedes secondary porosity formation. C) Strontium isotopic ratios and composition of various anhydrite burial salts compared with formation waters and Zechstein salts. There is a general trend of increasing isotopic ratio with increasing Sr content. D) Schematic section showing major sources of sulphur, strontium and carbon along with their migration from above and below the Rotliegende and subsequent mixing that precipitated anhydrite cements and fracture fill (modified form Sullivan et al., 1994).

According to Sullivan et al. (1994) the late stage anhydrite indicates an unusual, short-lived, catastrophic event that juxtaposed the Zechstein evaporites and the sandstones. The resulting geometry facilitated cross formational fluid flow via fractures that were tectonically induced during basin inversion. At that time pore fluids probably escaped from the overlying Zechstein evaporites and underlying Carboniferous mudstones into the lower pressures of the Rotliegende regional aquifer (Figure 4.28d). Mixing of the two sources precipitated dolomite and anhydrite cements that in turn produce a Na-K-Cl formation water. This pore water was lower in Ca+Mg than the overlying Zechstein which in turn supplied most of the S and some of the Sr to the anhydrite cements. Like the isotopic evidence, the chemical evidence shows the burial anhydrite precipitated from extremely saline formation waters, which were derived by deep subsurface mixing of Zechstein evaporite brines with meteoric waters. Even today the Zechstein brines in the Leman Field are extremely saline (165,530 - 238,531 mg/l), with cation/chloride ratios indicating ongoing mixing of Zechstein marine evaporite brines with fossil meteoric water.

Onshore in northern Germany the Rotliegende gas reservoirs also contain coarse authigenic poikilotopic anhydrite cements, particularly in the vicinity of faults. Cross-formational fluid flow of Zechstein fluids and fault-related fluid flow was also responsible for this anhydrite, especially during times of fault activity (Gaupp et al, 1993). This can be clearly seen in the isotopically distinct values of burial anhydrite occurring in horst blocks when compared with similar authigenic and nodular anhydrite from other wells (Figure 4.29). The anhydrite cement in those Rotliegende sands atop narrow horst blocks clearly show an affinity for Zechstein sulphates. In contrast, anhydrites in the other wells, where uplift has not juxtaposed Rotliegende sands with Zechstein fluids, retain a typical early Rotliegende time isotopic signature. Once again this underlines the importance of active faulting in facilitating brine mixing, which then precipitates authigenic minerals in the deep burial setting.

Late stage anhydrite and halite cements also characterise deeper burial cements in the aeolian and fluvial feldspathic sandstones of the Upper Jurassic Norphlet Formation and other Smackover sandstones of the Gulf Coast (McBride et al., 1987; Dworkin and Land, 1994). Coarsely crystalline anhydrite, along with quartz and ferroan dolomite, precipitated as intergranular cements. Anhydrite averages 3% by volume (maximum 21%) in Norphlet wells in Mississippi and 1.3% in Alabama. It occurs as scattered crystals, spherical and poikilotopic patches several cm across, it can completely cement sand laminae. All the anhydrite cements are patchily distributed in the Norphlet reservoirs and, as in the Rotliegende, tend to be better developed in the coarser sand laminae.

Oxygen isotopic values from the anhydrite averages 15.6 ‰ (SMOW). According to Dworkin and Land (1994) this burial anhydrite precipitated in near oxygen (and strontium) isotopic equilibrium with ambient burial fluids at temperatures of between 90°C and 100°C at depths ≈2.3 km (Figure 4.30a). Authigenic halite cements, along with late stage carbonates, followed at temperatures up to 120°C (<3km burial). Subsequent dissolution of the cements, along with some labile detrital grains, created secondary porosity that makes up at least half of the economic porosity in the Norphlet in this region. Potassium-

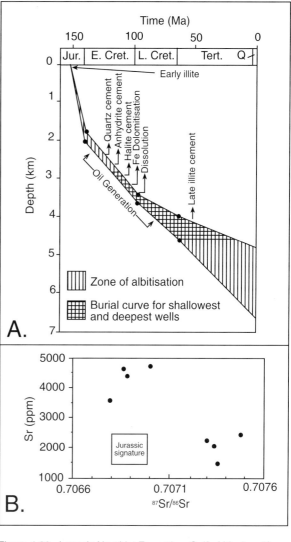

Figure 4.30. Jurassic Norphlet Formation, Gulf of Mexico. A) Burial curve showing the onset of major burial-related diagenetic events including dissolution derived burial anhydrite (modified from McBride et al., 1987). B) Sr concentration versus Sr isotopic composition of anhydrite cements in east Texas basin. Shows two separate groupings of the cement: an early reflux related group with a Jurassic isotope signature and a later burial dissolution related group. Box represents expected chemistry if the anhydrite was a direct precipitate of Jurassic seawater (modified from Dworkin and Land, 1994).

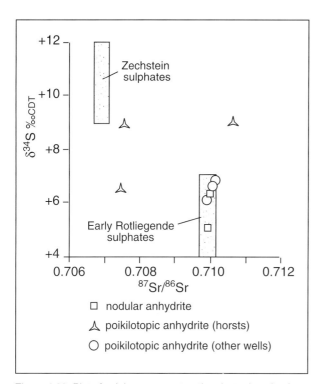

Figure 4.29. Plot of sulphur versus strontium isotopic ratios for coarse anhydrite cements from the German Rotliegende. It clearly shows the affinity of anhydrite cements in horst blocks for Zechstein sources rather than an early Rotliegende signature (after Gaupp et al., 1993).

145

CHAPTER 4

rich brines, generated by dissolution of nearby bitterns, tended to stabilise K feldspar in the Norphlet so that authigenic K-spar can even coat quartz grains, as well as its more commonplace occurrence as overgrowths on detrital K-spar grain substrata.

Dworkin and Land (1994) argued that the expected isotopic composition and trace-element concentration of a marine brine-reflux-derived anhydrite cement in Smackover Formation quartzose sands should be: $\delta^{34}S_{CDT} = +16‰$, $\delta^{18}O_{SMOW} = +14‰$, $^{87}Sr/^{86}Sr = 0.7069$, $Sr = 1500\text{-}2500$ ppm, $Ba < 20$ ppm, and $Mg > 200$ ppm (the box in Figure 4.30b). When geochemical analyses of Smackover anhydrite cements are compared with the predicted composition, it is apparent that most of the coarsely crystalline anhydrite is not a marine cement. Rather, two groupings or generations of anhydrite cement can be distinguished using a strontium isotope to strontium content crossplot (Figure 4.30b). An early cement, that may have been derived from slightly modified Late Jurassic sea water (possibly via reflux), is characterised by its Jurassic isotopic signature and high strontium content (>3,200 ppm). A second group of cements, that represent the bulk of coarsely crystalline cements in the Smackover sandstones, has a depleted strontium content (<3,000 ppm) and an elevated isotopic ratio. It may have precipitated later in the burial history or may represent burial recrystallisation of the first cement. It probably formed at slightly higher temperatures (which would have maintained relatively low strontium concentrations) and in the presence of pore fluids that had radiogenic Sr isotopic compositions. Such radiogenic pore fluids were probably derived from waters that were involved in the burial dissolution of feldspars.

As well as in sandstones, dissolution-derived sulphate-rich basinal brines also precipitate anhydrite burial salts in Smackover carbonates of the Gulf of Mexico (Figure 4.31; Heydari and Moore, 1989; Heydari, 1997). It precipitated as orthorhombic laths and blades that replaced both allochem grains and calcite cements. The anhydrite and the calcite formed prior to reaching the oil window and post stylolitisation. The arrival of liquid hydrocarbons and the precipitation of solid bitumens temporarily retarded further precipitation of calcite, anhydrite and dolomite burial cements. Later, the anhydrite was dissolved and replaced by a more finely crystalline calcite spar and elemental sulphur via thermochemical sulphate reduction processes (Chapter 9). This occurred in the higher temperatures of the gas window.

Dissolution-derived late burial anhydrite is also widespread in the Lower Cretaceous carbonates of the Pearsall and

Lower Glen Rose Formations, South Texas (Woronick and Land, 1985). As in the Smackover, the anhydrite formed late, after the development of saddle dolomite cements. It crystallised mostly as bladed to fibrous anhydrite laths some 0.1 - 6.0 mm in length. Much of the anhydrite fills preexisting intergranular pore space in the carbonates but it can also replace fossils, matrix material and earlier calcite cements. Once it had formed it was subject to corrosion and replacement by calcite. $\delta^{18}O_{SMOW}$ values for the anhydrite range from 19.5-22.9‰, values that are much more enriched than the Cretaceous seawater sulphate (+15‰). Woronick and Land (1985) interpret these enriched values as proof that none of the subsurface anhydrite in the Pearsall and the Glen Rose (including some associated nodular anhydrite) was precipitated directly from Cretaceous seawater. They argue that even the nodular anhydrites were recrystallised in the subsurface. This is supported by $\delta^{34}S_{CDT}$ values of the anhydrite and the modern Edwards Formation brines, which lie between 20 and 26‰. Directly precipitated Cretaceous marine evaporites have $\delta^{34}S_{CDT}$ values of less than +16‰.

They envision a basinward flowing meteoric aquifer, driven by hydrostatic flow, that was essentially saturated with respect to calcite, dolomite and $CaSO_4$. Periodically the meteoric waters mixed with upwelling hotter basinal

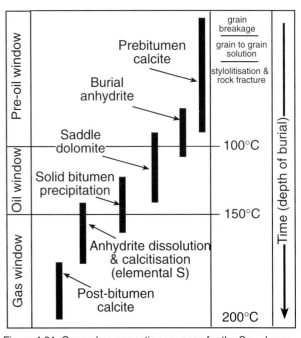

Figure 4.31. General paragenetic sequence for the Smackover Formation in the northern Gulf of Mexico showing approximate temperature range of various events (after Heydari and Moore, 1989; Heydari, 1997).

146

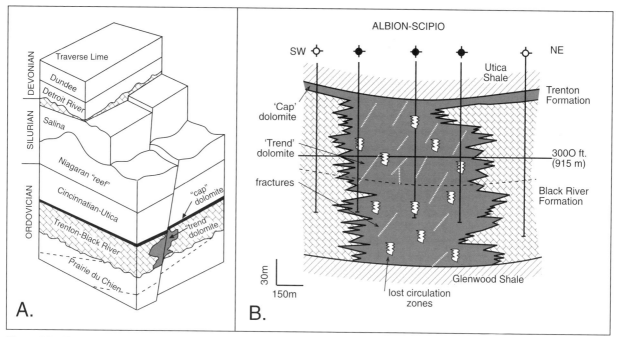

Figure 4.32. Fault-focused late dolomitisation in the Scipio-Albion trend, Michigan (after Hurley and Budros, 1990). A) Structural stratigraphic interpretation of the Albion-Scipio trend. The fault/fracture set that controls dolomitization is pre Devonian. B) Structural cross section of the Scipio-Albion Field. Note that the field is located in a synclinal trend (not usual for a hydrocarbon trap) and there is a rapid lateral transition from a regional nonreservoir limestone into a permeable reservoir dolomite that also hosts late stage anhydrite and other authigenic burial salts. Lost circulation and fracture zones are common indicators of secondary dolomite-hosted porosity.

fluids that were derived from more deeply buried and dissolving Jurassic salts. These basinal fluids contained abundant carbon dioxide, calcium, strontium (radiogenic) and barium, together with small amounts of lead, zinc and hydrocarbons. Mixing of the two waters caused the precipitation of burial anhydrite as well as other burial cements, including saddle dolomite and metal sulphides.

Other carbonate platform sequences that entrain burial anhydrite cements derived from evaporite dissolution and upflow of basinal waters include the Devonian Keg River Formation of NW Alberta (Aulstead and Spencer, 1985) and carbonate reservoirs in the Middle Ordovician Trenton Group, SW Ontario (Middleton et al., 1993). The latter host oil fields like Scipio-Albion, classic examples of fault-controlled permeability development associated with late stage burial dolomitisation (Hurley and Budros, 1990). The emplacement of the burial dolomite was followed by precipitation of anhydrite burial salts along with late stage calcite, minor sulphides, fluorite and baryte cements (Figure 4.32). Primary fluid inclusions in fracture-related saddle dolomite cements in the "trend" dolomite are saline (24 to 41 wt % NaCl eq.) and homogenization temperatures range between 100°C and 220°C (Middleton et al., 1993). These temperatures are considerably higher than those

likely to have been generated during peak burial of the hosting limestone sequence (≈70°C). Late stage calcite cements associated with degradation of burial anhydrites have fluid inclusion homogenization temperatures ranging from 70°C to 170°C, and were precipitated by somewhat less saline brines (16 to 28 wt % NaCl eq.) than the dolomites. They may be the result of fluids derived from thermochemical sulphate reduction migrating up the permeable pathway created by the fault-aligned dolomite. Pattern drilling of the region for oil exploration has once again demonstrated the importance of faulting in controlling emplacement patterns of most burial cements.

Evaporite pseudomorphs

This chapter has shown how evaporites can be replaced by silica, calcite or dolomite at any time in their burial and uplift history. So far, I have focused on the more commonplace evaporites that are pseudomorphed or grow in the subsurface —namely, gypsum, anhydrite and halite. These three salts constitute more than 95% of modern and ancient precipitates or their pseudomorphs. However, there are pseudomorphs that preserve the outlines of other evaporite salts, such as trona, glauberite, thernadite and

147

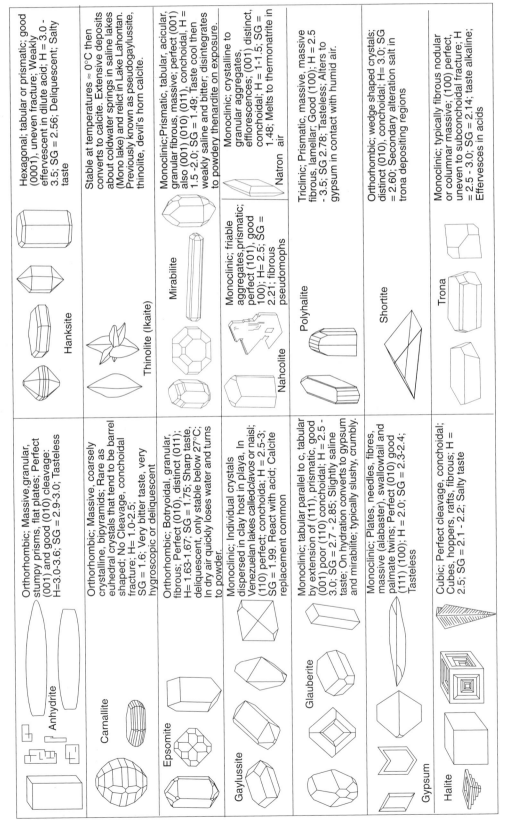

Figure 4.33. Characteristic shapes of some of the various depositional evaporite salts and their physical properties.

other less common precipitates. They are also the group of salts that make up a large group of mineable deposits (Chapter 7). Some of these less-common salts have crystal shapes that can be recognised if their euhedral forms are well preserved by carbonate or siliceous replacements. Figure 4.33 illustrates some of the more typical shapes of these less common salts along with the physical properties of the actual salts.

Of course the preservation of a crystal form is often accompanied by enlargement or partial collapse of the void during the replacement process. This complicates the process of recognition. Some of these salts are also more likely to alter to calcite with burial as they already contain elevated levels of the bicarbonate ion when they are deposited. This group includes trona, hanksite, shortite, pirssonite (e.g. Southgate et al., 1989). When preserved as isolated replacement crystals in a mudstone matrix other calcitic pseudomorphs, such as ikaite and gaylussite, have lenticular forms which are easily confused with the lenticular form of gypsum. In the past these salts have been commonly confused one with the other.

Ikaite in a saline lake sequence is typically preserved as thinolitic calcites (syn. jarrowites, devil's horn calcites, glendonites) about the margins of cold water saline lakes. Historically, many of these thinolitic calcites were interpreted as some sort of evaporite crust and called pseudo-gaylussite. We now know that ikaite pseudomorphs are much more likely to indicate coldwater carbonate seeps. These lenticular calcites, that are easily confused with gypsum and gaylussite pseudomorphs, are actually pseudomorphs of calcium carbonate hexahydrite (ikaite). This hydrated form of calcite is a cold water variety that is only stable at temperatures around 0°C. Its formation typically has little to do with an evaporite precursor, and it is often recovered in modern Antarctic bottom sediment or in glacial outwash fans (Shearman and Smith, 1985; Larsen, 1994; Riccioni et al., 1996). In one of ikaite's best documented saline occurrences it is a winter precipitate around coldwater tufa springs in Mono Lake, California, and probably also first formed the thinolitic platforms of Pleistocene Pyramid Lake, Nevada. Once the water temperature about a spring mound rises above 3-4°C the ikaite dehydrates to calcite, but the calcite often preserves the characteristic lenticular and twinned form of its precursor.

Summary

Evaporites are soluble in most subsurface situations. Even preserved evaporite beds show varying degrees of dissolution about their edges. The longer an evaporite bed remains in the subsurface its preservation it is not a question of whether an evaporite will dissolve, but when. This is why there are few preserved beds of evaporite salts older than Neoproterozoic. As thick evaporite units dissolve they leave behind characteristic breccias and other textures and replacements that allow the recognition of former extensive salt beds. The breccias can be interstratal beds or trans-stratal chimneys, distinguishing the two forms differentiates between blanket or more focused inflows. Dissolution can be driven by basinal fluids escaping from below or by fluids infiltrating from above. Salt units may not dissolve until they have flowed into diapirs and welds (halokinetic breccias) or have acted as décollement surfaces in fold and thrust belts (rauhwacke).

Other evaporites, especially those isolated crystals and nodules, originally deposited in a more permeable nonevaporite matrix, show a strong tendency to be replaced by silica or carbonate while still retaining the outline of the original salts.

All evaporite salt units will ultimately dissolve, many disappear in the diagenetic realm while others may be preserved until they attain metamorphic grades (Chapter 6) or interact with a magmatic body (Chapter 9).

Chapter 5
Salt tectonics

Introduction

Salt flow is an important mechanism in controlling the formation of complex traps for hydrocarbons and metals. Evaporites and associated flowage are responsible for trapping more than 60% of the hydrocarbon reserves of the Middle East (Edgell, 1996). Salt flow produces some of the most complex and visually impressive deformation features to be seen on the earth's surface. Some even have religious significance; Lot's wife on the edge of the Dead Sea is a 12m-high column of salt lying at the foot of the much larger Mt Sedom (Usdum). It is the dissolutional remnant of a gypsum-capped outcropping diapir composed of Miocene salt, which also makes up Mount Sedom. It was noted by Fulcher of Chatres who accompanied the crusader Baldwin I across the Dead Sea Valley in December 1100 AD. Mountain is somewhat of a misnomer when describing Mt Sedom, it is certainly 8 km across but only rises some 30-46 m above the floor of the Dead Sea Valley. Such low relief is typical of many outcropping salt structures, the only exceptions are the actively flowing namakiers (salt mountains) of Iran, which rise up to 1000m above the surrounding landscape.

The word diapir comes from the Greek *diaperin*, meaning "to thrust-through" and was coined by L. Marzic in 1902 to describe salt features of the Romanian Alps. However, it is preferable to define salt diapirs as salt structures having discordant contacts with the encasing sedimentary strata (overburden) rather than to uncritically encompass notions of active piercement or shouldering aside of thick piles of adjacent sediments. Diapirs are not just composed of salt (halite) and other evaporite salts, but can be composed of shale, serpentinite, and any other material that possess a lower relative density than its overburden. Many of the principles that describe salt diapirism can be applied to molten granites rising through the mantle (Talbot, 1992a). This chapter will concentrate on salt structures; halokinetic breccias have already been discussed in Chapter 4, while the economic aspects of the allochthon-metal association are discussed in Chapters 8 and 9.

Salt tectonics is a general term that encompasses notions of salt flow, trans-stratal salt movement, salt pillowing and diapirism. Salt tectonics (syn. halotectonics) refers to tectonic deformation involving halite or other evaporites as a substratum or source layer. Halokinesis is a form of salt tectonics in which flow of salt is powered by gravity, that is by the release of gravity potential without significant lateral tectonic forces (Trusheim, 1957, 1960). Concepts of salt tectonics have undergone a major revision in the last decade, due mainly to advances in 3D seismic acquisition and processing, especially 3D prestack depth imaging. The greatest surprise came from the recognition that allochthonous salt layers are widespread in classic halokinetic basins (Figure 5.1). Allochthonous salt structurally overlies parts of its (stratigraphically younger) overburden. For example, spectacular allochthonous salt

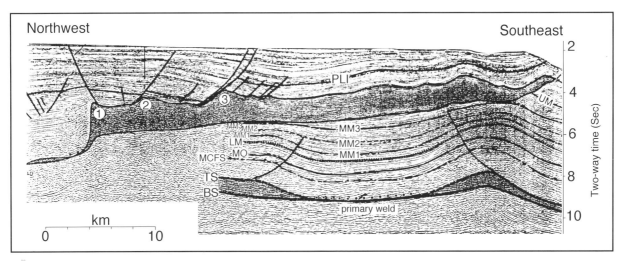

Figure 5.1. Interpreted seismic profile through allochthonous salt sheets that build the Sigsbee Escarpment (to southeast), base of continental slope off Louisiana, Gulf of Mexico (after Wu, 1993; Jackson and Vendeville, 1994). Note how supra-allochthonous faults have initiated diapiric upwelling at 1, 2 and 3. The allochthon is underlain by shortened strata of the Mississippi Fan Foldbelt. BS = base of Jurassic Louann Salt; TS = Top of salt; MCFS = Middle Cretaceous flooding surface equivalent; MO = Middle Oligocene; LM = Lower Miocene; MM1 = Lower Middle Miocene; MM2 = Middle Miocene; MM3 = Upper Middle Miocene; UM = Upper Miocene; PLI = Pliocene. Note how Jurassic salt has climbed the stratigraphy and now overlies and seals Neogene deepwater sediment.

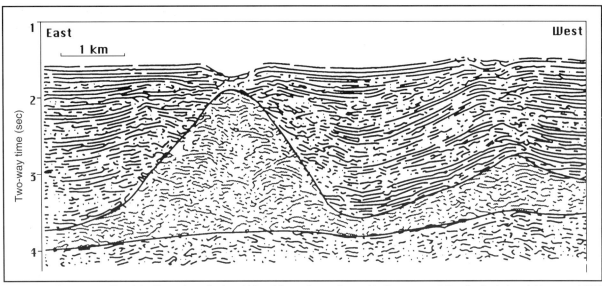

Figure 5.2. Seismic profile of the Xenophon diapirs of Messinian salt, southeast Mediterranean (after Garfunkel and Almagor, 1987).

sheets in the northern Gulf of Mexico form a line of five or so huge asymmetric tongues of Jurassic salt that now overlie and seal Pliocene and Pleistocene sediments. They coalesce behind the Sigsbee Scarp to form a salt tongue canopy that can be considered an analogue to structural nappes in orogenic belts (Worrall and Snelson, 1989). Allochthonous salt bodies climb the stratigraphy via a series of tiers or canopies, leaving evidence of passage along salt welds or overburden touchdowns, much like a garden slug leaves behind a slime trail. Prior to our improved seismic imaging most of these allochthonous sheets were thought to be firmly rooted in their mother salt beds (i.e. at the level of primary weld in Figure 5.1). The recognition that salt sheets can be allochthonous over large areas in a sedimentary basin opens a whole new realm of possible structural salt/sediment configurations and a range of potential subsalt hydrocarbon traps (Montgomery and Moore, 1997).

Tectonic setting of halokinesis

Salt structure provinces on the world-scale are most often found in active rifts or along previously rifted continental margins. Rifts are areas of regional extension, sediment dumping and high initial geothermal gradients, all features that also encourage salt flow. Thick beds of nonmarine and marine evaporites tend to accumulate in the incipient to early oceanic phase of a rift. This is seen today where juvenile continental margins in arid climatic settings, such as the Red Sea, are floored by thick layers of bedded salt that may also be rising into salt structures. Earlier stages

of older rifts are now found as buried halokinetic remnants, such as in the proto-Atlantic rift where salt structures define the eastern margin of the Americas and the western margin of Africa (Jackson and Vendeville, 1994; Evans, 1978; Kinsman, 1975a,b).

As a rift widens, the basin floor cools, subsides and is covered by layers of sediment. Bedded evaporites of the early rift undergo extension and burial by several kilometres of shelf sediment. Extension with associated differential loading initiates movement and diapirism of the previously bedded salt. Diapir growth in continental margin sediments then persists episodically for tens of millions of years, as long as extension, subsidence, differential loading and burial continue to affect the thick, mobile salt units. Depending on the thickness and lateral extent of the mother salt bed, salt spines can climb ten or more kilometres of section and in so doing create a number of highly effective structural and stratigraphic traps for hydrocarbons and metals.

As well as in active rift zones, thick bedded salt units can be laid down during continental collision (see Chapter 3). For example, the Upper Miocene (Messinian) evaporites of the Mediterranean were laid down as Africa collided with Eurasia. Subsequent transpression and extension remobilised this bedded salt so that extensive diapiric provinces exist now in and about the Mediterranean (Figure 5.2; Garfunkel and Almagor, 1987). Compression can also overprint early extensional salt provinces; such a regime has clearly overprinted extensional salt structures in the Triassic salt of Spain and the French Maritime Alps (Roca et al., 1996; Dardeau and Graciansky, 1990).

Salt units in older halokinetic provinces are more and more likely to dissolve, so that many of the oldest salt structures no longer retain salt. Fossil (Neoproterozoic) salt structures outcrop along the rift axis of the Adelaide Geosyncline in South Australia, and the Amadeus Basin in central Australia (Figure 4.10). Structures were active in the Neoproterozoic and the Cambrian, with some remaining episodically active into the Triassic (Lemon, 1985, 1988). Salt has long since been flushed from the nearsurface structures, so that they are now outlined by chaotic carbonate and siliceous breccias.

The amount of information on halokinesis is huge and the volume of relevant literature has more than quadrupled since 1970. The question for me in writing this chapter was how to review all this without writing a chapter that was as long as another book. I decided that rather than attempt a weighty review of all the literature back to the turn of the century I should summarise much of the relevant data from recent symposia and summary papers that were published in the last few years. I have drawn much of the material that follows from what I consider are keynote references. For an excellent perspective, the volumes edited by Jackson et al (1995) and Alsop et al. (1996) are "must reads". For keynote papers the interested reader is referred to papers by Jackson et al. (1994a,b), Duval et al. (1992), Talbot (1993), Worrall and Snelson (1989), Wu et al. (1990a,b). For seismic interpretation of salt structures there are excellent books by Jenyon (1986) and Nely (1994; this is an translation into English of Nely's 1989 book). Both works were written in the 1980s, and some of the structures would now be interpreted a little differently using the paradigms of the "brittle era", but the quality of the material and the quantity of seismic presented means both are still very relevant.

Historical perspective

The history of salt tectonics can be divided into three parts (Jackson, 1995): the pioneering era, the fluid era, and the brittle era. The *pioneering era* (1856 - 1933) is characterised by a search for a general hypothesis of salt diapirism. The early part of this era was dominated by notions of igneous activity, residual islands, *in situ* crystallisation, osmotic pressures, and expansive crystallisation. Then data from oil exploration began to confine speculation and the effects of buoyancy versus orogeny began to be debated, contacts were studied,

salt glaciers recognised and the concepts of active intrusion, downbuilding and differential loading were proposed as driving mechanisms (Figure 5.3).

Many of the early hypotheses now seem rather bizarre; but by the end of this period a consensus was arising that active piercement, the pushing aside of overburden by a rising salt plume, was an important mechanism, and perhaps dominant over downbuilding in the emplacement of most salt structures. As it rises through its overburden an active diapir forces its roof upward and sideways, thereby solving any room problems (Nelson, 1991). In contrast a downbuilding mechanism (nowadays also called passive diapirism) encapsulates the notion that the crest of the salt structure remains at or near the surface while sediments accumulate discordantly about it. The diapir grows taller because its base is sinking relative to the surface. No overburden need be displaced, so there is no room problem.

Then followed the *fluid era* (1933 - 1989). Rayleigh-Taylor instability models dominated in which a dense fluid overburden sinks into a less dense salt layer displacing it upward (Jackson et al., 1995). Density contrasts, viscosity contrasts and dominant structure wavelengths were emphasised, while strength variation and overburden faulting were largely ignored as possible controls to salt flow. In this period palinspastic reconstructions were attempted; salt upwelling below thin overburdens was recognised; internal structures in mined diapirs were mapped and interpreted; peripheral sinks, turtle structures, and diapir families were recognised; flow laws from experiments on dry salt were conducted and applied to generate notions of "hot" salt extrusion.

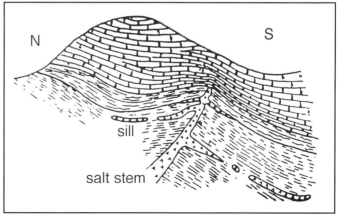

Figure 5.3. Ehrmann's (1922) figure where the broad lateral flanges of Triassic salt of the main salt stem are interpreted as intrusive sills, Gulf of Bejaia (Bougie) Kabylle, Algeria. Now the same features are interpreted as remnant salt allochthons and glaciers.

The 1970s revealed the basic driving force of salt allochthons, intrasalt minibasins, finite strains in diapirs, the possibility of thermal convection in salt, direct measurement of salt glacial flow stimulated by rainfall, and the internal structure of convecting evaporites and salt glaciers. The 1980s revealed salt rollers, subtle traps, flow laws for damp salt, salt canopies, and mushroom diapirs. Modelling explored effects of regional stresses on domal faults, spoke circulation and combined Rayleigh-Taylor instability and thermal convection. Toward the end of this period, downbuilding or passive diapirism, rather than active diapirism came to be considered the most important process in driving salt flow

The *brittle era* blossomed in 1989, although it had its beginnings in the 1947 discovery that a diapir stops rising if its roof becomes too thick (Jackson et al., 1995). Stimulated by sandbox experiments and computerised reconstructions, the onset of the brittle era has yielded regional detachments and evacuation surfaces along vanished salt allochthons (salt welds and fault welds), raft tectonics, shallow spreading and segmentation of salt sheets. The early 1990s revealed rules of section balancing for salt tectonics, salt flats and salt ramps, reactive piercement as a diapiric initiator resulting from tectonic differential loading, cryptic thin-skinned extension, influence of sedimentation rate on the geometry of passive diapirs and extrusions, the importance of critical overburden thickness to the viability of active diapirs, fault segmented sheets, counter regional fault systems, subsiding diapirs, extensional turtle structure anticlines, and mock turtle structures.

The brittle era is the time frame in which reactive diapirism came to be recognised as one of the most important ways to initiate salt flow. Reactive diapirism requires regional extension, created by rifting or thin-skin gravity spreading and gliding. The process of regional extension in itself generates space for the formation of the salt structure. It was also the period where the three modes of diapirism (active, reactive and passive) were for the first time integrated into the study of salt tectonics, which we will explore later in the chapter.

But, as in all scientific revolutions, the pendulum of opinion is still swinging. In the period 1990-1995 the paradigms of extension in controlling diapirism became so popular that other equally valid mechanisms of diapirism, such as differential loading and plastic responses in the overburden were largely ignored or downgraded. The pendulum has now started to swing back and differential loading, associated with prograding shelf wedges, is once

again being recognised as a mechanism that can both initiate and drive basin-scale diapirism (Ge et al., 1997). In this chapter we will consider both as equally valid mechanisms, but first we need to discuss the physics of salt.

Physics of salt systems

Density, viscosity, strength, strain and buoyancy

Bedded halite is an unusual sediment in that intercrystalline porosity is lost by 50-100m burial (Figure 1.14). It is near incompressible to depths of 6-8 km, where with further burial and entry into greenschist depths it can undergo massive recrystallisation and dissolution along with a slight decrease in density due to thermal expansion (Chapter 6). Halite's density of 2.2 gm/cc remains near constant throughout the diagenetic realm. In contrast, burial compaction in most other sediments leads to a progressive loss of porosity with an associated increase in density.

Most nonevaporite sediments are deposited with initial densities less than that of salt, but, with compaction and water loss, their densities increase to become greater than that of salt by depths of a kilometre or more. This means that salt buried beneath nonevaporitic overburden to depths in excess of a kilometre or more has positive buoyancy. A possible exception is salt buried beneath cemented carbonates. Reef carbonates and dolomites can have densities equal to, or greater than, salt from the time they are deposited. The depth of density crossover, where the densities of salt and its overburden are equal, is sometimes called the level of neutral buoyancy.

Few other rocks are as close to truly viscous as natural rock salt. Laboratory measurements show wet carnallite may have a viscosity as low as 10^8 Pa.s, while dry halite has a viscosity around 10^{18} Pa.s. The viscosity of salt in the Gulf of Mexico is 10^{13} to 10^{16} Pa.s, depending on moisture content and temperature. The adjacent overburden of clays and sands is less viscous by four to five orders of magnitude. Davison et al. (1996a) estimates the ratio, sedimentary rock:evaporite rock viscosity, ranges from 50 to 10^4. Different parts of a diapir can deform at different rates and by different mechanisms and so effective subsurface viscosity values for diapiric salt vary with position in the diapir and with time.

SALT TECTONICS

Type of flow	Strain rate (s^{-1})	Speed (mm a^{-1})	Speed
Lava flow	10^{-5} to 10^{-4}	5 x 10^{11} to 3 x 10^{13}	1 to 60 km hr^{-1}
Ice glacier	10^{-10} to 5 x 10^{-8}	3 x 10^{5} to 2 x 10^{7}	1 to 60 m day^{-1}
Salt glacier	10^{-11} to 2 x 10^{-9}	2 x 10^{3} to 2 x 10^{6}	10 to 100 km Ma^{-1}
Mantle currents	10^{-14} to 10^{-15}	10 to 10^{3}	2 m a^{-1} to 5 m day^{-1}
Salt tongue spreading (<30km wide)	8 x 10^{-15} to 10^{-11}	2 to 20	2 to 20 km Ma^{-1}
Salt tongue spreading (>30km wide)	3 x 10^{-16} to 10^{-15}	0.5 to 3	0.5 to 3 km Ma^{-1}
Salt diapir rise	2 x 10^{-16} to 8 x 10^{-11}	1 x 10^{-2} to 2	10 m to 2 km Ma^{-1}

Table 5.1. Representative strain and flow rates (after Jackson and Vendeville, 1994).

Wet salt at rest approaches Newtonian viscous behaviour, unless it is atypically coarse grained. Using a viscous fluid model, salt diapirism and its relationship to its overburden can be likened to Rayleigh-Taylor instability with a viscous substratum (salt) and an overlying denser viscous fluid (overburden; e.g. discussions in Koyi, 1991; Talbot, 1992a,b). Under such a fluid-fluid model, diapirism is spontaneously initiated by small irregularities in the fluid-fluid interface that then amplify with time. The only requirement for Rayleigh-Taylor instability in a fluid-fluid system is density inversion. Using this model the initiation of diapirism does not require any external trigger such as regional extension or differential loading. Under this fluid-fluid scenario, regional extension thins the fluid overburden and the source layer, so reducing the overall thickness. The resulting diapirs are smaller, more closely spaced, and more slowly rising than those in unthinned counterparts (Koyi, 1991).

In contrast, proponents of the "brittle school" tend to model salt as a pseudo-fluid at subsurface flow rates, but typically consider its sedimentary overburden to show brittle responses to stress (e.g. Jackson and Vendeville, 1994). In reality, the strength response of a buried salt bed and its overburden will depend on the strain rate. Salt can fault at high strain rates, but this rate is much higher than experienced in many typical subsurface situations (Jackson and Talbot, 1994). Most adherents of the brittle school contend that regional extension, and perhaps differential loading or gravity sliding, will localise, initiate and promote diapirism of salt and shale in most geological situations. Such models are firmly rooted in the observations of rock mechanics and relative strengths of rock salt compared to other materials (Weijermars et al., 1993).

Data on experimental deformation of virtually all sediments (apart from some evaporite salts) show them to deform brittlely rather that by creep at shallow crustal depths (<8 km). This encompasses much of the realm of salt tectonics. Thus, subsurface nonevaporitic sediments, including indurated shales, show time-independent, pressure-dependent brittle behaviour and deform most readily by frictional slip along faults. This means that axioms of Taylor-Rayleigh instability, and the results of fluid-fluid experimentation, may not directly apply to many real rock counterparts. That is, relative density as a trigger and a subsequent control to halokinesis, is perhaps a less important consideration than relative strength of salt and its overburden in most subsurface situations.

Under typical subsurface conditions slightly damp halite deforms as a near-Newtonian or power-law fluid and is much weaker than its sedimentary overburden. When the creep and frictional strengths of wet and dry salt are compared with equivalent strengths of other sedimentary rocks, using a representative strain rate of 10^{-14} s^{-1}, it can be seen that even dry salt is much weaker than any overburden sedimentary rock, except perhaps for extremely shallow water-saturated muds (<0.5km deep; Figure 5.4). Damp salt, which deforms by solution-transfer creep is hundreds of times weaker than dry salt, which deforms by dislocation creep, the movement by slippage along lattice defects (Urai et al., 1986). Solution-transfer creep occurs in regimes of differential stress, where mineral grains are dissolved in regions of high stress and redeposited in regions of low stress so that the deforming salt behaves like a Newtonian fluid.

Representative strain rates and the speed of salt deformation are listed in Table 5.1. A strain value of 10^{-15} s^{-1} indicates

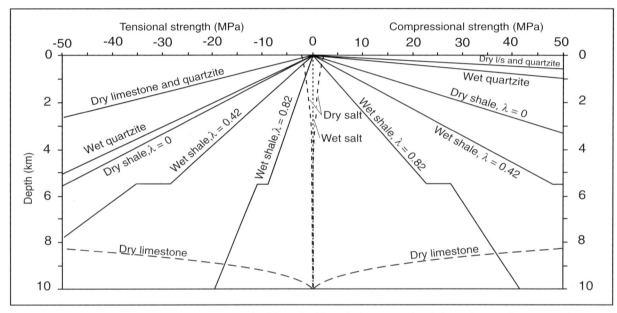

Figure 5.4. Comparison of creep and frictional strength of dry salt with equivalent strengths of dry and wet sedimentary rocks and sediments (after Jackson and Vendeville, 1994). Creep strength is the deviatoric stress required to drive salt flow at a representative strain rate of 10^{-14} s^{-1}. Pore pressure coefficient (λ) is the ratio of pore pressure to lithostatic pressure: λ=0 for dry rocks, λ=0.46 for hydrostatically pressured rocks and λ = 0.86 for overpressurred rocks. In contrast wet salt is so weak (strength of 0.01 MPa at this strain rate) that its creep curve plots virtually on the central axis.

that a rock stretches at a rate of 10^{-15} of its length per second and is a slow strain rate equivalent to rock deformation rates in active orogenic belts. A value of 10^{-9} is 1 million times faster and represents the strain rates in active salt glaciers in Iran (1.6 x 10^{-13} to 1.8 x 10^{-8} per second; Wenkert, 1979). Salt glaciers (namakiers) in Iran always flow faster after a rain; the Dashti glacier at Kuh-e-Namak flows at 0.5 m/day during the few weeks of the annual wet season, but flows little if any during the dry season when the glacier expands and contracts in response to changes in surface temperature.

Damp salt does not even require external water to act as a fluid. Urai et al. (1986) showed that water need not enter a salt body from the outside to be a factor in salt deformation and flow. They demonstrated that even the minute quantities of brine caught up as inclusions and intergranular films in confined natural salt have a profound effect on strength. Brine-inclusion-bearing samples showed marked weakening at low strain rates; a result not seen in earlier experiments using coarse-grained dry salt. More water entering natural salt allows it to flow even faster.

Confined rock salt layers with water contents over $\approx$0.01% can be considered to act as a very weak crystalline fluid (Talbot, 1993). Wet salt at rest is weaker than all sediments and sedimentary rocks, except for partly compacted wet muds within a few metres of the surface (Figure 5.4;

Weijermars et al., 1993). The extreme weakness of natural wet salt is seen in seismic onlap patterns of allochthonous salt sheets in the Gulf of Mexico (Figure 5.5; Nelson and Fairchild, 1989; Nelson, 1991). The salt sills were emplaced into low density shales at depths of less than 250-300m, with a mean depth of 150m. Shales deeper than that were too strong to be laterally intruded by the salt sheets. Were density contrast, rather than relative strength, the controlling factor then these salt sheets would have spread at depths $\approx$ 1400m. This is the level of neutral buoyancy where the shale and the salt densities are equal (Nelson, 1991). Salt weakness, or lack of strength, is thus more significant than buoyancy contrast in controlling the position of allochthonous salt sheets in earth-space.

The weakness of salt at geologic strain rates means the viscous forces in flowing salt cannot easily drag or stretch the overburden. Rather it acts as a lubricant that effectively decouples the overburden from the basement and makes it difficult for basement faults to directly influence deformation in the overburden. Salt's lack of strength also means a thick salt bed tends to act as a ductile "crack stopper". A buried layer of thick salt hinders faults and dykes propagating into higher or lower levels, and segments sediments into layers that deform independently or even by different processes. It can also help focus and entrap petroleum and mineral fluids against a salt seal at the tops

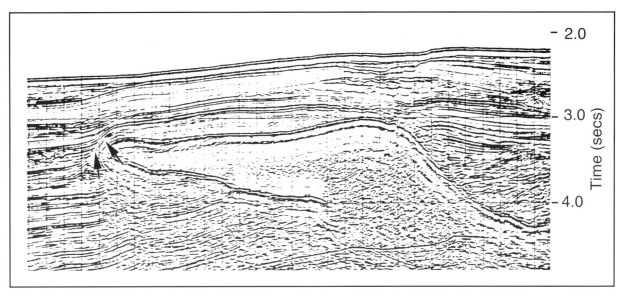

Figure 5.5. Salt allochthon, northern Gulf of Mexico (after Nelson, 1991). Onlap of frontal reflectors above a thin roof, averaging only 100m thick (arrows on left hand edge of salt tongue), indicates shallow emplacement of the allochthonous salt tongue.

of subsalt faults, even if the faulting postdates the deposition of the subsalt layer.

Lubricating effects of salt layers are clearly seen in the widths, taper angles, and internal structures of dynamic wedges at both converging and diverging plate boundaries By lubricating regional décollements, even thin layers of salt influence how huge areas shorten or extend due to lateral forces generated by gravity. It also leads to the complexities of raft tectonics, where rapidly opening ocean basins generate floundering blocks of brittle sediments (tens of kilometres across and kilometres thick) that glide down into the widening basin atop a layer of lubricating salt. For the same reason fold thrust belts in converging plate margins that are deforming over salt layers are much wider, with taper angles <2°; while those deforming over other rock layers are narrower, with taper angles nearing 8° (Davis and Engelder, 1987).

In reality, a simple division into brittle versus viscous fluid models is an oversimplification as salt and the adjacent sediments can, depending on the circumstances, exhibit either brittle or viscous responses (Davison et al., 1996a; Poliakov et al., 1996). Salt is sometimes brittle enough to fault or shear, and such displacements are not uncommon in salt mines. Likewise the overburden can show evidence of viscous behaviour. Sediments atop allochthonous salt sheets in the Mexican fold ridges in the Gulf of Mexico are folded not faulted. Seismic also shows many diapirs have sedimentary margins with smooth drag profiles (shale sheaths), rather than brittle faulted margins.

Pressure effects

Like all buried sediments, a salt source layer is pressurised by its own weight along with that of the overburden and results in a pressure head (Ψ). Piercing of a brittle sediment overburden can occur when the pressure head exerted by the salt creates a buoyancy force that exceeds the strength of the overburden. Figure 5.6 illustrates pressure head for three situations where the fluid layer (salt) is less dense than its brittle overburden (Jackson and Vendeville, 1994). Any net upward force is resisted by the strength of the overburden roof, which depends on roof thickness. Thin roofs are weak and easily pierced, whereas thick roofs are too strong to be pierced by pressurised salt alone and require regional extension to be diapirically breached. The thickness of sediment that can maintain sufficient strength to hold back a buoyant salt bed defines the piercement threshold layer (Figure 5.6). If sediment thickness is less than the piercement threshold, then a salt diapir can actively pierce or shoulder aside its overburden (shown by the arrow in Figure 5.6a). Where the floor of an erosional valley or sag thins the sediment layer to where the trough floor is below the piercement threshold, active piercement can occur once more (Figure 5.6b). Active piercement can also occur in situations where the roof is thinned by extension, a commonplace feature in actively spreading rift systems (Figure 5.6c).

In another series of physical models Vendeville and Jackson (1992a,b) made a number of interesting observations. Their analytical model assumed: (1) salt behaves as a

157

pressurised fluid at rest (internal viscous forces can be neglected), and (2) the sedimentary overburden deforms only by frictional sliding of a rigid diapir roof along faults. They observed:

• The taller or wider the ridge or plug, the more easily it pierces.
• The thinner the roof, the more easily it is pierced.
• Structures with linear planforms (ridges or troughs) are more easily pierced than those with circular planforms (plugs or cavities).
• Relief in the upper surface of the overburden (trough or cavity; Figure 5.6b,c) promotes piercement more than does the same relief in its lower surface (ridge or plug; Figure 5.6a).

Pressure forces below a surface trough or cavity are set up by the density ratio between overburden and air or water (typical ratios around 2 or higher). Conversely, pressure forces above salt ridges or plugs reflect the lower density ratio between overburden and salt, which is merely 1.1 to 1.2.

Thermal effects

In addition to pressure/strength interactions, variations in burial temperature gradients can also have some interesting effects on diapirism. In this context halite is an unusual mineral. At a depth of 5 km, under a geothermal gradient of 30°C/km, halite expands 2% due to heat and contracts 0.5% due to pressure. Under such a scenario the base of a thick salt unit becomes slightly less dense with increased temperatures associated with burial to create a temperature-induced density inversion. Thermally induced convective flow is theoretically possible in a salt layer thicker than 2.9 km under a geothermal gradient of 30°C per km (assuming a viscosity of $<10^{16}$ Pa.s; Jackson and Talbot, 1994). This mechanism may

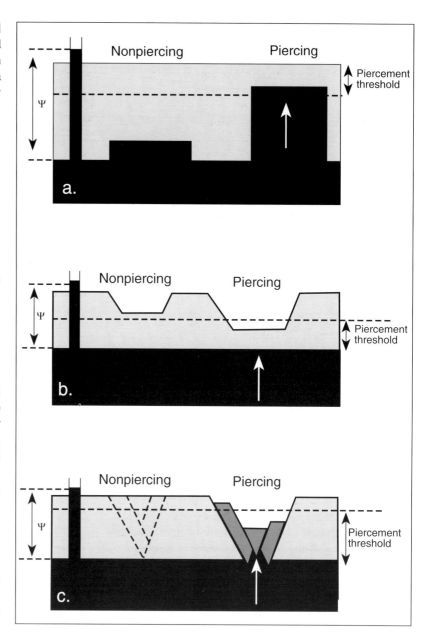

Figure 5.6. Conditions for active piercement of salt (black) overlain by rigid brittle overburden (grey). Where the overburden is thinner than the piercement threshold, the salt can break through (white arrows) under nonextensional conditions above salt ridges (a) or below topographic troughs (b). In (c) the non-stretched salt roof is too thick to permit diapirism but where it is thinned by extension to below the critical thickness a diapir can break through. Ψ is the pressure head of salt (after Jackson and Vendeville, 1994).

trigger diapirism in salt-filled rifts characterised by elevated nearsurface temperatures and high geothermal gradients (Talbot, 1978). The volcano-shaped salt mounds associated with hot springs in Ethiopia's Danakil Depression may be located above the tops of convecting plumes of salt.

Gussow (1968) argued that the outcropping Iranian salt glaciers were catastrophic extrusives emplaced hot, in a similar way to a lava flow. He thought that heated salt was intrusive and that the salt glaciers have now cooled and are no longer flowing. He was strongly influenced by early experimental work on dislocation creep using coarse-grained dry salt that had indicated salt could only flow at temperatures above 200°C. We now know that most confined salt is wet and capable of flow at earth surface temperatures (see above). Later work on the Iranian glaciers by Kent (1979, 1987) and Talbot (1979) indicated that the salt is still flowing today and that hot intrusion was not a major mechanism of salt emplacement in this or in other sedimentary basins. Temperature does become an important consideration once bedded salt enters greenschist facies conditions, as is seen in the diapiric remnants of the Damara Orogen in Namibia (Chapter 6).

Flow textures and rates

Average halite grain size in deformed recrystallised salt ranges between 5 and 10 mm in Zechstein salt in the North Sea and 15 to 20 mm in Gulf Coast salt (Davison et al., 1996a; Talbot and Jackson, 1987a). Both pressure solution creep and dislocation creep create the deformed and recrystallised halite textures that indicate mobile salt. As a salt crystal flows it stretches and deforms, leaving behind a memory of the stress regime in the various structures and crystal fabrics preserved in the diapir (Talbot, 1992a and references therein). Like many other crystalline rocks, such as gneisses and mylonites, the grains in flowing salt have such short strain memories that they rarely strain beyond length to width ratios of 4 or 5 before new crystals are created by the effects of pressure solution. These fine-scale grain-shape fabrics in confined flowing salt record only the late stages of the strain history. The actual strains, resulting from superposed deformations, are recorded by larger strain markers–primary layering or bedding, nodules, inclusions, and veins– all of which have much longer strain memories (Talbot and Jackson, 1987a,b).

Mapping of these internal strain markers along with modelling experiments have clearly shown internal flow in deforming salt is not homogeneous (Burliga, 1996; Davison et al., 1996a; Sans et al., 1996a,b; Smith, 1996; Talbot and Alavi, 1996). Foliations in deformed salt in a diapir are often axial to folds that develop in the primary layering and to the flow folds that map local gradients in flow rates along internal stream lines and surfaces in the moving salt (Talbot, 1979, 1992b). If these layers approach mechanical passivity then the flowing viscous "fluid"

picks up internal flow folds wherever the salt slows and thickens to pass obstructions. Beyond the obstruction the flow accelerates and internal folds tighten and almost disappear in the narrowing flow before they refold at the next obstruction. If the internal markers are strain active, the slow (competent) layers fold while fast (incompetent) layers mullion as they shorten within the slowing flow. Shears zones will form in a deforming salt body and will first grow in the least competent (usually potassic) horizons within the flowing salt body. Between the shear zones, less deformed halite layers are carried laterally and upward into the diapir stem.

Davison et al. (1996a) show that fold wavelengths are of the order of 20-30 km in salt layers in the North Sea, with an original thickness of more than one kilometre and an overburden that is more than one kilometre thick. The spacing of the folds controls the amount of salt that will drain into any particular structure. With a circular drainage envelope of 10 km radius a 1.5 km-thick salt layer can supply of the order of $470 \, km^3$ of salt to a single diapir. This is much greater than the volume of salt held in a typical North Sea diapir, which is around 25-75 km^3. This implies there has been substantial salt loss by dissolution. The rate of salt supply is not constant throughout the drainage area as the rate of radial flow increases closer to the diapir. According to Davison et al. (1996a) the constriction associated with flow into the neck creates strain rates within the neck up to 80 times greater than in the source layer. Thus vertical flow in the centre of a diapiric neck produces extreme prolate strain textures that are in turn nullified at the top of the diapir where vertical flattening takes place (Figure 5.7).

Any interbedded sediments within the salt source layer strengthen a salt unit and initially create drag zones in the flowing salt, and so create anisotropic strength and strain effects in the layer. This drag is often sufficient to generate small wavelength sheath folds. Once diapirism is more advanced then any intervening interbeds (up to 40m thick) are brecciated or boudinaged so salt can then flow across former intrasalt layers allowing the whole region to behave like pure salt. Hence during early pillowing, prior to piercement, the source layer with its interbeds will show high strength and high flow anisotropy. Once the system attains the diapir stage the strength is drastically reduced by boudinage and break-up of the stronger sedimentary rock layers. The style of boudinage depends on the degree of induration of the entrained sedimentary layers. Where the sediments are well indurated the break-up of intrasalt beds tends to create angular clasts, where less consolidated the beds tend to stream or flow plastically like the adjacent

159

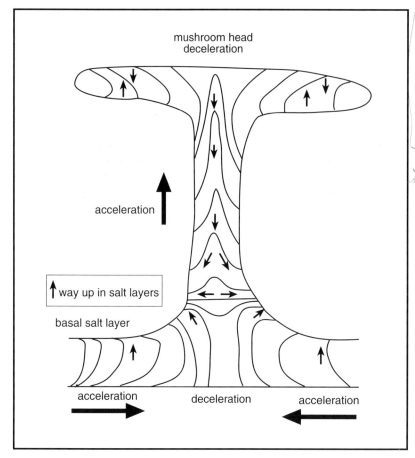

mushroom head
deceleration

acceleration

↑ way up in salt layers

basal salt layer

acceleration   deceleration   acceleration

Figure 5.7. Schematic velocity fields in a flowing diapir derived from physical models. The deformed outlines of originally vertical markers are shown, as well as bed facing directions (after Davison et al., 1996a).

salt. Boudinaged and aligned angular clasts and blocks can become a substantial component of any later halokinetic dissolution breccia (see Chapter 4).

Salt tectonics

Salt structures are triggered and driven by a variety of mechanisms. Studies since 1990 have tended to emphasise the role of regional extension (Duval et al., 1992; Vendeville and Jackson, 1992a,b; Demercian et al., 1993; Jackson and Vendeville, 1994). Studies conducted in the 1980s tend to place more significance on the triggering and driving effects of differential loading (Trusheim, 1960; Dailly, 1976; Seni and Jackson, 1983a,b; Jackson and Talbot, 1986, 1991; Worall and Snelson, 1989; Nelson, 1991). Recently, differential loading has come to be physically modelled in the same way as extension. As was postulated

for the past two decades, it is an equally valid mechanism as extension for both triggering and driving salt flow (Ge et al., 1997). In natural settings, both mechanisms work in conjunction, we will now discuss them in terms of their influence on salt flow and salt structure geometries.

Diapirism under extension

Under regional extension diapirs "pierce" the overburden in three ways: reactively, actively and passively (Figure 5.8; Jackson et al., 1994 a,b). A reactive diapir pierces during extensional faulting by filling the space created by divergence of overburden fault blocks. Extension typically reflects regional scale tectonic processes – as during gravity spreading or gliding in an opening rift system. In this scenario the extensional faulting is not caused by the diapir, the diapir merely reacts to regional stretching as salt moves into space created by the extension. Hence the name reactive diapirism. Reactive diapirism is an effective mechanism for initiating diapirism in many situations, regardless of the prior thickness, lithology or density of the overburden (Figure 5.8; Vendeville and Jackson, 1992a, b). In such settings there is always a low ratio between salt pressure (Ψ) and brittle strength of the overburden. Reactive structures can be found either beneath thin overburdens starting to extend, or in the deeply buried salt beds in regions undergoing extension.

An active diapir, by definition, lifts and shoves aside its overburden. This style of forceful intrusion is only possible if the overburden roof is relatively thin and the salt pressure exceeds brittle strength in the overburden (Figure 5.8). For this to happen Jackson et al. (1994b) argue the roof must be less than 20% of the entire thickness of the overburden flanking the diapir. However, once initiated a diapir can actively pierce each new increment of sediment that buries it. The repetition of active piercement episodes, which shed successive episodes of thin cover sediment, means that over time an active diapir can pierce large cumulative thicknesses of sediment.

As a diapir emerges and flows onto the earth's surface it becomes a passive diapir. As the stem grows taller, the diapir crest maintains its position near the surface, while the adjacent sediments, along with the source layer, subside - this is the process of downbuilding first documented by Barton (1933). Sediments accumulate around the diapir stem, at the same time that the diapir crest is a topographic bulge covered by a thin veneer of sediment or caprock. This thin sediment skin is continually breached by erosion or by the divergent flow of underlying salt at the diapir crest or by salt breakout. Thus sediment cannot accumulate atop the crest of a passive diapir. This lack of vertical constraint means there is no "room" problem, flanking strata are not rotated and faulted as they were in active diapirism. Faulting and folding are negligible in the sediments around passive diapirs apart from a narrow drag zone along their contacts (Figure 5.8; Jackson et al., 1994a,b).

If a passive pluglike stock is later stretched during ongoing regional extension, the adjacent country rocks may also stretch, creating normal faults next to the diapir. But more typically the stretching is taken up in the much weaker salt body, so that passive elongated salt walls under extension are widened via ductile flow of salt. The much stronger flanking overburden may remain grossly intact and virtually devoid of normal faults at the scale of seismic resolution. An apparent absence of normal faults in sediments adjacent to a passive diapir wall means that even large amounts of regional extension may remain hidden within the expanding and deforming salt wall. This cryptic extension can be inferred by accurate reconstruction of cross sections backward in time. But this process is far from straightforward and requires additional constraints, such as independent data on the original length of the section or original thickness of salt (Schultz-Ela, 1992).

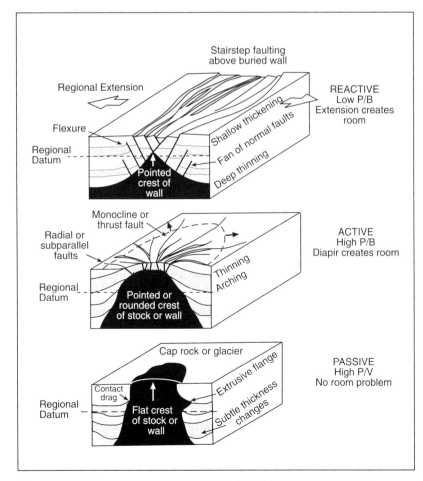

Figure 5.8. Three piercement modes for diapirs (black) and their characteristic structures. Regional datum (dashed line) is the base of the stippled layer. P, V and B refer to stresses due to salt pressure, salt viscosity and overburden brittle strength, respectively (after Jackson et al. 1994a).

Evolution of extensional diapirs

With moderate rates of extension a salt diapir will evolve from reactive to active to passive (Figures 5.8, 5.9). Slow reactive piercement is the first stage whereby in response to local thinning salt moves into overlying grabens and half grabens. The top of the diapir floats at its equilibrium level in response to pressure in the salt source layer. If regional extension ceases before the onset of active diapirism so does diapirism.

The next stage is rapid active diapirism, a stage that is largely independent of extensional forces. It can only occur when salt pressure at the crest of the diapir is sufficient to lift its roof. The third stage of diapirism, the passive stage, begins once salt reaches the landsurface. Without ongoing adjacent sedimentation to confine the spreading salt a passive diapir cannot rise, but it can widen if extension continues.

This sequence can be altered by changes in rates of extension, depletion of the source layer or changes in sedimentation rate, and so various

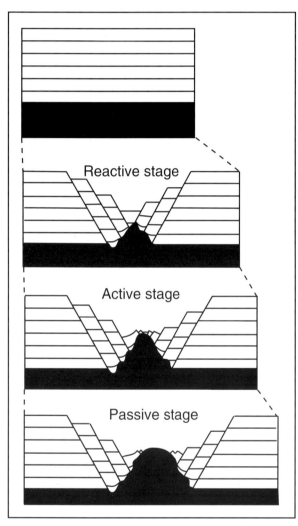

Figure 5.9 . Three stages of diapiric piercement through synkinematic overburden during thin-skinned extension. Extension is accommodated by both graben-related faulting and diapir widening (after Vendeville and Jackson, 1992a).

evolutional paths are possible (Jackson et al., 1994a):

• Although a diapir may appear to be initiated passively it most likely was reactive or active during its earliest stages. If the overburden was thin and unevenly deposited the reactive and active stages are very brief, and the strata faulted during these very early stages could be too thin to be resolved seismically.

• A diapir that appears to have been passive may have passed through numerous cycles. Each cycle could have comprised an episode of passive growth during slow sedimentation, followed by burial during a period of rapid sedimentation and terminated by a

brief active salt breakout. Each active stage removes the overburden roof by doming, local extension, entrainment, slumping and dissolution collapse.

• An active diapir can partially break through its roof but never emerge, as: 1) The source layer may become depleted, 2) the salt source may be channelled off (pirated) by an emergent segment of an adjacent salt wall, or 3) without erosion or differential loading an active diapir only partially penetrates all but the thinnest roofs before reaching equilibrium (Schultz-Ela et al., 1993).

Sedimentation and geometry

The interaction of upwelling salt and accumulating sediment on the diapir's flanks controls the geometry of a passive or downbuilding diapir. If the rate of salt rise is less than the sediment accumulation rate, then a passive diapir narrows upward as onlapping sediments encroach onto the diapir (R/A < 1; Figure 5.10a,b). Eventually all diapirs are buried by a slowing and eventual cessation in the rate of salt supply. This typically happens once the mother salt (source layer) is significantly thinned or depleted by salt withdrawal and overburden touchdown. Conversely, when a large volume of salt is supplied to the growing structure, the diapir rises faster than the surrounding sediment can aggrade. The resulting laterally unsupported salt ridge repeatedly overflows its unsupported margins, thus widening upward over time (R/A>1; Figure 5.10a,b).

Sedimentation rate also controls the overall planform (at-surface shape) of a passive diapir (Jackson et al., 1994a). A diapir's rate of salt flux increases with increasing width. At the diapir's widest part, salt flow rates are high relative to sedimentation rates. Conversely, contact drag at the extremities affects the narrow ends of an elliptical diapir more than at its wider parts. This slows the rate of rise at either end so creating relatively high aggradation rates. Thus salt spreads at the widest part of a diapir and retreats from its elliptical ends. Accordingly, an originally elongate salt wall evolves toward a string of pluglike passive stocks rising from the deeply buried reactive salt ridge. The number of exposed plugs declines as they exhaust their supply and become buried. The gross flow rate of salt in passive diapirs reflects the balance between driving forces, retarding forces and salt supply.

If, during extension, the sedimentation rate is high compared to the rate of regional extension, a reactive diapir evolves into a squat deeply buried salt roller bounded by a listric (upward steepening) fault. Conversely if the sedimentation rate is low compared to the rate of regional extension, a reactive diapiric ridge evolves into a tall salt wall or string of salt plugs that show steep or overhanging discordant contacts and shallowly buried crests (Vendeville and Jackson, 1992a). Likewise sedimentation rate also controls the form of advance of an allochthonous salt sheet (Figure 5.10c). Salt flats indicate times of low sedimentation, while salt ramps indicate higher rates of sedimentation.

Diapirism by differential loading

In addition to basin extension generating salt structures that pass from reactive to active to passive, the progradation of thick sedimentary wedges atop a salt bed can create lateral forces that also drive salt flow. This happens much in the same way that the vertical loading created by

pushing your thumb down on a tube of toothpaste drives the flow of paste from the tube. According to Worall and Snelson (1989) current salt flow in the Gulf of Mexico reflects sedimentary progradation driving the salt and other fluids into the basin, aided by gravity sliding and spreading.

In a series of scaled physical models Ge et al. (1997) illustrate the importance of progradation as a trigger for salt tectonics and formation of allochthonous sheets. They considered two end members. One group of experiments modelled salt with an underlying flat top to the basement (Figure 5.11), while another group considered a basement with a stepped top, as would occur in a block faulted basement terrane (Figure 5.12). Regional extension and contraction were excluded in their models.

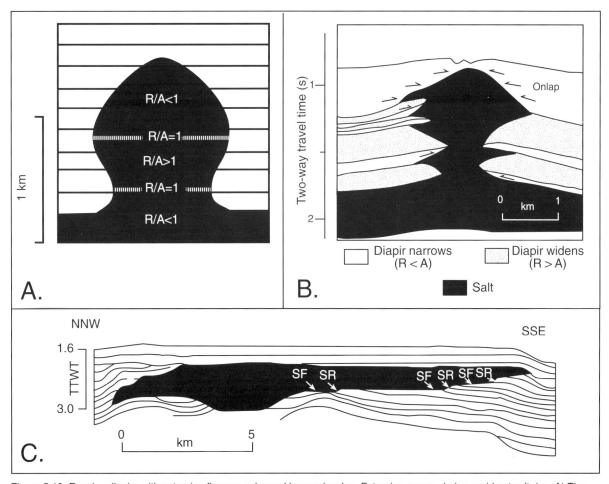

Figure 5.10. Passive diapirs with extrusive flanges onlapped by overburden. Extrusion occurs during rapid net salt rise. A) The shape of the diapir is controlled by the ratio of net diapiric rise R (incorporating the effects of dissolution), to aggradation rate (vertical accumulation of sediment). B) North Sea Diapir (after Jenyon, 1986). C) Mitchell dome, Louisiana, high R/A causes salt flats (SF), low R/A causes salt ramps (SR) (after Vendeville and Jackson, 1993).

The prograding wedges above a tabular, buoyant salt layer that was underlain by a flat basement top, expelled the salt basinward in a predictable sequence of structures. From proximal to distal they were: (1) sigmoidally distorted initially planar wedges, (2) relict salt pillows and salt welds, (3) basinward-dipping expulsion rollover and crestal graben, (4) rollover syncline, (5) landward-facing salt-cored monocline, and (6) distal inflated salt layer (Figure 5.11a). This deformation zone amplified and advanced basinward together with the prograding front. Complete evacuation of the salt left a basin-propagating salt weld below the prograding wedge; incomplete evacuation left salt pillows as residual highs, but no diapiric salt structures were formed.

In contrast, progradation over a buoyant salt layer that was underlain by landward-facing basement steps (Figure 5.12a), initially formed a broad anticline where salt flow was restricted across each basement step (Figure 5.12b). Early on, differential loading by the prograding wedge expelled the salt basinward in an identical fashion to that which occurs in a basin with no basement steps. Then the lateral escape of the salt was altered as it was slowed across the friction point created by the basement step. The salt backed up to form a broad asymmetric salt cored anticline. Distal aggradation continued to pin the anticline and enhance differential loading. Next the anticline actively pierced its crest, which had been thinned by faulting and erosion (Figure 5.12c). Thereafter, the diapir grew passively, locally sourcing allochthonous salt sheets. This deformation cycle repeated over each basement step so that the age, amplitude, complexity and maturity of salt-related structures decreased basinward (Figure 5.12d-f).

As each allochthonous salt sheet was buried and evacuated by sediment loading, arcuate peripheral normal faults formed along the sheet's trailing edge, detached wrench faults formed along its lateral edges, and active piercement at its leading edge allowed the sheet to break out and climb stratigraphic levels. This process formed a multi-tiered complex of salt sheets that migrated basinward with time. Ge et al.'s (1997) modelling clearly shows that immense landward-dipping (counter-regional) pseudofaults can arise entirely by salt expulsion rather than regional extension. Their physical models clearly show that progradation alone can generate salt anticlines that mature into salt walls, stocks, glaciers or linked, multi-tiered complexes of evacuated sheets and salt welds The prograding sediment

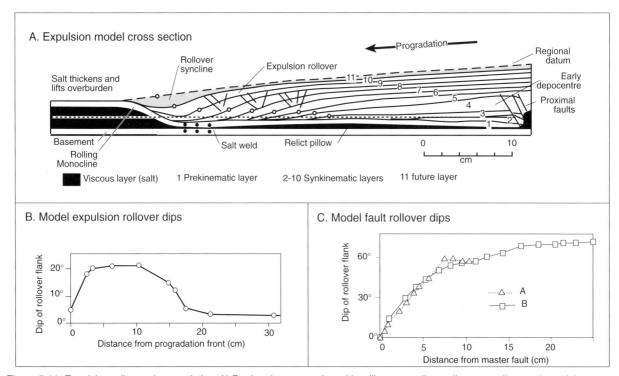

Figure 5.11. Expulsion rollover characteristics. A) Regional cross section with rolling monocline, rollover syncline, and expulsion rollover all migrating basinward with the basinward progression of the progradation front. Dashed horizontal line represents the top on an initially tabular salt body. Open black circles show positions of data points plotted in Figure 5.12b. B) Rollover plots of maximum dip associated with model illustrated in 5.12a. C) Rollover plot of model experiments A from experiment model 188 listed in Ge et al. (1997), B from experiment E30 in McClay (1990) (modified from Ge et al., 1997).

wedge expels salt basinward regardless of density relationships. In such situations there is a relationship between underlying basement discontinuities and the location of salt diapirs and allochthons. A result that is not obvious in purely extensional regimes where the zones of diapirism tend to begin in zones of overburden thinning that bear no relationship to subsalt discontinuities.

In reality salt structures can probably form both under extensional and progradational settings. Extension is probably the dominant mechanism for triggering diapirism in deeply buried salt bodies, while progradation is commonplace as a driving mechanism whenever salt is near the surface. This can happen syndepositionally with little or no burial of the salt, or it can occur much later once deeply buried salt has once again reached the surface in the

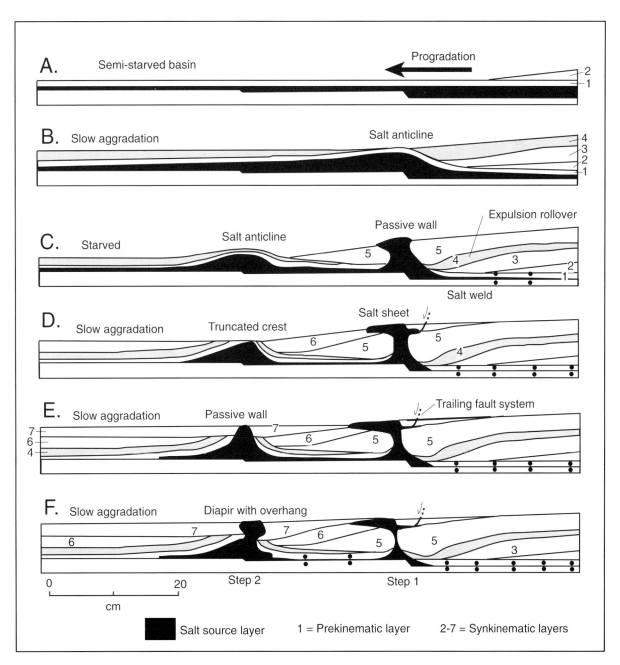

Figure 5.12. Evolution of salt structures during progradation atop a salt basin that is underlain by a stepped basement (modified from Ge et al., 1997).

form of a passive diapir. It also means that diapirs, initially formed by early progradation of a continental shelf or platform, can be later reactivated or overprinted during extensional or compressional regimes.

This leads to an obvious problem in distinguishing between expulsion rollover generated by progradation and a fault rollover in the hanging wall of a listric growth fault generated during extension. Ge et al. (1997) clearly show how dip analysis of the rollover plot can be used to help differentiate the two styles (Figure 5.11b, c). They plot the maximum dip of each sediment wedge in the expulsion rollover (open circles in 5.11a, b) versus distance from the present progradation front. Maximum dip was measured on the basinward dipping flank in the rollover syncline. The rollover plot is shaped like an asymmetric hill with dips increasing rapidly away from the front to a maximum and then decreasing less rapidly. The maximum dip corresponds to where the thickest salt was expelled. The oldest wedges preserved the initial depositional slope (5° in modelled system) and planar shapes, even though they were folded and then unfolded by the passage of the structural monocline that rolled through them as the salt evacuation front passed by. The plot of a fault rollover in the hanging wall of a simple listric fault is quite different and shows no decrease in dip away from the present progradation front (Figure 5.11c). Dips of the hanging wall cut off abruptly and then gradually increase to reach a maximum value away from the surface trace of the listric fault. The constant maximum dip away from the fault corresponds to flattening of the underlying fault. This type of analysis can only be carried out on synkinematic strata, which are distinguished from prekinematic strata as they show local lateral variations in thickness. If a listric fault has multiple flats and ramps, the prekinematic strata generate a tail of data points that are not diagnostic and should not be used. See Ge et al. (1997) for a full explanation of this methodology and its application.

Once salt is extruded onto the seafloor as a salt glacier, it can flow in any direction depending on the local slope; it can even flow landward into an adjoining rollover syncline (rim syncline). Over time the breakout sites, where the salt climbs the section and breaches to the surface, will ultimately shift basinward as salt is squeezed by the prograding basin fill. Regionally, the site of active diapirism in the Gulf of Mexico has been shifting basinward since the Jurassic deposition of the Louann Salt. The salt in the breakout may have moved more tens of km laterally and climbed as a series of tiers through kilometres of stratigraphy. As each newly emplaced salt sheet is loaded by prograding sediment, it carries its overburden basinward

by gravity spreading. While a sheet evacuates into a new stratigraphic level it also generates halokinetic profiles atop it. The tops of these diapirs in older, more deeply buried tiers, were what was interpreted as deeply connected primary salt structures in older (pre 1980s) seismic reconstructions.

Salt sheets or allochthons

Allochthonous salt sheets overlie younger strata as flowing salt beds that are typically no longer connected with the mother salt bed, yet they have areas measured in hundreds of square kilometres and thicknesses up to several kilometres. In contrast, a salt structure that extends continuously down several kilometres to the primary source layer is defined as autochthonous salt. Whether one views diapirism in terms of extension or in terms of differential loading, our increasing recognition of allochthonous salt sheets as multiple tiers within sedimentary fill of a halokinetic basin has created a Kuhnian revolution in our understanding of salt flow.

Allochthonous salt sheets are now known to characterise the present day continental margins of offshore Brazil, the Gulf of Mexico, West Africa, the North Sea and the Red Sea. Their recognition in petroleum provinces has opened up possibilities of prodigious subsalt hydrocarbon reservoirs, particularly in the continental slope and rise of the Gulf of Mexico, where a vast complex of salt allochthons increases in size and degree of coalescence westward (Montgomery and Moore, 1997; Liro, 1992; Wu, 1993). Many of the at-surface salt sheets can be seen to have partly or completely coalesced into composite diapirs, called canopies (Jackson et al., 1990). The leading edge of a canopy complex that now forms the Sigsbee Scarp at the base of the continental slope in the Gulf of Mexico has climbed 8 km vertically and up to 80 km laterally (Figure 5.1).

Major emplacement mechanisms for allochthons are varied and include: recumbent folding, deep intrusion, shallow intrusion and surface extrusion (Figure 5.13; Schultz-Ela and Jackson, 1996). In all categories, a massive seaward flow of salt is envisaged as a major part of the emplacement process. Ewing and Antoine (1966) envisaged the allochthonous Sigsbee Scarp was the basinward flowing edge of an autochthonous diapir (Figure 5.13a). We now know their original interpretation, limited by the state of the seismic art at that time, was not correct in terms of its vertical connectivity to the mother salt bed (Figure 5.1). Amery (1969) discovered that the Sigsbee Scarp was the leading tip of a wedge of allochthonous salt. Jackson and Cornelius (1987) modelled allochthonous salt sheets as flowing salt bodies ramping up to the nearsurface from a buried sub-salt step. The wedge then spread rapidly under gravity as a recumbent fold anticline. The whole system advanced like a tank track with a core of allochthonous salt and a deformed sedimentary overburden (Figure 5.13b).

D'Onfro (1988) modelled allochthonous salt sheets as dykes and sills that were intruded in areas of local compression, such as the toe regions of growth faults (Figure 5.13c). Podladchikov et al. (1993) modelled asymmetric salt sheets as intruding laterally from a diapiric salt mass at depths of approximately 1 km depth. This level, according to the authors, lies just above an abrupt thousand-fold downward increase in the viscosity in what otherwise are uniformly viscous sediments (Figure 5.13d). Amery (1969,1978) envisaged shallow allochthonous intrusion of the spreading salt layer as occurring beneath a thin sediment roof that was no more than 200 - 300 m thick. He visualised the salt as coming from an autochthonous diapir source. Nelson (1991) stated that most salt sheets had been emplaced shallow at depths of not more than 100m (Figure 5.13e). Talbot (1993) pointed out the resemblance of salt sheet profiles to viscous fountains spreading laterally into unconsolidated surface muds (Figure 5.13f).

Humphris (1978) was the first to postulate extrusion at the front of a vast thrust nappe that was riding basinward on the allochthonous sheet. Jackson and Cornelius (1987) hypothesised that extrusion was possible if progradation was slow. Vendeville and Jackson (1992a) modelled allochthonous salt sheets as forming under little or no roof. Their concept of extrusive salt sheets on the sea floor under little or no sediment cover is now the most widely accepted mechanism for allochthon emplace-ment (Figure 5.13f; Fletcher et al., 1995; McGuinness and Hossack, 1993; Schultz-Ela and Jackson, 1996; Schuster, 1995).

Salt sheets form by the geologically rapid widening of the crest of passive diapirs (Jackson et al., 1994b; Fletcher et al., 1995; Ge et al., 1997). As sediments are deposited around an extruding salt sheet the salt glacier climbs up across the newly deposited sediment. Thus a salt glacier

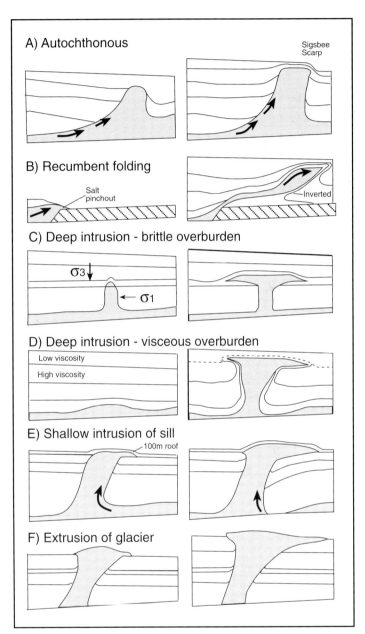

Figure 5.13. Hypotheses for the origin of allochthonous salt sheets. See text for explanation (modified from Schultz-Ela and Jackson, 1996).

climbs over progressively younger sediment as it moves away from its feeder. The slope at the base of the salt sheet is thus a function of the relative rate of sedimentation. Rapid widening and a flat basal slope (salt flat) are most likely to occur during unusually slow sedimentation (Figure 5.10c). A more steeply inclined base (salt ramp) indicates a slow advance or more rapid sedimentation. As salt glaciers spread they also laterally transport and deform any roof.

During times of surface extrusion a salt sheet forms a bathymetric high with several hundred metres of relief above the surrounding sea floor (Fletcher et al., 1995). A feedback occurs between the rate at which salt is pumped to the surface and the rate at which seawater dissolves the salt. The rate of salt pumping is thought to be many times greater than the sedimentation rate during peak growth rates. Salt can be protected from dissolution by a combination of: 1) sedimentary veneer, 2) an insoluble residue (or caprock) of anhydrite/gypsum, and 3) an overlying layer of saturated pore brines where circulation of sea water is locally restricted. If continuously deposited,

a low-permeability sedimentary roof, alone, can protect underlying salt, even if the roof is continuously breached by deformation and erosion.

Ancient equivalents of salt glaciers have not been widely recognised in the rock record. Now, with our improving understanding of salt tectonics and the significance of widespread salt evacuation, ancient counterparts are beginning to be documented. For example, Vila et al. (1996a,b) showed that the large deformed saline masses of Triassic rocks in the Ben Gasseur-El Kef anticline area of Northern Africa ($\approx$ 165 km^2) are former allochthonous sheets, now sandwiched in middle Albian sediments. They have, both underneath and on top, two originally horizontal sedimentary boundaries. Several Albian reefs located along the upper limit of the salt mass infer that a large lenticular saliferous body of Triassic salt (area $\approx$ 250 km^2) was emplaced on the Albian seafloor as a large submarine "salt glacier". Emplacement occurred on at the continental slope during a time of active basin rifting. Their interpretation required the removal, by section straightening, of two Tertiary folding events.

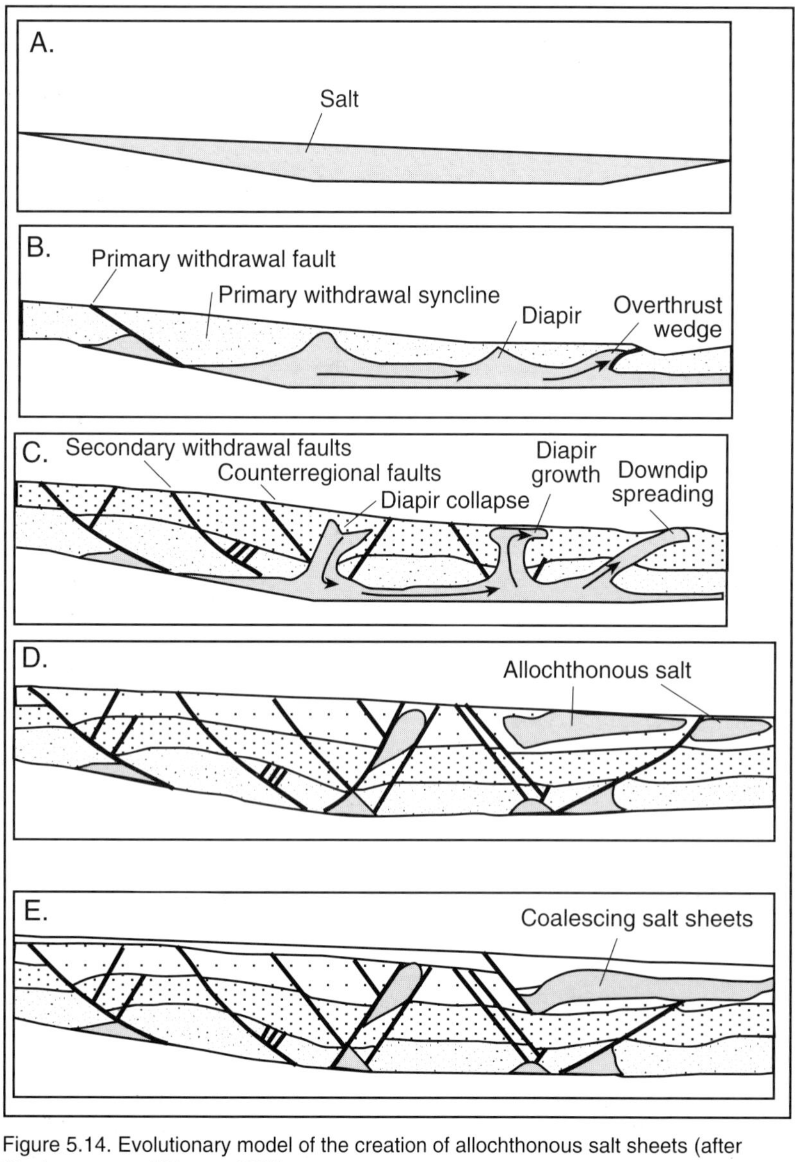

Figure 5.14. Evolutionary model of the creation of allochthonous salt sheets (after Montgomery and Moore, 1997).

Deformation of salt sheets

The emplacement of allochthonous salt sheets in the Gulf of Mexico clearly illustrates how salt features grow and deflate as the basin loads with sediment (Montgomery and Moore, 1997). There are two major phases of emplacement in the Gulf of Mexico: 1) initial loading of a Jurassic salt mass, which generates zones of salt withdrawal and diapir formation, followed by downdip gravitational spreading into sheets and tongues (Figure 5.14a-c); and 2) a later phase or phases of suprasalt loading, which creates complex growth faults of both regional and counter-regional faults (Figure 5.14c-e). This second stage occurs in levels of the stratigraphy that lie well above the primary salt level. It is associated with the spreading and/or coalescence of a number of salt sheets that are in turn evacuated by later sediment loading (Figure 5.15). The salt sheet is often segmented into smaller second generation structures as extension, extrusion and overburden loading cause the top and bottom contacts of the salt sheet to come closer and closer together until it becomes a salt weld.

A salt weld forms when the salt sheet or primary bed is so thinned by withdrawal that salt is no longer resolvable on seismic profiles (Figure 5.15; Ge et al., 1997; Jackson et al., 1994a,b; Schuster, 1995). The weld can still retain some salt or consist of a brecciated insoluble residue. Salt welds are also described as overburden touchdowns, or simply as touchdowns. A salt weld is usually, but not always, marked by a structural discordance. Another distinctive feature is the common presence of a structural inversion above the salt weld, which often forms some sort of broad anticlinal structure (e.g. turtle and mock-turtle structures described later).

A primary weld joins strata originally separated by gently dipping autochthonous bedded salt. Secondary welds join strata originally separated by steep sided salt diapirs (walls or stocks). Tertiary welds join strata originally separated by gently dipping allochthonous salt sheets (Jackson and Cramez, 1989). Each side of the weld can slip relative to the other, as it can in a more steeply inclined fault that is lubricated by salt. The latter structure is termed a fault weld. A fault weld is a fault surface or fault zone joining strata originally separated by autochthonous or allochthonous salt. Thus a fault weld is a salt weld along which there has been faulting or shearing.

Intrasalt minibasins above tabular salt allochthons typically entrain sediment fill that indicates evolution from slope basins to shelf basins. They are typically bound by arcuate growth faults formed largely by downslope salt withdrawal rather than by extension (Figures 5.15, 5.16). The sheet itself is segmented into discrete salt structures separated by salt welds and fault welds. Initially the allochthon acts as

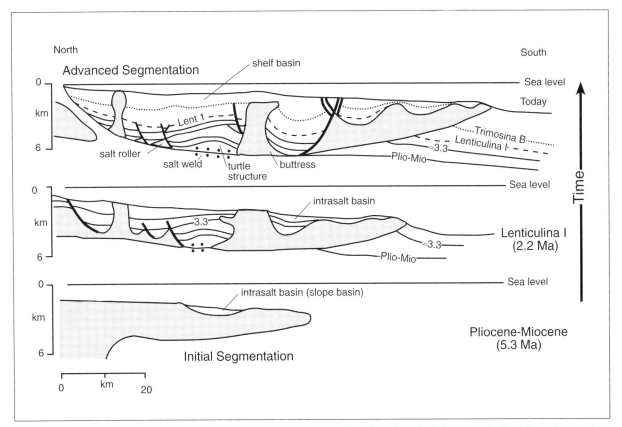

Figure 5.15. Schematic of evolution of salt structures driven by burial and loading of an allochthonous salt sheet. Based on eastern Green Canyon, Gulf of Mexico. At 5.3 Ma the salt sheet was extruded. From 5.3 to 3.3 Ma the ongoing basinward extension of the sheet promoted differential loading within half grabens. From 3.3 to 2.2 Ma the largest intrasalt basin displaced all the salt below it to create a salt weld and a structural inversion that was created by a turtle structure. Subsequently the sheet was further deformed as it continued to flow basinward and climb the stratigraphy (modified from Seni and Jackson, 1992).

a deflating cushion that accommodates sediment deposited in intrasalt basins. After time, loading or extension, the original salt sheet transforms into thin but broad welds, which then act as major subhorizontal detachments for deep growth faults whose slip creates space for further sedimentation. Second cycle salt structures rise from these deformed sheets. Substantial thicknesses of sediment can accumulate in intrasalt basins; in the Green Canyon region of the Gulf of Mexico the intrasalt basins contain more than 5 km of Neogene and Quaternary sediment fill (Figure 5.15).

Salt sheets can spread laterally as they deform until they eventually coalesce with neighbouring salt sheets. Collectively they create a salt canopy, a formation readily observed in the northeastern Gulf of Mexico (Figure 5.14; Talbot, 1993; Wu, 1993). There individual lobate sheets coalesce westward into an extensive salt canopy whose leading edge is the Sigsbee Scarp. Adjoining sheets merge into canopies along salt sutures. The formation of a more or less continuous salt canopy entails the upward and lateral movement of vast quantities of salt. When the canopy roof is thin, the adjoining sheets are salt glaciers. When a thick roof confines the adjoining sheets, the roof contracts into a cuspate syncline associated with small thrust faults and kinks (Jackson and Vendeville, 1994b).

Another type of syncline forms between sheets in a completely different way, where a once continuous salt sheet is segmented by intrasalt minibasins. This syncline can also contain thrusts, but they are the result of slumping from the steepening sides of the minibasin. These synclines steepen over time but do not contract laterally. The two types of syncline also differ in plan view. They are cuspate above salt sutures, versus bowl-shaped in minibasins above segmenting salt sheets.

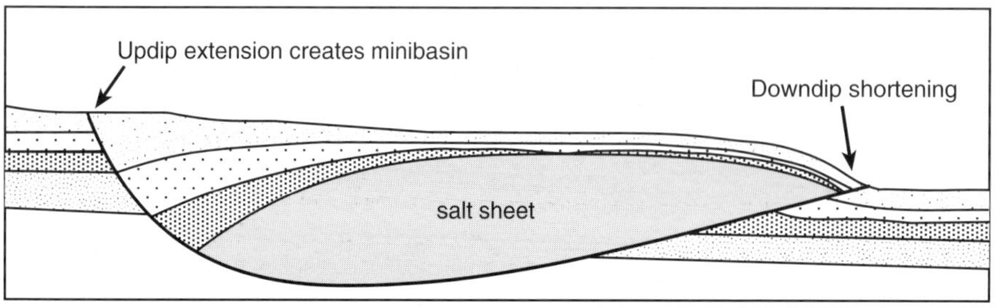

Welds, regional detachments and growth faults

Many regional growth faults sole out onto residual fault welds many tens of kilometres in length. The link between growth faults and salt welds is likely to be an inherited relationship (Jackson et al., 1994b; Fletcher et al., 1995). A thick salt sheet that was present before welding also serves to decouple faults above and below it. This decoupling causes overlying faults to sole out into the sheet. The sheet then evolves into a weld via removal of the salt.

The burial of a salt sheet beneath loading sediment typically follows the pinch-off of the feeder diapir (Fletcher et al., 1995). This allows the former sea floor high created by the glacier to relax, especially at the updip end of the sheet. Sediment can then prograde across the sheet beginning at its updip end (Figure 5.16). Even during burial of its updip end, a salt sheet typically retains several hundred metres of relief on the downdip end. Thus the downdip sheet continues to advance as a composite salt-sediment glacier. The sediments atop the glacier are deformed along with the flowing salt. The movement can create extensional breakaway zones at the updip end. These breakaways may initiate many of the fault bounded intrasalt minibasins found at the updip end of salt sheets. Ongoing slippage creates a growth fault rollover anticline and salt continues to extrude at the downdip end.

This same salt can recycle multiple times into ever higher stratigraphic levels. Each time salt is transferred, a largely welded smear of salt with vestigial salt structures is left behind at the lower levels. Each recycling also means substantial volumes of salt are lost via dissolution. Thus the salt cannot recycle indefinitely and the driving mechanism is lost once a salt weld or overburden touchdown is created beneath the intrasalt basins.

Once a salt sheet is isolated from its feeder stock it advances further only by cannibalising its trailing edge (Figure 5.16; Fletcher et al., 1995; Schuster, 1995; Jackson and Vendeville, 1994b). If the salt is inefficiently withdrawn from the trailing edge, the salt becomes sequestered into residual smaller bodies and the overall structure is called a fault-segmented sheet. In the Gulf of Mexico this is sometimes also called a "roho" system (Figure 5.17a).

Figure 5.16. Schematic of a composite salt-sediment glacier. Sediments on top of the salt sheet are carried along with the underlying salt, producing a pull-apart at the updip end and contractional structures at the downdip toe (after Fletcher et al., 1995).

Roho systems are characterised by major, listric, down-to-the-basin growth faults, which in the Gulf of Mexico sole into intra-Tertiary salt evacuation surfaces (tertiary salt welds). During its progradation, the rear (landward) margin of the salt sheet gradually segments and deep withdrawal structures are enhanced. At shallower levels, vestigial salt forms a train of sequestered salt walls formed by the partitioning of a formerly continuous sheet of salt. Each residual salt structure is typically in the footwall of a normal fault or is separated from its neighbour by a syncline. Each residual wall is generally overlain by a splay of extensional faults, which are commonly listric.

If salt is more efficiently withdrawn from the deflating trailing margin then a stepped counter-regional system results (Figure 5.17b). Strata above the deflating salt sheet subside to form a shallow flat step along the weld of vanished salt. The step resembles a landward dipping growth fault. However, the step can be merely a pseudo-fault, like the contiguous deeper step created by upward evacuation of the original diapir. In three dimensions the shallow step fault is the axis of a partly evacuated sheet that is shaped like a tongue with curled up lateral margins. Along the deepest axis of the tongue lies the deepest, most evacuated part of the salt sheet below the thickest overburden.

These two structures —the fault-segmented sheet (Roho) and the stepped counter regional system— are end members of an evacuation series where the typical structures are hybrids. It is possible that as the salt climbs the stratigraphy the same salt body can create both roho and counter-regional systems (Figure 5.17c). Initial salt emplacement in both systems occurs as salt fountains and glaciers located at or just below the seafloor. Downbuilding is the dominant process driving salt growth (Rowan, 1995). Once salt is near the surface, reactive and active diapirism is rare and confined mostly to the landward edges of salt sheets where the overburden is extending.

Extension, falling diapirs and turtle structures

Once a salt bed or salt sheet is emplaced, salt-lubricated extension can initiate and maintain diapirism throughout the reactive, active and passive stages. While connected to

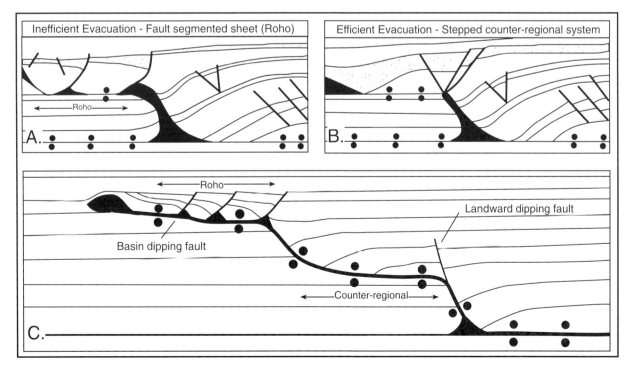

Figure 5.17. Efficiency of evacuation is reflected in the suprasalt structuring. A) With inefficient salt evacuation a fault segmented (Roho) system forms. B) With efficient evacuation a counter-regional system is created. C) Shows how the two systems can be created by the same salt body. Circles show position of salt weld. See text for explanation, black circles define salt welds (in part after Jackson and Vendeville, 1994b; Rowan, 1995).

an adequate salt supply passive diapirs can widen between separating blocks of overburden while still maintaining their crest at or near the surface (Vendeville and Jackson, 1992b). That is, once it attains its maximum height, a reactive/passive diapir can continue to widen by viscous flow. Salt inflow fills an ever-widening gap between the faulted extending blocks of overburden that are gliding apart on the salt source layer (Figure 5.18 a, b). Eventually the salt supply becomes restricted and ultimately depleted so that the diapir crest sags even as the diapir body continues to widen (Figure 5.18c).

Turtle-structure anticlines with crestal keystone grabens are another result of source-layer depletion. These anticlines were conventionally attributed to the evolution of a pillow as it contracted into a diapir (see Warren 1989 for discussion), but it is now known they form during the falling diapir stage (Figure 5.18c-e). Once the salt supply to a diapir is restricted, more and more salt from the surrounds is withdrawn into the falling diapir (Figure 5.18c). As this occurs the region of overburden touchdown forms and then widens beneath the rim syncline

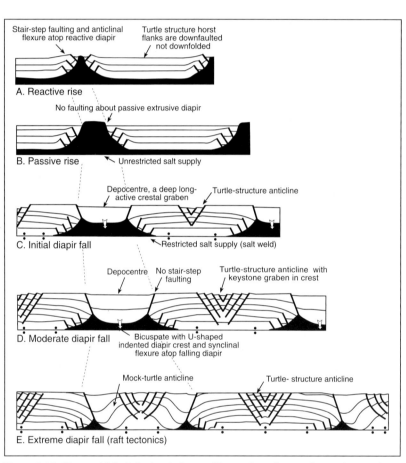

Figure 5.18. Rise and fall of diapirs during synkinematic sedimentation and extension. Dashed line implies the various stages are not necessarily separate in time, but can be found along a single salt wall (modified in part from Vendeville and Jackson, 1992b).

sediments. Over time, the ongoing widening of this zone of salt withdrawal and touchdown causes the sediments of the rim syncline to invert into a turtle structure anticline, often with a keystone graben in the anticlinal crest (Figure 5.18d). The touchdown and inversion of the rim syncline sediment pile can create a topographic high on the seafloor. Sediments deposited atop the anticlinal crest may be higher energy sands and carbonate build-ups, as has occurred in the reefal carbonate reservoirs that characterise buildups in the James Limestone in the Gulf of Mexico. Such depositionally enhanced potential reservoir intervals can be thought of as self-reservoiring, self-closing and self-draining (see Warren 1989, for discussion).

As the diapir crest continues to sag, the crest inverts from a topographic bulge into a subsiding depotrough or graben, which is rapidly filled with sediment. This creates a sediment thick atop the diapir (Figure 5.18c). Slippage along the bounding faults of this deepening graben continues, but no new faults generally form within the

graben. Such a crestal graben lacks the stair-step faults that characterise the smaller and shallower crestal graben of a reactive diapir. Adjoining the indentation are hornlike cusps of salt below each bounding fault of the graben.

As extension continues after all salt supply to a passive stage diapir is cutoff, overburden touchdown occurs in the position of the former passive diapir. The falling diapir overburden segments into two flanking triangular relics flanked by glide blocks composed of older overburden (Figure 5.18e). The diapiric relics are now separated by deep, wide depocentres filled with younger sediments. If extension continues, the flanks of these diapiric relics subside until they too touchdown. Thus ongoing extension and salt depletion causes even the young depocentre to subside until it too eventually inverts into a crestally faulted arch, called a mock-turtle anticline.

But what about the third dimension along the strike of a salt wall or diapir crest. In many diapiric settings, such as the North Sea and the Gulf of Mexico, salt walls pass along strike from a reactive rising section of the wall to a falling section of the wall marked by a thickening crestal graben. In other words stages a-e in Figure 5.18 can occur along strike within a single salt wall (joined by dashed line in figure). Salt tends to move into the rising (reactive-passive) sections of the wall and away from subsiding depressions below the crestal grabens, where subsidence is accommodating regional extension. As salt flows into the intervening culminations along the wall they evolve into increasingly pluglike, rounded diapirs. If extension continues, these rising diapirs, which are commonly emergent, continue to widen and accommodate the regional extension without faulting. With sufficient extension, these diapirs ultimately exhaust their salt supply so that they too must subside and also become buried.

Prior to the work of Vendeville and Jackson on falling diapirs, many of the crestal structures produced by diapir fall were attributed to salt dissolution or forceful active intrusion at the crest; all three possibilities should be evaluated in any areas of crestal grabens or remnant salt cusps/rollers. The long standing bias toward salt dissolution causing crestal grabens is partly because criteria for interpreting the origin of crestal grabens are usually neither clearly stated nor consistently applied (Vendeville and Jackson, 1992b).

Falling diapirs and raft tectonics

In some areas of falling diapirs the overburden becomes stretched to two or three times its original length by normal faulting, but the sub-salt basement is not involved. This prodigious extension of the overburden over a non-deforming basement is enabled by a weak intervening ductile layer or décollement of salt. This is an extreme form of thin-skinned extension or gravity sliding where rigid sediment blocks slip downslope atop a décollement layer of evaporite or shale and is described as raft tectonics (Spathopoulos, 1996). Rafts are allochthonous fault blocks that are so far separated that they are no longer in mutual contact (Duval et al., 1992). With less extension the fault blocks are called pre-rafts. Pre-rafts remain in mutual contact across the fault so that a hanging wall still rests on its original footwalls. In contrast rafts are so far separated that hanging walls no longer rest on their original footwalls. The downslope rafting movement produces listric faults and sediment-filled grabens atop the deflating salt. These sediment-filled grabens can be substantial structures; the Quenguela Graben in the Kwanza Basin measures 20 km wide and 90 km long, the nearby Gaivota Graben is 15-20 km wide and 225 km long (Figure 5.19). Similar large

rafts characterise blocks of Zechstein sediment in the NW Permian Basin of the North Sea (Clark et al, 1998). In this case, the blocks of sediment are composed of platform anhydrite sliding into the basin atop a halite lubricated sole.

Sedimentation sequences deposited between rafts of strata sliding downslope are now well understood. With initial extension between two overburden blocks, a salt wall first rises passively to fill the increasing gap (Figure 5.19a). Once the supply of salt can no longer keep up with demand, the diapir starts to fall. The crest of the sagging diapir becomes a subsiding trough that fills rapidly with sediment

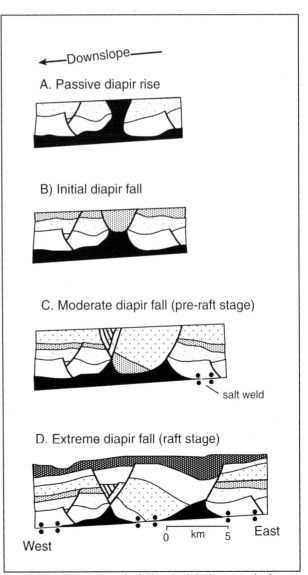

Figure 5.19. Restoration of a falling diapir in the zone of raft tectonics of the Kwanza Basin, Angola (after Schultz-Ela, 1992 and based on a seismic line in Duval et al., 1992).

173

(Figure 5.19b-c). While stretching and décollement are active the thickness of the evaporite layer is reduced both by downslope gravity drainage of the salt and ongoing dissolution. Once most of the salt is removed from under the sliding blocks, they ground or weld on the presalt layers and the downslope movement of the blocks stop (Figure 5.19d).

Stretching or extension that drives raft tectonics can usually be attributed to gravity gliding or gravity spreading. Gravity gliding is the translation of fault blocks down a gentle slope, driven by the downslope shear stress component of gravity on a tilted mass. Gravity spreading is the vertical collapse and lateral spreading of a rock mass under gravity. Both mechanisms require the creation of substantial lateral space into which the overburden can expand during stretching. This space is generated by: 1) formation of a downdip fold or thrust belt, 2) down dip displacement of autochthonous salt into allochthonous salt sheets that climb the stratigraphy, and 3) generation of new oceanic crust by a spreading ridge with allochthonous salt allowing overlying blocks to slide (Duval et al., 1992).

Thus salt basins characterised by raft tectonics can usually be divided into two different stress regimes (Spathopoulos, 1996): an updip extensional regime located under the shelf and upper slope where salt décollement has facilitated the

break-up of overlying sediments into raft blocks, and a downdip compressional or contractional regime typically under the lower continental slope where the up dip extension is taken up in a fold or thrust province.

Rafting requires a specific stress field. The regional σ_1 maximum effective stress is subvertical; σ_3 is subhorizontal and oriented downslope; σ_2 is oriented parallel to the regional strike. The same regional stress field also account for the grabens, growth faults, depocentres, rafts and salt walls in a halokinetic basin (Jackson and Talbot, 1991).

Contractional salt tectonics

Contractional salt tectonism is produced either by convergent (including transpressive) plate tectonics or is restricted to the narrow seaward portions of divergent continental margins where downslope gravity-sliding stacks and thrusts blocks against each other (Figure 5.20). Contractional salt tectonics require either the minimum principal stress (σ_3) to be vertical (creating shortening) or the intermediate principal stress (σ_2) to be vertical (resulting in trans-pression; Jackson and Talbot, 1994).

Contractional salt basins can be classified into compression belts dominated by regional gravity sliding, or into regions dominated by thin skinned compression or by basin

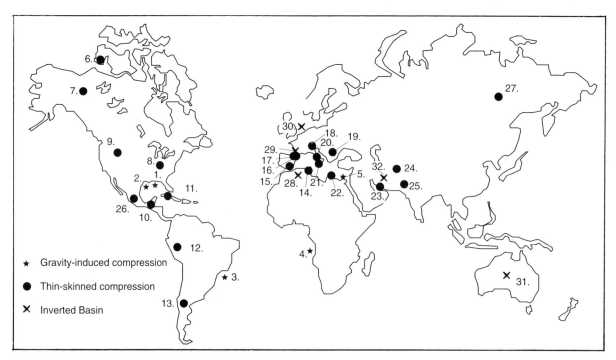

Figure 5.20. Worldwide occurrences of compressive deformation that involves salt sequences (after Davis and Engelder, 1985; Letouzey et al., 1995). Numbers refer to localities listed in Table 5.2.

inversion (Figure 5.20). The resulting regional shortening can be thin skinned (only the cover contracts) or it can involve the basement. There are three main structural settings within salt basins where compression is the dominant mechanism acting on the salt and adjacent sediment (Figure 5.21; Letouzey et al., 1995):

- shortening at the toe of prograding and gravity-gliding systems (thin-skinned; Figure 5.21a),
- thin-skinned deformation in fold and thrust belts (Figure 5.21b),
- intracratonic inverted basins with or without wrench components (thick-skinned with basement involved in the compression; Figure 5.21c).

Shortening with gravity gliding

Rapidly prograding shelf-slope systems with regions of upslope extension are not uncommon on passive continental margins (Letouzey et al., 1995). Shortening typically occurs at the toes of such systems. In cases of extreme extension, raft tectonics occur in the upslope position. Compressional foot of slope systems that involve salt décollement include: the Perdido and Mississippi Fan foldbelts in the Gulf of Mexico (Schuster, 1995), the Campos and Santos Basins of offshore Brazil (Cobbold et al., 1995), offshore west Africa (Duval et al., 1992), the eastern Mediterranean adjacent to the Nile Deep Sea fan (Sage and Letouzey, 1990) and offshore Yemen (Heaton et al., 1995). In all these regions contraction is driven by gravity gliding associated with progressive tilting of the margin in combination with gravity spreading of the salt layer generated by differential loading associated with the prograding shelf margin (Figure 5.21a). The salt in all the areas, with the possible exception of the Mediterranean, was precipitated in early post-rift time so that consequent thermal subsidence generates a basinward tilt of the margin (less than 1°-3°). The tilt is a contributing factor to the décollement but is not the driving mechanism. That is

Shortening related to gravity gliding at the toe of massively prograding systems		
1. Mississippi Fan	Gulf of Mexico	Middle Jurassic
2. Perdido	Gulf of Mexico	Middle Jurassic
3. Campos, Santos	Brazil	Lower Cretaceous
4. West Africa	Angola, Gabon	Lower Cretaceous
5. Nile delta	Mediterranean	Upper Miocene
Thin skinned deformation in fold and thrust belts		
6. Parry Islands	Canada	Ordovician
7. Franklins	Canada	Cambrian
8. Appalachians	USA	Silurian
9. Wyoming	USA	Upper Jurassic
10. Guatemala	Central America	Middle Jurassic or Cretaceous?
11. Cuba	Caribbean	Middle Jurassic
12. Santiago	Peru	Permian
13. Neuquen	Argentina	Jurassic or Cretaceous
14. Rif-Tellian	Algeria, Morocco and Tunisia	Triassic
15. Betics	Spain	Triassic
16. Iberics	Spain	Triassic
17. Pyrenees	France and Spain	Triassic
18. Southern Alps	France	Triassic
19. Carpathians	Romania	Lower Miocene
20. Apennines	Italy	Triassic
21. Messinian	Eastern Mediterranean	Upper Miocene
22. Mediterranean Ridge	Eastern Mediterranean	Upper Miocene
23. Zagros	Iran	Infracambrian
24. Tadjiks	CEI	Jurassic
25. Salt Range	Pakistan	Infracambrian
26. Sierra Madre Oriental	Mexico	Jurassic
27. Verkhoyansks	CEI	Cretaceous
Intracratonic inverted basins		
28. Atlas	Algeria, Morocco, Tunisia	Triassic
29. Aquitaine Basin	France	Triassic
30. Southern North Sea		Permian
31. Amadeus Basin	Australia	Neoproterozoic
32. Central Iran	Middle East	Eocene-Miocene

Table 5.2. Occurrences of salt involved in compression. Numbers refer to localities on Figure 5.20 (after Letouzey et al., 1995).

supplied by differential loading tied to the rapid progradation of the shelf margin.

Because the extension domains tends to lie in a proximal position within a sedimentary basin they are, to date, better studied. Structures characteristic of the extensional domain include: salt rollers in the footwalls of listric normal growth faults, salt walls with triangular cross sections lying between intersecting conjugate normal faults, turtle anticlines and salt welds (Cobbold et al., 1995). Contractional or compressional domains are less well studied as they tend to lie in deeper water far out into the

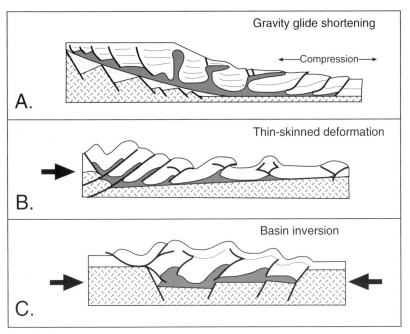

Figure 5.21. Compression tectonics. a) Gravity glide shortening at the toe of the continental slope. The basement is not involved in the deformation. b) Thin skinned deformation in front of the fold and thrust belt where subduction is the driving mechanism in the inner fold belt. C) Intracratonic inverted basin. Driving mechanism is regional contraction, which is orthogonal or oblique to the preexisting graben. Basement, salt and overburden are all shortened together (after Letouzey et al., 1995).

basin at the foot of a continental slope or prograding delta front. These offshore areas are covered by waters that are too deep to be of interest in petroleum exploration. Typical features include salt-cored contractional growth folds and salt-cored thrusts.

The style of compression is largely controlled by the extent of the salt layer across the toe of the slope system (Letouzey et al., 1995). If the salt extends well out into the basin and beyond the zone of compression, the system tends to form a fold belt dominated by salt-cored buckle folds with little or no evidence of thrusting. If, on the other hand, the salt pinches out near the toe of the slope, the friction coefficient at the base of the contracting sediment pile is greatly increased and a thrust-faulted zone forms (Figure 5.22). The degree of shortening in such thrust terrains can be substantial. In the deep water Santos Basin, Brazil, the downdip section is shortened by up to 50% of its original length (Cobbold et al., 1995). Many of these thrust blocks show typical contractional salt thrust assemblages of sediment wedges in their footwall and chronologically older (Aptian) salt in their hanging walls.

As widening salt-floored rift basins are often subject to later closure (Wilson cycle), compressional deformation is often superimposed on originally extensional halokinetic margins, often with spectacular results. The allochthonous salt acts as a lubricant to the subsequent thrust sheet as it is pushed back in a landward direction. For example, the halokinetic Triassic section of the external domain of the Betic Cordillera, Spain, is rootless and thus allochthonous (Flinch et al., 1996). The frontal accretionary wedge of the Guadalquivir allochthon consists dominantly of Triassic evaporites and red beds that form a gypsum-cemented melange with Upper Cretaceous-Palaeogene deepwater sedimentary rocks. Throughout the unit, Jurassic rocks are absent. The widespread Triassic evaporites of the Guadalquivir allochthon were originally emplaced as gravitational allochthonous masses or salt sheets at the foot of the Betic continental slope. This occurred in a passive-margin setting, much like the present day emplacement of allochthonous sheets of Jurassic salt atop Neogene sediments in the Texas-Louisiana Gulf Coast or the Santos Basin of Brazil. Thus, on the Betic passive margin, these allochthonous evaporites of Triassic salt were first emplaced along the foot of slope during the Late Cretaceous-Palaeogene. Later, during Neogene time and the formation of the Gibraltar Arc, these evaporites were overthrust as an accretionary wedge to form the Guadalquivir allochthon. This explains the lack of Jurassic sediments within the melange of Guadalquivir allochthon. They were thrust back up onto the former passive Betic margin, a lateral distance of some 200 km, from an area that lacked Jurassic deposits. The Guadalquivir allochthon ended up atop sediment that was originally located well landward of the Guadalquivir prism.

Thin-skinned fold and thrust belts

Thin-skinned fold and thrust belts form in mountain chains where the basement is not involved in the deformation, and evaporite layers are preferred detachment horizons (Figure 5.21b; Letouzey et al., 1995). Their structures are predominantly oriented parallel to the intermediate stress direction and perpendicular to the maximum shortening. Examples of thin-skinned fold and thrust belt include: the Parry Islands Foldbelt in Arctic Canada (Harrison, 1995), the European Jura (Philippe, 1994), the Spanish Alps (Flinch et al., 1996), and the Zagros Foldbelt of Iraq and Iran (Talbot and Alavi, 1996).

The pressure on a fluid salt décollement horizon undergoing compression promotes decoupling of the overburden and so has the following associations:

- folding is more common than thrusting;
- folds and thrusts lack consistent vergances, being commonly expressed as box folds and backthrusts;
- the cross sectional taper is typically low angle, which results in extremely broad fold and thrust belts (e.g. Davis and Engelder, 1987; De Celles and Mitra, 1995);
- complex three dimensional edge effects form at the lateral boundary of the salt substratum (Yeats and Lillie, 1991).

The style of deformation is dependent largely on the friction coefficient acting along the detachment level. The lower the basal friction level, the further the thrusts propagate out and so the thinner the cross section of deformation and the lower the taper angle (Davis and Engelder, 1985, 1987). Preexisting faults and salt welds in a halokinetic prism on a passive margin can be a major zone of weakness in the development of subsequent compressional structures. Hence the very close association

between preexisting faults and the current elongated salt structures of the southern North Sea.

Angular folds (box folds, chevron folds and mitre folds) predominate in a buckling thin-skinned thrust belt, reflecting internal deformation of anisotropic multilayers above the thin, ductile salt substratum. If the deforming salt substratum is thicker, the multilayers tend to buckle as a single unit into large sinusoidal folds, with consequent loss of fold angularity.

In physical models of a buckling compressional system the viscous salt-analogue layer does not breakout of any of the thickening anticlinal crests (Vendeville and Jackson, 1992b; Letouzey et al., 1995). Rather the silicone gel either thickens in the anticlinal cores or is dragged along the thrust planes, but it always stays in normal stratigraphic contact with the overlying thrust sheet. The same observations are also true in many natural salt cored thrust settings (e.g. Salt Ranges of Pakistan; Baker et al., 1988). These observation imply that salt undergoing compression finds it difficult to pierce a thick brittle overburden. However there are other natural situations, such as the Miocene salts in the Carpathians of Romania, where older salt appears to have been injected along thrust planes into younger sediments (Paraschiv and Olteanu, 1970). Similar salt-cored transtratal features occur in Triassic salts in alpine

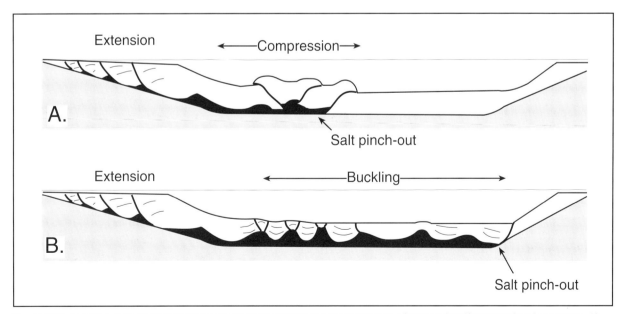

Figure 5.22. Location of basinward pinchout of salt layer controls the structural style of contractional/compressional structures. A) Where the salt layer does not extend out far past the foot of the continental slope, the frictional resistance increases at the base of the sedimentary pile and interferes with further basinward movement. This can create an imbricate thrust zone on the updip side of the salt pinchout. B) Where the salt layer extends across the basin and well out from the foot of the continental slope, the contraction is taken up in a series of buckle folds with crests that tend to lack thrust zones (after Letouzey et al., 1995).

thrust belts in southern France and Spain. The French examples have now been reconstructed using balanced section techniques and are seen to be reactivated diapirs formed first in an earlier extensional phase that was then caught up in the thrust belt (Dardeau and Graciansky, 1990).

Physical modelling has now led to a quandary as to whether the buckling and the thickening of an anticlinal salt core under compression can build up sufficient pressure to initiate diapirism. Lateral shortening during buckling should encourage salt upwelling by thickening the salt substratum in the anticlinal core, while the associated folding should create the structural relief necessary for buoyancy. But, using brittle overburdens, the experiments of Vendeville and Jackson (1992a,b) clearly show that buckling alone does not initiate diapirism nor lead to emergence. Vendeville and Jackson (1993) found that thrusting can initiate asymmetric diapirism. But it is a much less effective mechanism than extension and it does

not lead to salt emergence on the land surface without erosional thinning of the anticlinal crest to where the salt pressure exceeds the piercement threshold of the overburden (Figure 5.21b,c). In contrast, using "two fluid" models Koyi (1988) concluded salt diapirs can form and completely pierce overlying anticlines during buckling. In most natural situations, where the overburden tends to behave in a brittle fashion under geologically relevant deformation rates, salt may only become emergent when the crests of rising structural anticlines are faulted and erosionally breached.

For example, Zagros folds are box-shaped and symmetric with the main detachment surface in Precambrian Hormuz salt. Miocene evaporites of the Fars/Gachsaran Formations can act as secondary detachments, often where pairs of shallow thrusts are rooted with opposite vergence toward the crest of the anticline. Hormuz salt emerges as namakiers in the Zagros where salt cored anticlinal crests are cut by the Kazerun and Mangarak fault zones (Figure 5.23;

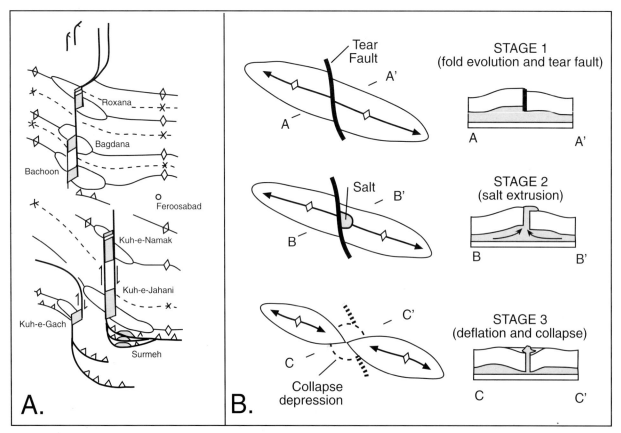

Figure 5.23. The importance of faulting as a control on salt extrusion in compressional buckling regimes. A) Current kinematics of Mangarak fault zone, Iran, showing the location of extrusions (grey) in zones where fault zone crosses salt-cored anticlinal culminations, shown as elliptical areas (after Talbot and Alavi, 1996). B) Schematic based on Algerian examples showing how salt extrusions occur where tear faults intersect salt-cored anticlinal culminations (modified from Letouzey et al., 1995).

Talbot and Alavi, 1996). Such a system of extensional tear faults across salt-cored compressional anticlines also explains emergent salt in the Atlas ranges of Algeria (Vially et al., 1994).

The amount of shortening in thin-skinned salt-effected thrust provinces ranges from mild inversion of an extensional system (such as "pillow" structures in the North Sea; Stewart and Coward, 1995) to fully developed salt-cored fold-and-thrust belts in the various Alpine chains of Europe, North Africa, southwest Asia and central Australia. Inversion in much of the southern North Sea and Germany is the foreland expression of severe Alpine shortening further south.

Inverted basins

Graben inversion is common in intracratonic basins (Figures 5.21c, 5.24; Letouzey et al., 1995). Under regional shortening that is orthogonal or oblique to preexisting structures, the inversion typically squeezes the hanging wall blocks upward. The basement, salt and overlying sediments are all shortened. Inversion also occurs in orogenic belts along former passive margins where rift-stage evaporites are also inverted. Examples of salt-entraining inverted basins include the Permian in the southern and central North Sea (Hooper et al., 1995; Stewart and Coward, 1995; Eggink et al., 1996), the Triassic of the Atlas fold belt of north Africa (Vially et al., 1994), the Neoproterozoic of the Amadeus Basin in central Australia (Stewart et al., 1991), and the Eocene-Miocene of central Iran (Jackson et al., 1990).

Where the horizontal compressive stresses are near orthogonal to preexisting extensional structures and where these are bound by high angle normal faults, much of the inversion is accommodated by forced folding rather than slip along the former faults (Letouzey et al., 1990). The new low-angle reverse faults are firmly rooted in the décollement and originate at the crest of the upthrown basement block,

and so passively transport the former high-angle normal fault planes. This initiates short-cut geometries. If the horizontal compression is oblique to the former normal fault, a strike-slip component is induced that can reactivate former high angle normal faults. In addition to conventional inversion features, such as forced folds and short cuts over basement faults, strike slip will also form *en echelon* synthetic faults in the hanging walls of preexisting basement faults (Letouzey et al., 1995). Such *en echelon*

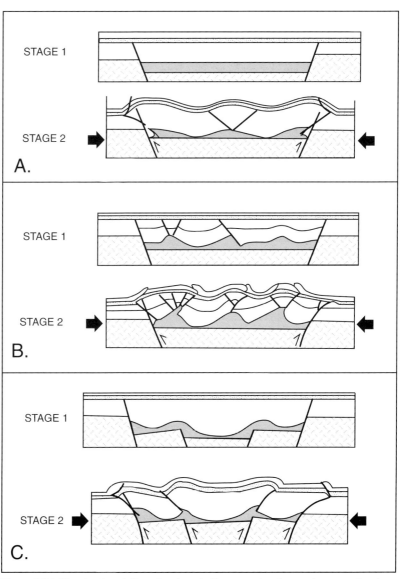

Figure 5.24. Structural evolution of an inverted basement graben varies according to structures present in preexisting graben (salt is grey). A) Folds and short-cut faults develop in homogeneous strata above inverted basement faults. Fold hinges in the graben are perpendicular to the regional compression. B) Thrusts and folds are localised by preexisting salt structures. C) Thrusts and folds are localised by reactivation of basement faults (after Letouzey et al., 1995).

179

systems define the basin edge of the Sahara flexure in Algeria (Vially et al., 1994).

The work of Letouzey et al. (1995) suggests that precontractional salt tectonics strongly influences the location of subsequent thrusts and folds (Figure 5.24). After salt deposition, renewed extension on underlying basement faults causes salt to flow toward footwall crests, especially to any salt ridges above footwall blocks (Vendeville and Jackson, 1992a,b; Nalpas and Brun, 1993). During subsequent contraction and inversion, folds and thrusts are initiated where thickness variations are already present in the overburden (Figure 5.24a,c). This is clearly illustrated by thrust faults and structures that fringe the Broad Forties basin (Hayward and Graham, 1989). Thinning of the overburden over salt ridges or salt walls also localises thrusts and folds even when they are not related to basement normal faults (Figure 5.24b). Normal-fault offsets on the top of salt can also localise thrusts (Figure 5.24b).

Sediments interact with flowing salt

Sediments adjacent to halokinetic structures either lie above or below the source horizon, which in turn may be composed of allochthonous or autochthonous salt. When flow salt approaches the land surface in either continental or offshore positions it creates topographic relief in the landscape (Table 5.3). Such topography means the flow of salt influences the style of deposition in adjacent sediments. Historically most of the sedimentology conducted on such sediments dealt mostly with suprasalt sedimentation (see summary in Warren, 1989). Conclusions as to the position of the sands and porosity belts in relation to salt structures

are still valid, but our understanding of mechanisms driving salt flow need updating. With the increasing recognition of allochthonous salt the sedimentology of subsalt systems has become more important. We will look now at both suprasalt and subsalt sedimentation.

Once flowing salt approaches a sedimentation surface it strongly influences the style and texture of nonevaporite sediments that are deposited adjacent or atop it. Reactive diapirism typically forms a graben or depression atop the diapir. Active diapirism and passive diapirism are both marked by topographic highs about the crest of the salt body. Salt flow rates are in turn influenced by loading or sedimentation rates. This interaction creates a classic feedback system where the geometry of a growing salt body is influenced by, and feeds back into, the depositional rate/water depth of the adjacent sedimentation setting. Once rising salt is sufficiently close to the surface to influence its topography, sedimentary feedback will typify all styles of halokinesis. At the scale of seismic resolution this is recorded in changes in bed thickness (e.g. formation of intrasalt/intersalt basins) as well as onlap, offlap and changes in bed amplitude and attitude as the salt body is approached. At the facies scale, it is recorded in changing styles of shallowing or deepening of the sequence, the development or the drowning of reefs and shoals, syndepositional bed disturbance, slumping and faulting, as well as the formation of local unconformities and karsting.

Style of sedimentation can be broken down into three groupings at the broad scale of seismically resolvable sediment geometries: pre-kinematic, synkinematic and postkinematic (Figure 5.25; Jackson and Talbot, 1994). The prekinematic layer records sedimentation before salt movement or any other deformation. A prekinematic sediment layer is an interval of strata whose initial stratigraphic thickness is constant above a salt structure or its adjacent rim. That is, the prekinematic sediment layer is isopachous or its depositional thickness varies only as much as would be typical for the same sedimentation style in a region where no salt was present.

In contrast, the synkinematic layers record sedimentation during salt flow or any other deformation and typically

OFFSHORE		
Abu Thama, Arabian Gulf	60m	Purser, 1973
Cabo St Tome, Brazil	97-144m	Demercian et al., 1993
Allochthons, Gulf of Mexico	130-260m	Liro, 1992; Wu et al., 1990a,b; Rezak, 1985
Offshore Mississippi Delta	100-240m	Jackson et al., 1995
CONTINENTAL		
Al Salif, Yemen	45m	Davison et al., 1996b
Jabal al Milh, Red Sea	45m	Davison et al., 1996b
Mt Sedom, Dead Sea	200m	Frumkin,1994
Saharan Atlas, Algeria	300m	Kulke,1978
Zagros Chain, Iran	300-1500m	Kent, 1987; Talbot and Alavi, 1996

Table 5.3. Landscape relief around salt structures.

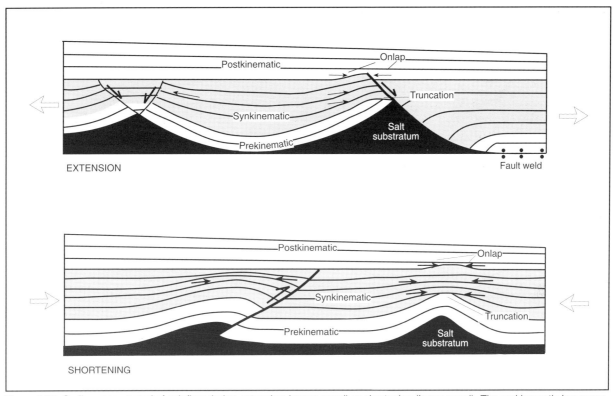

Figure 5.25. Sedimentary record of salt flow during extension (upper panel) or shortening (lower panel). The prekinematic layer was deposited before salt movement began. The synkinematic layer was deposited during salt flow and includes many internal onlap or truncation surfaces. The postkinematic layer was deposited after salt flow ceased. Its base may onlap or truncate originally exposed surfaces in the synkinematic layer (after Jackson and Talbot, 1994).

overlie the prekinematic layer. It shows local stratigraphic thickening or thinning and water-depth related facies variation above structures or parts of structures. Synkinematic layers thicken above withdrawal basins or locally thin against rapidly rising structures. For example, in the East Texas Basin in the Gulf Coast, the percentage of stratigraphic or seismic-stratigraphic thinning of synkinematic strata ranges from 10 to 100% in units influenced by the flow of Jurassic Louann Salt (Seni and Jackson, 1983a,b). Changes in bed thickness are frequently recorded by onlap or truncation at all levels of the synkinematic layer (Figures 5.25, 5.26).

The interval of strata deposited by sedimentation after salt flow, or any other deformation, has ceased is termed the postkinematic layer. Basal postkinematic strata can onlap an underlying, uneven deformed surface but show no thickness changes that can be directly ascribed to deformation induced by local salt-flow. The postkinematic layer typically overlies the synkinematic layer.

Truncations are particularly useful in distinguishing the three packages. Removal and termination of dipping

deformed strata by subaerial or subaqueous erosion forms a truncation or sequence boundary. Truncations cutting across anticlines or fault blocks indicate that the structures had surface relief at the time of erosion (Figure 5.26). Undeformed truncations mark the base of the postkinematic layer. Where a truncation is itself folded or faulted, it marks an episode of deformation of the synkinematic layer.

Terms have also been coined to describe synkinematic intrasalt basins that fill with sediment as they load and subside into relatively thick allochthonous or autochthonous salt (syn. minibasin or intersalt basin). An older and more specific term for such sedimentary bodies is a depotrough. In its strictest definition it describes an intrasalt basin filled with deepwater, mass flow or turbiditic deposits that uplap a diapir margin (Spindler, 1977). A depopod is an intrasalt basin with shallow water deltaic deposits draping over or against the flanking salt diapir. In many prograding shelf situations a depotrough passes upward into a depopod. There should be no assumption of predominant water depth in an intrasalt basin, although in the literature there is a strong bias toward deeper water. This bias reflects the

181

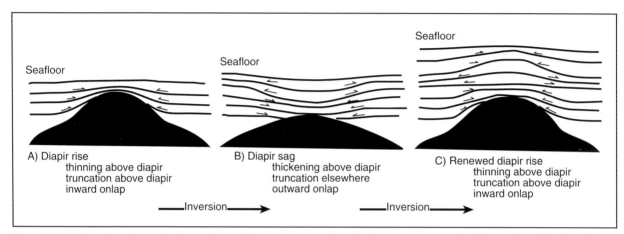

Figure 5.26. Schematic of two episodes of structural inversion caused by subsidence and renewed uplift of a diapir. The inversions are recorded by changes in bed thickness and onlap directions indicated by barbed arrows (after Jackson and Talbot, 1994; Nelson and Fairchild, 1991).

excellent, but geographically localised, work that has been widely published on the offshore Gulf of Mexico. It has led some authors to think mostly in terms of deeper water sedimentation models for all intrasalt basins. Outside of the Gulf of Mexico, modern and ancient depopods are often filled by other facies; such as deepwater chalks in the North Sea, or by shallow water lagoonal mudstones, grainstone shoals and reefs as well as sabkhas and saltern evaporites in the Middle East (e.g. Davison et al., 1996b). Once the rate of salt withdrawal slows or the salt body is cut off from its salt source an intrasalt basin fills with sediment and the vertical trend in the total infill package is almost invariably one of shallowing upward.

Diapir stage and sedimentation

Circum-salt faulting in any extensional regime is intimately associated with the first two of the three evolutionary stages of diapirism: reactive, active and passive. During the reactive (piercement) growth stage a diapir is triggered by faulting created by regional extension (Jackson and Vendeville, 1994a). The hanging wall of an initial fault then sinks into the source layer until resisted by increasing pressure forces in the salt source layer and bending resistance in the overburden. New faults form repeatedly nearer the axis of the graben. The dwindling central fault block sinks, while the diapir rises below it regardless of overburden density. Progressively smaller fault blocks are supported by fluid pressure of the salt at progressively higher levels flanking the triangular diapir. Syndiapir sedimentation keeps diapirs in the reactive phase longer by continually filling the graben. Reactive diapirism is controlled by the rate of regional extension; whenever

regional extension ceases, a reactive diapir stops growing. In contrast both active and passive diapirs are strongly influenced by sedimentation rates and the deposition of sediment wedges (Ge et al., 1997).

At the time a reactive diapir is forming there is an active depression or graben atop the extending structure (Figure 5.27). There are three depositional settings that can infill that depression: deepwater slope and basin, marine shelf, and continental. If the extensional graben or depression atop the reactive diapir ridge formed in a continental situation, then any surface waters or rivers will tend to occupy that depression. In hydrologically open settings, a series of stacked fluvial channels may infill the depression. In a more restricted to hydrologically closed scenario the graben may fill with of a series of elongate saline lakes and mudflats. If the graben depression forms in a marine shelf setting it may fill with deeper water lagoonal muds while shoal water sediments and reefs occupy the very shallow waters atop the edges of the faulted graben margin. If the reactive stage creates a graben depression in deeper waters on the upper part of the continental slope, then a series of growth faults amid thickened deltaic sand traps may occur and help load the system. Conversely the graben may create a deep seafloor depression that stacks successive episodes of submarine channel and fan sands.

Once a reactive diapir becomes sufficiently tall and its roof sufficiently thin, then pressure at the diapir crest impels the system into the active stage. The diapir lifts its roof, as it rotates, breaks through and shoulders aside its overburden. Erosion then disperses the fragmented roof materials as a series of monomict debris aprons composed of sediments that formerly overlay the diapir crest. Throughout this time, and during the following passive stage, the area above and about the diapir or salt wall is a topographic high. In the continental realm, fluvial channel sands may stack into the depression created by any adjacent salt withdrawal sink. In the marine realm, the high often lifts the sea floor above wavebase and into shallow euphotic waters suitable for carbonate growth. In this scenario muds are winnowed form seafloor sediments while sands and gravels tend to accumulate. In the transitional continental marine setting, the high energy shorezone tends to lap up onto the edge of this high and often reworks and winnows the debris apron of the active stage into well sorted shorezone sands. Later in the passive stage the shorezone also picks up and redeposits exotic blocks derived by the dissolution of extruded salt. In some Neoproterozoic salt allochthons in Northern Australia the transition from active to passive can be seen in a change of the debris apron constituents from monomict to polymict breccias (pers. obs).

In the Gulf of Mexico, a reef system is actively accumulating coarse biogenic sediment atop such a high in the Flower Gardens Bank. At the same time the surrounding upper slope siliciclastics are still accumulating as deep water muds. Carbonate buildups atop the Flower Garden Banks are there only because of the shoaling effects of salt movement. In the East Flower Garden Bank a brine lake in the central part of the structure was created and fed by salt dissolution (Rezak and Bright, 1981). Collapse foundering over the crest of the structure has generated the graben that houses the lake. Healthy growing corals surround the lake, except where a canyon cuts across the bank to the deep open seafloor. Halokinesis beneath the Flower Garden Banks has converted an otherwise muddy siliciclastic platform into a region suitable for carbonate deposition. Shoaling has lifted the seafloor into a clear, well-lit environment and created a potential carbonate reservoir in what is otherwise a muddy, siliciclastic province. Brine seeps atop salt allochthons, located in much deeper waters at the foot of the continental slope, have allowed chemosynthetic mussel communities to flourish at water depths that are a kilometre or more below the base of the euphotic zone (Chapter 9).

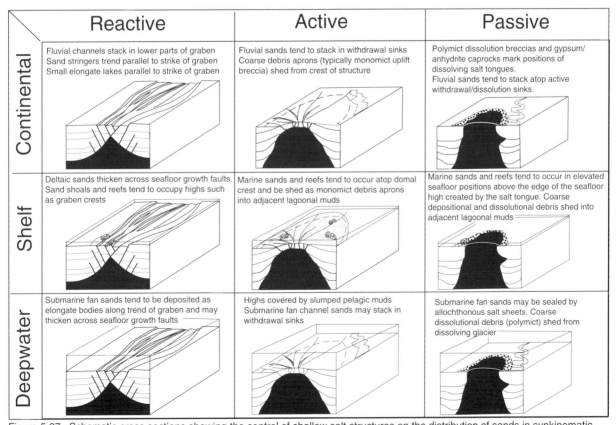

Figure 5.27. Schematic cross sections showing the control of shallow salt structures on the distribution of sands in synkinematic sediments. Sand distribution changes with the stage of salt structure and water depth in the depositional setting (modified in part from Warren, 1989). This sequence of schematics is drawn using a diapir, but similar depth-related sediment patterns also form about the seafloor highs created by nearsurface salt allochthons or atop turtle structure anticlines in interdiapir areas.

Once salt is emergent at the land surface or on the seafloor the diapir has evolved into its passive stage (Figure 5.27). At this stage the salt is extrusive and any thin sediment roof that may periodically develop atop the salt has little mechanical influence and is destroyed by ongoing active breakout. Dissolution of the salt sheet on the seafloor creates anhydrite/gypsum carapaces and may also supply exotic blocks that have been dragged up from greater depths. A salt-induced high on the seafloor can develop substantial relief over the surrounding region. In shoal-water tropical seafloor settings a rich biota of bioherms and associated grainstone flats can form. In the deeper water settings of the Gulf of Mexico the areas atop extrusive salt glaciers (salt fountains) have a relief measured in hundreds of metres above the surrounding seafloor (Fletcher et al., 1995). Such deep seafloor highs channel turbiditic sands into the relative lows along the allochthon edge. While the salt is replenished, the area about the salt glacier remains as a topographic high. The maintenance of the salt at or near the sedimentation surface also means that both the active and the passive stages are characterised by many unconformities and thinning of strata in their immediate vicinity.

Above the onshore portion of the active Al Salif diapir, Yemen, are three-thousand-year-old coral reefs that now occur some 18 m above sea level (Davison et al., 1996b) This active stage diapir has lifted its roof and continues to do so. The nearby Jabal al Milh diapir is the much reduced remnant of a formerly active extrusive structure. In both scenarios the highs atop the structures strongly influenced the facies distribution of nearby carbonate and evaporite sediments. Offshore from this region are a number of islands and grainstone shoals; they indicate sea floor shoaling atop other diapirs and salt walls in the same Miocene salt.

Topographic lows form above peripheral sinks associated with areas of salt withdrawal. The formation of lows in zones of peripheral sinks are often areas of preferential sedimentation, that is, sediment thicks. The loading associated with the sinks drives the salt flow via feedback associated with differential loading. The lows persist until the salt is completely evacuated and a salt weld forms. Salt-flow induced highs and lows in the sedimentation surface persist until the diapiric ridge or crest is cut off from its supply by the formation of welds or touchdowns. Then the diapir and allochthon crests are buried by sediments whose thickness continues to aggrade until it has exceeded the piercement potential.

Sedimentation associated with salt allochthons

Suprasalt clastic sedimentation

Alexander and Flemings (1995) documented the spatial and temporal distributions of siliciclastic sands within an evolving Pliocene-Pleistocene salt-withdrawal shelf minibasin (intrasalt basin). The Eugene Island Block 330 field, a giant oil and gas field in offshore Louisiana, is contained within the minibasin and is located immediately landward of the deeper water Green Canyon area. Based on their stratigraphic and structural analyses they conclude their facies model can also be applied to other minibasins.

The minibasin, or depopod, evolved through three phases: prodelta, proximal deltaic, and fluvial (Figure 5.28). In the prodelta phase, bathyal and outer neritic shales and turbidites loaded and mobilised the underlying salt sheet (Figure 5.28a). During the proximal deltaic phase, salt continued to withdraw from beneath the minibasin, and lowstand shelf margin deltas remained focused at a regional growth fault zone on the northern margin of the minibasin (Figure 5.28b). Sediment accumulation and fault slip rates were high, as thick sequences of deltaic sands were deposited adjacent to the fault system. During the final fluvial phase, salt withdrawal waned; consequently, the creation of accommodation space within the minibasin ceased (Figure 5.28c). The basin infilled and, during lowstands, deltaic systems prograded southward. Unconformities developed in the minibasin during these lowstands. During transgressions, thick packages of shallow-water deltaic and fluvial sands (capped by shales) were deposited on top of the unconformities.

Suprasalt carbonate marine shelf sediments

The multiplicity of depositional settings above salt structures in the Arabian Gulf gives an insight into the geometries of carbonate sands and potential reservoirs about ancient salt structures (Warren, 1989). Seafloor highs surrounded by deep water do not show the effects of widespread current reworking. These structures shed shallower water sediments centripetally into deep sub-wave base depths. Facies patterns tend to form concentric rings, with the best sorted sands or reefs at or near the crest of the structure. For an ancient counterpart that is a possible oil or gas reservoir, one would drill on or near the crest of the structure as identified in the seismic. Intermediate highs are surrounded by waters of moderate to shallow depths. Now, sediment shed from the carbonate high can be reworked by tides and other currents, both across the high and into the shoal water areas surrounding the high. Often, this type of structure generates a down current sediment tail, creating a strong asymmetry to any reef or sand buildup on the structure itself. The sediment tails can be quite thick; off Dalma Island in the Arabian Gulf the modern downwind grainstone tail attains thicknesses of 20 m (Figure 5.29). Drilling for relict primary porosity on the ancient equivalent of this type of structure would involve drilling both on or near the crest of the structure as well as expecting and drilling sand trends that cut across the structural crest. Twenty-metre-thick sands encased in muds and sabkha evaporites are not beyond the resolution of modern high-frequency seismic.

Sediment tails about Arabian Gulf salt structures often trend in the same direction as the prevailing winds of the region (Figure 5.29). This is a useful predictive tool when exploring for sediment tails atop ancient salt structures in shelf settings. Once the position and trend of an ancient grainstone unit is established about an individual salt structure, grainstone orientation and position of potential reservoirs about other time equivalent diapirs in the region can be predicted. In coastal settings in the Arabian Gulf the sediment trend is often modified by coastal transport processes. Leeward sediment tails often extend from the high to the shoreline (Yas Island), generating a progressive shallowing between the high and the mainland. Studies of the peritidal sand tail that extends from Yas Island onto the mainland at Jebel Dhanna show that shallowing increases the velocity of tidal currents. This causes the sequence to shoal upward from bioclastic grainstones and wackestones into oolite shoals that trend at an oblique angle to the trend of the sediment tail (Warren, 1989).

Subsalt sediments

Subsalt sediments have been studied mostly in the Gulf of Mexico, where more than 30 wells have been drilled with subsalt targets. This has resulted in 8 discoveries, at least 3 of which are commercial and 3 have reserves of 100 million bbl oil equivalent or more (Montgomery and Moore, 1997). The reservoirs are sands of Miocene-Pleistocene age. Most of the sands were deposited in deepwater-slope submarine fan channel and levee settings, often with channel orientations strongly influenced by the topography generated by nearby salt allochthons. Once deposited the sands were then overridden and sealed by these allochthonous sheets. Fans typically stacked into intrasalt basin lows and along basin margins. Their position can be closely related to the seafloor generated by the nearby allochthonous salt sheets. Over time the seafloor facies evolve with the alternating episodes of salt movement and dissolutional collapse to create complex vertical successions.

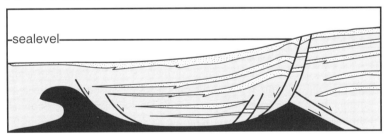

A. Prodelta

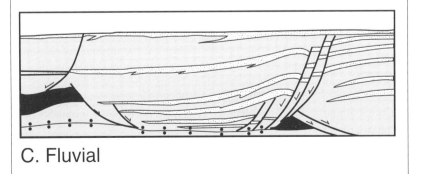

B. Proximal deltaic

C. Fluvial

Figure 5.28. Schematic of three-phase depositional evolution of intrasalt basin fill in the 330 Block, Gulf of Mexico (Alexander and Flemings, 1995). Evolution began prior to 2.8 Ma and continues today.

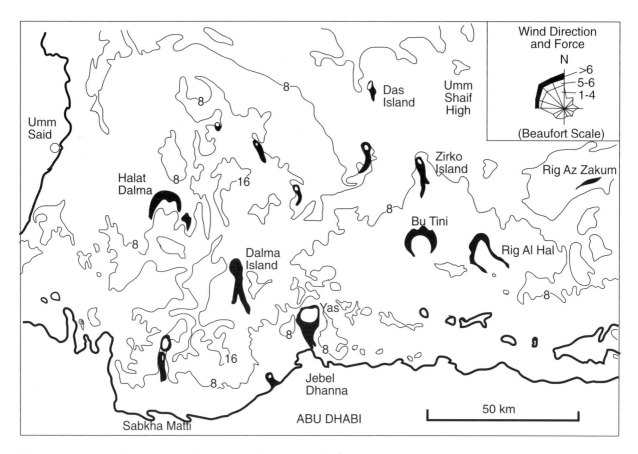

Figure 5.29. Relationship between wind direction and orientation of Holocene sediment tails (black) in shallow waters of the southern Arabian Gulf. Bathymetry in fathoms (after Purser, 1973; Warren, 1989).

Complications of shale diapirism

Both salt and shale form diapiric structures that, without drilling, can appear quite similar in seismic sections. In a benchmark paper Morley and Guerin (1996) discuss the similarities and differences present in salt and shale diapiric provinces and outline ways to distinguish the two styles of diapirism. Salt's mobility is a fundamental material property; so that structures in a mobile salt belt continue to evolve until the salt diapir is completely dissolved or ongoing salt withdrawal produces sufficient overburden touchdowns to stop further flow. Shale, on the other hand, only flows while it is overpressured. Whenever the main shale body depressurizes it freezes. While a shale is mobile it can generate a progression of kinematic structures similar to those generated by the reactive and active phases of salt diapirism (Figure 5.30a).

Halokinetic flow in a salt province builds diachronous levels and tiers of allochthons all the way up through the basin stratigraphy to the seafloor. Salt flows and evolves at higher levels in the stratigraphy, long after the original salt level has become a primary weld. Movement of a shale diapir tends to be much more episodic and related to particular overpressuring events within the shale-dominated section of a prograding deltaic continental margin (e.g. load driven compaction or thermobaric release of water or hydrocarbon generation). Once a mobile and chaotic shale mass approaches the surface and the pressure buildup is released via mud volcanoes, the shale freezes. For this reason shallow allochthons and widespread raft tectonics atop thin shallow décollement layers are much less common in mobile shale belts. Laterally limited shale allochthons do occur in the Niger Delta, where they form imbricate thrusts that pass down dip into gravity flows and slumped blocks of shale derived from the exposed mobile shale.

Another significant difference between salt and shale diapiric provinces lies in the permanence of the contact between the flowing unit and adjacent strata. While flow in halokinetic terranes only takes place within the salt unit, shale mobility can spread out into adjacent beds as the overpressured interval spreads. While overpressured, a mobile shale interval can cannibalise adjacent shale strata and so cut across time/bed boundaries (Figure 5.30b).

Morley and Guerin (1996) noted how such differences in mechanical behaviour lead to differences in structural style between shale and salt deformation, although many basic aspects of the gravitational tectonics are the same in both provinces. They note: 1) Prekinematic and synkinematic structures occur in both provinces, but much of the prekinematic deformation in mobile shale deltas, such as the Niger Delta of Africa and the Baram Delta of Brunei, are destroyed by aggrading cannibalisation of the overpressured shale. Not recognising this cannibalising effect can lead to the inappropriate placement of the lower portions of growth faults during seismic interpretation of shale diapiric belts (C. Morley pers. comm.). 2) Fault-dominated depopods occur above the flowing unit in both salt and shale mobile belts. For example, in the Niger Delta, the interval of overpressured shales beneath the actively accumulating delta prism varies from relatively thin intervals to massive chaotic zones some 4-6 km thick. The maximum depopod thickness is the thickness of the mobile shale plus the thickness of the overlying pregrowth fault strata (i.e. some 4-8 km). Such thicknesses are similar to the salt-induced depopods in the Gulf of Mexico. 3) Classic diapir geometries are common to both salt and shale provinces, as are normal fault haloes that are created during reactive diapirism (Figure 5.30a). Salt has the potential to completely evacuate as it evolves into salt welds or overburden touchdowns. Mobile shale deformation may also create local touchdown areas, but complete shale collapse is uncommon and large volumes of immobile dewatered shale are usually left after diapirism has ceased.

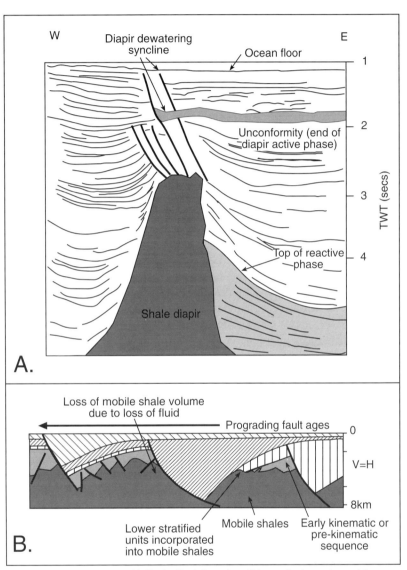

Figure 5.30. Shale diapirs. A) Shale diapir from the offshore diapir province of the Niger Delta, Africa. Adjacent sediments show synkinematic evidence of reactive and active phases of shale diapirism. The reflectors in the shaded region show thickening toward the shale diapir stem, suggesting expansion into an active growth fault during the reactive phase. B) Idealised evolution of growth faults associated with thick mobile shales (after Morley and Guerin, 1996).

Summary

With the increasing use of 3D seismic for oil exploration in Phanerozoic basins (e.g. Gulf of Mexico, North Sea, Red Sea) it has become obvious that previous notions of predominantly vertically movement of salt from a mother salt bed into thumb-like diapir appendages are wrong. We now know that huge salt sheets (allochthons) climb the stratigraphy as a series of predominantly subhorizontal sheets or tiers that are continually moving and squeezing as they are loaded by ongoing sedimentation. The end result is a feedback system where flowage create lows that fill with sediment, while the filling of the low with sediment triggers further salt flow. The upper side of the deforming salt sheet can flow into more classic appendage-like styles of diapirs. Prior to the advent of forward-modelling techniques in seismic processing and the gathering of 3D data networks we could not image beneath these salt bodies and so tended to interpret them as the primary level of the salt body. We now know this is not the case, and that most deforming salt beds flow into allochthons, which become rootless once the connection to the primary source or mother salt bed is converted to a salt weld or a fault weld. A salt weld forms as salt migrates or dissolves away to allow the former upper and lower contacts of the salt to come together to form a salt weld. A fault weld forms where salt that was once within or lubricating a fault or shear plane is lost and the two sides of the fault come together. When a thrust plane is lubricated by salt that is subsequently lost by migration or dissolution, the plane may convert to a characteristic nearsurface dissolution breccia called rauhwacke (Chapter 4).

The end result of salt allochthons climbing through an aggrading shelf sequence is a layered and faulted stratigraphy with former sheets left as salt welds and fault welds, rather like a slug leaves a slime trail as it moves through a pile of debris. The amount of salt left behind in the basin stratigraphy depends on the efficiency of salt evacuation. When the evacuation of the allochthonous salt sheet is efficient, all that is left behind is a series of stepped counter-regional fault welds and associated landward dipping faults. When the evacuation is less efficient the result is a series of salt rollers along the salt weld and an overlying fault-segmented sedimentary sheet cut by dominantly basin-dipping growth faults (sometimes called a roho system)

Wherever a salt allochthon is near the seafloor it can load, deform, coalesce or disaggregate as it continues its climb through the aggrading stratigraphy. Sediment loading atop a deforming allochthon creates fault-defined suprasalt basins or depopods. Depopod fill shows a characteristic upward shoaling pattern from initial deeper-water laminated shales and mud to shallow water sands and carbonates. Processes of ongoing halokinesis and depopod formation cease once the salt allochthon is sufficiently thinned, dissolved or buried. Thus, throughout the sedimentation history of a passive margin that is underlain by early thick salt beds there are many intervals where thick sheets of flowing allochthonous salt are present just beneath the deep sea floor and sandwiched into what are otherwise normal deeper marine bottom sediments.

Salt allochthons are triggered and driven by a variety of flow mechanisms, with extension and differential loading being the two dominant controls. We now know that extensional basins, characterised by thick beds of salts that were precipitated in the early stages of rift opening, are likely to have regional faulting and structural geometries controlled largely by salt tectonics. While salt is flowing, the sedimentation system is in a continual state of feedback; so that patterns of suprasalt sediment thickening are often controlled by the style of salt withdrawal beneath the various depopods. As an opening rift becomes a passive margin, allochthonous salt climbs through the various portions of an otherwise marine platform margin stratigraphy. As a salt allochthon continues to be thinned by ongoing extension, along with dissolutional loss to convecting basinal fluids, the deforming salt layer can become a décollement layer. This in turn facilitates gravity gliding and ongoing accommodation of sediment accretion between what were adjacent fault-bound blocks in a process known as raft tectonics.

The close association of salt allochthons, regional fault sets, growth faults and simultaneous décollement has only been appreciated in the last decade. The economic significance of salt allochthon dissolution acting to focus and feed metalliferous chloride-rich brines into active growth faults and venting these solutions to the seafloor has only begun to be documented and is discussed in detail in Chapter 9.

Salt tectonics is also active in compressional terranes, which occur at the foot of a gravity glide sheet basinward of the continental slope; but is especially obvious in ancient foreland fold and thrust belts where weak subsurface salt acted a décollement layer. The resulting anticlines are narrow and salt cored while the intervening synclines are broad. Most of the thrusts are blind and backthrusts typically occur in association with forward directed thrusts. Much of the salt in the cores of the anticlines is strongly folded and diapiric; although the notion that compressional diapirs can breach their overburden, without being erosionally thinned or disrupted by tear faults, has yet to be proven. There is a growing body of evidence that a substantial number of the compressional diapirs are actually reactivated extensional structures.

Chapter 6
Meta-evaporites

Introduction

Evaporites survive into the metamorphic realm, but are altered, recrystallised or transformed into new minerals. Consider "the before and after" situation as a sequence of evaporite entraining sediments is buried and passes into the metamorphic realm. Under a likely "before" regional metamorphism scenario, the buried evaporite body is hundreds of metres thick, tens to hundreds of kilometres wide, perhaps with halokinetic geometries and an extensively tilted and faulted overburden. Halite typically makes up more than 60-80% of the total rock salt volume. This halite is likely to be interlayered with anhydrite, bitterns, magnesian carbonates and siliciclastic clays. "After" metamorphism, the end products of this once dominantly NaCl body of rock are a series of sodic and magnesian aluminosilicates, magnesites, calcsilicates and marbles. The immediate question is, where did all that salt and the associated volatiles (Cl, CO_3, SO_3) go?

Some of it may still reside in meta-evaporite indicator minerals, such as sodic or magnesian talcs, marialitic scapolites or tourmalinites. The remainder went elsewhere, perhaps as it generated metalliferous volatile-entraining brines. What effect did the loss of this huge volume of metasomatic fluid have on surrounding lithologies? Where, when and why did it precipitate diagnostic meta-evaporite minerals? This chapter attempts to address some of these questions, while Chapter 9 addresses matters related to sulphides associated with meta-evaporites.

It must be noted that the relevant literature base dealing with the "after" is sparse when compared with the mainstream literature in metamorphic petrology, and is at times contradictory. Much of the metamorphic literature and the associated structural literature in meta-evaporite terrains also seems to assume that all sedimentary protoliths were flat bedded and little altered. This does not reflect the sedimentological reality that, prior to the onset of metamorphism in evaporitic basins:

• Steeply dipping blocks and folds were commonplace in the brittle overburden and many of these structures were decoupled from basement features.
• Widespread breccias defined the position of former salt units (dissolutional, halokinetic and fault plane breccias).
• Prior to the onset of metamorphism, pervasive alteration haloes had already formed with nonstratiform geometries. They outlined hydrological conduits associated with salt sheets and welds, diapirs and faults.
• Magnesium was replacing adjacent sediments almost from the time the evaporites were first deposited (e.g. reflux dolomites, zeolites).
• Sodic and potassic alteration haloes in adjacent finer-grained sediments were widespread, as were burial cements and replacements within coarser-grained nonevaporite units.

The onset of metamorphism occurs once the thermobaric regime, and associated diagenetic realm, passes with depth and increasing temperature and pressure into the zone of metamorphism. The temperature separating the diagenetic from the metamorphic regime is ≈200°C (Figure 6.1). Conceptually, it is difficult to separate the high temperature end of the zone of diagenesis from the zone of metamorphism. In the more advanced stages of burial

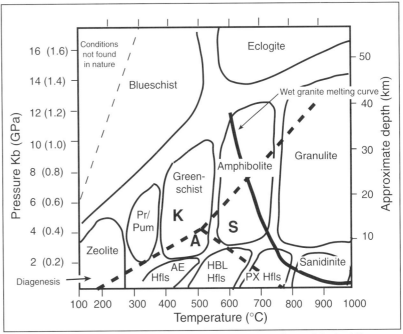

Figure 6.1. Metamorphic facies; K = kyanite, S = sillimanite, A = andalusite, Hfls = hornfels, Pr/Pum = prehnite-pumpellyite, AE = albite-epidote, HBL = hornblende, PX = pyroxene.

CHAPTER 6

diagenesis many clays become unstable as they convert to interlayered mixtures of chlorite and illite, often associated with kaolinite. These mineralogical changes in clays occur in both the late diagenetic and the early metamorphic realms and are associated with the ongoing loss of pore fluids, including generation of volatiles such as CO_2, H_2S and CH_4. Hence the rock matrix chemistries cannot be interacting isochemically at this time; there must be a flow of fluids and ionic components in and out of the system.

An increasing degree of illite crystallinity is often used by metamorphic petrologists to help divide the early stages of metamorphism. Some use the term anchizone to describe a subsurface region where chlorite-illite diagenesis

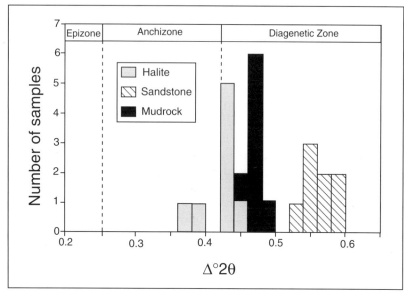

Figure 6.2. Illite crystallinity (Δ °2θ) of the <2μm fraction of samples from Hall in Tyrol. Note that different lithologies from the same region show distinct crystallinities (after Spötl, 1992).

is transitional into the metamorphic realm, and the term epizone for the succeeding lowest grades of metamorphism in which illite is largely replaced by white mica. Very fine-grained white micas of epizonal slaty rocks are known as sericites, with the dominant constituent typically phengite, a variety of muscovite.

Where studies of illite crystallinity have been conducted on evaporite samples the outcome, while fitting the general paradigm of increasing illite crystallinity, is a little more complicated. Spötl (1992) studied clay mineral assemblages in Upper Permian evaporites (Alpine Haselgebirge Formation) of the Northern Calcareous Alps in Austria. Using fresh samples from a salt mine from Hall in Tyrol, he found that illite crystallinity was strongly dependent on the host lithology (Figure 6.2). Illite from muddy halite was the best crystallised, and indicates an incipient anchizonal overprint (mean = 0.42 ± 0.02). Illites from intercalated mudrocks and sandstones were less crystalline and averaged higher Δ °2θ values of 0.47± 0.01 and 0.56 ± 0.03, respectively. Thus illites in these lithologies are significantly less crystallised compared with intercalated muddy halite samples and indicate nonmetamorphic (diagenetic) conditions. Spötl postulates that the greater crystallinity in the muddy halites reflects either higher potassium content and hence greater supersaturation with respect to illite in the high saline pore brines of the evaporites or a thermal anomaly associated with some sort of nearby diapiric ascent.

Instead of clay crystallinity, the diagenetic to metamorphic transition is sometimes tied to the subsurface interval where sedimentary porosity and permeability is lost (Phillips et al., 1994). That is, while sedimentary porosity is significant, diagenetic fluids are likely to dominate. Once compaction and authigenic cementation occludes this porosity to negligible levels, there is little capacity to retain either diagenetic fluids or significant permeabilities and so the strata pass into the realm of metamorphism. This definition is based on the assumption that in metasediments, unless there is evidence to the contrary, one generally assumes that with the loss of diagenetic permeability, the system can be approximated as a near isochemical closed system. Evolving mineral phases then reflect successive equilibria, and are tied to changes in temperature and pressure, with little change in the molecular components that make up the whole rock chemistry. This is, after all, the basis for the application of phase diagrams in metamorphic petrology. More recent work is showing that most metamorphic systems do have some degree of fluid throughput, especially those involving metasediments and decarbonation reactions (Yardley and Lloyd, 1995). The loss of volatiles from metamorphosing evaporites described in the first paragraph means an assumption of a closed system is not typical for areas adjacent to, or interbedded with, thick successions of evaporites that are entering the realm of metamorphism.

Porosity and permeability loss is directly related to increasing burial in all sediments, except for some evaporite

190

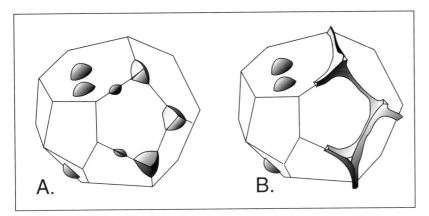

Figure 6.3. Effect of dihedral angle on pore connectivity in texturally equilibrated monomineralic and isotopic polycrystalline mosaic halite. Stipple shows position of fluid phase within the intercrystalline porosity. A) Isolated porosity for dihedral angle > 60°. B) Connected porosity for dihedral angle < 60° (after Lewis and Holness, 1996).

deeply buried halite. Most of the inclusions in chevron halite and other inclusion-rich cloudy primary salts are due to entrained brine inclusions and not mineral matter. Figure 6.4 plots the weight loss of various types of halite during heating. It clearly shows cloudy halite contains up to 5 times more water (0.2-0.5 wt%) than clear coarsely crystalline halite. An analysis of all fluids released during heating shows carbon dioxide and hydrogen contents are much lower than the water volumes: $CO_2/H_2O < 0.01$ and $H_2/H_2O < 0.005$. Organic compounds, with CH_4, are always present (<0.05% H_2O), and are twice as abundant in cloudy samples. There are also traces

salts where permeability is lost very early but regained at greater burial depths. In bedded halite, all effective porosity and permeability in a halite bed is lost by 100 m burial at temperatures that are less than 50°C. Anhydrite beds are tight by 500-1,000 metres depth. More deeply buried halite may regain permeability during halite recrystallisation due to a change in the halite-brine dihedral angle at depth (Lewis and Holness, 1996). Once this recrystallisation occurs, mosaic halite loses its ability to act as an aquitard and may instead act as a permeable conduit for escaping formation waters (Figure 6.3). According to Lewis and Holness, the depth at which the recrystallisation occurs may begin as shallow as a few kilometres. But their pressure bomb laboratory-based experiments largely ignore the ability of natural salt to creep and self-seal by diffusion-controlled pressure solution (Chapter 5). Even if the changing dihedral angles alter and open up permeability there is no guarantee that subsequent flowage associated with pressure solution will not reanneal these new pores. The ability of salt to continue to act as a highly effective hydrocarbon seal to depths of 8-10 km means, in my opinion, that salt does not become a relative aquifer until depths of 10 km or more, and probably at burial temperatures that mean the sequence is entering the greenschist facies field.

A release of entrained water at temperatures > 300-400°C is another factor that influences the textures of

of nitrogen and, in some samples, hydrogen sulphide and sulphur dioxide (Zimmermann and Moretto, 1996).

Nitrogen is released below 240°C and is probably of atmospheric origin. About 90% of the water is expelled from 280 to 420°C, corresponding to the decrepitation of inclusions (Figure 6.4). Likewise, CO_2 liberation occurs mainly between these temperatures, but above 420° C an additional part comes from carbonate dissociation through heating. Hydrogen, whose content is less than 0.5% that of H_2O, is extracted continuously from 200 to 460 °C. Most of the hydrogen originates from dissociation of organic matter present in the halite crystals and from water reduction during heating (Zimmermann and Moretto, 1996).

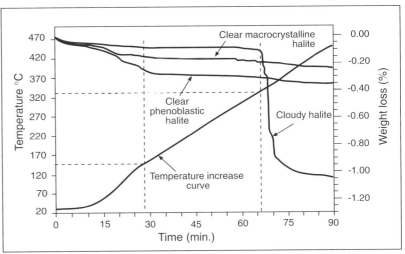

Figure 6.4. Weight loss of three pure halite samples upon heating from ambient temperature to 470 °C. Most of the weight loss corresponds to water loss from entrained inclusions (after Zimmermann and Moretto, 1996).

Both the textural transition and the dehydration of primary salt may explain why massive stratiform crystalline halite is rarely observed at metamorphic grades higher than early greenschist facies, even though the associated temperatures are well below its melting point of 750-800°C. As it recrystallises the increasingly permeable halite bed is also flushed by waters from adjacent devolatising and altering meta-sediments that further aid in removing and altering halite. Through its dissolution and metamorphism, halite generates zones of extensive sodic metasomatism in subjacent meta-sediments. The same chloride-rich fluids may also flush metals from adjacent beds and then carry them into contact with reducing intervals, such as graphitic schists, where metal sulphides or oxides crystallise in suitably prepared ground or pressure shadows (Chapter 9).

In contrast, a thick anhydrite does not dissolve as readily, and recrystallised anhydrite is known to survive into high grades of metamorphism (Butchins and Mason, 1973; Serdyuchenko, 1975; Markl and Piazolo, 1998). For example, hornfelsic anhydrite, interbedded with anhydrite-cemented shale breccias formed by evaporite solution collapse, is found at 1 km depth in the Magmamax 11 well in the central part the Salton Sea geothermal system (McKibben et al., 1988a,b). Thick beds of recrystallised sedimentary anhydrite occur in amphibolite grade metasediments at Balmat, New York (Whelan et al., 1990) and granulite grade metasediments in Central Dronning Maud Land in East Antarctica (Markl and Piazolo, 1998). Nonevaporitic anhydrite can also form as a hydrothermal precipitate from heated seawater, and will discussed fully in Chapter 9. The current chapter concentrates on metasedimentary anhydrite.

In summary, ongoing progressive regional metamorphism of an evaporitic sequence leads to a removal of water, CO_2 and other volatile components. The greatest expulsion of rock mass occurs in the lower grades of metamorphism. The H_2O, CO_2, $Cl°$, CH_4, etc. then migrates, as the fluid carries away large quantities of dissolved ions, particularly Na^+. Fluid released during devolatilisation of meta-evaporites is typically saline and chloride-rich with an

Process (indicator mineral)	Example	Reference
Scapolitization (scapolite) $(Na,Ca)_4[(Al,Si)_4O_8]_3(Cl,CO_3)$	Betic Cordilleras, southern Spain	Torres (1978) Gómez-Pugnaire et al. (1994)
	Corella and Staveley Formations, Mt Isa Block, Australia	Oliver (1995) Stewart (1994)
	Seve Nappe Complex, northern Sweden	Svenningsen (1994)
	Granulite complex, Dronning Maud Land, Antarctica	Markl and Piazolo (1998)
	Zambesi Orogenic Belt, Zambia	Hanson et al. (1994)
Albitization (albite) $NaAlSi_3O_8$	Olary Block, Australia	Cook (1992) Cook and Ashley (1992)
	Proterozoic Paranoa Group, Goias, Brazil	Giuliani et al. (1993)
Stratiform anorthositization (plagioclase) $(Na,Ca)Al(Si,Al)Si_2O_8$	Grenville Series, New York	Gresens (1978)
	Oaxacan granulite, Mexico	Ortega-Gutierrez (1984)
Tourmalinitization (tourmaline) $Na(Mg,Fe)_3Al_6(BO_3)_3$ $(Si_6O_{18})(OH)_4$	Barberton Greenstones, South Africa	Byerly and Palmer (1991)
	Willyama Supergroup, Broken Hill Block, Australia	Slack et al. (1989) Plimer (1988)
	Liaong borates, China	Peng and Palmer (1995)
Lapis Lazuli $(Na,Ca)_8(AlSiO_4)_6(SO_4,S,Cl)_2$	Baffin Island, Canada Edwards, New York	Hogarth and Griffith (1978) Hogarth (1979)
Columbian emerald (tourmaline)	Columbia, S. America	Giuliani et al. (1995)

Table 6.1. Examples of some meta-evaporite indicator minerals.

innate ability to leach and transport metals as chloride complexes. The resulting metamorphic minerals, dominated by sodic phases, create zones of widespread scapolitisation or albitisation (Kwak, 1977; Moine et al., 1981; Rozen, 1979).

Even in the zone of metamorphism, rock matrix chemistries cannot be interacting isochemically: there must be a flow of fluids and ionic components in and out of the system to explain the complete loss and alteration of the evaporite salts. The presence of a likely meta-evaporitic mineral assemblage, such as NaCl-scapolites with dravite-uvite tourmalines, does not, by itself, indicate an undeniable evaporite protolith. Ideally, there should be further evidence to support a meta-evaporitic association such as:

• strict stratigraphic/halokinetic control on the occurrence of meta-evaporite mineral phases and haloes (Hietanen, 1967; Sharma, 1981);
• the presence of recognisable salt pseudomorphs, such as hoppers, in less metamorphosed stratigraphic equivalents (Grotzinger, 1986);
• hypersaline fluid inclusions in the assemblage (Roedder, 1984);

• the occurrence of abundant associated Mg- and Cl-rich biotite (Moine et al., 1981).

Let us now look at some of the more common evaporite indicator minerals.

Common meta-evaporite minerals

As a general rule, metamorphic minerals with an evaporite protolith tend to be enriched in sodium or magnesian phases and can retain high levels of volatiles (Table 6.1). The sodium tends to come from the dissolution of salts, such as halite or trona; while the magnesium tends to be remobilised from earlier diagenetic minerals, such as reflux dolomites and magnesium-rich evaporitic clays. Boron in tourmalinites may have come from a colemanite/ulexite lacustrine precursor. Once a salt protolith is totally removed, the palaeoevaporite evidence in the metasediments is largely restricted to characteristic mineralogic associations along with occasional evaporite pseudomorphs. For example, evidence of early stages of a sodic transformation can be seen in the sodium phlogopites and sodian aluminian talcs in the metapelites of the Tell

Atlas in Algeria (Schreyer et al. 1980). Evaporitic sulphates are pseudomorphed in the NaCl-scapolite-dominated sequences of the Cordilleras Beticas of Spain (Gómez-Pugnaire et al. 1994). Rocks of higher temperature and pressure facies, such as the massive stratiform anorthosites in the Grenville Precambrian Province of North America, have been interpreted as possible meta-evaporites (Gresens, 1978), as have the anhydrite-containing Precambrian calcsilicates in the Oaxacan granulite complex in southern Mexico (Ortega-Gutierrez, 1984) and pervasive scapolites in the Zambesian orogenic belt of Zambia (Hanson et al., 1994).

Scapolite

The term scapolite describes a solid solution series. The sodium-chloride-rich end member is called marialite [$Na_4(Al, Si)_{12}O_{24}Cl$] and the calcium-carbonate-rich end member is meionite [$Ca_4(Si, Al)_{12}O_{24}(CO_3, SO_4)$]. The sulphate ion shown in the formula is typically barely more than a trace, even in meta-evaporites; but is found in far greater percentages than the occasional fluorine or hydroxide interlopers in the scapolite structure. The structure of scapolite is similar to some feldspathoids in that it is composed of large open spaces in the framework of silicate and aluminium tetrahedrons. These open spaces

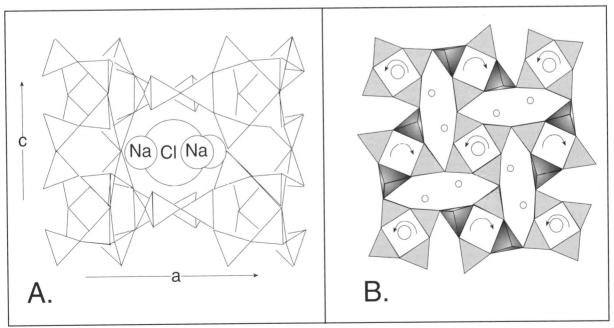

Figure 6.5. Scapolite structure. A) The [100] projection of the structure of marialite showing the cage-like structure and the linkage of the (Si,Na)—O tetrahedra parallel to the c axis. B) [001] projection of the crystal structure of scapolite. Small circles represent cations (Na), larger circles represent anions (Cl). The arrows indicate the sense of rotation of the four membered tetraheda rings with increasing pressure (modified from Comodi et al., 1990).

are large enough to cage very large ionic groups. The scapolite structure can be described in terms of four membered tetrahedral rings [(Si,Al)—O], linked together to form an open chain framework that is parallel to the c axis. A large subspherical opening is located at the fourfold axis and is surrounded by four oval shaped channels running parallel to the c axis [001]. This large cage is occupied by the Cl, CO_3, and SO_4 anions, whereas the channels host Na, Ca, and minor K cations (Figure 6.5a).

The cage-like lattice structure means that scapolite acts as a source or sink for a number of volatile species in the metamorphic realm (Cl⁻, CO_3^{2-} and SO_4^{2-}). Once the metasedimentary sequence passes beyond the early greenschist stage, scapolite takes over from halite as a potential source for volatiles. As the metamorphic grade increases, the pressure increases, and the lattice compresses by rotating the tetrahedra (Figure 6.5b; Comodi et al., 1990). As this happens the Cl content of the scapolite decreases as the lattice releases Cl.

In general, for each metamorphic grade, the variation in the amount of Cl in the scapolite suggests a local control. In a scapolite-biotite assemblage, elevated levels of Cl indicate a high NaCl activity at the time of their crystallisation from a likely feldspathic precursor (Orville, 1975; Ellis, 1978; Vanko and Bishop, 1982; Mora and Valley, 1989). The NaCl in the scapolite may have been derived from sedimentary halite, or introduced from an outside source such as Cl-rich hydrothermal or sedimentary brines. The distribution of scapolite in the field is often the best evidence for the origin of the scapolite (Hietanen, 1967). Scapolites derived from hydrothermal brines tend to be haloes around intrusives, while basinal chlorine tends to form replacement fronts associated with faults and other conduits. Conversely, where scapolite is distributed as fine-scale interbedded units of scapolite-bearing and scapolite-free layers that differ only in their Cl content, it argues strongly for an *in situ* salt precursor. This section will consider only those scapolites interpreted as meta-evaporites; for seawater hydrothermal systems a good starting point is Jiang et al. (1994) and for skarn deposits, Pan et al. (1994).

Whatever the ultimate ion source, most scapolites form via the metamorphic alteration of plagioclase feldspars. The entire scapolite series can be considered to be analogous to the plagioclase series (Figure 6.6a). If the formula of marialite is written as $3(Na(Al, Si)_4O_8)NaCl$, it is clear how well it matches the formula of the sodium-rich plagioclase, albite, $NaAlSi_3O_8$. A similar look at meionite's formula, $3(Ca(Al,Si)_4O_8)CaCO_3$, shows that it too is near three

times the formula of anorthite, $CaAl_2Si_2O_8$. Reinhardt (1992) noted the formation of scapolite in meta-evaporitic terrains of the Mary Kathleen region, Australia, can be expressed by end member reactions noted in Orville (1975):

$$3Albite + NaCl <—> Marialite$$
$$3Anorthite + CaCO_3 <—> Meionite$$
$$Anorthite + CaSO_4 <—> SO_3 \text{-Meionite.}$$

The addition of the extra sodium chloride or calcium carbonate occurs during metamorphism, as does a substantial alteration of the original lattice structure. Although nearly pure albite and anorthite specimens are sometimes found in nature, pure forms of meionite and marialite are unknown. Natural scapolites have compositional limits between 17.4% and 87.3% of the meionitic component (Me) calculated as:

$$Me = [Ca+Mg+Fe+Mn+Ti]:[Ca+Na+K+Mg+Fe+Mn+Ti]$$

Marialite corresponds to the compositions Me_{0-20} and meionite to Me_{80-100}. Intermediates Me_{20-50} and Me_{50-80} are known as dipyre and mizzonite, respectively.

Minerals of the scapolite group have received considerable attention in metamorphic petrology due to their rather wide field of stability, which extends from late diagenetic to high temperature and high pressure crustal conditions. An understanding of variations in NaCl-scapolite (marialite) and $CaCO_3$-scapolite (meionite) mineralogy is considered useful when studying the relationship between scapolite-plagioclase-halite reactions, metamorphic conditions, and the nature of the fluid phase.

Ellis (1978) considered the equilibrium between scapolite and a fluid, for a scapolite of fixed EqAn content, from exchange of the salt components:

$$NaCl_{(scapolite)} + CaCO_{3(fluid)} = CaCO_{3(scapolite)} + NaCl_{(fluid)}$$

at 750°C and 4 kbar:

$$K_D = [(X^{Scap}_{CaCO3})(X^{fluid}_{NaCl})] / [(X^{Scap}_{NaCl})(X^{Calcite}_{CaCO3})$$

where X^j_i is the mole fraction of NaCl or $CaCO_3$ (i) in the respective phases (j). This he fitted to an expression

$$lnK_D = -0.0028 (X_{Al})^{-5.5580}$$

where X_{Al} is the atomic Al/(Al+Si) ratio in scapolite (Figure 6.6b). He went on to suggest that if the mixing is ideal, the activity of NaCl in fluids coexisting with scapolite grown at temperatures and pressures other than those of the experiment can be calculated from this equation for K_D

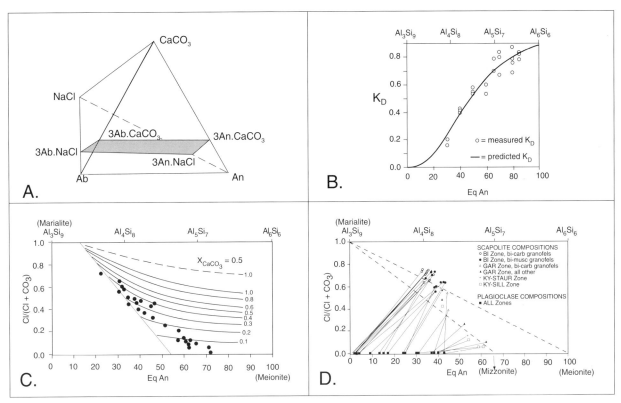

Figure 6.6. A) Compositional space for marialite-meionite. The shaded region is the theoretically possible plane of scapolite compositions. B) Measured values of K_D for the reaction $NaCl_{(scap)} + CaCO_{3(fl)} = CaCO_{3(scap)} + NaCl_{(fl)}$ at 750°C and 4000 bar plotted against EqAn. The predicted line $LnK_D = (-0.0028)[Al/Al + Si]^{-5.5580}$ is shown for comparison. C) The correlation between scapolite composition and fluid composition at 750 °C and 4000 bar predicted by the equation $LnK_D = (-0.0028)[Al/Al + Si]^{-5.5580}$. Lines of equal $NaCl/(NaCl + H_2O)$ in coexisting fluid are shown. The compositions of natural scapolites are shown as black dots. D) Compositions of coexisting scapolite and plagioclase in the Wallace Formation, Idaho. Plagioclase compositions are plotted along the horizontal axis. Dashed lines are solid solutions between marialite, meionite and mizzonite. In general, Cl in scapolite decreases with increasing metamorphic grade; but compositional variations within each grade suggest local control of fluid and scapolite compositions (B and C after Ellis, 1978; D after Mora and Valley, 1989).

(Figure 6.6c). This implies that Cl-poor natural scapolites coexisted with fluids low in NaCl and that regional occurrences of Cl-rich scapolites are likely to represent metamorphosed evaporite sequences.

Experimental results of Ellis (1978) have been used by Mora and Valley (1989) and Oliver et al. (1992) in studies of meta-evaporite paragenesis to estimate the salinity of the metamorphic fluids based on the composition of Cl-bearing scapolite. Mora and Valley found that abundant Cl-rich scapolite and biotite in the Wallace Formation, Idaho, indicate high Cl activities during greenschist-through amphibolite-facies regional metamorphism of metasedimentary rocks (Figure 6.6d). In general, Cl in scapolite decreases with increasing metamorphic grade; but compositional variations within each grade suggest local control of fluid and scapolite compositions probably related to the original distribution of halite in the sequence.

The abundance of Cl-rich phases decreases abruptly with the appearance of zoisite-bearing assemblages, suggesting infiltration of low-Cl H_2O at higher metamophic grades. The occurrence of buffered activity gradients in Cl, as indicated by scapolite and biotite compositions, also suggests that metamorphic fluid flow was highly channelised and aqueous fluid-rock interaction was limited. Similar channelisation defines scapolite distribution in the meta-evaporitic Corella Formation in the Mt Isa inlier (see later).

In applying postulates to their Cl-rich scapolites, Mora and Valley (1989) recognised limitations in the work of Ellis (1978), such as too high a temperature in the experiment for many meta-evaporites, and the high degree of uncertainty in terms of the nature and composition of the $NaCl-H_2O-CO_2$ fluid including the degree of immiscibility.

Generally, at temperatures up to 400-500°C, the NaCl in scapolites may be derived directly from solid halite, that is, associated with the reaction:

halite + plagioclase + calcite --> scapolite.

At higher temperatures the production of H_2O and CO_2 from devolatilisation reactions in adjacent metasediments tends to dissolve the remaining solid halite (Skippen and Trommsdorf, 1986; Oliver et al., 1992). Scapolite growing in palaeo-evaporitic metapelites at temperatures above 400 - 500 °C is probably not in equilibrium with halite but with a halite-undersaturated NaCl-bearing fluid. Some dispersed halite can remain up to the melting point of halite at 750-800°C. For example, solid inclusions of halite have been found in several samples of calcsilicate rocks from Switzerland and northern Italy that have been metamorphosed to lower amphibolite grade with the development of talc and tremolite. Idiomorphic crystals and dendrites of halite have been analysed and show a range of compositions in the system NaCl-KCl, which indicate temperature maximums greater than 500°C and pressures of 2 kbar or higher (Trommsdorff et al., 1985).

Gómez-Pugnaire et al. (1994) found that the existing well-studied relationships and assumptions of the alliance between scapolite, calcite and plagioclase (e.g. Aitken, 1983) were not the explanation of the meta-evaporitic scapolites of the Cóbdar region of Spain. There biotite, not albite, was involved in scapolite formation. This Permo-Triassic pelite-carbonate rock series (with intercalated metabasitic rocks) in the Cordilleras Beticas, Spain, was metamorphosed during the Alpine metamorphism with peak metamorphism at eclogite facies with high pressures (P_{min} near 18-22 kbar) and relatively low temperatures (≈650°C). These evaporitic metapelites have a whole rock composition characterised by high Mg/(Mg + Ca) ratios > 0.7, variable alkaline and Sr, Ba, contents, but are mostly K_2O rich (< 8.8 wt%). The F (< 2600 ppm), Cl (< 3600 ppm), and P_2O_5 (< 0.24 wt%) contents are also high. The pelitic member of this series is a fine grained biotite rock, where kyanite-phengite-talc-biotite aggregates in pseudomorphs developed in the high pressure stage. Albite-rich plagioclase was formed when the rocks crossed the albite stability curve in the early stages of the uplift. Scapolite, rich in NaCl [Ca/(Ca + Na) ≈ 24-40 mol%] and poor in SO_4, with Cl/(Cl + CO_3) ratios between 0.6 and 0.8, formed as porphyroblasts, sometimes replacing up to 60% of the rock in the later decompressional stage of metamorphism (between 10 and 5 kbar, near 600°C).

The Spanish scapolites are rich in Na and poor in CO_3. Plagioclase is not consumed by scapolitisation; but biotite,

the most important matrix material, almost completely disappears in areas of scapolite blastesis. Gómez-Pugnaire et al. (1994) conclude that scapolite formed from a biotite precursor by:

Al-biotite + $CaCO_3$ + NaCl + SiO_2 = Al-poor biotite + scapolite + $MgCO_3$ + KCl + $MgCl_2$ + H_2O

The rocks still retain well preserved sedimentary indicators of former evaporites, such as pseudomorphs after sulphate evaporites, which are now composed of talc, kyanite-phengite-talc-biotite, and quartz. In addition, there are relicts of baryte, anhydrite, NaCl and KCl still present in the rock matrix. All the metasedimentary evidence indicates a salt-clay mixture as the protolith, which was dominated by illite, chlorite, talc, and halite. Calculated fluid composition in equilibrium with scapolite indicates varying salt concentrations in the fluid, and the distribution of Cl and F in biotite and apatite also indicates varying fluid compositions.

Tourmaline

Tourmaline is a metamorphic mineral that forms in a wide range of nonevaporitic as well as evaporitic settings. It constitutes a complex series of aluminium borosilicates with the general formula: $XY_3Al_6Si_6O_{18}(BO_3)_3(OH,F,O)_4$. X represents large cations such as calcium, sodium and/or rarely potassium, while the Y cation represents smaller cations such as iron (+2 and/or +3), magnesium, aluminium, lithium and/or manganese (Table 6.2). The aluminium in the Al_6 site can be substituted for on a limited basis by iron and more rarely chromium or vanadium. Fluorine and oxygen substitute for hydroxide in a few of the members and then only to a limited extent. For a full discussion of the various petrogenetic indicators in tourmalines read Henry and Guidotti (1985).

Three solid solutions characterise the most common tourmalines in nature (Figure 6.7a,b):

- between the Fe-rich (schorl) and the Mg-rich (dravite) tourmalines,
- between schorl and the Li-rich (elbaite) tourmalines, and
- between dravite (Na>Ca) and uvite (Ca>Na) tourmalines.

The dravite-uvite series, is typical of tourmalines with meta-evaporite and/or metacarbonate affinities and has end member compositions: dravite [$NaMg_3Al_6Si_6O_{18}(BO_3)_3(OH)_4$], and uvite [$Ca(Mg,Fe)_3MgAl_5Si_6O_{18}$

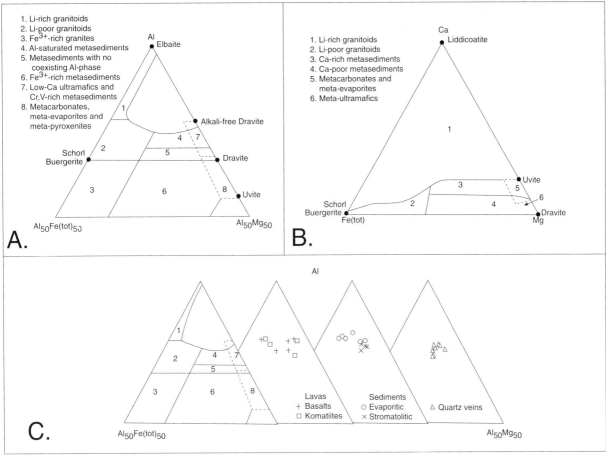

Figure 6.7. Tourmaline composition. A) Al-Fe$_{(Tot)}$-Mg diagram in molecular proportions for tourmalines showing compositional range of various rock types, note overlap of fields 4 and 5 with field 7 (after Henry and Guidotti, 1985). B) Ca-Fe$_{(tot)}$-Mg diagram in molecular proportions for tourmalines showing compositional range of various rock types, note overlap of field 4 with field 6 (after Henry and Guidotti, 1985). C) Compositional variations of various Barberton rocks using plot of A for comparison (after Byerly and Palmer, 1991).

End Member	Formula
Schorl	NaFe^{2+}Al$_6$(BO$_3$)$_3$Si$_6$O$_{18}$(OH)$_4$
Dravite	NaMg$_3$Al$_6$(BO$_3$)$_3$Si$_6$O$_{18}$(OH)$_4$
Tsiliasite	NaMn$_3$Al$_6$(BO$_3$)$_3$Si$_6$O$_{18}$(OH)$_4$
Elbaite	Na(Li,Al)$_3$Al$_6$(BO$_3$)$_3$Si$_6$O$_{18}$(OH)$_4$
Uvite	CaMg$_3$(MgAl$_5$)(BO$_3$)$_3$Si$_6$O$_{18}$(OH)$_4$
Liddicoatite	Ca(Li,Al)$_3$Al$_6$(BO$_3$)$_3$Si$_6$O$_{18}$(OH)$_4$
Buergerite	NaFe^{3+}$_3$(BO$_3$)$_3$Si$_6$O$_{22}$F
Ferridravite	NaMg$_3$Fe^{3+}$_6$(BO$_3$)$_3$Si$_6$O$_{18}$(OH)$_4$

Table 6.2. Common end members of the tourmaline series.

(BO$_3$)$_3$(OH)$_4$]. Brown and Ayuso (1985) and Henry and Guidotti (1985) also showed that tourmalines associated with meta-evaporites and metacarbonates in Precambrian sequences in North America are Mg-rich members of the schorl-dravite series with significant Ca-uvite components.

The restriction of boron-rich minerals such as tourmaline to particular stratiform units in a metamorphosed sequence is used by some authors to indicate likely meta-evaporite horizons (Abraham et al., 1972). However, even in meta-evaporitic settings, the boron ultimately captured in stratiform tourmalines may have come from:

• diagenesis or regional metamorphism of originally bedded borate mineral deposits (see below; borates in eastern Liaoning, China: Peng and Palmer , 1995);
• from fluid circulation and hydrothermal activity in areas underlain by dissolving continental evaporites, such as early rift basin activity (see below; Willyama complex: Slack et al., 1993); and
• from ongoing submarine fumarolic and hydrothermal activity (Byerly and Palmer, 1991).

Type 1 may well pass to type 2 and then 3 as a continental rift basin evolves into a passive continental margin.

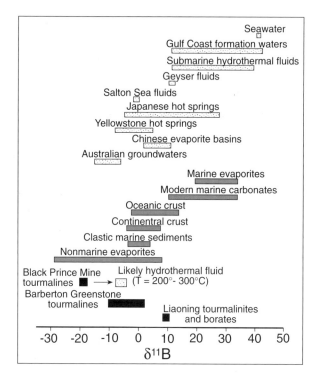

Figure 6.8. Boron isotopic composition of tourmalines in the Black Prince Mine, Broken Hill District, Australia, the Barberton greenstone belt, South Africa, and the Lianong region, China, (solid black boxes) compared with various boron sources. Stippled boxes indicate fluids, hatching indicate solids (after Slack et al., 1989; Byerly and Palmer, 1991; Peng and Palmer, 1995).

The utility of endmember diagrams and compositional field divisions is only as good as our understanding of the chemical system that they represent. The tourmalines in sediments and lavas from the Archaean Barberton greenstone belt in South Africa show similar compositional variations (Figure 6.7d; Byerly and Palmer, 1991). Using a Al-Fe(tot)-Mg plot puts many of the samples in the field of immiscibility between dravitic and elbaitic tourmalines (Figure 6.7c). The remainder plot in the Al-saturated metasedimentary field and are well away from the uvite compositional field. Only the Barberton tourmalines associated with stromatolites and quartz veins show a degree of Mg enrichment.

In my opinion, this quandary underlines a difficulty that seems to pervade the metamorphic literature, namely simplistic notions of the homogeneity of the sedimentary protolith. Not all borate evaporites contain high levels of calcium, and most mined borate salts contain little or no magnesium (see borates in Chapter 7). This means the magnesium component in the protolith to meta-evaporitic tourmalines comes mostly from the host sediments that

entrain the borate salts. There can be only a very minor contribution from the borax salts. If the host matrix contains dolomites and other carbonate minerals and the borate salts are dominated by colemanite or ulexite (major ore minerals in Turkey) there will be a high level of Mg and Ca in the system and dravite-uvite tourmalines will form. But, it is worth noting that there is not any calcium in lacustrine borates composed of borax or kernite (major salts in the US borate deposits). In many borate deposits the host volcanogenic matrices are aluminosilicate clays (such as smectite) with some zeolites. Analcime for example contains no Mg (see Chapter 7). Without evaporation of appropriate low salinity bicarbonate-rich inflow waters in the depositional basin the levels of magnesium and calcium carbonates in the matrix may be quite low. When borate salts in such a systems start to metamorphose and supply ions that will precipitate tourmaline, there may be little or no Ca or even Mg supplied to the system. The result will be meta-evaporitic tourmalines that do not plot in field 4 of Figure 6.7b.

Perhaps the most reliable method to distinguish between possible origins of tourmalines is to measure their $\delta^{11}B$ isotopic signature. Palmer (1991) describes how boron isotopic signatures vary in a predictable manner. Tourmalinites associated with marine evaporites have a high $\delta^{11}B$ signature near +10‰, tourmalinites associated with mafic volcanics have a signature near -2‰, tourmalinites from continental margin sediments have values near -15‰, while tourmalinites derived from nonmarine evaporites have very low isotopic signatures near -20‰. Figure 6.8 shows how Slack et al. (1989) determined a likely nonmarine evaporite source for the tourmalinites of the Black Prince Mine in the Broken Hill District, while samples from the Barberton Greenstone Belt indicate a mixed volcanic and sedimentary protolith.

Sodian phyllosilicates and talc

In Triassic meta-evaporites from Derrag, in the Tell Atlas of Algeria, sodian phlogopite occurs naturally as an unusual meta-evaporitic product in a sugary dolomite matrix that also hosts other magnesian phyllosilicates, as well as large porphyroblasts of albite, dravite-uvite tourmalines, quartz, scapolite, magnesite, rutile and pyrite (Schreyer et al., 1980). The phlogopite has a microprobe-determined chemical formula that is close to ideal $(NaMg_3[AlSi_3O_{10}](OH)_2$ with some possible addition of lithium. Crystals are invariably coated with thin rims of the more usual potassium phlogopite and these rims probably prevented the retrograde hydration of the sodian phlogopite. It coexists with an enigmatic talc-containing phase called kulkeite.

Kulkeite is an unusual sodian phyllosilicate mineral in this Triassic sequence, which appears to be associated with the meta-evaporites. It is an ordered 1:1 chlorite/talc mixed-layer mineral that forms platy, colourless, porphyroblastic, single crystals in its low-grade sugary dolomite host (Schreyer et al., 1982). From microprobe analysis kulkeite's empirical formula is $(Na_{0.38} K_{0.01} Ca_{0.01})(Mg_{8.02} Al_{0.99})[Al_{1.43} Si_{6.67} O_{20}](OH)_{10}$ with some variation in Na, Si and tetrahedral Al.

Both the sodian phlogopite and the kulkeite formed during low grade metamorphism of an evaporite protolith at temperatures that did not exceed 400°C. According to Schreyer et al. (1980, 1982) these meta-evaporitic aluminosilicates are the first known natural examples of sodian phlogopite and the first documentation of an aluminosilicate showing a regular mixed interlayering between chlorite and talc. Both occur in dolomites interbedded with clays and thick salts that were originally buried to depths varying from 2 to 10 km and then returned to the surface by halokinesis. The close proximity of clastic silicate rocks with dolomite and salt (mostly halite but with some K, Mg salts), as well as deep burial prior to exhumation, are thought to have been responsible for this association.

Sodic talc- kyanite is another likely meta-evaporite indicator (Schreyer, 1976; Schreyer and Abraham, 1976). Once again the fluid-driven chemistries of meta-evaporitic sequences do not fit easily into a conventional analysis of regional metamorphism. Meta-evaporites in the Sar e Sang region of Afghanistan exhibit mosaic equilibria across small volumes (in the cm^3 range) within a talc-kyanite schist (whiteschist) host. The microscale mineral variations are characterised by variations in mineral assemblages conventionally attributed to vastly different pressure/temperature conditions during regional metamorphism.

On the basis of petrographic and microprobe studies, these assemblages are attributed to three consecutive stages of metamorphism of a chemically exceptional rock with a composition that falls largely into the model system MgO-Al_2O_3-SiO_2-H_2O (Schreyer and Abraham, 1976). Stage 1, typified by Mg chlorite-quartz -talc and some paragonite, was followed during stage 2 by talc-kyanite, Mg gedrite-quartz and the growth of large dravites. Microprobe analyses of the phases gedrite and talc indicate variable degrees of sodium incorporation into these phases according to the substitution NaAl—>Si. In stage 3, pure Mg cordierite formed with or without corundum and/or talc, and the kyanite was partly converted into sillimanite. Pressure and temperature during this final stage of metamorphism was near 5-6 kb and 640°C.

Schreyer and Abraham (1976) concluded that chemical variations in the metamorphic fluids were generated by progressive metamorphism and mobilisation of an evaporite deposit. Relict anhydrite and gypsum still occur in the Sar e Sang area. Whiteschists and the associated lapis lazuli deposits of the region are part of a highly metamorphosed evaporitic succession (see the later discussion of lapis lazuli in this chapter). Salts have largely vanished due to ongoing melting and volatilisation. The preservation of the three stage succession of mineral assemblages, across such small scales and yet related to each other through isochemical reactions, means that the main factors governing the metamorphic history of this whiteschist were compositional changes of the coexisting fluids with time. Under this scenario any pressure-temperature variations were subordinate and the chemistry of the fluids evolved as the evaporites underwent metasomatic alteration.

The sedimentary pelitic layers of this evaporitic sequence first underwent a period of metamorphism in which fluid pressures approached lithostatic (stage 1). Subsequently at higher metamorphic grades, with the beginning of mobilisation of the salts, the metamorphic fluids became increasingly enriched in ions such as Na^+, Mg^{2+}, Cl^-, SO_4^{2-}, BO_3^{3-}, etc., so that water fugacity dropped considerably. This period is represented by stage 2 of the whiteschist metamorphism and was characterised by strong metasomatism that led, for example, to the growth of dravite. The physical and chemical character of stage 3 is less clearly defined. Kyanite/sillimanite inversion requires an increase in temperature or a decrease in pressure, or both; but changes in the composition of a coexisting gas phase may have played an additional role in the formation of cordierite.

In summary, this meta-evaporite-derived assemblage may in a single thin section entrain mineral assemblages that conventionally would be assigned to the greenschist facies, the hornfels facies, and to a high pressure regime. The assemblages are in effect a mosaic equilibria that reflects changes in fluid composition generated from a metamorphosing evaporite pile over time and only to a lesser degree by regional evolution of total temperature and pressure. Once again evaporites generate unusual responses compared to the general responses of metasediments.

Talc ($Mg_6[Si_8O_{20}][OH]_4$) also occurs as an authigenic component in evaporites that have yet to experience metamorphic grade burial temperatures and pressures. Talc has been recorded in Permian evaporites from New Mexico (Bailey, 1949), in Carboniferous evaporites in Nova Scotia (Evans, 1970) and from the Eskdale boreholes in NE Yorkshire (Scrivener and Sanderson, 1982). In Yorkshire it occurs as minor quantities of isolated flakes in the halite-rich parts of four evaporite breccia beds that also contain minor anhydrite, dolomite and quartz. In the Yorkshire deposit, the mineral is always enclosed by halite, but in New Mexico it is also found in the cleavage cracks of sylvinite. Both of these localities are associated with magnesium-bearing salts such as carnallite and kieserite. In the case of the Yorkshire deposits, the nature of the sediments enclosing the evaporites and in the breccia

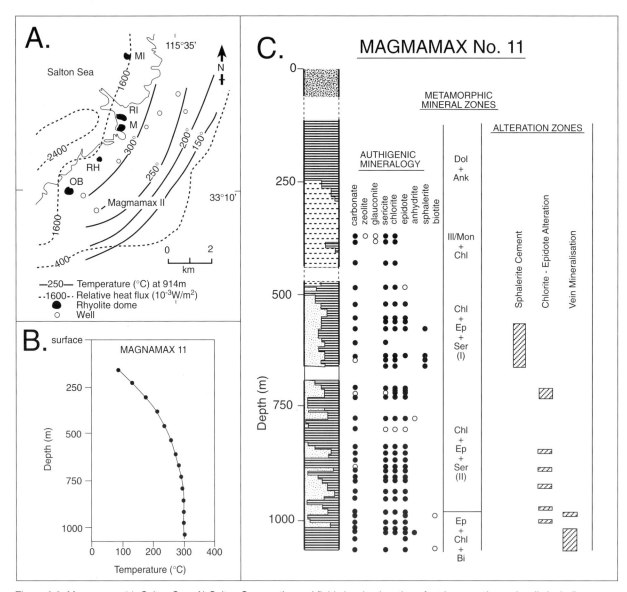

Figure 6.9. Magmamax 11, Salton Sea. A) Salton Sea geothermal field showing location of various geothermal wells including Magmamax 11 (solid dots). Also shown are the isotherms at 914m (3000 ft) and contours of relative heat flux. Black areas indicate Quaternary rhyolite domes extruded onto the lacustrine sediments (OB = Obsidian Butte, Rh = Red Hill, RI = Red Island, MI = Mullet Island. B) Downhole temperatures from Magmamax 11. Bottom hole temperature is 305°C. C) Sediment types and alteration zones in Magmamax 11 well. Solid circles indicate major authigenic phases; open circles, minor phases. Abbreviations: dol = dolomite, ank = ankerite, ill = illite, mon = montmorillonite, chl = chlorite, ep = epidote, ser = sericite, bi = biotite, I = epidote detectable by XRD only, II = epidote detectable optically (after McKibben et al., 1988a, b).

suggests a salt flat or sabkha environment of deposition. There is no evidence that the talc is depositional and Scrivener and Sanderson (op. cit.) suggest that postdepositional hypersaline conditions permitted the diagenetic reaction of dolomite with silica to form talc.

Meta-evaporites examples

The preceding discussion clearly shows that meta-evaporite interpretations are typically not based on the presence of actual salts but on the recognition of alteration products related to the mobilisation, volatilisation and dissolution of the precursor salts. What follows is a discussion of a number of meta-evaporitic terranes. The order of presentation is not chronological but rather in terms of the strength or the reliability of the meta-evaporite interpretation. Ancient meta-evaporite localities range from archetypical occurrences, with excellent preservation of the evaporite salts or their pseudomorphs, along with widespread sodic and potassic aluminosilicates, through to inferred areas with little more evidence for meta-evaporites than the presence of occasional sodic, magnesian and chloride-rich mineral phases.

Anhydrite in greenschist conditions,

Salton Sea

The Salton Trough is a continental rift zone, forming the transition from the divergent tectonics of the East Pacific Rise to the strike-slip tectonics of the San Andreas Fault System (Figure 6.9a; McKibben, et al., 1988a,b; Osborn, 1989). Infilling of the rift by the bi-polar Colorado River delta for the past 4 Ma has isolated the northern part of the trough, forming the closed Salton Basin in an orographic desert. High evaporation rates combined with an extremely variable hydrologic budget have resulted in the episodic formation of saline lakes and lacustrine evaporites. Rapid subsidence and rift-related intrusions at depth subject the Pleistocene and younger sediments within the Salton Sea geothermal system (SSGS) to temperatures up to 365°C and flushing by hypersaline brines with TDS up to 26 wt% at depths of 1-3 km. The geothermal system is associated with an arc of five Quaternary rhyolite domes (Figure 6.9a,b). The high salinity of the brines reflects dissolution of lacustrine evaporites (McKibben et al., 1988a, b).

Sediments are metamorphosed mainly to greenschist facies mineral assemblages: actinolite-quartz-epidote-biotite-andradite (Figure 6.9c; Muffler and White, 1969). Cores recovered from a depth of 3 km contain two-amphibole assemblages marking the transition from greenschist to amphibolite facies. Calcic amphiboles with up to 2.7 wt.% Cl occur in metasandstones, metabasites and veins at depths between 3100 to 3180 m where the temperatures are in excess of 350°C (Enami et al., 1992). There the Cl-bearing amphiboles range in composition from hastingsitic (Cl>1 wt.%) to actinolitic (Cl<0.5 wt.%). These amphiboles were formed by reactions involving high-salinity geothermal fluids, with 15.4 to 19.7 wt.% total dissolved Cl. Coexisting phases include quartz, plagioclase, K-feldspar, epidote, clinopyroxene, apatite, and titanite.

Stratiform calcium sulphate structures observed in drill cores in the SSGS were deposited in various playa-sabkha subenvironments: laminated to massively bedded sulphates were deposited subaqueously in perennial and ephemeral saline lakes; gypsum (now pseudomorphed by anhydrite) was precipitated within fluid-saturated sediments in lake-fringing saline mudflats; and nodular sulphates were formed in the capillary zone of surrounding dry mudflats (McKibben et al., 1988a). Gypsum is the only primary evaporitic sulphate found in the SSGS, it undergoes dehydration to anhydrite at 80-105° C and depths of 200-250m. Fluid inclusions in gypsum contain 4-12 wt.% equivalent NaCl fluids (Osborn, 1989). Gypsum dehydration apparently dilutes these fluids, forming inclusions in anhydrite that contain <4 wt.% equivalent NaCl fluids. Fluid inclusion homogenization temperatures closely parallel the thermal profile measured in boreholes, implying continual recrystallisation of anhydrite to maintain thermal equilibrium with the prevailing geothermal gradient. Continuous recrystallisation allows heterogeneous mixing of relatively dilute fluids from inclusions with more concentrated geothermal brines, producing inclusions containing fluids ranging from 13-25 wt.% equivalent NaCl. These coexist spatially with 25 wt.% equivalent NaCl geothermal brines.

The result of increasing temperature is a hornfelsic to granofelsic recrystallisation texture in the anhydrites. Bedded anhydrites are still present as a 0.6 m bed at a depth of 1,000 m and a temperature of 305°C in the Magmamax 11 well. Most of the anhydrite is made up of interlocking granular anhedral grains, which alternate with marly and pelitic laminae that still entrain small anhydrite nodules and clusters (Figure 6.9c). Original small-scale sedimentary growth banding and other depositional textures in the original evaporites were largely obliterated by metamorphic recrystallisation. Shales in a 0.8 m thick section immediately below this anhydrite are extensively brecciated, forming angular fragments cemented by clean, coarse anhydrite.

The angular nature of the shale breccia fragments, their sharp contact with the anhydrite cement and the lack of fine matrix between them implies that they were brecciated after lithification of the original mudstone. The breccia is interpreted as after halite dissolution (McKibben et al., 1988a). Dissolution at depth of former halite beds to supply much of the deeper high salinity brine is reinforced by the work of Bozkurt (1989). He recognised halite relics in calcite at 1240 m depth and interpreted this halite as a remnant of once more-extensive beds.

Below the breccia zone in the Magmamax 11 well is a 4.7m thick sequence of thin interbedded anhydrite and shale. The layers of anhydrite are composed of a mosaic of interlocking grains. Isolated coarse crystals of anhydrite occur scattered throughout the shale layers, possibly representing porphyroblasts or pseudomorphs formed in place of former gypsum crystals.

Metaplaya sequences, Damara Orogen (Namibia)

Evaporitic metaplaya strata, up to 1000 m thick, occur in the Neoproterozoic rift sediments of the Duruchaus Formation in the Damara Orogen, Namibia (Behr and Horn, 1982; Schmidt-Mumm et al., 1987). Deposition occurred some 950-650 Ma, while orogeny and overthrusting occurred 650-450 Ma. Morphologically well-preserved pseudomorphs of Na-carbonates (shortite, trona, gaylussite), borates (borax, colemanite, ulexite) and sulphates (thernadite, gypsum, baryte) indicate a playa metasedimentary precursor, as do laterally extensive albitites and tourmalinites (Figure 6.10). All of the pseudomorphs are now composed of microcrystalline albite with varying amounts of dolomite, calcite, tourmaline, rutile and chloritoids.

Ferroan dolomite is the most important carbonate phase in the region and at least five generations of dolomite can be distinguished in the area (Behr et al., 1983). The oldest is micritic dolomite, which is subsequently overprinted by successive coarser-grained dolospars. The composition of the dolomites varies markedly. Groups of $Ca_{90}Mg_{10}$, $Ca_{70}Mg_{30}$ and $Ca_{55}Mg_{45}$ were identified that also contained 5% to more than 10% FeO. Calcite occurs only locally and appears to be a product of dedolomitisation or hydrothermal alteration associated with the later stages of dolomitisation

A high NaCl content in the protolith is indicated by numerous horizons dominated by Cl-scapolite with thicknesses up to several metres (Figure 6.10). Scapolite is best developed in dark grey dolomitic mudstones in the upper part of the Duruchaus Formation (Facies E). It is associated with albite, phlogopite and quartz; dolomite exceeds albite, and phlogopite can comprise up to 30% of the total rock. Individual scapolite crystals are subhedral, up to 2 cm long and can make up to 80% by volume in some layers. The scapolitic dolomites still preserve mud cracks, depositional laminae and ripples. The scapolite typically replaces calcite and phlogopite and is often associated with newly formed sparry dolomite. Average make-up is SiO_2 52-54%, Al_2O_3 21-25%; Na_2O 8-9%, CaO 9-10%; SO_3 0.7%; Cl 2.5%. Meionite content is calculated as 45-50%. Behr et al (1983) postulate the reaction:

$$phlogopite + albite + NaCl(aq) + calcite \rightarrow scapolite + dolomite$$

Both the chlorine and the sodium were supplied by the metamorphism of calcareous shales containing varying amounts of halite, and were not from a magmatic source.

Laminated albite-dolomite rocks (albitites or albitolites) are also commonplace and are characterised by albite crystals up to 5 cm long (Facies D). Albite can make up more than 80% of the rock by volume, and dolomite proportions range up to 40% of the rock volume. Quartz, minor biotite and metamorphic muscovite are the other rock-forming minerals. Lamination is often distorted and brecciation is also widespread, possibly indicating the removal of former evaporites. The albite is usually crypto- to microcrystalline (10-30 µm) and is almost pure (CaO≈0.3%; K_2O≈0.1%). Occasional larger albite crystals occur as lath-like forms (up to 500µm) in brecciated zones. Only simple albite twins have formed and Roc-Tourne twins, so common in authigenic albites, were not found in the region.

Albite is thought to have formed during late diagenesis and the onset of metamorphism in the evaporite successions from: 1) clay minerals, 2) evaporites and 3) the alteration of tuffs (Behr et al., 1983). Albite can form in the presences of saline highly alkaline waters in the following fashion:

$$illite + NaCl_{(aq)} + SiO_{2(aq)} \rightarrow NaAlSi_3O_8 + HCl_{(aq)} + H_2O + KCl_{(aq)}$$

The liberated HCl was buffered by carbonates in the matrix as the protolith was replaced by albite, a similar evolution to that documented in Triassic salts of north Africa (Kulke, 1978). This style of albitisation is largely restricted to the pelitic sediments of the Duruchaus Formation, and the albites typically includes relics of sericite.

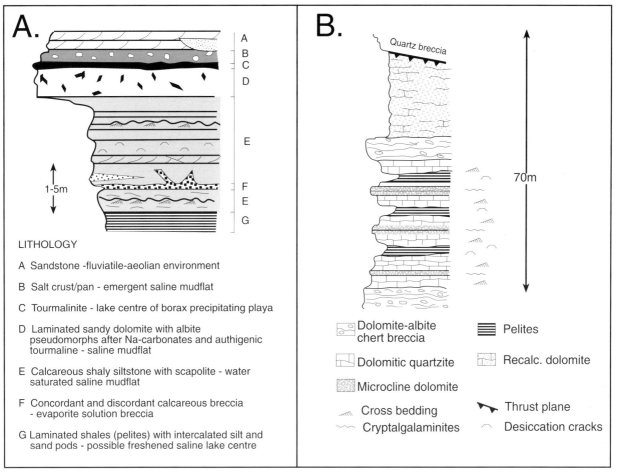

Figure 6.10. Stratigraphy of the metasediments in the Duruchaus Formation, Namibia. A) Idealised vertical sequence of the metaplaya sequence Geelkop Dome, Damara showing distribution of meta-evaporite indicators in the profile. B) Generalised section of the Naos dolomite sequence showing position of microcline dolomite and breccias, which indicate former evaporite horizons (after Behr et al., 1983).

Primary sodic salts in alkaline playas, such as gaylussite and trona, typically transform during dewatering to pirssonite and then to shortite. As burial proceeds and temperatures increase, the shortite typically transforms to albite via:

$$Na_2Ca_2(CO_3)_2 + 2Al_3OOH_{(aq)} + SiO_{2(aq)} --> 2NaAlSi_3O_8 + 2CaCO_3 + H_2O + CO_2$$

Carbonate liberated by this reaction is typically preserved as calcite and dolomite within albite pseudomorphs of the former sodium salts.

Albitites with a tuff precursor typically form dm-thick albite layers and tend to be associated with volcanics on the southern Damara margin. The pervasive recrystallisation associated with this transformation means relict shard textures or analcime and other zeolites are no longer present. However, this form of albitite typically retains elevated levels of detrital rutile and biotite. The proposed reaction is:

tuffitic glass + Duruchaus brines $--_{(pH>9)}-->$ Na-Al-Si(gel) $-->$ zeolites $--_{(+Na)}-->$ analcite $--_{(90°C; +SiO2)}-->$ albite

Microcline is another conspicuous mineral in Duruchaus sediments and is especially obvious in the Naos dolomite sequence that occurs toward the top of the Duruchaus Formation (Figure 6.10b). Microcline is present in a number of forms: 1) as microcrystalline anhedral crystals in the albite-phlogopite-quartz matrix of the pelites; 2) as rosette-like pseudomorphs after gypsum in sabkha facies sediments —these crystals are up to 3 cm long and are often arranged perpendicular to bedding planes in cryptalgal laminites; 3) as planar, up to 1 cm long crystals in dolomite layers of calcareous shales; 4) as crystal linings in quartz-filled

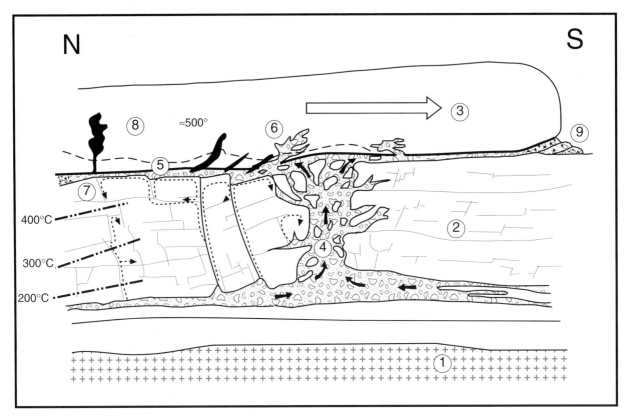

Figure 6.11. Processes associated with the hot Kudis Nappe overriding Duruchaus Formation. 1 = basement; 2 = Duruchaus Formation; 3 Kudis Nappe; 4 = discordant breccia; 5 = carbonate-quartz breccia and sole dolomite; 6 = breccia injected into faults and shear planes of Kudis rocks; 7 = zone of strong hydrothermal metasomatic reactions liberated by dewatering of evaporite horizons; 8 = pipes plugs and veins filled by quartz precipitates derived from highly saline hydrothermal solutions; 9 = breccias generated by overriding nappes (after Schmidt-Mumm et al., 1987).

shear fractures; and 5) as hydrothermally formed stratiform veins opened by hydrofracturing and the intrusion of a carbonate mush. Most of the microcline has an adularian habit with individual crystals having a dirty core and clear rims. It is interpreted by Behr et al. (1983) to have formed via hydrothermal crystallisation at the end of deformation stage.

Also within this meta-evaporite sequence are stratiform tourmalinites up to several metres thick that can be traced over areas measured in tens of kilometres and extending continuously for more than 20 km (Facies C). Electron microprobe analysis shows that they are dravites with a significantly high Mg content [average $Fe/(FeO + MgO) = 0.28$]. The sedimentary precursor of these tourmalinites is seen in their well-preserved finely laminated primary bedding along with preserved dewatering and crossbedded structures. Some of the tourmalinite is made up of fine alternating laminae of tourmaline and albite. These units are interpreted to indicate times of hot spring activity and were associated with times of widespread borate deposition

in the playa (see Chapter 7). Tourmalinites are of high economic significance in the area as they are commonly associated with stratabound accumulations of Cu, Au, Ti, U and megaquartz.

Metamorphism of the playa sediments, as deduced from the biotite isograd of the intercalated pelitic siltstones, and temperature-depth estimates indicated by the CO_2 and H_2O-NaCl isochore intersections, attained temperatures of 550°C at depths of 5 to 8 km burial (upper greenschist facies). During this rise of temperature the large fluid volume in the enclosing dolosparite caused, by aquathermal pressuring, a mechanical destruction of the grains and a liberation of the fluid phases (Behr and Horn, 1982). Additional fluids were liberated by the dehydration of the Na-carbonates, Na-silicates and Na-sulphate minerals that were concentrated in particular layers. This abnormal fluid pressure broke up the palaeoevaporite layers to form a highly mobile cataclastic sludge. Preserved fragments within the sludge range from several cubic micrometres to metres.

This process of palaeoevaporite fluidisation seems to have been initiated by the rapid overthrusting of nappes composed of hot (≈540°C) metasediments that were squeezed out of the Khomas Trough during the orogeny. Nappe emplacement formed a hot lid some 5 km thick atop the cooler meta-playa sediments (Figure 6.11). The underlying mobile carbonate sludge, enriched with palaeoevaporite pseudomorphs, was partly squeezed into the thrust plane of the Khomas Nappe. Three sequential stages of brecciation were distinguished (Behr et al., 1983; Schmidt-Mumm et al., 1987):

1) A stratiform breccia formed by dewatering of water-rich evaporites in layers now enriched in palaeo-evaporitic pseudomorphs.
2) With increasing mobilisation of the saline fluids a stronger style of brecciation occurred. It was associated with aquathermal pressuring due to abnormal pore pressures and it created strongly and mostly internal hydrofractured concordant units.
3) At the final stage a very ductile discordant breccia containing xenoliths of the wall rock formed. It intruded the wallrock along faults, fissures and bedding planes.

A further breccia evolved from the type-(c) breccia that, along with diapiric ascent of the breccia mush, was intruded into the contact zones with the basement nappes. By acting as a lubricant it re-initialised or further enhanced the southwest-directed movement of the nappes (Figure 6.11). The associated flushing of large volumes of hypersaline and alkaline solutions resulted in a strong hydrothermal and metasomatic alteration of the country rock along with mineralisation of large quartz-carbonate bodies. Single quartz crystals within these mineralisations can have crystal faces up to 50 m across and are intergrown with large dolomite crystals up to a metre long.

This region is an interesting meta-evaporite terrane for a number of reasons: it provides excellent evidence of how sodic metasomatism is controlled by former evaporites; it offers outcropping examples of crosscutting evaporite derived breccias; and it shows how dravite-tourmalines and albitites can be directly related to former borate, sodium sulphate and halite salts in lacustrine settings. What is yet to be considered is how much of the metasomatism, brecciation and diapirism was related to replacement and haloes set up during the rift stage then overprinted and reworked during the subsequent orogeny. This distinction has obvious economic implications in terms of localities of prepared ground for sulphide emplacement.

Borates of eastern Liaoning, China

Palaeoproterozoic boron deposits in eastern Liaoning, China, are now considered to be meta-evaporites that have formed with little or no fluid input from the granites of the region (Peng and Palmer, 1995; Jiang et al., 1997). The borate minerals of economic interest are suanite ($Mg_2B_2O_5$), szaebelyite or ascharite ($Mg_2B_2O_5.H_2O$), and ludwigite [(Fe, $Mg)_4Fe_2B_2O_7$]. The borates are hosted within crystalline Mg-silicate and Mg-carbonate rocks, but exhibit laminar textures implying a sedimentary precursor.

A generalized stratigraphy (from the base up) in of the Liaoning area is comprised of (Figure 6.12; Peng and

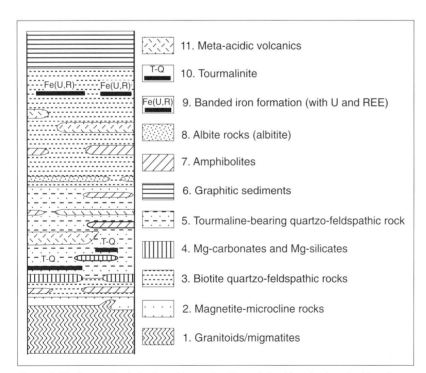

11. Meta-acidic volcanics

10. Tourmalinite

9. Banded iron formation (with U and REE)

8. Albite rocks (albitite)

7. Amphibolites

6. Graphitic sediments

5. Tourmaline-bearing quartzo-feldspathic rock

4. Mg-carbonates and Mg-silicates

3. Biotite quartzo-feldspathic rocks

2. Magnetite-microcline rocks

1. Granitoids/migmatites

Figure 6.12. Generalised stratigraphic section through the Liaoning borate deposits (after Peng and Palmer, 1995). Various proportions of the numerically listed lithologies (1 - 10) make up the metamorphic facies types I (lowest) through VI (highest). See text for discussion of these units.

Palmer, 1995): I) Magnetite-microcline rocks (that have undergone varying degrees of metamorphism) - this unit forms the base of the borate-bearing sequence, it is dominated by microcline and contains up to 5% magnetite. II) Biotite-quartz-feldspar (largely microcline) rocks, which are locally termed leptites or leptynites, where biotite is more abundant. Tourmaline may be present locally, but only in trace amounts. III) Mg-silicates (that have undergone varying degrees of serpentinisation) intercalated with magnesite marble - these rocks are the exclusive hosts to the borate ores and range in thickness from 15 - 160m. In some areas the borates are present as breccias within otherwise bedded rocks and many of the blocks are laminated. Phlogopite is present in both marbles and Mg-silicate rocks and can be up to 10% by volume of the rock. IV) Tourmaline-bearing biotite-quartz-feldspar rocks that are otherwise very similar to the footwall leptites and leptynites - except for abundant tourmaline this lithology has the same mineralogy as the underlying leptites and leptynites. The abundant tourmaline tends to overlie borates and is used as an indicator of nearby ore. V) Albite rocks may contain up to 90% albite (albitites). They overlie the tourmaline-bearing unit and define the upper limit of the boron-bearing sequence. VI) Graphitic gneisses - an unconformity may separate this unit from the borate bearing sequences. It is sometimes finely laminated and is interpreted as having a sedimentary protolith. Amphibolites are intercalated throughout the sequence but rarely exceed 10m in thickness and consist almost entirely of hornblende and albite.

The borate-bearing rocks comprise much of the Liaohe Group and are overlain by graphitic gneisses with intercalations of dolomitic marbles of the Gaojiayu Formation. The thicknesses of the units vary considerably between different areas as the whole sequence has experienced intense folding and faulting, such that the true nature of the original stratigraphic column is often obscured at individual sites.

The Liaoning borates experienced several postdepositional metamorphic episodes (Figure 6.13):

- *Stage I* The first of these prograde metamorphic events was associated with the widening of the rift basin, which led to increased geothermal gradients.
- *Stage II* This resulted in expulsion of Si-rich brines from the underlying sediments to form Mg-silicates by the alteration of lacustrine magnesite. In some places this alteration was via vigorous focused flow, which led to the development of fluidised breccias. Elsewhere in the deposits, where the fluid flow was

more diffuse, this led to the replacement of carbonate in the bedded carbonate-borate evaporites on a lit-par-lit basis.
- *Stage III* At the height of metamorphism felsic veins cut through the deposits producing tourmaline-quartz rims to the veins. Subsequent deformation led to intense faulting and folding of the ore bodies.

Parental fluids for the borate and carbonate evaporites may have been originally derived from the evaporation of surface waters fed by geothermal springs associated with local volcanic activity. They have strong affinities both with the Tertiary borate deposits of Turkey (see Chapter 7) and the metasediments of the Damara Orogen (Jiang et al., 1997). Previously the Liaoning deposits were considered to have formed as skarns or as metamorphosed volcanogenic exhalative deposits. A skarn origin is now known to be incompatible with their boron isotopic signature (Figure 6.8; Peng and Palmer, 1995). A direct volcanogenic exhalative model is unlikely, as borate minerals are simply too soluble to precipitate directly into a standing fresh to marine water body fed by un-evaporated hydrothermal fluids. A concentration phase associated with solar evaporation is needed to explain the borate protolith.

Liaoning borate ore bodies differ from economic borate deposits elsewhere in the world in two important respects. First, the deposits are Precambrian, whereas most of the world's other major borate deposits are located in Neogene (or younger) sediments. Second, with the exception of a few small deposits located in skarns in Russia, all other major economic boron deposits in the world are sedimentary evaporites. The largest such economic deposit is the Tertiary lacustrine borates of the Bigadic deposit, west Turkey (Helvaci, 1995), where the major ore phases are colemanite ($CaB_3O_4(OH)_3.H_2O$) and ulexite ($NaCaB_5O_9.8H_2O$) deposited as nodules and veins associated with laminated and volcaniclastic (tuffaceous) sediments of a perennial alkaline saline lake (Chapter 7).

Grenville Complex, St Lawrence County (USA)

Northwest of New York, beneath the Adirondack Lowlands, is a 5-6 km thick Mesoproterozoic Grenville Series of metasediments and metavolcanics (Whelan et al., 1990). It is made up of the Upper and Lower Marbles separated by some 750 m of the Popple Hill Gneiss and underlain by the Hyde School Gneiss (Figure 6.14a). Amphibolite to granulite facies metamorphism of the Grenville Series

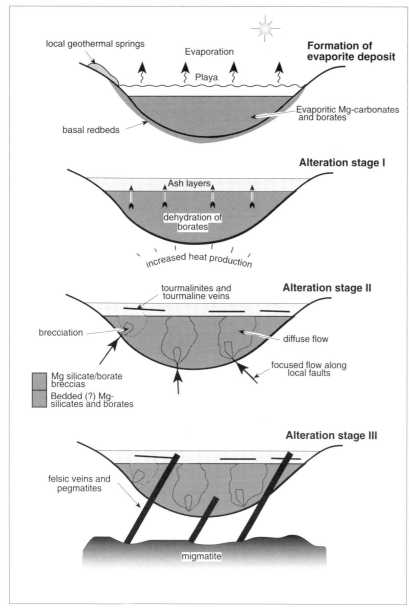

Figure 6.13. Alteration model of lacustrine borates, Precambrian Liaoning borate deposits, China (after Peng and Palmer, 1995).

sections that are more than 50% recrystallised anhydrite. Detailed mine mapping and drilling allows the Upper Marble to be subdivided into 16 units, numbered from oldest to youngest (Figure 6.14). In general, the sequence is composed of alternating fairly pure dolomites and quartzose calcsilicate-rich dolomitic marble. Units 1, 3, 5, 7, 9 and 12 are typically >95% dolomite (Whelan et al., 1990). Lenses and layers of anhydrite occur in unit 6 and 10 through 14, and may be up to 50% of the unit by volume. Brownish in unit 6, elsewhere the anhydrite is usually pale to deep lavender. Quartz and diopside may be present in the lenses, along with minor tremolite, talc, calcite, phlogopite and pyrite. The thickest lenses occur near the base of unit 6 (15-20m) and unit 11 (40-60 m) Plastic deformation has thickened the lens in unit 11 to more than 300 m in the hinge zone of the Fowler syncline (Unit 11a in Figure 6.14b). The anhydrite lenses in units 10, 12, 13 and 14 are 1-15 m thick.

$\delta^{34}S_{anhy}$ values in the oldest anhydrite lense (Unit 6) average 8.2 ‰; while progressively younger lenses average 19.6, 27.2, 26.8 and 19.1‰ (units 10, 11a 12 and 13/14, respectively; Whelan et al., 1990). The thickest and isotopically heaviest lens (unit 11a) displays a coincident up section increase in $\delta^{34}S_{anhy}$ (24.1-30.2‰) and $\delta^{18}O_{anhy}$ (19.8-22.5‰) values. These increases and the overall ^{34}S enrichment indicate the evaporitic anhydrite precipitation was concurrent with large scale bacterial sulphate reduction in the depositional basin. A similar increase in $\delta^{34}S$ values of biogenic pyrite (total range, -31.6 to 12.9‰) and decrease in $\delta^{13}C$ values (total range, -3.7 to 5.1‰) of the marble carbonate are also consistent with large-scale bacterial reduction. Whelan et al. (1990) interpret the isotopic data as indicating a protolith of biogenic-sulphide rich sediments. The anhydrite lenses were deposited in a restricted basin, which was also the site of accumulation of highly saline brines and extensive bacterial sulphate reduction. The low $\delta^{34}S$ and high $\delta^{18}O$ in

occurred about 1150 Ma, with peak conditions near the town of Balmat reaching 6.6 kb and 625°C. There are four phases of deformation in the region. The Grenville Series is host to some unusual iron-sulphide-rich metasediments with some intervals of more than 10m thick exceeding 70% pyrite. There are also some small-scale Pb-Zn mines in the region.

The Upper Marble also contains a number of indicators of meta-evaporites, including magnesian tourmalines and local scapolites. But probably the most impressive are

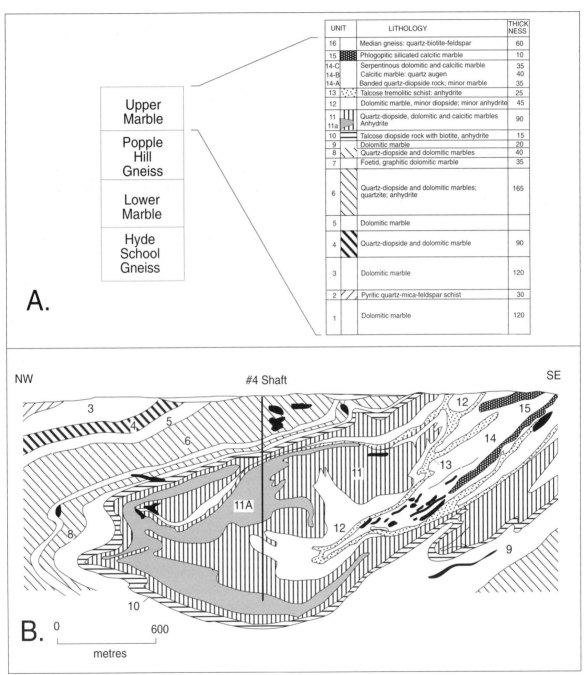

UNIT		LITHOLOGY	THICKNESS
16		Median gneiss: quartz-biotite-feldspar	60
15		Phlogopitic silicated calcitic marble	10
14-C		Serpentinous dolomitic and calcitic marble	35
14-B		Calcitic marble: quartz augen	40
14-A		Banded quartz-diopside rock; minor marble	35
13		Talcose tremolitic schist: anhydrite	25
12		Dolomitic marble, minor diopside; minor anhydrite	45
11 11a		Quartz-diopside, dolomitic and calcitic marbles Anhydrite	90
10		Talcose diopside rock with biotite, anhydrite	15
9		Dolomitic marble	20
8		Quartz-diopside and dolomitic marbles	40
7		Foetid, graphitic dolomitic marble	35
6		Quartz-diopside and dolomitic marbles; quartzite; anhydrite	165
5		Dolomitic marble	
4		Quartz-diopside and dolomitic marble	90
3		Dolomitic marble	120
2		Pyritic quartz-mica-feldspar schist	30
1		Dolomitic marble	120

Upper Marble

Popple Hill Gneiss

Lower Marble

Hyde School Gneiss

A.

B.

NW   #4 Shaft   SE

0   600
metres

Figure 6.14. Geology of the Upper Marble. A) Stratigraphy of the Upper Marble. B) Cross section of the Upper Marble, also showing location of the many Pb-Zn ore bodies in the region (black). Where unit 11 is more than 50% anhydrite it is shaded grey (after Whelan et al., 1990).

unit 6 are interpreted as episodes of oxidation of biogenic H_2S in the surface waters. They also infer a likely sedex setting for the precipitation of the widespread sulphides.

In a later study, Hauer (1995) argues that mineral composition, trace element and isotopic (O, C, and Sr) data show that the largely dolomitic rocks below the main anhydrite horizon (unit 11a of the Upper Marble) experienced open-system diagenesis, whereas the rocks above unit 11a experienced closed-system diagenesis. This diagenesis was accomplished penecontemporaneously by sea water-dominated fluids. Differences in the history of the Upper Marble below and above 11a are supported by differences in published Pb data for sphalerite in the Balmat Zn-Pb ores. Hauer (op. cit.) also concludes that most Upper Marble quartz originated as early diagenetic chert and that increases in aluminosilicate mineral abundances just below unit 11a are related to the tectonic events that caused the basin restriction and also drove anhydrite deposition.

Mt Isa Inlier, northern Australia

Several major sedimentary sequences in the Mt Isa Inlier contain relict evaporite pseudomorphs or meta-evaporites (Stewart and Blake, 1992; Stewart, 1994). The influence of evaporite interaction in the Mt Isa Inlier was ongoing through the diagenetic into the metamorphic realms. In the greenschist terrane in the vicinity of the Mt Isa mine many evaporite pseudomorphs and other hypersaline indicators, such as tepees, stromatolites and dissolution breccias, are still recognisable in the sediments that host the Mt Isa ores (Neudert and Russell, 1981; Neudert, 1986). In the more metamorphosed amphibolite facies of the Mary Kathleen Fold Belt in the central part of the Mount Isa Inlier most evaporites are remobilised within the zone of metamorphism and only meta-evaporite indicators, such as scapolites, albites and tourmalines, remain (Phillips et al., 1994; Oliver et al., 1994).

Mary Kathleen Fold Belt

The Mary Kathleen Fold Belt is composed of: 1) meta-evaporites and metacarbonates of the Corella Formation (deposited 1780-1750 Ma), and 2) metagranites and metadolerites (Figure 6.15a; Oliver and Wall, 1987). These metamorphic rocks range in grade from upper greenschist to upper amphibolite facies and are intruded by batholithic late- to post-tectonic granitoids. Evaporites and carbonates in these metasediments acted as a source of saline fluids during a protracted series of tectonothermal events.

An evaporitic origin for the metasediments of Corella Formation was first postulated by Ramsay and Davidson (1970) and has been supported by subsequent work. High fluid salinities in the Corella Formation are indicated by Cl-rich biotite, amphibole, and scapolite (maximum Cl contents: 0.60 wt%, 1.24 wt%, 3.67 wt%, respectively). Some SO_3 is present in almost all samples implying the former presence of $CaSO_4$. Reinhardt (1992) showed that the Corella Formation in the Rosebud Syncline now consists of interlayered Al- and Mg-rich metapelites, Ca-rich rocks, feldspathic schist, and scapolite schists that were originally deposited as pelite-carbonate- evaporite sequences in a Mid-Proterozoic ensialic rift basin. Widespread evaporite deposition took place in the final stage of rift sedimentation within a hypersaline shallow-water environment. The compositional variation of scapolite in the Rosebud Syncline on a layer-to-layer scale excludes large scale infiltration of externally derived fluids in the metasediments, but suggests variable proportions of evaporitic minerals in different layers and small scale variations of fluid salinity during metamorphism.

With burial, the Corella Formation interacted with distinctly saline fluids under conditions close to those of peak metamorphism. This occurred first during an early phase of extension (1760-1730 Ma), which resulted in intrusion of voluminous granitic and doleritic magmas into the carbonate-evaporite-dominated Corella Formation at $\approx$ 5-10 km depths (Oliver et al., 1994; Oliver 1995). Widespread high-temperature metasomatism ensued, involving scapolitisation in dolerite, formation of albite-scapolite shear zones in granite, exo- and endoskarn formation, and a zone of K-Na-Ca alteration in the lowermost Corella Formation. Granites and dolerites were altered to an unusual Na-Ca-rich bulk composition, reflecting high-temperature infiltration of highly saline, chemically reactive externally derived fluids. The alteration products and their distribution suggest not only reaction of magmatic/aqueous fluids with the country rocks but also extensive halite dissolution and recirculation of saline fluids back into the intrusive bodies. The bulk of fluid flow occurred at high temperatures (500-700° C), and within the metasediments the major element and isotopic fronts were generally not smoothed out by the effects of temperature gradients.

Oliver and Wall (1987) divided the lithologies of the Mary Kathleen Fold Belt into two distinct groups: a low fluid/rock (F/R) group and a high F/R group. The low F/R group is comprised of large areas of pristine calcsilicate metasediments showing evidence for internal buffering of fluid composition, as shown by $\delta^{18}O$ and $\delta^{13}C$ signatures typical of metamorphosed marine carbonates. In these low F/R metasediments the progress of calcsilicate reactions, and scapolite compositions vary greatly between adjacent layers. In contrast, the high F/R group is an assemblage developed in pretectonic intrusive rocks, as well as in metasomatic zones of calcsilicates around the margins of the intrusives (Figure 6.15a). This group is characterised by scapolitised metadolerites, albitised calcsilicates, and large calcite pods and veins with ^{18}O- and ^{13}C-depleted values. This group indicates localized throughput of predominantly externally derived hypersaline fluid, at high F/R ratios. It accompanied the metamorphic peak of 560-630°C and 3-4 kbar.

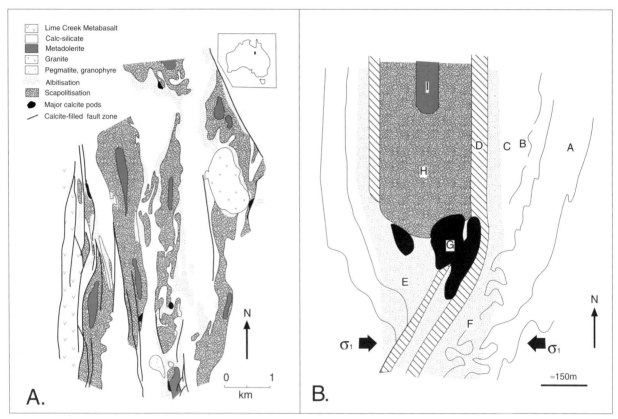

Figure 6.15. Geology of Mary Kathleen Fold Belt. A) Portion of the fold belt showing distribution of pristine calcsilicate (white) relative to albitised equivalents, shear zones, and scapolitised metadolerites. Inset shows locality of fold belt. B) Model of influence exerted by competent meta-intrusive units on localisation of fluid flow. Unit A is pristine calcsilicates; B is zone of boudinage and minor veining in zones of high differential stress; C is albitised and brecciated calcsilicates (failure adjacent to shear zone); D is albite-calcite shear zones; E is strain-shadow zone - abundant calcite veins and albitite breccia; F is refolded brecciated folds in albitised calcsilicates; G is large calcite pods in tensile and shear zones; H is scapolitised orthoamphibolite; I is scapolite poor metadolerite (after Oliver and Wall, 1987).

Scapolitic layers in the fold belt comprise approximately 40% of the metasedimentary sequence (low F/R group; Unit A and B in Figure 6.15b) and scapolitic modes are typically in the range 5 to 35%, with average chlorine contents ≈1.5%. Plots of Mary Kathleen Fold Belt metasediments compared with plots from the experimental work of Ellis (1978) indicate that fluids in equilibrium with the chloride-rich scapolites must have been hypersaline, particularly in calcite-bearing metasedimentary assemblages, with >30 mol% NaCl equivalent (Figure 6.16; Oliver and Wall, 1987). Abundant three-phase halite-bearing primary fluid inclusions in scapolite, amphibole and quartz are further evidence of synmetamorphic hypersaline conditions. The Cl content in units A and B (low F/R) varies widely, even in individual outcrops, indicating small-scale salinity gradients associated with the original Cl content of the metasedimentary protolith are still retained in the sequence (Figure 6.16). Similarly, the highly variable Cl content in the biotites in units A and

B suggests a protolith differential in the low F/R group. The original proportion of halite was at least 15% of the sedimentary sequence and probably much higher if the effects of diagenetic dissolution and the likely inefficiencies of halite incorporation into scapolite are considered.

Major competency contrasts between meta-intrusive and calcsilicate metasedimentary rock types influenced stress and strain patterns around the igneous bodies, which in turn controlled fracture permeability and the focusing of the metamorphic fluid (Figure 6.15). The relatively brittle behaviour of the meta-intrusive rocks favoured the development of small-scale fracture permeability and consequent pervasive alteration of these bodies. Calcite vein systems (Units C and D) and intensely albitised calcsilicate breccias (Units E) are localized in shear zones and dilatant areas around the meta-intrusives, either where shear stresses were high or where all stresses were low, resulting in shear and tensile failure, respectively (Units F and G). Thus, permeability enhancement, and the development of major fluid pathways during regional metamorphism of the terrain, is systematically related to variations in the stress field accompanying deformation.

Analysis of oxygen isotopic data and the position of isotopic and geochemical fronts reveals time-integrated fluid fluxes of up to 2×10^4 m^3/m^2 for the metasomatism (Oliver et al.,1994). Although very high salinities (up to 50 mol% NaCl) were attained by evaporite dissolution, $\delta^{18}O$ values of most alteration products are in the range 7-12‰, reflecting a predominance of oxygen derived from an igneous fluid. The position and interrelationships of metasomatic and isotopic fronts indicate an earlier stage of infiltration dominated by fluid released from crystallizing granite (with $\delta^{18}O$ 10-12‰), and a later stage ($\delta^{18}O$ 7-9‰) in which fluid had already interacted with halite and a mixed mafic-felsic igneous source or was repeatedly circulated between these rock types during alteration. The data reflect only a minimal contribution from fluids produced by devolatilisation of the abundant carbonate-bearing rocks in the Corella Formation, and there are substantial areas of Corella Formation rocks that have escaped metasomatism during this phase of intrusion-related hydrothermal activity and during the subsequent regional metamorphic overprint.

Supported by a requirement that the metamorphic fluids dissolved large amounts of halite from the same sequence, and their structural observations, Oliver et al (1994) favour a model where fluid was preferentially channelled along specific permeable conduits, including former evaporite layers, before interaction with the now exposed altered rocks. Fluid was probably driven by both convective circulation and dilatancy-related deformation accompanying emplacement of magmas into a major crustal extensional décollement.

Regionally, the halites and subsequent scapolites of the Corella Formation evaporites played an ongoing role over more than 600 million years of metasomatic alteration in the Mary Kathleen Fold Belt. There were four phases of deformation-related hydrothermal activity in the region (D1-D4), which occurred over a time range from at least 1750 Ma to 1100 Ma, and all were associated with evaporite-derived saline fluids (Oliver, 1995). The Na–Cl brine

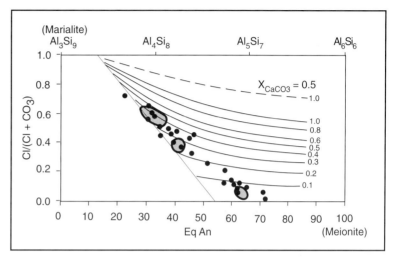

Figure 6.16. Compositions of scapolites (low F/R group) from a single outcrop (shaded zones) in the Mary Kathleen Fold Belt are plotted on the equivalent anorthite/fluid salinity plot. Black circles are data points from Ellis, 1978 (after Oliver and Wall, 1987).

source of much of these highly saline metasomatic fluids was halite in the Corella Formation, its equivalents, and subsequent scapolites.

In Phase 1, deformation, contact metamorphism, and metasomatism, accompanied the emplacement of the Wonga Batholith along with related mafic and felsic intrusives (1750-1740 Ma). This occurred not long after the deposition of the volcano-sedimentary package that included the evaporitic Corella Formation (Figure 6.17). It involved fluid pathways related to emplacement of intrusions in and above a major crustal subhorizontal extensional shear zone where a predominantly ductile lower plate -Wonga Belt- was flanked by a predominantly brittly deformed upper plate -Corella Formation. Intrusive rocks and contact aureoles formed discrete kilometre-scale bodies within the Corella Formation during Phase 1. As a consequence, fluid flow during Phases 2 and 3 (regional deformation and amphibolite-facies metamorphism) was localised around the boundaries of these very competent Phase 1 intrusions and contact metamorphic aureoles (Figure 6.15). Fluids infiltrating the rocks during Phases 2 and 3 were complex externally derived NaCl-CaCl$_2$-KCl-H$_2$O-CO$_2$ brines that may have undergone phase separation. This fluid flow culminated in the formation of widespread calcite vein systems (with copper) and the Mary Kathleen U-REE vein-style orebody. Mineralisation was sufficiently localised by the preexisting structural heterogeneity that many parts of the Corella Formation did not experience significant fluid flow at this

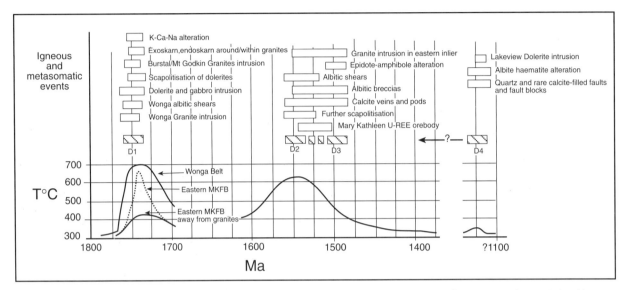

Figure 6.17. Hydrothermal history of the Mary Kathleen Fold Belt (MKFB), the time-temperature lines are poorly constrained by available geochronological data and are schematic (after Oliver, 1995).

time (perhaps a reflection of intrinsically lower permeability in meta-evaporite intervals not adjacent to the intrusive). Phase 4, post-metamorphic oxidising alteration (D4), was localised around major predominantly crosscutting strike-slip faults. These faults are very large with apparent strike-slip displacements of 5 to >40km and define the most obvious crosscutting regional geological features.

Staveley Formation

The Staveley Formation is a conformable evaporitic transition between the Corella Fm and the Marimo Shale (Stewart, 1991, 1994). Chalcopyrite (cobaltiferous)-pyrite mineralisation occurs as disseminations in pre- and syntectonic breccias in the Staveley Formation, and within disaggregated silica pseudomorphs (after thinly bedded evaporitic sulphate beds) that are associated with dolomitised limestone-calcilutites. Although this mineralisation may be partly diagenetic in origin, the 200m-thick meta-evaporitic packages are also host to a locally widespread (2 to 4 km long and 200 to >500 metres wide) and pervasive 'redrock' styled microcline-dolomite±haematite alteration. Within this gross alteration package there also seems to be a consistent association with narrow (10 to 100 metres wide) linear (stratabound) semi-massive magnetite-haematite-microcline and quartz-chalcedony replacement zones. The economic significance of mineralised haematitic intervals (ironstones) in evaporitic terrains is discussed further in Chapter 9.

Like much of the Corella Formation, the Staveley Formation, has undergone dissolution, palaeosolution

karsting (breccias), and tectonic brecciation such that the precursor lithological details have been progressively obliterated. The post-diagenetic to post-tectonic fluid pathway development and metasomatic alteration history of this important lithostructural package of rocks has been traced by Stewart (1994). He proposes that both the base of the Marimo Shale and parts of the underlying evaporitic-limestone dololutite have acted as conformable fluid pathways for hydrothermal and (perhaps) diagenetically derived, Cu-Au-(±Co±As)-K-Na-Ti-REE bearing fluids and brines.

Evidence for the evaporite protolith in this amphibolite grade sequence, where the actual salts have long since disappeared, is not just mineralogical but also textural. Stewart's (op. cit.) documented evaporite indicators in the Staveley Formation include:

1) Disaggregated chert nodules: 2 to 6 cm thick beds of 2 to 3 cm spheres comprising a collection of fragment-supported chert discoids, shingles, 'shards' and nodules with an ankerite feldspar matrix.
2) Enterolithic layers: chert nodule trains and contorted cherty nodule layers with overlying feldspathic dolomite laminae draped around nodules and slumped between them.
3) Beds of disseminated nodules and angular chert fragments with a sparry dolomite-quartz cement.
4) Cauliflower-textured and embayed nodules with contorted internal bands of cryptocrystalline (20 μm sized) chert.

5) Abundant hypersaline fluid inclusions (including halite crystals).

6) Displacive upward growth textures of nodules and discoids, alignment of nodule shards along low-angle cross bedding.

7) Lenticular and discoidal voids in shales. Shingle beds. Chickenwire or cottage cheese textures. Desiccation cracks. Soft sediment slumping.

8) Halite and scapolite pseudomorphs.

He shows how these shallow marine? evaporitic sulphates underwent Ca-Mg-Fe carbonate metasomatism and copper enrichment during diagenetic disintegration of the sulphate forms prior to general silicification.

To date workers in the Mary Kathleen Fold Belt have largely ignored the possibility that halokinetic horizons (including primary and secondary salt welds as well as fault welds) formed fluid conduits and channelways. My own, at this stage preliminary, work indicates halokinesis was a significant control on many of the faults and diagenetic mineral patterns in the Proterozoic rift basins (Chapter 9). At higher temperatures and regional metamorphic grades these early growth faults probably continued to act as conduits for what had evolved into meta-evaporitic fluids.

Meta-evaporites, Seve Nappes, Sarek Mts, Swedish Caledonides

Meta-evaporitic magnesite and scapolite, now hosted by diabase and amphibolite derived from diabase, occur in the Sarektjåkkå Nappe, Seve Nappe Complex, in the Caledonides of northern Sweden (Figure 6.18a,b: Svenningsen, 1994). The Sarektjåkkå Nappe consists of a sedimentary sequence, the Favoritkammen Group, that has been intruded by vast amounts of diabase as single or sheeted dykes so that it constitutes up to 70-80% of the nappe (Figure 6.19; Svenningsen, 1995). The diabase yields a Sm-Nd isochron of 573±74 Ma. The Sarektjåkkå Nappe largely escaped penetrative Caledonian deformation and still preserves recognisable igneous and sedimentary elements that are linked to the evolution of a pre-Caledonian rift margin. The sheeted dyke complex is interpreted to represent the early stage of seafloor spreading.

The magnesite and scapolite layers are part of the Spika Formation in the Favoritkammen Group. Locally the Spika Formation retains sedimentary structures, such as crosslamination in psammitic beds, dewatering pillars and soft sediment convolute folds in former calcarenite layers. More typically, the ductility of the carbonate bed and

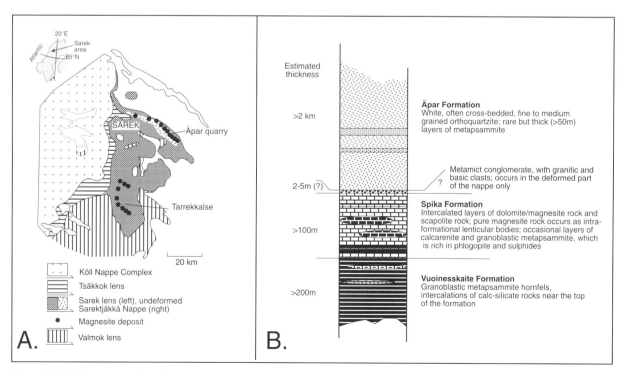

Figure 6.18. A) Tectonostratigraphic and location map of the Sarek National Park and adjacent areas. All the magnesite deposits occur in the Sarektjåkkå Nappe. B) Generalised stratigraphy of the Favoritkammen Group. The sequence is dismembered by enormous quantities of diabase dykes, and so thicknesses are no more than estimates (after Svenningsen, 1994).

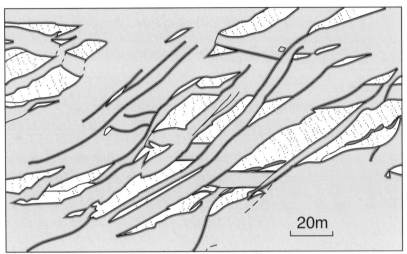

Figure 6.19. Typical sediment-dyke pattern from the central, undeformed part of the Sarektjåkkå Nappe (Favoritkammen Ridge east of Äpar quarry). Diabase dykes are stippled grey, with chilled margins. Vertical bedding attitude in sediment (Äpar Formation) is also indicated. Diabase is more than 70-80% of rock volume, yet bedding is still recognisable in metasediment blocks (after Svenningsen, 1994).

contrast between layers of differing compositions of scapolite dolomite and magnesite means that pervasive deformation and shearing has destroyed much of the primary structure.

Two horizons in the Äpar Magnesite Quarry contain matrix-supported intraformational breccias with intraclasts, which are interpreted as evaporite solution collapse breccias (Svenningsen, 1994). This is supported by the presence nearby of carbonate pseudomorphs after anhydrite nodules. The magnesite rock is conformable with the underlying calcsilicate layers and the contacts are gradual, indicating that the magnesite was not created by metasomatic alteration derived from the diabase. Rather, the rock reflects an original sedimentary magnesite deposited in a Coorong-like setting. The purest magnesite in Sarek is found in the Äpar Quarry, situated in a carbonate/ calcsilicate lens on the northern slope of the Äpar massif (Figure 6.18b). There it is composed of pure sparry white magnesite and has a visible area of 2500 m^2 and has the shape of a 120m by 40m flat lying horizontal cigar. The Äpar magnesite is hosted by massive grey diabase and is the only known deposit of pure magnesite in the region.

The Spika Formation was deposited in an evaporative setting of alternating influx of continental (magnesian carbonates) and marine water (halite and anhydrite/gypsum altered to scapolite during thermal metamorphism), possibly in rift-related basins in the Late Precambrian (Svenningsen, 1995). The overall stratigraphy of the Favoritkammen Group indicates a progressively deepening basin, consistent

with deposition in a developing rift basin. The diabase dykes that cut the Spika Formation caused regional thermal metamorphism, but were not responsible for the magnesite formation. The lithology of the Sarektjåkkå Nappe thus records the evolution of a rift into the formation of a passive margin. That passive margin was detached and thrust over the Baltic Shield during the Caledonian Orogeny.

Willyama Supergroup: Broken Hill and Olary regions

The Mesoproterozoic Willyama Supergroup straddles the border between far western New South Wales and eastern South Australia (Figure 6.20). It entrains Early to Middle Proterozoic metasedimentary and minor meta-igneous rocks that have a total stratigraphic thickness estimated at 7-9 km. It is subdivided into several structural blocks dominated by the Broken Hill and Euriowie Blocks in NSW and the Olary Block in South Australia (Stevens et al., 1988). Late Proterozoic (Adelaidean) sedimentary cover separates the Euriowie and Broken Hill Blocks, while the Broken Hill and Olary Blocks are separated by

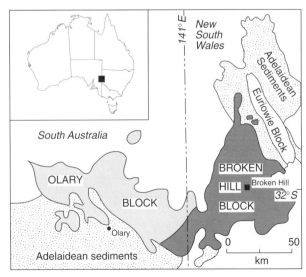

Figure 6.20. Location of the Willyama Block, made up of the Olary Block and the Broken Hill Block.

Cainozoic cover in the north and an inferred fault in the south.

The Willyama Block is intensely mineralised. In the Olary Block copper and uranium deposits are the most significant, with lesser iron, barium, cobalt, tungsten, gold and fluorite. In the Euriowie Block tin deposits are the most significant and common, while minor deposits of lithium, niobium, tantalum, copper and iron are also present. The Broken Hill Block is dominated by deposits of lead, zinc and silver with lesser tungsten, copper, iron, cobalt, gold, nickel, platinum, tin, beryllium, uranium and fluorite. The only deposit being worked at present is the 250 Mt of ore in Broken Hill lead-zinc-silver body.

The Willyama Supergroup metasediments are inferred to have been deposited in an ensialic intracontinental rift zone. Lithologies in the Olary Block indicate rift margin-related conditions, whereas the majority of lithologies in the Broken Hill Block indicate more central parts and deeper water parts of the progressively developing rift system (Willis et al., 1983; Plimer, 1984, 1985; Vernon and Williams, 1988; Slack et al., 1989; Cook and Ashley, 1992; Cook, 1993; Bierlein et al., 1995). Due to the intense deformation and high grade regional metamorphism in the region, any former evaporites have long since been destroyed.

Olary Block

Cook (1992) and Cook and Ashley (1992) demonstrated the importance of meta-evaporite associations within the Willyama Supergroup in the Olary Block. Their lithological, geochemical and fluid inclusion data from the Olary Block, west of the Broken Hill Block, suggests the presence of former evaporite rocks now metamorphosed to a variety of chemically and mineralogically unusual lithologies. The stratigraphic succession in the Olary Block displays a change in general composition from quartzofeldspathic at the base through calcsilicate-rich to pelitic at the top (Figure 6.21; Clarke et al., 1987; Forbes, 1991; Cook and Ashley, 1992).

The Composite Gneiss Suite is the basal unit and is made up of crudely layered coarse-grained quartz-feldspar-biotite±sillimanite±garnet gneiss. It grades into psammopelitic and pelitic schists, implying a sedimentary precursor; but there are zones exhibiting complete conversion, via *in situ* migmatites, into banded, migmatitic and locally massive leucocratic granitoids with intrusive characteristics (Figure 6.21; Cook and Ashley, 1992).

This is overlain by the laterally extensive Quartzofeldspathic Gneiss Suite, which is dominated by massive to layered quartz-albite and quartz-K-feldspar gneiss. These rocks have a felsic volcaniclastic precursor (Cook and Ashley, 1992) and are locally intercalated with pelitic

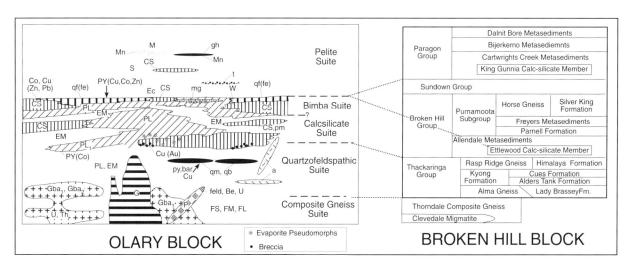

Figure 6.21. Mineralisation types and stratigraphic sequences in the Olary Block and Broken Hill Block comprising the Willyama Supergroup (after Cook and Ashley, 1992). Abbreviations; S - psammitic schist; M - psammopelitic schist; EM - psammopelitic and pelitic rocks; PL - plagioclase rock (albite-quartz rock); PL$_1$ - pyritic albite-quartz rock; FS - psammite-rich composite gneiss; FM - psammopelite-rich composite gneiss; FL - composite gneiss with leucocratic layers; CS - calcsilicate rock; PY - pyrite-rich rock; Ec - carbonaceous pelite; P - pegmatite; G - post-tectonic granite; Gba$_1$ - early syntectonic granite (pink porphyritic); Gba$_2$ - early syntectonic granite (sodic); a - amphibolite; qf(fe) - quartz-Fe sulphide rock (Bimba horizon); mg - magnetite-garnet rock; qm - quartz-magnetite rock; qb - quartz-baryte rock; gh - garnet-quartz rock; W - scheelite; t -tourmalinite; pm - piemontite-bearing rock.

mica schists, quartzites and calcsilicate-rich layers. Radiometric U-Pb dating of zircons gives ages of 1699 ± 10 Ma and 1703 ± 6 Ma for samples from a felsic metavolcanic rock and a comagmatic granitoid (Cook 1993). Minor rock types include quartzite, laterally discontinuous horizons of quartz-magnetite-haematite and quartz-magnetite-baryte rocks along with rare amphibolite and tourmalinite.

The overlying Calcsilicate Suite is characterised by calcsilicate-bearing albitites, grading into pure calcsilicate rocks. Unusual chemical sediments of possibly exhalative origin, such as scheelite-bearing rocks, haematite-rich banded iron formations and Mn-rich rocks, are common in this laterally extensive suite. Cook and Ashley (1992) proposed an evaporitic origin for the albite- and calcsilicate-rich metasediments on the basis of their textures, mineralogy and calcsilicate pseudomorphs after dolomite and gypsum. Rocks of sodic composition abound, represented by albite-quartz-rich rocks (+calcite, actinolite, epidote, garnet, clinopyroxene), and are inferred to represent metamorphosed analcite-rich strata of felsic volcaniclastic provenance, with variable proportions of evaporitic carbonates. Gypsum was recognised as pseudomorphic moulds and as epidote pseudomorphs in the rocks of the Calcsilicate Suite and in Bimba iron formation.

Whole rock analyses provide additional support for evaporitic conditions during deposition of some strata where substantial alkali enrichment is evident, other elements typically vary as a function of dilution by SiO_2 (Cook and Ashley, 1992). Pseudomorphs after carbonate were also observed in calcsilicate-bearing quartz-albite rocks of lower metamorphic grade and interpreted to be pseudomorphs after dolomite destroyed during prograde metamorphic reactions. The presence of halogen-rich minerals in a distinctive sodic mineral assemblage, in addition to hypersaline fluid inclusions, pseudomorphic voids and minerals after gypsum and carbonate, tourmalinites with isotopically light boron, all imply a nonmarine evaporite protolith.

A transition from generally magnetite-bearing to sulphide-bearing strata defines the passage from the Calcsilicate Suite into the overlying quartz-feldspar-calcsilicate-carbonate-sulphide-bearing Bimba Suite. The Bimba Suite is a thin (< 50 metres), laterally continuous unit that, due to the widespread occurrences of stratiform Fe-dominated sulphides, is the focus of metal exploration interest in the region. Oxidation of sulphides (dominantly pyrite and pyrrhotite), calcsilicates and carbonates has led to distinctly gossanous surface exposures typically associated with anomalous Cu, Zn, Pb, Co, As, Ag, Au, Mn, Ba, W, Mo, Bi and U.

The uppermost unit of the Willyama Supergroup metasediments in the Olary Block is the widely distributed Pelite Suite. It is made up of graphitic sulphide-bearing quartz-mica-garnet-andalusite-sillimanite-tourmaline schists, which are interbedded with layers of coarser-grained quartzofeldspathic metasiltstone and quartzite. Lenses of Mn-rich banded iron formations and garnet-quartz rocks occurring at the base or within the lower parts of the Pelite Suite imply continuing exhalative activity during deposition of this unit (Cook and Ashley, 1992).

In addition to recognisable metasediments, at least four phases of widespread emplacement of granitoids have been identified, ranging from pre- through syn- to post-high grade metamorphism in age (Cook and Ashley, 1992). Five deformational episodes (D_1-D_5) are recognised throughout the Olary Block (Berry et al. 1978; Clarke et al. 1987). Events D_1-D_3 only affected the Willyama Supergroup and some granitoid units. They occurred during the Middle Proterozoic Olarian Orogeny at about 1600-1580 Ma. Events D_4-D_5 also deformed the Adelaidean sequence during the Delamerian Orogeny (500 Ma; Preiss, 1988).

Broken Hill Block

As in the Olary Block, the presence of former widespread evaporites in the district can only be inferred on the basis of indirect lithologic, chemical, and isotopic data (Slack et al., 1993). Lithologic evidence for former evaporites includes the occurrence of bedded albitites in the Himalaya Formation at the top of the Thackaringa Group, which are believed to represent clastic sediments or tuffs that were largely altered to analcime in an alkaline lake setting (Figure 6.21; Plimer, 1977; Stevens et al., 1988). Other evidence for evaporites and/or nonmarine depositional settings comes from (Slack et al., 1993):

• the presence of abundant scapolite locally in the Ettlewood Calcsilicate Member at the base of the Broken Hill Group, which in correlative rocks of the adjacent Olary Block in the South Australia and contains up to 2.7% Cl (Cook, 1992; Cook and Ashley, 1992);
• abundant magnetite disseminated in some quartzofeldspathic rocks of the Thackaringa Group suggesting protoliths of oxidised red beds;
• quartz-muscovite nodules in the Rockwell area (Himalaya Formation) that have crystal forms suggesting pseudomorphic replacement of gypsum by silica in nonmarine evaporitic environments.

The former presence of evaporites is also suggested by the occurrence of chemically unusual halide minerals in the main Pb-Zn-Ag lodes at Broken Hill (Slack et al., 1993). The oxidised ores of the main lodes are famous for containing an unusual suite of secondary halides including: ioargyrite [AgI], bromargyrite [AgBr], marshite [CuI], miersite [(Ag,Cu)I], and embolite [Ag(Cl,Br)]. These halides were reportedly concentrated at the base of the gossan, above the supergene ores and formed following the weathering of the main ore loads in the Permian and/or Tertiary. Most workers have assumed that the source of the Cl, Br, and I within oxidised ores in such desert environments is from exotic groundwaters or seawater-derived salt spray, with little if any contribution from the protolith. Similar iodine and nitrate accumulations in the soils of the hyperarid Atacama Desert, South America, are discussed in Chapter 7. Slack et al., (1993) argue that in the case of Broken Hill a local bedrock or protolith source is equally plausible, if not compelling. They also contend that much of the sulphur in the Pb-Zn-Ag ores may also have come from nonmarine evaporites.

Tourmaline-rich rocks are widespread minor lithologies within the Early Proterozoic Willyama Supergroup in the Broken Hill district (Figure 6.8). Tourmaline concentrations occur in stratabound and local stratiform tourmalinites, clastic metasedimentary rocks, quartz-gahnite lode rocks, stratiform Pb-Zn-Ag sulphide ores, garnet quartzites, stratabound scheelite deposits, quartz-tourmaline nodules, discordant quartz veins, and granitic pegmatites. Most of the tourmaline-rich rocks are within the Broken Hill Group that hosts the main Pb-Zn-Ag ores.

Combined field and geochemical data indicate that the tourmalinites represent normal clastic sediments that were metasomatically altered by boron-rich hydrothermal fluids at or below the sediment-water interface. In the model of Slack et al. (1993), the hydrothermal system(s) acquired abundant boron by leaching evaporitic borates within the Thackaringa Group, the stratigraphic sequence that underlies the Broken Hill Group and hosts most of the tourmaline concentrations. They suggest that evaporites of the Thackaringa Group provided a source of readily extractable boron for formation of the tourmalinites and also acted as a source for the fluoride, sulphur, and perhaps the carbonate in the main lodes; such evaporites may have been critical for increased metal chloride complexing and transport necessary for deposition of the high-grade Pb-Zn-Ag ores.

Plimer (1994) interprets scheelite in the Willyama Supergroup in the Broken Hill region as occurring within a sequence of metamorphosed intracontinental rift sediments, volcanics, and hot spring precipitates (Figure 6.22). Scheelite ($CaWO_4$; an ore of tungsten) occurs within regional calcsilicate rocks, which are enriched in F and Zn

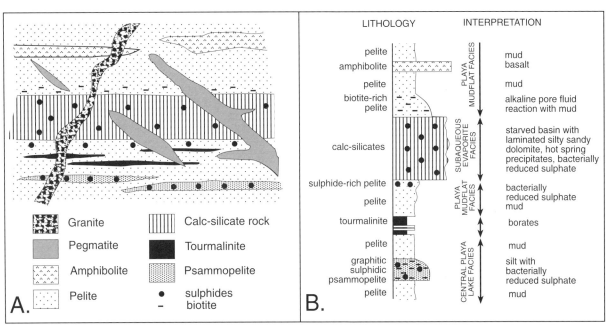

Figure 6.22. A) Schematic of stratigraphic relationships between the stratiform scheelite-bearing calcsilicate rocks and other lithologies in the Lakes Nob region of the Broken Hill Block. B) Palaeolimnological interpretation of the observed sequence in the Allendale metasediments and the Ettlewood Member at Lakes Nob (after Plimer, 1994).

within a sequence of carbonaceous-sulphidic metabolite, tourmalinite, and amphibolite. It overlies a felsic gneiss-metasediment sequence, part of which is interpreted as a meta-evaporite. The distribution of scheelite is unrelated to syntectonic pegmatite or post-tectonic granitic rocks. The scheelite underwent only minor remobilisation during tectonism. Plimer (op. cit.) interprets the calcsilicate rocks as dolomitic limey sediment that formed part of a playa-lake sequence into which B, W, Zn, and F-rich alkaline hot springs debouched (Figure 6.22). Coeval sudden deepening, an increased geothermal gradient, deposition of the Broken Hill Pb-Zn-Ag sulphide orebodies, and siliceous exhalite deposition all occurred after deposition of the evaporite sequence. Thus components for the highly saline Broken Hill Pb-Zn-Ag ore fluid may have been leached from the underlying evaporites and redeposited higher in the stratigraphy.

Pine Creek Inlier, Northern Territory

The Pine Creek Inlier is a major mineral province in the Northern Territory, Australia, and covers an area of some 66,000 km^2. It entrains Archaean and Palaeoproterozoic rocks surrounded by subhorizontal strata of several sedimentary basins that range in age from Mesoproterozoic to Mesozoic. It contains the Alligator River, Rum Jungle and South Alligator Valley uranium provinces.

Evaporite pseudomorphs after gypsum (lenticular) and halite (cubic) were noted in the coarsely crystalline magnesite and dolomite rocks in the Rum Jungle and Alligator Rivers uranium fields Northern Territory; in the Celia and Coomalie Dolomites at Rum Jungle, and the Cahill Formation in the Alligator Rivers region of the Pine Creek Geosyncline (Crick and Muir, 1980; Muir 1987). .

The rocks are now marbles, but they still retain many sedimentary features including algal bioherms (Figure 6.23a). Stratiform and domal stromatolites are common, with some domes reaching 1.5 m in height with a similar diameter; *conophyton* bioherms are also well developed. The stromatolites may also preserve evaporite casts after gypsum and are interbedded with nonstromatolitic carbonates that contain crossbeds, ripples, scour surfaces, intraclast breccias and tepee structures. Possible evaporite casts in the magnesite marbles are characterised by blade shapes and discoidal shapes up to 1 cm long. Cruciform and swallowtail twin forms are also present in the magnesite marbles. Orthorhombic shapes and cubes after halite are also abundant.

However, Bone (1983) in a study of inclusions in the discoidal forms of magnesite from Rum Jungle concluded that the discoids and the cube shapes were a high temperature form of magnesite that formed above 150°C and not evaporite pseudomorphs. Muir (1987) countered this argument with the observation that identical discoidal

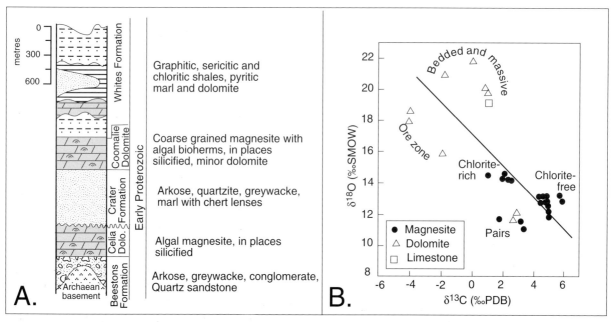

Figure 6.23. Magnesites in the Rum Jungle Uranium Field, Northern Territory, Australia. A) Stratigraphic column. B) Carbon and oxygen isotope crossplots of magnesites and coeval dolomites (after Aharon, 1988).

pseudomorphs after gypsum can be found in the Amelia Dolomite in the McArthur Basin in strata that have never been subjected to temperatures of more than 100°C. Further evidence for an evaporite association is adduced from the co-presence of lutecite, a variety of length-slow quartz that pseudomorphs anhydrite.

Work by Aharon (1988) on the same magnesites showed that they are replacement fronts. Magnesite beds up to 700m thick from the Lower Proterozoic rocks of the Rum Jungle Uranium Field yielded $\delta^{18}O$ and $\delta^{13}C$ compositions of +11.2‰ to + 14.6‰ (SMOW) and +1.0‰ to +5.8‰ (PDB), respectively (Figure 6.23b). Dolomites in the same region yield $\delta^{18}O$- and $\delta^{13}C$-values of + 11.7‰ to + 21.9‰ and - 4.0‰ to + 2.9‰, respectively, and show an ^{18}O enrichment gradient away from the magnesite. Fluid-inclusion homogenization temperatures, used in conjunction with the isotope determinations and an approximate magnesite-water equation, indicate that magnesites were formed in the temperature range of 100-200° C from fluids that became progressively enriched in ^{18}O. He argues the results support an epigenetic-metasomatic replacement model of dolomite by magnesite under the influence of invasive hot Mg-rich basinal fluids.

My own work on Proterozoic sparry magnesite supports the conclusions of both Aharon and Bone (see Chapter 9). That is, the discoidal magnesites are a diagenetic

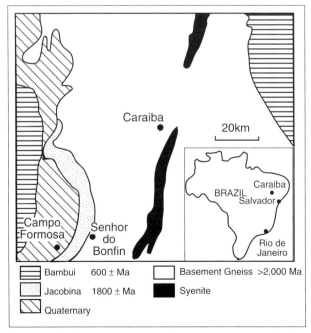

Figure 6.24. Geological setting of the meta-evaporites in the vicinity of the Caraíba Copper Deposit in the Archaean Basement Gneiss (after Leake et al., 1979).

replacement in a nonevaporitic platform carbonate host. But, my own work also implies that these sparry magnesites are ultimately tied to evaporites. They do not indicate the depositional association postulated by Crick and Muir, (1980). Rather, they are related to deep subsurface dissolution of halokinetic evaporites that then supplied metalliferous chloride brines into fault conduits. In my opinion these sparry magnesites are hydrothermal pinolitic structures that were fed via fault conduits, many of which were created or reactivated by the flow of underlying salt as the basin underwent extension. Similar pinolitic carbonate cements and replacements associated with faults, tied to rifting episodes, occur in other Precambrian basins in Northern Australia and the Flinders Ranges, South Australia. There are also strong similarities with Phanerozoic peridiapiric metasomatic cements in North Africa and Europe.

Sulphide-anhydrite Archaean carbonates, Brazil

Archaean carbonates host sulphide-sulphate phases in high-grade metamorphics in the Caraíba gneiss-granulite complex in central Bahia State, Brazil (Figure 6.24; Leake et al., 1979). These metasediments (marbles, calcsilicates, metapelites) are crosscut by younger granitic gneisses (≈2.0 Ga) and by K-rich zones. The anhydrite-entraining metasedimentary sequence is dominated by calcsilicates with varying mineral assemblages (Sighinolfi, et al., 1980):

1) diopside (≈90%), green spinel, apatite and sphene;
2) diopside, calcite microcline;
3) diopside, anhydrite microcline;
4) diopside, anhydrite, calcite, forsterite.

Assemblages (2)-(4) contain varying amounts of phlogopitic marble and scapolite occurs locally. The anhydrite-bearing assemblages occur as layers (0.05 - 0.5 m thick) within the metasedimentary sequence. The anhydrite forms large crystals in the layers or forms fine-grained aggregates that fill cracks and microfractures. Textural evidence shows the anhydrite growth was associated with formation of phlogopitic mica (Sighinolfi, et al., 1980).

Almost all the rocks in the Caraíba area have been metasomatised and deformed, so that determining the ultimate origin of the anhydrite is beset with difficulty. The textural evidence is inconclusive. Anhydrite distribution in the layers appears sometimes to have been little affected by metamorphism, but there is also extensive anhydrite crystallisation along cracks and fractures. The abundance of the anhydrite-mica association indicates its active participation in metasomatism but does not directly infer an evaporite protolith. The possibility of hydrothermal anhydrite has not been addressed.

Interestingly, many different coloured tourmalines are found in the meta-evaporites of Bahia State. Pink and green colours are particularly popular. In 1989, miners discovered gem tourmaline unlike any that they had ever seen before. The new type of tourmaline, which soon became known as Paraiba tourmaline, comes in incredibly vivid blues and greens. The demand and excitement for this new material grew in the early 1990s and better specimens soon fetched more than US$10,000 per carat.

Scapolitic meta-evaporites, Rajasthan, India

Electron-microprobe analyses of coexisting phases from a scapolite-garnet-epidote-calcite-plagioclase-amphibole-pyroxene (sphene-haematite-magnetite) rock of the Aravalli group (Early Precambrian) in Karera, district Bhilwara, Rajasthan, reveal that chlorine is an important constituent of both scapolite ($Me_{71.3}$) and amphibole (Sharma, 1981). The amphibole also contains an unusually high K_2O (3.7 wt.%), and is a chlor-potassium hastingsite. The epidote contains 41% pistacite and shows complete substitution of Al by Fe^{3+} in the M_2 site and to some extent also in M_1 and/or M_3 sites. The garnet is also rich in ferric iron and has a molecular composition $Pyr_{23}Alm_{8.5}Gro_{13}And_{54.5}Sp_{1.0}$. The pyroxene is dominantly a hedenbergite.

Phase relations and textural as well as geological criteria exclude metasomatic processes from a magmatic source

and favour equilibrium recrystallisation of the scapolite-bearing assemblage with the chlorine derived from an evaporite component of the Aravalli metasediments. Geothermometry based on the fractionation of Na and Ca between scapolite and plagioclase yields metamorphic temperatures of around 700°C, which are in agreement with those obtained by other mineral equilibria in the associated pelitic assemblages (Sharma, 1981).

Precious stones as meta-evaporites

Base and precious metals are not the only ore deposits formed in association with meta-evaporites. Some precious stones can also have meta-evaporite affinities. For example, lapis lazuli from the Precambrian of Baffin Island, Canada, and from Edwards, New York, are interpreted as meta-evaporites, as are the lapis lazuli deposits at Sar e Sang in the Kokcha valley, Afghanistan. First mined 6000 years ago, the Sar e Sang lapis was transported to Egypt and present day Iraq and later to Europe where it was used in jewellery and for ornamental stone. Europeans even ground down the rock into an expensive powdered pigment for paints called "ultramarine".

Lapis lazuli (literally "blue rock") is mostly composed of lazurite. Lazurite belongs to the sodalite group of feldspathoid minerals (Table 6.3). Feldspathoids have chemistries that are close to those of the alkali feldspars, but are poor in silica. If quartz were present at the time of formation it would react with any feldspathoids to form a feldspar. Natural lazurite contains both sulphide and sulphate sulphur in addition to calcium and sodium, it is classified as a sulphide-bearing haüyne. This sulphur gives lapis its characteristically intense blue colour. Other members of the sodalite group include sodalite and nosean. Sodalite is the most sodium-rich member of the sodalite group and differs from the other minerals of the group in that its lattice retains chlorine. Interestingly, sodalite can be created in the laboratory by heating muscovite or kaolinite with NaCl at temperatures of 500°C or more.

In the literature the commonly accepted origin of lazurite is through contact metamorphism and metasomatism of dolomitic limestone. Although not noted in most textbooks, such a metasedimentary system also requires a source of sodium, chlorine and sulphur; an obvious source would be interbedded evaporites. For example, the lapis deposits on Baffin Island and in Edwards, New York, were produced by high grade metamorphism of an evaporite-marble protolith (Hogarth and Griffin, 1978). The anhydrites preserved near Balmat are also part of this sequence and

Sodalite Group	
Sodalite	$Na_8(Al_6Si_6O_{24})Cl_2$
Nosean	$Na_8(Al_6Si_6O_{24})SO_4$
Haüyne	$(Na,Ca)_{4-8}[Al_6Si_6O_{24}](SO_4,S)_{1-2}$

Table 6.3. Sodalite group composition.

were discussed earlier in this chapter. On Baffin Island the two main lapis lazuli lenses lie at the structural top of two sequences of dolomitic marble, the thicker one being approximately 150 m across. The elongation of the lenses parallels the local layering and foliation and show a well-developed layering parallel to the regional foliation implying a sedimentary protolith to the deposits.

Some Colombian emeralds are also meta-evaporitic formed via fault-focused hypersaline basinal solutions released during the subsurface alteration of evaporites (Giuliani et al., 1995). Fluids trapped by emerald, dolomite and pyrite in the Colombian emerald deposits consist predominantly of Na-Ca brines with some KCl. The Na-Ca-K chemistry of the brines provides strong evidence for an evaporitic origin of the parent hydrothermal fluids. The $\delta^{34}S$ values of H_2S in solution in equilibrium with pyrite from six emerald deposits range from 14.8 to 19.4‰, whereas sedimentary pyrite from the enclosing black shales yields a $\delta^{34}S$ of -2.4‰. The narrow range in $\delta^{34}S_{H2S}$ between the different deposits suggests a uniform and probably unique source for the sulphide-sulphur. Elevated $\delta^{34}S_{H2S}$ values suggest non-participation of magmatic or Early Cretaceous black-shale sulphur sources. Salt diapirs occur in the emeraldiferous areas and the most likely explanation for high $\delta^{34}S$ involves the reduction of sedimentary marine evaporitic sulphates. This unique emerald-deposit type is a mesothermal deposit (300°C) produced through thermochemical reduction of diapir-derived sulphate-rich brines to hydrogen sulphide by interaction with organic-rich strata.

A similar model is used to explain emeralds hosted in the Red Pine Shale within the South Flank fault zone of the Uinta Mountains of Utah (Keith et al., 1996). The host rock is a 1 km thick, black (organic-bearing) Proterozoic shale sequence with thin interbedded arkoses. Abundant pyrite, baryte, vein quartz, blocky and fibrous calcite veins, along with bleached shale and green mica, are present along the fault zone, and within the Cr-rich (up to 200 ppm) Red Pine Shale. There are no granitoid or intermediate intrusions within 50 km of the area, and no geophysical evidence to indicate that the alteration is associated with concealed igneous activity (Keith et al., 1996).

The isotopic composition of pyrite from the South Flank fault zone ($\delta^{34}S$ = +4.6 to +4.2‰) is similar to the isotopic signature of sulphur from nearby Tertiary oil field brines ($\approx$ +5‰). These brines are present near the base of the Uinta Basin succession, within the Wasatch Formation and the highly evaporitic Green River Formation. The deepest portion of the asymmetric Uinta basin is bounded

by normal and reverse faults that were active during uplift of the Uinta Mountains. Consequently, the hydrothermal fluid, responsible for alteration of the Red Pine Shale, may be similar to that forming the Colombian emeralds. According to Keith et al. (1996), their data suggests that an external hydrothermal fluid (a highly saline, probably Tertiary age basinal brine) mixed with other subsurface waters along the fault zone. As these basinal fluids reacted with feldspar or calcite cements, they formed the sparse emerald and other alteration products now found in their Proterozoic host shales.

Summary

Meta-evaporites are not simple rock systems, their interpretation involves more than just recognising sodic or magnesian metamophic minerals. This must be done in conjunction with a full description of the geometries of these phases and the recognition that the evaporites and their subsequent minerals are altered many times during their metamorphic history, often in ways that do not fit the classic paradigms of metamorphic petrology. The following questions should always be asked of potential meta-evaporite sequences:

- Are the minerals distributed in a fashion that reflects a stratiform or a diapiric precursor?
- Are breccias present that contain nodules and other textures indicative of former evaporites?
- Do the breccias reflect a diapiric precursor or was the brecciation a later fluidisation event, perhaps associated with compression and overthrusting centred on evaporite -rich horizons that were undergoing metamorphism?

Evaporites undergoing metamorphism release large volumes of volatile-rich fluid, which are typically saline chloride solutions with elevated levels of sodium and magnesium. Many of these fluids have the capacity to mobilise and flush metals from adjacent strata that often include metalliferous metabasites. Thus meta-evaporites or their residues are often part of a metallisation series in a metamorphic terrain. The processes and styles of these evaporite-metallogeny associations are discussed further in Chapters 8 and 9.

221

Chapter 7
Evaporites as a mineral resource

Introduction

Sodium chloride (halite), the most common evaporite salt, is used in some form by virtually every person in the world. Most people think of this salt in terms of food processing, or in colder climes in terms of road de-icing. These are in fact lesser usages of halite, its main use is as a feedstock in the chemical industry. There are more than 14,000 reported usages of halite and it, along with other salts, has long played a very important role in human affairs. About 4,700 years ago the Peng-Tzao-Kan-Mu was published in China, it is probably the earliest known treatise on pharmacology. A major portion was devoted to a discussion of more than 40 kinds of evaporite salts, including descriptions of two methods of extracting salts from brine and how to recover it in usable forms. In ancient Greece the slave trade involved exchange of salt for slaves and gave rise to the expression, "not worth his salt". Special salt rations paid to early Roman soldiers were known as "salarium argentum", the forerunner of the English word "salary". There are more than 30 references to salt in the Bible. Salt also had military significance. For instance, thousands of Napoleon's troops died during his retreat from Moscow because their wounds would not heal as a result of a lack of salt.

The same diversity of usage is also true of salts other than halite (Table 7.1). Potash salts are used mostly as agricultural fertiliser; while the sodium salts, halite-trona-glauberite, are major feedstocks in the formation of a wide range of industrial chemicals. Gypsum is used in plaster and cement manufacture throughout the world. This chapter first focuses on the potash salts and then on the other industrial salts and the products of alteration associated with evaporite salts, such as sulphur and zeolites; emphasis is on depositional and diagenetic modes of formation. Halite and gypsum (anhydrite), the two most commonplace salts, are not considered in the same detail as the other evaporite resources; their origins were discussed in much greater detail in earlier chapters.

The annual production figures and tables for the various salts in this chapter were compiled from the raw data sheets in various pdf-format documents held on the US Geological Survey web site <http://minerals.er.usgs.gov/minerals/pubs/commodity/>. Other very useful sources that were used extensively in the compilation of this chapter were the excellent books on industrial minerals by LeFond (1983), Harben and Bates (1990), and Harben and Kuzvart (1996).

Potash evaporites

Natural potash evaporites are part of the bittern series precipitated at the surface or in the shallow subsurface at the higher concentration end of the evaporation series. Today the deposition of potash minerals at the earth's surface is a relatively rare occurrence. The extremely high solubility of potash means it can only accumulate in highly restricted, some would say highly continental, settings. Where it does occur in the modern playas of Qaidam in China and in the Danakil Depression of Africa, carnallite, not sylvite, is the precipitated salt. This has led some to postulate that carnallite is the primary potash phase, while sylvite is a secondary diagenetic mineral formed by incongruent dissolution of carnallite. Others have argued that ancient sylvite was sometimes a primary precipitate deposited by the cooling of highly saline surface or nearsurface brines (Lowenstein and Spencer, 1990).

Phanerozoic potash salts comprise two distinct groups: $MgSO_4$-rich and $MgSO_4$-poor. Potash salts have not been recovered from Precambrian sediments. Potash deposits rich in $MgSO_4$ are composed of some combination of gypsum, anhydrite, polyhalite, kieserite, kainite carnallite and bischofite. Potash deposits free of, or poor in, $MgSO_4$ are dominated by some combination of halite, carnallite, and sylvite. This latter group constitutes the majority of the potash found in Phanerozoic sediments. The hydrogeochemical significance of the two groups was discussed in Chapter 2; this chapter emphasises the geology of potash salts.

Potash is the generic term for a variety of ore-bearing minerals, ores and refined products, all containing the element potassium in water soluble form (Table 7.2). It was originally applied to potassium in carbonate-potassium hydroxide crystals, which were recovered in iron "pots" from washings of wood (or other plant) "ashes." Commercial potash can be potassium chloride (KCl or muriate of potash [MOP]), potassium sulphate (K_2SO_4 or sulphate of potash [SOP]), potassium-magnesium sulphate ($K_2SO_4MgSO_4$ or sulphate of potash magnesia), potassium nitrate (KNO_3 or saltpetre), or mixed sodium-potassium nitrate ($NaNO_3 + KNO_3$ or Chilean saltpetre). The most common naturally occurring potash minerals are carnallite and sylvite, with sylvite the most economically important. Other common potassium-entraining evaporite ores include: kainite, kieserite, langbeinite, leonite and polyhalite.

Type	Mineralogy	Source	Uses
Ammonium chloride	NH_4Cl	Brine processing	Fertiliser, industrial additive.
Ammonium sulphate	$(NH_4)_2SO_4$	Brine processing	Fertiliser.
Borate	B_2O_3	Borax (tincal), Boracite, Colemanite, Hydroboracite, Kernite, Ulexite	Glass, ceramics, agricultural chemicals, pharmaceuticals, flux cleansing agent, water softener, preservative and fire retardant.
Bromine Sodium bromide	Br_2 $NaBr$	Brine processing	Additive.
Calcium sulphate	$CaSO_4$	Gypsum, secondary after anhydrite	Construction industry, plaster, cement retardant, soil enhancer.
Iodine	I_2	Brine processing	Animal feed supplements, catalysts, inks and colorants, pharmaceutical, photographic equipment, sanitary and industrial disinfectant, and stabilisers.
Lithium	Li	Brine processing, Hectorite in saline clays	Production of ceramics, glass, and primary aluminium, battery technology, pharmaceutical.
Magnesium carbonate Magnesium oxide Magnesium sulphate	$MgCO_3$ MgO $MgSO_4$	Brine processing, Magnesite, Electolytic processing	Majority used for refractories. A constituent in aluminium-base alloys, used for packaging, transportation, and other applications. Remainder is consumed in agricultural, chemical, construction, environmental, and industrial applications.
Potash	KCl KNO_3 K_2SO_4	Sylvite, Sylvinite, Brine processing	Fertiliser, chemical industry.
Sodium carbonate	Na_2CO_3	Trona, Solvay process, Brine processing	Glass manufacture, chemicals, soap and detergents, pulp and paper manufacture, water treatment, and flue gas desulphurisation.
Sodium chloride	$NaCl$	Halite, Brine processing	Food processing, chemical and industrial products, chloralkali industry, swimming pool additive, stockfeed additives.
Sodium sulphate	Na_2SO_4	Mirabilite, Thernadite, Glauberite, Brine processing	Detergent powder, glass, pulp and paper.
Strontium	Sr	Celestite, Strontianite	Colour television tubes, ceramic magnets, colouring agent in pyrotechnics and signals.
Sulphur	S_2	Biogenic sulphur, Frasch process	Agricultural chemicals, chemicals manufacture, metal mining, and petroleum refining.

Table 7.1. Evaporite salts as a mineral resource.

In 1997 there were 15 potash producing countries, including small volumes from the Ukraine and Azerbaijan. The North Americas produce 42% of the world's total potash production, eastern Europe and Russia some 24%, western Europe 22%, and the Dead Sea 11% (Figure 7.1). Of total world production, just ten companies control 90%, and the five largest companies control more than 75%. Canada, the world's largest producer holds 36% of the world's potash reserves, and recovers potash by conventional underground mining of bedded sylvinite deposits. Production is centred on the Devonian Elk Point Basin in Saskatchewan, with lesser production in the Maritime Provinces and Quebec. Devonian potash is mined in the Pripyat depression of Belarus. Permian potash is mined from the northern margin of the Delaware Basin in New Mexico, USA; from the Zechstein of Europe (Germany, UK and the Netherlands); and from the Upper Kama Basin in the Russian Federation in the Solikamsk region west of the Urals. Oligocene

Mineral	Formula	K_2O Equiv.	Comments
Chlorides			
Sylvite	KCl	63.17	Principal ore mineral
Carnallite	$MgCl_2.KCl.6H_2O$	16.95	Ore mineral and contaminant
Kainite	$4MgSO_4.4KCl.11H_2O$	19.26	Important ore mineral
Sulphates			
Polyhalite	$2CaSO_4.MgSO_4.K_2SO_4.2H_2O$	15.62	Ore contaminant
Langbeinite	$2MgSO_4.K_2SO_4$	22.69	Important ore mineral
Leonite	$MgSO_4.K_2SO_4.4H_2O$	25.68	Ore contaminant
Schoenite (picromerite)	$MgSO_4.K_2SO_4.6H_2O$	23.39	Accessory
Glaserite (aphthitalite)	$K_2SO_4.(Na,K)SO_4$	42.51	Accessory
Syngenite	$CaSO_4.K_2SO_4.H_2O$	28.68	Accessory
Associated minerals			
Halite	$NaCl$	0	Principal ore contaminant
Anhydrite	$CaSO_4$	0	Common ore contaminant
Bischofite	$2MgCl_2.12H_2O$	0	Accessory contaminant
Bloedite (astrakanite)	$Na_2SO_4.MgSO_4.2H_2O$	0	Accessory
Loewite	$2MgSO_4.2Na_2SO_4.5H2O$	0	Accessory
Vanthoffite	$MgSO_4.3Na_2SO_4$	0	Accessory
Kieserite	$MgSO_4.H_2O$	0	Common ore contaminant
Hexahydrite	$MgSO_4.6H_2O$	0	Accessory
Epsomite	$MgSO_4.7H_2O$	0	Accessory
Ores			
Sylvinite	$KCl+NaCl$	10-35	Canada, USA, Russia, Brazil, Congo, Thailand
Hartsalz	$KCl + NaCl + CaSO_4 + (MgSO_4.H_2O)$	10-20	Germany
Carnallitite	$MgCl_2.KCl.6H_2O + NaCl$	10-16	Germany, Spain, Thailand
Langbeinitite	$2MgSO_4.K_2SO_4 + NaCl$	7-12	USA, Russia
Mischsalz	Hartsalz + Carnallite	8-20	Germany
Kainitite	$4MgSO_4.4KCl.11H_2O+NaCl$	13-18	Italy

Table 7.2. Potassium salts.

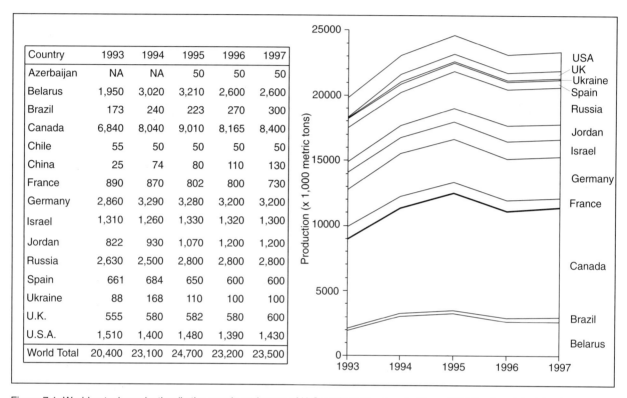

Country	1993	1994	1995	1996	1997
Azerbaijan	NA	NA	50	50	50
Belarus	1,950	3,020	3,210	2,600	2,600
Brazil	173	240	223	270	300
Canada	6,840	8,040	9,010	8,165	8,400
Chile	55	50	50	50	50
China	25	74	80	110	130
France	890	870	802	800	730
Germany	2,860	3,290	3,280	3,200	3,200
Israel	1,310	1,260	1,330	1,320	1,300
Jordan	822	930	1,070	1,200	1,200
Russia	2,630	2,500	2,800	2,800	2,800
Spain	661	684	650	600	600
Ukraine	88	168	110	100	100
U.K.	555	580	582	580	600
U.S.A.	1,510	1,400	1,480	1,390	1,430
World Total	20,400	23,100	24,700	23,200	23,500

Figure 7.1. World potash production (in thousand metric tons of K_2O equivalent - compiled from USGS online data tables).

potash is extracted from the Rhine graben in France, while Miocene potash is mined in the Stebnik area in the western Ukraine. In Spain, in the northern part of the Ebro Basin near the border with France, potash is produced from Eocene-Oligocene lake sediments. In Great Salt Lake, Utah, underground bedded potash is brought to the surface by solution mining of Tertiary playa deposits and potash is recovered from the brine by solar evaporation and flotation. Israel uses similar technology to produce large volumes of potash from brines recovered from Miocene evaporites buried beneath the floor of the southern Dead Sea. In California, potash and coproducts, borax pentahydrate, soda ash and saltcake, are recovered from subsurface playa brines using mechanical evaporation. Sylvite is also a byproduct of nitrate processing in the Atacama Desert of Chile, South America.

The fertiliser industry uses more than 90% of produced potash, and the chemical industry uses the remainder in the manufacture of soaps and detergents, glass, ceramics, synthetic rubber and numerous industrial chemicals. About 70% of the potash is produced as potassium chloride and so sylvite is the preferred ore mineral in most mines. Potassium sulphate and potassium magnesium sulphate, required by certain crops and soils, comprise about 25% of

potash production. Retailers typically sell potash, or potash blended with other fertilisers, in dry or liquid form. Potash is a source of soluble potassium, which is one of the three primary plant nutrients; the others are fixed and soluble phosphorus and nitrogen. Mankind's use of these three nutrients, in commercial forms, started the "Green Revolution" in agriculture.

The importance of potash to the world economy means it is worthwhile understanding how potash salts accumulate naturally in a variety of depositional and diagenetic settings. First we shall look at some Quaternary deposits and then some ancient counterparts with a view to understanding what depositional and diagenetic processes concentrate the potash salts to economic levels.

Quaternary potash

Natural Quaternary potash accumulations are rare: there are only two well documented examples of substantial modern potash accumulations, namely, the Qaidam Basin in China and the Danakil Depression in Ethiopia. Other examples where carnallite or sylvite are forming today are either as minor effloresences and thin crusts in continental lakes, playas and soils, or as artificial products, either

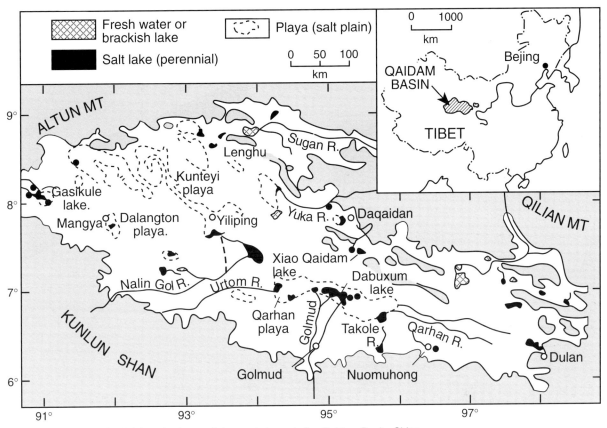

Figure 7.2. Location of salt lakes, freshwater lakes and playas in the Qaidam Basin, China.

brought about by enhanced solar pond evaporation of seawater as at Lake McLeod in Australia or by evaporation of $CaCl_2$-rich basinal waters that have been pumped to the surface as in the Dead Sea in the Middle East. Sylvite also occurs naturally as a widespread pedogenic salt in the hyperarid regions of the Atacama Desert in Chile.

Lake Qaidam

The high plateau region of northern Qinghai-Xizang (Tibet) in western China contains large closed basins with thick fills of Mesozoic and Cainozoic nonmarine sediments (Figure 7.2). Its modern at-surface playas are subject to intense evaporation and are characterised by hypersaline brines as well as accumulations of potash and borate evaporites. Lake Qaidam is of special geological interest as it is one of the few localities in the world where significant quantities of potash salts are precipitating (Yang et al., 1995; Vengosh et al., 1995; Casas et al., 1992, Kezao and Bowler, 1985).

The Qaidam Basin is a large intermontane sedimentary and structural basin located on the northern margin of the Qinghai-Tibet Plateau; with an area of 120,000 km², it has

an extent that approaches that of some ancient continental 'saline giants'. It encompasses a large, tectonically active system that is totally isolated from the ocean and filled with clastic and evaporite sediments (Figure 7.2). The Kunlun Mountains bound the basin to the south, the Altun mountains to the west and the Qilian Mountains (that reach 5,000 m above sea level) bind the northern portion of the basin. The basin centre has an average elevation of 2,800m and contains numerous shallow to ephemeral saline lakes and dry saline pans.

The Qarhan playa or salt plain, with an area of some 6,000 km², is the largest saltflat in the Qaidam Basin and is underlain by halite (Figure 7.2). It is a broad flat plain floored by a layered halite crust with a permanent groundwater brine located 0 to 1.3m below the surface. The surface crust consists of a chaotic mixture of fine-grained halite crystals and mud, with a rugged, pitted upper surface (Schubel and Lowenstein, 1997). Vadose diagenetic features, such as dissolution pits, cavities and pendant cements, occur where the salt crust lies above the water table. Interbedded salts and siliciclastic sediments underlie the crust to reach thicknesses of up to 75m, and over the last 50,000 years have formed the Qarhan Salt plain (Kezao

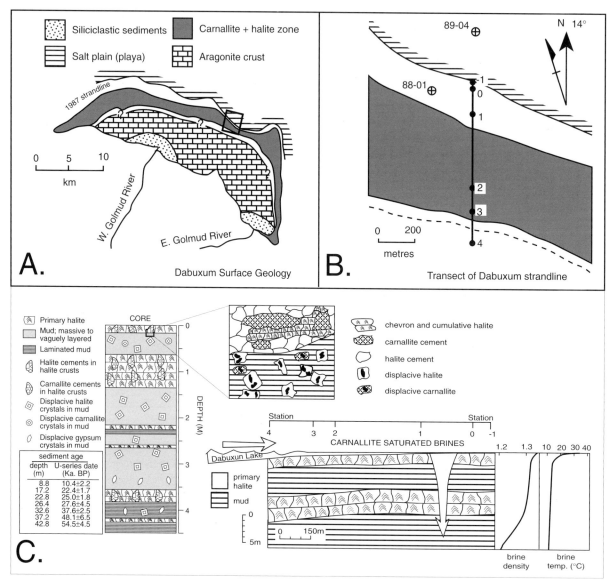

Figure 7.3. Geology of carnallite in Lake Dabuxum, Qaidam Basin. A) Distribution of surficial sediment types also showing the position of enlarged region seen in B. B) Strandline geology showing position of coring traverse seen in C. C) Cross section showing how process of brine cooling forms carnallite cement in the karstic voids in subsurface crusts and the resultant cement textures (after Casas et al., 1992).

and Bowler, 1985). Carnallite occurs in the upper 13m of the Qarhan playa sequence.

The compositions of fluid inclusions in primary (chevron) halite in the various crusts, represent preserved lake brines and indicate relatively wet conditions throughout most of the Late Pleistocene (Yang et al., 1995). Oxygen isotope signatures of inclusions record episodic freshening and concentration during the formation of the various bedded salt units. Desiccation and salt deposition occurred a

number of times: 1) during a short-lived event at about 50,000 yr. BP, 2) from about 17,000 to 8,000 yr. BP, and 3) from about 2,000 yr. BP to the modern (Figure 7.3). Modern halite crusts contain the most concentrated inclusion brines, indicating that the modern may be the most desiccated period recorded in the last 50,000 years. Associated diagenetic halite fluid inclusions (clear halite), formed from shallow groundwater brines, confirm the primary halite climatic record. The occurrence of carnallite-saturated brines in the diagenetic halite fluid inclusions in the top 12 m of sediments also imply a diagenetic, not depositional origin of preserved carnallites.

Today the Qarhan salt plain (playa) encloses 10 shallow perennial saline lakes fed mostly from the Kunlun Mountains. The largest lake on the Qarhan Salt Plain is Dabuxum Lake; an area where carnallite recently accumulated along its strandline (Casas et al., 1992; Casas, 1992). The greatest volume of water entering the lake comes from the Golmud River (Figure 7.3a). Cold springs, emerging from a linear karst zone some 10 km to the north of the strandline, also supply solutes to the lake. Thus, evaporation of a mixture of river water and spring water creates the carnallite-saturated brines of the Dabuxum Lake system (Lowenstein et al. 1989).

Surface sediments of Dabuxum Lake are zoned (Figure 7.3a,b). The southern part of the lake, closest to the Golmud River delta, is underlain by fine grained siliciclastic sediments that pass lakeward into a centimetre-thick aragonite crust that floors much of the lake. A small island in the centre of Dabuxum Lake in made up of mud and silt laminae coated by microcrystalline aragonite and mud chip breccias cemented with aragonite. Halite crusts floor most of the northern subaqueous parts of Dabuxum Lake.

Surface efflorescent crusts of halite and carnallite accumulated along the margin of Dabuxum Lake from 1987-1989 with carnallite reaching thicknesses of up to 0.3 m along a strandline zone approximately 1 km wide. The lake area at that time was approximately 250 km^2 and waters were less than 1m deep (Casas et al., 1992). The carnallite crust was brine-saturated, poorly cemented and had a mushy consistency. It was dominated by equigranular, anhedral crystals up to 5 mm in diameter. Thin sub-centimetre layers of very fine grained (sub-millimetre) carnallite and siliciclastic mud occurred within the carnallite mush. Individual carnallite crystals in the mush were often linked with textures identical to rafts or films of fine-grained carnallite, which could be seen forming on the surface of the very shallow brines near the lake shore. Rafts were being blown onto the strandline or left on the lake floor as the drying brine sheets retreated from the strandline. Strandline carnallites underwent extensive syndepositional diagenetic modification depending on daily temperature and weather conditions. In a single two-week period dominated by fine sunny weather with daily temperatures up to 35°C, carnallite precipitated in the afternoon as rafts, or as centimetre-size, needle shaped skeletal crystals, or as efflorescent puffs of sub-millimetre crystals on evaporitic mudflats wherever the brine level was below the surface.

Light rains and strong southwest winds are common in the summer in Qaidam Basin. During such storms Dabuxum lake waters are blown toward and pond on the north shore (Figure 7.3b). For example, vigorous winds on 6 and 10 August, 1988 shifted the northern lake shore several hundred metres (Casas et al., 1992). These open lake waters were at or slightly below halite saturation and considerably below carnallite saturation. After being covered by these waters the carnallite crust was modified and partly dissolved. During subsequent warm days the crust was re-established. In a subsequent large flood during July to September 1990 the Dabuxum Lake area expanded from 250 km^2 to 800 km^2, while the associated freshening completely dissolved the carnallite crust. Thus strandline carnallite on the north shore of Lake Dabuxum is an ephemeral sediment with little or no preservation potential.

Today it appears that carnallite is precipitated and preserved only in the upper 13 m of the 45m thick interbedded clay-halite succession that underlies the Lake Dabuxum strandline (Figure 7.3c). These subsurface carnallites are diagenetic void-filling cements and displacive crystals, which indicate carnallite saturation has a strong temperature control. As dense carnallite-saturated lake brines sink into the underlying sediment they cool and so precipitate carnallite in preexisting karst holes and pits in the halite bed (Figure 7.3c; Casas et al., 1992). The voids were created by the lowering of the water table and periodic freshening of the lake waters prior to the accumulation of the present salt crust. This style of secondary nearsurface carnallite accumulating as early cements in the active phreatic zone is analogous to much of the secondary carnallite and sylvite seen in ancient potash deposits in Thailand, Canada and the USA.

Danakil Depression, Ethiopia

The Danakil Depression, located in the axial zone of the Afar rift near the confluence of the East African, Red Sea and Carlsberg rifts, is an area of intense hydrothermal activity related to rift magmatism and deep brine cycling (Holwerda and Hutchison, 1968; Hardie, 1990). Located some 50 to 80 km inland, it runs parallel to the Red Sea coast, is some 185 km long, 65 km wide, with a basin floor as much as 110 metres below sea level (Figure 7.4). A shallow volcano-tectonic barrier prevents continuous direct surface inflow of seawater into the depression.

More than 970 metres of halite-dominated-evaporites have accumulated in the Danakil Depression, along with gypsum, anhydrite and shale interbeds (Holwerda and Hutchison,

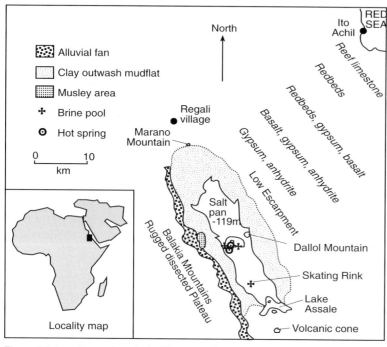

Figure 7.4. Location and geological features of the Danakil Depression (after Holwerda and Hutchison, 1968).

1968; Augustithis, 1980). Evaporites still outcrop extensively in its northern portion, and probably underlie the entire depression, but lava flows and alluvium cover the salt beds in the southern two thirds of the depression.

This km-thick Quaternary subcropping halite contains one and perhaps two dipping potash beds in the Musley area (Figure 7.4). The upper potash bed has been cored in a number of wells in the depression, while the lower bed has been interpreted from gamma logs. The upper potash interval is between 15 and 40 metres thick; it consists of a basal member rich in kainite (4-13m thick), an intermediate carnallite member (3-25 m thick), and an uppermost sylvite member (0-10 m thick). The first two are primary precipitates, while the sylvite zone formed by selective surface or nearsurface leaching of $MgCl_2$ from the carnallite by meteoric waters (Holwerda and Hutchison, 1968). Due to its secondary origin, the proportion of sylvite in the sylvite member decreases as the proportion of carnallite increases, along with kieserite, polyhalite and kainite.

Beneath the sylvite member, the carnallite member contains a varied suite of non-commercial potash minerals that in addition to carnallite include, kieserite, kainite and polyhalite along with minor amounts of sylvite. Minor anhydrite is common, while bischofite and rinneite may occur locally along with rust-red iron staining. Sylvite is

more abundant near the top of the carnallite member and its proportion decreases downward. Kainite is the reverse. The sylvite member and the carnallite member also show an inverse thickness relationship. Bedding in the carnallite member is commonly contorted with folded and brecciated horizons interpreted as slumps. The base of the carnallite member is defined as where carnallite forms isolated patches in the kainite before disappearing entirely. The kainite member is composed of nearly pure, fine-grained, dense, relatively hard, amber-coloured kainite with $\approx 25\%$ admixed halite. Texture and colour are characteristic, but even more so are the distinct "ruler-straight" bedding planes in the kainite, a striking contrast to the wavy, contorted and irregular bedding planes of the carnallite and sylvite members. The preponderance of $MgSO_4$ salts in these Pleistocene evaporites means they most likely formed by the evaporation of seawater.

This is in marked contrast to the present hydrology of the Danakil Depression (Hardie, 1990). Today hot springs, rich in basinal $CaCl_2$, supply and maintain the active brine pools and lakes of the Depression. Lake Giulietti, with an area of 70 km^2 and a floor some 100m below sea level, is located within the active volcanic field of the northern Danakil. Both the inflowing hot springs (40 - 50°C) and the brines of the lake itself are rich in $CaCl_2$ and poor in $MgSO_4$. Extensive concentration of these lake waters will eventually lead to deposition of an $MgSO_4$-poor series of salts dominated by sylvite, carnallite, halite and $CaCl_2$ salts. This will be in marked contrast to the thick underlying $MgSO_4$-rich Pleistocene evaporites. It implies that seawater was much more important as a brine source in the Pleistocene than today.

Interaction between hot basinal groundwaters and volcanic bedrock is the likely explanation of the present inflow chemistry. Dissolution of the earlier highly soluble potash evaporites in preference to the less soluble Mg-salts also plays a part. Other hot brine springs in this geothermally active area around the Dallol salt pan are rich in $FeCl_2$, sulphur and manganese, clearly indicating a hydrothermal interaction of saline groundwaters with volcanics.

Inland chotts and coastal sabkhas in north Africa

Chott el Djerid in southern Tunisia is an endoheic playa with ephemeral carnallitite crusts accumulating in the most central portions of the saline pan facies (Bryant et al., 1994). It has an area of approximately 5,360 km^2 in a drainage basin with an area of 10,500 km^2. Mean annual rainfall for the area is 80-140 mm, the mean annual temperature is 21°C and evaporation, which is highest between May and September, has a mean annual value of 1500 mm.

Chott el Djerid lies in the zone of chotts where a series of regional aquifers emerge to form a region of discharge playas along the northeast extremity of the Bas Saharan Artesian Basin (Figure 7.5a). The latter covers most of the Algerian and Tunisian Sahara, extends into Morocco and Libya and encloses the whole of the Grand Erg Oriental. The zone of chotts lies along major discharge areas of the 'Complexe Terminal' aquifer system, which is made up of permeable Turonian dolomites and Upper Senonian limestones and intercalated evaporites, along with some Mio-Pliocene sands and conglomerates (Figure 7.5b). Discharging continental ground-waters seep into the chotts beneath the edges of thin Quaternary clay aquitards that also form the bottoms to the playas. The same clay bed facilitates temporary ponding of waters supplied by winter floods onto the central playa facies.

A flood in January 1990 and its aftermath showed that the chott undergoes four evolutionary stages as it desiccates (Bryant et al., 1994): (1) initial flooding, (2) evaporative concentration of lake waters, (3) the movement of concentrated thin brine sheets over the playa surface as

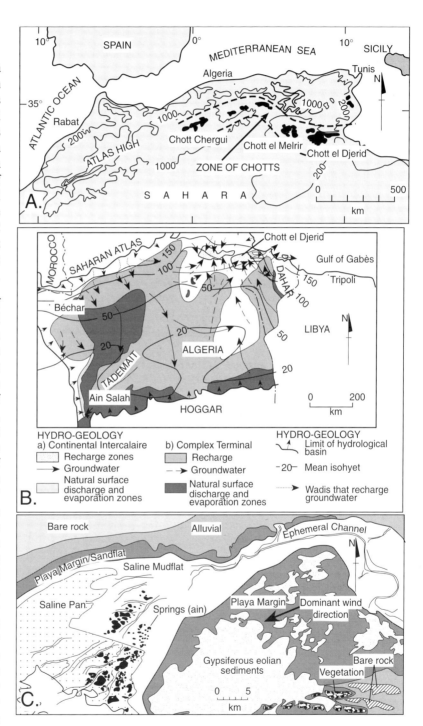

Figure 7.5. Chott el Djerid. A) Locality showing Zone of Chotts as regional groundwater outflow zone. B) Regional hydrology. C) Facies on northeastern area of Chott el Djerid (after Bryant et al., 1994).

a result of wind action, and (4) total desiccation of the lake by September 1990, some nine months after the initial flood. Surface waters and seeps in the chott have a consistent

231

Ca–SO$_4$–Cl rich and HCO$_3$–CO$_3$ poor chemistry, reflecting the recycling of older (Pleistocene - Cretaceous) evaporites within the catchment area. As the lake desiccates and shrinks, these waters concentrate to produce an Na–Mg–K–Cl–SO$_4$ continental brine with ionic proportions similar to those of modern seawater. During desiccation, saturation was reached first with respect to gypsum and then with halite and ultimately with potash. After complete desiccation, the main mineral phases in the crystal mush on the floor of the saline pan are gypsum and halite, with lesser ephemeral carnallite present mostly as cm-scale layers within small haematite-stained areas near the centre of the pan (Figure 7.5c).

Another modern area of ephemeral carnallite formation in North Africa is in the modern mudflats of Sabkha el Melah and the smaller Sabkha el Mehabeul. Both mudflats are situated on the northern edge of the Djeffara coastal plain near the town of Zarzis, Algeria. Sabkha el Mehabeul is the more isolated of the sabkhas; it is now a true endorheic basin as it has not been connected at surface to the nearby Mediterranean since the Upper Pleistocene (Stengele and Smykatz-Kloss, 1995). The larger Sabkha el Melah is still connected with the open sea by a channel that opens in times of flooding and turbulent seas. Evaporites in the salt crusts and in the upper 0.5 metres of the sediments are concentrically zoned. This zonation is more evident in Sabkha el Melah where gypsum and anhydrite occur at the rim of the sabkha depression; increasing amounts of halite occur towards the centre of the ephemeral saline pan facies. Traces of mirabilite and carnallite occur in the shallow subsurface sediments (< 1m depth), but only in the innermost zones of the ephemeral saline pan facies. The separation of the Sabkha el Mehabeul from the ocean since the Upper Pleistocene has enabled a more advanced stage of diagenetic maturity in this sabkha fill. Anhydrite predominates over gypsum and the concentric zonation of the pan minerals is much less obvious. Notably, in both these coastal sabkhas the sole potash mineral, carnallite, is no more than a minor mineral phase with little preservation potential. This lack of preserved potash beds in modern near-coastal depressions characterises all Holocene coastal sabkhas and salinas worldwide. It clearly illustrates that in terms of marine-fed potash deposits, so common in many ancient evaporite basins, the present marine coastal setting is a poor key to the widespread evaporites of the past (Chapter 3).

Sylvite in the Amadeus Basin, Australia

To the south of Alice Springs in the Amadeus Basin of Central Australia there are a number of small saline playas

(areas < 50 km^2) with ephemeral sylvite/glauberite crusts (Arakel and Hong Jun, 1994). Unlike these smaller playas, there are no documented sylvite crusts in the larger playas of the region, such as Lake Amadeus (area ≈ 750 km^2). The absence of potash precipitates in the larger playas reflects the nature of the inflow hydrologies in the region (Jankowski and Jacobson, 1990). Both types of playa experience the same arid climate; median and mean annual rainfall in the region are 190 mm and 220 mm, respectively, mostly falling in summer (Dec.-Feb.); while annual evaporation is ≈ 3300 mm. All the playas are ultimately supplied by a combination of surface runoff and seepage from a large regional groundwater outflow system that discharges into this chain of playa lakes, which is some 500 km long.

The difference in the mineralogy of the surface salt crusts lies in the depth to which the various playa groundwater systems tap into the regional hydrology. Throughout the artesian discharge zone, the various playa salt crusts form by capillary evaporation. All playa sediments contain highly concentrated brines; typically these are sodium-chloride rich waters with appreciable magnesium and sulphate and very low concentrations of calcium and bicarbonate. The larger playas in the chain, exemplified by Lake Amadeus, have dual shallow and deep groundwater inflow paths. The smaller playas, exemplified by Spring Lake and the Na$_2$SO$_4$ lakes of the Karinga Creek drainage system, have only shallow inflow paths. Brines in the larger playas are continually diluted by mixing with upwelling deep groundwaters. Thus, pore saturation with respect to gypsum/halite is obtained rapidly in the capillary zone of the mudflats of Spring Lake, a small hydrologically isolated playa. In contrast, halite saturation does not occur in the mudflats of large hydrologically open discharge playas, such as Lake Amadeus, where the inflow seepages are fed in part by the fresher waters of regional discharge aquifers. Subcropping groundwaters in both lake Amadeus and Spring Lake are undersaturated with respect to glauberite and sylvite; but owing to the inherently higher local salinities and a lake surface located above the regional groundwater discharge level, both minerals can accumulate as ephemeral efflorescent crusts about the edges of Spring Lake. The continual recharge of artesian waters beneath all the playa depressions in the region means there is little longterm preservation of buried salts that are more saline than gypsum (Chapter 2).

Potash from brine wells

Potash is not only mined, it is also produced by brine processing. Typically the brine is extracted by pumping, often in combination with solution mining from a potash horizon some hundreds of metres below the surface. Recovered brine is then pumped into ponds where it is further concentrated by solar evaporation. Often the mother potash interval was not suitable for extractive mining due to its structural complexity or a propensity for water flood. Potash units can also be solution mined at much greater depths than are feasible using conven-tional methods.

For example, potash is produced by the solar concentration of brines extracted from beneath the margins of the southern Dead Sea (Figure 7.6). The Dead Sea Basin, with a floor some 400 m below sea level is the deepest part of a pull-apart valley that lies on a transform fault that separates the Arabian and the Palestinian Plates. The brine level in the Dead Sea has been falling for many years and the Southern Basin is now a permanent saline pan (Warren, 1989). The Dead Sea is a modern example of a saline lake characterised by $CaCl_2$-rich saline brines held within a narrow fault-bounded basin.

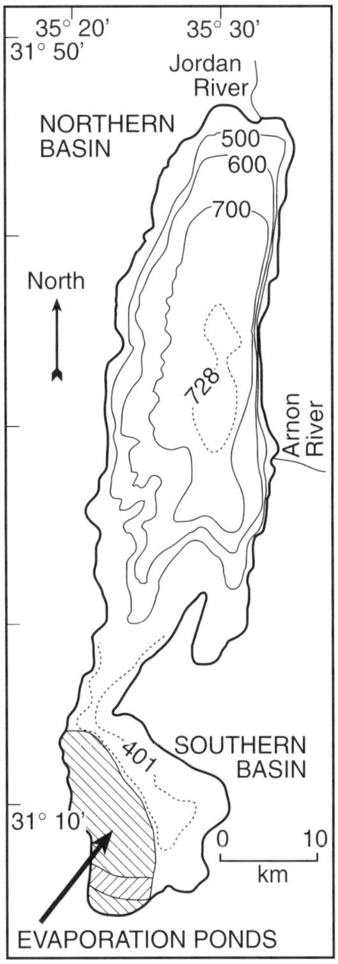

Exploitation of the Dead Sea brines takes place at the southwest end of the Dead Sea, by the Dead Sea Works Ltd, at Sedom on the Israeli side, and by the Arab Potash Company at Ghor al Safi on the Jordanian side. At Sedom the potash is extracted from evap-orative pans as a carnallite slurry (mixed with sodium and magnesium chloride) created by sequential evaporation. Brine is first pumped into

Figure 7.6. Evaporation pans at the southern end of the Dead Sea. The water surface in the Dead Sea is around 401 m below sea level; the Southern Basin is covered only by a thin brine sheet, while the deeper brines of the Northern Basin attain depths of more than 300m (seafloor isobaths are in metres below sea level).

evaporation pans atop the natural saltflat surface of the Southern Basin, and then allowed to concentrate to carnallite saturation while moving through a succession of progressively more concentrated ponds (Figure 7.6). The area of the concentration pans is some 130 km^2, within a total area of 1,000 km^2 of the Dead Sea floor. Potassium chloride was originally produced from the slurry via hot leaching or flotation, but is now produced more eco-nomically in a cold crystallisation plant. Output in Israel currently exceeds 1.3 million tonnes per year (Figure 7.1). In Jordan, the Arab Potash Company produces nearly a million tonnes per year.

Brine extraction for potash with subsequent solar pan concentration also takes place on the edges of Great Salt Lake and the Bonneville Salt Flats in Utah. On the east and west shores of Great Salt Lake near Ogden, Utah, brines are concentrated in the more than 160 km^2 of solar ponds within a total area of 4,100 km^2 of the Great Salt Lake. During normal summer conditions, 90% of the halite present in the brine is precipitated with little contamination by other salts. Mirabilite is a winter precipitate in the ponds (see later). Once potassium salts begin to precipitate from the saturated brine they form a complex mixture of double salts, such as kainite, schoenite and carnallite, while magnesium precipitates as bischofite, a highly hydrated magnesium chloride. This salt sequence is produced from a continental brine, but has a seawater-like geochemistry.

Formerly the solids were harvested using front-end loaders and trucks, and transported to the processing plants. Recently the Behrens Trench was completed, a 35 km underwater canal on the floor of the Great Salt Lake that transports saturated brine from the western shore solar ponds across the lake to its processing facilities on the eastern shore. The processing plant produces sulphate of potash from this brine as well as solid magnesium chloride. By 2000 it is predicted that the Great Salt Lake plant will produce more that 550,000 tonnes of sulphate of potash. The salt workings about the strandline of Great Salt Lake also illustrate one of the major problems in solar pond extraction atop saline mudflats. Record floods in 1984 filled the ponds with fresh water and destroyed much of the infrastructure. Harvesting did not recommence until the early 1990s.

Potash is also produced by solution mining of Pennsylvanian potash evaporites in the Cane Creek anticline in the northern part of the Paradox Basin in Utah near the town of Moab. Extracted brines are evaporated and flotated. The mother potash bed is a 3m-thick sylvinite bed some 1200 m below the surface. The Utah operation began in 1972, when an

233

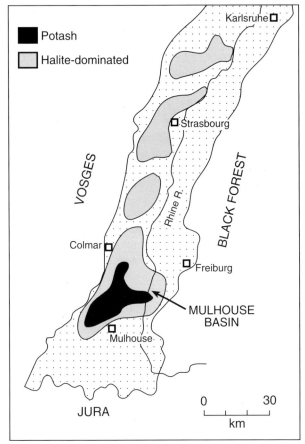

Figure 7.7. Mulhouse Basin, Rhine Graben, showing how the potash dominant section is the southernmost of a number of bedded halite deposits in the graben.

unsuccessful conventional mine was flooded and converted to the current solution mining- evaporation processing. The former mine was not successful owing to the high degree of deformation of the ore horizon. Solution mining for potash is also profitable in Saskatchewan, where solution mining operations have been active since 1964, and in NE Thailand since the late 1980s.

Ancient potash

Rhine Graben

Oligocene evaporites in the Mulhouse Basin in the Rhine Graben contain sylvite with subordinate carnallite, but lack the $MgSO_4$ salts characteristic of the evaporation of modern seawater (Figure 7.7). The Rhine Graben is a Tertiary rift some 150 km long and 10-25 km wide that straddles the Franco-German border as it runs from Basle in Switzerland to Frankfurt in Germany. The Mulhouse

Basin, with an area of 400 km^2, is the southernmost of a number of Oligocene evaporite basins in the graben. Total fill of Oligocene lacustrine evaporites is some 1,700m thick and is dominated by anhydrites, halites and mudstones. The sequence is underlain by Eocene continental mudstones with lacustrine fossils and local anhydrite (Lowenstein and Spencer, 1990; Hardie, 1990; Gely, et al., 1993).

The Oligocene halite section in the Mulhouse Basin includes two thin, but important, potash zones: the Couche inferieure (Ci, 3.9m thick), and Couche superieure (Cs, 1.6m thick; Figure 7.8). Both potash beds are made up of stacked, thin, parallel-sided cm-dm-thick beds (averaging 8 cm thickness), which are in turn constructed of couplets composed of grey-coloured halite overlain by red-coloured sylvite. Each couplet has a sharp base that separates the basal halite from the sylvite cap of the underlying bed. In some cases the separation is also marked by a bituminous parting. The bottommost halite in each dm-thick bed consists of halite aggregates with cumulate textures that pass upward into large, but delicate, primary chevrons and cornets. Clusters of this chevron halite swell upward to create a very hummocky boundary with the overlying sylvite. The sylvite member of the couplet consists of granular aggregates of small clear halite cubes and rounded grains of red sylvite (with some euhedral sylvite hoppers) infilling the swales in the underlying hummocky halite. The sylvite layer is usually thick enough to bury the highest protuberances of the halite, so that the top of each sylvite layer, and the top of the couplet, is flat. Dissolutional pipes and intercrystalline cavities are noticeably absent, although some chevrons show rounded coigns. Intercalated marker beds formed during times of brine pool freshening are composed of a finely laminated bituminous shale with dolomite and anhydrite.

The sylvite-halite couplets appear to record unaltered settle-out and bottom-growth features of a primary chemical sediment that accumulated in a shallow perennial surface brine pool. The sylvite layers, like the halite layers, are interpreted as primary subaqueous deposits (Lowenstein and Spencer, 1990). Based on their crystal size, the close association with cumulate halites in the sylvite layers, and the manner in which they mantle underlying chevron halites, sylvites are interpreted as precipitates first formed at the air-brine surface (or within the upper brine) that then sank to the bottom to form well sorted accumulations. Similar cumulate sylvite deposits are found on the floor of modern Lake Dabuxum in China (see earlier). The mosaic textural overprint of the Mulhouse sylvite layers was probably produced by postdepositional modification of the crystal boundaries, much in the same way as mosaic

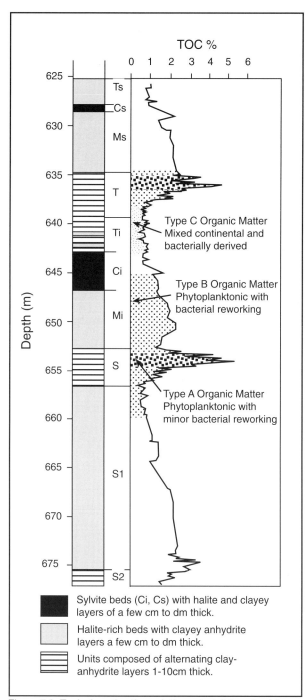

Figure 7.8. Typical section through the potash intervals of the Upper Salt. The log also illustrates the amount and style of total organic carbon (TOC%) extracted from marly layers in the various units. Curve is smoothed with a moving three point average (after Gely et al., 1993).

halite is formed by recrystallisation during burial (see Chapter 1). Inclusion studies in both the sylvite and the halites average 63°C, suggesting solar heating of surface

brines as precipitation took place. Similar high at-surface brine temperatures are not unusual in many modern brine pools, especially those subject to density stratification and heliothermometry.

TOC values in the sylvite interval also imply preservation of a primary sylvite deposit. TOC levels fluctuate in a rhythmic manner related to the salinity of the depositional brines (Figure 7.8; Gely et al., 1993). The highest TOC values are found near the top of the clay-anhydrite layers and the lowest values are recorded near the top of the halite-rich beds where the sylvite (and the highest brine concentrations) would have occurred. The organic matter is mainly of algal origin (A and B groups), while a third category of organic material (C group) is derived from a mixture of continental inflow and *in situ* bacterial productivity. The sylvite bed (Ci) occurs at the transition from Type B to Type C, indicating a change of brine chemistry from the hypersaline brines of the halite/sylvite pans to the freshened waters that precipitated the marls.

Khorat Plateau, Thailand

Potash in the halite-dominated Cretaceous Maha Sarakham Formation is preserved within two basins: the northern Sakon Nakhon Basin (which also extends into Laos) and the southern Khorat Basin. These basins underlie much of the Khorat Plateau in NE Thailand and are separated by the highlands of the Phu Phan Range (Figure 7.9a). It is possible that at one time there existed a single giant evaporite basin encompassing at least the area of the present day Sakon Nakhon and Khorat Basins. These deposits represent one of the last saline giants preserved in SE Asia and suggest an extreme arid period during Cretaceous time. The sequence lacks $MgSO_4$ salts, indicating a likely continental/basinal brine source. Most of the current margins of the basin are dissolutional, as are the upper and lower contacts of the salt units (Chapter 4).

In those parts unaffected by evaporite dissolution the Maha Sarakham is composed of three salt beds –Lower, Middle and Upper Salt members– each overlain by a clastic mudstones –Lower Middle and Upper Clastic members– (Figure 7.9b). Where the Maha Sarakham has not been thinned, either by dissolution or salt flow, it is 250-300m thick. In areas of salt dissolution it thins to <100m, while in halokinetic regions it can be more than a kilometre thick. Evaporite minerals include halite; the sulphates gypsum and anhydrite, along with a variety of halite-hosted potash phases dominated by carnallite and sylvite with accessory tachyhydrite. Portions of the bedded halite display intricate depositional, dissolutional and diagenetic features that rank amongst the most pristine I have seen preserved in an ancient evaporite.

Middle and Upper Salt members are generally flat lying and correlatable from well to well. In contrast, seismic and geomorphologic analysis shows that the Lower Salt of the Maha Sarakham has flowed so that thicknesses are much more variable from well to well (Figures 4.12, 7.10). Textures in the Lower Salt Member range from well preserved depositional textures in the upper part immediately below the potash entraining intervals to flowage features lower in the section. Petrographic evidence for salt flowage includes overburden-controlled elongate flattening of crystals and pervasive recrystallisation that has destroyed original bedding features. In wells that pass deeper into the Lower Salt Member there is often a transition down core from pristine depositional textures into flowage textures, implying that flow into the pillows was fed mostly from the deeper parts of the Lower Salt Member. This is also consistent with the upper portions of the salt unit still being deposited while the lower parts of the same unit were flowing into pillows.

At the regional scale, the halokinetic features follow curvilinear trends typically subparallel to the present basin margin. Some pillows in the Lower Salt Member, especially those associated with sylvinite along the western margin of the basin, were probably syndepositionally initiated, as they are truncated by sediments of the Lower Clastic Member (Figures 7.10, 7.11). Other salt flow features are probably active today, driven by laterally shifting surface river systems, including the Mekong and its tributaries. For example, $CaSO_4$ caprocks outcrop today in Laos in the erosional river banks in one of the major tributaries to the Mekong (pers. obs). As a modern growing salt structure approaches the land surface and enters the zone of active phreatic flow the rising salt is breached and altered by fresher groundwaters. With time this creates first a topographic high then, as the salt breaches, it transforms into a dissolution collapse graben often marked by a lake or a surface depression in the modern landscape.

Initiation of the salt flow in the Maha Sarakham is probably diachronous and driven by sediment loading from various depocentres moving across the salt bed. In the Cretaceous the depocentres were tied to the deposition of the Lower

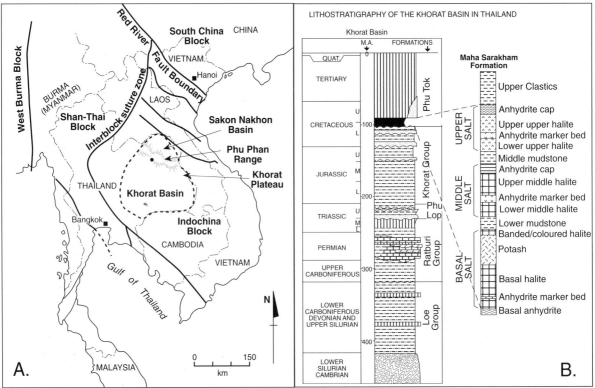

Figure 7.9. Potash in Thailand. A) Locality map showing location and tectonic setting of the Khorat Plateau, the Khorat Basin and the Sakon Nakhon Basin. B) Stratigraphy of the Khorat Basin and the Maha Sarakham Formation, NE Thailand (after Mouret, 1994; Suwanich, 1986).

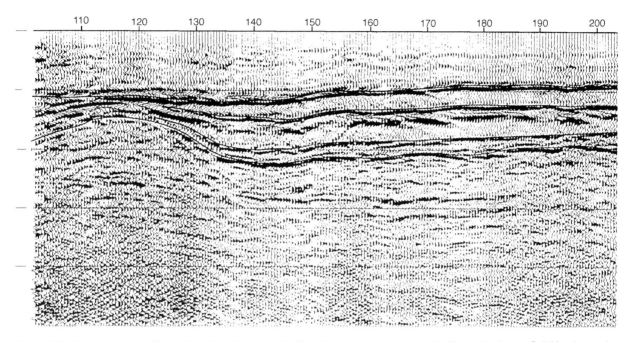

Figure 7.10. Seismic image of Maha Sarakham Fm, from the Khon Kaen area showing a salt pillow in the Lower Salt Member and the syndepositional reduction in sedimentary thickness in units above this structure (seismic line supplied by CRA-RTZ).

Clastic Member; today they are tied to the fluviatile deposits of the Mekong and its tributaries. Such feedback systems between depositional loading (sediment thicks) and salt buildups (pillows and anticlines) characterise the western margin of the Maha Sarakham in the vicinity of Khon Kaen (Figures 7.10, 7.11). As well as being seen seismically, the relationship between surface landform and the underlying structuring of the salt can be imaged using currently available topographic maps, aerial photography and satellite imagery.

Like much of the halite in the Maha Sarakham, the petrography, depositional style and preservation quality of potash in the Lower Salt Member has been complicated by dissolution and halokinesis (Hite and Japakasetr, 1979; Sessler, 1990; El Tabakh et al., 1995; Utha-aroon et al., 1995). In terms of the potash minerals, carnallite dominates over sylvite throughout the basin, while traces of tachyhydrite are also found preserved as enclosed euhedral to subhedral crystals in halite.

Potash is widely dispersed as minor shows across the basin. The only well studied and significant potash-rich zones are in the upper section of the Lower Salt Member along the western margin of the Khorat Plateau. The potash interval is dominated by carnallitite (up to 20-30m thick), which forms a widespread stratiform unit along the western margins of both the Khorat and Sakon Nakon basins. It is locally capped by lesser sylvinite (< 6 m) and

covered by a bed of colour-banded red and grey halite (up to 6 m thick). This colour-banded halite preserves pristine depositional textures and is in turn overlain by the Lower Clastic Member.

A complete interval in the potash-entraining region at the top of the Lower Salt Member is divisible into:

(4) uppermost colour banded halite ($\approx$ 6m),
(3) sylvinite zone (0-6m),
(2) carnallitite zone (15-30m,
(1) lower zone of massive to bedded halite with traces of carnallite (50-300m).

The lower zone (1), is dominated by massive, colourless, inclusion-free halite, which is made up of stacks of beds 0.2-0.5 m thick. The upper portions of this halite interval retain pristine depositional textures, including aligned bottom-nucleated growth forms, clay dissolution stringers and microkarst pits. This halite zone grades down into clean recrystallised mosaic halite, which may stack into intervals that are hundreds of metres thick. Often this clean halite is dominated by biaxially flattened or necked and recrystallised lensoidal textures indicative of salt flow. Minor anhydrite stingers and nodules ($\approx$1 cm thick) cross-cut the core axis often with steep inclinations that also suggest flowage. This form of anhydrite is microcrystalline and lacks any of the original gypsum ghost structures so well preserved in the regional marker beds.

The carnallitite zone (2) is typically 15-25 m thick. Three types of halite are found in this zone: (a) inclusion-rich, cloudy, chevron halite; (b) clear, recrystallised, coarse-grained halite; and (c) granular, fine-grained halite. The first typifies preserved bottom-nucleated beds, the second typifies massive halite beds with poorly defined upper and lower contacts and suggests textures associated with macro and micro karstification, while the third type is found as disseminated halite in carnallite-filled voids. Anhydrite and other insoluble residues are scarce in the carnallitite zone. Halite that hosts carnallite is often crosscut by dissolution surfaces, clay stringers, collapsed halite beds and karst cavities coated with microcrystalline to coarsely crystalline carnallite.

Many of the carnallite-filled voids are subvertical and narrow downward. This implies a syndepositional vadose karst prior to the precipitation of the carnallite cement. Textures in this zone are identical to those found in the bedded potash deposits of the Qaidam Basin and indicate the water table was subject to large oscillations (Figure 7.3). During deposition the carnallitite interval in the Maha Sarakham was subject to brine freshening, subaerial exposure and water table lowering related to climatic changes and tectonic tilting in the depositional basin.

The sylvinite zone (a mixture of halite and sylvite) directly overlies the carnallitite zone and is up to 6 m thick. It is not always present. Its contact zone is transitional with the underlying carnallitite zone, although sylvite crystal-clusters in the transition zone typically show sharp non-horizontal edges. Sylvite is present in its halite host as: a) individual and small crystal clusters of up to 1 cm size, making up amoeboid-shaped forms; b) coarsely crystalline massive beds up to 5 cm thick; c) even layered, thick beds of up to 4 cm thick; d) irregular medium to coarsely crystalline patches within massive halite beds and sometimes rimming preexisting carnallite crystals; and e) massive sylvite beds up to 25 cm across. In almost all cases, dark grey clayey matrix surrounds the larger sylvite crystals and/or alternates with the subhorizontal sylvite beds. Petrographically, sylvite crystals exhibit euhedral crystalline forms either as individual crystals or as clusters of crystals.

At the transition between the carnallitite and sylvinite zones some dissolution cavities are 25 cm across (in core) and filled with single clear sylvite crystals. Higher in the sylvinite interval the sylvite tends to occur as stratiform primary layers showing subhorizontal bedding, and ranges from 10 cm to 1 m thick. Individual primary sylvite crystals can reach up to 2 cm in size in the primary layers.

Macroscopically, sylvite layers are not homogeneous and several types of sylvite may coexist within single layers along with fine clastic components and halite.

Halite beds alternating with sylvite layers range in thickness from 1 cm to 15 cm. Where not destroyed by the processes of dissolution and recrystallisation, these halite beds can still exhibit relict bedding and occasionally are milky white from inclusions in the preserved chevrons. But most of the halite in the sylvinite zone is clear and coarsely recrystallised, with geometries suggesting repeated episodes of dissolution followed by precipitation as passive void-fill cements. Traces of anhydrite are present throughout the sylvinite zone, along with traces of tachyhydrite. Anhydritic cm-diameter nodules are locally present as are white nodules dominated by microcrystalline borate minerals (boracite and tyreskite).

More than one sylvite-rich interval may occur in the sylvinite zone. Workers in the region who have used a strict stratigraphic approach have often reported intervals with more than one sylvite layer as fault-repeated sections. Such an approach underscores the limitations of using a non-process based stratigraphy in an evaporite succession (Chapter 3). What is occurring is not associated with a fault-repeat of a single sylvite bed, but the vagaries of an anastomose sylvite-filled karst system being intersected a number of times in a single core. Many of the so-called repeat beds show classic insoluble-rich geopetal linings.

Colour banded halite (4) constitutes the uppermost part of the Lower Salt Member. It is coarsely laminated and retains pristine depositional textures (mostly chevrons) with little or no evidence of recrystallisation. There is often a sharp upper dissolution contact with the overlying Lower Clastic Member, defined by a diagenetic horizon made up of anhydritic dissolution residues (see Chapter 4).

The origin of the richest sylvite zone (3) is related in part to the action of actively refluxing, possibly freshened brines working on an actively rising (pillowing) carnallite/halite precursor. For example, in the Khon Kaen region (Figures 7.10, 7.11), the Lower Salt Member has flowed into a pillow structure and the overlying strata are truncated beneath a Quaternary alluvial cover. The synclinal structure associated with the rim of this pillow is also a zone where sylvinite thickens at the expense of the underlying carnallitite (Figure 7.11a). This sylvinite is capped by the Lower Clastic Member, which is in turn capped by the Middle Halite Member. Core from the Middle Halite Member in this region shows predominantly undisturbed subhorizontal bedding and still retains perfectly preserved primary textures, such as chevrons and halite intercrystal karst, along with a regional dm-thick marker bed made up of subaqueous swallowtail gypsum pseudomorphed by halite/anhydrite (pers. obs. of core).

Facies geometries suggest that sylvite deposition along the western margin of the Khorat Plateau was: a) post the deposition of the carnallitite; b) prior to deposition of the Lower Clastic Member; c) fed from lateral inflow, probably driven by a potentiometric head generated by relative uplift the adjacent growing salt pillow; d) related to the cannibalisation of the uplifted carnallitite bed atop a growing salt pillow; e) precipitated early, either as pond-bottom nucleates and foundered raft cumulates or as downward-fluxing karst-filling cements in the rising margins of the

brine pool; and f) then covered by subaqueous colour-banded halite, fed via the dissolution of halite from a now carnallitite-depleted pillow crest (Figure 7.11b). Sylvite precipitation was driven both by solar evaporation on the pond floor (as in the Mulhouse Basin) and by subsurface cooling of downward flushing potassium-saturated brines (as in the Qaidam Basin).

In terms of potash exploration in the region, it is important to note that the thickest potentially exploitable sylvinite fairways lie in rim synclines adjacent to breached salt pillows. They do not form laterally extensive horizons and are not akin to the Devonian sequences of Canada discussed in the next section. These pillows were deforming the carnallite zone and uplifting previously formed carnallite beds into a periodically freshened hydrological setting, while sylvite was precipitating on the pan floor and in karstic voids in the halite about the pan rim. It was the superimposed pillow hydrology that was karstifying the halite at the same time that the upper portions of the carnallite member were dissolving and supplying K-rich brines to the pool floor. The raising of the sedimentation surface, via pillowing, explains why some of the sylvite formed as clear coarse-crystal fills in large karst holes in the halite, while other sylvite was a primary precipitate.

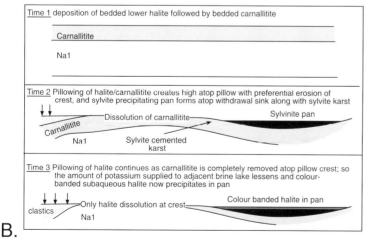

Figure 7.11. Sylvite controls in the Maha Sarakham Formation along the western margin of the Khorat Plateau near Khon Kaen, Thailand. A) Generalised schematic based on well data in the region. B) Sequential development of pillow and the erosion of crest facies controlling the formation of sylvite then colour banded halite in rim syncline depressions (see text for full explanation). Schematic based on author's research and field work in the region.

Permian potash in Europe and North America

The Zechstein Basin stretches from northern Britain, southeast across the North Sea through the Netherlands, Denmark, Germany and Poland to the edge of the Hercynian massifs (Harz, Rhine and Bohemian mountains). Potash extracted from the Permian Zechstein evaporites provide 20% of the world's potash. The Zechstein succession was deposited in a huge drawdown seaway and is today made up of four evaporite cycles: the Z1-Werra Series, Z2-Stassfurt Series; Z3-Leine Series and Z4-Aller Series (Figures 3.24, 3.25). Each has a general sequence of clastics, carbonates, one or two phases of sodium and potassium

Zone	Mineralogy	Potential
11	Mostly carnallite, minor sylvite and leonite	Noncommercial to date
10	Sylvite	2nd best in district
9	Carnallite, kieserite, minor sylvite	Noncommerical to date
8	Sylvite	Moderate reserves, future potential
7	Sylvite	Moderate reserves
6	Carnallite, kieserite, minor sylvite	Noncommercial to date
5	Sylvite and langbeinite	Moderate reserves
4	Langbeinite and sylvite	Main source of langbeinite
3	Sylvite	3rd in sylvite producton
2	Carnallite, kieserite, minor sylvite	Noncommercial to date
1	Sylvite	Formerly the major sylvite producing zone

Table 7.3. Mineralogy and potential of potash ore zones in New Mexico (after Griswold, 1982).

salts, regressive halite or anhydrite, and carbonate. Two potash cycles in the Z1 are exploited in the Werra -Fulda Basin southeast of Kassel. Two potash beds in the Z2 and Z3 are mined near Hannover, while potash is extracted from the Aller Series in the Netherlands and NE England (Boulby Mine).

Potash is also mined from the Carlsbad Potash District in the Permian Basin of New Mexico (Figure 7.12a, Table 7.3; Lowenstein, 1987). Potash minerals occur in eleven ore zones numbered 1 through 11 (in ascending order) in a 215 m thick cyclic interval in the middle of the Salado Formation. Zone 1 is the richest, but is now largely mined out. Production is now mostly from zones 3, 5, 7 and 10. The major ore minerals are sylvite and langbeinite, with accessory leonite, kainite, carnallite, polyhalite, kieserite, bloedite, halite and anhydrite. A typical ore is 60% halite, 30% sylvite, 5% langbeinite, and 2% of polyhalite and insolubles. Langbeinite is produced from zone 4, and mixed sylvite-langbeinite from zone 5. Several mines have been closed in the last decade, and it is unlikely that they will be reopened.

The ore zone in New Mexico lies above and immediately basinward of the older Permian basin margin, as defined by the position of the underlying Capitan Reef (Figure 7.12a). This location corresponds to the spill point of the aggrading basin-filling evaporite cycles that make up the basinwide evaporites of the Castile and Salado formations. The significance of fill and spill hydrology as loci of changes in mineralogy, as discussed in Chapter 3, is clearly illustrated by this facies position.

Two types of lithologic cycles dominate the ore zones in the Carlsbad Potash District (Figure 7.12b; Lowenstein, 1987). Type I cycles begin with mixed siliciclastic carbonate (magnesite) mudstone, followed by anhydrite polyhalite with gypsum crystal pseudomorphs, overlain by halite and

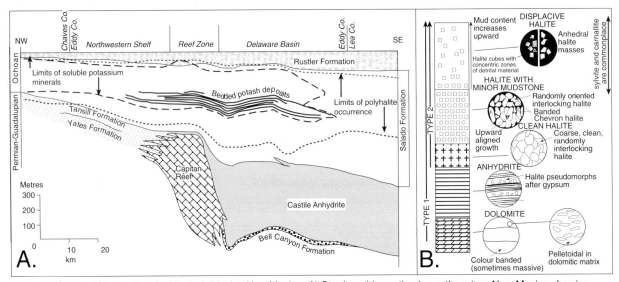

Figure 7.12. Potash in the Carlsbad Potash District, New Mexico. A) Stratigraphic section in southeastern New Mexico showing distribution of bedded potash about basin margin (after Bates, 1969). B) Idealised halite-dominated depositional cycles that host the ore beds (after Lowenstein, 1987).

capped by muddy halite. Type II cycles begin with halite that grades transitionally upsection into muddy halite and pseudomorphs after gypsum, which retain subaqueous bottom-aligned growth habits or mechanically deposited sedimentary structures (ripples, cross beds, etc.). In the anhydrite-polyhalite parts of the cycles there are many pseudomorphs of primary gypsum textures that are now replaced by anhydrite, polyhalite, halite or sylvite. Sylvite and carnallite are most common in the muddy halite parts of the cycles where they occur as millimetre- to centimetre-sized anhedral crystals. Both the blood-red sylvite and the purple carnallite colour come from included micrometre-sized plates of haematite. The crystals may be internally zoned; with dark haematite-rich bands developed at the crystal rims, and progressively lighter bands, depleted in haematite, in the cores of the crystals. Ore zones are composed of stacked halite beds up to 1 metre thick with up to 50% sylvite and carnallite. Thin sections of the ore show poikilitic textures, with several halite crystals encased in single carnallite or sylvite crystals. The sylvite and

carnallite occupy areas equivalent in size and shape to the void spaces in modern saline pan halites and are interpreted as pore-filling cements (Lowenstein and Spencer, 1990).

Given that halite beds lose porosity by 30-50m depth, it is also likely that the potash was emplaced early during shallow burial. Once again, textures and processes of formation are analogous to those in the Quaternary potash deposits of the Qaidam Basin, China. Like these more recent analogues, this Permian potash precipitated in voids in halite beds via cooling of a warm refluxing brine. This mechanism explains the irregular distribution of the potash salts, their general lack of bedding, and why they cross the stratigraphic boundaries of the primary halite beds.

West Canadian Basin (Devonian)

The Middle Devonian Prairie Evaporite Formation is a potash-entraining halite sequence that was deposited in the Elk Point Basin, an early intracratonic phase of the Western

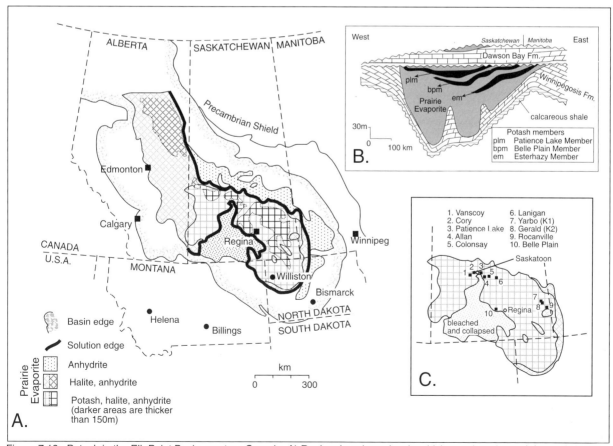

Figure 7.13. Potash in the Elk Point Basin, western Canada. A) Regional geology showing thickest salt and potash intervals in the Prairie Evaporite and its dissolution edge. B) Cross section of basin showing distribution of potash members. C) Map of potash mines in the basin (compiled and modified from Worsley and Fuzesy, 1979; Fuzesy, 1982; Boys, 1993).

241

Canada Sedimentary Basin (Chipley and Kyser, 1989). At that time the Elk Point Basin covered most of Alberta, the southern half of Saskatchewan, southwest Manitoba and extended south into the USA (Figure 7.13a). The basin was bound to the north and west by a series of ridges and arches; but due to subsequent erosion, the true eastern extent is unknown. The Pacific coast was near the present Alberta-British Columbia border, and the Elk Point Basin was centred at approximately 10°S latitude. The isolation of the basin resulted in the deposition of a drawdown sequence of evaporites, including the Prairie Evaporite Formation.

The Prairie Evaporite Formation lies atop the irregular topography of the carbonates of the Winnipegosis Formation (Figure 7.13b). Above the Prairie Evaporite Formation is a series of cyclic Devonian limestones, dolomites and evaporites that make up the Dawson Bay, Souris River, Duperow and Nisku Formations. Lying unconformably on these deposits are the Lower Cretaceous sands of the Manville Group, which are further overlain by younger Cretaceous shales and capped by Quaternary glacial sediments. The Prairie Evaporite Formation is continuous over a large portion of southern Saskatchewan, except where removed by dissolution.

Potash geology

Potash deposits mined in Saskatchewan are found within the upper 70 m of the Middle Devonian Prairie Evaporite Formation at depths of more than 300 metres. Within the Prairie Evaporite there are four main potash-bearing members, in ascending stratigraphic order they are: Esterhazy, White Bear, Belle Plain and Patience Lake Members (Figure 7.13b). The potash members are composed of halite, sylvite, sylvinite, and carnallitite (Fuzesy, 1983). The Prairie Evaporite Formation does not contain any $MgSO_4$ minerals (kieserite, polyhalite) that would be expected to form by the evaporation of modern seawater.

This region of Canada contains the world' largest reserves of exploitable potash and, as of 1998, Canada supplies more than 30% of the world's annual potash production (Figure 7.1). The Prairie Evaporite is 200 m thick in the potash mining district in Saskatoon and only 140 m thick in the Rocanville area. The Prairie Evaporite typically thins southwards; although local thickening occurs where carnallite, not sylvite, is the dominant potash mineral (Worsley and Fuzesy, 1979). The Patience Lake Member is mined at the Cory, Allan and Lanigan mines, and the

Esterhazy Member is mined at Rocanville (Figure 7.13c). Carnallite is uncommon in the Cory and Allan mines.

A typical sylvinite ore zone in the Patience Lake Member can be divided into six units, based on potash rock-types and clay seams (Figure 7.14, M1-M6; Boys, 1993). The units are mappable and can be correlated throughout the PCS mine with varying degrees of success dependent on partial or complete loss of section from dissolution. Potash deposition appears to have been cyclic, expressed in the repetitive distribution of haematite and other insoluble minerals.

Desiccation polygons, desiccation cracks, microkarst pits and chevron halite crystals indicate that the Patience Lake Member that entrains the potash ore was deposited in a shallow-brine, salt-pan environment.

The clay seams form thin stratigraphic layers throughout the potash ore zone(s) of the Prairie Evaporite, as well as disseminated intervals, and constitute about 6% of the ore

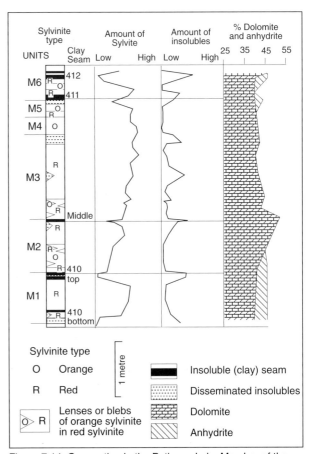

Figure 7.14. Ore section in the Patience Lake Member of the Prairie Evaporite Formation in the PCS Cory potash mine (after Boys, 1993).

as mined. The insoluble minerals found in the PCS Cory samples are, in approximate order of decreasing abundance: dolomite, clays [illite, chlorite (including swelling-chlorite/chlorite), and septechlorite], quartz, anhydrite, haematite, and goethite. The clay minerals make up about one-third of the total insolubles: other minor components include: potassium feldspar, hydrocarbons, and sporadic non-diagnostic palynomorphs (Boys, 1993).

The potash salts probably first formed as secondary precipitates just beneath the sediment surface and were modified to varying degrees by ongoing fluid flushing in the burial environment. The cyclic depositional distribution of disseminated insolubles was possibly due to a combination of source proximity and the strength of winds. Possible intrapotash disconformities, created by the dissolution of overlying potash-bearing beds, may be indicated by an abundance of residual haematite in clay seams. Except in, and near, dissolution and collapse features, the secondary redistribution of insolubles, other than iron oxides, is insignificant.

Bedded halite away from the ore zones generally retains primary depositional textures typical of halite precipitation in shallow ephemeral saline pans (Brodylo and Spencer, 1987). Crystalline growth fabrics, mainly vertically elongate chevrons, are found in 50-90% of the halite from many intervals in the Prairie Evaporite. Many of the chevrons are truncated by irregular patches of clear halite that formed as early diagenetic cements in syndepositional karst.

Halite associated with the potash salt lacks well defined primary textures and is atypical when compared with the well-preserved depositional textures in the non-ore sequences. From the regional petrology and the lower than expected Br levels in halite in the Prairie Evaporite Formation, Schwerdtner (1964), Wardlaw and Watson (1966) and Wardlaw (1968) postulate a series of recrystallisation events forming sylvite as a result of periodic flushing by hypersaline solutions. This is supported by intergrowth and overgrowth textures (McIntosh and Wardlaw, 1968), large scale collapse and dissolution features (Gendzwill, 1978), radiometric ages (Baadsgaard, 1987) and palaeomagnetic orientations of the diagenetic haematite linings associated with the emplacement of the potash (Koehler et al., 1997).

Dating of clear halite crystals associated with ore shows that the exceptionally coarse and pure secondary halites in pods in the potash mining horizons may have precipitated as soon as early burial or have formed as late as Pliocene-Pleistocene times. Even today, alteration appears to be an ongoing process perhaps related to the encroachment of the current dissolution edge.

Halite-sylvite (sylvinite) rocks show two end member textures, the most common is a polygonal mosaic texture with individual crystals ranging from millimetres to centimetres and sylvite grain boundaries outlined by concentrations of blood-red halite. The other end member texture is a framework of euhedral and subhedral halite cubes enclosed by anhedral crystals of sylvite. This is very similar to ore textures in the Salado Formation of New Mexico; which formed by passive precipitation in karstic voids.

Petrographically the halite-carnallite (carnallitite) rocks display three distinct textures. Most halite-carnallite rocks contain isolated centimetre-sized cubes of halite enclosed by poikilitic carnallite crystals. The halite cubes are typically clear, with occasional cloudy crystal cores that retain patches of syndepositional growth textures (Lowenstein and Spencer, 1990). The second texture is coarsely crystalline halite-carnallite with equigranular, polygonal mosaic textures. The third type, which is very rare, contains interlayered halite and carnallite with a primary bedding style similar to that in the Rhine Graben. In zones where halite overlies anhydrite, most of the halite is clear with only the occasional crystal showing fluid inclusion banding.

Clay minerals, which tend to occur as long continuous seams or markers in the potash zones, are mainly composed of detrital chlorite and illite and authigenic septechlorite, montmorillonite and sepiolite (Mossman et al., 1982; Boys, 1990). Of the two chlorites, septechlorite is the more thermally stable. The septechlorite, sepiolite and vermiculite very likely originated as direct products of evaporation or early diagenesis under hypersaline conditions. Absence of the otherwise ubiquitous septechlorite from Second Red Beds west of the zero-edge of the evaporite basin supports this concept.

The fluids

Analysis of fluids currently leaking into the Canadian potash mines, and into collapse anomalies in the Prairie Evaporite, further supports the notion that recrystallising fluids had access to the evaporite minerals throughout the burial history of the Prairie Formation (Chipley, 1995; Koehler et al., 1997). Rb-Sr isotope systematics of halite and sylvite indicate that the Prairie Evaporite was variably recrystallised during fluid events at 371, 284, 241, 185,

147, 85, 35 and 0.5 Ma. These ages are more recent than original deposition and correspond to ages of tectonic events that affected the western margin of North America.

Chemical compositions of inclusion fluids in the Prairie Evaporite, as determined by their thermometric properties, reveal at least two distinct fluids: a Na-K-Mg-Ca-Cl brine, variably saturated with respect to sylvite and carnallite; and a Na-K-Cl brine that is the result of fluid-rock interaction and did not result from simple evaporation of seawater to the sylvite facies (Chipley, 1995). This is supported by δD and $\delta^{18}O$ values of inclusion fluids in halite and sylvite, which range from -146 to 0‰ and from -17.6 to -3.0‰, respectively. Most of the values are different from those of evaporated seawater, which has δD and $\delta^{18}O$ values near 0‰. Furthermore, the δD and $\delta^{18}O$ values of inclusion fluids are probably not the result of precipitation of the evaporite minerals from a brine that was a mixture of seawater and meteoric water. The low latitude position of the basin during the Middle Devonian (15° from the equator), the required lack of meteoric water to precipitate basinwide evaporites, and the expected δD and $\delta^{18}O$ values of any meteoric water in such a setting, make this an unlikely explanation. Rather, the δD and $\delta^{18}O$ values of inclusion fluids in the halites reflect ambient and evolving brine chemistries as the various growth layers were intermittently trapped during the subsurface evolution of the Prairie Formation in the Western Canada Sedimentary Basin. They also suggest that nonmarine pore water has been a major component of the basinal brines throughout much of the recrystallisation history (Chipley, 1995).

Other significant potash deposits

Canadian Maritimes (Mississippian of Nova Scotia and New Brunswick)

The Carboniferous geology of the Maritime Provinces and Quebec is characterised by a series of northeast trending fault-bound grabens or half grabens, containing sediments and evaporites up to 10 km thick (Roberts and Williams, 1993). Of particular interest in this sediment package is the Windsor Group. In Nova Scotia and New Brunswick, the Windsor Group (Visean) hosts $MgSO_4$-poor potash evaporites in several continental intermontane basins that were created by extension driven post-Acadian strike-slip faulting (papers in Beaumont and Tankard, 1987). At Pugwash, the potash deposit consists of halite, sylvite and carnallite with minor rinneite and polyhalite. Deformation is characteristic of much of the ore at Pugwash; where much of the potash sequence has been syntectonically

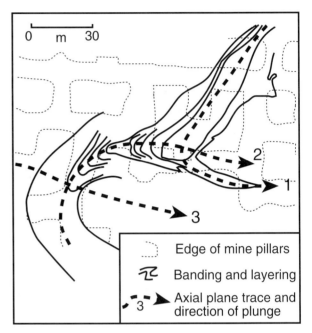

Figure 7.15. Refolded fold in sylvite-halite, Pugwash Mine, Nova Scotia (after Evans, 1967).

remobilised during gravity sliding, thrusting and folding in intervals centred on the evaporites. Figure 7.15 shows a refolded fold: the axial plane of the initial antiform (axial trace 1) was folded and the nose (axial trace 2) of the resultant synform was folded (axial trace 3). The axial planes of each of the folds have the same axial orientations as the axial plane traces of the other two implying a high degree of plasticity in the potash ore zone during deformation (Evans, 1967). Such high ductility characterises halokinetic potash worldwide. Sylvite and carnallite are two of only a few minerals that are even more ductile than halite.

The Potash Member (15-45m) of the Cassidy Lake Formation in the Windsor Group of the Moncton sub-basin consists of cm-scale beds of halite and halite+sylvite; while the Upper Salt Member is composed of interlayered thin beds and laminae of argillaceous halite, sylvinite and claystone-anhydrite (Anderle et al., 1979). Borates are common in particular horizons: for example, danburite occurs in the Basal Halite; while boracite, hydroboracite, szaebelyite, hilgardite and ulexite are found in the Upper Salt Member of the Windsor Group.

Potash is not the only evaporite-related commodity of interest in the region. Windsor Group sediments, and gypsum-anhydrite units of equivalent age in Newfoundland, host hydrothermal Pb, Zn and Ba mineralisation. The significance of this metal-evaporite association is discussed in Chapter 8.

Trans-Atlantic potash: west Africa and Sergipe Basin, Brazil

Cretaceous (Aptian) evaporites were deposited in a series of growing rifts created as Africa separated from South America. Today the remnants of these beds are to be found on opposite sides of the Atlantic in the Sergipe-Alagoas and Gabon-Congo Basins. In the Sergipe Basin the evaporite sequence (Ibura member of the Muribeca Formation) is a $MgSO_4$-free potash deposit composed of stacked cycles of halite, carnallite and sylvite up to 800m thick. Several of these cycles carry thick units (tens of metres thick) of the highly soluble $CaCl_2$ mineral tachyhydrite (Figure 7.16; Meister and Aurich, 1972; Wardlaw, 1972; Wardlaw and Nicholls, 1972; Borchert, 1977). The Sergipe deposits are characterised by thick cycles with a characteristic upward trend: halite (100%) —> halite (35-90%) + carnallite (10-65%) —> tachyhydrite (5-100%). Wardlaw (1972) showed that these deposits are primary as they contain bottom-growth features and primary bedding in the halite-carnallite units. In the upper part of the evaporite sequence, a sylvinite bed sits atop a carnallite +halite bed and is overlain by a halite bed; a mineralogical evolution that is similar to that of the Thai potash succession.

A similar cyclic potash sequence, also free of $MgSO_4$ salts and carrying thick tachyhydrite units, occurs in the Congo and Angola (de Ruiter, 1979). Within the Congo Basin the most extensive and thickest potash deposits occur in the coastal Kouilou region. There, ten evaporite intervals are recognised with a cumulative thickness of potash beds of more than 100m. Carnallite is the predominant potash mineral, while a few of the cycles also contain bischofite

and tachyhydrite, and some of the carnallite has converted to sylvite. The region may be suitable for solution mining of potash and magnesium. Further inland, for nine years in the 1970s, the equivalents of these evaporite cycles were conventionally mined for sylvite in the Holle mine area. Production, at rates of 360,000 to 450,000 tonne/year, was from the uppermost sylvinite layers. Geological and water problems prevented the mine from ever reaching its target production level of 720,000 tonne/yr. Operations ceased in 1978 when an aquifer was penetrated and the mine was flooded.

Neither the Sergipe nor the Congo evaporite mineralogies can be derived from the evaporation of modern day sea water. Yet both deposits retain bottom nucleated crystal fabrics that show they are primary. Hardie (1990) argues they are both examples of hydrothermally sourced lacustrine rift successions (see Chapter 2).

Moroccan Meseta (Triassic)

Nonmarine redbeds and evaporites fill many Late Triassic (Keuper) fault basins in the Moroccan Meseta (van Houten, 1977; Clement and Holser, 1988; Lorenz, 1988). Thick basalt flows (up to 500m) are interbedded with these saline deposits. In four of the sub-basins the evaporites interbedded with basalts also carry potash salts. No detailed petrography of these units has yet been published. The potash intervals consist essentially of halite, carnallite, sylvite and rinneite, with less that 1% of other minerals (anhydrite, kieserite, bischofite, douglasite; Salvan, 1972). Once again this is a $MgSO_4$-depleted potash sequence.

According to Hardie (1990) these evaporites, interbedded with nonmarine sediments and basaltic lava flows, are another candidate for deposition from upwelling $CaCl_2$ waters of hydrothermal origin in a rift setting.

So how does potash form?

Our discussions so far show there are two groups of processes used to explain ancient potash salts. One sees the potash salts, carnallite and sylvite, as primary precipitates and syndepositional, early diagenetic replacements, while the other visualises sylvite as forming later

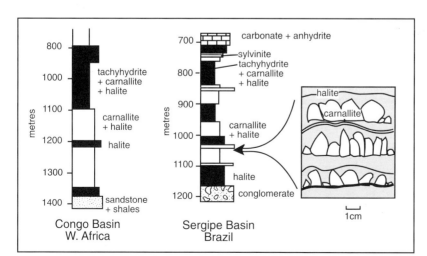

Figure 7.16. Similarities in potash sequences from the Sergipe and Congo Basins (after Hardie, 1990). Inset on the right shows primary layering and bottom-growth features of the Sergipe halite-carnallite subfacies (after Wardlaw, 1972).

245

in burial via ongoing incongruent dissolution of carnallite. Both views are valid but, if used exclusively, are perhaps a little too dogmatic. My own view is that owing to its high solubility the various textures found in ancient potash salts reflect variegated origins that are dependent on how many times in its burial history an evolving brine chemistry came into contact with the potash beds. In other words, the various forms and textures of potash may dissolve, recrystallise and backreact with each other from the time it is first precipitated until it is exploited. Textural and mineralogical evolution depends on how open the hydrology of the system is at various stages during its burial evolution.

Similar, but less intense, textural evolution can also be seen in the burial history of the less soluble salts. For example, $CaSO_4$ can flip-flop from gypsum to anhydrite depending on temperature and pore fluid salinity. This means that, as in gypsum/anhydrite sequences, there will be primary and secondary forms of both carnallite and sylvite that can alternate during deposition, during burial and during any deep meteoric flushing.

In Thailand, both primary precipitation and diagenetic backreactions seem to have worked simultaneously with hydrologies that were first driven by salt flow, and then by regional gravity drive of throughflushing fluids whose flow was focused along nearby clastic aquifers. The Canadian examples show how carnallite/sylvite has continued to evolve with the evolution of basinal and deeply flushing meteoric waters as they came into contact with the partially dissolving salt units.

Whatever model, or combination of models, is used to explain the origin of the major potash salts, the most important economic corollary is that, wherever a potash bed is in contact with mobile fluids, these highly soluble salts are continually subject to dissolution, mobilisation and reprecipitation. When trying to understand the distribution of potash in a halite-dominated sequence the ability of the potash salts to flow, dissolve, alter and reprecipitate must be accounted for in any predictive exploration model or development programme. Its ability to dissolve also explains variations in that most important feature of any mine — ore grade.

So what controls potash quality? Salt anomalies and potash leaching

Worldwide, the problem areas in most potash mines can be related to thinning or disappearing ore seams in zones that always show evidence of dissolution and solution collapse (Boys, 1990, 1993; Woods, 1979). Increased water inflows

in Saskatchewan potash mines can be consistently linked to salt anomalies, that is, to areas of little or no potash within the potash ore bed (Gendzwill and Martin, 1996). Worldwide, other than intersecting old open well bores through the ore zone, most episodes of flooding, including the complete losses of working mines, stem from uncontrolled flooding associated with anomalous non-ore zones.

For example, potash ore beds and three associated salt anomalies were studied by Boys (1993) in the PCS Cory potash mine in the Patience Lake Member of the Prairie Evaporite Formation. He defined up to five postburial facies that are present in the vicinity of the salt anomalies, all are related to an influx of water that is undersaturated with respect to potash (Figure 7.17a). In his model, collapse structures act as conduits for overlying formation waters and drive pervasive leaching and recrystallisation of the Prairie Evaporite Formation while also forming salt anomalies. The same fluids that cause widespread recrystallisation and possibly enrichment of the potash salts at the outer edge of the salt anomaly, can also dissolve huge cavities in the more central parts of the anomalies. The effect of the leaching is highly variable and the timing of major leaching events is still poorly constrained. Boys (1993) postulates at least two major influx events in the PCS mine, possibly driven by uplift and tectonism: one occurred in the Late Devonian, the other in the Cretaceous.

Recrystallised bedded sylvinite is one end member of a continuum of secondary leach features that form via interaction of bedded potash with crossflowing groundwaters (Figure 7.17a). Where undersaturated groundwater continually interacts with a sylvinite bed it ultimately dissolves all the salts to leave behind only insoluble residues in a solution collapse breccia (facies 1). This potash-free interval is one end member of a facies continuum in the PCS mine (facies 1 through 5). At the other end of the groundwater-potash interaction spectrum are beds that are composed of completely recrystallised and often potash-enriched sylvinite (facies 5). These lie adjacent to leached beds where recrystallised halite (facies 4) passes laterally into collapse breccias with blocks of anhydrite and halite (facies 3 and 2; solution residues made up of the less soluble salts), which in turn pass laterally into the most mature dissolution breccias of facies 1, where no evaporite salts remain.

Similar solution cavity-fill features creating salt anomalies occur in the Boulby Potash Mine in the UK (Figure 7.17b). There they consist of a lense of coarse pure

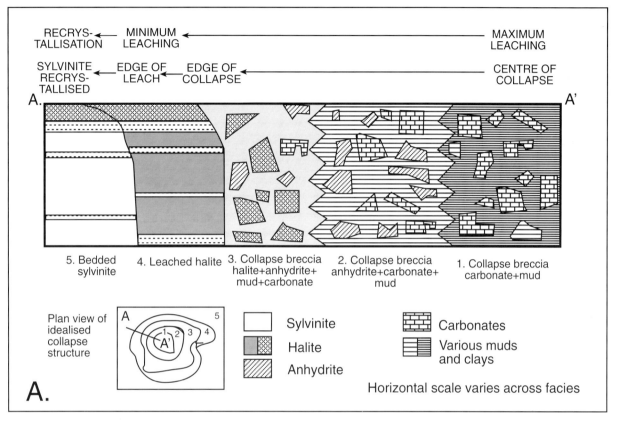

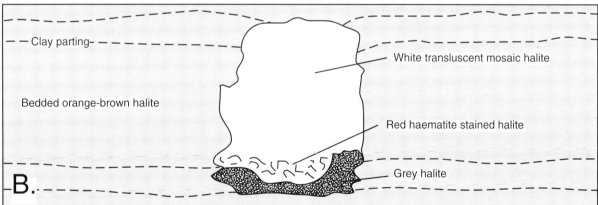

Figure 7.17. Potash solution features. A) Cross section A-A' through half of an idealised collapse structure in the Prairie Evaporite Formation (after Boys, 1993). B) Recemented solution cavity in the Boulby Mine, UK. This is probably equivalent to facies 3 in 7.17a (after Woods, 1979).

translucent halite, which is up to 2 metres across, and typically overlies a band of halite encrusted with red haematite (Woods , 1979). The amount of anhydrite is greater than would accumulate as a simple insoluble residue at the base of the cavity. According to Woods (1979), the cavities formed in the subsurface from $CaSO_4$-saturated brines that were involved in dissolving the carnallite and sylvite. The resulting brines became supersaturated with

respect to $CaSO_4$ causing halite to precipitate and accumulate on the cavity floor.

I have observed similar $CaSO_4$-floored dm-scale cavities in core from the Maha Sarakham potash intervals in Thailand. It seems that phreatic dissolution is commonplace in many potash intervals worldwide. The infill, however, is not always halite; and in Thailand, in the heavily recrystallised sylvinite zones, some of these anhydrite floored cavities have been filled by coarse-grained clear sylvite made up of single crystals more than 30 cm across.

Salt anomalies (barren zones in potash ore) also form crosscutting intervals in the Permian potash ores in New Mexico (Linn and Adams, 1966). Locally called "salt horses", these crosscutting irregular zones range from 0.3 to 100 metres in width and 3 - 200 m in length. Beds in salt horses are thinner than the equivalent beds in ore. The contacts between salt horses and ore are sharp. The colour of the clay changes from grey in ore to brown in salt horses, perhaps indicating more oxygenated throughflushing waters. Pods and lenses of langbeinite, leonite, kainite, recrystallised halite, and recrystallised sylvite occur in ore near the salt horses. The disseminated polyhalite of the marker beds is locally concentrated into intergranular seams and pods. According to Linn and Adams (op. cit.) the NaCl-saturated brines entered the ore zone from below and formed salt horses by the selective leaching of sylvite. In most other examples of salt anomalies, workers have concluded unsaturated fluids entered from above. Fluid entry from below may not be unreasonable in the New Mexico situation, as the evaporites along the western margin of the Delaware Basin sit atop an aquifer that entrains deeply circulating meteoric waters (Figure 2.19).

Effects of the leaching that produced the salt anomalies and loss of potash in Canada or elsewhere, range from weak to strong, from selectively preserving delicate laminae and chevron textures, to deforming and destroying salt beds (Boys, 1990). Good preservation of iron oxides in halite may indicate that the leaching was weak or of short duration. In many leach anomalies, salinities increase downward, possibly because fluids exit downward, and NaCl-saturated fluids tend to follow the chemical gradient provided by potash beds.

Identifying proximity to a salt anomaly is obviously important in terms of ore quality and mine safety (Boys, 1990). Geological indicators of proximity to a halite anomaly include: a change in insoluble seam colour from greenish-grey to mottled brown and greenish-grey (probably corresponds to increasingly oxidised fluids), unusual local increases in ore grade, large patches of sylvite-poor potash crosscutting units near the top of the ore zone, and drops in topography >10m of the marker seams across an anomaly. For example, within 5 m of a large salt anomaly, Boys (1990) found large blebs (>200 cm^2 on the mine wall) of sylvite-poor potash that crosscut

the units of an incomplete potash cycle near the top of the main ore zone. Anhydrite, a less soluble salt, is also more common in a salt anomaly than in the adjacent ore. Once into the anomaly, possible indicators of a nearby major collapse feature include: stretched clay seams; folded beds; small collapse features (1-20 m scale); and split clay seams, with salt appearing to be injected into the seam.

In a study of "barren bodies" in the Subiza Mine, Navarra, Spain, Cendon et al. (1998) recognised a syndepositional mechanism of "salt horse" formation. It was controlled by brine pool stratification, not the various diagenetic processes discussed so far. This Subiza potash deposit contains a 100 m thick Upper Eocene succession of alternating claystone and evaporites (sulphate, halite, and sylvite). The evaporites accumulated in an elongated basin that is one of the depocentres within the 250 km long South Pyreneean foreland basin. Slope instability along the margin of the basin, perhaps promoted by tectonism, created mass wasting of the evaporite beds. This formed subaqueous mounds 0.5-2 m high and tens of metres across. As evaporation progressed, a stratified brine system formed and encroached over the mounds. Halite precipitated at the air-brine interface and sank to the bottom of the basin along with terrigenous clays. Sylvite, however, precipitated only from the lower more dense brine. As many mound crests extended into the upper brine, the sylvite could not precipitate over these subaqueous highs. With progressive accumulation, the lower brine ultimately covered the mounds as sylvite beds overlapped the mound tops.

This model, however, is problematic in terms of its hydrological restraints: it requires a hydrological system that encapsulates sylvite saturation and precipitation in the lower brine simultaneous with halite precipitation in the upper brine. As discussed in Chapter 3, such a system is extremely difficult to maintain over time frames that allow metre-thick beds of evaporites to accumulate. A more likely hydrology is perhaps a desiccating and concentrating single brine body, which reached halite saturation at the "lake full" stages while the mounds were still covered with brine. It did not reach sylvite saturation until ongoing drying of the brine lake had lowered the level to where the crests of the mounds were exposed. Whichever model is accepted, what is most important in terms of potash geology is that breached syndepositional salt horses or barren zones are less likely to possess subsurface hydrologies that will flood mines, as compared with horsts induced by subsurface leaching.

Economic evaporite salts

Borates

Boron is a highly dispersed element in nature - averaging 3 ppm in the earth's crust and 4.6 ppm in seawater. The few occurrences worldwide where the element is sufficiently concentrated to be economic involve local volcanic activity as a source of boron. They also involve a standing body of water to accumulate the boron salts leached from the volcanics and evaporation to concentrate the solution to the point of precipitation of the various boron-bearing minerals. A protective layer (often of other impervious evaporite salts) forms a carapace that prevents subsequent redissolution. Sodium and calcium borate salts are the major economic minerals (Table 7.4). Water content of the various borates tends to decrease with diagenesis and to increase with exposure and weathering. Borax is the primary mineral in many deposits, but with burial and dewatering it converts to tincalconite and kernite. If kernite is exposed to weathering or throughflowing groundwater it reconverts to borax.

MINERAL	COMPOSITION	B_2O_3 wt%	REMARKS
SODIUM BORATES			
Borax (tincal)	$Na_2B_4O_7.10H_2O$	36.5	Major ore mineral in USA and Turkey.
Kernite	$Na_2B_4O_7.4H_2O$	51.0	Major mineral often converts to borax.
Tincalconite	$Na_2B_4O_7.5H_2O$	47.8	Intermediate or accessory mineral.
SODIUM-CALCIUM BORATES			
Ulexite	$NaCaB_5O_9.8H_2O$	43.0	Major ore mineral particularly in S. America
Probertite	$NaCaB_5O_9.5H_2O$	49.6	Secondary/accessory mineral
CALCIUM BORATES			
Colemanite	$Ca_2B_5O_{11}.5H_2O$	50.8	Major ore mineral, particularly in Turkey. Secondary after inyoite
Meyerhoffite	$Ca_2B_5O_{11}.7H_2O$	46.7	Intermediate ore mineral, rarely survives
Priceite (pandermite)	$Ca_2B_4O_{10}.7H_2O$	49.8	Ore mineral in Bigadic, elsewhere minor
Inyoite	$Ca_2B_6O_{11}.13H_2O$	37.6	Minor ore mineral
OTHER BORATES			
Sassolite	$B(OH)_2$	56.4	Natural boric acid once extracted in Italy
Hydroboracite	$CaMgB_6O_{11}.6H_2O$	50.5	Secondary mineral
Szaibelyite (Ascharite)	$MgBO_2(OH)$	41.4	Ore in Russia and China (meta-evaporite)
Boracite	$Mg_3B_7O_{13}.Cl$	62.2	Associated with potash especially in Europe
Howlite	$H_5Ca_2SiB_5O_{14}$	44.5	Accessory mineral
Bakerite	$Ca_4B_4(BO_4)(SiO_4)_3(OH)_3.H_2O$	42.1	Accessory mineral
Tunellite	$SrB_6O_9(OH)_2.3H_2O$	55.1	Accessory mineral
Tertschite	$Ca_4B_{10}O_{19}.2H_2O$	65.1	Accessory mineral
Kurnakovite	$Mg_2B_5O_{11}.15H_2O$	37.3	Accessory mineral

Table 7.4. Boron minerals in various exploited deposits.

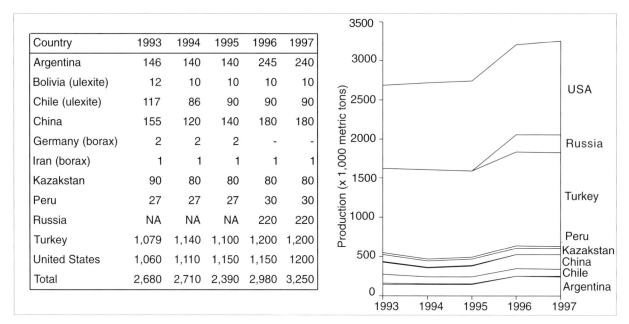

Figure 7.18. Borate production (thousand metric tons of boric oxide B_2O_2 equivalent - compiled from USGS online data tables).

Country	1993	1994	1995	1996	1997
Argentina	146	140	140	245	240
Bolivia (ulexite)	12	10	10	10	10
Chile (ulexite)	117	86	90	90	90
China	155	120	140	180	180
Germany (borax)	2	2	2	-	-
Iran (borax)	1	1	1	1	1
Kazakstan	90	80	80	80	80
Peru	27	27	27	30	30
Russia	NA	NA	NA	220	220
Turkey	1,079	1,140	1,100	1,200	1,200
United States	1,060	1,110	1,150	1,150	1200
Total	2,680	2,710	2,390	2,980	3,250

More than half of the borate produced worldwide is used in the manufacture of glass products (including fibre): borates are also used in the manufacture of cleaning agents, flame retardants and as an additive to agricultural fertilisers. Use of sodium borate is not just a modern process, it was used in mummification in Babylonian times and as a kiln additive to braze or harden precious metals. It was also traded along the Tibet to Europe trade route travelled by Marco Polo.

Borax is the preferred sodium borate in the chemical industry, as it crushes freely and dissolves readily in water with a rate of solubility that increases with increasing temperature. In contrast, kernite, which is present in large tonnages in the USA and Argentina, is less suitable despite its high B_2O_3 content. Kernite has excellent cleavage, but crushes into fibres that tend to mat and clog the handling equipment. It also is much less soluble in water than borax. Colemanite is the preferred calcium bearing borate, it is only slowly dissolved in water, but is highly soluble in acid solutions. Szaebelyite, a magnesium borate, was the principal source of borate in the former Soviet Union and is the main ore mineral in the Proterozoic meta-evaporite ores of the Liaoning Province, China (Figure 6.12). "Tincal" is the old name for borax and is sometimes still used to distinguish the naturally occurring ore mineral from the chemically identical manufactured salt.

Today, Turkey is the largest producer of borate ore and the second largest producer of boron compounds after the USA (Figure 7.18). Borates have been mined in Turkey for the last fifteen centuries. Ulexite is mined at Bigadic; colemanite at Bigadic, Emet, and Kestelek; and tincal at Kirka. The company also refines tincal and colemanite ores, and concentrates them in the Kirka and Bandirma plants.

The Bigadic borates are extracted from what is the largest colemanite/ulexite deposit in the world, with high-grade colemanite and ulexite ores running 30% and 29% B_2O_3, respectively. Ores originally precipitated within Neogene perennial saline lake sediments located in a series of northeast-southwest-trending elongate basins (Helvaci, 1995). The Bigadic, Emet and Kirka lacustrine basins of western Turkey are hosted in Tibet-type graben structures developed during the Miocene collision along the Izmir-Ankara suture zone complex. The volcano-sedimentary sequence in the deposits consists of (from bottom to top) basement volcanics, lower limestone, lower tuff, lower borate zone, upper tuff, upper borate zone and olivine basalt (Figure 7.19). The borate deposits formed under arid conditions in perennial saline lakes fed by hydrothermal springs associated with local volcanic activity. The deposits are interbedded with tuffs, clays and limestones.

Borate minerals at Bigadic form two ore zones, now separated by thick tuff beds that have been transformed to montmorillonite, chlorite, and zeolites (mainly heulandite) during diagenesis (Gundogdu et al., 1996). Colemanite and ulexite predominate in both zones: but other borates, including howlite, proberite, and hydroboracite are present in the lower borate zone; whereas inyoite, meyerhofferite, pandermite, tertschite (?), hydroboracite, howlite, tunellite, and rivadavite are found in the upper borate zone (Helvaci, 1995). Calcite, anhydrite, gypsum, celestite, K feldspar, analcime, heulandite, clinoptilolite, quartz, opal-CT, montmorillonite, chlorite, and illite are also found in the Bigadic deposit.

Nodular-shaped colemanite and ulexite minerals predominate in both borate zones. Colemanite and ulexite construct alternating horizons with sharp boundaries. The transformation of one mineral to another has not been observed. Because these minerals are readily dissolved, secondary pure and transparent colemanite and ulexite are often encountered in cavities of nodules and cracks. Some nearsurface colemanite and ulexite are so weathered they are now completely replaced by calcite.

Colemanite nodules in both ore zones probably formed as a secondary precipitate in concentrated pore brines within unconsolidated gypsiferous sediments located just below the sediment-water interface. Nodules continued to grow as the sediments were compacted to give septarian-like fracture fills. Later generations of colemanite and ulexite are found in vugs and veins and as fibrous margins of early formed nodules. It is unlikely that the colemanite formed by dehydration of inyoite and/or by replacement of ulexite after burial. Diagenetic changes to the main borates include the partial replacement of colemanite by howlite and hydroboracite, and replacement of ulexite by tunellite.

Proberite bands are found in some ulexite horizons, especially in the lower borate zone. It formed in the same chemical environment as ulexite, and according to Helvaci (1995) indicates short periods of more extreme desiccation and possibly subaerial exposure within the lakes. Euhedral tunellite formed during dissolution and recrystallisation of some Sr-rich ulexite horizons. In the Bigadic deposits, hydroboracite formed by replacement of colemanite, with Mg^{2+} ions supplied from adjacent tuffs and clays by ion exchange. Howlite grew in clay bands alternating with thin colemanite bands and coincided with periods of relatively high Si concentrations. Diagenetic processes also produced small howlite nodules embedded in unconsolidated colemanite nodules. The initial solutions that formed the

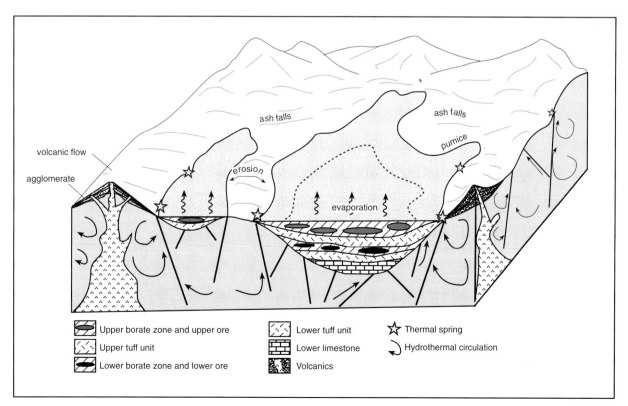

Figure 7.19. Deposition of the Bigadic Deposit (after Helvaci, 1995). Deposition occurred under arid to semi-arid conditions within separate or interconnected perennial saline lakes located in an active volcanic area with many lakes fed by thermal springs.

251

alkaline perennial saline lake(s) were low in Cl^- and SO_4^{2-}, and high in boron and Ca^{2+}, with subordinate Na^+. The boron isotope data from the Turkish deposits are consistent with colemanite being precipitated from a more acidic brine than ulexite, and with borax being precipitated from a brine that was more basic (Helvaci, 1995; Palmer and Helvaci, 1997).

Detailed analysis of Ca-sulphate-borate textures in a number of Turkish deposits by Orti et al. (1998) have shown that many borates formed as early diagenetic precipitates, preferentially replacing basin-centre gypsum laminites and debris flows. Replacement occurred in early burial as evaporitic brines mixed with upwelling boron-rich hydrothermal solutions, typically beneath a deeper water gypsum-dominated lake bottom. With further burial the anhydritisation of gypsum began (depths < 250m). At the same time, anhydrite also replaced some of the earlier formed borates (priceite, bakerite and howlite), along with associated celestite. Subsequent uplift and exposure of some parts of the sequence rehydrated nearsurface anhydrite to secondary gypsum and partially transformed priceite and howlite into secondary calcite.

The United States is the world's largest producer of boron compounds and the second largest producer of borate ore, mostly extracted from saline brines and playa deposits such as Searles Lake in California (Smith, 1979) or their Tertiary counterparts at Boron and Death Valley (Kisler and Smith, 1983). The Boron deposit near Kramer, California, is about 1600 m long, 800 m wide by 100 m thick, and consists of a lenticular central mass of borax/kernite and interbedded clays. Ores are part of the shale member of the Kramer beds, which include an arkose member above and rest on the Saddleback Basalt. Laterally and vertically away from the borax/kernite core, the borate facies changes to ulexite and colemanite, and clays become more important (Kyle, 1991). The deposit is believed to have formed in a shallow Mid-Miocene lake, fed by Na- and B-rich thermal springs, a result of the latter stages of volcanism. Saline sodium-borate bearing thermal spring water, being denser, sank to the lake bottom, cooled, and precipitated borax. In Death Valley, the ore is contained in the lower 150m of the 2,100m thick Pliocene Furnace Creek Formation. There the Billie orebody is a 1,100m long lens of ulexite, proberite and colemanite, interbedded and surrounded by limey lacustrine mudstone and shale. The orebody has an average thickness of 45 - 55 m and a width of 210m, with a dip of 20-30° to the southeast.

Well to the south of these deposits in a 880-km-long stretch of the Andes along the common borders of Argentina, Bolivia, Chile and Peru, there are more than forty borate deposits hosted in Cenozoic volcanogenic-sedimentary rocks. Most are small aprons or cones of borax and ulexite near volcanic vents, with reserves measured in thousands of tonnes. But in the large endoheic basins, where the brines can pond and concentrate, there are beds of ulexite, borax and inyoite in salars ranging up to hundreds of square kilometres in area, with reserves measured in millions of tonnes. In the Pleistocene Blanca Lila Formation near Salar Pastos Grandes in the Argentinian Andes, ulexite was the first evaporite to precipitate followed by gypsum then halite (Vandervoort, 1997). In contrast, in Neogene salar deposits near the Salar Hombre Muerto, Alonso (1991) documented a facies pattern of marginal sulphates (gypsum), transitional borate facies and a basin-centre halite unit. Alonso (op. cit.) also documented a similar facies distribution within the Tincalyu ores of the Neogene Sijes Formation, Argentina.

In plan view most of the South American salars have a central zone, consisting of a relatively thick halite crust, surrounded by a thin marginal sulphate zone consisting of a variety of sulphate minerals, as well as halite and ulexite (Ericksen, 1993). Salars that are filled mostly with gypsum lack such a zonation. The zoning may be symmetrical (bull's-eye pattern) with sulphate surrounding a central halite core, or asymmetrical with the halite zone crowded against one side of the salar. This is the situation in Salar de Atacama where tectonic tilting has displaced the lowest part of the brine sink to the southern end of the playa (Figure 7.26). Both the halite and sulphate zones are typically underlain by gypsum and this in turn by siliciclastics. Alternating arid and pluvial climatic episodes, since the mid-Miocene, means the playas are filled by alternating salt beds and siliciclastics. Drilling near the centre of Salar de Uyuni penetrated 121m of alternating massive rock salt and lacustrine clastics. Salar Grande contains massive rock salt to depths of 160 m while Salar de Atacama is filled with massive rock salt to a depth of more than 390 m. The deepest hole in this salar was still in rock salt when the hole was abandoned.

Most South American salar deposits have undergone relatively little deformation since they were first emplaced. As a result, borates are mined in a number of open pit operations in salars of the Andean Altiplano (Alonso et al., 1991). In Argentina, borax is mined at Tincalyu, hydroboracite at Sijes, and ulexite from two dry lake beds, Salar Cauchari and Salar Diablillos, all in the Salta province. In Bolivia, ulexite is extracted from mines in the Salar de Uyuni. In Chile, borates are extracted from Salar de Surire; the largest ulexite deposit in the world. Ulexite reserves are

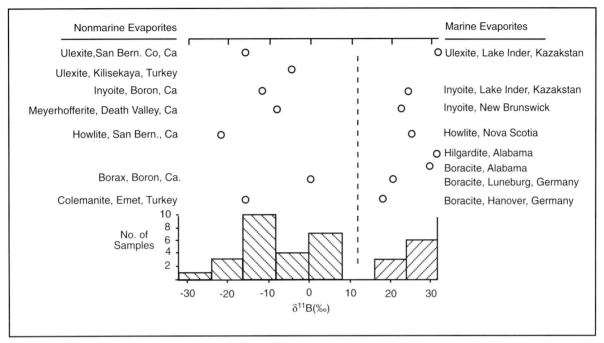

Figure 7.20. Isotopic analysis of evaporitic borates showing clearly separate fields of marine and nonmarine samples (dashed line). Upper part of figure shows analytical determinations from various deposits. Lower part shows a histogram when these values are grouped (after Swihart et al., 1986).

reported to be 1.5 million tonnes of 35% boron oxide. The salar is located at 4,250 metres altitude within the border of the Monumento Natural de Surire national park of Chile.

In China, borates occur in the salt lakes on the Qinghai-Xizang (Tibet) Plateau, where typical lake waters contain 85 mg/l of boron oxide (Figure 7.2; Sun and Li, 1993).. The maximum value of boron oxide in the brine is ≈ 1,500 mg/l. However, the largest economic borate reserves in China occur not in Quaternary/Tertiary lakes but in the Precambrian meta-evaporites of Liaoning Province (Figure 6.12). Reported reserves are 44 million tonnes of boromagnesite with 8.4% boron oxide content. The deposits in the Liaoning Province represent 64% of the country's total boron resources and 90% of these deposits are associated with banded iron formations.

In the Inder district of Kazakstan (former USSR), north of the Caspian sea, borates form bedded lenses that are up to 3 metres thick and associated with 100+ km^2 crestal area of a Permian salt dome. Commercial ores are magnesian borates, rather than the more commonplace sodian-calcian borates. The ores occur along fractures near the top of the dome, where they replace the halite or associated clay and gypsum. Boron is also extracted from the brines of nearby Lake Inder. This salt lake has a surface salt crust along with

about 30m of brine saturated salt, which contains potassium chlorides and bromides as well as borax (Kisler and Smith, 1983). Unlike the other economic borate deposits, this deposit is sourced by a dissolving ancient marine evaporite, hence the association with the magnesian salts. Its marine origin is clearly seen in its boron isotope signature, which is much more enriched in [11]B than its nonmarine counterparts (Figure 7.20; Swihart et al., 1986). The clear separation of nonmarine and marine borates using boron isotopes is the basis for the separation of the various meta-evaporitic tourmalines that were discussed in Chapter 6 (Figure 6.8).

Gypsum and halite (rock salt)

The depositional, diagenetic and geological settings of halite and gypsum/anhydrite were discussed in detail in Chapters 1 and 3, and will not be discussed further in this chapter. Only those aspects relevant to economic application and production are presented.

Gypsum

Annual world production of gypsum exceeds 100 million metric tonnes, with fifteen countries producing more than 80% of the world's annual production (Figure 7.21).

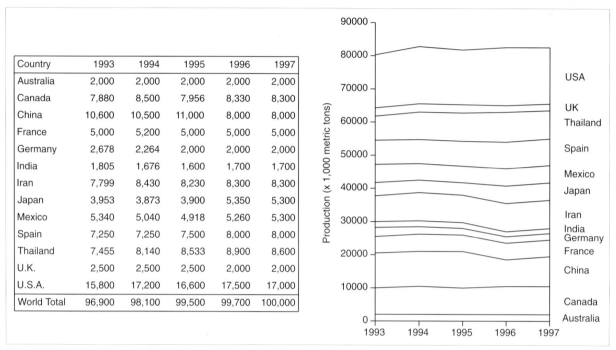

Country	1993	1994	1995	1996	1997
Australia	2,000	2,000	2,000	2,000	2,000
Canada	7,880	8,500	7,956	8,330	8,300
China	10,600	10,500	11,000	8,000	8,000
France	5,000	5,200	5,000	5,000	5,000
Germany	2,678	2,264	2,000	2,000	2,000
India	1,805	1,676	1,600	1,700	1,700
Iran	7,799	8,430	8,230	8,300	8,300
Japan	3,953	3,873	3,900	5,350	5,300
Mexico	5,340	5,040	4,918	5,260	5,300
Spain	7,250	7,250	7,500	8,000	8,000
Thailand	7,455	8,140	8,533	8,900	8,600
U.K.	2,500	2,500	2,500	2,000	2,000
U.S.A.	15,800	17,200	16,600	17,500	17,000
World Total	96,900	98,100	99,500	99,700	100,000

Figure 7.21. World gypsum production (in 1,000s metric tons) of nations producing more than 1,500,000 metric tons per year (compiled from USGS online data).

Gypsum's main usage is in the construction industry. Its utility stems from its hydrated character. When gypsum is calcined at 160°C it loses 1.5 moles of its combined water to form calcium sulphate hemihydrate ($CaSO_4 1/2H_2O$), commonly known as Plaster of Paris. When mixed with water it can be spread, cast or moulded into the desired form, which then sets to a rocklike hardness. When spread between sheets of heavy paper and allowed to set it makes the plasterboard or wallboard used worldwide in the "dry wall" industry. Other uses include uncalcined finely crushed gypsum as a retarding agent in cement, as a soil fertiliser and conditioner, and as a filler in a diverse range of products including paint, paper and toothpaste.

Production is either from quarrying or mining. In North America and Europe, gypsum is mined at relatively shallow depths where ancient buried anhydrite beds have reconverted to gypsum during uplift and erosion. The Maritimes region of eastern Canada and the USA produces the most gypsum from a single area in the world, with more than 6 million metric tonnes extracted each year from Mississippian strata of the Windsor Group. There is also substantial US production from Tertiary playa deposits in the Salton Trough, California. France is Europe's leading gypsum producer, with the main production from Lower Eocene beds in the Paris Basin and Upper Eocene-Oligocene beds in the Rhone Basin. In Germany, the greater volume

of gypsum is mined from the rehydrated crests of near surface salt structures of Permian Zechstein salt. Secondary targets are the Middle Muschelkalk and minor production from the Keuper. In the UK, the most important beds are Permo-Triassic in age, except for those in Sussex, which are Jurassic. Large volumes of high-grade Holocene gypsum are quarried, processed and shipped from Holocene coastal lakes in southern (>1,000,000 tonnes/year) and western Australia (300,000-500,000 tonnes/year). This production supplies the construction industry of Australia and other parts of southeast Asia. Small amounts of coarsely crushed gypsum are used as aggregate to pave unsealed minor roads in many places in southern Australia. Under the region's semi-arid to arid climate it tends to compact and set into a hard flexible crust.

Rock salt (halite)

Almost every country in the world exploits ancient rock salt (halite) deposits or deposits from solar evaporation operations of various sizes. The largest production comes from 15 nations, which produce more than 80% of the world's total production (Figure 7.22). Important commercial deposits are found in the Silurian of northeastern North America; the Devonian of western Canada (also a major source of potash); the Permian of

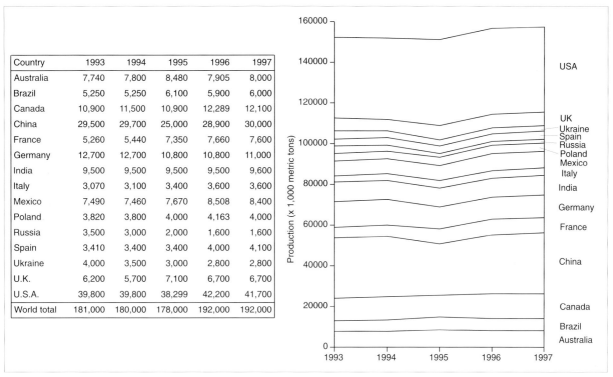

Country	1993	1994	1995	1996	1997
Australia	7,740	7,800	8,480	7,905	8,000
Brazil	5,250	5,250	6,100	5,900	6,000
Canada	10,900	11,500	10,900	12,289	12,100
China	29,500	29,700	25,000	28,900	30,000
France	5,260	5,440	7,350	7,660	7,600
Germany	12,700	12,700	10,800	10,800	11,000
India	9,500	9,500	9,500	9,500	9,600
Italy	3,070	3,100	3,400	3,600	3,600
Mexico	7,490	7,460	7,670	8,508	8,400
Poland	3,820	3,800	4,000	4,163	4,000
Russia	3,500	3,000	2,000	1,600	1,600
Spain	3,410	3,400	3,400	4,000	4,100
Ukraine	4,000	3,500	3,000	2,800	2,800
U.K.	6,200	5,700	7,100	6,700	6,700
U.S.A.	39,800	39,800	38,299	42,200	41,700
World total	181,000	180,000	178,000	192,000	192,000

Figure 7.22. Salt (NaCl) production from countries producing >2,000,000 metric tons (values in thousand metric tons - compiled from USGS online data tables).

Russia, western Europe, the midwest of North America, the Jurassic of the US Gulf Coast and the Miocene of the Middle East. In Europe and North America the salt is recovered either by hard rock mining (mostly room and pillar method) or by solution mining where water is pumped down a well to dissolve the salt and then recovered as brine. In warmer climates, such as Australia, Spain and Mexico, large volumes of salt are recovered by solar evaporation of sea water or less commonly of inland playa waters.

World resources of salt are practically unlimited and production levels have remained static since the early 1990s. The majority of salt usage worldwide is in the chemical industry, some 58% of total production, where it supplies sodium and chlorine chemicals that are used to manufacture a multiplicity of products. The principal basic chemicals, chlorine and sodium hydroxide (caustic soda) are produced from brine via electrolysis. Sodium chloride is reacted with limestone in the Solvay process to produce sodium carbonate (soda ash), and with sulphuric acid to produce hydrochloric acid and sodium sulphate (salt cake). Salt usage in the northern hemisphere increases in the winter months when salt is used to de-ice frozen roadways (≈13% of world production). The chloralkali industry also uses substantial quantities of salt. Most of its output is of

Country	1993	1994	1995	1996	1997
Azerbaijan	500	400	350	300	300
Chile	5,550	5,600	5,000	5,000	5,600
China	500	500	500	500	500
Indonesia	14	15	80	80	80
Japan	6,490	6,400	6,200	5,500	5,500
Russia	180	160	160	150	150
Turkmenistan	500	251	250	260	260
United States	1,940	1,430	1,220	1,270	1,330
Total	15,700	14,800	16,500	13,100	13,700

Figure 7.23. World iodine production (in thousand kilograms of elemental iodine - compiled from USGS online data tables).

polyvinyl chloride and vinyl chloride monomer manufacture; chlorine is another product. Salt is also used as a food additive (19% of world production) as well as in a variety of other processes ranging from dye manufacture to textile processing.

Iodine, bromine and lithium

Iodine is either recovered from saline brines associated with natural gas production (Japan - Kanto gasfield), oilfield brines (e.g. Woodward, Oklahoma) or mined from lakes and saline soils (Atacama Desert - Chile). Subsurface brines now provide 80% of iodine produced each year. Japan today controls over 40% of the world capacity, Chile 41% and the USA about 10% (Figure 7.23). Chile is now the largest producer of iodine in the world and the sole producer of commercial quantities of iodine from a non-brine source where iodine is recovered as a product in the processing of Chilean nitrate deposits (e.g. at Coya Sur, in northern Chile- see nitrates in following section). The ultimate source of the iodine is thought to be leaching of Tertiary volcanics (Ericksen, 1993). The iodine salts occur in caliche ores, largely as the minerals lautarite (CaI_2O_6) and dietzite ($7CaI_2O_6 \cdot 8CaCrO_4$), and is leached from the nitrate ore using an alkaline solution during the extraction of potassium nitrate and sodium sulphate. Overburden is less than 1 metre. Processing of the quarried ore involves crushing, leaching with fresh water, thickening, filtration, and iodine precipitation. Many of the iodine plants use solar evaporation ponds to concentrate (thicken) the leachate.

Downstream uses of iodine include animal feed supplements, catalysts, inks and colorants, pharmaceutical, photographic equipment, sanitary and industrial disinfectant and stabilisers. A worldwide effort is now underway to eliminate iodine-deficiency disorders by fortifying world salt supply. Iodine deficiency is the Third World's leading cause of mental health problems in the form of severe retardation, deaf-mutism and partial paralysis, along with more subtle problems such as clumsiness, lethargy, and reduced learning capacity. Iodine is an essential part of thyroid hormone, a substance that contributes to brain development during foetal life and later metabolism.

Bromine is recovered from modern brines in Searles Lake and the Dead Sea, as well as from formation waters in the Gulf of Mexico and the Michigan Basin, and seawater in Japan and Spain (Table 7.5, Figure 7.24). For example, a brine averaging 5000 ppm Br is recovered from the upper

Reynolds Oolite Member of the Upper Jurassic Smackover Formation in Union Co, southern Arkansas (Kyle, 1991). The brine with its elevated Br content is believed to have been derived from the natural leaching of the underlying Louann Salt. Since 1957 bromine has been produced at Beersheva, Israel, as a byproduct from bitterns associated with potash, chlorine and caustic soda production. There solar evaporation concentrates brines pumped from Miocene evaporites buried beneath the floor of the Southern Basin of the Dead Sea (Figure 7.6). After the potash slurry is removed, the waste bitterns are processed with chlorine to recover bromine. The bromine-free bitterns are then further processed to recover magnesium. The Dead Sea feed-brine for bromine recovery contains 11,000-12,000 ppm of bromine and is one of the richest bromine feeds in commercial operation. The brines reserves are also huge, with the Dead Sea brine containing more than 1 billion tonnes of bromine. In Germany, bromine is produced as a byproduct of potash mined from the Zechstein salt.

Country	1993	1994	1995	1996	1997
Azerbaijan	4,000	3,000	2,000	2,000	2,000
China	24,600	31,400	30,000	30,000	31,000
France	2,290	2,190	2,260	2,200	2,200
India	1,400	1,400	1,500	1,500	1,500
Israel	135,000	135,000	135,000	135,000	135,000
Italy	300	300	300	300	300
Japan	15,000	15,000	15,000	15,000	15,000
Spain	200	200	200	200	100
Turkmenistan	10,000	8,000	7,000	7,000	7,000
Ukraine	5,000	4,000	3,500	3,000	3,000
United Kingdom	27,423	28,000	28,000	28,000	28,000
United States	177,000	195,042	218,000	227,000	250,000
World Total	396,000	412,000	432,000	450,000	470,000

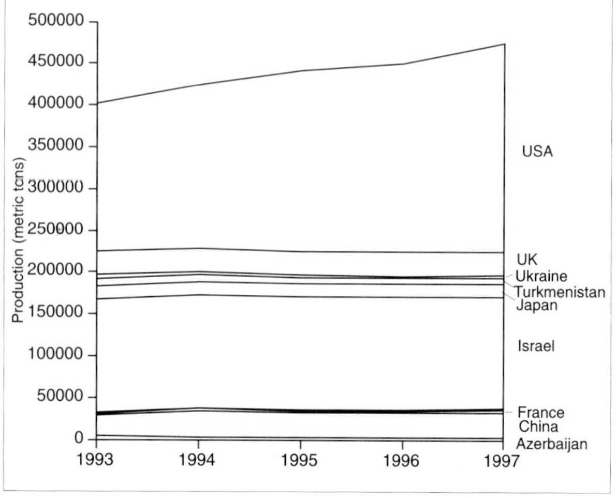

Figure 7.24. World bromine production (in metric tons of bromine content - compiled from USGS online data tables).

Country and Company	Location	Capacity	Brine Source
AZERBAIJAN			
Neftechala Bromine Plant	Baku	5,000	Underground brines
CHINA			
Laizhou Bromine Works	Shandong	30,000	Underground brines
FRANCE			
Atochem	Port-de-Bouc	13,600	Seawater
Mines de Potasse d'Alsace S.A.	Mulhouse	2,300	Bitterns of mined potash
INDIA			
a)Hindustan Salts Ltd.	Jaipur	1,500	Seawater bitterns from salt
b)Mettur Chemicals	Mettur Dam		production.
c)Tata Chemicals	Mithapur		
ISRAEL			
Dead Sea Bromine Co. Ltd.	Sodom	190,000	Bitterns of potash production from surface brines.
ITALY			
Societa Azionaria Industrial Bromo Italiana	Margherita di Savoia	900	Seawater bitterns from salt production.
JAPAN			
Toyo Soda Manufacturing Co. Ltd.	Tokuyama	20,000	Seawater
SPAIN			
Spain:Derivados del Etilo S.A.	Villaricos	900	Seawater
TURKMENISTAN			
Nebitag Iodine Plant	Vyska	3,200	Underground mines
Cheicken Chemical Plant	Balkan	6,400	Underground mines
UKRAINE			
Perekopsky Bromine Plant	Krasnoperckopsk	3,000	Underground mines
UNITED KINGDOM			
Associated Octel Co. Ltd.	Amlwch	30,000	Seawater
USA			
Arkansas Chemicals Inc	Union Co., Ark.	25,000	Well brines
Albemarle Corp.	Columbia Co., Ark.	140,200	Well brines
Great Lakes Chemical Corp.	Union Co., Ark.	59,000	Well brines
Dow Chemical Co.	Mason Co., Mich.	25,000	

Table 7.5. Bromine production methods in 1997 (in metric tons of bromine content - compiled from USGS online data tables).

Since 1973, when the United States produced 71% of the world supply, its market share has decreased, a result of environmental constraints and the emergence of Israel as the world's second largest producer. Today, the United States produces 53% of the world's bromine, Israel, 29%; China, 7%; the United Kingdom, 6%; and other countries, 5% (Figure 7.24).

Traditionally, the major market for bromine was in the production of ethylene dibromide; a lead scavenger in gasoline antiknock compounds. However, this market has been on a rapid decline in recent years as more and more countries adopt the catalytic converter and lead-free gasoline. Flame retardants are now the largest end use and account for more than 30% of the US bromine market. Other uses include intermediates in chemical manufacture of dyes, fragrances and pharmaceuticals; it is also used in photographic compounds, as a bleach, as an additive in drilling fluids, and in the production of methyl bromide, a pre-plant soil fumigant.

In 1997, for the first time, both of the lithium carbonate brine operations at the Salar de Atacama, Chile, produced lithium throughout the year and were

Country	Source	1993	1994	1995	1996	1997
Argentina	Spodumene and amblygonite	300	400	400	400	400
Australia	Spodumene	52,900	45,987	81,841	117,944	88,399
Brazil	Concentrates	1,600	1,600	1,600	1,600	1,600
Canada	Spodumene	18,900	20,000	41,900	54,000	50,000
Chile	Carbonate from subsurface brine	10,369	10,439	12,943	14,180	22,000
China	Carbonate	8,248	9,050	12,800	15,000	15,500
Namibia	Concentrates, chiefly petalite	742	1,861	2,611	2,081	2,000
Portugal	Lepidolite	13,289	11,352	10,000	8,740	9,000
Russia	Minerals not specified	3,000	2,000	2,000	2,000	2,000
United States	Spodumene and subsurface brine	Not released	Not released	Not released	Not released	Not released
Zimbabwe	Minerals not specified	18,064	25,279	33,498	30,929	35,000

Table 7.6. World lithium production (metric tons) and sources (compiled from USGS online data tables)

257

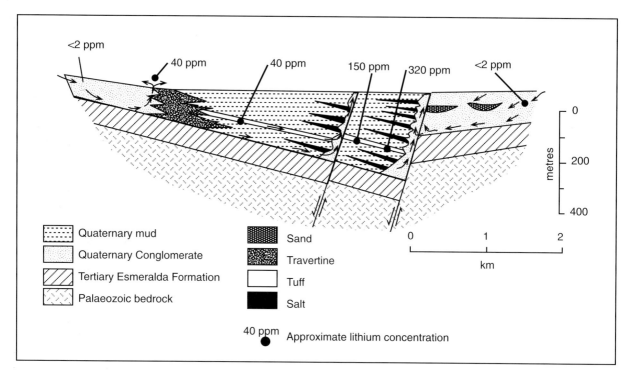

Figure 7.25. Generalised cross section of Clayton Valley Playa, Nevada. Shows the relative position of the major tuff-bed aquifer, inferred directions of groundwater flow and representative lithium concentrations (ppm) in the various aquifers and springs (after Davis et al., 1986).

no longer seasonal operations. And so, after decades as the world's leading producer of lithium and its compounds, the United States was surpassed in 1997 by Chile (annual online USGS commodity summaries - lithium). In the same year an additional lithium carbonate brine operation began producing in Argentina, further shifting lithium production dominance to South America. The United States is still a major producer and is still the world's largest consumer of lithium, but US lithium companies do not release annual production figures (Table 7.6). Australia, Canada and Zimbabwe are other important exporters of lithium ore concentrates, which are extracted from large volumes of spodumene pegmatites.

Lithium occurs in more than 140 minerals, but economic production is restricted mostly to lithium-rich brines or spodumene pegmatites. Spodumene concentrates are typically processed via heating and treatment with sulphuric acid to yield lithium carbonate, but chemical processing methodology is energy intensive and subject to environmental risks. Production of lithium carbonate from a brine source is much less energy intensive. Only the geology of the saline brine sources are discussed in detail in this chapter, the interested reader is referred to Kunasz (1983) for a full discussion of all lithium raw materials.

Subsurface lithium in playa sediments typically occurs either as a soluble brine component, or within lithium-bearing clays such as hectorite —$[Na_{0.33}(Mg,Li)_3 Si_4O_{10}(F,OH)_2]$— a clay mineral of the montmorillonite group where the replacement of aluminium by magnesium and lithium is essentially complete. Such clays are the result of alkaline brines flushing through volcaniclastic playa sediments.

Lithium carbonate is produced from playa brines in Clayton Valley, Nevada. Historically, the area has produced about one-third of the US lithium requirements (Davis et al., 1986). The playa has an area of 50 km² and an elevation of 1400 m. It lies in the rain-shadow of the Sierra Nevada, with an annual rainfall ≈130 mm and an evaporation rate of ≈1380 mm. Its near surface sediments consist of a mixture of clays (smectite, illite, chlorite, kaolin) and salts (halite and gypsum) and widespread pedogenic calcite. Lithium in the brines is thought to be derived by the weathering and leaching of the Tertiary Esmeralda Fm and Quaternary ash-fall tuffs (Figure 7.25). The lithium content of the playa sediments is highest on the eastern side of the playa adjacent to the outcropping marls of the Esmeralda Fm. Before it is leached the lithium is held in the clay fraction of the playa sediments and is probably part of the clay structure (hectorite is present in the playa clays).

Brines are pumped from playa sediments in areas where extensive saline alteration of lithic clays has released lithium. The highest lithium concentrations in the brines occur within smectites of the Miocene Esmeralda Formation and within tuff aquifers. The Esmeralda Formation is also exposed along the eastern side of the playa where subsurface brines contain up to 200,000 ppm NaCl. This brine carries high levels of lithium gained by the dissolution of subsurface playa salts.

Lithium-rich brines, averaging 0.023% lithium, are pumped from the ground via a number of gravel-packed wells and progress through a series of evaporation ponds. Over the course of 12 to 18 months of solar evaporation the lithium concentration increases to 6,000 ppm. When the lithium chloride level reaches optimum concentration, the liquid is pumped to a recovery plant and treated with soda ash, so precipitating lithium carbonate. The carbonate is then removed through filtration, dried, and shipped.

Chile has emerged in the last decade as the most important lithium producer, largely through the exploitation of Salar de Atacama, Chile (Figure 7.26). The salar lies on the Tropic of Capricorn at an altitude of 2,300m in the Desierto de Atacama, some 200 km inland from Antofagasto. In its more central portions this salt encrusted playa contains a massive halite unit (nucleus), which is more than 900 m thick, with an area ≈ 1,100 km^2. Fringing saline muds, with an area ≈ 2,000 km^2, surround this nucleus.

The current salt crust atop this halite nucleus is filled with a sodium chloride interstitial brine that is rich in Mg, K, Li, and B (Alonso and Risacher, 1996). Lithium contents of the pore brines range from 200-300 ppm in the marginal zone, some 500 - 1,600 ppm in the intermediate zone and 1,510-6,400 ppm in the salt nucleus. This central zone averages 4,000 ppm lithium. The main inflows to the salar drain volcanic formations of the Andean Highlands to the east side of the basin. Salts dissolved in inflow waters have a double origin. Weathering of volcanic rocks supplies K, Li, Mg, B and, to a lesser extent, Na and Ca. Leaching of ancient evaporites beneath the volcanic formations provides additional amounts of Na, Ca, Cl, SO$_4$ to the most saline inflow waters. The mass-balance of the upper nucleus of the salar centre shows a strong excess of NaCl with respect to the bittern solutes Mg, K, Li, B. According to Alonso and Risacher (1996) this suggests that the nucleus did not originate from evaporation of inflow waters similar to the present ones. Rather the excess of NaCl is due to NaCl-rich

inflow waters that formerly drained the Cordillera de la Sal, a Tertiary-age evaporitic ridge along the western rim of the present-day salar.

The average sedimentation rate of halite in the lake centre is ≈ 0.1 mm/year, based on the age of an ignimbrite interbedded with the salt. This slow aggradation rate implies a climatic setting of long dry periods and inactivity that alternated with short wet periods during which large amounts of water, and so large amounts of salt, accumulated in the basin centre. The lack of peripheral lacustrine deposits and the high purity of the salt also suggest that the main salt unit is not the remnant of an ancient deep saline lake, but originated mostly from evaporation of waters supplied by subsurface and subterranean saline seeps.

Brines are pumped from 30 metre deep holes in the salt into solar evaporation ponds. In the ponds they are concentrated

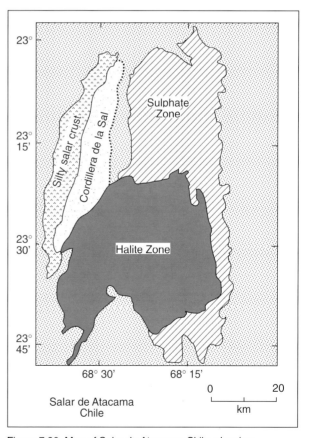

Figure 7.26. Map of Salar de Atacama, Chile, showing asymmetrical zoning of surface mineralogies. Tectonic tilting of the basing floor has crowded the halite zone up against the southwestern margin of the playa. The Cordillera de Sal is an intensely folded sequence of Tertiary continental saline clastic sediments, which contain domal structures with cores of coarsely crystalline rock salt (after Ericksen 1993).

by a factor of 25, generating a final brine strength of 4.3% Li. During evaporation processing the ion ratios are constantly monitored and adjusted to avoid the precipitation of lithium potassium sulphate. The high initial lithium content of the brines and the extremely arid setting means that only 90 hectares of evaporation ponds are required, only 5% of the area required at Clayton Valley, Nevada.

Because lithium is electrochemically reactive and has unique properties, there are many commercial lithium products including the rapidly expanding battery technologies. Most lithium compounds are consumed in the production of ceramics, glass, and primary aluminium.

Nitrates

Nitrogen is abundant on and near the earth's surface where it makes up 78% of the world's atmosphere. It, along with potassium and phosphorus, make up the three essential plant nutrients. Natural deposits of nitrate are rare because of the extreme solubility of nitrates in water; synthetic sources of nitrate today far outweigh the natural sources, with natural sources today supplying less than 0.3% of the world's nitrogen needs.

Although nitrates have been found in several of the world's deserts including the Namibian Desert in southern Africa and Death Valley in California, they are only abundant and in mineable quantities in the Atacama Desert, Chile (Ericksen, 1981, 1983; Searl and Rankin, 1993). Deposits occur in a narrow zone between the low-lying Pampa del Tamarugal to the east and the hills of the Coastal Range to the west (Figure 7.27a). This area contains nitrate-rich soils, as well as surfaces encrusted with nitrate-bearing soils, saline-cemented regolith, and nitrates in closed basin playas or salars. Commercial concentrations of nitrate salts occur on or in weathered rock types ranging from granite to limestone and shales. Nitrate deposits can be found in all topographic positions, from hilltop to hillside to playa and valley floor. The richest deposits tend to be on the lower slopes of hills marginal salars and clay playas (Ericksen, 1993).

Chilean ores have an average composition of: 7-10% $NaNO_3$; 4-10% NaCl; 10-30% Na_2SO_4; and 2-7% Mg, Ca, K, Br, and I (Table 7.7; Ericksen, 1983; Searl and Rankin, 1993). Highest purity nitrate ore occurs as stratiform pedogenic seams about 20 cm thick, located some 3-7 m below the desert surface. The ore, locally called caliche, is part of a sequence of pedogenic layers; with coba at the base, through caliche and costra, to chuca as the overburden. The nitrate-entraining caliche zone is typically 1-3 m thick (Figure 7.27b). It can be alluvial caliche, where the saline minerals, including the nitrates, cement the regolith; or it can be bedrock caliche, where the economic salts form impregnations, irregular masses, veins and fracture-fills in porous or fractured bedrock.

These nitrate deposits are unique. The ultimate source of the nitrate is not clear and transport mechanisms are still under study, but the fixing process is thought to be organic. Ericksen (1993) argues that cyanobacteria fix nitrogen on the surface of moist soils, in or near, ephemeral lakes and playas; or in soils of other areas that are moistened by the frequent and heavy winter fogs. The hyperarid Neogene climate of the Atacama Desert is what allows the highly soluble nitrate salts to build to economic concentrations in the soils. Average annual rainfall is less than a centimetre, and as much as twenty years can pass between rainstorms. This desert is among the driest in the world and similar conditions have prevailed since the middle Miocene (Alonso et al., 1991). This has allowed the gradual buildup of large volumes of pedogenic saline materials in the soils, including the exploited nitrates (Table 7.7). The precipitating parental fluids were derived from a variety of meteoric sources: westward flowing Andean groundwater, coastal fogs, occasional rainfall and Andean-derived surface floodwaters.

The dominant nitrate is mostly nitratite (=soda nitre or Chilean salt petre). A number of other unusual, highly soluble salts, have also accumulated along with the nitrates and include: iodates, perchloriates, chromates and dichromates, as well as the more familiar borates, sulphates and chlorides. It is the only known deposit in the world that contains natural perchlorate. The salts formed at the more saline end of a phreatic precipitation sequence made up of progressively more soluble evaporitic minerals: silicates (zeolites); calcite; Ca-, Na-, (K-) and Mg-sulphates; Na- and K-Mg-nitrate-sulphates; nitratine; and small amounts

Mineral	Composition	Characteristics	Occurrence
Nitratite (soda nitre)	$NaNO_3$	rhombohedral cleavage	main ore mineral in Atacama Desert
Nitre (salt petre)	KNO_3	acicular	secondary salt
Darapskite	$NaSO_4 \cdot NaNO_3 \cdot H_2O$	monoclinic	minor salt
Humberstonite	$K_3Na_7Mg_2(SO_4)6(NO_3)_2 \cdot 6H_2O$	prismatic	minor salt

Table 7.7. Typical nitrate minerals in the Atacama Desert, Chile.

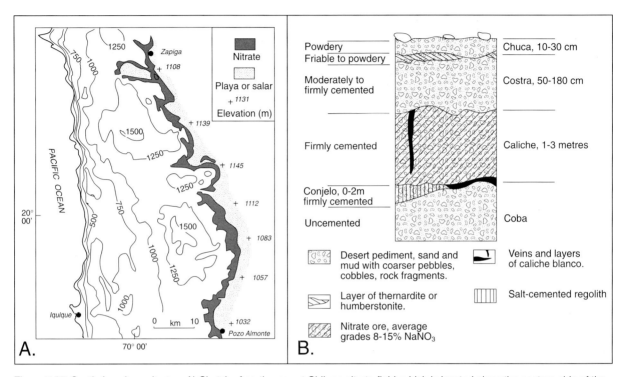

Figure 7.27. South American nitrates. A) Sketch of northernmost Chilean nitrate field, which is located along the eastern side of the Coastal Range, and marginal to salars and clay playas on the western side of Pampa del Tamarugal (after Ericksen, 1993). B) Typical regolith profile through alluvial-type nitrate deposits in a Chilean nitrate field (after Ericksen, 1981).

of iodates, Na- and K-Mg-iodate-sulphates, chromates, borates and perchlorates (Pueyo et al., 1998). Two specific precipitation trends can be distinguished within this more general regional sequence. A Na-trend characterised by an association of glauberite-darapskite-(hectorfloresite), and a K-Mg-Na-trend related to bloedite-polyhalite-humberstonite-niter-(fuenzalidaite).

Parental brine source for the deposits is the hypersaline discharge portion of a regional phreatic system, which Pueyo et al. (1998) consider to be older than 6 Ma. At that time the main physiographic features were similar to today, but the climate was more humid. Evaporation was involved in the nitrate paragenesis; it is the only mechanism effective enough to saturate a brine with these very soluble supersaline minerals. Nevertheless, the nitrate ores do not show typical displacive nodular and enterolithic structures, rather, the salts tend to passively infill veins, fractures or preexisting matrix porosity.

The unusual mineralogy of the nitrate ore reflects the extreme chemical evolution of the precipitating brines through multiple episodes of salt precipitation and remobilisation during transport to the nitrate horizons. The formation of high purity nitrate ore appears to be the result of multiple fractionation episodes driven by dissolution,

reprecipitation and recrystallisation in the desert soil profile and nearsurface lithologies. Millennia of these processes have separated the highly soluble nitrate salts from less soluble salts in the overlying profile. Salts have largely accumulated through passive growth, but some of the host silicate and carbonate lithologies have also undergone a small degree of active salt replacement.

Magnesite and magnesia

Magnesium is the eighth most abundant element in the earth's crust and is widely distributed in numerous minerals. Among the more commercially important minerals are magnesite, brucite, dolomite, olivine, talc and serpentine, as well as the magnesium sulphate and chloride salts recovered from natural brines and seawater. World resources of magnesite, dolomite and magnesium-bearing evaporite minerals are enormous. Magnesium-bearing brines are estimated to constitute a resource measured in billions of tonnes, and magnesium can be recovered from seawater at places along world coastlines where salinity is high. Much of the currently produced magnesium is recovered by electrolytic processes. The majority of magnesium compounds are used for refractories. The aluminium industry is the largest consumer of magnesium, where it is a constituent in aluminium-base alloys used for packaging, transportation, and other applications. The remainder is consumed in agricultural, chemical, construction, environmental, and industrial applications.

There are two forms of magnesite that are commercially exploited, crystalline magnesite and cryptocrystalline magnesite. Much of the world's crystalline magnesite formed as a hydrothermal alteration of carbonate shelf sediments in evaporite entraining basins. These deposits form the greater majority of the world's commercially exploited deposits, their genesis is considered in a discussion of sparry ferroan carbonates in Chapter 9. There are also forms of cryptocrystalline magnesite that formed as hydrothermal alterations. Hydrothermal forms lie outside the sedimentological focus of this chapter (see Lefond, 1983, for full discussion). The following discussion considers only those forms of cryptocrystalline magnesite that formed as evaporitic carbonate.

Worldwide most of these deposits are too small and impure to be commercially viable. The huge lacustrine deposits in the vicinity of Rockhampton, Australia, are an exception. Extensive drilling in the 1980s proved several large magnesite deposits in the area; Marlborough, Kunwarara, Merimal and Yaamba. Of these, the most commercially viable, but not yet in full production, is the Kunwarara deposit. It lies some 60 km northwest of Rockhampton with an area of 31 km^2, an average thickness of 7.75 metres and is covered by an average thickness of 4.45 m of overburden (Burban, 1990; Hill, 1993). The deposit contains some 550 million tonnes, including 260 million tonnes in a nearsurface "blanket" of 40-45% cryptocrystalline lacustrine magnesite. Most of the magnesite occurs as mm-metre diameter nodules and lumps in a matrix of soft magnesitic mudstone that is underlain by dolomitic mudstone, sandstone or gravel. The deposit is thought to have formed via secondary alteration of magnesian-rich lacustrine sediments.

The best analogue for the deposit is thought to be the magnesite deposits of Salda Lake in southern Turkey (Schmid, 1987). Based on this comparison, the following conditions are thought to have formed the Kunwarara deposits:

• Deep weathering of primary stockwork-bearing serpentine rocks and erosion by heavy rains.
• Water and groundwater transport of eroded magnesian materials into a lacustrine depression.
• Deposition and crystallisation of magnesite muds in an ephemeral mudflat under evaporative capillary conditions. Agglomeration of material into cryptocrystalline nodules.
• Periodic waves and sheet flooding during nodule crystallisation, which carry away non-crystallised impurities such as SiO_2, Al_2O_3 and CaO.
• These impurities are deposited in adjacent shallow lakes, which, with a lowering of lake levels, then act as matrix for later magnesite nodule formation.
• Newly formed magnesite zones are covered by soil and compacted. A silica-rich skin forms around the nodules during this process, while ongoing pedogenic magnesite crystallisation continues to raise the level of magnesium in the nodules.

Magnesia is also produced from surface evaporitic and subsurface basinal brines in the USA, the Netherlands, Israel and Mexico. In the Netherlands the brine is formed by solution mining of magnesium/potassium-bearing salts (carnallite, bischofite and kieserite) in the Zechstein section. The halokinetic thickening of the exploited section is thought to be related to the greater ductility of the bittern salts during salt flow. Another brine source is the Dead Sea brines. The brines contain 42,430 ppm of magnesium, which when evaporated and processed produce caustic-calcined and dead-burned magnesia, along with small amounts of magnesian chloride.

Sodium carbonate

The bulk of the world's supply of sodium carbonate (soda ash) is synthetic, created by the Solvay process, which utilises halite and limestone as raw materials with an ammonia catalyst. Major producers using the Solvay method include China, Germany, India and Russia (Figure 7.28). Until the 1960s, soda ash in the USA was almost totally derived from Solvay plants near industrial markets in the country's northeast. Today the position in the USA is reversed: trona is mined from the Eocene Green River Basin, Wyoming, and processed from lacustrine brines in Searles Lake, California. America today produces over 10 million tonnes of trona per year and is an exporter of product. Lesser production is extracted from Quaternary Lake Magadi in the East African rift ($\approx$250,000 metric tonnes/year) and from alkaline lakes in Mexico. Sodium carbonate end usage is mostly in glass manufacture followed by chemicals, soap and detergents, pulp and paper manufacture, water treatment, and flue gas desulphurisation.

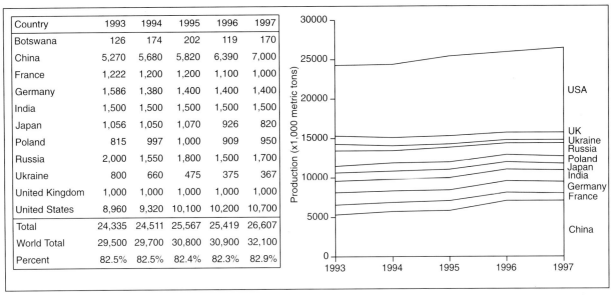

Country	1993	1994	1995	1996	1997
Botswana	126	174	202	119	170
China	5,270	5,680	5,820	6,390	7,000
France	1,222	1,200	1,200	1,100	1,000
Germany	1,586	1,380	1,400	1,400	1,400
India	1,500	1,500	1,500	1,500	1,500
Japan	1,056	1,050	1,070	926	820
Poland	815	997	1,000	909	950
Russia	2,000	1,550	1,800	1,500	1,700
Ukraine	800	660	475	375	367
United Kingdom	1,000	1,000	1,000	1,000	1,000
United States	8,960	9,320	10,100	10,200	10,700
Total	24,335	24,511	25,567	25,419	26,607
World Total	29,500	29,700	30,800	30,900	32,100
Percent	82.5%	82.5%	82.4%	82.3%	82.9%

Figure 7.28. World soda ash production (in thousand metric tons - compiled from USGS online data tables).

A number of naturally occurring minerals contain Na_2CO_3, but only trona is economically mined or quarried at this time (Table 7.8). Sodium bicarbonate, sodium sulphate, potassium chloride, potassium sulphate, borax, and other minerals are produced as coproducts from sodium carbonate production from brines in Searles Lake, California. Sodium bicarbonate, sodium sulphite, sodium tripolyphosphate, and chemical caustic soda are manufactured as coproducts during trona processing at several of the Wyoming soda ash plants. The Wyoming operation is huge, for example, General Chemical's Green River Basin plant has known trona reserves of more than 500 million tonnes of high grade ore and currently maintains more than 200 km of underground workings.

Green River trona precipitated as nodules and pavements in a saline lake within the intermontane foreland basin known as Lake Gosiute within Green River Basin of southwestern Wyoming (Figure 7.29). The depression originated in the Palaeocene and continued to accumulate sediment well into the Miocene. During its maximum extent it covered more than 38,000 km^2. During drier periods the lake shrank as saline sodium bicarbonate minerals (mostly trona) were deposited either as bottom nucleates in shallow brine lakes or an displacive intrasediment nodules. The interbedded dolomitic oil shales represent deposition from relatively dilute water during expanded lake phases. The trona beds were deposited in the interior part of the basin during times when surface waters were extremely shallow to ephemeral with chemistries akin to those in modern Lake Magadi (Surdam and Wolfbauer, 1974; Fischer and Roberts, 1991).

Trona ore in the Green River Basin is restricted to the Eocene Wilkins Peak Member. This unit ranges from 110 to 410 m thick; it is made up of dolomitic limestone, mudstone, thin oil shales and numerous beds of salts, including trona. There are at least 42 trona beds up to 14 metres thick in the member, currently at depths ranging from 120m to over 1,000m. Almost all the trona beds are flat lying, with dips typically less than a degree. Faults are uncommon. The trona beds are remarkable for their widespread purity. Sulphate minerals are essentially absent, and other than halite, the chief impurities are dolomitic marl stringers and vertical seams of mudstone. In the

Mineral	Formula	Na_2CO_3(%)
Trona	$NaHCO_3.Na_2CO_3.2H_2O$	70.4
Thermonatrite	$NaCO_3.H_2O$	85.5
Nahcolite	$NaHCO_3$	63.1
Bradleyite	$Na_2PO_4. MgCO_3$	47.1
Pirssonite	$CaCO_3.Na_2CO_3.2H_2O$	43.8
Northupite	$Na_2CO_3.NaCl.MgCO_3$	40.6
Tychite	$2MgCO_3.2Na_2CO_3.Na_2SO_4$	42.6
Natron	$Na_2CO_3.10H_2O$	37.1
Dawsonite	$NaAl(CO_3)(OH)_2$	35.8
Gaylussite	$CaCO_3.Na_2CO_3.5H_2O$	35.8
Shortite	$Na_2CO_3. 2CaCO_3$	34.6
Burkeite	$Na_2CO_3.2Na_2SO_4$	27.2
Hanksite	$2Na_2CO_3.9Na_2SO_4.KCl$	13.6

Table 7.8. Minerals containing sodium carbonate.

263

southern part of the basin some pure trona beds grade laterally into mixed halite and trona.

The lack of sulphate in the trona reflects bacteriogenic sulphate reduction in the lake brines, whereby bicarbonate was formed in the ratio of two moles of bicarbonate for each mole of sulphate that was reduced (Chapters 2 and 8). Associated hydrolysis of detrital silicate minerals and glassy volcanic ash (zeolites) also contributed additional dissolved bicarbonate, sodium, and other ions to the lake brines.

In the Green River Basin, about 47 billion metric tons of identified soda ash resources could be recovered from the 56 billion tonnes of bedded trona, and the 47 billion tonnes of interbedded or intermixed trona and halite, which are in beds more than 1.2 metres thick. About 34 billion tonnes of reserve-base soda ash can be obtained from the 36 billion tonnes of halite-free trona and the 25 billion tonnes of interbedded or intermixed trona and halite that are in beds more than 1.8 metres thick (USGS on-line data sheet - Soda ash).

Outside of the USA the largest production of natural soda ash is from Lake Magadi, located some 120 km southwest of Nairobi, Kenya. The lake is situated in the lowest part of the East African Rift Valley and has a total area of 100 km^2 (Eugster, 1980). It contains more than 30 billion metric tons of trona, which are believed to have formed in the last 9,000 years in a trona bed up to 40 metres thick. Thermal springs, driven by the area's volcanic activity, and sourced from a deep actively circulating groundwater reservoir, are the major source of sodium in the lake. Trona still forms each year as a bottom-nucleated crystal pavement in the more central areas of Lake Magadi. The annual crop is estimated at 1.5 million tonnes/year, although the bulk of this crop ($\approx$75%) is recycled by dissolution each wet season.

The trona-depositing part of the lake desiccates in the dry season, subjecting the upper part of the trona pavement to subaerial exposure. The lake experiences two rainy seasons, so that the texture of the trona deposit reflects the marked seasonality of the lake climate. The longer rainy season (March–April) usually provides enough runoff to flood the

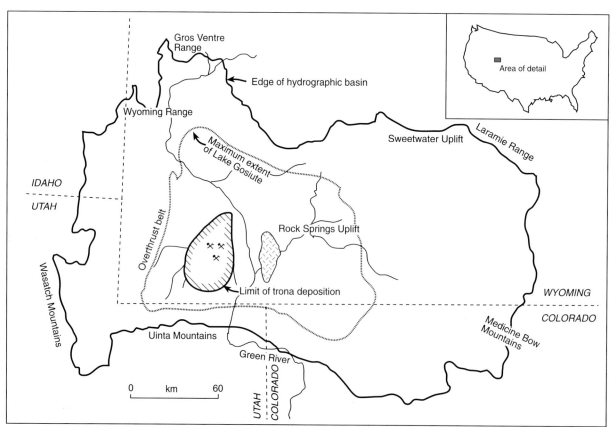

Figure 7.29. Inferred hydrographic basin and maximum extent of Lake Gosiute (Eocene) with area of trona deposition and location of mines in the basin.

trona-forming areas of the lake with water up to 1 m deep. Trona begins to crystallise when the brine is evaporatively drawndown to where it is no more than a few centimetres deep (Warren 1989). Trona than forms in two modes: as rafts (cumulates) and as bottom-nucleated, upward-pointing cm-scale blades. Bottom crystals first nucleate atop the pavement on a flat dissolution surface created by freshening, which was brought on by flooding in the previous wet season. Repetition of the precipitation-dissolution cycle over a number of years creates a layered salt bed dominated by bottom-nucleated trona splays that are truncated by multiple subhorizontal dissolution surfaces. This trona texture is analogous to the bottom-nucleated textures of halite and gypsum described in Chapter 1.

Even as it forms, the trona pavement is rock hard, composed of an interlocking meshwork of growth-aligned trona crystals. Its inherent strength allows it to support expansion polygons or pressure ridges (petees), up to several tens of metres in diameter. They cut up the trona into a series of overthrust saucers, with small brine pools in the centre of each polygon. Pressure ridges are the result of pavement overthrusting created by the sideways expansion of aligned trona crystals jostling for space in the growing crystal pavement.

Once brines atop the pavement dry up, the brine level still stays close to the surface. During exceptionally dry periods it drops only 1 or 2 cm into the trona layer. Thus, a stable brine curtain in Lake Magadi maintains salinities near or above trona saturation year round, so that trona currently extends over 75 km² of the lake floor and is between 7 and 40 metres thick.

Trona in Lake Magadi has been mined since 1914 and is currently quarried from the pavement using a dredging barge, which floats on the lake brines that pond in its wake. If for any reason the dredging barge breaks down for more than a day or two, the trona crystallizing about the base of the barge freezes the barge in place. Trona has to be physically chipped from around the barge before operations can continue. Replenishment of trona in the lake is so rapid that at present removal rates the Magadi trona is a renewable resource. Once extracted the trona is crushed, slurried and piped to the treatment plant where it is washed, screened, dewatered and calcined to produce soda ash.

Sodium sulphate

Sodium sulphate (salt cake or Glauber's salt) is a relatively common component of salts precipitated in many saline lakes and playas (Table 7.9). Thenardite and mirabilite — the anhydrous and decahydrate single-sodium sulphate salts— are commercially important, as are glauberite and bloedite. Thenardite is a colourless to white mineral with a specific gravity of 2.67 and a hardness of 2.5 - 3.0. It is stable below 33°C and is extremely hygroscopic below 25°C, forming the decahydrate mirabilite. This mineral contains 55.9% water of crystallisation and forms opaque

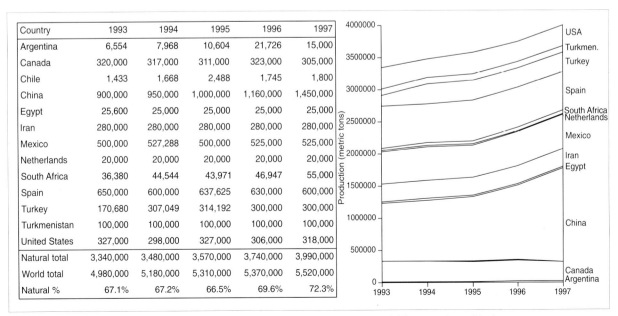

Country	1993	1994	1995	1996	1997
Argentina	6,554	7,968	10,604	21,726	15,000
Canada	320,000	317,000	311,000	323,000	305,000
Chile	1,433	1,668	2,488	1,745	1,800
China	900,000	950,000	1,000,000	1,160,000	1,450,000
Egypt	25,600	25,000	25,000	25,000	25,000
Iran	280,000	280,000	280,000	280,000	280,000
Mexico	500,000	527,288	500,000	525,000	525,000
Netherlands	20,000	20,000	20,000	20,000	20,000
South Africa	36,380	44,544	43,971	46,947	55,000
Spain	650,000	600,000	637,625	630,000	600,000
Turkey	170,680	307,049	314,192	300,000	300,000
Turkmenistan	100,000	100,000	100,000	100,000	100,000
United States	327,000	298,000	327,000	306,000	318,000
Natural total	3,340,000	3,480,000	3,570,000	3,740,000	3,990,000
World total	4,980,000	5,180,000	5,310,000	5,370,000	5,520,000
Natural %	67.1%	67.2%	66.5%	69.6%	72.3%

Figure 7.30. World production of sodium sulphate (in metric tons - compiled from USGS online data tables).

to colourless needle-like crystals sometimes referred to as Glauber's salt. Solubility temperatures are generally lower in the presence of other minerals. The temperature drop in winter and/or autumn, and even the cooling on summer nights in higher altitudes, are sufficient to allow mirabilite to crystallise on the pan floor. In some places, the crystals can redissolve into the lake brine with the next passage to warmer temperatures. This could be on the next day or with the passage of winter into spring.

Natural sources of sodium sulphate make up over 70% of total world production, the balance is a byproduct of various chemical and waste recovery processes.. Spain and Mexico are the world's highest tonnage miners of sodium sulphate, followed by a group comprising the USA, Canada, Turkey and Iran (Figure 7.30). The bulk of sodium sulphate usage is in the manufacture of glass and powdered detergents. It is also used in the pulp and paper industries. Commercial sources of sodium sulphate are mostly composed of crystalline mirabilite, or occur as brines in playas and saline lakes within cold or arid saline

Mineral	Formula	Na_2SO_4 (%)
Thernadite	Na_2SO_4	100
Hanksite	$2Na_2CO_3.9Na_2SO_4.KCl$	81.7
Sulphohalite	$2Na_2SO_4.NaCl$	73.9
Glauberite	$Na_2SO_4.CaSO_4$	51.1
Loweite	$2NaSO_4.2MgSO_4.5H_2O$	46.3
Ferronatrite	$3NaSO_4.Fe_2(SO_4)_3.6H_2O$	45.6
Mirabilite	$Na_2SO_4.10H_2O$	44.1
Bloedite	$Na_2SO_4.MgSO_4.4H_2O$	27.2
Tychite	$2NaCO_3.2MgCO_3.Na_2SO_4$	42.5
Aphthitalite	$(Na,K)_2SO_4$	21-38
Tamarugite	$Na_2SO_4.Al_2(SO_4)_3.12H_2O$	20.3
Mendozite	$Na_2SO_4.Al(SO_4)_3.12H_2O$	15.5

Table 7.9. Minerals containing sodium sulphate.

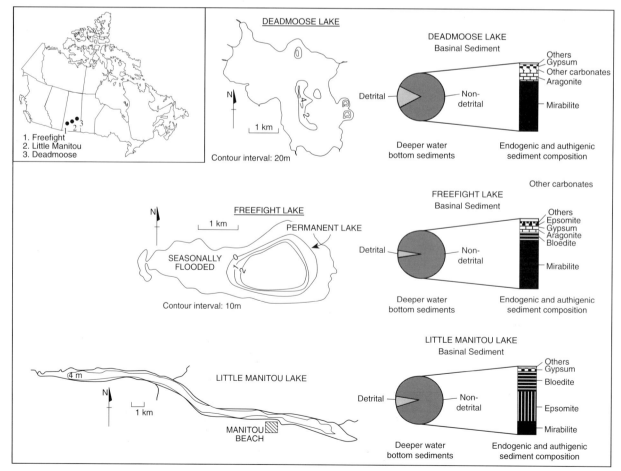

Figure 7.31. Sodium sulphate lakes, Canada, showing bathymetry and bottom sediment composition (after Last, 1994).

valleys lacking an outlet. Sodium sulphate can also be obtained as a byproduct from the production of ascorbic acid, boric acid, cellulose, chromium chemicals, lithium carbonate, rayon, resorcinol, and silica pigments.

Canadian production comes from the winter cooling/ freezing of playa brines in southern Saskatchewan and southeastern Alberta where mirabilite accumulates in winter within low lying saline lake depressions in glacial drift. Most of the lakes are small (< 100km^2), shallow (>3-20m) and occupy the lowest portions of hydrologically closed depressions (Last, 1989, 1994). For example, Deadmoose, Little Manitou, and Freefight Lakes are lacustrine basins in which soluble and sparingly soluble evaporite minerals are accumulating today, even in the relatively deep offshore areas of the lakes (Figure 7.31). Both Deadmoose and Freefight Lakes are hypersaline and meromictic, with strongly anoxic bottom waters. Little Manitou is also hypersaline, but seasonal density and chemical stratification usually breaks down by early to midsummer. The salts are generally coarsely crystalline, massive to thickly bedded, and interfinger laterally with laminated organic-rich clays and poorly sorted carbonate-clastic debris from nearshore areas. There are four distinct styles of salt occurrence in these lakes: a) crusts and hardgrounds, b) massive and bedded, c) spring deposits, 4) subsurface and groundwater related accumulations (Last, 1989). Even in the shallowest basin of the three, Little Manitou Lake, the bottom salts show no indication of widespread dissolution or recycling. Thus, relatively thick sections of subaqueous evaporites have accumulated and are being preserved in these basins (Last, 1994).

As to the origin of the sodium in the Canadian lakes there is still no consensus (Last 1989). The geographic coincidence of the lakes to the subsurface solution edge of the Devonian Prairie Evaporite led Grossman (1968) to suggest dissolution-derived resurging groundwaters was the source. Others have argued that weathering and leaching of the glacial drift is the likely source (Wallick, 1981).

Most of the commercial Canadian deposits are in Saskatchewan, the largest lake in production is Ingebright Lake, in southeast Saskatchewan, where the salt bed thickness averages 7 metres and in places is more than 40 metres. Reserves exceed 8 million tonnes. Traditional extraction involves pumping of surface brines into storage reservoirs during the summer and then allowing the lower autumn and winter temperatures to precipitate mirabilite. The mirabilite is then trucked to the processing plant, where water of crystallisation is removed, and then dried to salt cake in a rotary kiln.

Mexico's major source is Laguna del Rey, a desert playa containing an extensive glauberite bed with minor bloedite and other sodium and magnesium salts including halite. Minor mirabilite is present as a cement between the glauberite crystals. The salt body is lenticular in cross section, has a maximum thickness of 35 metres, measures 4 by 10 kilometres and contains some 360 million tonnes of sodium sulphate and 77 million tonnes of magnesium sulphate.

Glauberite is also a primary precipitate in the modern playas of northern Mali, central Africa. Nearsurface glauberite in these lakes is backreacting with nearsurface brines, via a bassanite intermediary phase, to form a gypsum cap to the deposit (Mees, 1998). Previously, this gypsum layer was considered a depositional unit.

Modern saline lakes, similar to those in Canada, are the major source of sodium sulphate in the former USSR. Examples include Selenga Lake in the eastern Transbaikal, Lakes Azhbulat, Ebeity and Tengis in Kazakhstan, and the Batalpashinsk and Tambukan Lakes between the Black and Caspian Seas. The enormous potential of these deposits can be illustrated in Lake Azhbulat where some 40 million tonnes of mirabilite is deposited each winter. On the Kulunda Steppe, western Siberia, lacustrine deposits consist of sodium chloride in the upper layers and mirabilite below. The largest lake in the region, Lake Kuchuk, southwest of Novosibirsk, covers some 150 square kilometres. It contains an estimated 600 million tonnes of sodium sulphate in brines and underlying strata; each winter between 380,000 and 640,000 tonnes of mirabilite

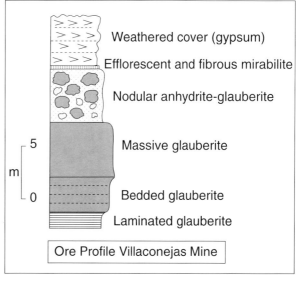

Figure 7.32. Glauberite profile Villaconejas Mine, Spain.

precipitates out of solution. Sodium sulphate also occurs in the Karabogaz Gulf on the eastern side of the Caspian Sea.

Ancient bedded lacustrine deposits containing sodium sulphate are non-commercial in the USA at the present time. Most US production comes from brine processing in Searles Lake, California. This brine contains about 450 million metric tons of sodium sulphate resource, representing about 35% of the lake brine. In Utah, about 12% of the dissolved salts in the Great Salt Lake is sodium sulphate, representing about 400 million tonnes of resource. There, below the lake bottom near Promontory Point, a naturally occurring irregular, 21-metre-thick mirabilite deposit is interlayered with clay beds 4.5 to 9.1 metres thick.

Tertiary sediments are the major source of sodium sulphate in Spain. Compared with other deposits worldwide, the Spanish deposits are unusual in that deeply buried solid thenardite or glauberite is mined and a brine or a Quaternary deposit is not exploited. For example, the "El Castillar" mine in Toledo Province yields thenardite from an ore averaging 67% Na_2SO_4, 19% $CaSO_4$ and 12% clay and marl. The ore bed thickness averages 9 metres and lies atop a 13 metre-thick halite bed. Glauberite is extracted at Cerezo de Riotiròn, Burgos Province, where the deposit is practically horizontal, undisturbed and made up of glauberite in four beds hosted in Miocene lacustrine marls.

The salts have a primary paragenesis of glauberite + anhydrite + dolomite ± chalcedony; the secondary minerals related to hydration and incongruent dissolution of glauberite are mainly gypsum and calcite (Menduina et al., 1984; Salvany and Orti, 1994). Glauberite types are: 1) massive glauberite within fining-upward sequences containing nodular anhydrite; 2) millimetre-scale glauberite layers alternating with dolomicrites; 3) single glauberite crystals dispersed in dolomicritic matrix (Figure 7.32).

A 'perennial saline lake' model with extensive saline mudflats explains the deposit (Figure 7.33; Salvany and Orti, 1994). Solutes were derived from the phreatic dissolution of Mesozoic evaporites, mainly Ca-sulphates and halite, in the nearby mountains (Pyrenees and Iberian Ridge). The precipitation sequence was: carbonate, gypsum-anhydrite, glauberite, halite-polyhalite. Because of dissolution by less concentrated (meteoric) brines fed into the basin margin and by occasional storm run off, backreactions related to salinity fluctuations were commonplace (see Chapter 2). Glauberite, anhydrite and polyhalite are early diagenetic minerals, which formed via backreaction with a solid precursor (typically gypsum)

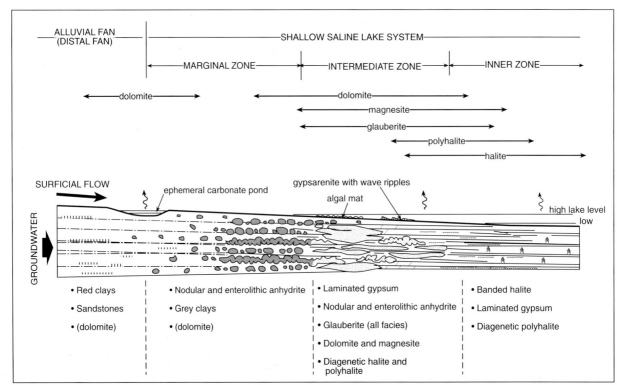

Figure 7.33. Sedimentological model for deposition of Spanish glauberite deposits (after Salvany and Orti, 1994).

Country	1993	1994	1995	1996	1997
Canada	8,450	9,140	9,010	9,010	9,200
China	6,030	6,030	5,430	5,470	5,200
France	1,260	1,100	1,170	1,200	1,100
Germany	1,170	1,240	1,110	1,110	1,100
Iran	800	880	890	890	890
Iraq	800	800	475	475	475
Japan	2,920	2,900	2,810	2,800	2,800
Mexico	1,670	2,920	2,880	2,890	2,900
Poland	2,130	2,380	2,660	1,770	1,700
Russia	2,000	1,830	2,200	4,000	4,000
Saudi Arabia	2,400	2,300	2,200	2,000	2,000
South Africa	575	524	459	469	470
Spain	687	702	756	752	700
United States	11,000	11,500	11,800	11,800	11,900
Other countries	7,840	7,430	7,526	7,550	10,200
Grand total	51,300	53,700	53,200	52,400	54,000

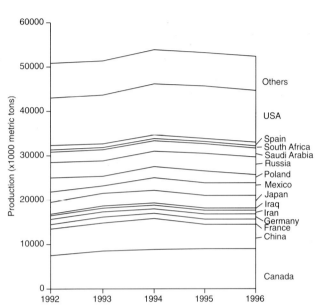

Figure 7.34. World production of sulphur (thousand metric tons - compiled from USGS online data tables).

during early diagenesis. The brines that drove this early alteration reacted with the solid precursor in two ways: 1) reaction of the concentrated brine with the solid phase precipitated during episodes of lowered salinity (e.g. anhydrite altered to glauberite; glauberite altered to polyhalite), and 2) reaction of dilute brines with solid phases precipitated in stages of higher salinity (e.g. glauberite altered to anhydrite, polyhalite altered to glauberite or anhydrite).

Burial led to further transformation of gypsum to anhydrite, and during exposure all preexisting sulphates (anhydrite, glauberite, polyhalite) were transformed into gypsum near the surface and formed the weathered cover or carapace seen atop many ore profiles (Figure 7.32).

Sulphur

Sulphur may be found alone as native or elemental sulphur, or as a gregarious element in combination with many other elements, including metal sulphides and evaporitic sulphates. Worldwide resources of elemental sulphur amount to about 5 billion tonnes. It occurs in evaporite and volcanic deposits or as sulphur associated with natural gas, petroleum, tar sands, and metal sulphides. Poland and the United States are the only countries that each year produce more than a million tonnes of native sulphur from evaporite related deposits, using either the Frasch hot water method or conventional mining (Figures 7.34, 7.35). Small quantities of native sulphur are mined in Asia, Europe, and North and South America. The use of pyrite as a sulphur

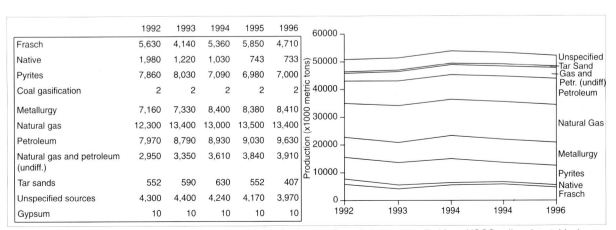

	1992	1993	1994	1995	1996
Frasch	5,630	4,140	5,360	5,850	4,710
Native	1,980	1,220	1,030	743	733
Pyrites	7,860	8,030	7,090	6,980	7,000
Coal gasification	2	2	2	2	2
Metallurgy	7,160	7,330	8,400	8,380	8,410
Natural gas	12,300	13,400	13,000	13,500	13,400
Petroleum	7,970	8,790	8,930	9,030	9,630
Natural gas and petroleum (undiff.)	2,950	3,350	3,610	3,840	3,910
Tar sands	552	590	630	552	407
Unspecified sources	4,300	4,400	4,240	4,170	3,970
Gypsum	10	10	10	10	10

Figure 7.35. Processes of sulphur manufacture - worldwide (in thousand metric tons - compiled from USGS online data tables).

source has significantly decreased in the last two decades. China, South Africa, and Spain are now the only countries in the top 15 sulphur producers whose prime sulphur source is pyrite; together these three account for more than 80% of all pyrites-based production. Chemically recovered elemental sulphur is the predominant sulphur source in Canada, France, Germany, Iran, Russia, Saudi Arabia, and the United States, and an important sulphur source in Japan and Mexico. Elemental sulphur can also be recovered during petroleum refining or at natural gas processing plants and coking plants (Figure 7.35).

In addition to the known native sulphur resource, the amount of sulphur in gypsum and anhydrite is almost limitless. Some 600 billion tonnes are contained in coal, oil shale, and shale rich in organic matter, but low-cost recovery methods have not yet been developed. Agricultural chemicals (primarily fertilisers) comprise the greater bulk of sulphur demand, with chemicals manufacture, metal mining, and petroleum refining using the remainder.

Native sulphur is naturally produced in diapir caprocks by the bacterial degradation (sulphate reduction) of anhydrite and gypsum in the presence of organic matter or hydrocarbons at temperatures of up to 110°C. This process occurs in both bedded and diapiric precursors (Aref, 1998b). In the sulphur-producing diapirs of the Gulf Coast and Mexico, the CaSO$_4$ precursor is created by dissolution of halite in the rising salt stem that leaves behind a residual carapace of CaSO$_4$ (Figure 7.36a; Ruckmick et al., 1979). The dominant species of bacteria acting on this CaSO$_4$ changes with temperature; *Desulphovibrio desulphrican* is active at normal shallow subsurface temperatures, *Desulphovibrio orientis* is active at moderate temperatures and *Clostridium nigrificans* dominates in hot conditions. All these bacteria consume organics or hydrocarbons as a source of energy, but use sulphur instead of oxygen as a hydrogen acceptor and so produce hydrogen sulphide, calcite and water as byproducts. In situations where the

H$_2$S accumulates for long periods in the subsurface it is oxidised to colloidal sulphur (see Chapter 8 for a full discussion of bacterially mediated sulphate reduction).

Native sulphur occurs in the Upper Permian Castile and Salado Formations on the western margin of the Delaware Basin, west Texas (Figure 7.36b). There masses of biogenic calcite are created by bacterially induced replacement of gypsum so that sulphur ore resides in pores and vugs of this biogenic limestone. In its unaltered state the Castile Formation is made up of bedded laminar anhydrite. Its underside acts as a seal and a focus for hydrocarbons seeping up from the oil kitchen in the deeper parts of the Delaware Basin. The ground about the edge of the Delaware Basin is naturally prepared for sulphur formation by a series of faults, which will act as foci for the creation of porous dissolution collapse breccias, that crosscut the anhydrite beds. This creates chimneys of brecciated

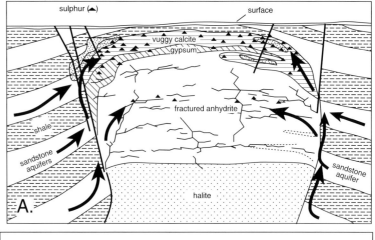

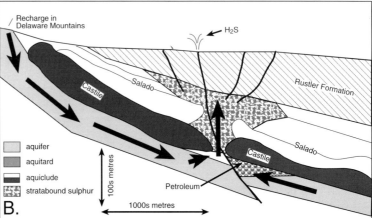

Figure 7.36. Conceptual models for the origin of evaporite-related elemental sulphur. A) Genetic model for diapir caprocks. B) Genetic model for stratabound sulphur deposits (after Ruckmick et al., 1979).

anhydrite (see Chapter 4 for a full discussion of mechanism). Once the dissolution breccia/karst system is created it focuses the ascent of petroleum-bearing waters from the underlaying carrier beds (Bell Canyon Sandstone). The hydrocarbon-bearing waters flush through the breccia and the bacteria converting the anhydrite to limestone and depositing the sulphur via bacterially mediated sulphate reduction. Thus the location of the sulphur in this style of deposit is structurally controlled. Deposits are generally small, although the Duval Mine in Culberson Co., Texas, had original reserves of more than 55 million tonnes between depths of 70 and 380 metres (Davis and Kirkland, 1979).

Sulphur was mined in Poland as early as the fifteenth century, and although declining, annual sulphur production in Poland still exceeds 2 million tonnes. Native sulphur deposits today occur beneath fluvio-glacial outwash in the Vistula Valley, some 190 km south of Warsaw, and in the area around Lublin. All deposits lie within Middle Tortonian gypsum. They form part of the 3,000 m thick fill of Miocene sediments that accumulated during the Alpine Orogeny in the foredeep north of the Carpathian Front. The sulphur is biogenic and formed in a fashion very similar to that hosted in the bedded deposit of the Castile Formation, west Texas (Pawlowski et al., 1979). The main difference is that the Polish gypsum host was probably never converted to anhydrite during burial.

Sulphur in Poland occurs in several modes (Pawlowski et al., 1979). The most commercially important is in a fine-grained poorly lithified limestone (marl) that contains disseminated microcrystalline sulphur. The sulphur makes up 15% to 80% of the rock mass and averages 30%. It is pale coloured and may occur in rounded concretionary masses, in thin irregular stringers, or in thick masses of high purity. A second mode is as a coarser recrystallised form, it is bright yellow and typically occurs as crystals lining vugs and along seams. Masses of this coarsely crystalline form commonly grade into the microcrystalline forms. Two other non-commercial occurrences are as finely divided sulphur in primary bottom-nucleated gypsum crystals and as cements in sandstone beds. Both styles contain less than 15% sulphur and in a form that is difficult to concentrate. The Polish deposits are either extracted by open-pit mining or using the Frasch hot-water/steam method, in which the native sulphur is melted underground and brought to the surface by compressed air. This is the same method used to extract much of the sulphur in the Gulf of Mexico and west Texas.

Zeolites

Zeolites are hydrated aluminosilicates of the alkaline and alkaline earth metals. Natural zeolites form by the reaction of pore waters with materials such as volcanic glass, poorly crystalline clay, plagioclase or silica. The formation of evaporitic zeolites is favoured by elevated ratios of Mg and H to Na, K and Ca, conditions that characterise groundwaters in hypersaline basins; but not all zeolites form in evaporite basins. The particular zeolite that forms depends on the composition of the original rock as well as chemical activity, pressure, temperature and partial pressure of water in the fluid-filled pore spaces.

Some zeolites can begin to form days after emplacement of the precursor lithology and the transformation is typically rapid. More than 30 different zeolite minerals have been identified in Cenozoic rocks, only 7 in Mesozoic strata, 2 in the Early Palaeozoic, 4 in the Late Palaeozoic, one in the Neoproterozoic and none in the Archaean (Iijima, 1980).

Zeolites as a group comprise a number of crystalline, hydrated aluminosilicates made up of an infinitely extending three-dimensional network of $(Si,Al)O_4$ tetrahedra where some of the quadrivalent silicon has been replaced by trivalent alumina. This creates a charge imbalance that is rectified by the addition of monovalent or divalent cations such as Na^+, K^+, Ca^{++}, or Mg^{++}. There are also highly variable secondary building units in the zeolite silica framework. They are composed of single rings of 4, 5, 6, 8, 10 and 12 tetrahedra; double rings of 4, 6, 8 tetrahedra; or larger symmetrical polyhedra. Variations in these secondary building units is the basis for classification of zeolites into seven basic groups (Table 7.10).

Essentially a zeolite has three basic components —the aluminosilicate framework, the interconnected void spaces in the framework that contain metals, and the water molecules present as an occluded phase (Flanigen, 1984). This gives an empirical formula for any zeolite:

$$(M^+_2,M^{++})Al_2O_3 . gSiO_2 . zH_2O$$

where M^+ is usually Na or K, M^{++} is Mg, Ca, or Fe. Occasionally Li, Sr or Ba can substitute for M^+ or M^{++}.

This rather unusual lattice structure gives zeolites some useful characteristics (Breck 1974):

• high degree of hydration;
• low density and large void volume when dehydrated;

271

	Typical unit cell content	Void volume	Specific Gravity	Ion exch. (meq./gm)
Group 1: Single 4-ring (SAR)				
Analcime	$Na_{16}[(AlO_2)_{16}(SiO_2)_{22}].16H_2O$	0.18	2.24-2.29	4.54
Phillipsite	$(K,Na)_{10}[(AlO_2)(SiO_2)_{22}].20H_2O$	0.31	2.15-2.20	3.87
Laumonite	$Ca_4[(AlO_2)_6(SiO_2)_{46}].16H_2O$	0.34	2.20-2.30	4.25
Group 2: Single 6-ring (S6R)				
Erionite	$(Ca, Mg,K_2,Na_2)_{4.5}[(AlO_2)_2(SiO_2)_{27}].27H_2O$	0.35	2.02-2.08	3.12
Group 3: Double 4-ring (D4R)				
A (Linde)	$Na_{12}[(AlO_2)_{12}].27H_2O$	0.47	1.99	3.12
Group 4: Double 6-ring (D6R)				
Chabazite	$Ca_2[(AlO_2)_4(SiO_2)_8].13H_2O$	0.47	2.05-2.10	3.81
Group 5: Complex 4-1, T_5O_{10} unit				
Natrolite	$Na_{16}[(AlO_2)_{16}(SiO_2)_{24}].16H_2O$	0.23	2.20-2.26	5.26
Group 6: Complex 5-1, T_8O_{16} unit				
Mordenite	$Na_8[(AlO_2)_8(SiO_2)_{40}].24H_2O$	0.28	2.12-2.15	2.29
Group 7: Complex 4-4-1, $T_{10}O_{20}$ unit				
Clinoptilolite	$Na_6[(AlO_2)_6(SiO_2)_{20}].24H_2O$	0.34	2.16	2.54

Table 7.10. Properties of some important zeolites (after Breck, 1974).

• a stable crystal structure in many zeolites when they are dehydrated;
• cation exchange properties;
• uniform molecular sized channels in many dehydrated crystals;
• variable physical properties such as electrical conductivity;
• adsorption of gases and vapours;
• catalytic properties.

Some of these properties, in particular adsorption, molecular sieving, ion-exchange character and catalytic properties are very useful in industrial applications.

There are seven major geological settings where zeolites form (Table 7.11). Many zeolites form from altered volcaniclastics, as such rocks are characterised by an elevated initial content of high-temperature glassy material. Volcanic glass is thermodynamically unstable at the temperature and pressure conditions at or near the earth's surface, and so quickly alters to a zeolite complex. The Japanese extract such hydrothermally formed zeolites

from the volcanics of onshore Japan, but these deposits are non-evaporitic and not considered further. The interested reader is referred to Hay (1981) for a full discussion of all geological settings. In the context of evaporites we need consider only the saline lake setting and groundwater flushing of the immediate sedimentary and volcaniclastic surrounds.

Saline alkaline lakes with pH levels as high as 9.5 are environments where abundant zeolites can precipitate in time-frames measured in thousands of years. Most commercial zeolite development in this setting is in the Basin and Range province of the United State (Surdam and Sheppard, 1978) and the Andes (Hartley et al., 1991). The Basin and Range region is characterised by an arid climate, an abundance of volcanics, and a tectonic setting that has produced many hydrologically closed basins with alkaline lakes and pore waters; similar conditions occur in the Chilean examples. The San Simon valley in Arizona is one such area (Sheppard et al., 1978). The valley is underlain by 700 m of Pliocene to Holocene flat-lying fluviatile and lacustrine rocks. Within what is known as the "older alluvial fill" is a zeolite tuff, which is part of the tuff marker bed in the Green Lake Beds. The most common zeolites in this succession are chabazite, erionite and clinoptilolite, with small quantities of analcime in the southern part of the deposit. Parent materials include volcanic glass, biogenic silica, clays, plagioclase and quartz. Non-zeolite minerals in the tuff include smectite clays, calcite, halite, gypsum and/or thenardite. Chabazite, the main commercial mineral, constitutes up to 80% of the bed.

The zeolites crystallised after the accumulation of a volcanic ash in the lake, as throughflushing alkaline pore waters altered the ash. Factors controlling the intensity of alteration included lake bottom topography, depth and cation content of the saline-alkaline lake water and proximity to postdepositional erosion surfaces (Eyde and Eyde, 1987). Zeolitic alteration was complete when an extensive system of younger palaeochannels deeply eroded the Green Lake Beds leaving only a few remnants of the original tuff marker and the "high grade zeolite bed". Both the channels and the Green Lake Beds are overlain by a section of halite-bearing brown mudstone known as the Brown Lake Beds.

Throughout burial, and on into the zone of metamorphism, zeolites continue to react with evolving pore fluids, and new minerals are formed by the replacement of initially formed zeolites. Ongoing alteration generates analcime and even feldspar. This is reflected in a common transition from unaltered volcaniclastic glass in the outer parts of the saline lake system through a series of alteration alkaline silicic zeolites to analcime to potassic and sodic feldspar in the lake centre (Eugster, 1986; Hay et al., 1991). It also explains the abundance of albitised metasediments and metavolcanics in meta-evaporitic settings (Chapter 6) and the loss of natural zeolites with increasing geological age. Zeolites in the mudflats of alkaline saline lakes, such as Lake Magadi, are commonly associated with magadiite chert nodules (Chapter 4).

Summary

Evaporite salts that are mineable resources occur in a wide range of depositional and environmental settings. Ages of exploitable potash deposits range from Quaternary to Devonian; no economic potash deposits are known to occur in Precambrian strata. The substantial age range probably reflects the impervious nature of salt once it is buried. Even though the various potash salts are highly soluble, undersaturated solutions cannot reach them until a thick protective wrap of tight halite has been etched away.

Larger potash deposits, such as in the Zechstein and Prairie evaporite units, formed as part of the evaporite fill of basinwide settings (Table 7.12). This probably reflects the very high salinities that are needed before any brine, be it marine or nonmarine, reaches carnallite/sylvite saturation. The highly isolated conditions that can be attained in "saline giants" tend to favour the accumulation of potash salts in such settings (Chapter 3). There is no modern day equivalent to a seawater-fed drawdown basin such as the Zechstein. The closest modern approximation, in terms of isolation, is the highly arid continental salt lakes of Qaidam in inland China. Potash salts in these Chinese deposits are also illustrative of the importance of brine evolution in attaining appropriate brine chemistries to precipitate potash.

The mineralogy in the Chinese deposits underscores the difficulties in using the presence of $MgSO_4$ salts as an indicator of a seawater feed to the basin. Some lakes in the Qaidam Basin lack these salts, while nearby lakes contain them; it is all dependent on the balance in the lake brines between

Mode of occurrence	Temp (°C)	Zeolite species
• Alkaline saline lake and percolating groundwater (in basic tephra)	20 - 50	Ph, Cp, Ch, Er, Mo, Gi, Fa, Go, Na, An, (He)
• Percolating groundwater (in acid tephra)	20-50	Ph, Cp, Er, Mo, Fe, Th, Me
• Deep sea sediments	4-50	Ph, Cp, (An)
• Shallow burial diagenesis (low-temp. hydrothermal)	25-100	Sc, He, St
• Deep burial diagenesis (mod. temp. hydrothermal)	100	La, An
• Low grade metamorphic (high temp. hydrothermal)	>200	Wa, Yu, An
• Magmatic primary	high	An

Ph: phillipsite; Cp: clinoptilolite; Ch: chabazite; Er: erionite; Mo: mordenite; Gi: gismondite; Fa: Faujasite; Go: gonnardite; An: analcime; Na: natrolite; He: heulandite; Th: thomsonite; Me: mesolite; Sc: scolecite; St: stilbite; La: laumontite; Wa: wairakite; Yu: yugawaralite.

Table 7.11. Geological occurrence of some natural zeolites (after Iijama, 1980).

273

Location	Age	Mineable	Tectonic setting	Depositional Setting	MgSO$_4$	Main potash minerals	Style and origin of potash
Qaidam Depression, China	Quaternary	Yes	Foreland basin	Continental playa	Yes	Carnallite, lesser sylvite	Void filling cements and displacive crystals via cooling of syndepositional sinking brines.
Danakil Depression, Ethiopia	Quaternary	Possibly	Rift aulocogen	Continental playa	Yes	Primary carnallite, kainite, secondary sylvite	Seaward seepage into Pleistocene subsealevel rift depression.
Chott el Djerid, North Africa	Quaternary	No	Endoheic depression	Continental discharge playa depression	No	Ephemeral halite-carnallite crusts	Brine evaporation in area of regional artesian aquifer discharge.
Small playas in Amadeus Basin, Australia	Quaternary	No	Endoheic depression	Continental discharge playa depression	No	Ephemeral halite-sylvite crusts	Brine evaporation in small playas atop, but not connected to, zone of artesian aquifer discharge (only rare and minor occurrences of potash salts).
Dead Sea Depression, Middle East	Quaternary	Yes	Continental transform or "pull-apart" basin	Artificial brine recovery and pan evaporation	No	Carnallite slurry with sodium and magnesium chloride	Sequential evaporation of brine pumped from pores and dissolution cavities in subsurface Miocene evaporites.
Solfera Fm., Caltanissetta Basin, Sicily (Messinian)	Late Miocene	Yes	Foreland basin	Continental depression fed by seawater seepage	Yes	Syndepositional carnallite and early secondary sylvite	Syndepositional carnallite and secondary sylvite via brine cooling within karsted halite host. Deposited in early stage of a basinwide seaway drawdown.
Mulhouse Basin, France	Oligocene	Yes	Continental rift graben	Continental saltern, bedded	No	Primary carnallite and sylvite	Interlayered primary potash and halite cm-scale couplets with settle-out/bottom growth textures.
Maha Sarakham Fm., NE Thailand	Cretaceous	Possibly	Continental foreland basin	Continental saltern, now halokinetic in part	No	Syndepositional carnallite and early secondary sylvite, also tachyhydrite	Syndepositional carnallite and secondary sylvite via brine cooling within karsted halite host.
Transatlantic potash basins, West Africa and Brazil	Cretaceous	Possibly	Continental rift graben	Continental salterns, now halokinetic in part	No	Primary carnallite and sylvite, also tachyhydrite	Bottom growth textures in both halite and halite layers in cm-scale stacked couplets.
Moroccan Meseta, Northern Africa	Late Triassic (Keuper)	Possibly	Continental rift graben	Continental salterns, now halokinetic in part	Rare	Not known - no detailed petrography available	Potash/halite evaporites interbedded with redbeds and basaltic lava flows.
Windsor Group, Canadian Maritime Provinces	Carboniferous (Visean)	Yes	Continental rift graben	Continental salterns, now halokinetic in part	No	Syndepositional carnallite and sylvite within primary halite. Associated borates	No detailed petrography yet published. Some thickening of potash units on salt pillow margins.
Z$_1$, Z$_2$ and Z$_3$ evaporite intervals in Zechstein Basin, Europe	Permian	Yes	Basinwide marine evaporite	Mostly salterns, occasional deeper water, now halokinetic in part	Yes	Syndepositional carnallite and sylvite within primary halite	Carnallite as void fill during early brine reflux. Incongruent dissolution of carnallite to form sylvite. Potash units thickened by salt flow.
Salado Fm., Delaware Basin, New Mexico, USA	Permian	Yes	Basinwide marine evaporite	Mostly salterns, occasional deeper water, now halokinetic in part	Yes	Syndepositional carnallite and sylvite within primary halite	Carnallite and sylvite precipitated as void fills during early brine reflux.
Prairie Evaporite Fm., Elk Point Basin, Canada	Devonian	Yes	Basinwide evaporite	Mostly salterns, occasional deeper water	No	Syndepositional carnallite and sylvite within primary halite	Syndepositional potash textures, often overprinted by later alteration events related to tectonically driven episodes of fluid flushing by basinal brines or deeply circulating meteoric waters.

Table 7.12. Summary of major potash evaporite occurrences as discussed in this text.

spring and river inflow, salt dissolution and backreactions (Vengosh et al., 1995; also Chapter 2). Some authors have distinguished/classified particular saline giants as marine or nonmarine on the basis of the presence or absence of these salts, but as was discussed in Chapter 2, this is all predicated on the constancy of ionic proportions in Phanerozoic seawater. If an absence of $MgSO_4$ salts indicates a nonmarine feed, then evaporites of the Elk Point Basin, the largest potash salt resource in the world, are nonmarine deposits. However, there is a growing body of fluid-inclusion evidence that suggests that during the Phanerozoic the chemical composition of marine brines oscillated between the Na-K-Mg-Ca-Cl type and the Na-K-Mg-Cl-SO4 type (Kovalevich et al., 1998).

Aside from drawdown saline giants, the other major depositional setting for economic accumulations of potash salts are active continental rift grabens. Some, such as the Danakil Depression, have abundant $MgSO_4$ salts whose presence has been used to argue for a seawater spring feed. Others such as the Mulhouse Graben in the Rhine Graben probably had nonmarine feeds. In both cases, the greater volume of inflowing ions within the depression was via brine springs, rather than surface runoff. In both the saline giants and the continental rifts there was a substantial input of basinal brines along fault conduits, which crosscut thick sediment columns beneath the subsiding floor of the depression. As most basinal brines are typically Ca-Cl fluids, tachyhydrite is a relatively common associated mineral in such hydrothermally fed potash deposits. It is commonplace in the Transatlantic potash basins and in the continental saltern that hosts the potash interval of the Maha Sarakham in Thailand.

Potash enrichment in a thick halite host often occurs at an intrasalt unconformity or disconformity. This reflects an episode of brine fractionation and recrystallisation related either to episodes of exposure and subaerial leaching/concentration, or to episodes of subsurface flushing of the most soluble salts followed by their reprecipitation.

In all mined potash deposits the two major recovered salts are sylvite and carnallite. Historically sylvite was the preferred ore mineral; it was often simply crushed, prior to mixing and supply to the fertiliser industry. As our ability to manipulate brine chemistry in slurry feeds has improved, carnallite feeds are coming more into use.

Without exception, nonpotash salt resources formed in continental lacustrine or playa settings. They owe their existence to the unusual ionic compositions that come from groundwater leaching, under highly arid conditions,

in an appropriate bedrock terrane. Phanerozoic seawater simply never had the ionic proportions necessary to precipitate borates and nitrates, or the carbonates and sulphates of sodium. This chemical evolution can only happen when seawater is mixed with nonmarine brines or, more typically, when nonmarine waters discharge into an arid, highly isolated depression in the landsurface. The effect of isolation and concentration in producing unusual brine chemistries is often enhanced by the presence of nearby mountains. Mountains act both to form a rain shadow zone and to create topographic heads, which drive deep circulation of meteoric groundwaters. In rift basins, the creation of nearby mountains may also be tied to an active tectonic regime and possible volcanism, which tends to create enhanced hydrothermal circulation.

Borates, for example, form by the evaporative concentration of spring inflow, which has passed through an active volcanogenic terrain. Likewise, the trona beds of Lake Magadi reflect groundwater and hydrothermal/basin brine leaching of their rift-volcanic surrounds. Interestingly, there are even igneous rocks composed of molten salts near where the Magadi trona beds are accumulating (see natrocarbonatites in Chapter 9). In other situations a degree of hyperaridity can be created in the rain shadow of a mountain range. This situation is most obvious in the creation of the salars of South America, which lie in intermontane depressions with floors that are thousands of metres above sea level. Rain shadow effects in combination with offshore upwelling explain the hyperarid deserts of Atacama. There is so little rainfall in this region that highly soluble salts, including economic levels of nitrates and iodates, accumulate in the regolith.

The high solubility of the nonpotash salt resources also explains the limited age range of this group of economic deposits. Almost all the exploited deposits have ages that are Miocene or younger (Table 7.13). Their lacustrine depositional setting and their high solubility mean that such salt deposits are rarely preserved in rocks that have pre-Tertiary ages. The Proterozoic borates of Liaoning Province are an obvious exception. The impervious nature of this succession, its metavolcaniclastic host, and the formation of a silicified envelope around the borates, have allowed these salts to be preserved. As a general rule in a frontier region, those arid basins with post Mid-Tertiary ages should be considered more prospective than any older basins.

Location	Age	Tectonic setting	Depositional Setting	Main minerals	Style and origin of mineralisation
BORATES					
Bigadic, Turkey	Miocene	Tibet-style grabens in collison arc	Volcanogenic playa/mudflats	Colemanite, ulexite	Precipitated as secondary nodules and fissure-fills within unconsolidated sediments, just below the sediment-water interface.
Boron, California	Miocene	Basin and Range	Volcanogenic saline lake/mudflats	Borax/kernite core facies with more distal ulexite, proberite and colemanite	Lacustrine precipitates with secondary overprints.
Salars, South America	Miocene	Intermontane grabens	Saline lakes and mudflats	Borax, ulexite, hydroboracite and inyoites	Lacustrine beds in salars fed by volcanogenic springs.
Meta-evaporites, Liaoning Province, China	Proterozoic	Amphibolite-grade metamorphics	Saline playa and mudflats	Boromagnesite	Metamorphosed lacustrine beds associated with banded iron formations and volcanics.
NITRATES					
Atacama Desert, South America	Neogene	Intermontane depressions	Hyperarid desert regolith	Pedogenic nitratite with associated iodates perchlorates, bromates and chlorides	Hyperarid conditions have prevailed since the mid-Miocene allowing buildup of highly soluble salts in the desert regolith.
SODIUM CARBONATES					
Lake Magadi, East African Rift, Tanzania	Holocene	Continental rift graben	Volcanogenic perennial saline lake	Cumulates and bottom-nucleated trona blades	Trona accretes each year as bottom growths atop 15m thick trona beds, which are sometimes deformed into pressure ridges (petees).
Wilkins Peak Member, Green River Basin, USA	Eocene	Foreland basin	Arid lacustrine interval in Tertiary basin fill	Trona beds and nodules with associated oil shales and dolomites	Subaqueous bottom growth on the floor of saline lakes and intrasediment nodules beneath the surface of desiccated mudflats.
SODIUM SULPHATE					
Continental playas in lowlands of Alberta and Saskatchewan	Holocene	Spring-fed hypersaline depressions on Northern Great Plains of Canada	Hypersaline, meromictic, perennial continental salt lakes	Coarsely crystalline, massive to thickly bedded mirabilite with lesser epsomite, bloedite, gypsum, magnesite and aragonite	Cooling of freestanding lake brine causes subaqueous bottom and nodular mirabilite to form.
Tertiary lacustrine deposits, Toledo Province, Spain	Miocene	Continental foreland basin to nearby Pyrenees and Iberian Ridge	Hypersaline, mudflats of continental salt lakes	Glauberite with lesser anhydrite, gypsum chalcedony, dolomite and magnesite	Beneath the surface of hypersaline lacustrine mudflats, glauberite, anhydrite and polyhalite are early secondary minerals, created by brine-mediated backreactions with gypsum.

Table 7.13. Summary of nonpotash salt resources discussed in this chapter.

Chapter 8
Evaporite-metal associations: lower temperature and diagenetic

Introduction

I hope to show in this chapter and the following chapter that most subsurface evaporites ultimately dissolve and, through their ongoing dissolution and alteration, create conditions suitable for metal enrichment and entrapment. Salt beds are merely the solid part of a large ionic recycling system, dissolved metals are another part. Halite-dominated sequences tend to supply chloride ions to the brine system, while dissolving gypsum or anhydrite beds can supply sulphur. When the chemistries of the dissolving salt beds and the metal carriers interact so that redox fronts are set up, an ore deposit can form. Thus, in base and precious metal exploration in evaporitic terranes, we are ultimately searching for those parts of the subsurface ionic cycling system where economic levels of metals have accumulated.

In most sedimentary/metasedimentary fluid flow systems the mineralisation event is part of an ongoing burial history preserved in the textures of the rock matrix. One should see the role of evaporites and metal sulphides as each contributing its part to this larger scale "mineral systems" approach. Ore deposits should be recognised as no more than part of the fluid evolution and burial story. This means one must integrate the ore paragenesis with the regional geology, sedimentology, diagenetic-metamorphic-igneous facies, fluid flow conduits and structural evolution so that a useful model can be garnered from a detailed analysis of the rock matrix. Unfortunately, much economic geology still concentrates on the detailed ore paragenesis of the sulphides and the resulting localised ore deposit models are typically not capable of prediction in a regional sedimentary framework.

For clarity, I have restricted my discussion in this chapter to those ore deposits where the evaporite presence is unequivocal, either through the remaining presence of evaporite salts, or the presence of more than one indicator of former evaporites. Hence, most of the examples in this chapter tend to occur in sedimentary (low temperature) host rocks. In the next chapter I will discuss more speculative aspects of the metal-evaporite association including examples where the host matrix tends to show metasedimentary and igneous affinities and the former presence of evaporites may be indicated by nothing more than an unusual, but widespread, mineral phase. Once an evaporite presence is established in or adjacent to an ore deposit it becomes a useful indicator to the next step, generating a workable exploration and targeting model.

Evaporites in the subsurface?

Before, or as, evaporites disappear, they can transport, seal, trap, focus and precipitate ore from metalliferous basinal fluids. Thus a subsurface evaporite unit acts as:

• Brine and ion source - As the salt dissolves it acts as a brine or volatile source feeding into the basin hydrology. It can supply economically important ions that, depending on the dissolving mineralogy, may be chloride-rich or a source of sulphur.
• Seal and trap - Thick evaporite units act as a seal to the free escape of basinal, hydrothermal, hybrid or upwelling meteoric waters. Given the right geometry, an evaporite unit can also act as a fluid trap that, until it is breached, will pond escaping metalliferous basinal waters and hydrocarbons.
• Focus - breach points or zones in the evaporite bed created by flowage (halokinesis), décollement, or faulting tend to focus the escape of pressured basinal waters into the zones above or sometimes even below the breach. Hot metalliferous basinal water escaping into nearsurface and surface settings then can create brine ponds and replacement haloes about, or near, the trap position or along the escape conduit.

Even after the salts have gone, the zone of dissolution residues can act as an aquifer or focus for ongoing metalliferous brine flow. Recognising the process and timing under which an evaporite disappeared is often the most important step in understanding the evaporite/base metal association. This is especially true of Proterozoic and Archaean base metal deposits where no actual salts remain, only indicators of former evaporites.

Ore fluid chemistry ("carriers")

Evaporites through their ongoing nearsurface and, more importantly, subsurface dissolution/alteration are important contributors to the major ion chemistry of a basinal brine. The chemical make-up of the brine depends on the mineralogy of the dissolving sequence. Dissolving thick beds of halite (the most common basinal evaporite) generate chloride-rich brines, while leached anhydrite/gypsum successions are likely to develop more sulphate- or sulphide-enriched waters (depending on the oxidation state at the site of dissolution). Associated brine reflux during dissolution can drive circulation cells as well as leach and transport metals within and through the adjacent sediments. Ongoing reaction of this brine with the nonevaporite matrix of other nearby beds can also modify overall brine chemistry (Chapter 2).

This can be seen in halogen water chemistry (Cl, Br, F, I) of Phanerozoic sedimentary basins (Worden, 1996). Key geological parameters that influence chlorine and bromine (and possibly fluorine) concentrations in the basinal waters are, in order of importance: 1) the presence of salt in the basin, 2) the age of the aquifer unit, and 3) the kerogen type within any significant organic-rich rocks in the basin. Worden concluded the presence of salt in a basin meant that any connate water contributing to the basin water chemistry would be hypersaline. More importantly, he also showed that most high chlorinity subsurface waters are not connate (i.e. not preserved seawater), but are generated though pore water-salt bed interaction during burial. For example, the chloride-dominated deep brines in the Canning Basin and beneath the modern Gulf of Mexico are the result of dissolution of thick buried halite beds (Ferguson et al., 1992, 1993; Land 1995a,b).

Worden also found that brines in Tertiary-age basins typically have much lower chlorine and bromine concentrations than pore waters in Mesozoic or Palaeozoic rocks. This age separation may simply reflect the different amount of time the basinal waters had to interact with any buried salt beds in the basin. Dominance of type II marine kerogen in a basin leads also to a higher bromine concentration in the basinal waters. According to Worden this may reflect the dominance of marine influences in a type II entraining basin, which in turn is more likely to lead to greater volumes of salt deposition compared with a basin dominated by terrestrial/meteoric influences. Iodine concentrations in subsurface waters are independent of all these parameters. Other geological parameters, such as depth of burial, temperature, basin-forming mechanism and reservoir lithology, exerted no recognisable influence upon halogen concentrations.

Basinal brines derived from dissolving evaporites can act as potential metal "carriers" or metal "fixers". In order to qualify as a potential ore carrying fluid, a basinal fluid must contain on the order of 1 mg/l or more of the dissolved metal in question (Figure 8.1; Eugster, 1989; Hanor, 1994). Most basinal brines that achieve Pb, Zn or Cu values at or above this level have pore salinities that are typically in excess of 200,000 mg/l, with their entrained base metals carried as chloride complexes. These waters also have high chloride concentrations, lowered pH and most occur in basins that contain, or once contained, thick dissolving salt beds.

The capacity of chloride-rich basinal brines to become ore forming fluids is well illustrated by the Na-Ca-Cl oil field brines in the Cheleken region of Turkmenistan (Lebedev, 1972). These brines contain up to 10 mg/l Zn+Pb and deposit native lead, sphalerite, galena and pyrite in the well pipes and holding tanks at rates analogous to ore formation. Some Na-Ca-Cl oil field brines in Jurassic and Cretaceous formations in the Salt Dome Basin of central Mississippi, USA, have total dissolved solid contents up to 350,000 mg/l, densities that exceed 1.2 g/ml, and contain up to several hundred mg/l of Pb + Zn (Figure 8.1; Carpenter et al., 1974; Kharaka et al., 1987). Both these highly metalliferous brines occur in regions underlain by thick dissolving evaporites.

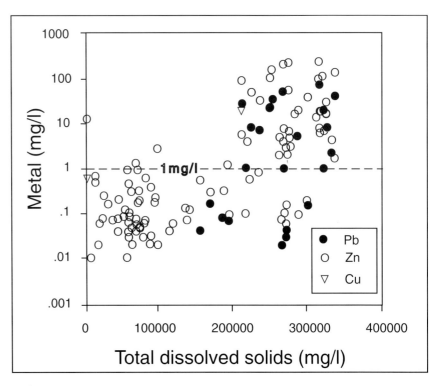

Figure 8.1. Metal content of subsurface waters worldwide (after Hanor, 1994).

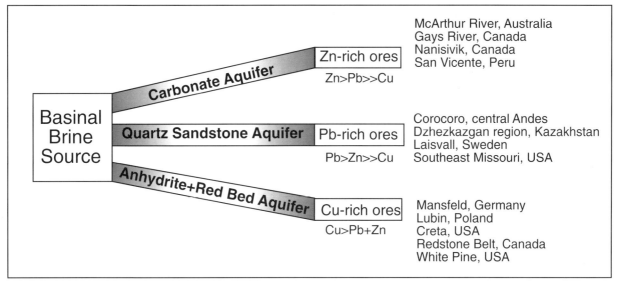

Figure 8.2. Influence of flushed matrix mineralogy on the base metal precipitated (after Sverjensky, 1989).

Sphalerite and galena scale is forming today in the pipes of producing geothermal wells in the Asal Rift in the Republic of Djibouti, Africa (Damore et al., 1998). Sulphide precipitation starts at the flash level, located at 850 m depth where temperatures are close to 260°C, and extends deeper into the main reservoir. The main geothermal reservoir is located at depths between 1000 and 1300 m with temperatures between 260 and 360°C. Seawater seepage, with input from dissolving salt beds, is the main brine source in the field. As a result of sulphide precipitation, about 90% of H_2S is removed from the original fluid before it is discharged at the surface. Brines produced in the Asal thermal field all have salinities in excess of 100,000 ppm, with a low gas content of the order of 0.6 mmol/mol.

Much of the metal content in these sedimentary basins has an intrabasinal nonmagmatic source. Some researchers propose that the metals in basinal brines are derived either from the alteration and leaching of feldspars and clays to supply most of the iron, lead, barium and strontium, or from the alteration of carbonates to supply most of the zinc (Land and Prezbindowski, 1981; Sverjensky, 1986; Kharaka et al., 1987). Others workers conclude that leaching of iron-oxide-hydroxides from a redbed parent can source many metals (Carpenter et al., 1974). Copper-rich mafic rocks, such as basalt, or copper-rich mafic minerals (such as hornblende, pyroxene, biotite, oxides and sulphides) in felsic rocks are other ideal copper sources when incorporated as detritus in first-cycle redbeds (Kirkham, 1989). Flushing of metals sequestered into organic-rich evaporitic mudstones are yet another possible source for base-metals. Dissolving evaporite beds themselves have

also been evoked as possible metal sources. Thiede and Cameron (1978) argued that Cu remains in solution in a brine during the evaporation process; whereas Pb and Zn enter solid precipitates, such as gypsum and anhydrite. Later dissolution of such evaporites generates a metal source enriched in Pb and Zn and depleted in Cu.

Sverjensky (1989) further refined earlier brine transport models by demonstrating that significant water-rock interaction takes place as warm, basinal chloride-rich brines traverse and interact with various aquifers. He concluded that, depending on the dominant lithology of the aquifer traversed by the brines, a single basinal brine may evolve chemically to become the ore-forming fluid for a wide range of sediment-hosted base-metal sulphide deposits. Migration through carbonate, quartz sandstone, or anhydrite + redbed aquifers could result in the formation of Zn-, Pb-, or Cu-rich deposits, respectively (Figure 8.2). As a starting point he considered a typical chloride-enriched basinal brine that was initially saturated with galena, chalcopyrite, muscovite, kaolinite and quartz, but undersaturated with respect to sphalerite, with up to 1 mg/l Pb, 0.1 mg/l Cu and 5 mg/l Zn. Passage of this fluid through a carbonate-cemented aquifer produces a metalliferous brine characterised by high Zn/Pb and (Zn+Pb)/Cu ratios. Transport of the same fluids through a quartz-cemented sandstone exhausts the buffering capacity of the aquifer relatively quickly so that any mineral accumulations precipitated from the resulting waters are galena-rich. Maintenance of the oxidation state in a typical basinal brine near the haematite-magnetite buffer prevents copper mobilisation (solubility of Cu in a typical basinal brine is ≈0.07mg/l). However, if the basinal brine migrates though a red bed succession that contains haematite and/or anhydrite, the oxidation state and the SO_4^{2-}/H_2S ratio of the fluid are increased by reactions such as:

$$CaSO_4 = Ca^{++} + SO_4^{2-}$$

$$4Fe_2O_3 + H_2S + 14\,H^+ = 8\,Fe^{2+} + SO_4^{2-} + 8H_2O$$

The migration of this more oxidising fluid through a red bed can then scavenge the necessary Cu, Zn and Pb to form a mineralising fluid with elevated Cu content (Cu > Pb + Zn). Redox conditions of $\log f_{O_2} > -46$ at 125°C are necessary in the brine to mobilise copper as cuprous chloride complexes. This explains the common association of copper sulphides with reduction haloes adjacent to redbeds (see later).

So far we have discussed slightly acidic metal-rich chloride brines. However, a high salinity in a brine, by itself, does not guarantee a high dissolved metal content, nor are metals always carried in the subsurface as chloride complexes in these acidic brines. A large number of basinal brines and formation waters are low in metals and high in dissolved sulphides. Figure 8.1 clearly shows that many highly saline basinal waters in sedimentary basins worldwide contain <1 ppm metal. Such waters become metal-rich and sulphide-poor only when they flush metalliferous sediment, such as redbeds, metal-rich shales or hydrothermally altered basalts (Ellis, 1968; Coveney, 1989). Metal leaching and chloride transport from such metal sources involves desorption of loosely bound metals, release of metals on mineral recrystallisation or replacement, or release of metals to the brine from metal-organic complexes through thermal alteration or thermal destruction of such complexes (Carpenter, et al., 1974; Gize, et al., 1991).

Acidification is an essential step in the acquisition of metals during dissolution of aquifer minerals. Chloride-rich, high salinity basinal fluids may become acidified by pyrite oxidation, by decomposition of organic matter, or by the precipitation of Mg silicates, such as smectite or chlorite. Organic acids may also play an important role in leaching the metals from metalliferous source beds (Raiswell and Al-Biatty, 1992). In contrast, Eugster (1985,1989) argued that not all metalliferous brines are acidic. In a Green River style of ore deposit (in which Eugster includes the Mt Isa and Dugald River

deposits of Australia) the brines were alkaline with pH values of up to 11. Metals were carried not as chloride-complexes, but as hydroxy- or carbonate-complexes. Such non-chloride waters are generated today in saline continental rift lakes where sodium bicarbonate evaporites dominate, there is no primary gypsum, and the lake evolves by the evaporation of source waters that drain adjacent basic igneous or metamorphic terranes (see Chapter 2 for a discussion of nonmarine brine evolution).

Sulphur redox reactions ("fixers")

Redox fronts are mostly subsurface interfaces where base metals can precipitate. Such fronts are typically created in the subsurface by the presence of H_2S and organic matter on the reducing anoxic side of the interface and metalliferous waters on the other side (Figure 8.3). Much of the H_2S in a sedimentary basin is generated by sulphate reduction, although it can also be generated from magmatic sources. Sulphate reduction requires the presence of both the sulphate ion and organic matter. Hence, the syndepositional association of evaporitic sediments and organic matter is fortuitous when considering H_2S generation during shallow early burial (Warren 1986, 1996). Connate seawater also contains abundant dissolved sulphate and its importance, along with hydrothermal anhydrite, in forming volcanogenic hosted massive sulphide (VHMS) deposits will be further discussed in the next chapter. For now, we shall consider the evaporite/sulphate-reduction association in the sedimentary/diagenetic realm. In this context it is important to note that most, but not all, redox interfaces are subsurface hydrological features. This chapter concentrates on mineralisation associated with such subsurface interfaces. The next chapter includes some less commonly cited examples of redox mineralisation, where the formation

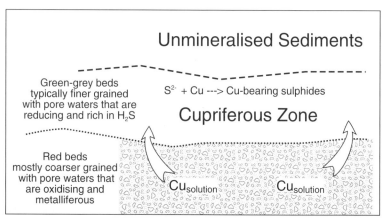

Figure 8.3. "Fixing" of copper sulphides at a redox interface (after Brown, 1993).

of a dense bottom brine on the seafloor moves the redox interface into ocean waters located tens of metres above the deep seafloor.

The reduction process can be bacterially mediated through the activities of sulphate-reducing bacteria (bacterial sulphate reduction-BSR) or inorganically mediated under conditions of increased temperature and pressure (thermochemical sulphate reduction -TSR). Both scenarios require the presence of preserved organic matter or its more evolved subsurface equivalent in the form of hydrocarbons (oil or gas).

Bacterial sulphate reduction (BSR)

During bacterial sulphate reduction, bacteria produce H_2S at temperatures less than 80-110°C (Figure 8.4; Trudinger et al., 1985; Hill, 1995; Riciputi et al., 1996; Aref, 1998b). For example, the strictly anaerobic, sulphur-reducing bacteria (*Desulfo*-x) produces isotopically light H_2S by metabolising organic matter (or hydrocarbons) with sulphate as the oxidising agent:

$$Ca^{2+} + 2SO_4^{2-} + 2CH_4 + 2H^+ \text{ --> } 2\ H_2S + CaCO_3 + 3H_2O + CO_2$$

In this equation, CH_4 represents a host of possible organic compounds or hydrocarbons and SO_4 is dissolved sulphate either from shallow phreatic refluxing brines or from dissolved sulphate evaporites such as gypsum, glauberite, or anhydrite. $CaCO_3$ represents bioepigenetic limestone products that often precipitate as replacements of precursor evaporite sulphates (e.g. calcitic caprocks of the Gulf Coast diapirs and the Castile mounds of west Texas). At pH<6-7 the H_2S is undissociated; at pH>6-7, H_2S dissociates so that $H_2S \longrightarrow H^+ + HS^-$ (Hill, 1995).

Bacterial sulphate reduction in and beneath dense anoxic brines and shallow subsurface brines explains the abundance of early pyrite in the sediments of many density-stratified brine lakes and seaways and its preponderance in many Archaean oceanic sediments. In the presence of soluble ferrous iron, the sulphide produced during sulphate reduction will immediately precipitate as metastable iron monosulphides, such as mackinawite, greigite and amorphous FeS. These intermediates are kinetically favoured over the direct precipitation of pyrite due to its much lower solubility product. Later they will transform to framboidal pyrite in the shallow subsurface. With time this framboidal form is in turn overgrown by euhedral pyrite. Evidence that early syngenetic biogenic pyrite is subsequently replaced via epigenesis can be seen in many bacterially influenced base metal laminite deposits, such as Bou Grine in north Africa, McArthur River in Australia and Lubin in Poland (Sawlowicz, 1992).

Sulphate-reducing bacteria are ecologically diverse and tend to develop wherever sulphate is present along with a supply of organic matter sufficient to create anaerobic conditions. Their growth is generally restricted to waters with pH values of 5 - 9. The creation of microniches means there are many additional situations where the bacterial reducers are flourishing locally while the overall environment exhibits pH levels that are unfavourable to bacterial growth.

The upper temperature limit for biogenic sulphate reduction is considered by many to be 80-85°C (Trudinger et al., 1985). But under suitably pressurized conditions it may extend into higher temperatures. Recent work in deepsea hydrothermal vent sediments of the Guaymas Basin in the Gulf of California has revealed archaebacterially mediated sulphate reduction at temperatures up to 110°C, with an optimal rate at 103-106°C (Huber et al. 1989; Jørgensen et al., 1992).

In the pore fluid of an actively dewatering sedimentary basin, the sulphate-reducing bacteria commonly thrive within, or near, any water-oil contact down to depths of a few kilometres. These are the aerobic/anaerobic (redox) transition zones between reduced paraffinic crude oils and oxygenated groundwaters. Locally anoxic conditions mean the aerobic bacteria provide nutrients to the sulphate-reducing anaerobes. For example, such anoxic interfaces tend to form near the dissolving margins of salt allochthons or diapir caps, where the outflow of reduced basinal waters is focused by the edge of the dissolving salt plumes (see Chapter 9). The metabolic activities of sulphate-reducing bacteria can also help precipitate metal sulphides at such interfaces (Bechtel et al., 1998).

Modern sulphate-reducing bacteria also flourish syndepositionally in anoxic water bodies within, or immediately beneath, lacustrine, estuarine, marine and hypersaline environments; there they also contribute to the formation of early pyrite framboids (Demaison and Moore, 1980). This geographic isolation of the reducing biota to restricted anoxic bottoms was not so in the Precambrian; under the reducing conditions that typified the Archaean oceans, the whole water column was populated by sulphate-reducing bacteria (Logan et al., 1995). Through their metabolism, they controlled the style and amount of organic matter preserved in bottom sediments. The widespread maintenance of an oxygen-stratified oceanic column in the

Proterozoic ocean, with a sulphate-reducing bacterial population in the bottom waters of the open ocean and at times on the shelf, played a major role in oceanic circulation patterns and the widespread preservation of pyritic laminites.

Some have even argued that the flowering of the calcareous macrobiota, and the marked increase in animal diversity that defines the end of the Proterozoic, was brought about by the lowering of the oceanic oxic/anoxic boundary into the sediment column. The lowering was driven by the evolution of organisms that produced faecal pellets. Such pellets rapidly remove organic matter to the ocean bottom, so moving the sulphate-reducing bacteria into these sediments, and allowing the encroachment of oxygen to take place throughout the whole water column (Logan et al., 1995).

Thermochemical sulphate reduction (TSR)

During thermochemical sulphate reduction, H_2S is produced as sulphate is inorganically reduced via reactions with hydrocarbons at temperatures in excess of 140°C (Figure 8.4; Heydari 1997). Until recently the efficiency of thermochemical sulphate reduction in producing H_2S was inferred from experimental evidence or levels of sour gas (H_2S) in oil and gas wells, rather than directly observed in any natural situations (Trudinger et al., 1985; Noth, 1997). This has now changed with the direct documentation of TSR-produced H_2S (sour gas) in reservoirs in the Permian Khuff Formation of Abu Dhabi (Worden et al., 1995, 1996; Worden and Smalley, 1996). In reservoirs hotter than 140°C, anhydrite has been partially replaced by calcite, and hydrocarbon gases have been partially or fully replaced by H_2S, that is, anhydrite and hydrocarbons have reacted to produce calcite and H_2S. Carbon and elemental sulphur isotope data from the gases and minerals show that the dominant general reaction is:

$$CaSO_4 + CH_4 \longrightarrow CaCO_3 + H_2S + H_2O$$

Gas chemistry and isotope data also show that C_2+ gases reacted preferentially with anhydrite by reactions of the type:

$$2\,CaSO_4 + C_2H_6 \longrightarrow 2CaCO_3 + H_2S + S + 2H_2O$$

Sulphur was generated by this reaction and is locally present, but was also consumed by the reaction:

$$4S + CH_4 + 2H_2O \longrightarrow CO_2 + 4H_2S$$

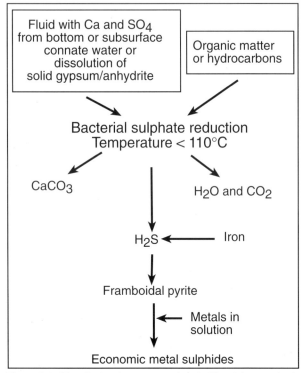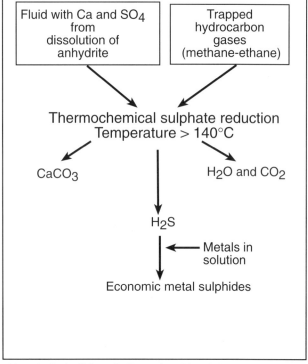

Figure 8.4. Chemical pathways associated with sulphate reduction. Note the lack of biogenically produced framboids during thermochemical sulphate reduction.

The frequently quoted and experimentally observed reaction between anhydrite and H_2S with CO_2 to produce calcite and sulphur:

$$CaSO_4 + 3H_2S + CO_2 \longrightarrow CaCO_3 + 4S + 3H_2O$$

has been shown, by gas chemistry, calcite $\delta^{13}C$ and sulphur $\delta^{34}S$ data, to be insignificant in the Khuff Formation. Rather, the direct reaction between hydrocarbons and anhydrite occurs in solution. It takes place within residual pore waters that are initially dominated by carbonate dissolved from the marine carbonate matrix. The first-formed replacive calcite thus contains carbon that was derived principally from a marine dolomite matrix with a $\delta^{13}C$ signature of 0 to +4 ‰. Continuing reaction leads to the progressive domination of the water by TSR-derived carbonate with a minimum $\delta^{13}C$ of about 31‰.

Thermochemical sulphate reduction also produces substantial volumes of very low salinity subsurface water. The salinity of formation water in evaporite lithologies undergoing TSR is, therefore, not necessarily high. In the Khuff reservoir the water salinity and isotope data show that the original formation water was diluted between four and five times by water from TSR (Worden et al., 1996). A typical Khuff gas reservoir rock volume suggests that initial formation water volumes can only be increased by about three times as a result of TSR. The extreme local dilution shown by the water salinity and $\delta^{18}O$ data in the Khuff must, therefore, reflect transiently imperfect mixing between TSR water and original formation water. Dissipation of this water into surrounding lithologies may aid further dissolution of adjacent evaporites.

Sulphate reduction and metallogeny

H_2S (sour gas) is a major product of both BSR and TSR (Figure 8.4). Wherever it comes into contact with a metalliferous brine this H_2S can then be involved in a number of metal sulphide forming reactions that often also involve alteration of the adjacent rock matrix. Sulphide products range from the cool shallow precipitation of syngenetic framboidal pyrite to the formation of hot, hydrothermal sulphides. For example, under a Mississippi Valley type scenario, a metal sulphide phase typically co-precipitates with hydrothermal or saddle dolomite at temperatures in the range of 60 - 180°C (Hill, 1995):

$$H_2S + CO_2 + MeCl^+ + Mg^{2+} + 2CaCO_3 + H_2O =$$

$$MeS + Ca^{2+} + CaMg(CO_3)_2 + HCO_3^- + Cl^- + 3H^+$$

When H_2S accumulates at a nearsurface redox interface, where oxygen is present and metalliferous brines are absent, it can form accumulations of native sulphur, rather than metal sulphides, via the reactions:

$$2H_2S + O_2 = 2S° + 2H_2O \ (pH < 6\text{-}7)$$

$$HS^- + O_2 + H^+ = S° + H_2O \ (pH > 6\text{-}7 \text{ to } 9)$$

Elemental sulphur forms and persists near the redox interface at Eh ≈ -0.2 to +0.2. This is thought to be the mechanism that generated the economic evaporite-associated sulphur deposits in hydrocarbon leakage haloes within the Gulf of Mexico diapir caprocks and the Castile mounds of west Texas (Figure 7.36a, b). The presence of abundant free sulphur (S°) in peridiapiric Pb-Zn deposits such as Bou Grine is used by some workers to argue that in such situations the supply of metals into a sulphate/sulphide reduced environment was the reaction-limiting step (Bechtel et al., 1996).

When considering the association of sulphate evaporites with mineralisation it is important to realise that it is not the sulphate evaporites that are directly responsible for the precipitation of metal sulphides. Rather, evaporite dissolution or reaction places sulphate/sulphide in a fluid medium so that it can react with hydrocarbons. This sulphate-entraining solution is reduced in the presence of organic matter or hydrocarbons to form H_2S. It is the H_2S in solution that then precipitates the metal sulphides at or near a redox interface, a zone where slightly more oxidised metalliferous chloride brines can interact with a H_2S buildup on the reduced side of the interface. Thus, the ore textures in areas where sulphate evaporites are dissolving and metal sulphides are precipitating are most likely to be open space or porosity-fill textures. Ore textures and the degree of layering will depend on a number of factors including: the size of the dissolution front (interpore versus laminar versus vug versus solution-collapse cavern), the rate of fluid crossflow at the interface, the degree of stability of the redox interface, and the distance the sulphate ion travels before it is reduced (e.g. Warren and Kempton, 1997).

The impervious nature of any subsurface evaporite unit, even as it dissolves, means that until it is breached it is an excellent seal to migrating hydrocarbons (both liquid and gaseous). Thus the organics required for the sulphate reduction process may not only be locally derived from organic-rich sediments, they may have seeped and flowed into the interval as a hydrocarbon charge and so be trapped directly beneath an evaporite seal. The requirement for organics in the sulphate reduction process and the ability of evaporite beds to act as a seal explain, at least in part, the ubiquitous association of sulphate evaporites, hydro-carbons/organic matter, and diagenetically created base metal deposits.

Ores formed by TSR and BSR can have near identical textures, both are typically pore or vug-filling cements or evaporite replacements (Machel et al., 1995). One way to help differentiate the origin of the precursor H_2S is via sulphur isotope analysis of ore sulphides that, if possible, is tied to $\delta^{13}C$ and $\delta^{18}O$ analysis of the various gangue carbonate cements and matrix that bracket the ore minerals. H_2S derived via bacterially mediated sulphate reduction tends to be isotopically light and so tends to exhibit more negative δ values (Figure 8.5). This is true of both the sulphur in the sulphides and of the carbon in any late stage dolomite or calcite cements. Unlike carbon and sulphur isotopes, oxygen is more thermally sensitive below 200°C and is better used as a geothermometer (Emery and Robinson, 1993).

H_2S from thermochemical reduction is not biologically fractionated, and so $\delta^{34}S$ values of the ore sulphide tend to reflect the isotopic signature of its evaporite precursor (Worden et al., 1997). Limited experimental work on deeply buried anhydrite undergoing thermochemical reduction suggests that the sulphur isotopic composition of the derived H_2S is isotopically similar to, or a few per mille lighter than, the precursor anhydrite (Krouse et al., 1988). Thus $\delta^{34}S$ and $\delta^{13}C$ values under TSR are much higher (more positive), while the $\delta^{18}O$ reflects elevated temperatures of late stage diagenetic spar formation (more negative).

Isotopic distinction is not as clear cut as it first appears; sulphur in the ore need not have come from the reduction of evaporitic sulphate, it may also come from organically derived sulphur in oil trapped in the host rock. Under this scenario the ore has low $\delta^{34}S$ values that are similar to any associated hydrocarbons (Kesler et al., 1994). The range of values for these hydrocarbon systems is similar to that from bacterially mediated sulphate reduction. This overlap underlines the need for matrix characterisation, which ties

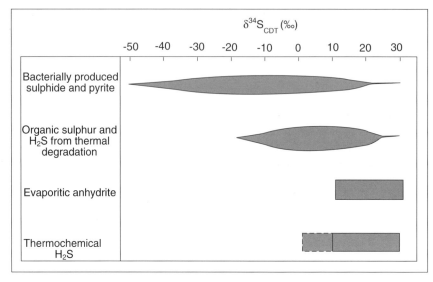

Figure 8.5. Typical ranges for naturally occurring sulphur isotopes under thermochemical and bacterial sulphate reduction (after Emery and Robinson, 1993).

stable isotope determinations to detailed petrographic and sedimentological logs of the ore host. For example, if the ore sulphur was locally sourced, then calcite may have completely replaced the precursor anhydrite. This calcite should be sampled for carbon and oxygen isotope analyses to see if its genesis parallels that of any ore sulphides. In addition, even though sulphate salts no longer remain, the "salt that was" textures should still be present.

So far our discussions of the metal-evaporite association have established:

• Evaporites as they are deposited can include substantial volumes of organic-rich hydrocarbon-prone laminites with reduced anoxic pore waters (Chapter 1).
• As they dissolve, thick halite beds can generate substantial volumes of chloride-rich basinal waters (Chapter 2). Depending on the mineralogy of the dominant basin aquifers these chloride brines can be enriched in Pb, Pb-Zn, or Cu (this chapter).
• Evaporite beds tend to focus the escape localities of metalliferous brines to particular areas in a sedimentary basin (e.g. above dissolutional or diapiric breaches that focus the escape of sub-salt basinal waters - Chapters 2 and 4).
• Subsurface sulphate salts can act as suppliers of sulphate ions to both BSR and TSR processes, and so generate subsurface redox fronts (this chapter).

Let us now take these ideas and apply them to various evaporite/base metal occurrences by looking in detail at a number of relevant examples, first in the low temperature diagenetic realm, and then in the igneous/hydrothermal and metamorphic realms of Chapter 9.

Diagenetic evaporite-metal association

Redox fronts about a dissolving subsurface salt bed play a fundamental role in localising and focusing base metal precipitation. The reducing conditions adjacent to and beneath a salt bed are created by the high salinity of the pore fluids and the likely occurrence of elevated levels of organics. Thus, the conditions suitable for the fixing of metal sulphides at redox interfaces can be present in an evaporite system from the time of deposition of the sediment host through to the metamorphic realm.

Bedded salt and diagenetic sediment-hosted copper

Many sediment-hosted stratiform copper deposits are closely associated with evaporites or indicators of former evaporites. Some ore deposits may still retain actual salts, especially gypsum or anhydrite, in close proximity to the ore. Such deposits include the Zambian and Redstone copperbelts, Creta, Boleo, Corocoro, Dzhezkazgan, Kupferschiefer (Lubin and Mansfeld regions) and Largientère. All these accumulations of metal are associated with the formation of a burial diagenetic redox front (Figure 8.3).

Redox fronts are commonplace haloes around dissolving subsurface salt beds that, when preserved, are indicated by a greenbed halo (reduced iron) adjacent to a dissolving salt bed or its dissolution breccia. When such a redox interface is set up between a reduced brine halo and slightly more oxygenated and acidic upwelling basinal waters it creates redox conditions suitable for base metal precipitation. For example, the underbelly and edges of a dissolving sequence of bedded evaporites tend to focus and deflect the escape of basinal or meteoric water along the underside of the evaporite. At the same time the ongoing dissolution maintains a stable and continually renewed source of reduced anoxic brine that survives until the evaporite bed is leached or breached. Dissolving sulphate evaporites, pyrite, or hydrocarbons near such redox fronts are possible sources of sulphur.

Stratiform copper/evaporite deposits span host lithologies that range from carbonates to siliciclastics. I am using the term "sediment-hosted stratiform copper" in the way defined by Kirkham (1989), that is, to encompass a group of Cu deposits that are not truly stratiform in that they may cut the stratigraphy either locally or regionally. Yet, locally, most of these deposits are not only stratabound (confined to a particular stratum), but are more or less concordant or peneconcordant with bedding. The problem with targeting this group of deposits, especially when they are associated with now dissolved evaporites, is that we are looking at metal precipitation mechanisms that are both diagenetic and dynamic, a reflection of a basin/brine hydrology that is no longer active. Accordingly, sulphide precipitates are concordant with permeability trends and chemical interfaces, but they are not always concordant with regional depositional bedding. Thus, to predict areas of likely stratiform sulphide occurrence within a basin, one must understand not just depositional geometries but also the subsequent diagenetic patterns. Diagenetic understanding comes from a detailed study of the history of the rock matrix and of basin architecture and not just a detailed sampling and study of the orebody.

Kupferschiefer/Rote Faule, Lubin region, Poland

Polish stratiform Cu-Ag deposits are hosted in Kupferschiefer shales or in adjacent aeolian sands (Weissliegende) that together make up the upper part of the Rotliegende sequence (Figure 8.6). The Lubin region has estimated ore reserves of 2,600 million tonnes at grades >2% Cu, 30-80g Ag/tonne, and 0.1g Au/tonne (Kirkham, 1989). Mineralisation typically occurs in reduced, evaporite-capped beds at the top of a thick continental volcanic and redbed sequence that was deposited in the Permian within a continental rift basin.

Stratiform Cu-Ag Kupferschiefer deposits were long considered classic examples of syngenetic mineralisation. However, detailed work in the last decade has shown that the Cu mineralisation cuts depositional boundaries. The richest intervals of mineralisation define low angle transgressive metal zones associated with the redox edge of the diagenetic Rote Faule units (Jowett et al., 1987; Jowett, 1987, 1989, 1992; Oszczepalski, 1989, 1994; Vaughan et al., 1989; Hammer et al., 1990; Kucha, 1990; Bechtel and Hoernes, 1993; Wodzicki and Piestrzynski, 1994; Bechtel et al., 1995; Heppenheimer et al., 1995; Speczik et al., 1995; Sun et al., 1995; Large et al., 1995; Large and Gize, 1996). The overlying evaporitic Zechstein sediments enhanced and focused the precipitation of C;, both by forming prepared ground at the time of deposition (pyritic and organic-rich evaporitic laminite host), and during diagenesis by creating a stable dissolution hydrology, which focused the greater volumes of economic mineralisation into a redox zone adjacent to the Rote Faule boundary.

Analysis of organic matter in the Kupferschiefer provides reliable evidence of: depositional conditions, salinity variations, the nature of organisms living in the water column, and the importance of bacterial activity in the sediment during deposition (Bechtel and Puettmann, 1997). The degree of methylation of 2-methyl-2-trimethyl-tridecylchromans (MTTC) indicates euryhaline to mesohaline (30-40‰) salinity during laminite

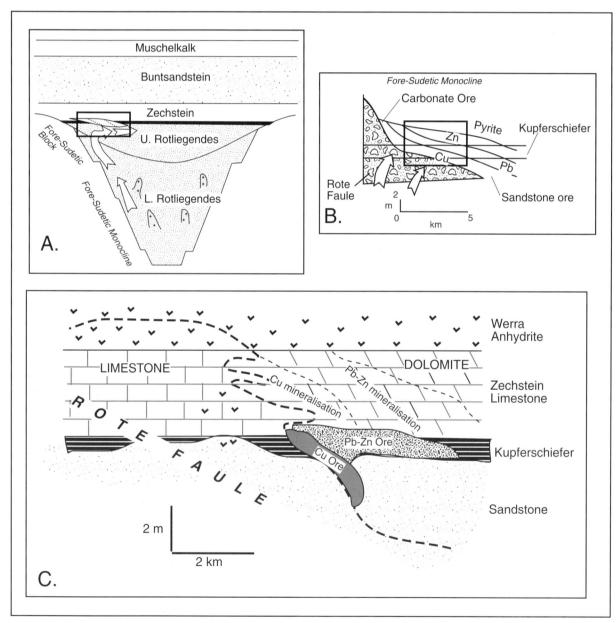

Figure 8.6. Rote Faule and its influence on mineralisation. A) Regional schematic showing focus of upwelling basinal waters toward basin edge. B) Enlargement of marginward area outlined in A, shows the mineralisation cross cuts the stratigraphy. C) Enlargement or area outlined in B showing relative position of respective ore zones. See Figure 8.7 for data detail (modified from Jowett, 1992; Vaughan et al., 1989).

sedimentation. An abundance of biomarkers, derived from green/purple sulphur bacteria, suggests H_2S saturation of the anoxic bottom waters and maximum water depths below 100 m. These organisms lived near the boundary between the photic zone and the anoxic (euxinic) bottom water at depths of 10 to 30 m. Primary production in the upper water column was dominated by photosynthetic cyanobacteria or green algae. In the sediment, sulphate reduction was driven by the availability of abundant sulphate and organic detritus settling from the overlying water column. Methanogenesis was active mostly during early Kupferschiefer deposition; it is indicated by the light carbon isotopic composition of the organic matter that originated from recycling of CO_2, which was generated by methane-oxidizing bacteria in the water column.

Thus the Kupferschiefer matrix was deposited mostly from suspension load in density-stratified subaqueous lacustrine/restricted marine-fed brine lakes and seaways with reduced anoxic bottom waters (Glennie, 1989a,b). Higher energy sediments were deposited in shallower shorezone areas about the basin edge or atop basement highs. The anoxic organic-rich laminates of the deeper water areas quickly became the hosts to syndepositional bacterially mediated pyrite framboids (Sawlowicz, 1992). Later this pyrite and the associated organics were replaced by base metal sulphides, which were further concentrated in regions adjacent to Rote Faule intervals.

In regions not overprinted by Rote Faule processes the high mineral content of the Kupferschiefer was attained via a three step process (Sun and Puettmann, 1997): 1) framboidal pyrite and pyrite precursors were precipitated by bacterial sulphate reduction during deposition of the sediment; 2) pyrite and pyrite precursors were largely replaced by mixed Cu/Fe minerals and by chalcocite during early diagenesis; and, 3) in those intervals with very high Cu contents (>8%), the amount of reduced sulphur from Fe-sulphide precursors was not sufficient for precipitation of Cu and other trace metals. In this part of the profile, thermochemical sulphate reduction (TSR) occurred after pyrite replacement, as is indicated by the presence of pyrobitumen and sparry calcite.

Rote Faule ("red rot") actually describes a non mineralised diagenetic zone and loosely encompasses barren red-coloured rocks typically found in the vicinity of high-grade ore. It describes an oxidised haematitic interval that occurs adjacent to reduced highly mineralised intervals. Levels of mineralisation in the Kupferschiefer adjacent to a Rote Faule interval are typically far higher than background levels in the Kupferschiefer shales not

influenced by Rote Faule development (Figure 8.6). Rote Faule is not formation specific and occurs in the Weissliegende, Kupferschiefer and Zechsteinkalk strata where various types of red colour have been created by the diagenetic formation of disseminated authigenic haematite and goethite. Pyrite, and the original organics of the Zechstein carbonate and the Kupferschiefer mudstones, are oxidised and largely removed in Rote Faule intervals (Oszczepalski, 1989).

The development of Rote Faule interval indicates the oxidised portion of a redox front. In Poland, it is located at the top of the thick redbed sequence, which during burial dewatering was flushed by upwelling, saline, slightly oxidised, basinal waters (Figure 8.6a). Ore sulphides occur on the far side of this subhorizontal oxidation-reduction front (Figure 8.6b; or, in the words of the poet Byron, "a gilded halo hovering around decay"). Thus Rote Faule zones outline redox interfaces centred on the outflow positions of hydrological conduits carrying ascending, slightly oxidising, basinal chloride brines, which also carried Cu, Pb and Zn. In the ore zones, the sulphides are arranged in three distinct haloes around "Rote Faule", in the order Cu, Pb and Zn (Figures 8.6b,c; 8.7). Thus, around and above the Rote Faule, the base metals are zoned laterally and vertically in successive mineralisation belts of chalcocite, bornite, chalcopyrite, galena and sphalerite.

Mineralisation in Kupferschiefer-style deposits transgresses depositional facies as it is tied to the Rote Faule/saline brine interface and not to the depositional pattern. Thus copper ore is not just hosted in the organic laminites of the Kupferschiefer and the Zechsteinkalk, it also occurs in parts of the Weissliegende aeolian sandstone, a unit that is exceptionally free of organics, laminites and pre-ore sulphides. To the dismay of many syngeneticists, this depositionally-clean aeolian sandstone hosts half the economic Cu-sulphides in the Lubin district (Figure 8.7). A likely ephemeral reductant in this sand was methane, another was authigenic anhydrite. Jowett (1992) argues that the widespread diagenetic anhydrite cements in the Weissliegende, along with the trapped methane, set up burial chemistries suitable for thermochemical sulphate reduction and the precipitation of Cu-sulphides in the Weissliegende host. The likely seal to this methane was the overlying Zechstein evaporites. It was also a likely source of the refluxing syndepositional brines, which first deposited the anhydrite cements (Chapter 4).

Throughout Europe, and in the offshore southern Zechstein Basin in North Sea, the Weissliegende is a significant gas reservoir. In Poland, the entrained gases are mainly methane and nitrogen and are often found in the same localities as significant sulphide concentrations (Peryt, 1989). These deposits cannot be mined unless vented by Alpine-age faults, as they have been in the Lubin district. As in the rest of Europe, this gas was sourced in the underlying Carboniferous coal beds and trapped beneath the Zechstein evaporite seal.

Geometries of mineralisation suggest that the underside of the Zechstein sediments acted as a longterm seal, which ponded upwelling metal-rich slightly oxidising basinal waters within the immediately underlying Kupferschiefer or its lateral equivalents in the Weissliegende. It also acted as a pressure seal that aided hydrofracturing within the maturing organic laminates of the Kupferschiefer, as evidenced by numerous sulphide-calcite veinlets in the Kupferschiefer. An inherent lack of permeability in widespread thick evaporites also meant the upper side of the Zechstein salt beds prevented downward percolation of fresh oxygenated waters, which would otherwise have destroyed the stability of the redox interface along the underside of the dissolving salt beds (the "block of ice" model for salt dissolution - see Chapters 2 and 4).

Metals that ultimately resided in the ore zone were leached from underlying Rotliegende volcanic detritus and carried as chloride complexes by escaping basinal brines. On the other side of the Rote Faule the dissolving Zechstein evaporites (including basinwide sulphates) atop the Kupferschiefer laminates acted both as a burial seal to escaping basinal brines (and gases) and as an ongoing supplier of anoxic hypersaline brines to the brine curtain. In unison these two sets of diagenetic hydrologic processes maintained a long term redox interface at the edge of the Rote Faule, which allowed substantial quantities of metal sulphides to accumulate in the various ore hosts.

Thus, before they were ultimately dissolved by flushing formation waters, the dissolving Zechstein evaporite beds maintained the Rote Faule interface between the more oxidising basinal redbed brines and the reducing dissolution brines that bathed Kupferschiefer laminites and Weissliegende sands. As the underbelly of Zechstein salts gradually dissolved, it created a brine halo that maintained a reservoir of anoxic reduced fluids in the laminates. It was this longterm maintenance of the interface between the anoxic hydrology of the Kupferschiefer laminates and the Rote Faule that created the worldclass ore deposits of the Kupferschiefer.

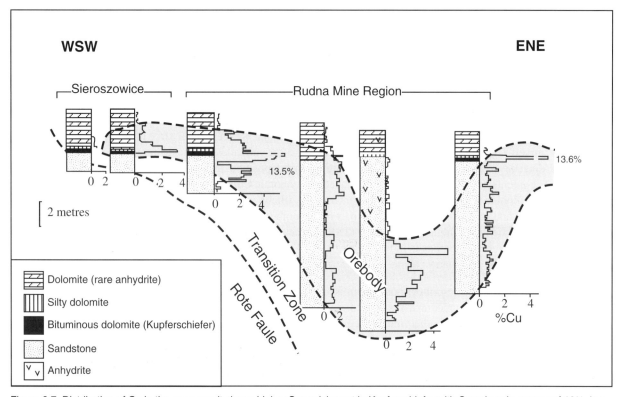

Figure 8.7. Distribution of Cu in the ore zone, it shows higher Cu enrichment in Kupferschiefer with Cu values in excess of 13%, but the greater thickness/volume of ore resides in the sandstone beneath an evaporite seal (after Wodzicki and Piestrzynski, 1994).

Redstone Copper Belt, Canada

Stratiform copper deposits of the Redstone Copper Belt, Northwest Territories, Canada, occur along a 250 km, northwest-trending, arcuate thrust belt of Neoproterozoic (Helikian) carbonates and redbeds of the Mackenzie Mountain Supergroup (Figure 8.8a; Chartrand et al., 1989; Ruelle, 1982; Brown, 1993). The copper deposits occur at the contact (Transition Zone) between the continental alluvial redbeds of the Redstone River Formation (0-1000 m thick) and the shallow marine/evaporitic rocks of the overlying Lower Coppercap Formation ($\approx$250 m thick). The Transition Zone host strata are composed of several blanket-like units of dolomitic silty, evaporitic cryptalgal-laminated carbonates that were deposited atop the fault-bound Coates Lake, Hayhook Lake and Keele River embayments. Beneath the Transition Zone sediments, in the Lower Redstone River Formation, Ruelle (1982) mentioned the presence of an unusual high Na-Mg Precambrian evaporite assemblage. It is made up of 55% glauberite [$Na_2Ca(SO_4)_2$], 20% magnesite, 15% talc and < 10% combined gypsum and anhydrite. At this time the significance of this sulphate-evaporite-rich assemblage in relation to the overlying sulphate evaporites and to any metal-transporting brines is neither well documented nor understood.

Sediments in the Transition Zone are made up of six shoaling peritidal cycles deposited along arid shorelines with sabkha affinities (Chartrand et al., 1989). Higher levels of mineralisation are typically found within the evaporitic carbonate portion of each depositional cycle. At the base of a typical cycle, red siltstone and mudstone grade up over a few decimetres into tan and green-coloured siltstone, exhibiting the same sedimentary textures and structures as their underlying red counterparts (Figure 8.8b). These terrigenous strata grade over a few centimetres into grey-green

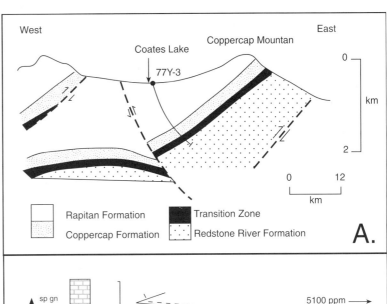

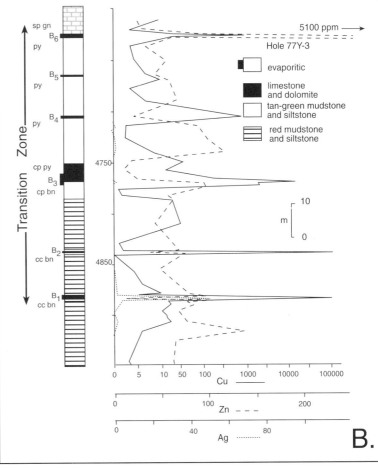

Figure 8.8. Redstone Copper Belt, Canada. A) E-W geological cross-section of Coates Lake region showing well 77Y-3. B) Stratigraphy and metal content of well 77Y-3 (after Chartrand et al., 1989). Mineral abbreviations as in Figure 8.9.

289

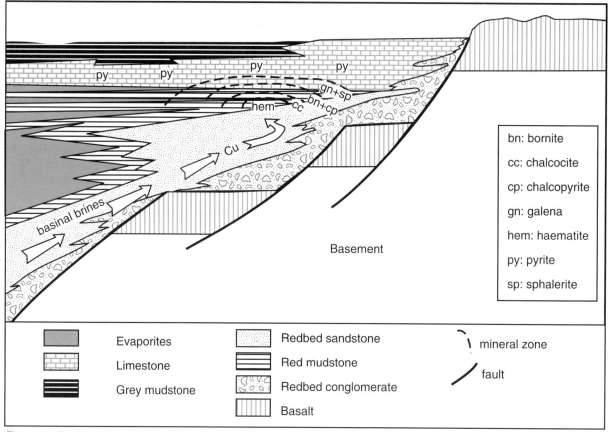

Figure 8.9. Diagrammatic section across a redbed-evaporite-rift margin showing mineral zonation of precipitated sulphides. Modelled from Redstone Copper Belt, Northwest Territories Canada (after Kirkham, 1989).

dolosiltite and dololutite overlain by microbial laminated carbonate, the principal host for the cupriferous sulphides. The dolosiltites and dololutites are capped by gypsum/anhydrite-rich units that are decimetres thick. Evaporitic units are overlain by tan-green siltstone and mudstone and, finally, by red siltstone. Beds B4-B7 are dominated by tan-green colours with most of the entrained mudstone and siltstone intervals containing more calcite and/or dolomite than their counterparts in beds B1-B3 (Figure 8.8b). The succession is crosscut by late-stage calcite and gypsum veins.

The largest and most economically promising Cu deposit in the Redstone Copper Belt is situated within the Coates Lake embayment and covers an area greater than 12 km². It occupies a bed 1 metre thick and contains drill-indicated reserves of 37 million tonnes, averaging 3.92% Cu and 0.33 oz. per tonne Ag (11.3 g/t). Mineralisation is best developed in the lower three carbonate units. For example, in drill hole 76Y-4, the richest Cu intersections are in bed 2a (0.15m @ 5.5% Cu) and bed 3 (1.16m @ 1.18% Cu),

both occurrences are tied to preserved beds of nodular anhydrite (Ruelle, 1982).

In well 77Y-3, the sulphide minerals are zoned vertically with bornite and chalcocite in the lower portion, passing up into a chalcopyrite-bornite interval, upwards into chalcopyrite-pyrite and finally into pyrite (Figure 8.8b). Sphalerite and galena occur at the uppermost part of the Transition Zone in contact with the overlying carbonates of the Coppercap Fm (Chartrand et al., 1989). The relatively high silver content coincides with high copper contents in the base of the Transition Zone and, according to Ruelle (1982), indicates the silver is entrained within bornite.

Like the Kupferschiefer, deposits in the Redstone Copper Belt show a pronounced mineral zonation, with the Cu:Fe ratio in the sulphides decreasing upsection through the Transition Zone and toward the embayment margins (Figure 8.8b, 8.9). The copper sulphides occur as partial to complete diagenetic replacements of anhydrite nodules in the cryptalgal-laminated units and as disseminations of anhedral to euhedral grains in the intertidal mud flat facies. Bladelike anhydrite crystals are also partially to completely replaced by chalcopyrite. Preserved anhydrite nodules are ellipsoidal in cross section and up to several centimetres in diameter, although most are less than 1 cm. Mineralisation of these nodule layers creates a laminar fenestral fabric to the mineralisation. Further away from the Cu mineralisation front, pyrite with chert replaces the anhydrite nodules (Ruelle, 1982).

Metals were leached and transported as metal-chloride complexes from volcaniclastic/redbeds that were derived from Cu-bearing mafic lavas in the Redstone River Formation (Figure 8.9). As these slightly acidic and weakly oxidised metalliferous basinal brines moved up and outward from the basin sediment pile, they were focused into the basin margins by the presence of overlying evaporite aquicludes. Near the margins, where the evaporite aquicludes were dissolved or had pinched out, the solutions seeped up section to encounter redox fronts associated with the organic-rich cryptalgalaminites of the Transition Zone. This interaction then fixed the metals on the reduced side of the redox front with a sulphide zonation identical to that found in the Kupferschiefer ore zones. In each of the cycles of the host stratigraphy the positions of oxidation-reducing fronts are still preserved as the red to grey-green colour transition. Haematite is still visible in the red-coloured intervals and the proportion of Cu in the grey-green intervals is much higher (Rose et al. 1986).

Creta, Oklahoma

The Creta deposit occurs in the upper part of the Flowerpot Shale in the Hollis-Hardeman Basin, Oklahoma. Similar cupriferous, but non-economic, sediments occur in several other locations within Upper Permian strata in Oklahoma, Kansas, Texas and New Mexico. Regionally some 900-3400 m of mixed clastic and chemical sediments of Ordovician to Permian age make up the Hollis-Hardeman basin fill, which, in turn, overlies a Precambrian to Cambrian basement made up of metasediments and metavolcanics. Locally, in the vicinity of the Creta deposit, sediment thickening is associated with movement on the Birch Fault, so that the sedimentary sequence is locally 3200 m thick and dominated by fine-grained clastics that show a gradual upward increase in the proportion of evaporites in to the Upper Permian section.

Mining of the Creta deposit between 1965 and 1975 removed 1.9 million tonnes of copper ore averaging 2% copper, making it the second largest shale-hosted ore deposit in the United States. Lead and zinc levels are negligible at Creta. According to Huyck and Chorey (1991), the mineralisation event at Creta was early diagenetic, as indicated by: 1) copper sulphide replacing large spores and framboidal pyrite, 2) lack of compaction of replaced spores relative to unreplaced spores, 3) enclosure of uncompacted mud and copper sulphides by early "matrix gypsum", 4) location of the ore bed within a thick sequence of fine-grained low permeability sediments.

Seven lithologic units occur in the upper 19 metres of the Flowerpot Shale and include the Prewitt Shale member, which hosts the Creta deposit (Figure 8.10). This interval is characterised by two common features: a) satinspar gypsum veins and b) variegated red-green layers. The veins both crosscut and parallel shale bedding and partings. They typically terminate within either nodular or massive gypsum beds or in gypsiferous green shale beds. In mineralised zones they commonly encapsulate shale intraclasts and copper minerals. The variegated layers and lenses are composed of discontinuous silty (60-80% silt) calcareous and gypsiferous red-brown layers. When the variegated layers are traced laterally they pass into adjacent gypsum, the layers obviously reflect dissolved gypsum or satinspar intervals.

The Prewitt Shale averages 2% Cu in the mine area, with a matrix composed of illite and chlorite and as much as 20% satinspar gypsum. It is made up of three layers: a lower blocky layer, a laminated layer and a discontinuous upper bed of impure gypsum (the "cap gypsum"). The lowest blocky layer is a claystone to mudstone; copper, when present, occurs only in the upper 10-13 cm and is concentrated in the mudstone (Figure 8.10a). The overlying laminated shale, which contains as much as 4.5% Cu, includes both gypsiferous and nongypsiferous clayshale. Copper is concentrated in the silty laminae both in outcrop (as malachite) and in thin section (as anilite). The cap gypsum bed directly overlies the laminated portion of the Prewitt Shale in the west, but pinches out in the east (Figure 8.10b). Where fully developed the cap gypsum bed (10-25 cm thick) is made up of a lower gypsum, a thin green clayshale, and an upper gypsum.

Copper mineralisation only occurs in association with concentrations of pyrite and spores. The main ore mineral is anilite ($Cu_{1.75}S$) and is concentrated along silty or sulphide rich layers in the laminated unit. In gypsiferous shale the opaques are concentrated along the edge of the satinspar veins and within clay blebs in the matrix gypsum. Anilite is also weakly developed throughout the nongypsiferous

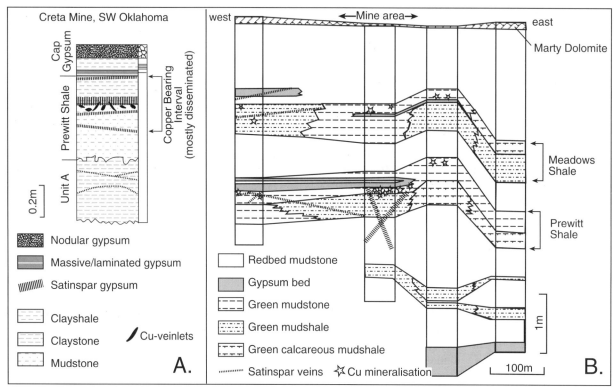

Figure 8.10. Creta copper shale deposit, Oklahoma (after Huyck and Chorey, 1991). A) Mineralised stratigraphy in the Creta Mine. B) Mine region stratigraphy showing intimate association between mineralisation and evaporite occurrence.

portions of the Prewitt Shale. The association of Cu mineralisation with evaporite occurrence in the Prewitt Shale is obvious, and reflects a late diagenetic mineralisation at a redox interface. In my opinion it is associated with an anoxic reduction front (red-green transition) that was generated and maintained by the dissolving adjacent evaporites.

Most published work on this deposit documents satinspar gypsum as well as the cap gypsum, but none (including the most comprehensive work by Huyck and Chorey, 1991) discuss its significance. The presence of satinspar gypsum indicates a combination of halite solution-collapse and gypsum rehydration. As such, the current forms of $CaSO_4$ in the mine area are tertiary evaporites, a result of uplift and rehydration. They were not responsible for the precipitation of the sulphides as they postdate that event. They do, however, indicate the former presence of a more substantial, but now dissolved, salt unit composed either of anhydrite, or perhaps a combination of halite and anhydrite. It was this former evaporite bed that was responsible for brine focusing and, through its dissolution, maintained a redox front that precipitated the copper sulphides.

Corocoro and other sandstone-hosted deposits of the Central Andes

Stratabound deposits of copper (+Ag) hosted by continental clastic sedimentary rocks occur in Central Andean intermontane basins and are known to postdate compressive deformation/uplift events in the region (Flint, 1989). The deposits are relatively small and include Negra Huanusha, central Peru (Permo-Triassic); Caleta Coloso, northern Chile (Lower Cretaceous); Corocoro, northwestern Bolivia (Miocene); San Bartolo, northern Chile (Oligo-Miocene); and Yasyamayo, northwestern Argentina (Miocene-Pliocene). The location of mineralisation is controlled by sedimentary facies and host-rock diagenesis. Deposits are irregular, elongate lenses of native metal, sulphides, and their oxidation products, and are restricted to alluvial fan and playa sandstones or conglomerate facies. The Corocoro area has produced the largest amount of copper, something like 6.4 million tonnes of copper at a grade of 5% (Kirkham, 1989).

Critical factors in ore genesis include (Flint, 1989): 1) the stratigraphic association of evaporites, organic-rich lacustrine mudstones, clastic reservoir rocks, and orogenic, igneous provenance areas for both basinfill sediments and metals; and 2) the intrabasinal evolution of metal-mobilising saline brines from the buried and dissolving lacustrine evaporites and mudstone-derived reducing fluids. The same diagenetic fluids also caused the dissolution of early, framework-supporting cement in the host clastic sedimentary rocks (Figure 8.11).

The Corocoro deposits in Bolivia are the largest of the deposits listed above and have been mined sporadically since they were first exploited by the local Indians prior to the Spanish invasion in the 16th century. Orebodies are elongate lensoid bodies confined to sandstones and conglomerates that invariably show bleaching or greying of the original redbed host. Ore minerals (including native copper) are secondary cement fills within secondary intergranular pores created by the dissolution of earlier carbonate and sulphate cements. Twelve grey sandstone beds, which were host to the long worked-out native copper ores, occur within a stratigraphic thickness of 60 m. Ores are stratabound, but not necessarily stratiform. The ore-hosting clastics lie within the Vetas Member of the Ramos Formation and were deposited as redbeds in braidplains or fluviodeltaic playas about the margins of

saline evaporitic lakes (Flint, 1989). Abundant evaporites are still present in the Ramos Member at Corocoro.

The playa depositional environment, in combination with the burial evolution of the adjacent evaporites, controlled the formation, transport and precipitation of the copper ore. Playa sandstones, sealed between impervious evaporitic mudstones, created the plumbing for focused metalliferous fluid migration toward the basin margin. It appears that the carbonaceous material at Corocoro was concentrated in the sandstones and conglomerates and not in the shalier members of the sedimentary sequence (Eugster, 1989). The organics entrained as plant matter in the sandstones, along with those migrating as hydrocarbons out of the basin, created locally reducing pore environments. This in combination with sulphur supplied as H_2S from the adjacent dissolving calcium sulphate beds, as well as from dissolving intergranular sulphate cements, precipitated copper in the newly created secondary porosity. The pore water chemistry and flow hydrology of this sandstone-hosted Cu system shows many affinities with diagenetic uranium-redox precipitating systems.

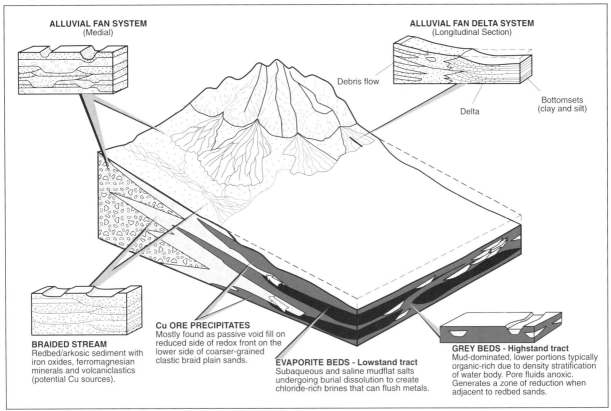

ALLUVIAL FAN SYSTEM
(Medial)

ALLUVIAL FAN DELTA SYSTEM
(Longitudinal Section)

Debris flow

Delta

Bottomsets
(clay and silt)

Cu ORE PRECIPITATES
Mostly found as passive void fill on reduced side of redox front on the lower side of coarser-grained clastic braid plain sands.

BRAIDED STREAM
Redbed/arkosic sediment with iron oxides, ferromagnesian minerals and volcaniclastics (potential Cu sources).

EVAPORITE BEDS - Lowstand tract
Subaqueous and saline mudflat salts undergoing burial dissolution to create chloride-rich brines that can flush metals.

GREY BEDS - Highstand tract
Mud-dominated, lower portions typically organic-rich due to density stratification of water body. Pore fluids anoxic. Generates a zone of reduction when adjacent to redbed sands.

Figure 8.11. Model for precipitation of Corocoro ores. Published with permission of JK Resources Pty Ltd.

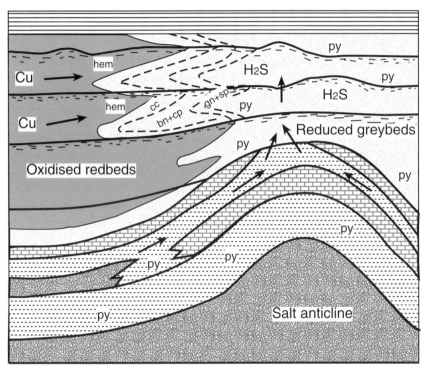

Figure 8.12. Schematic cross section of the Dzhezkazgan region showing Cu zonation at the redbed/greybed interface (See Figure 8.9 for key to mineral abbreviations; after Kirkham, 1989).

Dzhezkazgan region, Kazakhstan

Another sandstone-hosted Cu deposit is the large Dzhezkazgan copper deposit located at the northern edge of the Chu-Saysu basin, Kazakhstan. It is hosted in Middle to Late Carboniferous continental successions of conglomerate, sandstone, shale and evaporites (Gablina, 1981; Susura et al., 1986). Reserves estimated for this deposit, but based on questionable data, are 400 million tonnes with a copper grade of 1.5% (Kirkham, 1989). The depositional setting of the host was an intermontane closed-basin, which was made up of deltaic-lagoonal, alluvial and saline lake sediments. Hosts to the ore are siliciclastic redbeds that are laterally equivalent to bedded evaporites and underlain by Early Carboniferous organic-rich sediments (presumed source rock) and sealed by shaly carbonates. The main copper sulphide and pyrite orebodies occur in an anticlinal trap located along the flanks of a basement high, and sit atop a salt ridge (Figure 8.12). The host redbed sandstones become more evaporitic basinward and coarser-grained toward a pinchout along the basement highs. The mineralised sandstone is pyritic, with reduced pore fluids created by entrained hydrocarbons and high salinities. In contrast, the interbedded shales are red due to more oxidising pore fluids. These redbeds were in part derived from continental volcanics.

Gablina (1981) suggested that an early welt developed in the marine rocks and lowermost redbeds that overlie the salt anticline. Bed geometries illustrated in Figure 8.12 suggest this welt was induced by halotectonics. The breach allowed hydrocarbon-bearing basinal waters to migrate vertically into the overlying redbeds, where H₂S was produced by sulphate-reducing bacteria. Extensive precipitation of copper and other metals occurred at the regional redox front between reducing evaporite-fed brines and adjacent red beds. In this case the role of the dissolving catabaric evaporites was to supply reduced brines and to create a focus atop a halokinetically controlled salt anticline. The precipitation mechanism and the metal zonation is the same as in the Kupferschiefer model, but in this case the relative position of the evaporite and the redbeds are reversed. The main evaporite bed lies below the redbeds, which are the ultimate source of the metals. Although the mineralised beds are stratiform, this deposit is really a result of halokinesis; a similar, but less economically significant, analogue is to be found in the Lisbon Valley, USA, (discussed shortly).

Bedded salt and stratiform sediment-hosted Pb-Zn deposits

There are many sedimentological studies of Zn-Pb deposits that argue for, or allude to, the presence of evaporites as the probable source of sulphur in the metal sulphides (Fontboté and Gorzawski, 1990; Ghazban et al., 1990; Warren and Kempton, 1997). In such orebodies all that remains of the former evaporites are dissolution breccias or other indicators of vanished evaporites ("the salt that was"). Zn-Pb deposits associated with bedded evaporites generally occur in dolomitized shelf carbonates or sandstones that are adjacent to, or within, feeder faults. Such systems are sometimes located at or near the high energy depositional shelf edge, as defined by the rock matrix character of the host, or are located adjacent to the dissolution edge of the focusing evaporite. In many deposits, such as Gays River or San Vicente, the two features may coincide.

Many of the carbonate-hosted Pb-Zn deposits within this group have been be classified as Mississippi Valley Type deposits (MVT). Fluid inclusion studies indicate that most MVT sulphides are precipitated from highly saline basinal fluids at temperatures within the normal range for basinal brines (80 -150°C; Anderson and MacQueen, 1990). Not all MVT deposits, nor all stratiform Pb-Zn deposits, are associated with former evaporites.

Personally, I do not find much utility in the generalised MVT grouping when generating a viable exploration and targeting model for Pb-Zn in a metalliferous platform sequence. All it does is tie a wide range of Zn-Pb stratiform deposits to a general association of Pb-Zn precipitation from flowing basinal brines in sedimentary successions. Some MVT models have the fluids escaping from the basin during basin subsidence, others call upon deep meteoric circulation driven by basin collision and uplift of the basin margin, still others draw upon convection of basinal brines beneath evaporite seaways. The term MVT encompasses a generalised fluid flow concept not tied to any understanding of rock matrix character. Without a good understanding of the regional mechanisms that drive fluid flow and fluid focusing, the concept of a single MVT overgroup may be misleading. That is, in terms of predicting Pb-Zn ore bodies in carbonate platforms, our current understanding of what are termed MVT deposits is about as useful as an ashtray on a motorbike.

Facies analysis tied to an understanding of the history of a fluid flow pathway, as preserved in the rock matrix, allows us to do a much better job than this. My approach in this section is not to attempt to define an all encompassing Pb-Zn/MVT model. Rather, as in the preceding section on Cu deposits, I restrict my discussion to those Pb-Zn deposits that have an intimate association with an evaporite bed or a dissolution breccia. These matrix features can then be used to recognise and predict diagenetic zones of likely Pb-Zn accumulation at the various times when metalliferous basinal brines were moving through the stratigraphy. Buried salt creates a dynamic subsurface hydrology where sites of possible metal accumulation will change as foci of brine flow change with the structural and halokinetic evolution of the basin. Ore deposits can then be classified as pre-major salt flow and dissolution (i.e. bedded salt), syn-salt flow and post-salt flow or after complete salt dissolution. This approach generates specific paradigms, based on the recognition of "the evaporite that was", that can be used in combination with aeromagnetics and other geophysical techniques to generate targets in a regional framework.

Gays River, Nova Scotia

The Gays River Zn-Pb deposit of southern Nova Scotia, Canada, represents a well studied example of evaporite-associated Mississippi Valley-type mineralisation. It has reserves ca. 2.4 million tonnes at 8.6% Zn and 6.3% Pb of either massive or disseminated ore. The deposit is located in the Meguma Terrane of southern Nova Scotia and is hosted by Windsor Group (Visean) carbonate/evaporites that were deposited in several sub-basins of the larger Maritimes Basin of eastern Canada (Kontak and Jackson, 1995; Sangster et al., 1998).

In terms of ore focusing, it was the adjacent dissolving and sealing evaporite units, along with associated entrapped hydrocarbons, and not the adjacent permeable dolomitized carbonate buildup, that controlled the locality and the process of mineralisation. The massive ore adjacent to the sulphate evaporite bed is made up of fine-grained, beige coloured sphalerite and medium to coarse-grained galena. The massive ore is only found along the carbonate-evaporite contact (Figure 8.13). Disseminated ore adjacent to the massive ore and its evaporite seal is hosted by Visean-age, dolomitized carbonate rocks (bank and interbank facies) that are part of a series of carbonate banks (i.e. the Gays River Formation). The banks grew on fault-controlled palaeotopographic highs that were underlain by Lower Palaeozoic metaturbidites of the Meguma Group, sometimes with local intervening basal breccia units (Horton Group). The latter contain fragments (centimetre to metre scale) of Meguma Group lithologies within a dolostone matrix. The disseminated, lower-grade mineralisation infills the porosity in the dolomitised carbonate banks and today consists of yellow to orange, millimetre-scale sphalerite and euhedral galena on the walls of larger pores and cavities. Trace amounts of Fe and Cu sulphides as well as fluorite and baryte are also present.

The main ore-hosting authigenic dolomite replaced its limestone precursor prior to the mineralisation event (Savard, 1996). Dolomitization played a role in ground preparation, but was not directly tied to the Zn- and Pb-rich basinal fluids that mineralised the Gays River deposit. Fluid inclusion work on the coarse grained sphalerites, and rare earth element profiles in late stage carbonate cements associated with the mineralisation, indicate that the mineralising fluids were hot (<250°C), saline brines (>35 wt% NaCl) (Akande and Zentilli, 1984; Chi and Savard, 1995; Kontak and Jackson, 1995). [40]Ar/[39]Ar dating places

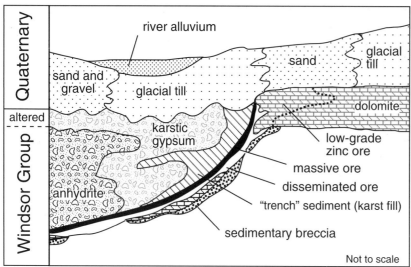

Figure 8.13. Gays River, Nova Scotia. Note the intimate association of the dissolved evaporite underbelly and edge with massive ore (after Kontak et al., 1994).

the mineralisation event at 300 Ma, while other isotopic data (Sr, C, O, Pb) indicate that the mineralising fluids were derived by fluid flushing of the underlying metaturbidites of the Meguma Group or of the derived detritus that resides in the Devonian-Carboniferous Horton Group (Kontak, 1998; Kontak et al., 1994).

Heroux et al. (1994, 1996) found that the mineralising brines also altered the clay mineralogy and organic maturity profiles of the adjacent carbonate/claystone matrix. The Gays River deposit is associated with smectite replacement of well-crystallised detrital illite and chlorite. This forms a halo around the main ore zone, while authigenic kaolinite and chlorite-corrensite typify the ore zone. Vitrinite reflectance (R_o) values and their absolute variability increase in the host rocks toward the ore zone (0.8 to 1.88%) and show a relatively low R_o within the ore zone. That is, the R_o exhibits an anomalous suppressive effect with respect to depth of burial at the mine site, suggesting a chemical alteration of the organic matter in the mineralised zone. Alternatively, as in other oil generating basins in the world, vitrinite suppression can indicate overpressuring or the occurrence of H-enriched liptinites. The highest R_o values are associated with mosaic (recrystallised) textures in the host carbonates and outline a thermal anomaly in the region, probably related to the focused escape of basinal brines. Their work suggests that rock matrix relationships between clay mineralogies, organic anomalies and the mineralisation at the mine can be used as sedimentological guides in the exploration for base-metal sulphides in this area and elsewhere.

Rock matrix relationships, and the position of the massive and disseminated ore zones (Figure 8.13), clearly show the evaporites of the region controlled the precipitation of the metal sulphides. The presence of massive ore at the carbonate-evaporite(anhydrite) contact shows the focusing effect of the impervious evaporites, and that the basal anhydrite along the underside of the thick evaporite package was probably the major source of sulphur. It is highly likely that the evaporites that once covered the Gays River carbonate buildup formed a regional seal at the time of ore emplacement, which allowed mineralising solutions to pond at the contact with the permeable dolomite host rocks beneath it. Then, the evaporite seal was breached. This allowed the local area to act as an ongoing focus for escaping metalliferous basinal fluids. The effect of this breaching clearly has a surface expression in the present landsurface and the lithofacies thicknesses and distribution above the deposit.

It was the evaporite-carbonate interface that focused the mineralisation, not the dolomitising of the matrix. The dolomitised carbonate's main role was that of passive host to the mixing of mineralising solutions, entrained hydrocarbons, and the thermochemically reduced sulphur derived from the dissolving evaporites. However, the prior burial-dolomitisation event, probably driven by less evolved escaping basinal brines, may well have enhanced the aquifer properties of the ore host in the vicinity of the evaporite aquiclude. Unfortunately, the mine is currently flooded and all the published sedimentological work to date on rock matrix in this region deals with everything except the evaporites and their evolution. To generate improved predictive exploration models, the evaporites at the mineralised contact need to be characterised in a basinal diagenetic context in as much, if not more, detail than the carbonates that host the lower ore grades.

Jubilee Zn-Pb deposit, Nova Scotia

The Jubilee Zn-Pb deposit, Nova Scotia, is located 2 km southeast of Little Narrows village, near the quarry of the Little Narrows Gypsum Co. (Fallara et al., 1998). Like Gays River, it is one of many lead and zinc prospects in

Early Carboniferous limestones of the Maritime region, eastern Canada (Armstrong et al., 1993). At Jubilee, a laminated limestone is overlain by transitional, interstratified limestone and anhydrite, with a massive anhydrite unit capping the sequence. The deposit extends over 2.75 km^2, and contains 0.9 million tonnes of 5.2% Zn and 1.4% Pb, and is hosted in the laminated limestones and breccias of Macumber Formation (basal Windsor Group). The Macumber Formation is the lateral equivalent to the Gays River Formation that hosts the Gays River deposit (see earlier). The Macumber Formation in the Jubilee deposit is made up of three main lithofacies: 1) a micritic lithofacies that locally contains ooids, oncolites and pellets, 2) a finely laminated lithofacies, and 3) a mineralised breccia facies (Fallara et al., 1998). The facies transition between the first two lithofacies is gradational and concordant, whereas the contact between the finely laminated lithofacies and the mineralised breccia is sharp. The breccia is conformable at the top of the Macumber Formation and never crosscuts it (Figure 8.14). The deposit is located near two faults interpreted by Hein et al. (1993) as normal synsedimentary faults. Armstrong et al. (1993) suggested the mineralisation was a low temperature sulphide system related to the former presence of hydrocarbons.

Overlying the breccia at the base of the Carrols Corner Formation is a transitional zone of interbedded limestone and evaporite. In the basal part of the transition zone some of the limestone is replaced by anhydrite. The overlying Carrols Corner Formation is a thick sequence of anhydrite and gypsum up to 300 m thick (shown in grey in Figure 8.14). The breccia that hosts the ore has been interpreted as: 1) an evaporite solution collapse breccia, 2) a tectonic breccia that was the result of normal faulting and magmatic intrusions that created a horst structure, and 3) a collapse breccia produced by the subsurface dissolution of evaporite during the circulation of the mineralising fluid. Recent work on breccias in the lower part of the Windsor Group by Lavoie et al., (1998) has shown that all three origins are correct. What previously was loosely called the Pembroke Breccia, and considered a single unit, does in part host the ore. The problem is that the Pembroke Breccia is a term that is used to describe all breccias in the region, when in reality it does not describe a single lithofacies or a formation specific unit. Rather, the term Pembroke Breccia has been applied to three temporally and genetically distinct breccia bodies. There is a pre-ore synsedimentary carbonate slope breccia, a syn-ore tectonic breccia related to the Ainslie Detachment, and a modern karstic post-ore evaporite solution collapse breccia related to the modern groundwater system. The formation of the syn-ore breccia was aided by pressure buildup and hydrofracturing beneath the regional evaporite cap; processes which also facilitated the formation of the regional subsalt detachment (Ainslie Detachment).

The main ore host is this syntectonic hydraulic breccia; it is a mineralised-matrix breccia is that contains pyrite, marcasite, sphalerite, galena and baryte, in addition to calcite. Its clasts have been extensively recrystallised and commonly retain finely disseminated sulphides. Anhydrite is absent, as is true sedimentary interclast matrix. Instead, the recrystallised breccia fragments are cemented by equant sparry, or locally radiaxial fibrous calcite, together with sulphides. Liquid hydrocarbons and solid bitumen are abundant, particularly within sulphide-rich domains, and also occur as inclusions within sphalerite and baryte. Stable isotope data clearly show the metals in the mineralised matrix breccia were derived from escaping basinal brines (Armstrong et al., 1993).

In contrast, the evaporite solution breccia that forms via surface-related dissolution of the evaporites is a classic rock matrix breccia; it contains clast-supported, chaotically distributed, centimetre-sized, subrounded to angular fragments of the Macumber Formation. Most of the clasts retain their original laminated texture and have not been extensively recrystallised. According to Hein et al. (1991), the clasts in the breccia fell into open spaces created by dissolving evaporite interbeds and interlaminae. The interclast matrix is made up of finely comminuted rock fragments of the overlying Macumber Formation. This rock matrix breccia can be present in cored intervals that also intersect mineralised intervals, hence the earlier confusion in separating the two styles of breccia.

The NNW-trending faults that cut the Jubilee deposit have been shown by Hein et al. (1991) to be synsedimentary and continued to be active solution conduits after they were overthrust by a thick evaporite seal. In the early stages, prior to the emplacement of the evaporite, the faults supplied nutrient-rich waters that vented on to the sea floor. This allowed enhanced carbonate sedimentation on the deep sea floor in the vicinity of the vents, where chemosynthetic communities of fossil tubes, serpulid and spirorbid worms, conularids, brachiopods, bryozoa and ostrocodes flourished (von Bitter et al., 1990). Later, during burial diagenesis, and after emplacement of the evaporite cap, the faults acted as conduits that carried metalliferous basinal fluids up into the base of the anhydrites where the sulphides then precipitated. This explains why the mineralised breccia unit, even though it is syn-ore

NW   SE

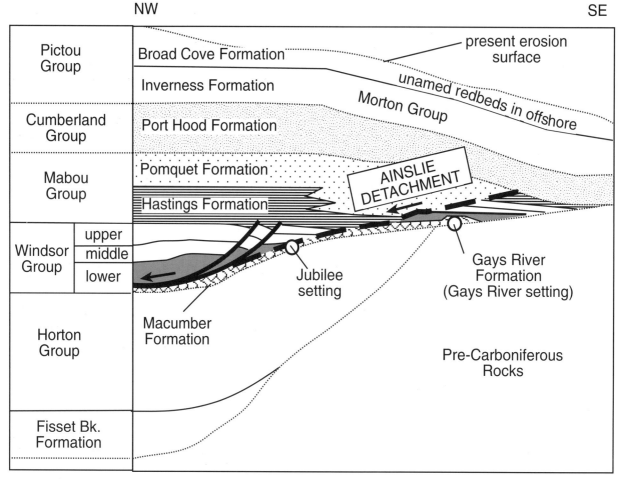

Figure 8.14. Schematic cross section showing the relationship between regional stratigraphy, mineralisation and the position of the Ainslie Detachment (after Lynch et al., 1998). Note the timing of the Ainslie Detachment is indicated by a gap in the Windsor Group stratigraphy, created by a former evaporite-lubricated gravity slide into the basin. This withdrawal sink was filled by sediment of the Mabou Group. Jubilee mineralisation occurs in the porous carbonates that make up the footwall of the detachment, a syntectonic mineralisation event. The facilitating evaporites of the Windsor Group (Carrols Corner Formation) are shown in grey. In contrast Gays River mineralisation occurs as massive ore adjacent to the evaporite seal of the lower Windsor Group and adjacent porous carbonates. In both cases the evaporite acted as a focus and seal to the hydrocarbons that facilitated thermochemical sulphate reduction, and at the same time supplied sulphate to the reaction. In the case of Jubilee, fluid escape was focused by the feather edge of the salt unit, in the Gays River it was trapped and focused by the overlying evaporite acting as a stratigraphic trap. Both styles of mineralisation are the result of an escaping metalliferous basinal fluid interacting with the lower dissolving portions of an evaporite bed.

tectonic breccia, is thickest and most heavily mineralised adjacent to the Jubilee Fault.

Gays River and Jubilee deposits are part of a larger regional scale Zn-Pb-baryte-evaporite association related to evaporite focused outflow interfaces during dewatering of the various sub-basins of the Maritimes Basin of eastern Canada (Lynch et al., 1998). Pb-Zn and baryte deposits generally occur at the Touraisian-Visean (Mississippian) clastic-carbonate-evaporite disconformity throughout central Nova Scotia, and overlie Cambrian to

Devonian metasedimentary and granitic rocks. In addition to Gays River and Jubilee, significant deposits include: the Southvale baryte occurrence, the Pembroke Pb deposit, the Smithfield Pb-Zn-Cu-Ba deposit, the middle Stewiacke baryte occurrence, the Brookfield baryte deposit, the Hilden baryte occurrence, the Walton-Magnet Cove Ba-Pb-Zn-Cu-Ag deposit and the Lake Fletcher baryte deposit. All these deposits occur on the southern margin of the >10 km thick Carboniferous Maritimes (Fundy-Magdalen) Basin. Regional isotope and inclusion studies, summarised by

Ravenhurst et al. (1987), show the deposits all formed from basinal brines, which were derived from >5 km depths in catabaric sediments that lay beneath the Visean evaporites.

The regional detachment surface, known as the Ainslie Detachment, was probably created by lateral "into-the-basin" gravity sliding or displacement along the sole of the regional salt thick, which characterises the lower Windsor Group (350 m+ evaporites of the Carrols Corner Formation). Its underside focused the flow of escaping metalliferous brines, which migrated laterally through the Tournisian clastic rocks and the immediately overlying tectonic breccia. Brines then escaped and mixed with shallower basinal fluids at breach points in the evaporite aquiclude about the basin margin (Figure 8.14; Lynch et al., 1998; Lynch and Keller, 1998). Thermochemical sulphate reduction was the main metal fixing reaction of the various deposits, with the sulphur/sulphate supplied by the adjacent evaporites and the organics from the migrating hydrocarbons.

The salt tectonic style that characterises the main offshore basin is similar to the raft tectonic styles that typify other salt-floored passive margins such as the Gulf of Mexico and offshore Brazil (see Chapter 5). The notion of raft tectonics creating breach points in a salt seal, and so allowing the escape/reduction of metalliferous chloride brines is discussed further in Chapter 9.

Baffin Island, Canada

The Zn-Pb-Ag ore at Nanisivik, near the north end of Baffin Island, consists of massive sulphides hosted in dolomite of the Middle-Upper Proterozoic Society Cliffs Formation (Figure 8.15a; Clayton and Thorpe, 1982). Textural and mineralogical variations divide the Main orebody into three vertical and four horizontal ore zones (Arne et al., 1991). The upper lens of the Main orebody can be further subdivided into six texturally distinct, laterally extensive mine units. Pyrite is the main sulphide mineral in the region, but in places sphalerite and galena are present in ore-grade proportions. Most of the known sulphides are in horizontal lenses about 10 m thick. The Main Ore Zone is 3,000 m long and, on average, about 100 m wide. This body is estimated to contain 11 million tonnes of sulphides, of which 6.5 million tonnes are ore-grade, averaging 12% Zn, 1.5% Pb and 50 g Ag/tonne (McNaughton and Smith 1986). There are also sulphides below the main lenses, either in horizontal lenses or in near-vertical veinlike structures. The host dolomite has been extensively block-faulted. Sulphide bodies pass laterally through some of these faults and appear to be confined to the upper half of the formation, which is mostly constructed of laminated algal dolomite that has been extensively brecciated and recrystallised. Crosscutting relationships shows the deposits are clearly diagenetic, not syngenetic.

The orebodies are characterised by well-banded ore textures interpreted by Arne et al. (1991) as progressive replacement of the carbonate wall rock. In contrast, Ghazban et al. (1990, 1992) note the obvious knife-sharp contacts between the massive pyrite and the adjacent host dolostone, and the presence of unaltered relict "fins" of the host dolostone. They use this relationship to argue for a more passive fill of karstic caves, with ore precipitating as the matrix was dissolving. There is a third possibility, namely the precipitation of ore in cavities adjacent to or within dissolving sulphate bodies, in a manner akin to that that characterises the Cadjebut ores where similar "fins" are present (Warren and Kempton, 1997). A study of the solution breccias in the Nanisivik deposit and listing of the presence or absence of evaporite indicators would help in resolving which of the three possible mechanisms is the most viable.

In this context it is interesting to note that the sphalerite - galena body is dominantly stratabound and lies within the dolostones of the Society Cliffs Formation, which dips 10°-20° to the north (Figure 8.15 b, c). Gypsum outcrops in the Society Cliffs Formation near Arctic Bay (25 km west of Nanisivik) and in the southeastern Borden Basin (Ghazban et al., 1990, 1992). There are also red beds associated with gypsum-bearing sabkha sequences within the lower Society Cliffs Formation on Bylot Island, some 150 km east of Nanisivik. One thin bed of gypsum (5-10 cm thick) was encountered at the same stratigraphic level as the ore zone in two boreholes immediately west of the main ore zone. In addition, gypsum (satin spar) commonly fills fractures in the dolostones surrounding the ore zones. Much of the Society Cliffs Formation is extensively brecciated, particularly in the western Borden Peninsula, with interfragment spaces filled with sparry dolomite and less frequently with sulphide minerals. The breccia zones appear to be stratigraphically controlled, they bear no close or obvious relationships to faults and folds within the formation (Olson, 1984). The association of secondary $CaSO_4$ and stratiform brecciation implies karstic brecciation associated with, and perhaps controlled by, evaporite solution collapse (Jackson and Iannelli, 1981).

Heavy $\delta^{34}S$ values of sulphides, ranging from 21.5 to 31.2‰, suggest complete reduction of marine evaporitic

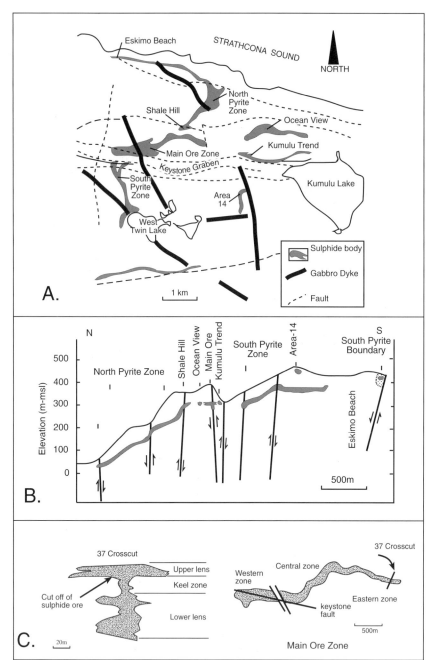

Figure 8.15. Nanisivik region, Baffin Island. A) Map of the major sulphide bodies and structural features in the Nanisivik region. B) Schematic N-S cross section showing major faults and sulphide bodies. C) Cross section of Main Ore Zone showing upper and lower lenses connected by keel zone at crosscut 37. Plan view of upper lens, Main Ore Zone (after Ghazban et al., 1990; Arne et al., 1991).

with total *in situ* thermochemical sulphate reduction (Ghazban et al., 1990). The likely reductant was organic matter, traces of which remain as pyrobitumen, although methane and volatile hydrocarbons may also have been involved. Striking evidence for the role of organic derivatives is found in $\delta^{13}C$ values of the white sparry ore dolomite, which varies between +6 and -12 ‰.

Two distinct stages of dolomitization can be dis-tinguished by their petrographic, cathodoluminescence, and isotopic characteristics: 1) massive dolomitization of precursor carbonates, and 2) late-stage cementation (Ghazban et al., 1992). The massive dolo-mites of the sabkha facies of the lower member of the Society Cliffs Formation and laminated algal stromatolitic to massive dolostones of the upper member are isotopically similar. Isotope signatures imply the later dolomite cement was precipitated from warmer pore fluids along with partial replacement of earlier dolomite. Dolomitising fluids responsible for later fracture and vug filling cements may have been released by compaction of the underlying Arctic Bay Shale, which also appear to have been the source of the base metals.

Whatever the fine-scale detail, overall ore formation at Nanisivik reflects processes involving *in situ* reduction of sulphate by hydrocarbons and minor H_2S at the site of ore deposition. The location and configuration of the orebodies was probably controlled by the former position of near horizontal fluid redox interfaces. The question to be resolved by future matrix characterisation is: was that interface there because of a now-dissolved evaporite seal?

sulphate, with no isotopic difference between sulphate and sulphides. Isotopic temperatures from intersulphide fractionations range from 90 to 270°C. Plots of fractionation versus $\delta^{34}S$ show that the sulphur source was heavy throughout sulphide precipitation, and was also consistent

San Vicente, Peru

The San Vicente Zn-Pb ore deposit is situated 300 km east of Lima, in central Peru, within the Upper Triassic-Lower Jurassic carbonate platform (Pucara Group) at the western margin of the Brazilian Shield (Fontboté and Gorzawski, 1990). Production during the past 20 years, plus present reserves, exceeds 12 million tonnes of ore assaying about 12% Zn and 1%Pb. Sphalerite and galena, the only ore minerals, occur as lens-shaped bodies generally parallel to the bedding. The three ore-bearing dolomites lie within the 1,400 m thick Pucará sequence (Figure 8.16). The San Vicente dolomite is the main ore-bearing unit and hosts the exploited part of the San Vicente mine. Other ore occurrences are located in the San Judas Dolomite and the Alfonso Dolomite. Individual ore lenses are in dolomitised tidal flat and lagoon facies characterised by cryptalgal laminae and evaporite moulds after former sulphate evaporites (including gypsum).

Ore lenses tend to occur adjacent to oolitic grainstones of former barrier-shoal facies. All three ore-hosting dolomite units were deposited in a peritidal carbonate platform that comprised: 1) the inner margin of the lagoon (tidal flat), 2) the lagoon *sensu stricto* and 3) the outer margin of the lagoon (barrier subenvironment).

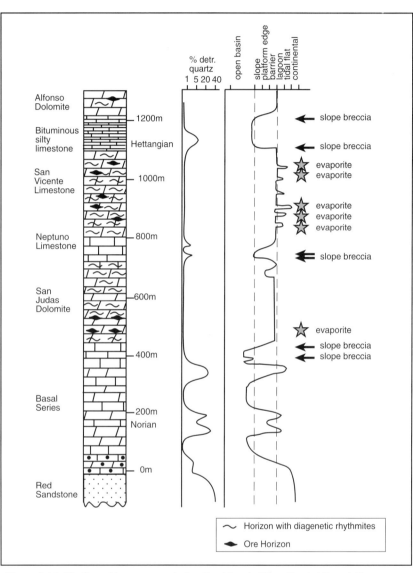

Figure 8.16. Stratigraphic sequence in the San Vicente Mine, Central Peru (after Fontboté and Gorzawski, 1990).

Strontium, carbon, oxygen, and sulphur isotope geochemistry was carried out on consecutive crystallisation generations in the San Vicente region (Spangenberg et al., 1996; Moritz et al., 1996). The results obtained display systematic trends that all show San Vicente lead-zinc deposit formed during burial diagenesis via the mixing of incoming hot saline, slightly acidic, radiogenic (Pb, Sr) basinal fluids and local sulphate-rich formation waters. The temperatures indicated by sulphur isotope geothermometry (75-92°C) are consistent with temperatures reached at a burial depth of ≈ 2 to 3 km. Such burial depths were probably reached by the end of the Jurassic. The ores were emplaced by the thermochemical reduction of evaporite-derived sulphates at or near the ore site and the introduction of a zinc- and lead-bearing basinal brine (Fontboté and Gorzawski, 1990).

Cadjebut region, Lennard Shelf, Australia

Bedded calcium sulphate evaporites, and their lateral dissolutional equivalents, have been cored in the Cadjebut mine area on the Lennard Shelf. Evaporite solution breccias (rock matrix breccias) are the lateral equivalents of the mineralised sequence in the Cadjebut mine (Figure 4.17). The deposit undoubtedly represents the *in situ* replacement of a number of evaporite beds by Pb-Zn sulphides on a one for one basis. The sulphate to sulphide replacement process occurred across cm- to metre- sized cavities, and so the banded and breccia replacement textures are layered, sometimes forming "zebra" textures. This cavity infill replacement process means the ore textures are not necessarily a direct mimicry of the precursor laminar to nodular evaporite textures (Tompkins et al., 1994 versus Warren and Kempton, 1997). One of the problems with interpreting sulphate textures in the Cadjebut region is that much of the remaining shallow anhydrite is now converting to porphyroblastic gypsum as it is flushed by modern meteoric waters. Some earlier workers failed to realise that the current gypsum textures in relict anhydrite beds are rehydration features (tertiary evaporites). Such textures should not be compared to possible primary textures, as the respective precipitative processes are totally unrelated. The timing and mode of formation of ore textures and breccias in the Cadjebut mine were described and discussed in Chapter 4, and so will not be repeated here. For a more complete discussion of Cadjebut and its surrounds the interested reader is referred to Arne (1996); Tompkins et al. (1994); Warren and Kempton (1997) and references therein.

Largentière, Ardèche, France

The Pb-Zn-Ag mineralisation at Largentière, France, is partly vein-type and partly stratabound (e.g. bed 5 ore; Bouladon, 1984). Mineralisation occurs mostly within Lower Triassic sandstones and arkoses (30-50 m thick), less frequently in about 20 m of overlying Middle Triassic dolomite, and rarely in Upper Triassic argillaceous sandstones. The mineralised intervals are broken down into sequentially numbered beds. All formations contain evaporites or their indicators The mineralised formations constitute the base of a Mesozoic cover that is transgressive onto the eastern border of the Cévennes basement. At Largentière the basement is made up of granite and, locally, of Carboniferous terranes overlain by Permian redbeds of pelite or sandstone.

Bed 5 ore formed as a cement within coarse-grained sandstones at the base of the Triassic section and is located less than 15 metres above the Permian contact. There is not always a single mineralised horizon, but stratiform character is the rule. Although local ore enrichments are related to shear zones, the general connection between

mineralisation and faults in bed 5 ore is not clear. About 10 metres above bed 5, the redbed sandstones change to dolomite-cemented greenbeds and green argillaceous, gypsum-bearing lagoonal siltstones appear (the reduced facies). The Pb-Zn mineralisation in beds 1, 2 and 3 (termed the "upper mineralised horizon") is consistently tied to the presence of gypsum within the mineralised intervals (Bouladon, 1984). In this upper interval, mineralisation is associated with faults, and "fissure ore" dominates in dolomitic sandstones and marls. In the Volpilliare region gypsum has been reprecipitated, along with galena in fibrous veins crosscutting the marls. Once again, the significance of satinspar texture in terms of former more widespread evaporite units is not mentioned in the existing literature.

Diapiric salt, diagenesis and mineralisation

For many years diapirs or salt domes have been viewed as potential targets for petroleum, salt and sulphur. Only in the last decade have they also come to be viewed as potential targets for Cu, Pb and Zn deposits (Wang et al., 1998). Basinal fluid circulation about peridiapiric systems in the Mesozoic sediments of Northern Africa has created metasomatic haloes of siderite and other ferroan carbonates that may act as indicator haloes for this style of deposit (see Chapter 9). Salt dome-associated mineralised systems can be divided into three types: 1) salt-dome (diapir) hosted deposits, where the host rock consists of anhydrite and bacteriogenic limestone; 2) peridiapiric deposits, where the bulk of the mineralisation resides in a laminar organic-rich sediment host adjacent to the diapir; and 3) fault-associated mineralisation that lies above the salt/halokinetic breccia, where the same faults that were created during synkinematic salt flow also channelled the mineralisation. All styles of mineralisation can occur in a single salt dome province, and the relationships can be preserved well into the metamorphic realm (Wang et al., 1998).

Caprock-hosted Pb-Zn deposits, Gulf Coast and north Africa

Small deposits of Pb-Zn with a diapir association were recognised in northern Africa more than 100 years ago, but it was not until the geology of diapir-hosted subeconomic deposits in the Gulf of Mexico (Kyle, 1991) and economic north African diapir- and peridiapir-hosted deposits, such as Bou Grine (Rouvier et al., 1985; Orgeval, 1994), were fully understood, that the significance of diapirism in controlling this style of Pb-Zn was fully appreciated. Mineralogically, the diapir-associated mineral deposits are similar to many other carbonate and shale-hosted mineral deposits, as they contain pyrite, marcasite, sphalerite, galena and baryte; many deposits also contain high concentrations of pyrrhotite and celestite.

Bacteriogenic limestones created during caprock formation are an important ore host in this style of deposit. However, diapir caprocks are no longer considered to reflect an exclusively meteoric setting, and so mineralisation models are still evolving for this style of deposit. The popular consensus of caprock genesis is shifting more to a subseafloor mode for most caprocks. All that is required to form the anhydrite portion of a caprock is that the top of the diapiric salt stem or sheet be bathed in a solution that is undersaturated with respect to halite (see Chapter 5). To form the calcite portion from this anhydrite/gypsum precursor requires the presence of organics/hydrocarbons and of sulphate-reducing bacteria in the region of the calcium sulphate cap. Such a situation exists subseafloor, as well as in zones of continental meteoric circulation. Active marine calcite caprock is forming today at deep seafloor sites atop shallow salt diapirs and allochthonous salt sheets in the present day Gulf of Mexico (Kennicutt et al., 1985; Roberts et al., 1990). These sites are coincident with petroleum seeps, so that the calcite cap, including that in entrained mollusc shells, has isotopic signatures indicating a crude petroleum source. Mineralising brines in such areas can be "exhaled" on to the seafloor to form brine pools in depressions adjacent to the same salt seeps. Given that methane and brine are seeping today into offshore sediments in active salt structure provinces, and that sulphide smokers accompany some of these seeps

(Martens et al., 1991), it is likely that diapir-hosted sulphides are also forming today beneath the brine pools adjacent to these seeps (Chapter 9).

Pb-Zn sulphides begin forming as soon as an anhydrite cap rock carapace encrusts a developing salt structure. This can happen during the earliest phases of salt flow and continue, perhaps episodically, as long as the structure's crest is bathed in nearsurface halite-undersaturated waters and oil- or methane-bearing fluids continue to circulate in the vicinity of the diapir. Some of the earliest minerals to form in the caprock are sulphides precipitated in equilibrium

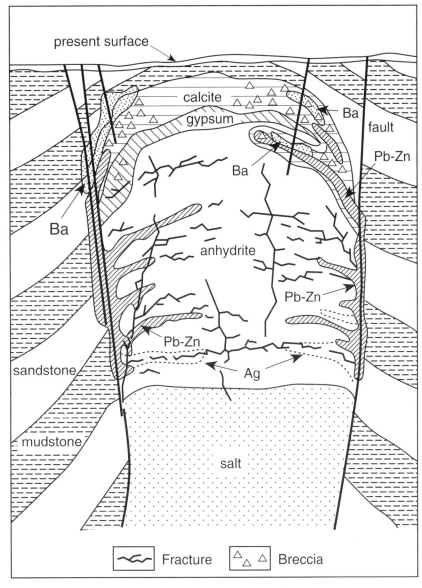

Figure 8.17. Patterns of mineralisation in relation to cap rock facies (after Price et al., 1983; Warren, 1997).

303

with anhydrite, where they often form laminar sulphide-anhydrite textures controlled by the morphology of the dissolution interface (Kyle and Posey, 1991). As crude oil and thermogenic methane migrates up a diapir's margin to come into contact with anhydrite/gypsum within a low temperature environment ($\approx 80°C$), biogenic calcite forms as a byproduct of bacterial sulphate reduction (Figure 8.17). Sulphides, baryte and celestite can form at the same time or postdate the calcite. Gypsum cap rock continues to form after anhydrite, and calcite caprock interfaces are well established, typically via the ongoing infiltration of undersaturated nearsurface waters that ultimately will rehydrate any remaining anhydrite (Werner et al., 1988).

Thus, at any particular site in a caprock deposit, iron sulphides typically form early; galena and sphalerites form later; and baryte, celestite and sulphur are generally the last phases to precipitate. Banded or colloform textures consisting of finely crystalline aggregates are the most common textures exhibited by the iron sulphides and sphalerite. These base metal sulphides occur as: inter-anhydrite crystal cements, open-space fills in anhydrite, as carbonate/sulphate replacements, and as open-space (vug/cavity) fills. Most of the anhydrite-hosted mineralisation formed during anhydrite caprock accumulation, whereas most of the sulphides hosted in biogenic calcite formed prior to or during calcite precipitation (Figure 8.17; Kyle and Price, 1986).

In general, an anhydrite caprock accretes or thickens by growth along its underbelly by a process of underplating (Figure 8.17). As diapiric salt dissolves from the uppermost portions of the underlying halite stem/sheet it leaves behind the entrained anhydrite as a residue. Calcitic caprock also forms first at its bottom, but it does so the progressive bacteriogenic replacement of the underlying calcium sulphate (Light and Posey, 1992). Metal sulphides that precipitate syngenetically within either the anhydrite or the calcite cap rocks show isotopic compositions and major element compositions that vary with depth. This range in isotopic values may record chemical changes in upward migrating metal-bearing brines or changes in local chemical conditions relating to bacteriogenic overprints during sulphide precipitation. The range in metal concentrations most likely reflects changes in brine composition due either to brine mixing or to evolution of brine composition.

A detailed record of episodic basin dewatering events is preserved in mineralised anhydrite-rich cap rocks of two well-documented Gulf Coast salt domes: the Hockley Dome and the Winnfield Dome (Hallager et al., 1990).

There the majority of the highgrade mineralisation occurs as massive laminar sulphide lenses. The lenses are typically vuggy, fractured and brecciated. At Winnfield, the layers consist mostly of iron sulphides, with vugs and fractures lined with sphalerite, galena, baryte and calcite. The sulphide precipitates had their origins in deep metal-rich basinal brines that were intermittently expelled from geopressured zones deeper in the stratigraphic section. The brines were channelled upward along escape structures bounding the diapir stems. A nearsurface overhanging caprock helped focus escaping fluid into the zone of salt dissolution between the top of salt and the overlying residual anhydrite cap rock. Iron, lead, and zinc sulphides saturations were exceeded in this zone, possibly in response to H_2S generation related to periodic dissolution and biogenic reduction of the caprock sulphates.

Episodes when metalliferous brine entered accreting diapir caprocks are recorded as relatively thin bands of sulphide sandwiched between thicker bands of anhydrite. Continued dissolution of salt and underplating of residual anhydrite caused the sulphide bands to be displaced upward relative to the base of the cap, leading to an inverted stratigraphic record of basin dewatering events. Palaeomagnetic data from the Winnfield salt dome suggest that sulphide-producing basin dewatering events and anhydrite caprock accumulation occurred between 157 and 145 Ma, and that widespread sulphide precipitation in the cap is no longer active.

Peridiapirically hosted stratiform Pb-Zn

As well as forming in the diapiric caprock, Pb-Zn deposits can be hosted in organic-rich pyritic and sideritic laminites located centripetal to salt sheets or halokinetic breccias, so making them stratiform peridiapiric ore deposits The highly reduced, organic-rich, nature of laminated peridiapiric sediments means that pyrite easily forms as a syndepositional precipitate (see Chapter 9). As this pyrite is buried deeper into the brine/hydrothermal system it is bathed and altered by escaping metalliferous basinal brines. Examples of this brine pool style of mineralisation range from small to medium scale Pb-Zn accumulations, such as Bou Grine in North Africa, to worldclass sulphide laminites, as are found today beneath the brine pools on the floor of the Red Sea. The Red Sea group of deposits are also tied to higher temperature volcanogenic hydrothermal/sedex mineralisation processes and so are discussed in the next chapter.

The Bou Grine Zn-Pb deposit is situated at the edge of the Lorbeus diapir, in the Domes region of the southern

SALT DIAPIR ASSOCIATED MINERAL SYSTEMS

Known deposits are of low tonnage (10-15 million tonnes) and modest grade (5-15% combined Pb-Zn) e.g. Bou Grine has 7.3 million tonnes of ore @ 2.4% Pb and 9.7% Zn, corresponding to 880,000 tonnes of metal.

PERIDIAPIRIC DEPOSITS	DIAPIR-CAPROCK HOSTED DEPOSITS
Bulk of the mineralisation resides in a laminar organic-rich sediment host that was deposited in seafloor depression (salt withdrawal sinks) adjacent to a seafloor high induced by diapirism (e.g. Lorbeus diapir adjacent to the Bou Grine deposit, Tunisia).	Bulk of the mineralisation resides in anhydrite and bacteriogenic limestone of the diagenetic capstone (e.g. Winnfield and Hockley domes in the Gulf of Mexico).
Synkinematic (salt flow) induced movement on the sea floor with: 1) local deposition of conglomerates and breccias 2) tilting of sediment panels by listric faults with internal syndepositional deformation 3) overlapping of faulted sedimentary bodies by progradation or truncation.	Crest of diapir subject to shallow sub-seafloor dissolution that created cap rock that in turn hosted sulphides.
Stratiform Pb-Zn mineralisation in calcareous laminites rich in organic matter. Minor sub-economic MVT and vein-style Pb-Zn occurrences.	Small massive to laminar sulphide lenses in zones of transition from diapir crest to sedimentary surrounds.
Sphalerite is present as diagenetic microcrystals that are disseminated along bedding planes or coalescent as flattened nodules; some later remobilisation into fissures and late fractures. Galena as euhedral diagenetic crystals along bedding planes and as scattered crystals in later fractures associated with calcite and schalenblende.	Gulf of Mexico examples: As crude oil and thermogenic methane come into contact with anhydrite in low temperature environments (below 70 - 80°C) bacterially mediated calcite forms. Colloform Pb & Zn sulphides, baryte and celestite are co-precipitates in massive laminar sulphide lenses. Host sediments are typically vuggy, fractured and brecciated. The layers in the lenses are mostly of iron sulphides (pyrite, marcasite), while the vugs and fractures are lined with sphalerite, galena, baryte and calcite.
Ores are hydrothermal precipitates in organic-rich laminites from brines supplied by compacting, dewatering basinal sediments with flow exits focused along salt diapir margins and crests.	Metal sulphides that are co-precipitates have isotopic compositions and major element compositions that vary with depth. This range in isotopic values records chemical changes in upward migrating metal-bearing brines and changes in local chemical conditions during sulphide precipitation.

Biodegradation of hydrocarbons and microbial sulphate reduction contribute to the formation of the high-grade mineralization at the contact with the salt dome cap rock. The metals probably were derived through leaching of deeper sedimentary sequences by hot hypersaline basinal brines, evolved by dissolution of salt at the flanks of the diapirs. These hot metalliferous brines are proposed to migrate up around the diapir, finally mixing with near-surface, sulfate-rich brines in the roof zone. When the fluids came in contact with the organic-rich sediments of the adjacent withdrawal sink, the dissolved sulphate was reduced by the sulphate-reducing bacteria. Hydrocarbons generated or reservoired in the diapir-associated sediments were utilised by sulphate reducers.

Organic matter in laminites of Bou Grine region show a high total organic carbon content (4-5% TOC), a low maturity (T_{max} = 423°C) and a marine phytoplankton origin. Organic geochemistry shows sulphate reduction in the laminate facies was bacterially mediated.

References: Bechtel et al., 1996; Charef and Sheppard, 1987,1991; Hallager et al., 1990; Kyle, 1991; Kyle and Posey, 1991; Rouvier et al., 1985; Orgeval, 1991, 1994.

Table 8.1. Properties of diapiric and peridiapiric Pb-Zn mineralisation.

Tunisian Atlas (Table 8.1; Figure 9.6; Rouvier et al., 1985; Charef and Sheppard, 1987,1991; Orgeval, 1991, 1994; Bechtel et al., 1996). The Domes region is oriented NE-SW, covers almost 8,000 km^2 and is characterised by a large number of Triassic diapirs that pierce Cretaceous cover rocks and are aligned parallel to the regional folding. The area formed as a trough during the Early Cretaceous, when it was a permanent depositional feature controlled by deep basement tectonics. It is covered by pelagic Cretaceous sediments and in places is overlain by Tertiary phosphates. All these sediments are pierced by diapirs and salt sheets made up of thick Triassic salts, along with entrained sands, carbonate lenses and mottled clays. The Cretaceous cover includes a major discontinuity in the upper Albian strata that divides it into two megasequences: 1) an Early Cretaceous to Late Albian succession, and 2) a terminal Albian to Palaeocene sequence. There are strong facies and thickness variations in the Cretaceous sediments located close to the diapirs, indicating that halokinesis was active in the Aptian and continued into the Tertiary.

Pb-Zn (Ba, Sr, F) mineralisation in the Domes area occurs mainly in Cretaceous cover rocks near the edges of the Triassic diapirs. There are four main mineralisation types: 1) lenticular Pb-Zn (Ba-Sr) Fe bodies in the zone of transition from the Triassic to the Cretaceous host sediment (caprocks); 2) stratiform Pb-Zn mineralisation in calcareous laminites, which are also rich in organic matter (e.g. Bahloul Formation); 3) Pb-Zn-Ba-F and Fe-Pb deposits, with features similar to Mississippi Valley Type deposits, in the Aptian reef formations that formed atop seafloor highs created by salt allochthons; and 4) vein type deposits.

Bou Grine mineralisation belongs mostly to the first and second groupings (Table 8.1). The Bou Grine deposit differs from subeconomic caprock sulphides in the Gulf of Mexico in that much of the Pb-Zn resides not in diapir cap rock but in the organic-rich limestone laminites of the Bahloul Fm adjacent to the diapir. As well, a semi-massive ore body (hanging wall orebody) cuts the Cretaceous series in the second megasequence in Xenomenia-Turonian Bahloul Formation at Bou Grine. The geologic reserves, with a cutoff value of 4% Pb + Zn, are currently estimated at 7.3 million tonnes of ore at 2.4% Pb and 9.7% Zn, equivalent to 880,000 tonnes of metal.

The zinciferous laminites of the Bou Grine deposit are made up of mineralised bioclastic clayey-carbonaceous calcimicrites whose major features are: 1) a regular, in places laminar, bedded structure implying low energy, sub wavebase bottom conditions; 2) numerous foraminifera and other bioclastic components showing clear marine affinities; 3) mineralisation that is in part affected by early diagenetic structures, especially ferron and calcitic cements. Sphalerite is present as microcrystals disseminated along the bedding planes or as coalescent flattened nodules; it also appears to have been remobilised as concentrations along fissures and later fractures. Galena occurs in euhedral crystals along the bedding planes and as scattered crystals in later fractures associated with calcite and schalenblende. Calcite is observed in fissures that were deformed during compaction, as well as in all the later fractures; it also pseudomorphs occasional sulphate crystals (baryte and anhydrite).

The early nature of mineralisation at Bou Grine is reflected in: 1) convoluted mineralised laminar beds deformed by intraformational slip, with flow structures; 2) compaction structures around rigid cores of pyrite nodules and galena crystals; 3) deformed infilled fissures. Sphalerite developed as: 1) early regularly disseminated microsphalerite crystals in a slightly compacted sediment; 2) as almost simultaneous concentrations along sedimentary structures, textures and discontinuities; 3) as later concentrations and deformations of the mineralisation due to compaction and subsequent fracturing; and 4) as diagenetic remobilised cements that were either passive infills or provoked by circulation of mineralising fluids.

Laminites of the Bou Grine region also show high total organic carbon contents (4-5% TOC), low organic maturity (T_{max} = 423°C) and a marine (Type II) phytoplankton origin (Bechtel et al., 1996). Hydrocarbon enrichment at particular levels suggests possible former hydrocarbon reservoirs in some of the sediments adjacent to the diapirs. Montacer et al. (1988) showed that sulphate reduction in the laminite facies was bacterially mediated, with evolution of the organic matter probably occurring by *in situ* maturation of the laminite source rock. Their isotope studies support the notion of a close relationship between the initial metal stock and the sedimentological/diagenetic setting. Sulphur isotopes also indicate development of mineralisation in a closed sulphur system. Along with Sr and Pb isotopes, they also show the deposit, and its aureole, formed under the influence of an organic-rich diagenetic palaeocirculation, with Pb derived from a single source.

The deposit's locality and host rock geometry is controlled by a feedback between salt flow and the synkinematic movement of the seafloor. Thus halokinesis and the associated allochthons control: 1) local deposition of conglomerates and breccias adjacent to the highs atop the growing salt allochthons, 2) local deposition of organic-rich pyritic laminites in areas of brine ponding on deepened seafloor, 3) tilting of panels of sediments by listric faults activated by the emplacement of allochthons and diapirs, 4) overlapping of sedimentary bodies by shoal/shelf progradation or truncation. Halokinesis also led to marked thickness variations of the various facies across the regions via the repetitious pulsation of the active diapir and its ongoing dissolution.

The Bou Grine region, from its beginnings, was a submarine slope at the edge of a structural high. Other than around a few local diapirs and salt allochthons, this area never seemed to have shoaled sufficiently for a true shallow water carbonate platform to have developed. The organic-rich sediments may well have accumulated in this region only in the deeper areas of rim synclines about the active diapirs and salt sheets, hence the localisation of the mineralisation to laminites in the peridiapiric areas. These organic-rich anoxic depressions probably also acted as early sinks for metal concentrations (Chapter 9).

The peridiapiric model

A model for peridiapiric Pb-Zn accumulations, based in part on the sulphide paragenesis in the Gulf Coast and in the north African deposits, is presented as Figure 8.18. Once near the seafloor, a flowing and dissolving salt

allochthon or diapir acts as a focus for migrating metalliferous fluids as: 1) sediment slippage and faulting driven by salt flow and dissolution about a diapir or allochthon edge provide an ongoing easy passage and focus for upwelling Fe-rich basinal fluids, and 2) hydrocarbon-saturated caprocks and adjacent organic-rich laminites provide favourable anoxic traps for inflowing metalliferous brine (Chapter 9).

Sulphide mineralisation in this system begins with the syndepositional, probably bacteriogenic, accumulation of pyrite framboids within an anoxic organic-rich sediment matrix beneath an anaerobic seafloor brine pool (Figure 8.18). The brine pool itself is also a result of the venting of these upwelling highly saline waters. As the pyritic organic-rich laminites are buried, they pass into a new diagenetic environment, which is bathed in hotter, deeper haloes of the same Fe-chloride brines, which are escaping at the surface of brine seeps and vents (see Chapter 9 for details of this hydrology and mineralisation chemistry). A deeper burial-imposed equilibrium then favours the replacement of the pyrite precursor by Pb-Zn sulphides. In effect, peridiapiric mineralisation is part of a fluid cycling system instigated during early diagenesis and continued into deeper burial under a plumbing that is controlled by basinal faults and tectonics that are in turn, driven by halokinesis.

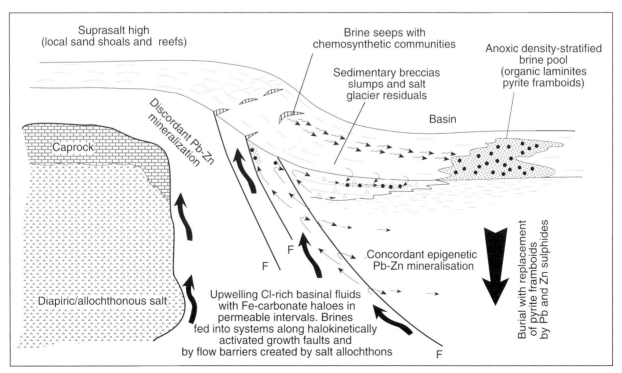

Figure 8.18. Model for development of peridiapiric mineralisation. Note the similarity of the hydrology to classic sedex style deposition (figure published with permission of JK Resources Pty Ltd).

Although Phanerozoic examples of this peridiapiric/allochthon style of Pb-Zn precipitation are largely small scale, this model of brine-controlled fault-fed evolving sulphide facies also has possible larger scale analogues in sedex deposits within Proterozoic metalliferous basins (Chapter 9). There, in contrast to the oxygenated deep ocean waters of the Phanerozoic, the lower oxygen levels in the Proterozoic oceans meant deeper ocean waters, and the floors of large intrashelf basins, tended to be capable of maintaining dysaerobic to anaerobic conditions over much wider areas for much longer times. This facilitated conditions suitable for preservation of laminated pyritic sediments adjacent to brine/allochthon outflows over much larger areas on the seafloor. As they were buried, these pyritic Proterozoic sediments were also replaced by Pb and Zn sulphides to form large laminite deposits such as HYC in Australia.

Pb-Zn-Cu Kipushi deposit, Africa: a metamorphosed allochthon

In order to emphasise the diapir focus of this section I will continue to restrict my discussion to those aspects of the ore geology that are related to meta-evaporites and in particular to mineralisation in the southern Shaba belt in the vicinity of Kipushi. For more comprehensive summaries of the general geology, and evaporite sedimentology of this very interesting meta-evaporite province, the interested reader is referred to Sweeney et al. (1991); Cailteux (1994); Cailteux et al. (1994) and Hanson et al. (1994).

The Neoproterozoic metallogenic province of Central Africa contains rich stratiform concentrations of Cu and Co in both the Shaba Province (Shaba cupriferous arc - Zaire) and Zambian copper belt (Cailteux, 1994). Most Shaba Province orebodies occur in carbonate host rocks, while most Zambian ore bodies are hosted in siliciclastics. Both provinces contain substantial evidence of meta-evaporites. For example, scapolite is widespread in amphibolite grade calcsilicate rocks and marbles in Katangan rocks of the copper belt and in the central provinces in Zambia. Ramsay and Ridgway (1977) point out that "scapolite is rarely found in such abundance and Zambia must be one of the major scapolite provinces in the world". Munyanyiwa (1990) argues for an evaporite precursor to this scapolite, he also notes that metasedimentary anhydrite is still present in parts of the Katangan succession in the copper belt.

The Kipushi Zn-Pb-Cu deposit is located in the northern extension of the central African Copper Belt in southeastern Zaire. It is one of the main producers of zinc, cadmium and germanium in Africa. It is a discordant ore body at the contact between a halokinetic breccia body and adjoining host rocks of the Upper Proterozoic Lower Kundelungu Group of the Katangan System (De Magnée and Francois, 1988). It is currently exposed at the surface on the northeast margin of a large plunging anticline (Figure 8.19a). The halokinetic breccia in the core of the anticline contains large blocks (up to 100s of metres across) of Roan Supergroup rocks of several lithologies, as well as large blocks of gabbro. Much of the breccia at Kipushi is made up of fragments of white sparry dolomite, which are thought to have originally been derived from the halokinetic and solution induced break-up of the Dipeta Dolomite. The gabbro (now completely amphibitolised) is thought to have originally been emplaced as a sill in the Dipeta Dolomite within the Roan Supergroup (Figure 8.19b).

Most workers accept there is an association with former evaporites and the Kipushi breccias which host, or are adjacent to, mineralisation. Roan Supergroup sediment once contained several evaporite intervals (De Magnée and Francois, 1988). Regionally these former salt intervals are defined by evaporite dissolution breccias. Such solution breccias occur in the lower part of the Katangan System and are interlayered with alternating sparry dolostones, black shales, and chlorito-talcose dolomitic siltstones (Figure 8.19). Prior to dissolution, the evaporites are thought to have acted as the mother salt bed for the Kipushi allochthon and other halokinetic features in the region. The upper part of the Katangan System is made up of the Lower and Upper Kundelungu Supergroup and is composed mostly of siliciclastics with some limestones, which are locally dolomitised.

Portions of the Katangan System are now intensely folded in the copper belt, with a disharmonic style reminiscent of the salt-cored European Jura. Cores of many anticlines in the copper belt are formed by megabreccias, containing blocks of Roan metasediment. Regionally, some of the larger clasts contain sufficient stratabound Cu-Co sulphides to be actively mined. The matrix between the blocks is a dolomitic metasiltstone, which also contains gypsum and anhydrite pseudomorphs (De Magnée and Francois, 1988). Regionally, the megabreccias cut the flank of many megabreccia cored anticlines and reach into Upper Kundelungu strata. This is the case at Kipushi (Figure 8.19a).

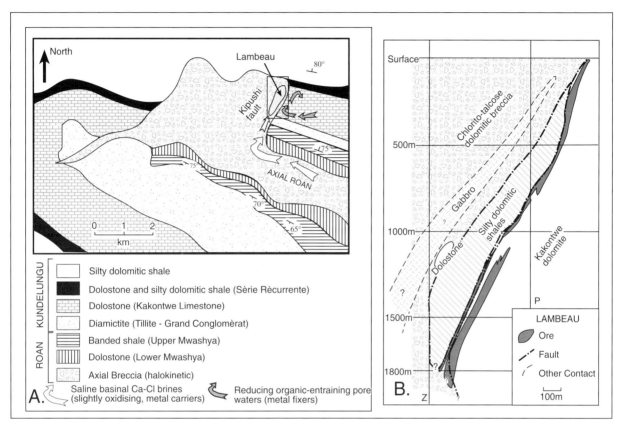

Figure 8.19. Pb-Zn-Cu Kipushi deposit. A) Geologic map of Kipushi anticline showing trans-stratal halokinetic nature of the axial breccia and location of ore body at its contact with adjacent sediment (Lambeau deposit - box shows position of enlargement in B. B) Detail of Kipushi ore body based on mine intersections (after De Magnée and Francois, 1988).

The megabreccia and its crosscutting relationships are very similar to outcropping halokinetic breccias in the Neoproterozoic Callana Group megabreccias of the Flinders Ranges, South Australia (Chapter 4). There, many of the breccia geometries reflect the falling or deflation stage of diapir evolution and are located along regional extensional faults (pers. observation). Breccias in the Katangan System also follow major faults, which De Magnée and Francois (1988) called extrusion faults. If analogies to the Flinders Ranges are valid, some of these "extrusion faults" may be examples of outcropping fault welds. Once a basin undergoes inversion such faults are often reactivated and can be easily misinterpreted as late-stage thrust faults.

The Kipushi structure is interpreted as a salt diapir/ allochthon that, after emplacement, was subject to a long period of selective dissolution of the more soluble salts, along with associated collapse brecciation. According to De Magnée and Francois (1988) and Chabu (1995), this dissolution process played an essential role in the base metal mineralisation, as it supplied and focused the brines

that leached and transported the heavy chalcophile metals. De Magnée and Francois (1988) went on to argue that sulphur and organic matter must also be present to form a mineral deposit, as well as a high lateral thermal gradient to set up the circulation cells and an appropriate permeability structure to channel the ascending hydrothermal solutions. At Kipushi these conditions were created by a huge slab of silty dolomitic shale (Kundelungu Group), which lies alongside the wall of the halokinetic breccia and forms the current roof of the ore body. De Magnée and Francois go on to draw parallels with peridiapiric mineralised laminites in north Africa.

However, De Magnée and Francois' paper was published in 1988, well before models of peridiapiric mineralisation were well understood and well before our understanding of diapirism underwent major changes as the interpretive paradigm moved from a two-fluid model to the fluid-brittle-overburden extension models that are in use today (see Chapter 5). The block of metasediment that now hosts the ore in the Kipushi structure could either be related to extensional collapse of the diapiric structure during the diapir deflation stage or, as De Magnée and Francois suggest, is an example of peridiapiric replacement akin to the deposits at Bou Grine.

Although the evaporite nature of the host and the focusing effect of the breccias is accepted by most workers, the actual timing of the mineralisation within the evaporitic host and its ultimate origin are still controversial. Many argue that the presence of folded, sheared and brecciated ores implies that the deposit predates the peak of the Katangan deformational event (650-500 Ma; Chabu 1995). Intiomale (1982) argued for ascending hydrothermal solutions with a magmatic origin. De Magnée and Francois (1988) related the mineralisation to basinal fluids produced by the dissolution of the salt diapir, which also produced the ore host. Unrug (1988) suggests the mineralising fluids were formed via metamorphic dewatering. Chabu and Boulégue (1992) and Chabu (1995) suggest the mineralisation occurred in evaporite solution cavities. They argue the karst formed contemporaneous with the mineralisation and predates the peak of Lufilian tectonism.

In a regional study, that included Kipushi, Cluzel (1986) noted that the Late Precambrian copper-bearing strata of the southern Shaba arc underwent metamorphism during the latest Proterozoic to Early Palaeozoic times. During the first phase, prior to compressional folding, a static recrystallisation was accompanied by several chemical changes, such as sodic, then K-Mg metasomatism. A second hydrothermal event followed the main folding phase, and he linked this to halokinetic structures in the region. According to Cluzel, it was essentially a retrograde event with oxidising and low-temperature (≈200°C) fluids that leached any remaining evaporites. It also formed retrograde clays such as kandites or smectites.

Walraven and Chabu (1994) showed there was little variation between Pb-isotopic compositions of sulphide minerals and the host rocks at Kipushi. They calculated that the model Pb-age data of galena from the deposit indicated a probable mineralisation age of 454 ±14 Ma. This is younger than the peak of the Lufilian tectonism and metamorphism noted by Cluzel (1986). They go on to suggest that the Kipushi mineralisation ultimately has a magmatic or volcanic source. If one believes model Pb-ages can be used to interpret genesis, then it suggests that the mineralisation was epigenetic and postdated halokinetic folding and brecciation at Kipushi. However, the geochemistry of ore-related phlogopite and chlorite in the Kipushi deposit indicates an emplacement of the orebody that predates, not postdates, the peak of the Lufilian

(Katangan) Pan African orogeny some 650-500 Ma (Chabu, 1995).

For the purposes of "on ground" targeting of an analogous deposit this distinction in terms of timing may be largely academic in defining greenfield sites for a preliminary sampling program. What is obvious from the relationship of mineralisation to the halokinetic breccia at Kipushi is that the breccia margin in contact with organic-rich sediments was the precipitational focus for the mineralising solutions. This occurred either as the residual salt was still dissolving, or once it had dissolved and scapolites were supplying the metal-entraining volatiles (Figure 8.19). An exact determination of the time of the mineralisation is not the fundamental control on the location of the deposit. The salt structure induced focus to fluid flow and its associated redox front is what is important.

Peridiapiric copper

Copper in the Lisbon Valley, Utah

Cu(±Ag) ores occur along extensional faults, controlled by salt anticlines, in redbed sediments of the Paradox Basin and in coal-bearing horizons of the overlying Cretaceous Dakota Sandstone in SE Utah and SW Colorado (Figure 8.20a). Deposits of this type occur at Salt Valley and Paradox Valley, with the largest deposits at Lisbon Valley (Morrison and Parry, 1986). Cu is found as vein fillings, sandstone pore fillings, and as replacement of coalified plant fossils. The deposits are clearly related to halokinetically induced faults, and have not been offset by post Middle Cretaceous movements (Figure 8.20b, c). The largest deposits, Big Indian and Blackbird, occur along the Lisbon Valley fault in coal-forming horizons of the Cretaceous Dakota Formation. Almost all the mineable copper ore occurs within several hundred metres of either the Lisbon Valley fault or major fracture zones.

All the deposits in the area are related by geologic setting as well as ore and gangue mineralogy (Morrison and Parry, 1986). Fluid inclusions, mineral chemistry, and C-O stable isotopes in calcite gangue associated with Cu ores at Lisbon Valley suggest they formed via subsurface mixing of two diagenetic waters. Fluid temperatures ranged from 72°C to 103°C , while salinities were 5 to 20 equiv. wt % NaCl during calcite deposition. Cu-bearing, saline, basinal fluids migrated upward along fault zones, and upon mixing with shallower reduced Ba-rich groundwater deposited ore and gangue minerals. Mineral precipitation was induced by a combination of dilution and reduction of upwelling basinal brines.

The metal-carrying fluid probably originated in the redbed aquifers of the Permian Cutler Formation. The brines may have equilibrated for long periods in these Cutler Formation aquifers and only been released during faulting and halokinetic activity driven by the Late Cretaceous - Early Tertiary Laramide orogeny. Then faults, such as the Lisbon Valley Fault, breached the redbed aquifer and so allowed pressurised, heated metalliferous fluids to escape upward along the faults until they encountered permeable units replete with reductants, such as occurred in the Dakota and Wingate Sandstones (Kirkham, 1989).

In my opinion, the role of salt has not been given the importance it deserves in the explanation of mineralisation in a Lisbon-Valley style deposit. The style of faulting that focuses the mineralisation is a response to movement on the underlying salt structure (Figure 8.20b,c). The dissolution of this underlying salt also supplied the chloride-rich brines that leached the metals from adjacent redbeds. It also created the thermal and brine density contrasts that drove the convection and metal enrichment of this brine. Without halokinesis and salt dissolution there would be no focusing into this structurally controlled deposit. Nor would there be any long term maintenance of the redox haloes about the faults that allowed this deposit to accumulate. Without the underlying halokinetic salt, economic levels of copper sulphides simply would not be there. As is to be expected, this cupriferous flow regime is

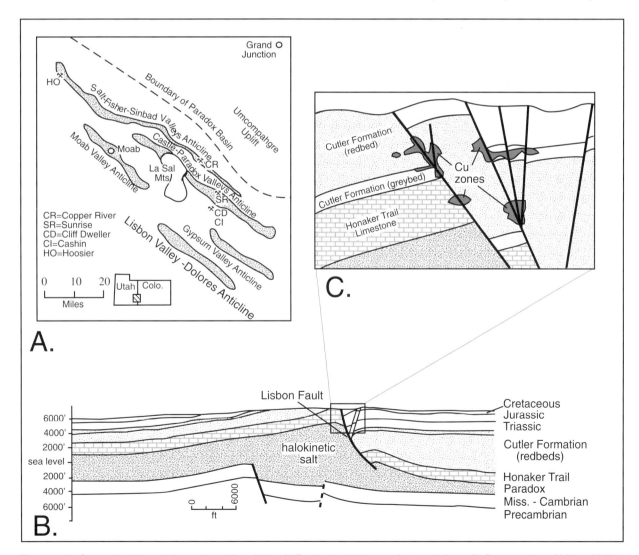

Figure 8.20. Copper in Lisbon Valley salt anticline, Utah. A) Regional distribution of salt anticlines, B) Cross section of Lisbon Valley salt anticline. C) Detail of crestal position of Lisbon Fault showing distribution of Cu ore (after Morrison and Parry, 1986; Kirkham, 1989).

311

very similar to, although less prolific than, those of the Dzhezkazgan region (Figure 5.12).

Dongchuan copper, China

Stratabound sediment-hosted copper deposits, associated with a widespread halokinetic breccia unit, are found in a 450 km long rift in the Yunnan Province of China (Huichi et al., 1991). The Dongchuan area is the largest of five copper mining areas in this north-trending rift, which contains sandstone, shale, and carbonate rocks of the Kunyang Group of Middle Proterozoic age. Deposits in the area range from 10 to 100 million tonnes of ore with ore grades of 1 to 1.5%. Three formations, about 1200 m thick, in the lower part of the group contain the copper deposits. They are from bottom to top: the Heishan Formation, mainly a carbonaceous slate; the Laoxue Formation, mainly a stromatolitic dolostone; and the Yinmin Formation, chiefly purple slate and sandstone. These formations were intruded by mafic igneous rocks. Below the Yinmin Formation is a thick breccia consisting of clasts of Yinmin-type shale and sandstone, diabase, and other rocks. The three ore-producing formations and the breccia are exposed in a band of steeply dipping strata extending across the Dongchuan area.

Most of the copper ore occurs in the lowermost dolostones of the Laoxue Formation. Vein type ore bodies account for only 3% of the total ore reserves in the area. Stratabound ores characteristically occur as disseminations of sulphides along stromatolitic algal layers. Locally, highgrade vein-type ores crosscut the dolostone. Both ore types contain bornite, chalcopyrite, chalcocite, and other minerals. Fluid inclusions in quartz veinlets contemporaneous with the copper mineralisation contain liquid, vapour and daughter minerals: $NaCl$, $CaCl_2$, KCl and $BaCl_2$. They homogenise between 200 and 280°C.

The breccia was produced by tectonically induced salt diapirs from which the salt was later removed by hydrothermal leaching. Copper ores are associated with the thermal anomaly that resulted from this salt diapirism and was contemporaneous with diabase intrusion (Huichi et al., 1991; Wu Moude and Li Xiji, 1981). Hot brines circulating upward along the flanks of the diapir leached copper from the oxidised Yinmin Formation and carried it to the base of the overlying Laoxue Formation. Sulphide-rich fluids were formed by reaction of evaporite sulphates with decayed organic matter in the dolostones. Copper sulphide deposition in the dolostones resulted from the mixing of these fluids. Continued circulation leached all of

the remaining salt from the diapir, and the overlying and surrounding rocks collapsed to form the breccia. Texturally, these halokinetic breccias are very similar to the diapiric breccias that are widely exposed in the Flinders Ranges, South Australia, and the Amadeus Basin of central Australia.

Another Proterozoic example of the focusing effects of the peridiapiric setting is clearly illustrated in Palaeoproterozoic Fe-Cu sulphide deposits, eastern Liaoning, Northeastern China (Wang et al., 1998). This area lies within the same region as the metaborate deposits described in Chapter 6. The sulphides were precipitated at two sites: as breccias, lenses, and veins around venting centres; and as stratiform pyrite-chalcopyrite, which are associated with anhydrite located distal to the salt domes. Within each mining district the proximal ores are associated with brecciated, albite- and tourmaline-rich metasediments. The sulphides in the proximal ores have $\delta^{34}S$ values of 8.9‰ to 12.7‰, whereas the distal ores have $\delta^{34}S$ values of 2.6‰ to 8.8‰ for the sulphides and 7.9‰ to 19.0‰ for the anhydrite; suggesting a sulphate-dominated sulphur source, with the pyrite $\delta^{34}S$ variations arising mainly from decreasing temperatures away from the centres of the deposits. According to Wang et al. (1998) the origin of the deposits is best described by a model in which sulphate-carbonate rocks were deposited in peridiapir sinks during salt diapirism. The salt domes then acted as foci for subsequent hydrothermal venting of chloride-rich metalliferous fluids and the adjacent peridiapir sediments became the centres of sulphide mineralization. The sulphides precipitated during interaction of the reduced, metal-bearing hydrothermal fluids with the *in situ* sulphates, which acted as the major *in situ* source of evaporitic sulphur.

Base metals, evaporites and diagenetic accumulations: a summary

Many low-temperature base metal deposits in evaporite-entraining sedimentary basins can be related to the presence of preserved evaporites or indicators of the former presence of evaporites, such as bedded and halokinetic breccias, chert nodules and other indicators of "the evaporite that was". The link is either direct, with metal accumulation at the site of the present or former evaporites; or indirect, from ascending hypersaline fluids supplied from a dissolving mother salt bed via a single or a series of feeder faults. Thus mineralisation can be subsalt, intrasalt, or suprasalt and the salt, or its breccia, can be bedded or halokinetic. In many cases the H_2S that causes the ore to precipitate at the redox interface is derived from bacteriogenic or thermochemical reduction of nearby evaporitic sulphate in combination with organics supplied by hydrocarbons ponded beneath an evaporite seal (Table 8.2).

The diagenetic evaporite-base metal association includes world class Cu deposits, such as the Kupferschiefer-style Lubin deposits of Poland and the large accumulations in the Dzhezkazgan region of Kazakhstan. The Lubin deposits are subsalt and occur where longterm dissolution of salt in conjunction with upwelling metalliferous basin brines created a stable redox front, now indicated by the facies of the Rote Faule. The Dzhezkazgan deposits (as well as smaller scale Lisbon-Valley style deposits) are suprasalt and formed where a dissolving halite-dominated salt structure maintained a structural focus to a regional redox interface. It also drove the circulation system whereby metalliferous saline brines leached adjacent redbeds. In both scenarios the resulting sulphides are zoned and

arranged in the order Cu, Pb, Zn as one moves away from the zone of salt-solution supplied brines. This redox zonation can be used as a regional pointer to both mineralisation and, more academically, to the position of a former salt bed. In the fault-fed suprasalt accumulations the feeder faults were typically created and maintained by the jiggling of brittle overburden blocks atop a moving and dissolving salt unit. A similar mechanism localises many of the caprock replacement haloes seen in the diapiric provinces of the Gulf of Mexico and Northern Africa.

Evaporite-associated Pb-Zn deposits, like Cu deposits, can be associated with both bedded and halokinetic salt units or their residues. Stratiform deposits such as Gays River, Cadjebut and Nanisivik, have formed immediately adjacent to or within the bedded salt body, with the bedded sulphate acting as a sulphur source. In diapiric deposits the Pb-Zn mineralisation can occur both within a caprock or adjacent to the salt structure as replacements of peridiapiric organic-rich pyritic sediments. In the latter case the

Deposit (Age of Host)	Reserve (million tonnes)	Cu (%)	Pb (%)	Zn (%)	Evaporite Role	Evaporite Association
COPPER DEPOSITS						
Kupferschiefer-style, Lubin, Poland (Permian)	≈2,000	>2	-	-	Subsalt, stratiform with brine focus	Anhydrite and halite in Zechstein hangingwall and anhydrite in Rotliegende footwall
Kupferschiefer-style, Mansfeld, Germany (Permian)	≈75	≈2.9		±1.8	Subsalt, stratiform with brine focus	Anhydrite and halite in Zechstein hangingwall and anhydrite in Rotliegende footwall
Redstone River, Canada (Neoproterozoic)	small	2.7	-	-	Intrasalt with sub salt focusing of upwelling brine	Algal limestone host with abundant nodular CaSO₄
Creta, Okla. USA (Permian)	1.9	2	-	-	Intrasalt bedded (breccia)	Redbed - evaporitic mudflat host with CaSO₄ in hangingwall and footwall. Former thick halite beds
Corocoro, Bolivia (Tertiary)	≥7	≈5	-	-	Intrasalt bedded (breccia)	Redbed/ greenbed host with gypsum and baryte in veinlets and disseminated nodules
Dzhezkazgan, Kazakhstan (Carboniferous)	≈400	1.5	0.1-1	-	Suprasalt diapiric, anticlinal focus to brine flow	Interbedded redbed and greybed host with CaSO₄ and halite in underlying salt anticline and lateral equivalents
Lisbon Valley, Utah, USA (Cretaceous)	small >0.15	1.4	-	-	Suprasalt	Ore in vein and pore fills adjacent to faults created by halokinesis
Dongchuan deposits, China (Mesoproterozoic)	10 - 100	1 to 1.5	-	-	Suprasalt or intrasalt breccia	Stratabound ore in cryptalgal laminites adjacent to diapiric breccia
LEAD-ZINC DEPOSITS						
Gays River, Nova Scotia (Carboniferous)	2.4	-	8.6	6.3	Intrasalt bedded with subsalt focusing of upwelling brine	Anhydrite both within main ore zone and as cement in disseminated ore
Jubilee, Nova Scotia (Carboniferous)	Not economic	N/A	N/A	N/A	Intrasalt (tectonic breccia), stratiform, subsalt focusing of brine	Interstratified fractured carbonate and anhydrite ore host, created by regional salt-induced gravity slide.
Nanisivik, Baffin Island (Proterozoic)	6.5	-	1.5	12	Intrasalt?, bedded	Host is a stratiform karst cavern: evaporites and pseudomorphs in equivalent Society Cliffs Formation
San Vicente, Peru (Trias.-Jurr.)	12	-	1	12	Intrasalt, bedded	Ore hosted in cryptalgal laminites with pseudomorphs after nodular CaSO₄ in barrier-lagoon host
Cadjebut, Australia (Devonian)	3.8	-	17% (Pb+Zn)		Intrasalt, bedded (breccia)	Bedded nodular anhydrite as lateral equivalent to each ore lense with intervening solution breccias
Largentière, France (Triassic)	9.6	-	0.7	3.7	Intrasalt	Disseminated CaSO₄ cements and nodules within evaporitic lagoonal beds
Gulf Coast, USA (Cretaceous)	Not economic	N/A	N/A	N/A	Suprasalt, diapiric capstone, brine focus from halokinesis	Anhydrite caprock undergoing bacteriogenic alteration in association with reservoired hydrocarbons
Bou Grine, Tunisia (Cretaceous)	7.3	-	2.4	9.7	Suprasalt, peridiapiric, brine focus from halokinesis	Ore in peridiapiric organic-rich pyritic sediments, adjacent to halite of allochthonous diapir sheets

Table 8.2. Properties of ore deposits formed in the diagenetic realm in association with evaporites.

conditions of bottom anoxia that allowed the preservation of pyrite were created by the presence of brine springs and seeps fed from the dissolution of nearby salt sheets and diapirs. The deposits in the peridiapiric group tend to be widespread; but, in Phanerozoic strata, the individual deposits are relatively small and mostly subeconomic. However, the fluid flow and brine pool mechanisms, as well as models of replacement haloes in bottom brine sediments, can be used to model larger counterparts within Proterozoic settings for which there are no modern analogues. This notion is developed further in the next chapter.

In all the deposits discussed in this chapter there are four common factors that can be used to recognise suitably prepared ground and so improve exploration models:

- A dissolving evaporite bed acts either as a supplier of chloride-rich basinal brines capable of leaching metals, or as a supplier of sulphur and organics that can fix metals.
- Where the dissolving bed is acting as a supplier of chloride-rich brines, there is a suitable nearby source of metals that can be leached by these basinal brines.
- There is a redox interface where these metalliferous chloride rich waters mix with anoxic waters within a pore fluid environment that is rich in organics and sulphate/sulphide/H_2S.
- There is a salt-induced focusing mechanism that allows for a stable, longterm maintenance of the redox front, e.g. the underbelly of the salt bed (subsalt deposits), a dissolution or halokinetically maintained fault activity, the overburden (suprasalt deposits), a stratiform intrabed evaporite dissolution front (intrasalt deposits).

Most important is the notion that the sulphide orebodies that form at the redox interfaces can remain long after the actual salt units have dissolved or evolved into meta-evaporites. To an explorationist trained in the recognition of former salts, the indicators remain visible and act as pointers to zones of structural or diagenetic foci for the mineral-depositing redox system. When such zones are recognised, and placed within the regional framework or basin architecture, then particular zones or lithologies of interest are quickly defined and suitable for more intense exploration.

Chapter 9
Evaporite-metal associations: brines, magma and metamorphism

Introduction

In this the final chapter I discuss the more enigmatic groups of metal-evaporite associations that range across the diagenetic to the metamorphic and magmatic realms. Unlike the deposits discussed in Chapter 8, preserved salts or solution breccias are not always a recognised part of the stratigraphy of the preserved mineralisation in this group. Rather, evidence for the association is less direct, or perhaps more distant from the site of mineralisation. It can be classified in terms of: i) a near ore association (including brines and solution breccias), ii) a geographic association, and iii) an isotopic association. As in Chapter 8, the significance of evaporites in this context is that as they dissolve, flow or "go elsewhere" they can establish a brine hydrology, or a chloride-rich fluid, or a volatile or sulphur-rich melt, which facilitates the accumulation of Pb, Zn, Cu and even Au.

Evidence of the importance of evaporites in some of these deposits can be interpreted equivocally; I have chosen for the purposes of this chapter to emphasise the likely role of evaporites. This makes the interpretive discussion far more speculative than in the previous chapter. Nonetheless, the circumstantial evidence for the role of evaporites in the mineralisation is strong, but in most cases is not well understood. In my opinion, a major reason for this lack of understanding lies in the difficulties inherent in recognising relic textures of former evaporites and brines, especially when the salts have long since vanished via processes of flowage, dissolution, melting and assimilation, and the original rock matrix has been drastically altered by the processes of melting and/or metamorphism.

Similarly, in low temperature sulphide deposits, such as the laminites of the Red Sea or some ancient sedex deposits, there is a lack of appreciation of how evaporites can dissolve in the diagenetic realm to create hydrothermal seeps and anoxic brine pools. Anoxic brine pools can focus the formation of pyrite framboids and sulphur-bearing organic-rich sediments, while the structurally defined feeder pathways to the pools can be outlined by haloes of ferroan carbonate cements and replacements, as well as zones of sodic alteration. These evaporite-solution-associated bottom brine sediments also act as "prepared ground" for the precipitation of stratiform base metal accumulations, while nearby flowing and dissolving salt units can create synprecipitational faults to focus the throughflow, trapping and venting of metalliferous chloride-rich brines.

Shales and allochthons

Until ten years ago most sedimentologists did not appreciate the importance of allochthonous salt sheets which; as they flow and dissolve, supply chloride-rich brines to many nonevaporitic seafloor systems. Chapter 5 discussed how these huge salt sheets (allochthons) climb the stratigraphy as a series of predominantly subhorizontal sheets or tiers that are continually moving and squeezing as they are loaded by ongoing sedimentation. Wherever a salt allochthon is near the seafloor it can load, deform, coalesce or disaggregate as it continues its climb through the aggrading stratigraphy (Figures 5.13, 5.17). Sediment loading atop a deforming allochthon creates fault-defined suprasalt basins or depopods. Depopod fill shows a characteristic upward shoaling pattern from initial deepwater laminated shales to shallow water sands and carbonates. Processes of ongoing halokinesis and depopod formation cease once the salt allochthon is sufficiently thinned, dissolved or buried. Thus, throughout the sedimentation history of a passive margin that is underlain by early thick salt beds there are many intervals preserved where thick sheets of flowing allochthonous salt were present just beneath the deep sea floor and sandwiched into what was otherwise normal deeper marine bottom sediments.

Breaks in the lateral continuity of shallowly buried salt sheets tend to focus the upward escape of saline metalliferous/hydrocarbon-bearing basinal and hydrothermal brines (Figures 2.11c, 9.1). Venting of these waters can create substantial anoxic brine pools atop seafloor deeps, with bottoms characterised by pyritic and sometimes metalliferous laminites. Most post-Mesozoic examples of these brine pools are relatively small, and the resulting metalliferous deposits are also small (e.g. Bou Grine, Algeria; Chapter 8). Occasionally the examples are much larger and, when formed in combination with active hydrothermal circulation cells in seafloor rifts can contribute to the creation of widespread metalliferous laminites (e.g. Atlantis II deep, discussed later).

As an allochthonous salt sheet lies just beneath the seafloor it dissolves from the edges inward and releases chloride-rich brine into its surrounds (Chapter 2). Any subsalt brine forms a dense chloride-rich plume that then sinks into the underlying sediments and so becomes caught up in subsalt convection cells (Figure 2.11c). As it convects it mixes with escaping basinal brines whose upward passage is

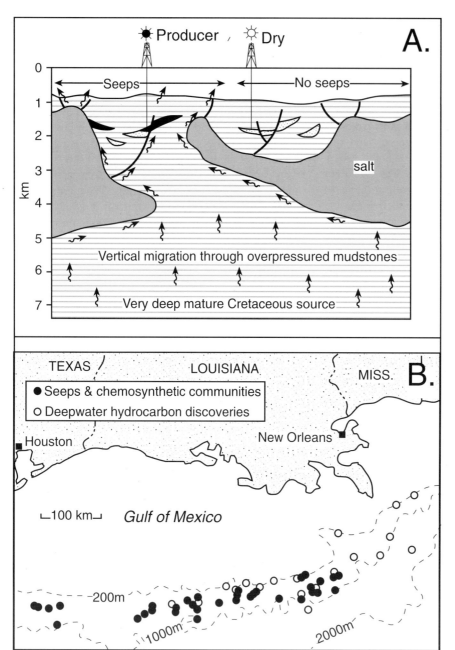

Figure 9.1. Brine seeps. A) Model of hydrocarbon migration and seepage in the deepwater Gulf of Mexico. Seepage indicates a change in the salt continuity in an intrasalt basin, but it does not indicate the presence of an effective migration pathway to a specific hydrocarbon trap. B) General association of seeps and hydrocarbon discoveries in the deepwater Gulf of Mexico (after Thrasher et al., 1996).

congruent seawater dissolution of the underlying Louann salt, and in part by hydrocarbon-rich basinal waters escaping from deeper within the sediment column (Figure 9.1b; Aharon *et al.* 1992). The venting brines are seven times saltier than ambient sea water and the geometry of their subsurface passage to the seafloor largely reflects breaks in the continuity of allochthonous salt sheets that underlie much of the continental slope of the Gulf of Mexico (Thrasher et al., 1996).

Venting of hydrocarbon-rich hypersaline brines onto the seafloor, both by lateral flow and by diffusion upward through the underlying clastic sediments, gives rise to drainage patterns on the seafloor that show an anastomose pattern similar to that of braided stream channels on land. Venting brines can either dissipate via mixing with oxygenated marine bottom waters or seep into closed depressions on the seafloor to form pools of density-stratified anoxic bottom brine. Brine transport pathways in those seepage drainages that lie beneath open ocean bottoms are floored by nonindurated red-orange iron-oxide deposits, which are similar to the iron-oxide deposits found about the edges of the Red Sea brine pools (Aharon et al. 1992). The ferric iron now in the seafloor oxides was originally extracted by flushing of buried sediments by salt-dissolution-driven convection of anoxic pore waters through the peridiapiric and salt sheet surrounds (Hanor, 1994). Once vented onto the oxygenated seafloor, the dissolved iron precipitates as haematite and other ferrous hydroxides on the floor of the drainage network. In this scenario the seafloor surface itself constitutes a redox front between the anoxic reducing

focused along the dissolving undersides and edges of the salt before it finally seeps and vents to the seafloor (Figure 9.1a). For example, brine-issuing vents occur at 1920m-deep seafloor on top of Green Knoll, an isolated salt diapir rising seaward of the Sigsbee Escarpment in the northern Gulf of Mexico. Vent brines are supplied in part by basin brines and oxygenated marine waters. Figure 9.2a and b, shows clearly that red iron oxides are stable in the oxygenated conditions that typify the open seafloor (see also Sheu and Presley, 1986a,b).

In contrast, under the more reduced conditions that typify highly saline subsurface waters and bottom waters of brine pools, the precipitated forms of iron are reduced phases dominated by pyrite, siderite and other ferroan carbonates. In such diagenetic settings the pH does not vary widely (Figure 9.2b), and the relationships between pyrite and siderite become clearer when activities of dissolved sulphur or dissolved bicarbonate are plotted against Eh. Pyrite is the stable phase in anoxic conditions where levels of dissolved sulphide are high (Figure 9.2c), while siderite and other ferroan carbonates are the stable phases under anoxic conditions where the activities of dissolved sulphide are very low and bicarbonate activities are high (Figure 9.2d). We shall shortly return to this chemistry when discussing dominant mineral phases accumulating on the floors of anoxic brine pools and in subsurface brines derived from the dissolution of buried halite units. The formation of iron silicates, such as chamosite and berthierine (the Fe_3O_4 silicates in Figure 9.2c,d), is favoured in anoxic environments where the activities of both sulphide and bicarbonate are very low. This has important implications for the formation of some types of banded iron deposit on the deep seafloors of the Archaean ocean, but further discussion lies outside the evaporites brief of this book.

Locally, anoxic brines will pond on the deep sea floor wherever vented brines can seep into closed seafloor depressions. The lows form via subsidence atop dissolving shallow allochthonous salt sheets or atop areas of salt withdrawal where the underlying salt is flowing into adjacent growing salt structures (allochthons and diapirs). The redox interface in such seafloor pools is located at the halocline and so may be located some metres or tens of metres above the seafloor.

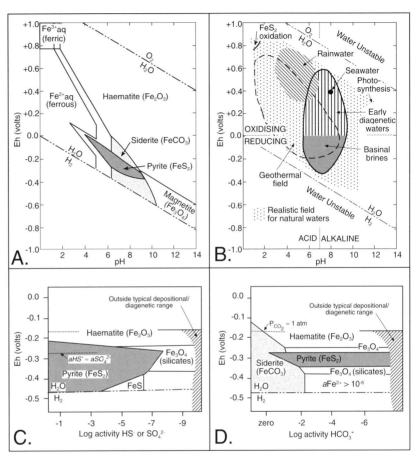

Figure 9.2. Iron stability: phase diagrams showing the stability of iron mineral phases with respect to A) & B) Eh and pH, C) Eh and bicarbonate activity and D) Eh and bicarbonate activity (after Garrels and Christ, 1965; Hesse, 1986; Taylor and Curtis, 1995).

For example, the Orca Basin in the Gulf of Mexico is a closed intraslope depression at a depth of 2,400 metres, some 600m below the surrounding seafloor (Figure 9.3a). Its bottom is filled by a 200m column of highly saline (259‰) anoxic brine that is more than a degree warmer than the overlying seawater column (Figure 9.3b). The pool is stable and has undergone no discernable change since it was first discovered in the 1970s. It is a closed dissolution depression fed by brines seeping from a nearby subsurface salt allochthon (Addy and Behrens, 1980). A large portion of the particulate matter settling into the basin is trapped at the salinity interface between the two water bodies. Tefry et al. (1984) noted that the particulate content was 20-60 µg/l above 2,100m and 200-400µg/l in the brine column below 2,250m. In the transition zone, the particulate content was up to 880 µg/l and contained up to 60% organic matter.

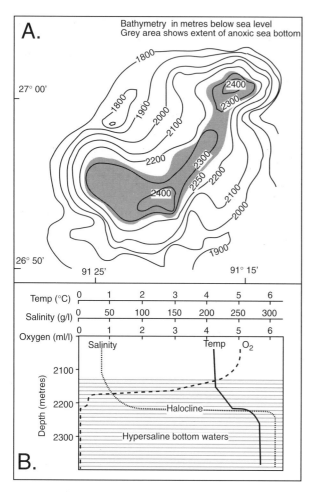

Figure 9.3. Orca Basin, northern Gulf of Mexico. A) Location and bathymetric profile (in metres below sealevel). Grey area outlines the extent of deep highly saline anoxic bottom waters created by the dissolution of nearby allochthonous salt. B) Temperature, salinity and dissolved oxygen profile in the basin (after Williams and Lerche, 1987).

A core from the bottom of the Orca brine pool captured laminated black organic-enriched pyritic mud from the seafloor to 485 cm depth and entrains 3 intralaminite turbidite beds of grey mud with a total thickness of 70cm (Addy and Behrens, 1980). This is underlain by grey mud from 485 cm to the bottom of the core at 1079 cm. The laminated black mud was deposited in a highly anoxic saline environment, while grey mud deposition took place in a more oxic setting. The major black-grey boundary at 485 cm depth has been radiocarbon dated at 7900 ± 170 years and represents the time when escaping brine began to pond in the Orca Basin depression. Within the dark anoxic laminates of the Orca Basin there are occasional mm- to cm-thick red layers where the iron minerals are dominated by haematite and other iron hydroxides and not

by pyrite. These layers represent episodes of enhanced mixing across the normally stable oxic-anoxic halocline and indicate the short-term destruction of bottom brine stratification.

Brine pools do not just form by allochthon solution on the floors of rifts and passive margins, they also form in the compressive terrain of the central and eastern Mediterranean where sheets of Miocene evaporites have been folded and thrust to shallow depths below the deep seafloor. The consequent dissolution is creating saline brine pools (Camerlenghi, 1990). Two of the better-studied marine regions with hypersaline deep bottom waters are the Tyro Basin and Bannock area. In the latter area the various brine-filled depressions or sub-basins create a closed outer moat around a central seafloor mound that is 10 km across (Figure 9.4a).

Both areas occur on the deepsea floor of the Mediterranean, and both areas entrain bottom waters that are more than ten times as saline as Mediterranean seawater. The chemical composition of the Tyro Basin bottom brine is related to the dissolution of a halite-dominated Miocene evaporite, while the chemical composition of the Bannock Basin (Libeccio Basin in the Bannock area) implies derivation from dissolving bittern salts (de Lange et al., 1990). Marine sediment fill in both brine basins ranges in age from Holocene to Late Pleistocene and is made up of turbidites interlayered with laminated intervals composed of alternating green gelatinous muds and grey reduced oozes. As in the Orca Basin, the organic content of these brine pool bottom sediments is much higher than is found in typical deep seafloor sediment (>1-3% TOC; Figure 9.5a).

The anoxic hypersaline brines of the Tyro and Bannock Basins are among the most sulphidic bodies of water in the marine environment, with H_2S concentrations consistently greater than 2-3 mmol (Figure 9.4c; Henneke et al., 1997). In contrast, there is little to no H_2S in the anoxic bottom brine of the Orca Basin. There the iron concentration is 2 ppm, a value more than 1000 times higher than in the overlying Gulf of Mexico seawater. Such high levels of reducible iron in the Orca Basin are thought to explain the lack of H_2S in the bottom brine and a preponderance of framboidal pyrite in the bottom sediments (Sheu, 1987). Both the Orca Basin and the brine pools on the floor of the Mediterranean show sulphate levels that are more than twice that of the overlying seawater.

Pyritic sulphur is the main phase of inorganic reduced sulphur in the Tyro and Bannock Basins, where it makes up

50-80% of the total sulphur pool. It is also present at the same levels in cores from the two basins, that is, around 250 µmol per gram dry weight (Figure 9.5b; Henneke et al., 1997). Humic sulphur accounts for 17-28% of the total sulphur pool in the Tyro Basin and for 10-43% in the Bannock Basin. Sulphur isotope data show negative $\delta^{34}S$ values for both pyritic sulphur, with $\delta^{34}S = -19‰$ to -39‰; and for humic sulphur, with $\delta^{34}S = -15‰$ to -30‰; indicating that both pyritic and humic sulphur originated from microbially produced H_2S within the brine column.

Usually, the density of the bottom brine in the Bannock region is high enough to support finely dispersed organic debris, allowing substantial bacterial mats to float at the seawater-brine interface (halocline) and form suspended deep mid-water bacterial mats (Erba, 1991). While the organics float on the oxic side of the halocline they are subject to oxidation and biodegradation. Owing to the high density of the bottom brines, moving this floating organic material to the floor of density-stratified brine pools appears at first to be a difficult proposition. Yet, organic contents in the anoxic bottom muds beneath these pools are as high as 4-5% organic carbon (Figure 9.5a). The propensity for organics to remain suspended at the halocline is overcome when mineral crystals, including pyrite, precipitate at the interface via diffusive brine mixing in combination with bacterial sulphate reduction. Newly formed crystals attach to organic matter floating at the halocline to create a combined density that carries the suspended pellicular organics to the pool floor. In such a brine pool scenario, the redox front that fixes pyritic sulphur occurs at the halocline,

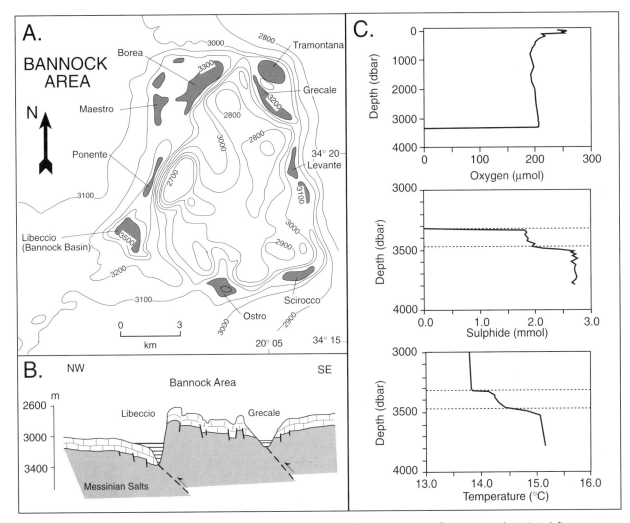

Figure 9.4. Brine pools on the Mediterranean sea floor. A) Bathometry of Bannock area, seafloor contours in metres (after Camerlenghi, 1990). B) Interpreted cross section based on seismic profiles (after Camerlenghi, 1990). C) Water column and brine column characteristics (oxygen, sulphide and temperature profiles; replotted from data tables in Bregant et al., 1990).

319

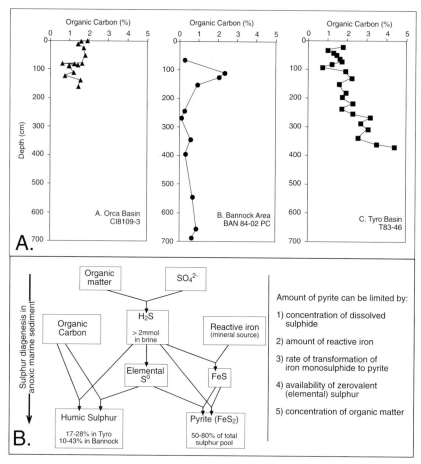

Figure 9.5. Organic characteristics of brine pool bottom sediment. A) Total organic carbon as % dry weight versus depth (plotted from Tyro and Orca data in Table 1 of Sheu, 1987, and Bannock data from Hennecke et al., 1997). B) Modes of occurrence of sulphur in brine pool bottom sediments (after Hennecke et al., 1997).

evaporite, but its seepage-fed genesis means it is a poor analogue for deepwater basinwide salt units (Chapter 3).

Neither the Tyro nor the Bannock Basin bottom sediments show a significant correlation between pyritic sulphur and organic carbon, suggesting predominantly syngenetic pyrite formation (Henneke et al., 1997). That is, both pyritic and humic sulphur preserved in the bottom sediments formed mostly by syngenesis, either in the water column or at the sediment-brine interface. Ongoing diagenetic processes within the bottom sediments form an additional 5% of the total pyrite. For example, van der Sloot et al. (1990) clearly showed that metal sulphides, as well as organics and other minerals, precipitated at the brine-seawater interface in the Tyro Basin. They found extremely high concentrations of Co (0.015%), Cu (1.35%) and Zn (0.28%) in suspended matter at the halocline. These high particulate Co, Cu and Zn concentrations corresponded to sharp increases in dissolved sulphide across the interface (a redox front), and indicate precipitation of metal sulphides at the interface (Figure 9.4c). Humic sulphur in the bottom sediments correlates with the pyritic sulphur distribution and is related to the amount of gelatinous pellicle derived from bacterial mats growing at the halocline between oxic seawater and bottom brine (Erba, 1991, Henneke et al., 1997).

Additionally, the degree of pyritization in the sediments (DOP ≈ 0.62) indicates that present-day pyrite formation is limited by the reactivity of Fe in the Bannock and Tyro Basins and not by the availability of organic matter, a process that limits pyrite formation in most normal marine settings. The degree of pyritization (DOP) is defined as [(pyritic iron)/(pyritic iron + reactive iron)]. Raiswell et al (1988) showed that DOP in ancient sediments can distinguish anoxic from normal marine sediments. Anoxic sediments show DOP values between 0.55 and 0.93, while normal marine sediments have DOP values less than 0.42.

and not at its more typical marine position located at or just below the sediment-seawater interface.

Organic debris then accumulates as preserved pellicle layers within the pyritic bottom muds (laminites). Pellicular debris is also carried to the bottom during the emplacement of turbidites when the halocline is disturbed by turbid overflow (Erba, 1991). Hence, pellicular layers are typically aligned parallel to lamination, or are folded parallel to the sandy bases of the turbidite flows, or line up parallel to deformed layers within slumped sediment layers. Individual pellicle layers are 0.5 to 3 mm thick and dark greenish-grey in colour. Similar pellicular layers cover the surface of, or are locked within, recent gypsum crystals recovered from bottom sediments of the Bannock area. This gypsum is growing today on the bottom of the Bannock Basin, atop areas about the brine pool margin that are directly underlain by dissolving Miocene evaporites (Corselli and Aghib, 1987). It is one of the few modern examples of a deepwater

The DOP levels in the Bannock and Tyro Basins confirm observations made in ancient anoxic sediments. Thus, although the Tyro and Bannock Basin brines differ in their major element chemistry, reflecting a different salt source, their reduced sulphur species chemistry appears to be similar, but significantly different from normal marine systems.

Pools of hypersaline, anoxic brines, underlain by laminated pyritic organic-rich sediments, are far more widespread in halokinetic terrains than just in the Orca, Bannock and Tyro Basins. For example, brine seeps and pools have now been documented atop and adjacent to numerous salt allochthons along the continental slope of the Gulf of Mexico. In the Green Canyon area, where shallow allochthonous salt sheets are dissolving just beneath the seafloor, organic-enriched pyritic sediments are also commonplace (Reilly et al., 1996). Similar methanogenic cool-seep systems are active today along the foot of the Florida Escarpment, they are focused and in part fed by subadjacent dissolving salt allochthons (Paull et al., 1992). Their abyssal chemosynthetic communities are supported by bacterial oxidation of reduced chemicals carried onto the seafloor by methanogenic brines that seep out through sediments at the base of the escarpment. These modern methanogenic seeps are surrounded by carbonate hardgrounds (fibrous calcite and aragonite crusts) and are associated with sediments rich in preserved organic carbon.

The escape of nutritive methane into these brine-filled pockmarks today supports a symbiont-containing mussel community that constructs dense biostromes around the pool rim. This biota is the cold water equivalent of the chemosynthetic seep communities flourishing in the vicinity of black smoker vents (MacDonald, 1992). Vagaries in the rate of brine supply mean many mussels can die in a short space of time, and, along with bacterial, algal and faecal residues, spread out as organic-rich debris atop the halocline. This proto-kerogen then sinks into the anoxic bottom, where it is preserved and protected from further biodegradation. The inherently unstable nature of the seafloor in the vicinity of active salt allochthons means parts of the biostrome are periodically killed "en masse" as sediment about brine pool edge collapses and slides into anoxic waters, carrying with it the chemosynthetic community. As well as further elevating levels of preserved organics in the brine pool bottom sediments, this process also creates potential fossil lagerstatte ("death communities"). Cretaceous counterparts to these methanogenic brine systems have been documented in the Canadian Archipelago where methane-rich mound seeps, dominated by fossiliferous fibrous calcite and aragonite

hardgrounds, grew atop faults related to salt diapirism in the Sverdrup Basin (Savard et al., 1996).

If the evaporite indicators in ancient counterparts of all these brine pool/seep systems are not recognised, then, once the associated allochthonous salt sheets have dissolved or moved on, an evaporitic association for the organic-rich pyritic brine pool sediments will become an enigma. Any remaining organic constituents are made up primarily of marine plankton, a benthic marine biota and bacteria from a normal marine setting. Sediments above and below the pyritic pellicular laminites are made up of normal marine deepwater deposits. There will be little geochemical evidence to show these organic-rich pyritic sediments were preserved in a hypersaline deep marine environment. Evidence of the significance of evaporites in creating such a sequence will only come from sedimentological and structural analysis of core or outcrop, and the recognition of halokinetic breccias, salt welds, chemosynthetic biostromes and marine-cemented hardgrounds within a deeper water marine sediment matrix.

Recognition of similar bottom brine systems is even more enigmatic in the Proterozoic, where elevated levels of methane and CO_2 were the norm for the world's oceans (Grotzinger and Knoll, 1995). In such systems, widespread fibrous calcite and aragonite cementstones/hardgrounds characterised the typical shallow tropical carbonate seafloors as well as the margins of much more localised brine seeps. Unlike today, the Proterozoic seafloor was dominated by splayed acicular cementstones that built dm-m-thick beds separated by shallow marine grainstones and rudstones. Such cementstone hardgrounds are widespread in outcropping carbonates of the Proterozoic basins of northern Australia (pers. obs.), but have been misconstrued by some previous workers as evaporite indicators (e.g. Coxco structures, supposedly after gypsum, in the McArthur Group of the McArthur Basin; see Figures 99, 109 in Jackson et al., 1987).

Ferroan cements (iron ores) tied to salt/allochthons

Within high temperature metamorphic or magmatic settings the migration of iron-rich hydrothermal and basinal waters is now accepted as an important process in the generation of many Fe-oxide (ironstone)-rich Cu-Au deposits (see later). What is not appreciated by many economic geologists is that iron is also highly mobile at lower temperatures (<250°C) that characterise the hypersaline basinal setting, especially in regions where Fe-chloride brines are created by subsalt density-driven convective flow through shales.

Surface expressions of the escape of these Fe-enriched basinal waters are clearly seen as the ferric hydroxides and pyrites so common about modern brine seepage vents (see above). The effects of their subsurface passage to the outflow vent are less understood. However, there is a definite burial diagenetic association between siderite and the hypersaline brines that decompose modern methane hydrates (Matsumoto, 1989). There is also a strong association between evaporitic basinal brines and various styles of ferroan carbonate cement/replacement (Table 9.1). Some of these replacement deposits are mined for iron ore in northern Africa, Spain and Austria. It is a style of pervasive *lit-par-lit* replacement, which is distinct from the better documented nodular siderites, so common as minor early cements in organic-rich early diagenetic settings (e.g. Middleton and Nelson, 1996).

Before any Fe-rich basinal brine can reach a sulphate/sulphide-rich sea-floor vent, the ascending hypersaline brines may precipitate metasomatic/hydrothermal ferroan carbonates in sediments adjacent to the faults and breached salt layers, which define the ascent pathways of basinal waters in evaporite entraining basins. As the temperature rises these ascending Fe-entraining brines are also capable of carrying much of the Pb and Zn that overprints and replace pyrite already present in the brine pool sediment (Figure 8.18). Once a salt allochthon has moved on, or dissolved, all that will be left as evidence of a former salt focus

	PERIDIAPIRIC METASOMATIC SIDERITES	VIETSCH-TYPE SPARRY MAGNESITES	HALL-TYPE SPARRY MAGNESITES
Geometry	Stratabound, irregularly shaped sideritic orebodies within marine platform carbonates.	Large stratabound lenses with pinolitic margins in less metamorphosed deposits.	Layered but 3D distribution not clear. Occurs in abandoned salt mine at Hall in Tyrol, Austria.
Ore Bodies	Ouenza ore bodies, over 4 km in length, occupy central position near SW closure of a major NE/SW striking anticline. Halokinesis contributed to the complicated faulted roof of the diapir that occupies the core of this anticlinal fold. Ouenza had geological reserves of 120 Mt prior to mining. Jerissa orebodies occur within a faulted circular dome structure some 2 km across. Likely this dome is the roof of a diapir structure. Jerissa contained some 45 Mt of geological reserve.	Magnesite reserves range from several to 50 Mt (Austrian deposits). Deposits in Austria, Brazil, Czechoslovakia, India, Korea, Manchuria, Russia, Spain and Tasmania, Australia.	Not economic.
Host lithology	Ouenza ore bodies occur within blocks of bioclastic or biohermal Aptian limestone host. Jerissa ore bodies are mainly within Aptian reef carbonates and their lateral equivalents.	Marine platform clastic/carbonate sediments, typically dolomite and limestones, also in black and grey shales, sandstones and conglomerates. Host ages range from Proterozoic to Palaeozoic. Austrian deposits in Devonian host.	Middle - Late Triassic evaporitic carbonates and shales. Coarsely crystalline pinolitic ferroan carbonate replacing dark-grey bituminous dolomicrites and recrystallised anhydrites.
Metamorphic grade	Africa - diagenetic in North Africa.	Mostly greenschist facies, but range from diagenetic (Asturreta, Spain) to amphibolite (Namdechon, Korea) and granulite grade (Snarum, Norway).	Anchimetamorphic grade. Hosted in "Haselbirge" facies; this is a syntectonic breccia consisting of angular fragments of black pelite, anhydrite, dolomite and polyhalite all in granoblastic anhydrite matrix.
Preserved sedimentary features	Ghost carbonate platform and basin textures, replaced intraclasts and nodules, preserved bedding. No obvious evaporite precursor was directly replaced in a "lit-par-lit" fashion but deposits were formed from upwelling hypersaline evaporite-related brines.	Ghost clastic textures, stromatolites, algal mats, bedding, ripples and cross beds. No obvious evaporite precursor was directly replaced in a "lit-par-lit" fashion but deposits were formed from hypersaline hydrothermal brines that typically occur in an evaporite-entraining basin.	Breccias generated by Alpine thrust tectonics. But ferroan carbonate mineralisation (Early Cretaceous) postdated brecciation related to Alpine thrust tectonics.
Diagenetic features	Earliest siderite is a fine-grained dark-grey to light-brown siderite that is typically recrystallised into a coarser spar.	Pinolites and rosettes of magnesite (including breunnerite), authigenic silica, chert nodules, magnesite or dolomite-filled fractures cutting sparry magnesite. Bi-polar growth of sparry magnesite.	Zoned sparry ferroan carbonate replacing alternating beds of micrometre sized dolomite and clay-rich anhydrite, fine-scale evidence of ongoing dissolution and reprecipitation. Later, outer zones of crystals tend to be more iron-rich.
Fluid conduits and origin	Fault-fed brines in halokinetic terrain. $FeCl_2$-rich brines from dissolving evaporites and adjacent flushed sediments ascend along extensional faults. These brines and other basin-derived fluids migrated upward adjacent to diapirs and through breaches in dissolving diapirs, salt sheets and their altered edges. Fluid flow is driven by high heat and pressure gradients. Solutions rendered acidic and saline via albitisation and other reactions associated with halite dissolution and circulation through nearby pelites.	Fault-fed fluids with varying origins explained as metasomatic/diagenetic replacement from subsurface brines in evaporite-entraining basins with fluids derived from: 1) flushing of redbed pelites. 2) rising fluids from cooler carbonates into hotter overthrust carbonate nappes. 3) rising hot basinal waters. 4) fluids from adjacent active volcanic systems.	Brine movement and brine sediment interaction confined to evaporite complex. Overlying Anisian carbonates are tight and unaffected by ferroan carbonate replacement events. Hall-type sparry magnesites are genetically linked to evaporites. They form by interaction of carbonate precursors with $MgCl_2$-rich brines during late diagenesis and anchimetamorphism. Early Cretaceous (post Alpine) anchizonal thermal event generated a fluid throughout within porous "Haselbirge" facies that precipitated metasomatic breunnerite.
References	Frimmel, 1988; Laube et al., 1995; Pohl et al., 1986; Thibieroz and Madre, 1976	Ilavsky et al., 1991; Kucha et al., 1995; Frimmel, 1988; Morteani and Neugebauer, 1990; Morteani et al., 1982; Pohl, 1990; Pohl and Siegl, 1986; Ribeiro de Almeida and Bergamin, 1993; Velasco et al., 1987	Pohl and Siegel, 1986; Spötl, 1988, 1989a,b

Table 9.1. Characteristics of various styles of ferroan carbonates (Mt = million tonnes).

may be iron-rich replacement haloes about salt welds and faults in what are otherwise normal marine sediments.

The predominance of ferroan carbonates as metasomatic precipitates, merely reflects the relative lack of sulphur in a chloride-rich basinal brine derived by halite solution and, the abundance of carbonate in the platform matrix adjacent to the fluid escape conduits. That is, siderite, ankerite and ferroan magnesite tend to form ferroan replacements and cements in the carbonate matrix, rather than authigenic pyrite, which tends to form from ferroan brines in a sulphide-rich matrix. In settings where sulphide is present, as it is today in the bottom waters of a marine brine pool or in the reduced humic sulphur of buried brine pool laminites, then authigenic pyrite (framboidal) is formed via brine mixing; precipitation occurs at the interface between the ascending chloride brines and marine phreatic waters or anoxic bottom brines (Figure 9.2c).

A modern example of the influence of ambient pore sulphate can be seen in the massive pyrite-calcite cemented sand beds about Gulf of Mexico diapirs, where abundant caprock anhydrite is undergoing bacterial sulphate reduction (McManus and Hanor, 1993). Where sulphur is absent, then siderite and the other ferroan carbonates are the stable mineral phases under similar reducing conditions (Figure 9.2d).

In Precambrian evaporitic basins, where there was even less dissolved sulphate in the subsurface basinal waters than there is in Phanerozoic counterparts, the precipitation of a Fe carbonate replacement was more widespread. Evaporite pseudomorph evidence shows that prior to 1.6-1.8Ga there was little or no sulphate evaporite accumulating in marine halite-dominated systems. This reflects much lower levels of Ca and SO_4 compared with Phanerozoic seawater and meant gypsum was not a widespread marine salt in evaporite beds older than Mesoproterozoic (Chapter 2). Free sulphate was even less likely to be present in subsurface pore waters derived from dissolving marine evaporites than it is today. Rather, the pore brines were chloride-rich solutions also characterised by elevated levels of sodium and bicarbonate ions. Such compositions, along with elevated levels of ferric iron, encouraged the subsurface precipitation of ferroan carbonates over the iron sulphides we see today, often as replacements of precursor limestones and dolomites.

In many Proterozoic and older evaporite-associated systems, where any salt beds have long since dissolved, these metasomatic ferroan carbonate haloes can be used to map ascension pathways of potentially mineralising basinal brines. The explorationist can use this as a regional tool to focus attention on what are likely to be more prospective areas in a basin for base metal sulphides. Mapping such indicators, in association with evaporite solution breccias and other evaporite indicators, can define areas where such brines vented and pooled beneath a sulphur-entraining seawater column.

However, owing to the highly metamorphosed nature of many exploited siderite terrains in Europe (e.g., Erzberg, Austria and Batere, France) and a lack of preserved salt beds other than breccias, many widespread metasomatic siderites and other ferroan carbonates have been attributed to the action of iron-rich magma-derived hydrothermal fluids (e.g. Mostler, 1984). This interpretation has led to the perception in the economic geology literature that most "sparry magnesites and siderites" are associated with magmatic activity. Such models largely ignore the likely effects that the now-dissolved salt allochthons had on focusing Fe-rich fluid flow during basin diagenesis.

Fe-carbonate replacement adjacent to Phanerozoic salt allochthons

Less metamorphosed, ferroan carbonate counterparts to the sparry systems outcrop in the Ouenza (NE Algeria) and Djebissa (Tunisia) siderite deposits of North Africa (Figure 9.6). These iron ore deposits are two of a number of iron carbonate deposits (siderite – ankerite) that outcrop along the Algerian-Tunisian border. Throughout the region the deposits are largely restricted to Cretaceous carbonate hosts, located immediately adjacent to outcropping Triassic diapirs. At the present land surface these Fe-carbonate deposits are partially to completely oxidised to haematite-goethite and are commonly associated with above background concentrations of Pb, Zn, Cu, Ba and F. Both Ouenza and Djebissa are located within 50 km of the peridiapiric Bou Grine Pb-Zn deposit discussed in Chapter 8. Prior to the onset of ore production, Ouenza had geological reserves of some 120-150 million tonnes, while Djebissa entrains some 45 million tonnes of ore. Both mines are important iron sources for their respective countries with Ouenza yielding ≈2.5 million tonnes/yr.

Mapping of the diapirs and their breccias over the last two decades has shown that they were emplaced as salt allochthons or seafloor salt glaciers on the Cretaceous and Tertiary seafloor. At that time mobile Triassic salt was actively flowing down into the basin and was capable of carrying suprasalt blocks of Cretaceous sediments, including reefs (Vila, 1995; Vila et al., 1996a,b). There were three main periods of halokinesis in the region;

Aptian, Vraconian and Miocene. The first two episodes were associated with regional extension and depositional loading, while the Miocene episode was associated with regional compression (Bouzenoune and Lécolle, 1997).

Ore bodies at Ouenza are over 4 km in length, and formed as diagenetic replacements within blocks of bioclastic or biohermal Aptian limestone, which occupy a central position near the SW closure of a major NE/SW striking salt cored anticline. Halokinetic processes contributed to the brittle fracturing of a very complicated roof to the diapir/allochthon that now occupies the core of this anticlinal fold (Thibieroz and Madre, 1976). These fractures focused the ascent of the replacing fluids, and are outlined by ferroan carbonate replacement haloes.

The Ouenza ore bodies are made up of stratiform and crosscutting geometries, both showing a strong structural control by faults, with the major faults striking subparallel to the anticlinal axis. Faults continued to be active after the first emplacement of the siderite, with movement driven by the ongoing flow and dissolution of the underlying and adjacent Triassic allochthon. Cloudy metasomatic siderite and ankerite replacements grade into adjacent fossiliferous limestones at the deposit boundaries. Pockets of calcite, baryte, quartz and minor sulphides (tetrahedrite, chalcopyrite and pyrite) are also found in these ferroan-carbonate-replaced areas. Later vein-type mineralisation with Pb, Zn, Cu and Ba in a calcitic, dolomitic quartz gangue occasionally cross-cut the sideritic ore bodies and their surrounding country rocks. Nearsurface siderite is extensively oxidised to goethite/haematite.

The iron mineralisation at Ouenza was emplaced mostly in the extensional phase and it tends to be hosted by Aptian neritic limestones, which contain rudists and foraminifera

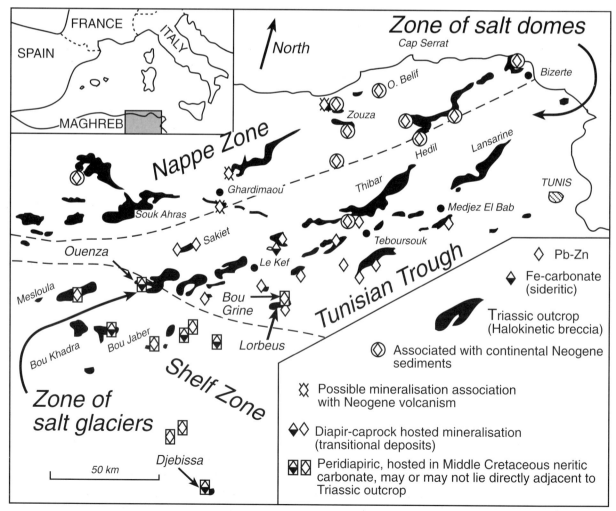

Figure 9.6. Regional geology of Tunisia. Illustrates the association between zones of outcropping siderite replacement, evaporites and Pb-Zn (after Pohl et al., 1986; Vila et al., 1996).

(Bouzenoune and Lécolle, 1997). Coral limestones of the same age are mostly barren. This may reflect an early loss of permeability by pervasive syndepositional marine cements, which then prevented the pervasive entry of Fe-rich pore waters. The main expression of the iron carbonate in the mine is as fine-grained siderite and ankerite, produced by the replacement of the host rock, particularly in those carbonates that contain rudist fragments. In addition, there are crosscutting veins of white sparry ankerite and blonde sparry siderite. The ores commonly enclose blocks of unreplaced host rocks and exhibit a brecciated texture, which encloses euhedral quartz. Sulphides, sulphosalts, baryte and fluorite form mostly as NE-SW oriented vein fills, which cross-cut the iron ore and the folded Aptian-Albian sequence arguing that the sulphide mineralisation is Neogene. It is commonly silicified.

Oxygen isotope values for the ferroan/mineralised portions of the recrystallised limestones at Ouenza are more negative than typical Cretaceous marine values, and more negative than the coral limestones, which tend not be replaced by ferroan carbonate (Figure 9.7; Bouzenoune and Lécolle, 1997). Siderite and ankerite tends to be even more negative in their $\delta^{18}O$ signature than the recrystallised limestone, with the ankerites more negative than the siderites. In general, the oxygen isotope signature of the host limestone becomes more negative as the contact with the iron carbonate interval is approached. The $\delta^{13}C$ values lie in the normal range for marine carbonates, so that Bouzenoune and Lécolle conclude that ferroan mineralisation had little effect on the carbon isotopic signature. The same mineralising fluids that introduced Fe to the system, also introduced minor amounts of Mn and Mg, while removing Sr. The ankerite and the siderite were both precipitated from the same fluid, which, according to Bouzenoune and Lécolle (1997) and Pohl et al. (1996), was a basinal fluid (120-150°C) whose escape was focused by the same structures that controlled the structure of the halokinetic flows.

At Djebissa, the ore bodies occur within a circular dome

structure some 2 km across. Traces of underlying evaporite rocks are found as injections into the faults. Based on regional geology, Pohl et al. (1986) postulate a diapir/allochthon roof once existed immediately below the deposit. As in Ouenza, deformation is dominated by faults and the major ore bodies are aligned along a SW-NE trending structure. Siderite ore bodies at Djebissa occur mainly as replacements in reef carbonates and their lateral equivalents. There are also minor stratiform, but predominantly crosscutting, siderite vein ores. Partial replacement of fossiliferous limestone is clearly seen about the edges of the Djebissa ore bodies, as well as in the ankerite and coarse sparry calcite haloes. Near these metasomatic ferroan margins the rock is often vuggy and brecciated. Bedding parallel stylolites can be traced from the limestones into the sideritic bodies, showing siderite replacement of the marine carbonate was postlithification.

A combination of halokinesis and hydrothermal fluid focusing of hot basinal brines, tied to syndiapiric faulting, explains this style of siderite replacement (Pohl et al., 1986; Bouzenoune and Lécolle, 1997). Ascending Fe-rich basinal brines were derived by reflux-driven convective flushing beneath deeply buried, flowing and dissolving Triassic salt beds. Driven by high heat flows and pressure gradients, ferroan chloride-rich basinal brines migrated upward, into and through breach points in the diapirs and salt sheets and along their tectonised margins.

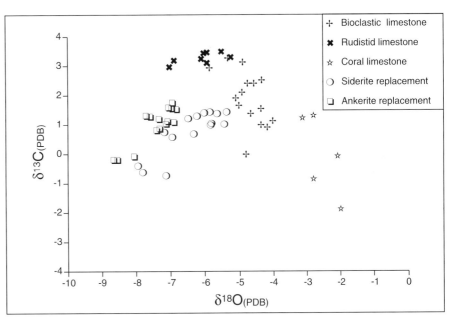

Figure 9.7. Crossplot of oxygen and carbon isotopes in the Ouenza iron deposit, NE Algeria. Clearly shows the more negative oxygen isotope signature of the replacement process (compiled from data in Table 3 and Table 4 in Bouzenoune and Lécolle, 1997).

Deeper in the basin sequence, the diagenetic reactions between refluxing brines, derived from deeply buried dissolving halite beds, and nearby fine-grained basin sediments, such as:

$$Muscovite + 6SiO_2 + 3NaCl \longrightarrow 3\ albite + KCl + 2HCl$$

along with organically mediated pyrite oxidation and other reactions discussed in Chapter 8, rendered these convecting brines acid and slightly oxidising. Their high chloride content enhanced their metal-carrying capacity. Iron (and lesser amounts of Pb and Zn) was leached, as these brines flushed adjacent redbed and shale aquifers. As the Fe-rich fluids rose to shallower depths, their ascent was focused by the edges and undersides of salt allochthons and along faults and joints in the fractured roof rocks. Extensional faults acted as the main fluid conduits atop the breached diapirs and salt sheets. Changes in Eh, pH and pressure/temperature conditions in the ascending brines facilitated the metasomatic replacement of the flushed carbonates by siderite and other ferron carbonates (Figure 9.2).

Recent work by Laube et al. (1995) has shown that convecting basinal waters, not magmatic waters of the earlier interpretations, precipitated the classic, but now more metamorphosed, sparry and laminar siderite ores of the Steirischer-Erzberg region of Europe. Carbon and oxygen isotope values in these siderites still retain the original diagenetic signature of the siderite replacements, even after multiple episodes of superimposed metamorphism. Ferroan basinal waters, driven by increased heat flows associated with Devonian rifting of the Noric terrane from Africa, ascended along active extensional faults to precipitate siderite as a replacement of the host carbonates and shales. A likely supply of iron was via convective flushing of hypersaline basinal brine through the underlying Ordovician volcaniclastics. Later regional metamorphism, related to collision tectonics in the Late Carboniferous (Hercynian), and later again during the Alpine Orogeny, caused intense recrystallisation and partial mobilisation of the various ferron carbonates, including earlier siderites, at temperatures in excess of 300°C.

Fe-carbonate replacement as Proterozoic evaporite indicators?

Sparry ferroan magnesites, siderites and other ferroan carbonates are commonplace hydrothermal precipitates from basinal brines in many Proterozoic basins of northern Australia (Table 9.2; see p. 337). In some regions, such as

the Pine Creek Inlier, intense metamorphism has largely overprinted their diagenetic origin (Chapter 6), while in other regions, such as in sub-greenschist facies sediments of the McArthur Basin, the diagenetic relationships are still clearly preserved.

My own field work in the McArthur Basin over the past three years has shown the "siderite marbles" of Jackson et al. (1987) have been incorrectly interpreted as gypsum pseudomorphs. In reality, they are sparry hydrothermal ferroan replacements of marine platform sediments. They occur along and adjacent to faults, fractures, salt welds and joints that were fed by subsurface $FeCl_2$-rich basinal brines. Away from their feeder faults, the "siderite marbles" show a tendency to become bedding parallel and then pass laterally into non-replaced sediments. This reflects preferential fluid entry along beds that were more permeable at the time the Fe-Cl rich basinal brines were ascending the feeder faults. Thus, coarser grained beds (grainstones and floatstones), along with joint planes, show a propensity to be replaced first and most effectively by ferroan carbonate. Less permeable micritic intervals tend not to be fully replaced and so retain only the early stages of ferroan carbonate precipitates, such as cm-sized isolated lozenges of ferroan carbonate that creates a pinolitic texture.

These sparry lozenge-shaped crystals originally formed as diagenetic replacements in dolomitic, open marine, intraclastic and peloidal wackestones, grainstones and rudstones. They do not, as reported in earlier work, define former beds composed of lenticular gypsum. A nonstratiform facies expression of the "siderite marbles" can be clearly seen when the edges, and not the permeability-parallel tops, of these stratiform ferroan bodies are walked out and mapped. The margins of the replaced regions are typically pinolitic and facies transgressive. The edges of a "sideritic marble" unit do not parallel bedding but, at the broad scale, follow faults or pass laterally into unaltered carbonate.

On the mesoscale, the replaced ferroan intervals can sometimes be seen to nucleate off joint planes (e.g. Figures 74, 75 in Jackson et al., 1987; Plate 1 in Pietsch et al, 1991). Similarly, bedding planes and chert nodule horizons in adjacent unaltered carbonates can be traced into the sparry ferroan carbonate replacement zones without any loss in vertical spacing of the bedding indicators. This clearly shows that the sparry replacement did not form as a replacement of a bedded evaporite unit nor of its solution breccia. Likewise, the chert nodules that remain in the "siderite marble" zones are smooth-walled nodules with normal marine affinities. There are no knobbly surfaced cauliflower cherts after anhydrite, gypsum or trona in any of the "siderite marble" units.

XRD analysis shows these "siderite marbles" of the McArthur Basin are not composed solely of siderite but of a range of ferroan carbonates dominated by sparry ferroan magnesites. In many sampled outcrops siderite was not present in the coarse spar facies. The mineralogy of the replacing sparry carbonates is in large part dependent on what lithology was replaced. Ferroan magnesites tend to dominate where platform dolomite was replaced. Siderite tends to occur where a calcitic precursor mineralogy dominated, or in more localised limestone regions that were flushed by large volumes of bathyphreatic $FeCl_2$ brine. In modern outcrop all these ferroan carbonate phases ultimately weather to fine powdery to massive goethite/ haematite, and so reliable analysis of the ferroan carbonate parent is best conducted on samples from core or mine workings.

Thin sections of the "siderite marbles" retain poikilitically enclosed ghost textures of peloidal carbonate grain fabrics, which were the dominant textures in the precursor platform carbonates. Lenticular outlines of individual ferroan crystals (e.g. Figure 74 in Jackson et al. 1987) do not mimic lenticular bird-beak gypsum or axe-head anhydrite (Chapter 1). Rather, the crystals show the characteristic blunt-ended or stubby lozenge-shaped outlines of pinolitic ferroan carbonates (especially sparry ferroan magnesite and siderite). The textural term "pinolitic" describes the two-dimensional similarity of aggregates of the lenticular lozenges to the patterning along the outside of a closed or partially closed pine-cone.

The old notion of a "sideritic marble" being a replacement of a syndepositional gypsum mush bed, means the presence of siderite lozenges were used incorrectly by some authors to map outcrop of particular formations in the McArthur Group, such as the Amelia Dolomite. Elsewhere, in other Proterozoic basins of the Northern Territory, the same lenticular texture in sparry ferroan carbonates has been applied inappropriately as an indicator of evaporite pseudomorphs. For example, Muir (1987) makes an early diagenetic lenticular gypsum interpretation of what are hydrothermal pinolitic structures in fault-fed metasomatic magnesites in the Pine Creek Inlier. Work by Bone (1983), and later by Aharon (1988), on the same Pine Creek magnesites correctly interprets them as indicators of epigenetic replacement of a dolomite precursor by throughflushing of hot basinal brines. Interestingly, Muir (p.8 opp. cit.) quotes the misinterpreted "sideritic marbles" of the Amelia Dolomite in the McArthur Basin as her main

argument that the Pine Creek magnesites are low temperature replacements of an earlier evaporite bed and so erroneously concludes the marbles cannot be hydrothermal replacements.

Sideritic marbles in the McArthur Basin, and other similar fabrics worldwide, are not formation or precursor specific but process specific; they indicate replacement of platform carbonates by ferruginous chloride-rich basinal waters supplied via hydrothermal feeder zones located near active faults or salt welds. The ultimate source of these chloride-rich waters was from convecting basinal brines, with their high levels of chloride supplied by thick dissolving halite units, which were buried deeper in the basin.

When occurrences of "siderite marbles" are plotted against the McArthur Group stratigraphy, the "marbles" are not found deeper in the stratigraphy than the Wollogorang Formation, implying the ultimate source of the ascending chloride-rich brines was at this level in the stratigraphy. In the McArthur Basin the primary salt weld was probably at the level of the Wollogorang Formation.

Multiple levels of evaporite dissolution breccias in the well DD91RC18 (Figure 9.8) confirm the former presence of a number of thick intervals of now dissolved thick salt beds in the Mesoproterozoic Wollogorang Formation. This core, and many Wollogorang breccia outcrops, show characteristic salt dissolution features, such as smooth breccia bases that are overlain by thin beds of insoluble residues, along with commonplace radial quartz micronodules and length-slow quartz euhedra preserved within the breccia matrix. The breccia beds pass up through rubble breccia into mosaic to crackle-brecciated bed tops (see Chapter 4).

When public-domain base-metal data are replotted against same scale sedimentological log, the utility of an applied sedimentological approach in the economic geology of the region is obvious (Figure 9.8). The crossplot clearly shows that the Pb and Zn anomalies are suprasalt and centred on dark organic-rich laminites (locally known as the "ovoid beds"), while the Cu peaks occur mostly in dissolution-effected breccia intervals. Cu migration through this sequence is clearly later diagenetic and postdates salt solution. In contrast, the Zn and Pb anomalies show a early diagenetic (possibly syngenetic) control, perhaps in subaqueous depressions that filled with anoxic bottom waters.

In summary, if ancient fine-grained or sparry ferroan replacements in sedimentary hosts are interpreted as possible palaeoplumbing indicators, rather than as pseudomorphed evaporites or as the products of magmatic waters, they become useful indicators of fluid conduits, which typically centre of extensional faults. They show that metalliferous $FeCl_2$ basinal brines have ascended from depth. The lowest occurrence of these cements in the stratigraphy may well define the level of the former thick salt, which supplied the high chloride content to the basinal brine. Such a mother salt bed also acted as the overpressured seal, which facilitated hydrothermal convection and metal enrichment of density-fed brines within the underlying shales (see Chapter 2). The flow and breaching of the same mother salt also facilitated the formation of various structures that acted as conduits for the escape of the Fe-rich basinal brine. During their subsequent ascent along focused conduits (faults, fractures, joints and salt welds), the brines leaked into adjacent permeable sediments to form characteristic pinolitic and massive replacements. After we take a look at the large scale metalliferous brine pool sediments on the modern deep seafloor of the Red Sea, we shall return to this notion of ferroan carbonates as palaeo-plumbing indicators in our discussion of evaporite-associated sedex deposits.

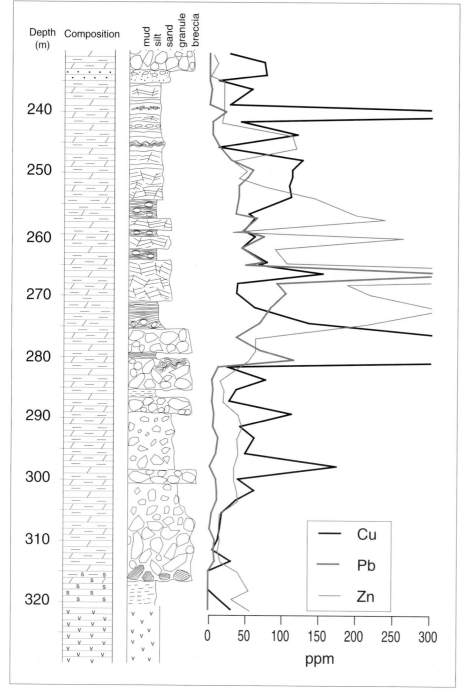

Figure 9.8. Sedimentological log of salt solution breccia interval in Proterozoic Wollogorang Formation, McArthur Basin, Australia (borehole DD91RC18). Metal log compiled from open file data tables held at Northern Territory Geological Survey.

Red Sea deeps

The Red Sea is a young intracratonic basin atop a region of the seafloor that is currently transitional from continental to oceanic rifting. Beneath the Red Sea there are now eighteen known brine pools or deeps, some of which are underlain

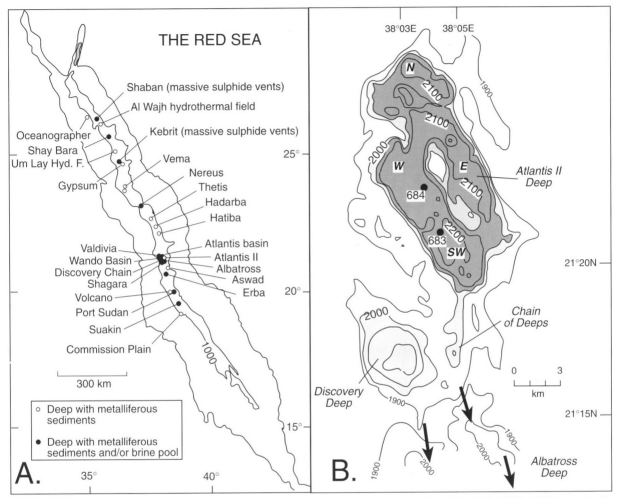

Figure 9.9. Red Sea deeps. A) Locality map showing different types of deeps (in part after Blum & Puchelt, 1991). B) Bathymetry of the Atlantis II deep, contours in metres (after Anschutz and Blanc, 1995).

by metalliferous sediments (Figure 9.9a; Blanc and Anschutz 1995, Blum and Puchelt, 1991). Economically, the most important brine pool is in the Atlantis II Deep; other smaller deeps with metalliferous muds include Commission Plain, Hatiba, Thetis, Nereus, Vema, Gypsum, Kebrit and Shaban Deeps. Along with the laminites of the Atlantis II Deep, the Kebrit and Shaban deeps are of metalliferous interest in that fragments of massive sulphide from hydrothermal chimney sulphides were recovered in bottom grab samples (Blum and Puchelt, 1991). All brine-filled deeps are located in depressions along the spreading axis, in the region of the median valley. Most of these axial troughs and deeps are also located where transverse faults, inferred either from bathymetric data or from continuation of continental fracture lines, cross the median rift valley in regions that are also characterised by nearby halokinetic Miocene salt.

Three conditions favour the formation of the brine-filled deeps in the region: 1) The Red Sea is a narrow depression only 200 km wide, with water circulation to the south limited by a shallow subsea sill near the connection with the Indian Ocean. 2) The slow rate of clastic or pelagic accretion along the axial part of the rift allows transverse faults to influence the topography of the seafloor. 3) Nearby dissolving halokinetic salt contributes huge volumes of hydrothermal and basinal brine to an irregular bottom topography it helped create (Blum and Puchelt, 1991). Many deeps are filled by permanently stagnant brine pools derived either directly from the dissolution of shallow evaporites or overspill from adjacent brine pools. Others deeps, filled by episodic hydrothermal brine seeps, may have been intermittently affected by hydrothermal convection within the bottom brine.

329

Atlantis II Deep, Red Sea

The Deep is roughly elliptical in shape, 14 km long, 5 km wide and is located in the most geothermally active part of the central Red Sea (Figure 9.9b). The bottom of the brine pool is 2,200 metres below sea level and is flanked on all sides by thick sequences of halokinetic Miocene evaporites and clastics (Figure 9.10a). Through subsurface dissolution and convective flushing, these Miocene evaporites supply ions to the seafloor and to high salinity brines caught up in hydrothermal flow cells beneath the salt units. Hot hybrid brines ultimately discharge onto the floor of the Deep, and settle into the main depression to form a two layered brine pool that is some 200 m thick and is underlain by hypersaline metalliferous muds (Figure 9.10b). For example, the southwest portion of the Atlantis II Deep contains 150 to 200 million tonnes of unconsolidated mud averaging 5-6 wt% Zn and 0.8 wt% Cu on a dry salt free basis. These tonnages rival some of the larger sediment-hosted deposits in the world and represent the only known deposit of Fe, Mn, Zn, Cu, Ag and Au of economic proportions currently forming on the sea bottom (Shanks, 1983; Backer and Lange; 1987).

The Deep was first discovered in 1966, it contains about 5 km^3 of brine, stratified into distinct dense and convective layers of differing temperature and chlorinity. The lower brine body has an area of 1.8 km^2 and is about 150 m thick atop a sea bottom as deep as 2,040 metres (Figures 9.9b, 9.10a). In 1977 the bottom brine had a temperature of 61.5°C, a chlorinity of 156‰, and contained no free oxygen. The overlying 50-m-thick brine layer of the Transition Zone was at 50°C with an 82‰ chlorinity. This partially oxygenated upper brine was a transitional zone of mixing between the lower brine and overlying normal seawater. Since then there has been a substantial increase in temperature of the brines (Anschutz and Blanc, 1996; Hartmann et al., 1998). By 1995 the temperature in the bottom brine of the SW basin was 71.7°C. The structure of the lower transition zone, between about 1990 m water depth and highly saline bottom brine, has also changed significantly; it now contains two convective layers with nearly constant temperatures (61° and 55°C, respectively). What is most significant is the long term stability of the lower brine, and its isolation from the overlying normal seawater. For the last 26 years, almost all additional heat and salt supplied to the Deep was confined to the depression and caused an increase in temperature , very little heat was dispersed into the overlying seawater (Anschutz and Blanc, 1996). The rate of heat input to the Deep was constant during this period, and amounted to 0.54 x 10^9 W. The salt input was also constant, and equalled 250-350 kg/s. The effect of this longterm brine pool stability has been the accumulation of some rather unusual sulphide-rich saline sediments.

In the last 28,000 years some 10 to 30 metres of the oxidic-silicatic-sulphidic laminites, along with hydrothermal anhydrites, have accumulated beneath the Atlantis brine pond and atop the basaltic basement (Figure 9.10a,b; Shanks and Bischoff, 1980; Pottorf and Barnes, 1983; Anschutz and Blanc, 1995). Metalliferous sediments beneath the floor of the deep are composed of stacked delicately banded mudstones with bright colours of red, yellow, green, purple, black or white indicating varying levels of oxidised or reduced iron and manganese. They are anhydritic and very fine grained, with 50-80% of the sediment less than 2μm in size. Intercrystalline pore brines constitute up to 95 wt% of the muds, with measured pore salinities as much as 26 wt% (Pottorf and Barnes, 1983).

The sulphide-rich layers are a metre to several metres thick and form laterally continuous beds that are several kilometres across. Sulphides are dominated by very fine-grained pyrrhotite, cubic cubanite, chalcopyrite, sphalerite, and pyrite, and are interlayered with iron-rich phyllosilicates (Zierenberg and Shanks, 1983). Sulphur isotope compositions and carbon-sulphur relations indicate that some of these sulfide layers have a hydrothermal component, whereas others were formed by bacterial sulphate reduction. Ongoing brine activity began in the western part of the Deep some 23,000 years ago with deposition of a lower and upper sulfide zone, and an amorphous silicate zone (Figure 9.10b). The metalliferous and nonmetalliferous sediments in the W basin accumulated at similar rates, averaging 150 kg/k.y./m^2, while metalliferous sediments in the SW basin accumulated at a higher rate of 700 kg/k.y./m^2 (Figure 9.10b; Anschutz and Blanc, 1995). The lowermost unit in the pile in the W basin consists mainly of detrital biogenic carbonates with occasional thin beds of red iron oxides (mostly fine-grained haematite) or dark interbeds entraining sulfide minerals.

Hydrothermal anhydrite in the Atlantis II sediments occurs both as at-surface nodular hydrothermal beds around areas where hot fluid discharges onto the sea floor and as vein fills beneath the sea floor (Degens and Ross 1969, Pottorf and Barnes 1983, Ramboz and Danis 1990, Monnin and Ramboz 1996). White nodular to massive anhydrite beds in the W basin are up to 20 cm thick and composed of 20-50 μm plates and laths of anhydrite, typically interlayered with sulphide and Fe-montmorillonite beds. The central

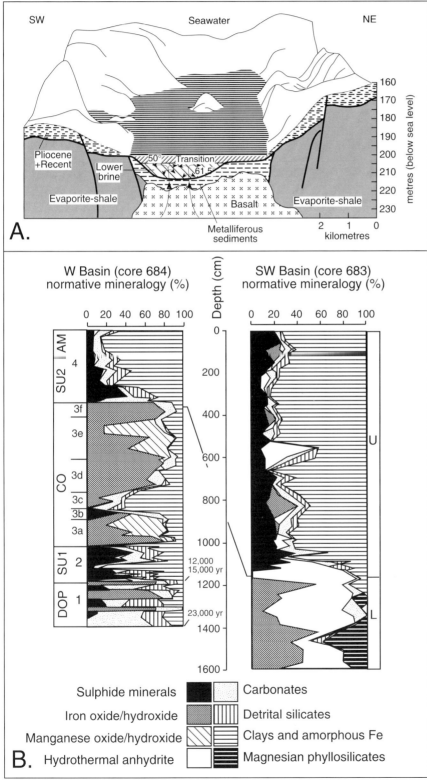

portion of individual anhydrite crystals in these beds can be composed of marcasite. The lowermost bedded unit in the SW basin contains much more nodular anhydrite, along with fragments of basalt toward its base. Its 4 metre+ anhydritic stratigraphy is not unlike that of nodular sekko oko ore in a Kuroko deposit, except that the underlying volcanics are basaltic rather than felsic (see later).

The anhydrite-filled veins that crosscut the laminites act as conduits by which hot, saline hydrothermal brines vent onto the floor of the Deep. Authigenic talc and smectite dominate in deeper, hotter vein fills, while shallower veins are rich in anhydrite cement (Zierenberg and Shanks, 1983). The vertical zoning of vein-mineral fill is related to cooling of the same ascending hydrothermal fluids, with stable isotope ratios in the various vein minerals indicating precipitation temperatures up to 300°C.

Because of anhydrite's retrograde solubility, it can form by a process as simple as heating circulating seawater to temperatures in excess of 150°C (see later). Pottorf and Barnes (1983) concluded that the bedded anhydrite of the Atlantis II Deep, like the vein fill, is a hydrothermal precipitate. Based on marcasite inclusions in the anhydrite units, it precipitated at temperatures down to 160°C or less. At some temperature between 60 and 160°C, probably close to 100-120°C, the fluid became undersaturated and hydrothermal anhydrite precipitation ceased. Thus, anhydrite distribution in

Figure 9.10. A) Schematic of Atlantis II Deep, Red Sea. Note the preponderance of dissolving evaporites adjacent to the floor of the deep. B) Mineralogical proportions in core from W and SW basins (after Anschutz & Blanc 1995).

331

the Atlantis II deep is related to the solution mixing and thermal anomalies associated with hydrothermal circulation.

The fact that Holocene sediments in the Atlantis II Deep contain sulphate minerals and that particulate anhydrite is still suspended in the lower brine body strongly suggests that anhydrite is stable at the bottom of the water column or is at least only dissolving slowly. These conclusions were clarified by Monnin and Ramboz (1996), who found that the Upper Convective Layer (UCL; or Transition Zone) of the Atlantis II hydrothermal system was undersaturated with respect to hydrothermal anhydrite throughout their study period, 1965-1985. The system reached anhydrite saturation in the lower brine only for short periods in 1966 and 1976.

What distinguishes the hydrothermal salts and sediments in the Atlantis II Deep from the much smaller hydrothermal accumulations in volcanogenic hosted massive sulphide (VHMS) deposits (see later) is that the hot metalliferous brines now found on the seafloor achieved their elevated salinities and acidic, saline chemistries via hybrid hydrothermal circulation of seawater both through seafloor basalts and adjacent dissolving halite intervals (the nearby halokinetic Miocene salt). Heat to drive this circulation probably came from the hot basaltic rocks in the slowly spreading axial zone, while most of the chloride and fluid focusing was contributed to circulating seawater by nearby dissolving salts. Thus a seawater/brine flushing of newly formed basalts as well as of Miocene evaporitic sediment both contribute metal to the brine pool. This is clearly seen in studies of lead isotopes in the brine pool laminites, which indicate that a substantial portion of the lead came from the Miocene evaporites and their clastic interbeds and not from the basalts (Dupre et al., 1988).

As in anoxic brine pools elsewhere in the world (see above), the reduced sulphur in the shallow pore brines and the bottom brine layer were mostly derived from bacteriogenic reactions using seawater sulphate and entrained organic matter. This created conditions suitable for widespread metal sulphide precipitation at the halocline, in the brine column, and in brine-saturated sediments beneath areas covered by stable anoxic H_2S-enriched brines. Brine pool stability and an associated lack of mixing with overlying seawater allowed metal sulphides to accumulate across large areas of the brine pool floor and not just near vent localities. Hence, the larger volume of the metal resource and the greater dispersal of metal sulphides in the Atlantis II Deep than in the much smaller near-vent precipitation sites typical of VHMS deposits, such as TAG

Mound, the smokers of the east Pacific rise, and in Kuroko- and Cyprus-style deposits (see later). It is the hydrological and hydrogeochemical effects of the dissolving halokinetic evaporites that distinguish this style of hydrothermal "sedex" precipitate from much smaller open marine VHMS deposits.

Kebrit and Shaban Deeps, Red Sea

Massive sulphides dredged from the Kebrit Deep (Arabic for sulphur) represent black smoker fragments, novel to most Red Sea brine pool deposits (Figure 9.9a; Blum and Puchelt, 1991). TV camera observations at 1366m depth in the Kebrit Deep, and just above the seawater/brine interface at 1370 m, imaged some 40+ inactive smokers. Brines below the interface in Kebrit Deep at a depth of 1373m have a temperature of 24.3°C and a salinity of 256‰. Above the interface the brine has a temperature of 21.5°C and a salinity of 41.9‰. According to Hartmann et al. (1998) there have been no significant changes in temperature, brine level or chlorinity over the last 23 years. The transition from high saline brine to normal seawater salinity probably extends over a depth of only a few decimetres or less. As in the Atlantis II Deep, the dissolution of nearby Miocene evaporites accounts for the very high salinities in the bottom brine layer.

Piston core sampling in the Kebrit Deep revealed stratified metal-enriched sediments in the upper portions of the sedimentary column, while box and dredge sampling recovered specimens of elemental sulphur, sulphide and sulphates. The recovered sulphides consist primarily of Fe- Zn- and Pb-bearing phases that are often tar and asphalt impregnated, and probably derived from hydrothermal cracking of organics held in adjacent sediments. The sphalerite typically forms banded colloform nodules. Concentration of Ni in discrete bravoite aggregates points to a basalt-brine leaching process as a source for at least some of the metals in the brine pool. Unlike the Atlantis II Deep, Cu-sulphides are virtually absent from the paragenesis. This, and the presence of gypsum rather than anhydrite interlayers in chimney fragments points to formation of metal sulphides at lower temperatures compared with the Atlantis II Deep. Sulphide layering in the chimneys shows classic zonation, with Zn-bearing phases preferentially lining the interiors of the chimneys while pyrite is dominant in the outer portions.

The bottom brine layers in the Shaban Deep in the northern part of the Red Sea are 200m thick, but the brines are only 2-3°C warmer than the normal bottom waters (Cocherie et al, 1994). An insignificant decrease in temperature (to about 24.1°C) has occurred in the upper brine since detection of the Shaban Deep in 1981 and no changes were found for the brine-seawater level (Hartmann et al., 1998). The bottom sediment in the Shaban Deep is somewhat unusual compared with other brine pool sediments in the Red Sea in that the sediment consists of biodetrital silicate mud (diatomaceous), with occasional volcanic glass fragments and only a thin (4 cm thick) hydrothermal sediment layer. This lithofacies is similar to the lowermost layer in the western part of the Atlantis II Deep that lies beneath the metalliferous sediments. This biogenic layer in the Shaban Deep overlies a tholeiitic ferrobasalt with transitional affinities, it too is different from the typical tholeiite basalt found further south. Thus, the Shaban Deep, the northernmost deep in the Red Sea rift, represents a very recent stage of oceanic opening. The lack of pervasive hydrothermal signature in the Shaban Deep compared with the Atlantis II Deep is also seen in both the Sr-Nd isotopes and the rare earth element signatures (Cocherie et al., 1994). Metal sulphides in Shaban Deep are less frequent than those in Kebrit, with framboidal pyrite being the dominant ore mineral.

Gypsum occurs in the bottom sediments of both deeps, both as euhedral prisms and as lenticular forms up to several centimetres across that often interpenetrate the sediment layering to construct classic "desert rosette" morphologies, especially, in the Kebrit Deep where gypsum is commonplace. Deeper drilling has yet to be undertaken in either of these deeps, but it is likely that this gypsum forms atop areas of subcropping and dissolving autochthonous salt in the same way morphologically identical gypsum forms atop areas of dissolving halokinetic salt in the deep bottom sediments of the Mediterranean brine pools and the Dead Sea (see earlier and Warren, 1989). Sulphur isotopes from gypsum in the rosettes in Kebrit Deep have values in the range +25.1‰ to +30.0‰. These values clearly exceed those of normal seawater-sulphate-derived gypsum and imply a significant input of bacteriogenically fractionated residual seawater sulphate. Some samples show intimate intergrowths of idiomorphic gypsum and spongy sulphide, pointing to coprecipitation of both minerals from the same hydrothermal fluid in the presence of possible bacterial mediators. Inclusion studies show rather low sulphide and sulphate precipitation temperatures of 110-130° C in the Kebrit brine pool and 100° C in the Shaban Deep. In either deep, the gypsum, like the anhydrite in the Atlantis II deep, is not a true solar-derived deepwater evaporite, but a hydrothermal or a burial salt.

Red Sea summary

Stable stagnant brine pools have developed in sea floor depressions along the medial ridge of the Red Sea in close association with adjacent thick dissolving halokinetic evaporites. Hot metalliferous brines focused by salt allochthons and escaping into the lower convective layers of these pools become widely dispersed across the extent of the deepwater brine pool (Atlantis II Deep). During active venting of metalliferous brines, the longterm brine pond facilitated stable anoxia at the sea bottom, as well as intra-brine-pool convection and the dispersal of metalliferous brine away from the immediate vent area. Diffusive mixing at the halocline atop the brine pool precipitated metal sulphides, mostly as pyrite framboids that then sank to the bottom carrying with them dispersed organic water. Thus, a stable bottom brine pool increases the area of prepared ground for any subsequent Fe, Pb, Zn and perhaps Cu emplacement. The leaching of adjacent thick halite units not only enhances the permanency of the brine pool, it also raises the metal-carrying capacity of hydrothermal brines and increases the reaction rate of any saline fluids flushing metals from the basaltic basement and suitable first cycle sediments. The presence of a dissolving salt aquiclude atop seafloor basalt also focuses reflux driven hydrothermal fluids into the outflow zones about the edges of the stable brine pools. Although not yet documented as a control on the location of the deeps on the Red Sea floor, it is likely that synkinematic salt gliding formed allochthonous salt sheets that aided restriction and brine focusing in the sea floor depressions. The Red Sea is, after all, a classic extension basin with well developed halokinetic sequences along its length (see papers in Jackson et al., 1995, especially Heaton et al., 1995).

Are some ancient sedex deposits a type of evaporite-associated exhalite?

Lydon (1996) defines a sedex deposit as "a sulphide deposit formed in a sedimentary basin by the submarine venting of hydrothermal fluids and whose principal ore minerals are sphalerite and galena". Goodfellow et al. (1993) added that the ore is typically a regularly layered sulphide interbedded with other hydrothermal products, such as chert or baryte, or host lithologies. A fault scarp with aprons of sedimentary fragmental lithologies (breccia flows, talus breccias and conglomerates) commonly defines the position of a nearby fault zone often marked by hydrothermal vents that feed hot brine into the deposit. The dominant sulphide mineral in most sedex deposits worldwide is pyrite, although in some deposits (e.g. Mt Isa and Sullivan) pyrrhotite is dominant. In addition to sphalerite and galena, copper (as chalcopyrite) can be economically important (Mt Isa, Rammelsberg). In the case of Mt Isa the copper mineralisation postdates the lead and zinc by more than 150 My. Goodfellow et al. (1993) and Turner (1992) go on to stress the importance of active rifting and bottom anoxia as two other favoured characteristics. Werner (1990) emphasised the importance of extensional faults as feeder conduits for the mineralising solutions into a sedex deposit.

After this consensus as to what defines sedex deposits, there followed a plethora of hypotheses interpreting the actual processes of sulphide precipitation and accumulation within a sedex deposit. In an excellent review, Lydon (1996) breaks out the various hypotheses into 4 main groups:

1) Brine pool/bottom-hugging brine model: bedded ores and distal hydrothermal facies are sediments deposited in a stagnant brine pool. The pool is the result of the ponding of hydrothermal vent fluids adjacent to a vent area or in the path of a bottom-hugging brine flowing downslope from a nearby vent. The advantages of this model are that it accounts for: a) finely laminated ore textures, b) sheetlike morphology of the bedded ore, c) the lateral continuity of individual laminae, d) a lack of bioturbation in Phanerozoic sedex deposits, e) preferential formation of many sedex deposits in second-order basins (seafloor depressions), and f) large average tonnages of some sedex deposits due to widespread dispersal of suitable precipitational fluids across the brine pool, where the density stratification under a stable halocline prevents early dispersal and dilution of the venting brine.

2) Buoyant plume model: When a venting plume created by the mixing of seawater and hydrothermal fluids has a temperature greater than 250°C, then such fluids vented into normal seawater remain buoyant for most of their cooling and precipitational history. Turner (1992) uses this property of the venting fluid to invoke a precipitation mechanism for sedex precipitates as occurring during times of widespread oceanic anoxia. The stratified anoxic water column is required to act as a longterm sulphide source and to prevent the rapid oxidation of

seafloor sulphides that would otherwise occur.

3) Clastic apron reworking model. The occurrence of reworked fragments of bedded ore in many sedex deposits has been used to infer that a portion of the laminar bedded ore is evidence for contemporaneous erosion and collapse of an elevated hydrothermal sulphide mound. According to Lydon (op. cit.) this process can explain part of the bedded ore in a sedex sequence, but not the total deposit.

4) Subsurface replacement model; This increasingly popular model for sedex deposits advocates early diagenetic/hydrothermal overprinting of an originally fine-grained bedded and probably pyritic laminite precursor. The control to the ore-sulphide replacement was the original permeability of the unconsolidated laminated sediments immediately prior to the lamina-by-lamina entry of the ore-precipitating fluids.

Mechanism (4) is most clearly seen in the unmetamorphosed deposits of the HYC in the McArthur Basin (Eldridge et al. 1993). HYC ore sulphides clearly postdate and replace an earlier framboidal pyrite. Associated sulphur isotope data require interpretations involving different sulphur sources for the early diagenetic pyrite (2 types) and the ore sulphides. A replacement model also explains mineralogical and chemical zonations around those sedex deposits with defined hydrothermal vents (e.g. Sullivan, Tom and Jason deposits in Canada; Rammelsberg in Europe). It is also consistent with observations of ongoing sulphide replacement in modern hydrothermal metalliferous analogues, such as in the Red Sea, Salton Sea and Middle Valley.

In my opinion none of these mechanisms, with the possible exception of (2), are mutually exclusive. The dominant mode of formation of most ancient sedex deposits was perhaps a combination of 1) and 4), with elements of 3) about the margins of many deposits.

Fluid inclusion and stable isotope data from ancient sedex ores worldwide show that the lead and zinc sulphides were emplaced from brines with temperatures ranging from 100-300°C and salinities of 9-26 eq. wt% NaCl (Turner, 1992). Most authors also accept that the ore-carrying fluid was sulphide-free brine. Experiments by Barrett and Anderson (1982) showed that at temperatures below 150°C it is impossible for H_2S-entraining chloride brines to carry enough lead to form an ore deposit. Unless the pH is near or less than 4, a chloride brine also cannot carry enough

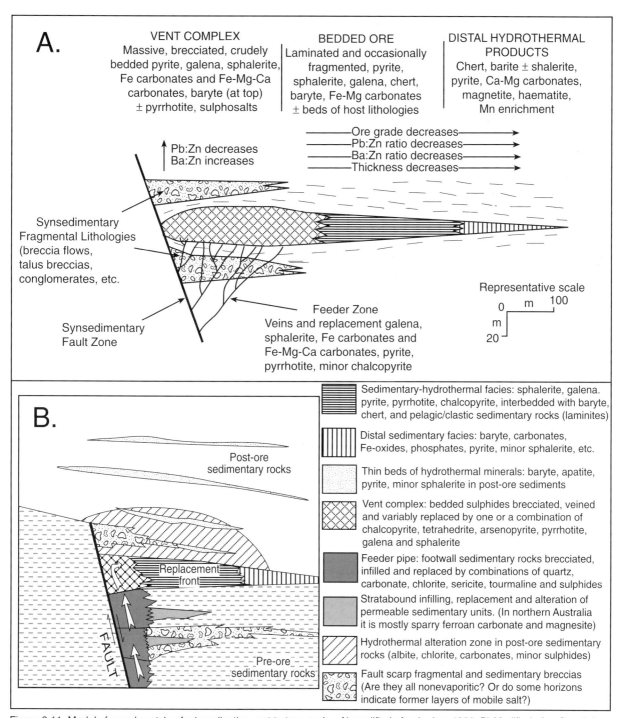

Figure 9.11. Models for sedex style of mineralisation, not to true scale. A) modified after Lydon, 1996. B) Modified after Goodfellow et al., 1993.

zinc to act as a mineralising fluid. Such low pH levels are not common in typical hydrothermal and basinal waters (Figure 9.2). Thus the reduced sulphur in a sedex deposit must be carried either as a component in a separate fluid or come from reduced seawater sulphide on or near the seafloor. Likewise, in those sedex deposits that contain stratiform baryte the sulphur isotope ratios of the baryte are typically close to that of seawater, indicating that much of the baryte was precipitated as venting hydrothermal barium combined with ambient marine sulphate (Lydon, 1996).

Figure 9.11 a and b shows typical examples of current models for sedex deposits (after Goodfellow et al., 1993: Lydon, 1996). Unlike the earlier sedex models of the 1970s and early 1980s that emphasised the exhalative (syngenetic) mode of mineral formation, the assumptions for emplacement mechanisms inherent in these models underline the significance of ongoing early intradeposit replacement of earlier pyrite by later metal sulphides via fault-fed flushing (mechanism 4). One of the shortcomings of current sedex models is that other than a nearby fault they do include any criteria from outside of the actual deposit, which help focus any regional exploration effort into particular intervals. In other words, you really need to intersect the margin of a deposit before the criteria are of any real use in further focusing the drilling and exploration effort. The same is also true of most current MVT models (see discussion in Chapter 8).

If one accepts the presence of copper in the bedded ore, then Atlantis II deep bottom sediments are a modern analogue of the surface and nearsurface mechanisms that form a sedex deposit. Its dependence on nearby thick dissolving allochthonous salt units to create a widespread stable brine pool, that in turn facilitates the preservation of widespread organic-rich metalliferous laminites, clearly distinguishes the Atlantis II deep deposit from much smaller open-marine base metal hydrothermal laminites and replacements (see later discussion of Kuroko, Cyprus and Besshi style deposits). Given the fundamental role of dissolving halokinetic evaporites in establishing and maintaining a bottom brine pool in the Atlantis II and other Red Sea deeps —a role that allowed metalliferous hydrothermal laminites to accumulate up to 15 km from the nearest active vent— the next question must be, how important were any nearby dissolving evaporites in creating anoxic intrashelf brine pools and so facilitating the formation of ancient sedex deposits of Northern Australia and elsewhere?

Sedex deposits, northern Australia

Table 9.2 lists characteristics of some major "sedex" deposits hosted in the Proterozoic sediments of Northern Australia. The ubiquity of evaporite indicators, including dissolution breccia beds in sequences below the ore deposits, begs the question, were bottom brine pools and convecting pore brines derived from dissolving subsurface evaporites significant in focusing Pb-Zn sulphide mineralisation in

deposits such as HYC, Lady Loretta, Century and Mount Isa? Or an even more trying question, was halokinesis, with its associated extensional faulting and raft tectonics, an important structural control on the precipitation of ferroan carbonates as well as Pb-Zn mineralisation in fault-defined seafloor depopods?

Historically, all breccias in proximity to the mineralised laminites that make up the ore in these deposits have been interpreted as sedimentary and fault scarp breccias. The possibility of salt allochthons controlling basin faults and focusing fluid flow has not been considered. At this stage I do not know the whole answer, but my fieldwork in northern Australia over the past few years offers some tantalising suggestions of possible halokinetic and brine pool influences.

Beneath the ore horizon, all the deposits listed in Table 9.2 show abundant evidence for the former presence of sedimentary evaporites in particular breccia horizons, and evaporitic breccias along fault planes (intrastratal and trans-stratal breccia). Some of these evaporite solution breccias occur in normal marine deepwater and platform hosts, and were probably emplaced as salt glaciers and allochthons that were sourced from a more deeply buried, now dissolved, mother salt bed. When the salt allochthon dissolved, it left behind subtle indicators of former salt within the breccia matrix, such as radial quartz micronodules, length-slow chalcedony and halite moulds. Generally, these breccia units are encased in deepwater normal marine sediments and very little sedimentary petrography has been done on them. Future work in such sedex regions should attempt to distinguish between salt allochthon breccias and other styles of sedimentary breccia.

The precursor salt beds and allochthonous sheets, through their flow and dissolution, would likely have contributed to the formation of chloride-rich metalliferous basinal brines in the region of a sedex deposit. Through fault-controlled venting they may well have created density-stratified bottom brine pools on the deepened seafloor (see Figure 9.1a,b).

Sedex deposits adjacent to structurally controlled Fe-carbonate haloes: McArthur Basin HYC deposit, Australia

In the Proterozoic McArthur Basin during Barney Creek time, the ore-host matrix of the HYC Pb-Zn deposit was deposited as laminated pyritic shales in local anoxic deeper water depressions.

It accumulated adjacent to major feeder faults now outlined by Fe-carbonate haloes (Table 9.2). These faults intersect underlying breccias that entrain indicators of former evaporite beds and allochthons. My mapping of feeder faults to these sulphide-rich laminites shows these hydrothermal haloes of diagenetic ferroan carbonate cements have replaced adjacent platform carbonate. Previously, these Fe-carbonate haloes have been

DEPOSIT & GRADE	LITHOLOGICAL AND MINERALOGICAL ASSOCIATIONS	EVAPORITE & BRINE STYLE	EVAPORITE SIGNIFICANCE	REFERENCES
Mt Isa (Pb-Zn-Cu-Ag) 77 Mt production 47 Mt reserve 6.8% Zn 5.9% Pb 148g/t Ag	• Formed in Leichhardt Fault Trough, a NW trending rift valley approximately 600 km long and 50-60 km wide. • Pb-Zn-Ag stratiform lenses accumulated in this fault-bounded basin following deposition of shallow water sediments and continental volcanics. Total sediment package = 16 km thick. • Intrabasinal sediment composition evolved from carbonate, to carbonate-sulphate, to carbonate-sulphate-Na- evaporites with the ore-hosting sediments formed during the main sulphate precipitating phase. • Current- deposited sulphate-rich sediments provide the sulphate source for formation of diagenetic pyrite that formed as a result of microbial sulphate-reduction. • Primary stratiform Pb-Zn mineralization occurred when metal-rich chloride brines inorganically reduced the remaining sulphate; subsequent high-temperature mineralization overprinting caused extensive sulphide replacement. • Silica-dolomite Cu ore formation postdates stratiform ore formation by two stages of deformation and 190 million years.	• Host sequence is interpreted as a saline lake complex formed in flood-plain and subaerial to subaqueous saline mudflat environments. • Sediments above and below the ore-hosting stratiform sequence contain stromatolites, flat pebble breccias, halite casts, dissolution breccias, gypsum/anhydrite pseudomorphs, chert nodules, large and small scale cross beds, tepees and pisolites.	• Sulphide- hosting strata contain pseudomorphs of continental evaporite minerals. • Rapid lateral variations in sediment thickness controlled by west and west-north-west trending growth faults, which also acted as channelways for ore-bearing fluids.	Neudert, 1986 Neudert and Russell, 1981 Swager, 1985 Perkins, 1984,1990 Derrick, 1982 Forrestal, 1990
HYC/McArthur River (Pb-Zn-Cu-Ag) 227 Mt reserve 9.2% Zn 4.1% Pb 0.2% Cu 41g/t Ag	• Located on eastern side of northerly trending Proterozoic rift valley, the Batten Trough, which is filled with 5,000m sediment. • Sphalerite and galena are ubiquitous ore minerals hosted by bituminous pyritic shales of the Barney Creek Formation. These shales formed as fills of subsidiary basins within the rift valley. • Stacked stratiform sulphide lenses separated by weakly mineralised sedimentary strata with rapid lateral sedimentary changes and abrupt changes in thickness of ore-hosting sequence. • Local conglomerates and breccias and slumped strata beneath massive sulphides suggest contemporaneous faulting and gravity sliding were major factors controlling sedimentation and discharge of metalliferous hydrothermal fluid onto basin floor and into permeable subsurface sediment. • Mineralised beds exhibit soft sediment deformation structures such as laminar beds that are deformed by intraformational slip with flow structures, compaction structures, microfaults, compactionally deformed infilled fissures. • At microscopic scale ore is made up of finely disseminated sphalerite and galena that is isotopically distinct from both types of the fine grained laminar pyrite that they replace. Co-occurrence of euhedral ankerite and prismatic quartz with the base metal sulphides implies dolomitic siltstone host underwent extensive recrystalisation during hydrothermal mineralisation, which included addition of Fe, $_2$, and Si as well as the more obvious Pb, Zn, Cu and possibly H_2S. • Ores are epigenetic/hydrothermal and not accumulations of fallout from hydrothermal pools or deeps nor are the ores biogenically produced. Replacement of pyritic beds by Pb and Zn sulphides was probably early, diagenetic and shallow.	• Chloride brines carry metals as part of a hydrothermal circulation cell. • Restricted marine and non marine euxinic conditions proposed for thinly laminated carbonaceous sulphide-hosting mudstones.	• No evaporite pseudomorphs have been documented in the ore-hosting shales. • Evaporite pseudomorphs and basinwide dissolution breccias in underlying and overlying strata. • Dissolution of these underlying evaporites supplied chloride and sulphate and the consequent brines carried metals into suitably prepared ground.	Muir, 1979, 1983, 1987 Eldridge et al., 1993 Jackson et al., 1987 Logan et al., 1990
Century (Pb-Zn-Ag) 118 Mt reserve 10.2% Zn 1.5% Pb 36g/t Ag	• Hosted by deeper water finer-grained siliciclastics. • Mineralisation is laminar sphalerite with lesser galena and rare pyrite in ore zones with a diagenetic pyrite halo around the ore body. • Sphalerite is unusually pure (62% Zn) and is associated with significant quantities of pyrobitumen. • Ferroan carbonate (≈70%Fe), not pyrite, is principal iron-bearing gangue. • Deposit was once a continuous body but is now broken into several blocks by faulting and tectonism. • Mineralisation is post nearsurface diagenesis and post stylolitisation and transgressive of depositional lithologies. • Mineralisation was related to a sulphate-bearing hydrothermal fluid that entered the organic rich host sequence at burial depths between 800 and 3000m.	• Thermochemical sulphate reduction took place in the host shales when they were bathed in metalliferous brines. • The most intense zones of mineralisation may represent palaeo oil-water contacts. • Onset of mineralisation coincided with basin inversion.	• Century deposit is hosted in the upper Lawn Hill Formation, which is the upper part of the McNamara Group. • Several formations in the McNamara Group contain documented evaporite pseudomorphs, they include: Gypsum pseudomorphs in the Torpedo Creek Quartzite, Paradise Creek Formation and the Esperanza Formation. Cauliflower cherts after anhydrite occur in the Lady Loretta and Paradise Creek Formations. • Dissolution of these precursor evaporites supplied ions to the mineralising brines.	Broadbent et al., 1996 Walker et al., 1978 Waltho and Andrews, 1993

Table 9.2. Characteristics of major Australian sedex deposits (Mt = million tonnes).

misidentified as replacement of gypsum beds (see earlier discussion of siderite marble).

It is true that former bedded and halokinetic evaporite intervals are widespread across the basin, but they only occur at two or three levels in the McArthur Group stratigraphy (Table 9.3). These former salt units are now indicated by solution breccias with chert micronodules, length-slow chalcedony and rare double-terminated euhedral quartz in the breccia matrix, as well as smooth lower and crackle upper contacts to the breccia unit. They do not occur in all formations of the McArthur Group, and so are far less commonplace than supposed evaporites indicated by the erroneous interpretation of "coxcos" and siderite "marbles". Interpretations of ubiquitous evaporite pseudomorphs throughout almost all the formations of the McArthur Group were based on misidentified botryoidal marine carbonate cements, or wrongly interpreted hydrothermal siderite. When such textures are correctly interpreted, then a depositional model for HYC becomes one of local intrashelf anoxic deeps within a typical Proterozoic platform marine carbonate (Figure 9.12a).

The stratigraphy of most wells in the McArthur Basin that pass through the equivalents of the Barney Creek Formation, host to HYC deposit, clearly show a deepening profile upward from the platform cementstones of the Teena Dolomite, and its equivalents, into the deepened waters of the Barney Creek mass flows and laminites (Figure 9.12b). Units in almost all wells through the Barney Creek Formation and its equivalents can be broken down into three facies belts (A, B, C).

Facies A is now dolomitised; it was originally deposited on the floor of a shallow normal-marine tropical ocean as cementstones interlayered with micritic intraclast packstones and rudstones. Terminations of many of the crystal outlines suggest it was dominated by grasslike cement layers and botryoids of aragonite and perhaps Mg-calcite that were subsequently dolomitised. This facies, with its abundant normal marine cements, is identical to other tropical carbonate platform sediments of the same age (e.g. Grotzinger 1995; Grotzinger and Knoll, 1995).

Facies B is a layered to laminated dolomitised siltstone with interbedded packstones. It is usually slumped or cross-laminated and was deposited as some form of intrabasin slope deposit. Its lack of cementstone layers implies it was deposited in deeper, but still oxygenated, shelf waters. It too is now extensively dolomitised, although the occasional sandy interval is still a limestone. The preservation of local limestone beds in this interval suggests

the dolomitising mechanism for the region was relatively early and driven by hydrological processes that were most active in platform sediments (e. g. brine reflux or mixing zone processes).

Facies C is a finely laminated organic-rich siltstone probably deposited in the deepest waters of the three facies. Its inherently high levels of organics and framboidal pyrite imply it accumulated under anoxic bottom conditions. This is the facies style that hosts the HYC ore.

The presence of an evaporite solution breccia (possible former allochthon) at the Myrtle-Mara contact immediately below the level of the Barney Creek Formation (Table 9.3), and evidence of a major thick regional mother salt bed at the level of the Wollogorang Formation (Figure 9.8), suggests an alternative "raft tectonics" model of formation for sedex deposits in the McArthur Basin and other sedex entraining basins in Northern Australia. This model can only be applied to sedimentary basins with evidence of former thick salt units below the ore level and an extensional tectonic setting at the time of mineralisation.

Analysis of underlying evaporite solution breccias and vertical changes in sedimentation style in the vicinity of the HYC deposit, and its equivalents in other nearby sediment thicks such as the Glyde sub-basin, suggests the various intraplatform deepwater pyritic thicks in the Barney Creek Formation and its equivalents are probably salt withdrawal depopods. They formed by salt tectonic focusing of escaping metalliferous basinal brines into local brine-filled anoxic deeps. The subsurface passage of the salt-solution-derived Fe-chloride rich brines that fed these pools is indicated by Fe-carbonate haloes along faults and glide shears, such as portions of the Emu fault zone in the vicinity of the HYC deposit.

However, as in North Africa, these ferroan carbonate haloes are not deposit or formation specific, and can be found over much wider areas than the immediate region of metalliferous laminites of the HYC deposit. The accumulation of substantial sulphides, especially pyrite framboids formed in the early stages of sulphide accumulation, requires the presence of anoxic bottom brines in the deeper portions of isolated depressions on the shelf floor.

Sedimentological analysis of public domain cores and well exposed outcrops in the vicinity of the HYC deposit suggests allochthonous salt sheets played a role in the sedimentation of the ore-hosting pyritic shales. Figure 9.13 is one possible halokinetic scenario to explain the regional setting of the HYC deposit. It uses a five stage

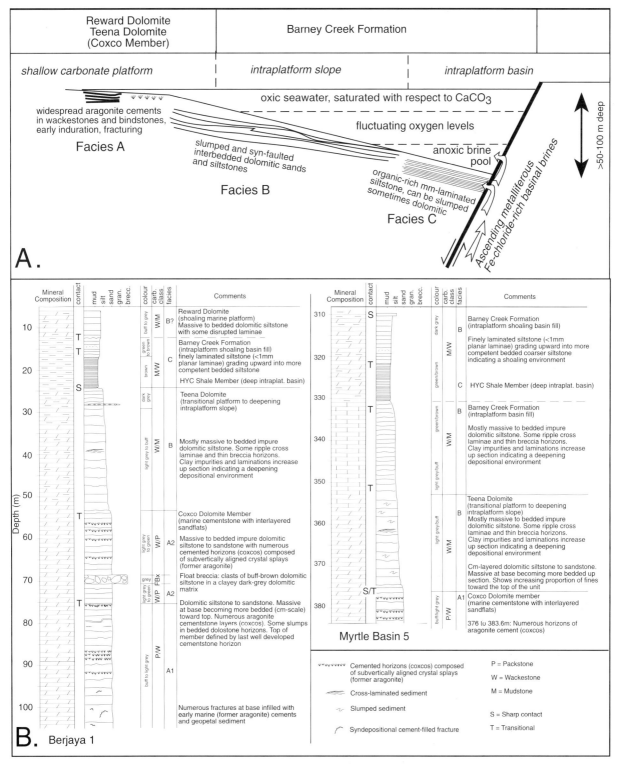

Figure 9.12. Depositional analysis of the Barney Creek Formation. A) Schematic model showing distribution of facies A, B, C in relation to water depth and anoxia. B) Vertical distribution of facies in wells Berjaya 1 and Myrtle Basin 5 (courtesy of JK Resources Pty Ltd).

Group	Subgroup	Formation	Members	Main Lithologies	Thickness (m)	Evaporites (from literature)	Interpretation (this study)
McArthur Group	Batten Subgroup	Looking Glass Formation (Amos Fm)	Donnegan Member / Hot Spring Member / Caranbirini Member	Chert; chert breccia; minor dolomite	25 - 230	None noted	Coarse dolospar in the Caranbirini Member is a possible late stage dolomitisation overprint
		Stretton Sandstone		Flaggy quartz sandstone	30 - 245	None noted	
		Yalco Formation		Chert; chert breccia; minor dolomite	130	C, LD, DBH	
		Lynott Formation		Dolomitic siltstone; chert; chert breccia; minor dolomite	210 - 760	LD, C , DBH?	
	Umbolooga Subgroup	Reward Dolomite		Dolomite with chert pellets; dolomitic siltstone	50 - 300	N, Co, DBH?	Non-evaporitic shallow carbonate platform
		Barney Creek Formation	HYC Pyritic Shale Mbr. / W-fold Shale Mbr. (Cooley Mbr)	Pyritic carbonaceous, dolomitic shale; dolomite; dolomitic breccia; HYC deposit	0 - 490	None Noted, Breccias DBH?	Infill of deeper intraplatform depressions on a shallow carbonate platform. Tectonism possibly associated with halokinesis.
		Teena Dolomite	Coxco Dolomite Member	Massive and laminated stromatolitic dolomite	57	H, N, Co, DBH	Non-evaporitic shallow carbonate platform
		Emmerugga Dolomite	Mitchell Yard Dolomite / Mara Dolomite Member	Massive and laminated stromatolitic dolomite	500	H, LD, N, DBH	Non-evaporitic shallow carbonate platform
		Myrtle Shale		Thin evaporitic red bed unit	30-100	H, DBH	Evaporite dissolution breccia at Mara/Myrtle contact.
		Leila Sandstone		Coarse-grained and crossbedded dolomitic sandstone	20-30	None noted	Transitional restricted terrigenous evaporitic platform (?lacustrine in part)
		Tooganinie Formation		Stromatolitic dolomite; dolomitic shale; quartz arenite; dolomitic siltstone	800	H, LD, SM	Non-evaporitic marine dolomites and cherts overprinted by coarsely crystalline diagenetic ferro-magnesite (not replacing gypsum)
		Tatoola Sandstone		Quartz arenite; dolomitic siltstone	140	LD	
		Amelia Dolomite		Laminated and stromatolitic dolomite, (?massive gypsum pseudomorphs in beds)	90 - 240	H, LD, C, DBH, SM	Ferro-magnesite haloes near fault zones incorrectly interpreted as SM. These are diagenetic ferroan carbonate facies that crosscut lithological contacts
		Mallapunyah Formation		Purple dolomitic siltstone; minor dolomite; quartz arenite	30 - 750	H, LD, C, DBH, SM	Diagenetic anhydrite nodules perhaps formed by postdepositional brine reflux (evaporites are possibly arid lacustrine, but not sabkha)
		Masterson Sandstone		Coarse braided sandstone, ferruginous and siliceous	30 - 350	H, LD, in uppermost beds	Postdepositional halite growth in an alluvial fan/fluvial sandstone. Indicates reflux rather than evaporitic at deposition

Stratigraphy after Pietsch et al. 1991
Evaporite occurrence in literature after Jackson et al., 1987)

KEY H = Halite casts; LD = Euhedral and discoidal gypsum and/or anhydrite pseudomorphs, not all are syndepositional; N = Acicular gypsum pseudomorphs; C = Cauliflower chert nodules; SM= Sideritic "marble" (after bedded gypsum?); DBH= Dissolution breccia horizons; Co = "Coxco" structures

SM is not siderite but ferro-magnesite and is a diagenetic hydrothermal precipitate. It is not a replacement of a gypsum mush bed. LD is mostly displacive and replacive late stage ferroan carbonate associated with "SM" haloes. N is a Proterozoic seafloor carbonate cement, no association with a gypsum precursor. Some DBH may be halokinetic.

Table 9.3. McArthur Group Stratigraphy showing evaporite distribution and horizons with misidentified evaporites (courtesy JK Resources Pty Ltd database).

model of a deforming salt allochthon in a salt entraining basin that is undergoing extensional deformation. It is based upon a typical extensional margin undergoing raft tectonics (see Chapter 5 for a full discussion of the relevant halokinetic and raft tectonics mechanisms that can form local deeps and depopods). Salt flow settings, other than the salt-hinged platform-edge raft tectonic model given in Figure 9.13, are also possible. Which of the various halokinetic deformation styles was the most likely requires a detailed sedimentological analysis of breccias, both beneath and within the ore deposit, to be tied to a backstripped regional structural analysis. This requires access to proprietary company-held core.

For the purposes of this discussion, I will only use the salt tectonic style shown in Figure 9.13, it does explain:

- underlying and syndeformational solution breccias of the deposit;
- ongoing soft sediment extensional faulting, which breaks up the continuity of individual ore layers in the HYC deposit;
- pyritic and organic-rich nature of the ore host;

- localised syndepositional thickening of the Barney Creek Formation within an otherwise shallower water setting (compare Time 3 scenario with depositional setting based on sedimentological analysis shown in Figure 9.12a);
- ferroan carbonate halo distribution about faults, especially along the intrashelf basin edge; and
- syndepositional extensional graben structuring seen in the regional structure of HYC the deposit.

Time 1 encapsulates a typical passive diapir scenario, where the upper portion of the diapir is near, or at, the seafloor. The underlying source salt bed or allochthon is intact and acting as a pressure seal to deep basinal brines. Ongoing dissolution along its underbelly is continually supplying dense chloride-rich basinal brines into active hydrothermal convection cells. These convecting brines are leaching metals from sediments and volcanics below the salt sheet. At the surface, periodic dissolution and slumping of the allochthon carapace is creating monomict and occasional polymict sedimentary breccias atop the surrounding seafloor (facies A).

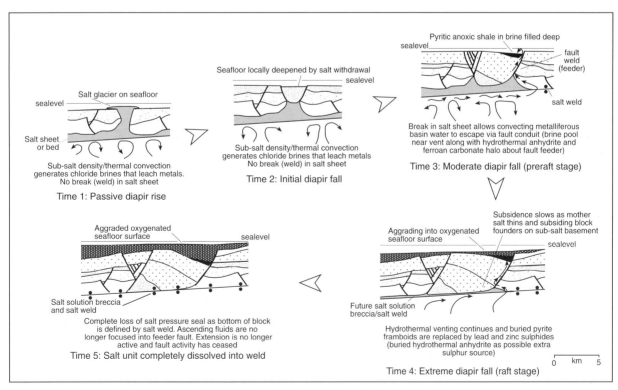

Figure 9.13. Possible halokinetic model that explains facies distribution, bottom brine stability and timing of mineralisation postdating formation of pyritic laminites (courtesy of JK Resources Pty Ltd).

Time 2 is the falling diapir stage, where ongoing extension causes the diapir crest to fall, with the formation of a local deep atop the falling diapir crest (the bicuspidate diapir stage). The deepening of the seafloor is indicated in core by the transition from the shoal water platform carbonates of facies A to the deeper water carbonate slope deposits of facies B. Hydrothermal circulation continues unabated beneath the source salt unit.

By time 3, ongoing basin extension has created a breach in the mother salt unit (a salt weld), which allows pressured hot Fe-rich basinal waters to escape across the weld. Their subsequent upward passage is focused into the still active zone of extension associated with the growth fault. The subsurface passage of these rising and cooling Fe-rich hydrothermal waters along the feeder fault creates replacement haloes of ferroan carbonate cement in the adjacent sediments. Cooling of these ascending waters means copper is not carried far above the level of the sealing salt bed and so does not make it to the seafloor (see later discussion and Figure 9.14). Where these brines vent and pond as a seafloor brine pool it facilitates the formation of pyritic shales in anoxic closed deeps (facies C). The anoxic acidic bottom conditions means that at times little or no carbonate matrix accumulates in these pelagic shale intervals.

By the end of time 3, the subsidence of the down-hinged structural block is slowing as the mother salt bed is thinning to where the bottom of the block begins to founder on the sub salt basement. Throughout time 3 and time 4, the dissolution feather-edge of the mother salt continues to focus the ascending hydrothermal brines into the main set of basin defining faults. The burial of the pyritic ore-host laminites means hotter ascending brines now have access to the pyritic laminites (Figure 8.18). The early diagenetic replacement of the pyrite by Zn and Pb sulphides begins.

As long as the withdrawal edge of the mother salt bed continues to focus ascending hydrothermal brines, the feeder fault continues to act a focus for the ongoing subsurface influx of metalliferous basinal brine into the now buried pyritic laminites. The elevated burial temperatures associated with ongoing brine flushing facilitate the nearsurface replacement of earlier formed pyrite by Pb and Zn sulphides (Figure 9.14). Any hydrothermal anhydrite initially formed near the brine vent will now redissolve and may act as an extra source of sulphur for these newly forming metal sulphides. Once the mother salt is completely dissolved or drained from the system, then its focusing effect is lost and the mineralisation process becomes inactive (time 5).

At the surface, this cessation of salt-induced subsidence, or a lessening in the rate of supply of brine, means the brine pool sediment surface can now aggrade back into oxic settings (facies C to B in Figure 9.12b). Ultimately, pyrite accumulation on the seafloor ceases as shallower, oxygenated seafloor sediments cover pyrite-rich laminites for the last time.

Other sedex deposits within Fe-carbonate haloes/hosts: Century and Lady Loretta deposits, Australia

The nearby Century deposit, Queensland, with its mineralisation hosted in deepwater sideritic siltstones, is another structurally and hydrothermally focused metal trap (Table 9.2; Waltho and Andrews, 1993; Broadbent et al., 1996). Ore hosting siderites were emplaced via predominantly subsurface seepage, followed by metalliferous basinal brines that were focused into anticline traps in their deepwater sediment hosts (Broadbent et al., 1996).

In the Lady Loretta deposit, the stratiform Pb-Zn ore lens is surrounded by an inner sideritic halo, followed by an outer ankerite/ferroan dolomite halo, which merges with low iron dolomitic sediments with regional background compositions (Large and McGoldrick, 1998). Carbonate within the inner siderite halo varies in composition from siderite to pistomesite ($Fe_{0.6}Mg_{0.1}CO_3$), whereas carbonate in the outer ankerite halo varies from ferroan dolomite to ankerite ($Ca_{0.5}Mg_{0.3}Fe_{0.2}CO_3$). As yet little information has been released on the structural setting of this deposit.

In my opinion, much of the basinal Fe-rich fluid flow in Century and Lady Loretta was focused into rollover anticlines, adjacent to active growth faults, created by "into the basin" salt flow and dissolution; much in the same way that salt flow controls rollover structures and up-dip faulting in the Gulf of Mexico and the Maritimes Basin of Canada (see Chapter 5). In formations that underlie the Lawn Hill Formation, the host to the Century deposit, and beneath Lady Loretta there are documented evaporite indicators. Examples include: pseudomorphs after gypsum crystals in the Torpedo Creek Quartzite, Paradise Creek Formation ($\approx$1670 Ma) and Esperanza Formation, and cauliflower cherts which are replacements of anhydrite nodules in the Lady Loretta and Paradise Creek Formations.

Evaporite-associated sedex model

Not all sedex deposits are evaporite associated. The discussion that follows concerns only those sedex deposits where abundant evidence of former widespread evaporites can be found in particular intervals below a potential ore horizon.

Diapirs and allochthonous salt sheets in all salt tectonic realms can act as focusing zones for metalliferous basinal fluids, from deposition until well into the diagenetic burial/ metamorphic realm as:

- impervious diapir and salt sheet margins provide an ongoing focus for upwelling basinal fluids that typically feed into suprasalt extensional faults,
- ongoing salt flow creates suprasalt structural folds, such as rollover anticlines, that affect adjacent organic and pyritic laminites and so provides favourable subsurface traps for metals and hydrocarbons.

That is, sulphide mineralisation in a halokinetic platform package can begin in a bottom brine pool via the syndepositional, probably bacteriogenic, precipitation of pyrite framboids in an anoxic organic-rich sediment matrix. It can continue into much greater depths as ongoing focused hydrothermal alteration processes overprint early formed sulphides via the passage of increasingly hotter chloride rich waters. Throughout, the feeder faults are kept active and permeable by ongoing collapse and extension generated by the deforming and dissolving salt unit that underlies the sulphide accumulation. As metalliferous laminites are buried they pass into diagenetic environments that are part of a hotter, deeper feeder zone for the same Fe-chloride brines that are escaping at the seafloor brine seeps. A deeper burial-imposed equilibrium favours the replacement of the pyrite precursor by Pb-Zn sulphides. In effect, mineralisation in salt-solution controlled basins is a fluid cycling system, instigated by early diagenetic facies controls, and continued into burial diagenetic situations. The plumbing is largely controlled by basinal faults, which in turn are generated by halokinesis and salt dissolution. Not only can suprasalt sedex deposits form in such a system; but subsalt "MVT" style deposits, such as Gays River and Jubilee, can also form as a response to the salt focused hydrology. Metal sulphides are especially likely to precipitate in subsalt areas where sulphate evaporites are present to act as reductants (Chapter 8).

If a dissolving salt bed or allochthon is creating an evaporite-associated sedex deposit, then the following features should be present in the rock matrix. This list passes from the general to the specific:

- General evidence for former salt beds in particular horizons and tectonic settings within the stratigraphy. Thick salt beds tend to accumulate at times of basinwide restriction associated either with early continental rifting or with continent-continent collision and transfer/tear faults. Thus in Precambrian sediments, where the former salt unit has long since gone, the solution breccias of the mother salt bed (primary weld) tend to occur atop or immediately below major sequence boundaries.
- Structural styles and faulting patterns that indicate gravity gliding and possible raft tectonics occur atop the salt bed/breccia (a major décollement). Many depopod-defining faults sole into major evaporite layers/breccias.
- Abundant evidence for fault-fed ferroan carbonate replacement haloes (feeder haloes may also entrain minor metal sulphides and baryte).
- Evidence of these large faults intersecting a pyritic/ sideritic interval and so lying adjacent to a possible brine pool or structurally closed deposit where the precipitating redox front was located.
- Evidence for a salt-solution breccia/former salt bed beneath deeper water pyritic deposits. If the salt bed was an allochthon, then the primary salt/fault welds, and the original salt bed, may lie much deeper in the stratigraphy. In this case, salt allochthons may be encased in normal marine sediments, which contain no evidence of an evaporite depositional setting. This is an extremely important determination, and the explorationist must be able to separate tectonic from sedimentary and salt solution breccias, even when the sandwiching lithologies are normal marine. Recognising indicators of vanished evaporites is very important.

Feeder faults to sedex systems may be reactivated during any subsequent time of extension or compression and basin inversion. Time frames for reactivation can range from less than ten to hundreds of million years. In such a scenario, the feeder faults may evolve into reverse faults, and associated halokinetic solution breccias can become preferred conduits for cupriferous hydrothermal waters.

Alternatively, as the solution breccia beds are more deeply buried, and they pass into a temperature range suitable for the emplacement of copper from hotter basinal brines, their higher permeability means they may become preferential conduits for cupriferous basinal brines (Figure 9.14). Such reactivation scenarios are possible explanations for the 100 million year plus gap at Mt Isa between the formation of the Pb-Zn laminites and the copper mineralisation hosted in a dolomitised metabreccia that still retains evidence of evaporite pseudomorphs.

A specific exploration paradigm for evaporite-associated sedex deposits in a basin of interest cannot be generalised from a single halokinetic model. Salt can "flow and go" at all times from early burial through to amphibolite facies, and can create multiple tiers of deformation and regional-scale faults as it does so. Thus, it is extremely important to define the style of breccia and its structural paragenesis, so that it then can be assimilated into a metal exploration model. Features such as evaporite dissolution breccias or stratabound pseudomorph horizons should be sought, even during the initial reconnaissance stages. As we have seen throughout this book, evaporite beds are important foci for fluid flow and in many cases for mineral precipitation. In regions where such prospective strata intersect faults at depths suitable for exploitation then a more detailed targeting program can come into play.

Some more pragmatic metal explorationists would say this approach is fine when you have outcrop, but what about when the area is largely covered by regolith. My answer is that in moderately mature exploration regions where sedex accumulations are likely, such as many parts of Australia, Canada, Europe, South Africa and the USA, enough relevant core is already held in state and federal government repositories. Much of that core is largely unlogged in terms of sedimentary and metasedimentary features. Even when prospective targets are cored beneath the regolith, many mining companies do little or no sedimentary or matrix analysis of the core outside of a general broad lithological log and sampling for sulphides. Detailed sedimentological logs are rarely compiled. From personal experience in the last few years I know that when existing public-domain core is sedimentologically logged, it defines useful information for an exploration programme.

Using public domain core in the McArthur Basin, one can determine (e.g. Figures 9.8, 9.12): 1) timing of Pb-Zn versus Cu mineralisation in relation to evaporite dissolution; 2) the presence in the stratigraphy of sedimentary horizons that are much more likely to be mineralised where they intersect or are adjacent to fault systems; and 3) the presence or absence of indicator sulphides and metasomatic haloes in pore cement stratigraphies, which can be used to determine where metalliferous basinal brines have moved through the strata and at what stage in the burial history they moved through.

Elsewhere in Australia, and worldwide, halokinetic scenarios can be envisaged as facilitating and focusing other metal sulphide laminites in sedex deposits that contain evidence of associated thick former evaporite units. Unlike our current sedex models, this type of deposit model, that is in part controlled by salt tectonics and the feedback it exerts on secondary basins and structural styles, always entrains matrix textures and features that are outside the actual ore deposit. Recognising these features is very important; they can be used to focus a regional geochemical/drilling programme toward particular horizons and structural styles that are more likely to be mineralised.

So far, we have discussed extensional settings, but salt tectonics can continue its influence into the structural styles of compressional terrains where salt, or its solution breccia, tends to occupy the cores of compressional anticlines. Then the breaches allowing fluid escape from beneath active allochthons and salt glaciers tend to be confined to areas atop regional transtensional tear faults that intersect compressional anticlines (Chapter 5). Fault conduits atop compressional allochthons can also influence the position of sea floor brine pools and their resultant mineralisation trends.

Evaporite-metal interaction in the metamorphic realm

There are only a few published papers that deal directly with sulphide/meta-evaporite relationships. A major problem when studying evaporites in the metamorphic realm is our lack of understanding of how evaporite salts evolve once the greenschist facies is approached, and how to differentiate the residual indicators and breccias from similar, but nonevaporitic, tectonic features (see Chapters 4 and 6). Another is how this metasomatic evolution can supply chloride-rich fluids that are capable of carrying metals long after the evaporites have evolved into daughter products such as marialitic scapolite. Yet another is the relative role of preexisting sulphide accumulations and how to differentiate their metamorphosed counterparts from much later mineralisation events in highly altered sequences.

For example, in some metamorphosed compressional terrains, such as the scapolitic Eastern Succession of the Cloncurry region, Queensland, the basinal brine conduits are still indicated by ferroan carbonate haloes, which form ironstone ridges at the surface (see later). Salt units or vent deposits have evolved into tourmalinites or albitites or marialitic scapolites, but the relative positions of the various lithofacies can still be recognised (Chapter 6). Let us now look at some examples of the meta-evaporite metal association.

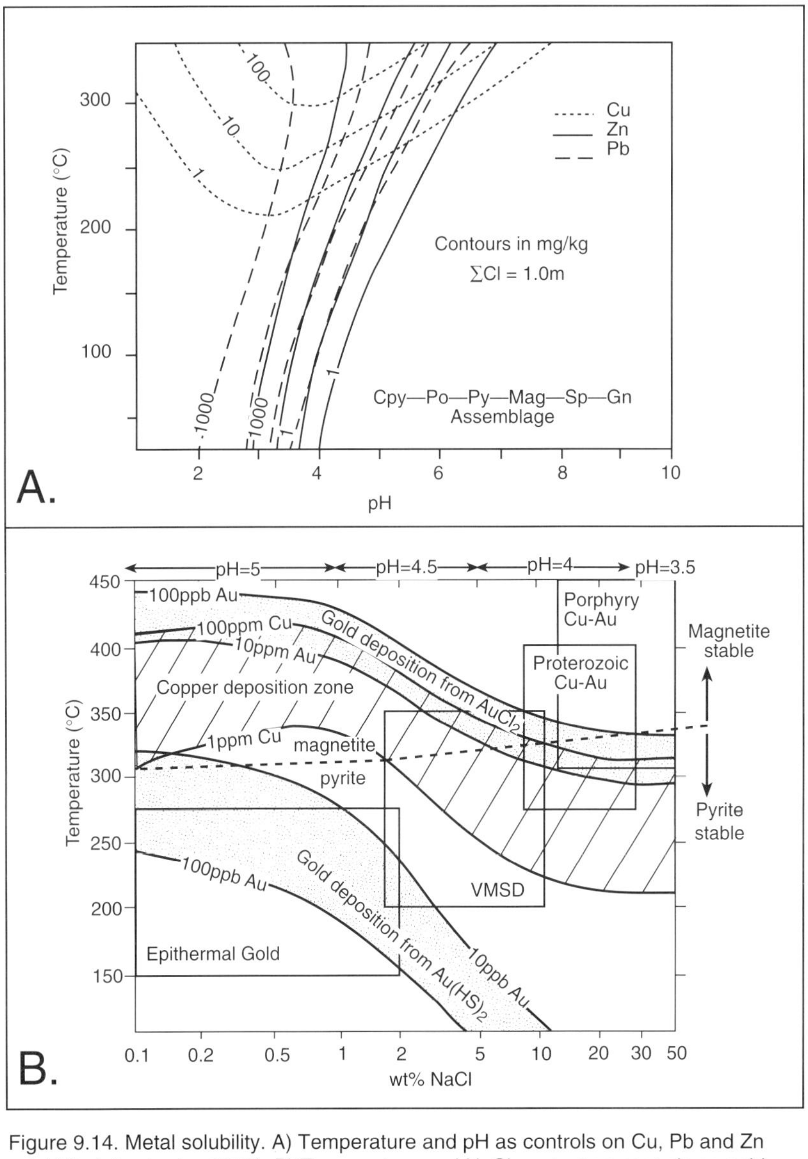

Cu-Au associations within Proterozoic meta-evaporites

Until now we have discussed the transport and precipitation of Pb, Zn and Cu in relatively low temperature diagenetic settings (T<200-300°C). But once the metamorphic and igneous realms are entered, temperatures of subsurface waters and chloride-brines rise above 200°C. Such fluids are increasingly capable of transporting and depositing copper and gold, as well as Pb and Zn, over considerable distances (Figure 9.14a,b). Such deposits typify a large group of Proterozoic Cu-Au deposits associated with meta-sediments in Australia, Scandinavia and elsewhere. All show characteristics that indicate hydrothermal transport of Cu and Au via hot chloride-rich brines and, by their relative lack of pyrite, are readily distinguished from the more typical pyritic epithermal and VHMS deposits of the Phanerozoic and Archaean (Davidson and Large, 1994). In these two styles of gold deposit the hydrothermal waters supplying the pyritic mesothermal deposit styles are typically reduced H_2O-CO_2-CH_4 waters with relatively low salinities. The same lower temperature hydrothermal waters are also capable of entraining abundant sulphur, and so deposit gold in equilibrium with pyrite or pyrrhotite and arsenopyrite (Figure 9.14b).

In contrast, Proterozoic Cu-Au systems commonly show a marked lack of iron sulphides, but are associated with iron oxides, significant quantities of Cu, along with other elements such as U, REE, Bi, Co and Mo, all metals that can be carried in high-salinity brines (Davidson and Large, 1994). Inclusion studies show these deposits formed from high salinity (15-35 wt. % NaCl equiv.), low total sulphur ($a_{ss} = 10^{-3}$ to 10^{-2}), high fO_2 hydrothermal fluids, in which metal transport was dominated by chloride-complexing (Figure 9.14b). They tend to form Proterozoic ore

Figure 9.14. Metal solubility. A) Temperature and pH as controls on Cu, Pb and Zn solubility (after Lydon 1988). B) Temperature and NaCl content as controls on gold transport (after Davidson and Large 1994).

345

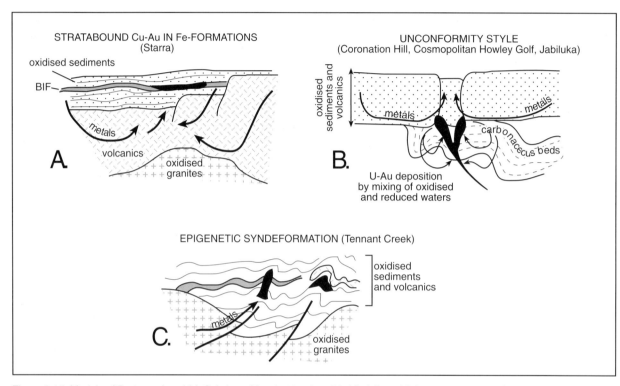

Figure 9.15. Models of Proterozoic gold (±Cu) deposition that involve chloride brines, high temperatures and evaporites (modified from Davidson and Large, 1994).

bodies within intracontinental basins that are frequently associated with anorogenic magmatism and indicators of former evaporite beds. Intrusions of oxidised fractionated granites, acted in some places to heat and focus basinal fluids, and in others were the ultimate source of metals. Mineralogies in such Proterozoic Cu-Au deposits can be separated into two temperature-dependent groups: a) Cu-Au-magnetite +/- haematite types formed at relatively high temperature (300-450°C), and b) Cu-U + Au-haematite types formed at 150-300°C. For a full discussion of the geology of the Cu-Au deposits the reader is referred to an excellent review by Davidson and Large (1994) of the Australian deposits, while Frietsch et al. (1997) provide an excellent summary of the Scandinavian deposits. Much of the following geological description is drawn from these references.

Proterozoic Cu-Au deposits, Australia

According to Davidson and Large (1994) there are four styles of these Proterozoic, iron-oxide, gold-producing deposits in Australia and all accumulated some 1.55-2.00 Ga. The groupings are:

• Type 1: Stratabound Au +/- Cu-bearing iron formations (152.4t Au) - exemplified by the Starra deposit.
• Type 2: Unconformity-style U +/- Cu/PGM/Au deposits (53t Au) - exemplified by the Jabiluka deposit.
• Type 3: Iron oxide-dominated Au + Cu mineralisation hosted within elements of ductile deformation (146.7t Au) - exemplified by the Tennant Creek deposits.
• Type 4: Iron oxide-dominated, breccia-hosted, Cu-U +/- Au hydrothermal replacement deposits spatially associated with felsic intrusions (273t Au) - exemplified by the Olympic Dam deposit.

Davidson and Large (op. cit) go on to postulate that the most effective method of metal fixing was via subsurface fluid mixing, which achieved a synchronous decrease in fO_2 and temperature. In some deposits, especially those in Scandinavia, the redox barrier that regulated the deposition of copper and gold was supplied by the presence of graphite, often in graphitic schists. In Australia, Type 1 through 3 deposits are found in association with well developed often stratabound ironstones, suggesting the same reducing environment that formed the ironstones was also capable of fixing the Cu and Au. Davidson et al. (1989) interpreted the ironstones in the Starra deposit as syngenetic exhalites (their 'magnetite-rich VHMS deposits'). Williams (1994) argues the different modes of occurrence of various ironstones indicate an epigenetic origin where late hypersaline fluids generates magnetite replacement bodies. In my opinion, both styles of formation can be encapsulated in the notion of FeCl-rich brines supplied by sequences flushed by brines fed by dissolving and altering halite-dominated evaporites.

Type 1, stratabound Au +/- Cu-bearing iron formation deposits, are exemplified by the Starra deposit, Queensland (Figure 9.15a; Davidson, 1994). The Starra ore deposit is hosted by the meta-evaporitic Staveley Formation (1720 Ma) of the Mary Kathleen Group, Mt Isa Eastern Succession. Starra is the major mine of the Selwyn Cu-Au Project consisting of 4 geographically separate outcropping lodes, totalling 5.3 million tonnes at 5.0 g/t Au and 1.98% Cu. Stratabound Cu-Au mineralisation occurs in a series of massive ironstone ridges, known as the Western Haematites, with a strike length of 6.5 km near the base of the Staveley Formation. The ironstones strike north-south and occur as discontinuous lenses, which cross cut the host rocks in places. Unmineralised varieties of ironstones crop out over a much larger area.

Within the ironstones, the Cu-Au mineralisation occurs at the lithological contact of a metamorphosed sandstone/ siltstone sequence with a succession of foliated chlorite-biotite-magnetite schists (Rotherham, 1997). Magnetite ironstone textures are generally massive in appearance, but relict deformed breccia and foliation textures are locally preserved. Sulphide ore textures are more brittle and consist of fine, pervasive brecciation of the ironstones and host rocks.

The ironstone hosted ore lenses are immediately underlain by variably sheared albite-magnetite-haematite-pyrite-bearing rocks, and overlain by mainly unaltered, less-sheared, meta-sediments. Davidson (op cit.) interprets the sodic metasomatism that formed the albite as a metasomatic halo fed by the metamorphic evolution of preexisting halite-dominated evaporites. Regionally, the sodic rocks in the Proterozoic Mt Isa Eastern Succession are increasingly being recognised as the legacy of volatile-derived fluids fed, at least in part, by evolving meta-evaporites. There were several episodes of sodium introduction prior to, during, and following the major

deformation. They were in part driven by hydrothermal fluid flow associated with granite intrusions and by halite/scapolite evolution (Oliver and Wall, 1987; Oliver et al., 1992; Williams et al., 1993; Williams, 1994).

Rotherham's (op. cit.) petrological and textural observations support a hydrothermal origin for both ironstone and mineralisation in the Eastern Succession. He recognises three dominant post-peak metamorphic paragenetic stages of alteration and mineralisation: 1) early widespread Na-Ca metasomatism consisting of albite-quartz-actinolite-scapolite, 2) localised K-Fe metasomatism consisting of biotite-magnetite-haematite-quartz-pyrite, and 3) the mineralising stage consisting of anhydrite-calcite-haematisation of magnetite-pyrite-chalcopyrite-gold and extreme chloritisation of biotite. The Na-dominant alteration occurs throughout the Eastern Fold Belt and for at least 50 km to the east and 150 km to the north of Starra. The ironstones are interpreted as products of localised Fe-metasomatism associated with shearing and brecciation of previously Na-metasomatised host rocks. Gold and copper mineralisation resulted from the interaction of magnetite, a late oxidising fluid, and an increase in pH. This produced variable haematisation of magnetite and caused the solubility of chloride-complexed gold to decrease dramatically.

Albite was abundant prior to deformation in the Starra deposit, and also prior to the main hydrothermal addition of Fe, S, Cu and Au (Davidson, 1994; Davidson and Large, 1994; Rotherham, 1997). Evidence for abundant early albite includes: 1) regional albite development at this stratigraphic level, 2) alteration of albite by ore-related sericite, and 3) inclusion of unfoliated inclusions of albite and haematite in ore-related pyrite. Davidson (1994) goes on to conclude that the fluids responsible for the sodic alteration at Starra, and for ore deposition, were saline, acid (pH = 3.9-6.0) and high temperature with relatively low levels of sulphur, with the gold carried as chloride complexes (Figure 9.15b). This, and the association with abundant meta-evaporite (mostly halite not sulphate) indicators in the region, suggests that dissolving halite at high temperatures (or chloride released during the pressure-induced lattice collapse of marialitic-scapolitic daughter products – see Chapter 6) were responsible for the chloride transport and perhaps focusing of the gold into the deposit.

Albite in the Starra region also shows protolith and structural controls. As well as being preserved as bedding parallel units, it is often developed around dolerites and within hangingwall shears. Davidson saw this as a possible problem in interpreting the sodic metasomatism as being related solely to an evaporite precursor. In my opinion, the nonstratiform fault- and dolerite-related geometries of some of the sodic metasomatism is not a problem when invoking a meta-evaporite fluid genesis. All it is reflecting is the typical reality that deforming salt is non stratiform. Once a salt body flows, it and its brine products will have climbed the stratigraphy as a number of salt slugs and tiers that generate both salt and fault welds. These features can act as ongoing planes of weakness and fluid conduits, whenever the succession undergoes renewed extension, compression or even igneous intrusion. With ongoing metasomatism the fault/shears will gain sodic haloes, supplied by the evolving salt or its daughter products. Even in the metamorphic realm, the sodic alteration, which is commonly associated with meta-evaporites, will show non-stratabound trends that can be related to the deformation of a flowing salt body as it is squeezed through the subsurface (e.g. Wang et al., 1998).

Type 2, unconformity-style, U±Cu/PGM/Au deposits are exemplified by two provinces in Northern Australia; the South Alligator Valley minerals field in the Pine Creek Inlier, and the Alligator Rivers mineral field (Figure 9.15b; Davidson and Large,1994). Mineralisation in both provinces is localised at an unconformity between overlying relatively undeformed oxidised siliciclastic sediments and strongly deformed carbonaceous graphitic units. Coronation Hill is a viable, but legally unmineable, South Alligator Valley deposit (Mernagh et al., 1994; Needham and De Ross, 1990), while exploited deposits like Jabiluka are well documented in the Alligator River minerals field.

Mineralization at Coronation Hill occurs below a major unconformity, which separates a deformed and metamorphosed basement (composed of igneous intrusions, volcaniclastic rocks, and calcareous and carbonaceous sedimentary units) from a cover sequence of little-deformed haematitic quartz sandstones, organic shales and sedimentary breccias. Geologic, fluid inclusion, stable isotope, and thermodynamic data indicate that mineralization at Coronation Hill was driven by reaction of oxidised and acid saline basin brines originating from the Palaeoproterozoic cover sequence (and its hydrosphere), with feldspathic and reducing rocks and fluids of the crystalline Archaean basement (Mernagh et al., 1994). Inclusion analysis shows the ore fluids were saline and calcium chloride dominated (ca. 26 wt% $CaCl_2$ equiv.) and mineralization occurred at around 140°C. These are the characteristics of hydrothermal basinal brines in evaporite-entraining basins (Chapter 2).

The lowest member of the Palaeoproterozoic cover sequence is the Koolpin Formation. It has been shown by Matthai and Henley (1996) to have been deposited in a terrigenous peritidal setting. Further south, the unit hosts the Cosmopolitan Howley Golf deposit. In some areas the Koolpin Formation contains textures indicating the former presence of evaporites as well as scapolites (J. Stewart, pers. comm.). It would be an interesting exercise to test, via matrix characterisation, if some of the "sedimentary breccias", including portions of the Kombolgie Unconformity that hosts the ore, are possible evaporite solution breccias. That is, did a now-dissolved, clean thick salt unit once reside at this unconformity and by its dissolution act as a focus for mineralisation? Under this scenario, any improved exploration targeting requires better definition of mineralised relationships to alteration haloes within faults, which also intersect or sole out into synkinematic breccias. Sparry carbonate replacement haloes, which occur around some feeder faults to the mineralisation in the region, should be redefined and mapped with a halokinetic emphasis. This will test any regional lithological/salt solution relationships.

In the Alligator Rivers mineral field a structural control to the mineralisation is less obvious, although mineralised reverse faults are present and several deposits are fault displaced. Regionally, the massive haematitic Kombolgie Sandstone (1648 ±219 Ma) and mafic Nungbalgarri Volcanics unconformably overlie polydeformed semipelitic to graphitic schist and amphibolite of the Lower Proterozoic Cahill Formation and the Myra Falls Metamorphics (Davidson and Large, 1994). The deepest mineralisation (mostly U) occurs as much as 500m below the unconformity. Primary ore at Jabiluka occurs as uraninite-filled veins, cements and disseminations in which gold and tellurides are present as inclusions and veins along with magnesian chlorite and anhydrite (Binns et al., 1980).

Sparry ferroan magnesite occurs in the Coomalie and Celia dolomites in the nearby Rum Jungle area. Disparate interpretations of halite pseudomorphs (Crick and Muir, 1980), versus high temperature recrystallised magnesite (Bone, 1983), versus hydrothermal replacement of precursor dolomite (Aharon, 1988) were discussed in Chapter 6. These, at first, disparate explanations, can be drawn into a much more coherent escaping-basinal-water model, centred on fluids derived as host strata pass from the diagenetic into the metamorphic realm. All it requires is that analogies are drawn with fault-fed hydrothermal sparry replacements, which occur atop halokinetic provinces in northern Africa and the McArthur Basin.

Such intervals show fluids capable of carrying metals have passed along faults outlined by sparry replacement haloes.

Type 3 deposits are exemplified by Tennant Creek style deposits, where iron-oxide -bearing Cu-Au pipes are the main ore bodies (Figure 9.15c; Davidson and Large, 1994; Rattenbury, 1994). The Tennant Creek goldfield is characterised by large massive haematite/magnetite + quartz + chlorite "ironstones", which host gold, copper and bismuth ores in metasediments of the rift-fill Warramunga Group. The ironstones formed during east-west folding, initially from high salinity basinal brines that later mixed with magmatic brines to form the Cu-Au-Bi mineralisation stage. Rattenbury interprets the folding as derived from a relatively thin skinned thrust system that drove the folding and deformation of the Warramunga Group, thrusting rocks up and over from the south toward the north resulting in a series of steep listric reverse fault splays and folds. The hypersaline ironstone forming fluids used the thrust faults as major conduits as well as the developing cleavage planes in the turbidites and other deepwater sediments of the Warramunga Group. The ironstones then formed where the cleavage-parallel conduits intersected iron-rich parts of the stratigraphy, and in numerous instances show a close spatial association with stratiform quartz-feldspar porphyry. The latter has recently been interpreted as being emplaced into the sediments of the Warramunga group while they were still unconsolidated and water-saturated (McPhie, 1993). Thus the ironstones formed early in the burial history of the basin sediment pile. It is tempting to argue such a mechanism may well be tied to ascending FeCl-rich basinal waters.

Based on fluid inclusion work, Zaw et al. (1994) favour a mostly basinal brine genesis for the chloride-rich solutions that first created the ironstones and then mineralised them. They also acknowledge that some magmatic contribution cannot be ruled out. Fluid inclusions in the ironstones have a salinity range of 10 to 30 NaCl equiv. wt % and homogenisation temperatures of 100 to 350°C, with a mode at 200 to 250°C. Salinity measurements on fluid inclusions in the mineralised ironstones gives a range of 10 to 50 NaCl equiv. wt %, with a mode of 35 NaCl equiv. wt%; while homogenisation temperatures range from 250 to 600°C, with a mode of 350°C. Similar elevated salinities in basinal brines are typically found in modern basins where evaporite beds are dissolving.

No evaporite solution breccias have yet been recognised in or atop the sediments of the Warramanga Group. If they are present, there are two likely positions in the stratigraphy for the primary weld. One likely position is near the base of that portion of the Warramanga sediments that indicates the first incursion of marine waters into the opening rift. Given the deepwater setting of the ore-host sequence, this tectonic setting means the dissolution breccia would likely lie below the mineralised part of the stratigraphy. The other setting likely to deposit widespread evaporites, which could have supplied ions and high salinity to the basinal brines, would be at or near the time of the collision event that created the thrust faulting. These beds would tend to lie stratigraphically above the mineralised intervals.

Type 4 deposits are exemplified by the Olympic Dam Breccia Complex, in the northeast part of the Gawler Craton, South Australia (Haynes et al., 1995). The geologic setting of this worldclass ore body is poorly defined as the breccia complex is buried beneath approximately 300m of Stuart Shelf cover rocks, and the nearest exposed rocks of the Gawler Craton are more than 100 km distant (Figure 9.16). The upper parts of the deposit and the surrounding rocks have long since been removed by Precambrian erosion. Any local geologic setting at the time of mineralisation must be inferred from the remnants of the upper parts of the complex, now preserved within the breccia complex, and from coeval rocks preserved elsewhere in the northeast part of the Gawler Craton. The geological setting that drove the mineralising system at Olympic Dam is even more enigmatic than the other three types, with divergent opinions as to its ultimate origin. Interpretation is further complicated by multiple brecciation and alteration events within the breccia body. Of prime importance to the genesis of this deposit is the relative significance of direct magmatic versus hypogene hydrothermal processes, and the relative input of nearsurface and basinal brines generated via dissolving evaporite units.

A two-fluid mixing hypothesis for ore formation at Olympic Dam was first proposed by D. W. Haynes (in Roberts and Hudson, 1983) and was based on the sulphide-pitchblende-haematite association of the ore body, as well as the ore textures and the overall sulphide zonation pattern. Reed and Haynes (in Haynes et al., 1995) numerically modelled a hypothetical Olympic Dam system where an oxidised 150°C water, derived from a synthetic arkose, interacted with a proto-ore assemblage consisting of siderite, pyrite, magnetite and chlorite. Wall and Valenta (1990) speculated that an analogous, but two-stage, mechanism was responsible for the genesis of both the Olympic Dam and the Tennant Creek orebodies. The modelling of Reed and Haynes implied that such a two stage mechanism was unlikely to result in ore formation at Olympic Dam, and

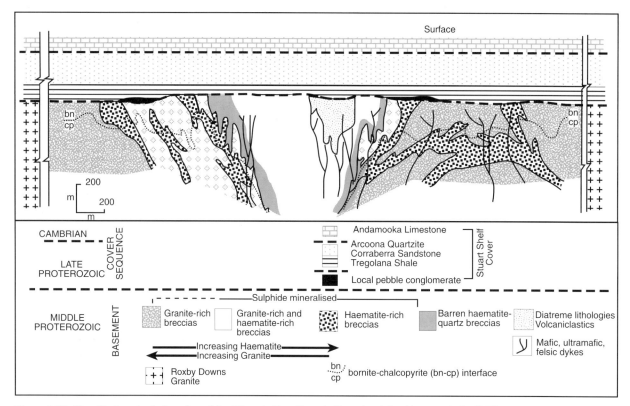

Figure 9.16. Simplified geological section of the Olympic Dam Breccia Complex (after Haynes et al., 1995).

was not consistent with textural and mineralogical characterisation of the orebody (Reeve et al., 1990) nor with its Sm/Nd and U/Pb isotopic signatures (Johnson and Cross, 1991).

Einaudi and Oreskes (1990) and Oreskes and Einaudi (1990) also presented a multistage mechanism for the development of the deposit. They proposed a primary hypogene origin for the iron and some of the copper and uranium in the deposit, but suggested that much of the ore-grade mineralisation was produced by a discrete (up to 200 My younger) second stage Cu-U overprint. They argued that the orebody had undergone a low temperature secondary enrichment as a result of uplift, unroofing and weathering at the Proterozoic surface. Reeve et al. (1990) described a model involving nearsurface cyclic interaction of hot ascending saline magmatic waters and cooler, more oxidised saline waters. They argued against the notion that the orebody had been extensively upgraded by any distinct low temperature weathering event, and supported a two fluid mixing model first presented in Roberts and Hudson (1983).

Oreskes and Einaudi (1992), with reference to fluid inclusion and isotopic data, refined their earlier hypothesis noting that two fluids were involved in ore genesis. They concluded that a deep-seated or magmatic fluid was responsible for the genesis of the magnetite-rich proto-ore. Water of superficial origin was then involved in the later stages of hydrothermal activity, which in turn formed the haematite breccias, and that sulphide deposition mainly occurred late relative to brecciation and deposition of haematite. They continued to speculate that the magnetite deposition may have preceded haematite deposition by some 150 My. In contrast, Haynes et al. (1995) concluded that their two-fluid fluid mixing model best explains the precipitation of the sulphides, magnetite and haematite and many associated minerals. They do not think the data support a discrete episode of sulphide enrichment related to later hypogene hydrothermal events, or to later weathering of the upper part of the orebody. The debate continues.

If both Haynes et al. and Reeve et al. are correct in their assertion of the presence of a saline water body in the caldera, which mixed with deeper magmatic water to form the deposit, then there may well be indications of former saline minerals and burial salts still preserved in the locally preserved pods of slumped and deformed caldera sediment fill. To my knowledge these textures have yet to be documented in the published literature base.

Other Proterozoic Cu-Au Fe-oxide/evaporite associations

As well as these Australian examples, there are other Proterozoic Cu-Au deposits that show strong affinities with hypersaline basinal waters and possibly with chloride-rich brines derived from dissolving evaporites. These include the Cu-Co-Au deposits of the Idaho Cobalt Belt, USA (Nash and Connor, 1993). There the deposits are stratabound and, in the Blackbird Mine itself, the mineralisation is locally stratiform. The ore host is the Mesoproterozoic Yellowjacket Formation, composed of dark-coloured, thin-bedded to laminated argillites and siltites, quartzites and minor marbles. These units have equivalents in the Belt Supergroup that are known to be meta-evaporitic. The mineralised Blackbird Member shows anomalously high levels of Cl, K and Fe in the biotites, as well as the presence of stratiform scapolite. Locally, the Fe-enriched sediments become ironstones. The Blackbird Member is also characterised by crosscutting growth faults, soft sediment disruption zones and likely hydrothermal seafloor vents. Nash and Connor (op. cit.) underline similarities in the mineralising system with the evaporite-associated Red Sea deeps and the Salton Sea hydrothermal sulphides.

Some of the best documented Cu-Au meta-evaporite/ volcanic associations are to be found in the Cu or Cu-Au deposits of Sweden. In the Kiruna-Tjåmotis-Arjeplog district of Northern Sweden, the iron oxide deposits and associated greenstone lithologies are extensively scapolitised (Parák, 1991). These Kiruna greenstones were formed close to 1.9 Ga. Their eruption and emplacement, along with associated evaporitic rift sedimentation, was closely followed by metamorphism and hydrothermal circulation driven by the volcanism.

Scapolitisation can often be seen around lava pillows as well as in syenite, gabbro and diabase. Scapolite is commonplace in the copper mineralised areas at Svappavaara and Nautanen where it is accompanied by apatite, tourmaline, magnetite and skarn. At Ultevis it occurs as well developed crystals up to several centimetres in size in coarse-grained skarn and as pegmatite in a rusty, graphite-bearing rock. The most impressive scapolite

occurrences occur in the metasedimentary pile, where it forms almost pure bands or layers in skarn, biotite-bearing limestone, biotite-bearing schist, and graphite bearing schist. In some cases scapolite in the metasediments makes up more than 50% of the rock volume. The strongly stratiform distribution of scapolites in the Kiruna district implies an evaporite precursor (Parák, 1991), while preserved salt casts in the metasediments in the Viscaria region support this interpretation (Godin and Lager, 1986).

The Viscaria copper deposit also reflects a strongly evaporitic protolith with mineralisation occurring in Proterozoic playa metasediments in a mafic volcanic terrane (Godin and Lager, 1986). Within a 700m section of predominantly tholeiitic basalts there are four mineralised horizons. The highest concentration of sulphide minerals occurs within graphitic schists and calcsilicates within the basalts. The graphitic units are interpreted as original organic-rich sediments deposited in the deeper parts of a fault-bounded anoxic lake and as the units that acted as reductants to the Cu carrying chloride brines (see Chapter 8 for a discussion of mechanisms). The calcsilicates were originally limestones deposited as shallower nearshore deposits. The original shallow water setting is also reflected in locally preserved sedimentary structures: mud cracks, ripple marks, shallow-water crossbeds, erosion surfaces, and breccia beds. Evaporite mineral pseudomorphs suggest the chert-albite units were formed in a saline lake as magadiite chert replacements of sodium silicate precursors (Godin and Lager, 1986). Some of the aligned growth fabrics, still preserved in the albite-chert layers, are reminiscent of bottom-nucleated trona precursors (pers. obs, 1986). Areas of scapolite and albite alteration may represent metamorphosed end-products of authigenic reaction haloes in the volcaniclastics that surrounded the lake. Sulphide minerals appear to have first replaced the organics relatively early in the diagenetic history of the lacustrine sediments. Parák (1991) suggests that the hydrothermal system of the greenstone belt, which was driving alteration in the subsurface, may also have supplied the ionic components to the depositing primary evaporites at much shallower burial depths.

The likelihood of an evaporite protolith for the Swedish Cu deposits is further reinforced by the presence of anhydrite lenses in the Kiirunavaara deposit, where re-examination of drill core from the deposit showed anhydrite in 26 holes, partly as lenses and partly as secondary fissure-filling (Parák 1991). Tuisku (1985) discussed the origin of scapolites in the schists of nearby central Finland; as in Kiruna, he concluded the chlorine for the scapolitisation

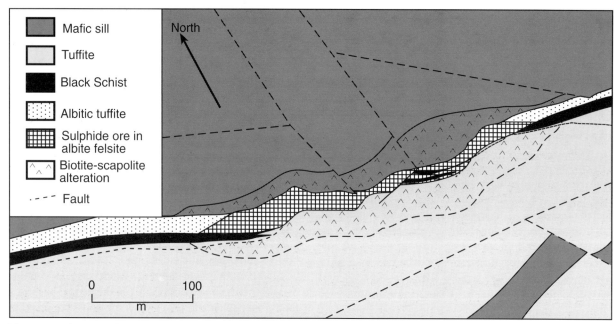

Figure 9.17. Geology of the Pahtohavare copper deposit, Norway (after Ettner et al., 1994).

was provided by dissolving precursor evaporites. "The chlorine of the basic igneous rocks was either introduced directly into the magma from evaporites or brines or into the cooling igneous rocks from later circulating brines" (p.169).

Evaporite-associated Au-Cu deposits also occur at Pahtohavare and Bidjovagge in the greenstone terranes of Norway (Lindblom et al., 1996; Ettner et al., 1994). The Pahtohavare deposit is located some 10 km south of the Viscaria Cu deposit, while Bidjovagge is located in greenstones some 200 km to the northeast. At Pahtohavare the gold mineralisation occurs as impregnations, epigenetic quartz-rich breccias and fracture fillings, along with pyrite, chalcopyrite and pyrrhotite. It is hosted in zones of fine-grained albitised rocks within tuffite, black graphitic schists and mafic sills (Figure 9.17). The graphite horizons in the albite-altered blackschist have been replaced completely by albite in the most mineralised sections. Albite-altered zones are surrounded symmetrically with barren biotite-scapolite-altered rocks. The scapolite is marialitic and occurs as porphyroblasts and irregular veinlets or networks. Coarse-grained carbonate veins are common both in the ores and surrounding altered rocks. Syngenetic magnetite occurs at several stratigraphic levels within the tuffite (Lindblom et al., 1996).

At Bidjovagge the deposit is hosted in the metavolcanic and metasedimentary Cas'kejas Formation, which is composed of amphibolites and mafic metavolcanics interbedded with minor thin beds of carbonate, graphitic schist, albitic felsite and metatuffite (Ettner et al., 1994). Originally, the host sequence is thought to have been a sequence of lava flows, ash deposits, black shales, limestones and shales deposited in a shallow early rift basin. A series of thick diabase sills intrude the Cas'kejas Formation and a zone of albitic felsite has developed between the contact of the diabase sills and graphitic schists. Mineralisation is of two types (Au-rich and Cu-rich), hosted in a series of mineralised lenses along a shear zone some 2.5 km long. This mineralised shear zone lies in an albitic felsite made up very fine-grained albite with impurities of rutile and carbonate. It is thought to be the product of sodic metasomatism of black shales, driven by the intrusion of the diabase sills and associated circulation of hot saline brines. The gold-rich ore is hosted within ductile to brittle structures in the shear, while the copper-rich ores occur within brittle structures in shears that crosscut the ductile fabric. A series of syenodioritic intrusions crosscut the shear texture in the albitic felsite and are cross cut by quartz-carbonate chalcopyrite veins. Based on crosscutting relationships these dykes are interpreted to have been intrude during the period between the gold-rich and copper-rich mineralisations.

Based on inclusion studies in both Pahtohavare and Bidjovagge, gold transport can be tied to highly saline NaCl fluids (30-45 wt%) that carried CO_2 and CH_4 at temperatures near 350°C. These are fluid characteristics very similar to those associated with the emplacement of the higher temperature Australian Cu-Au deposits. Ettner et al. (1994) relate gold precipitation at Bidjovagge to redox reactions between the fluid and the graphitic schists. Lindblom et al. (1996) suggest gold precipitation at Pahtohavare was the result of cooling of a chloride brine as it mixed with a lower salinity fluid during tectonic fracturing along with associated pressure fluctuations and CO_2 mixing and go on to suggest, by comparison with Olympic Dam (as per Oreskes and Einaudi, 1992), that the ore fluids were magmatically derived. They largely ignore the work of Haynes et al. (1995) showing that magmatic fluids at Olympic Dam may have mixed with saline evaporite-derived subsurface waters. They also ignore the possibility that the source of the chlorine and sodium in the inclusions, and the marialitic scapolites, and the albitites, could reflect a meta-evaporite association. This in a deposit located less than 10 km from the Viscaria Copper deposit and along the same greenstone trend as the classic meta-evaporite Viscaria locality of Godin and Lager (1986). Ettner et al (1994) are a little more circumspect in their interpretation of the origin of the Na-Cl brine, and suggest these highly saline chloride brines may have had either an igneous or a metasedimentary source.

The Santa Rita gold deposit in the Middle to Upper Proterozoic Paranoá Group in Goiás, Brazil, is another example of a possible evaporite/brine formed gold deposit. But, in this case, the solutions were at times also capable of precipitating pyrite within the quartz-carbonate veins that host the ore (Giuliani et al., 1993). Fluid inclusion studies of quartz from these veins identifies two fluids: (1) a highly saline CO_2-N_2-rich aqueous fluid with halite ± sylvite daughter minerals, and (2) a CO_2-N_2-rich aqueous fluid with moderate salinity. The two fluid types can occur in the same quartz domain and display great variation in the degree of filling and notable dispersion of the microthermometric data. On heating, all the inclusions decrepitate between 200 and 300° C. The simultaneous entrapment of compositionally variable fluids in the system H_2O-CO_2-N_2-NaCl-KCl is interpreted as heterogeneous trapping resulting from the mixing of high-salinity fluids (H_2O-NaCl-KCl system) with carbonic fluids (H_2O-CO_2 system). The latter moderate-salinity fluids were produced by devolatilisation of carbonate and phyllitic host rocks. The former, the high salinity brines, were derived by leaching of the evaporites that characterised the lower part of the Paranoá Group (Giuliani et al., 1993). Gold was initially transported as an $AuCl_2$-complex (T>300°C, low pH, moderate fO_2) much like the Proterozoic Cu-Au deposits of Australia. When temperatures fell below 290°C, the

"switch-over" process (of Large et al. 1989 - see Figure 9.14) lead to a predominance of $Au(HS)_2$ in the carrier fluid, so that pyrite was a coprecipitate at these lower temperatures. The oscillatory zoning in the As-Co-Ni-bearing pyrites also indicates episodic fluctuation of the fluid composition. Such changes in fluid character typify fluid-mixing systems subject to fluctuating levels of feed waters within an active hydrothermal system. The model proposed by Giuliani et al. encompasses a Proterozoic geothermal system that was characterised by a heated fluid flow regime along fault feeder structures that also crosscut the dissolving Paranoá evaporites. Today, the preserved equivalents of such brine-fed systems are likely to be represented by dissolution breccias and salt welds intersecting the faults rather than actual salt beds that have long since dissolved.

A similar fluid mixing system, involving highly saline evaporite-derived basinal brines and magmatic waters, is also suggested as the mechanism for mesothermal gold mineralisation in the Transvaal Sequence in South Africa (Anderson et al., 1992). In this case, the mixing occurred between a deep-seated homogenous CO_2-rich fluid of magmatic/metamorphic origin and a saline (subsurface/intraformational) basinal brine derived by the dissolution of evaporites in the sediments of the Chuniespoort and Pretoria groups of the Transvaal sequence, where Tyler (1986) documented widespread salt casts.

The common theme in all these Proterozoic Cu-Au deposits (except possibly Olympic Dam) is transport of copper and/or gold at relatively high temperatures within saline chloride complexes derived from the dissolution of deeply buried salt units or their metamorphosed successors. Precipitation was typically either at some sort of redox front associated with graphitic schists, to give a stratabound style of precipitation, or at some form of subsurface groundwater/fluid mixing zone, often with a more structurally focused mode of precipitation associated with ironstone precipitation or replacement. Those cases showing a strong structural control to the mixing zone, imply precipitation was coincident with, or later than, a major deformation event that had facilitated the formation of a more permeable pathway. Precipitation was often associated with some form of pressure shadow. In all cases a strong argument can be made that the high-temperature metal-carrying brine was derived by dissolving chloride-rich evaporites or pressure driven escape of volatiles from subsequent meta-evaporites. In some cases an argument can also be made for an evaporitic origin to the organic-rich-shale precursor of the graphitic schists, which later set up a metal-precipitating redox interface.

Most published papers to date have not made an argument for an evaporite-associated protolith to any of the ore-hosting ironstones. And yet, as was clearly shown earlier, refluxing brines beneath dissolving evaporites can build up high metal contents and then release ferroan chloride-rich basinal brines along structurally focused paths. As the ferroan chloride brines seep upward they precipitate ferroan carbonate haloes along the fault conduits, or in structurally focused traps or as parts of widespread stratiform iron-hydroxide and iron-sulphide rich sediments about the brine outflow zones. Could these ferroan replacements and structurally controlled ferroan haloes along regional basinal faults be the precursors to some of the stratiform and structurally aligned ironstones in strongly sodic metasedimentary terranes? Only when the rock matrix is characterised by geologists who understand possible evaporite and metal precursors, as well as high temperature magmatic systems, can exploration paradigms based on

hybrid fluids from magmatic and meta-evaporite sources be considered.

Magmatic - evaporite associations

Evaporite-derived waters can also be caught up in hydrothermal circulation cells driven by igneous intrusives. For example, saline fluids derived from Triassic evaporites played a role in polymetallic Sb-As-Au-W mineralisation hosted in the granodiorites at Dubrava in the Western Carpathians (Chovan et al., 1995). Evaporite-derived solutions also played a role in the Saint-Salvy vein-hosted Zn(+Ge) deposit on the southern margin of the late Variscan Sidobre Granite (Munoz et al., 1994).

LOCATION / DISTRICT	AGE / SETTING	SIZE* / METALS / ALTERATION	COMMENTS
CENOZOIC			
Red Sea Region (Afar, Atlantis II)	Holocene / Rift transform setting with basaltic volcanism and abundant evaporites	>100? (>157)/ FeOx + MnOx in sediments with modern Fe-rich hot springs or (Zn-Cu-Pb brines)/local sodic ± peralkaline alteration	Modern geothermal circulation through evaporitic beds leading to syngenetic FeOx ± Zn + Pb sulphides
Chilean Andes (El Laco, Magnetita Pedernales)	Neogene / Intermediate volcanic centres in closed basins	>1000? (>500) / FeOx in two districts / minor sodic ± peralkaline alteration, extensive gypsum	Arid climate; volcanoes are adjacent to high altitude salars containing halite
Mexican Altiplano, Cerro de Mercado, La Perla	Mid-Tertiary / continental arc felsic volcanic centres (incl. calderas)	>300 (>100)/FeOx ± REE, Cu in >20 occurrences / modest sodic ± peralkaline alteration, local stratiform FeOx with gypsum	Arid palaeoclimate, sulphate evaporites in volcanic basins, Jurassic evaporites
MESOZOIC			
Basin and Range province, USA (Humbolt complex; Cortez Mountains)	Jurassic / extensional or back-arc mafic or felsic volcano-plutonic complexes	>1000 (>75) / FeOx in > 40 occurrences, ± minor REE & Cu / regionally extensive sodic ± peralkaline ± minor potassic alteration	Arid palaeoclimate, evaporites in region; variable S (locally heavy): Volcanogenic massive sulphides (VMS) in same arc
Chilean coastal belt (El Romeral, Candelaria)	Jurassic - Early Cretaceous / extensional arc mafic-intermediate volcanoplutonic	>3000 (350) / FeOx ± major Cu(Au), minor Co, REE in > 100 occurrences / regionally extensive sodic ± minor potassic alteration	Arid palaeoclimate, Jurassic evaporites; variable S (locally heavy)
Peruvian coastal belt (Raul, Monterrosas, Marcona)	Early Cretaceous / extensional arc mafic-intermediate volcano-plutonic complexes	>1100 (>900) / FeOx ± major Cu in > 20 occurrences / regional sodic (±peralkaline?) alteration and skarn	Local redbeds & evaporites; evidence for evaporite/seawater source; VMS in same arc
Anhui and Hankow regions, east-central China (Fanchang?)	Jurassic / extensional back-arc mafic intermediate volcano plutonics	>1000 (>300) / FeOx ± minor Cu in > 25 occurrences / extensive sodic ± minor potassic alteration and skarn	Local redbeds & Triassic evaporites; evidence for evaporite/seawater source
Transvaal, RSA (Messina)	Jurassic (?) / mafic volcanics and intrusions associated with Limpopo aulocogen	>10 (>?) / Cu (≈3%) associated with FeOx / extensive sodic alteration	Controversial, but geochemistry and structure interpreted to indicate Jurassic evaporitic source
Mid-Atlantic (Cornwall, Grace)	Triassic-Jurassic / mafic sills in early Mesozoic basins	>350 (>100) / FeOx ± minor Cu, Co (±U) in > 100 occurrences / skarn and sodic alteration	Deposits in northern half of rift with redbeds and minor evaporites; evidence for evaporite source
Siberian Platform, Russia (Krasnoyarsk, Korshunovsk, Tagar)	Permo-Triassic / mafic flows and intrusions in flood basalt province	>3000 (>650) / FeOx + minor Cu, anhydrite and halite in >150 deposits / skarn and sodic alteration	Interaction of dolerites with Cambrian salts (>0.5 km thick); geochemical evidence for evaporite involvement

* In millions of tonnes; first number is total district resource, numbers in parentheses are tonnages for largest deposits in district.

Table 9.4. Characteristics of some Fe-oxide (-REE-Cu-Au-U) regions (after Barton and Johnson, 1996, and Geol. Soc. America Data Repository 9612).

Location / district	Age / Setting	Size* / Metals / Alteration	Comments
PALAEOZOIC			
Turgai province, Kazakhstan (Sarbai, Kachar)	Carboniferous / mafic-intermediate arc volcano-plutonic complex	>4000 (>1500) / FeOx + minor Cu, Co, MnOx / extensive sodic ± minor potassic alteration and skarn	Arid palaeoclimate, possible evaporites; setting analogous to Mesozoic Andes
Altai-Sayan, central Asia (Tuva, Abakan)	Mid Palaeozoic / volcano-plutonic complexes (mafic-felsic), redbeds	>2000 (>500) / FeOx ± significant REE, Cu, U in > 25 occurrences / extensive sodic alteration and skarn	Arid palaeoclimates; Devonian evaporites in region; geochemical evidence for evaporite involvement
Central Iran (Bafq, Gole Gohar, Hamadan)	Cambrian (and younger) / anorogenic felsic volcanic successions	>2500 (>1000) / FeOx ± Cu ± REE in stratiform and crosscutting bodies in > 25 occurrences/ sodic ± peralkaline ± minor potassic alteration	Extensive regional Cambrian and younger evaporites; distal MnOx and Pb-Zn
PROTEROZOIC			
Northwestern Canada (Great Bear, Wernecke Mountains)	Palaeo- and Meso-proterozoic / intermediate arc-like volcano-plutonic centres	>50? (>10?) / FeOx ± Cu, U, Ag, Co in > 25 occurrences/ regionally extensive sodic ± minor potassic alteration	No evaporites reported; zoned from central FeOx to outward Ag-Co-U mineralisation
Stuart Shelf, South Australia (Olympic Dam, Acropolis, Emmie Bluff)	Mesoproterozoic / extensional anorogenic(?) felsic-mafic(?) volcano-plutonic sequence	>3000 (>2000) / FeOx + major Cu, REE, U, Au in >5 deposits/ ± minor potassic alteration	No published evidence for evaporites in Mesoproterozoic
Northern Sweden (Kiruna, Svappavaara, Ekstromberg, Gallivare)	Mesoproterozoic / mafic-felsic volcanic sequence, possible extensional	>5000 (>2000) / FeOx + minor Cu, REE, Au, U in >20 deposits / regional sodic ± peralkaline ± minor potassic alteration	Evidence for rifting with evaporitic materials
Bayan Obo, Inner Mongolia	Mesoproterozoic(?) overprinted(?) by Palaeozoic / felsic sediments and dolomites, younger granitoids	>1000 / FeOx + major REE in stratabound orebodies / extensive sodic ± peralkaline ± minor potassic alteration	Controversial origin; features interpreted as syngenetic and epigenetic
Northeastern USA (Benson mines, Mineville, Dover)	Mesoproterozoic / intermediate-felsic anorogenic suite overprinted and intruded by Grenville granitoids	>1000 (200) / FeOx + REE ± minor Cu, U, Au in > 30 occurrences as discordant and stratabound orebodies / extensive sodic (± peralkaline) alteration and skarn	Evaporites present, some interpret FeOx to be related to Grenville event rather than Mesoproterozoic
Bihar-Orissa area, India (Bihar, Singhbhum)	Mesoproterozoic / mafic-intermediate igneous complex	>? (?) / FeOx ± zoned REE, Cu, U, in > 15 occurrences with discordant and stratabound ores / extensive sodic and minor potassic alteration	Strongly deformed and metamorphosed; local peralkaline alteration well developed zoning
Southeastern Missouri, USA (Pea Ridge, Boss - Bixby, Pilot Knob)	Mesoproterozoic / anorogenic felsic province	>1000 (>300) / FeOx ± Cu ± REE, U, Au in > 25 occurrences, minor sodic and potassic alteration known	No published evidence of evaporites; distal
* In millions of tonnes; first number is total district resource, numbers in parentheses are tonnages for largest deposits in district.			

Table 9.4. continued.

Currently our understanding of the effects of evaporites interacting with igneous melts, or their surface lavas, is minimal. Intense arguments as to the primacy of magmatic or evaporitic influences are often based on little more than conjecture. We do know that magma chambers can interact with evaporites in two main ways, either as a heat source that drives hydrothermal circulation or via heating and assimilation of any adjacent salt bodies. The resulting deposits encompass a wide range of styles from successions, which can be interpreted as skarns, to others best interpreted in the context of regional metamorphism. In regions such as the Mt Isa inlier and the Fe-oxide deposits of Scandinavia we have already discussed some of the complexities in separating meta-evaporite influences from meta-sedimentary and metavolcanic influences. Now we will look at a wider range of deposit styles where the evaporites act as: 1) a source of chloride and other ions; 2) as possible contributors to lavas composed of molten salts; and 3) as bodies of sediment that are assimilated into a magma chamber, so creating a rich body of magmatic sulphides.

Fe-oxides with evaporites as a hydrothermal chloride source

Barton and Johnson (1996) explore the notion of hybrid evaporite/magmatic hydrothermal chloride-rich fluids playing a role in the formation of many of the major Cu-Au Fe-oxide deposits worldwide. They carry this association out of the Proterozoic Cu-Au deposits into the Phanerozoic and the Archaean to document possible hydrothermal deposits in evaporite-associated settings ranging form rifts to collision belts. They go so far as to conclude that variation in magmatic composition is a secondary control when compared to variations in the hydrothermal chemistries contributed by dissolving chloride-rich evaporites.

According to Barton and Johnson all this group of deposits have abundant Fe oxides, as well as a characteristic suite of accessory elements, at least some sodic hydrothermal alteration, evidence for vigorous emplacement processes,

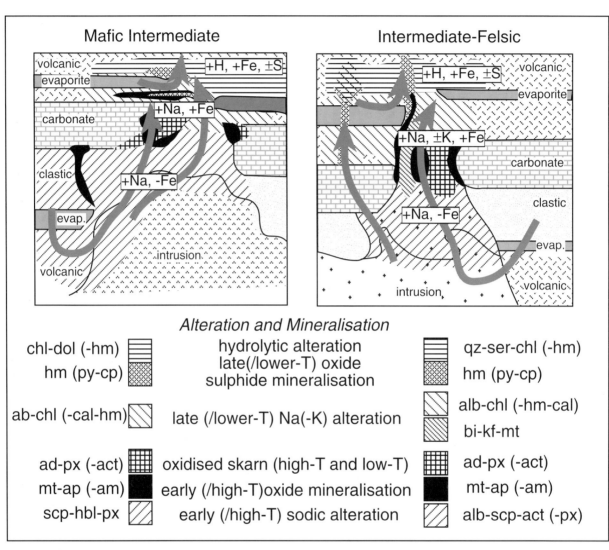

Figure 9.18. Model of igneous driven evaporitic fluids controlling large portions of the alteration zoning in both mafic and felsic systems. Abbreviations; act(actinolite), ad(andalusite), alb(albite), ap(apatite), bi(biotite), cal(calc-silicate), chl(chlorite), cp(chalcopyrite), hbl(hornblende), hm(haematite), kf(K-feldspar), mt(magnetite), px(pyroxene), py(pyrite), qz (quartz), scp(scapolite), ser(sericite) (based on Barton and Johnson, 1996).

and a genetic link to magmatism interactions in evaporitic terrains (Table 9.4). The circulation systems are typified by large volumes of Ti-poor Fe oxide (commonly >10^6 tonnes magnetite or haematite) with lesser amounts of phosphates (apatite, REE phosphates), Cu-Fe sulphides and sporadic Au, U, Ag and Co minerals. Proximal and deeper mineralisation consists of early magnetite ± apatite rocks, commonly with abundant sodic (albite ± scapolite ± hornblende) wallrock alteration. Superimposed or distal mineralisation consists of haematite ± Cu-Fe sulphides ± REE minerals with ± potassic alteration. Variants include deposits with substantial sodic alteration and Cu or U mineralisation, but subordinate Fe-oxide. Sodic alteration

occurs in nearly all districts, is typically voluminous, and a number of examples have local peralkaline assemblages containing NaFe^{3+} bearing amphibolites or pyroxenes. Calcareous host rocks contain calcic or magnesian skarn assemblages with hydrous distal assemblages (actinolite-chlorite-carbonate) that are superimposed on anhydrous assemblages (pyroxene + garnet). In form, these deposits may consist of stratabound massive Fe-oxides hosted in volcanic and sedimentary rocks, although most commonly they are discordant, variably brecciated tabular to irregular masses. The former have previously been interpreted as syngenetic replacement bodies, while the latter have been interpreted as veins, dykes and breccia fills associated with magmatism.

Igneous rocks are closely linked to these deposits, but there appears to be little correlation of the mineralisation to igneous composition or intrusive phases (Barton and Johnson, 1996). The ions in the circulating fluids are largely derived by reflux/dissolution of coeval or older evaporitic sources. Circulation is usually driven by the thermal effects of the magmatism. Igneous rocks often also serve as the source and host for mineralisation, making the total system a form of large-scale contact metamorphism. The dissolving evaporite provides much of the chloride necessary for metal transport as well as the Na observed in the widespread sodic haloes that characterise alteration in many of these deposits.

Products from the dissolution of thick halite beds are caught up in circulating basinal fluids. Owing to the chloride-rich nature of most dissolving salt beds, sulphur levels tend to be relatively low in the brine. Reaction of these fluids along hydrothermally driven circulation paths produces predictable patterns of alteration and mineralisation (Figure 9.18; Barton and Johnson, 1996). The patterns can be similar to those created by nonevaporitic igneous-related hydrothermal circulation cells (e.g. metamorphosed porphyry deposits). However, there are diagnostic differences in the proportions of indicator minerals in the alteration haloes and nearby meta-evaporites, such as scapolite and sodic amphiboles, along with characteristic solution breccias that distinguish these deposits. Voluminous sodic alteration forms independent of other factors of the magmatic history and in far greater volumes than any potassic alteration that forms later in the circulation history.

On the downwelling limb, sodic alteration reflects the high Na contents in relatively low temperature fluids (Figure 9.18; Barton and Johnson 1996). Alteration volumes are limited only by the amount of fluid that can circulate (heat balance) rather than by the S and Cl contents of causative intrusions and so contrast with the typical situation in many porphyry-type systems. Coincident with the sodic alteration, the downwelling fluid may leach components in which it is undersaturated (Fe, K, Ca, reduced S, REEs, SiO_2; Barton et al., 1991; Dilles and Proffett, 1995). If temperatures are relatively high, marialitic scapolite (scp) + hornblende (hbl) dominate the alteration halo. However, the presence of scapolite and hornblende is, by itself, not diagnostic as both minerals can also form in nonevaporitic intermediate and mafic systems. It is the total association that is the key to interpretation.

Albite (alb) + actinolite (act) + chlorite (chl) tend to form as co-precipitates in the more mafic systems (Figure 9.18).

Such peralkalic assemblages are typically present in many evaporite-associated Fe-oxide occurrences. They are hard to rationalise by ordinary igneous or magmatic hydrothermal mechanisms. In contrast, the extremely high Na^+/H^+ of a dissolving evaporitic source can drive reactions such as:

$$2NaCl_{aq} + Fe_2O_3 + SiO_2 + H_2O \longrightarrow 2NaFeSi_2O_6 + 2HCl$$

On the upwelling limb, cooling wallrock reaction and mixing of hot fluid with other basinal fluids creates distinctive mineral products dependent on rock type and flow path. Sodic alteration persists where fluid-to-rock ratios are high. Hydrolytic alteration [producing K-mica (ser) + quartz (qz) in felsic rocks, chlorite (chl) in mafic rocks] forms in response to cooling, and precipitation of metals from chloride complexes now takes place. In rocks with moderate K_2O contents, potassic alteration (secondary biotite (bi) + K-feldspar (kf)) can form by alkali exchange (Hitzman et al., 1992).

Evaporitic brines sourced from dissolving thick halites have exceptionally high Cl/S; thus metal content greatly exceeds the amount of sulphur available to precipitate them (Barton and Johnson, 1996). Partial reduction and incorporation of sulphides may modify any original sulphate-bearing fluids, but they generally have too little sulphur to precipitate sulphide minerals at higher temperatures associated with circulation close to the intrusive. Consequently, only the least soluble chalcophile elements (e.g., Cu) form sulphides along with the other elements that form oxides or oxysalts (Fe, REEs, U) or the native elements (e.g. Au). Because adequate S is lacking, Pb and Zn will be removed from the system, and Fe-sulphides will be sparse. In some circumstances, reduction of these fluids might precipitate distal Co-As-U. Early precipitation of magnetite (mt)/apatite (ap) is superseded by haematite (hm)±copper-iron sulphides (cp-py)± other REE minerals as H-O(-Cl-S) fluids become more oxidising and sulphidising with decreasing temperatures (Barton et al., 1991).

In summary, the evidence of former evaporites in these deposits is subtle and unless one seeks it out, will be largely lost in the milieu of hydrothermal alteration. The criteria for upgrading the possibility of an evaporite association in an Fe-oxide Cu-Au deposit include: 1) abundant sodic alteration, often combined with widespread elevated Cl content in some minerals (e.g. abundant marialitic scapolite, and increased Cl content in amphiboles and biotites); 2) laterally equivalent metasediments or sediments that retain evaporite pseudomorphs or dissolution breccias; 3) sediments that were deposited under arid/redbed conditions;

357

4) an abundance of transported iron with little evidence of sulphur in the transport system.

Molten salts:

Natrocarbonatite and brine

Natrocarbonatites are highly alkaline igneous rocks, exemplified by modern sodium carbonate lavas, which are extruded at Oldoinyo Lengai in Tanzania in the African Rift. With eruption temperatures between 500 and 600°C, they exhibit what are the lowest known magmatic eruption temperatures for modern lavas. Their salt mineralogy, and exceptionally high alkali content

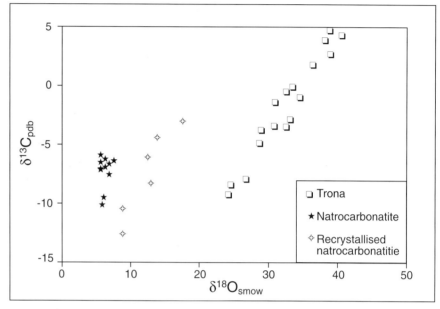

Figure 9.19. Carbon-oxygen crossplot of natrocarbonatites from Oldoinyo Lengai and trona from Lake Magadi. Recrystallised natrocarbonatites are either altered or recrystallised in the presence of meteoric water. Plot compiled from Bell and Dawson (1995) and data sources therein.

(total alkalies ≈ 40%), means they are unique amongst the various types of modern carbonatites. At Oldoinyo Lengai, Na_2O makes up 32% of the lava chemistry and the Na_2O/K_2O ratios hover around 4. This, and their proximity to the nearby trona deposits of Lake Natron and Lake Magadi, has led some authors to postulate formation via interaction between a silicate melt and either soda-rich subsurface trona beds (Milton, 1968, 1989) or associated saline basinal brines (Eugster, 1970, 1986).

A sedimentary influence in the magma chemistry is strongly contested by many in the igneous community who interpret these natrocarbonatites as the end result of a highly saline magmatic separation. For example, Bell and Dawson (1995) conclude the stable and radiogenic isotopic signatures of natrocarbonatites of Oldoinyo Lengai are different to those of sedimentary trona and similar to other magmatic rocks, including young carbonatites, from elsewhere in east Africa (Figure 9.19). Therein lies the problem. Bell and Dawson's isotopic data for trona comes from unaltered sedimentary trona, they ignore the highly reactive nature of buried evaporites within active hydrothermal hydrologies, which characterise rift systems (Chapter 2). Nor does their argument consider temperature-related fractionation effects that typify evolving stable isotope signatures of oxygen-entraining minerals undergoing burial diagenesis. How much is the oxygen isotope signature of trona lightened as it is buried in

settings where it undergoes dissolution/reprecipitation interactions with increasingly hotter throughflowing hydrothermal brines?

The fact that these sodium carbonate lavas at Oldoinyo Lengai, with their highly elevated sodium contents, are: 1) unique amongst all the world's carbonatites; 2) erupting today; and, 3) occur in proximity to one of the largest examples of modern trona accumulation (see Lake Magadi in Chapter 7); must at least argue for some influence of trona-derived brines. This is the model proposed by Eugster (1970, 1986) who envisaged a deep saline groundwater reservoir for the Magadi Basin. He interprets the sodium carbonate lavas of Oldoinyo Lengai, not so much as remobilised sedimentary trona, but as the product of the interaction of an evaporite-influenced basinal brine with alkaline silicate magmas.

There can be no doubt that the sodium and bicarbonate ions now held in trona salts of the Magadi and Natron Basins were derived, at least in large part, by surface weathering and groundwater leaching of the surrounding volcanic ashes and flows, including the natrocarbonatite lavas of Oldoinyo Lengai (Chapter 2). Even today, this leachate is carried into perennial trona lakes that occupy the lowermost parts of the two basins. There it is concentrated by solar evaporation to deposit beds of sedimentary trona. Ongoing extension and subsidence within the rift then caries this evaporite into the subsurface, where it is leached by an active hydrothermal hydrology. Under this scenario categoric statements of primacy in the intimate relationship between sedimentary trona, magma assimilation and natrocarbonatite is as useful as the question of the primacy of a chicken or an egg.

Dolerite intrusives and halite

When a mafic intrusive is emplaced in, or adjacent to, an evaporite unit, one of the most noticeable initial effects is melting and recrystallisation of the salt adjacent to the intrusive. Halite has a melting point ≈ 800°C, sylvite ≈ 780°C and anhydrite ≈ 1450°C. As many intrusives are emplaced at temperatures of 700° - 900°C, the likelihood of molten salt near an intrusive, with subsequent recrystallisation, is high. Although anhydrite's melting point is much greater than halite or sylvite, it tends to dissolve rather than melt in molten salt and so it too shows textures altered by the emplacement of an intrusive in a halite body.

Alteration aureoles about dolerite intrusives extend 1-10 metres into the salt in the Werra Potash zone of Germany, in the Yesso Hills in New Mexico, in Castle Valley in Utah and in the Irkutsk amphitheatre in Siberia (Sidle et al., 1985; Pavlov, 1970). In the 10m-wide alteration zone in lower Cambrian halite at Irkutsk, the halite crystals adjacent to the dolerite are 4-6 cm across. The anhydrite has been altered to rounded grains with characteristic fused edges, which indicate partial exsolution as the molten halite cooled (Pavlov, 1970). Sylvite also appears as separate crystal segregations in recrystallised salt, whereas prior to melting it was finely dispersed in brine inclusions throughout the sedimentary halite.

Magmatic Ni-Cu associations with evaporites as a sulphur source

Where an intrusive magma body comes into contact with a sulphate-entraining sediment body, the interaction can facilitate the precipitation of metals that may be in solution within the system. For example, when igneous trap intrusives interact with sediments containing hydrocarbon-water-sulphate-carbonate at temperatures of 100° to 700°C, the sulphates and carbonates in the skarn region decompose to form large volumes of hydrogen sulphide and carbon dioxide, respectively. At the same time mercaptans appear among the hydrocarbons (Kontorovich et al., 1996). The H_2S and mercaptans can then act as reductants to fix metals that may be in solution within the adjacent hydrothermal system.

The assimilation of skarn evaporites typically forms smaller sulphide deposits in the immediate vicinity of the emplaced magma. However, there are also a few larger magmatic metal accumulations where the assimilation of substantial volumes of sulphate evaporites contributed significant volumes of sulphur. Assimilation of solid sulphate evaporite units into a magma chamber is a different mechanism to that of dissolving evaporites contributing large volumes of chloride to the chemistry of the circulating hydrothermal and metamorphic fluids, as outlined in previous sections. The magma chemistry driving the direct assimilation of the sedimentary sulphur to form magmatic sulphides tends to be picritic or komatiitic.

The importance of komatiitic and picritic magmas in ore forming processes arises from the fact that they are among the few mafic/ultramafic magma types that are sulphur-undersaturated at the time of magma formation (Keays, 1995). As a general rule, and independent of whether sulphate evaporites were involved, the ore forming process is controlled by the availability of S, not Ni or Cu. Sulphur-saturation of a magma as it forms leads to marked depletion in the chalcophile metals. Komatiitic and picritic magmas are S-undersaturated because they are high temperature magmas, produced by large degrees of partial melting of upper mantle source regions, which were already depleted in S through earlier partial melting events. Most mafic magmas, produced by less than 25% partial melting, are S-saturated and hence depleted in the ore-forming chalcophile elements. In contrast, komatiitic and picritic magmas have their full complement of chalcophile metals. When they do become S-saturated, they form sulphide precipitates, which are strongly enriched in Ni, Cu, Au, PGE (platinum group elements), and other strongly chalcophile metals. The metals may accumulate to form ore deposits directly (e.g. magmatic Ni-Cu-PGE sulphides) or be dispersed in komatiitic/picritic rocks or their fractionation products, making these source rocks for hydrothermal Au or VMS deposits as discussed in the preceding sections.

Because the sedimentary sulphate is completely assimilated, the evidence for the evaporite contribution is subtle, largely isotopic, or perhaps tied to the presence of residual sedimentary evaporite units encasing the magma body. The high temperatures and near complete assimilation of the evaporites within the magma mean that this is a most enigmatic ("salt is elsewhere") style of deposit. Notions of evaporite assimilation for these deposits are usually only one of multiple possible explanations. For the purposes of this review, I have chosen to emphasise the evaporite connection in this style of deposit using a detailed description of the Noril'sk deposit in Siberia. Alternate explanations for the same deposit can be found in papers such as Wooden et al. (1992) and Lightfoot et al. (1997).

Sudbury, Noril'sk and Jinchaun are three of the largest currently mined magmatic Ni-Cu deposits. Although of less tonnage, the higher grade of the Noril'sk ore (555 million tonnes of 2.7 wt.% Ni) means that its Ni content is almost as much as Sudbury, the largest Ni deposit in the world (1648 million tons at 1.2 wt.% Ni). The Noril'sk ore contains 2 times as much Cu and 15 times as much platinum groups elements as at Sudbury (Naldrett, 1993). Jinchuan in China is the world's third largest Ni deposit with 515 Mt. at 1.06 wt.% of Ni (production + reserves) and forms the keel of a much larger body of magnesian basalt.

Unlike Noril'sk, most worldclass Ni deposits are Precambrian, most are Archaean and show no obvious association with evaporites, although some are associated with the assimilation of non-evaporitic S-rich sediments. Sudbury Ontario, for example, is a 1.7 Ga bolide impact crater (Reimold, 1993). Most Australian Ni deposits in the Agnew-Wiluna greenstone belt of Western Australia formed as hot (1600°C+) thermally erosive Archaean ultramafic lava rivers (komatiites) that were contaminated with sedimentary sulphide to create the rich nickel ores of the district (Hill et al., 1989; Evans et al., 1989). Such mineralising systems are tied to the more active tectonics, thinner crust and reducing atmosphere of the Archaean.

Noril'sk, Siberia

The anomalous Phanerozoic age of Noril'sk, compared with the Precambrian ages of other magmatic Ni-Cu deposits, and its relative enrichment in Ni and Cu compared with Sudbury and Jinchuan, may reflect the anomalously high volumes of sulphur contributed to the evolving magma chamber by the sulphate evaporites (Figure 9.20; Naldrett 1981, 1993, 1997). At Noril'sk, intense intracontinental rifting brought mafic magmas into contact with supracrustal sulphur from evaporitic sulphates. Noril'sk lies at the extreme northwestern margin of the Siberian Platform, an area that has been a stable cratonic block since the end of the Palaeozoic. To the north, the Khatanga Trough separates

the Siberian Platform from a second stable platform (the Taymyr Peninsula). To the east, the Yenisei Trough separates the Siberian Platform from the East European-Urals cratonic block. Thus, the area is a classic rift aulocogen. Naldrett and Macdonald (1980) noted that triple junctions, in particular the aulocogens that may result, are areas where substantial thicknesses of evaporite and mafic magmatism are natural consequences of the tectonic development. Thus, they are likely regions for magmatic Ni-Cu ore formation.

In the vicinity of the Noril'sk deposits, lower Palaeozoic marine argillaceous sediments are overlain by thick extensive Devonian evaporite deposits along with lower Carboniferous lagoonal and continental sediments that include a number of coal measures (Figure 9.20a). An extremely large volume (approximately one million cubic kilometres) of Permo-Triassic flood basalts, known as the Siberian Traps, now covers this sedimentary sequence. Sill-like tholeiitic intrusions, varying in composition from subalkaline dolerite to gabbro-dolerite were emplaced in the sediment pile at the same time as the basalts, and were part of the feeder system to the flood basalts. The ore-bearing gabbroic-dolerites are differentiated, with picrite and picritic dolerite overlain by more felsic differentiates. Individual sills may attain lengths of 12 km, widths of 2 km and thicknesses of 30 to 350 m. The Cu-Ni-platinoid mineralisation forms relatively persistent horizons, which in the lower portions of the intrusions are made up of segregations and accumulations of pyrrhotite, pentlandite and chalcopyrite.

The structure in the district is dominated by NNE-NE Permo-Triassic block faulting, which was coeval with the volcanic activity. Individual faults may be over 500 km in length with throws of up to a kilometre (Naldrett, 1997). Mineralised intrusions radiated outward and upward from intrusive centres as they penetrated the sedimentary sequence. Many of the intrusive centres are associated with prominent block faulting. The main Noril'sk-Kharaelakh fault occurs within the Siberian Platform, but is parallel to the main fault system that defines the boundary between the platform and the Yenisei Trough. This is the fault that guided the main upwelling magma body.

Sulphur isotope data show the mineralised intervals are anomalously heavy in ^{34}S (Figure 9.20b). These data are inconsistent with sulphur derived from the mantle, and are consistent with assimilation of an evaporite sulphate source (Godlevsky and Grinenko, 1963; Grinenko, 1985). As the magma rose through the sedimentary cover it penetrated and assimilated sulphur from extensive thick sedimentary

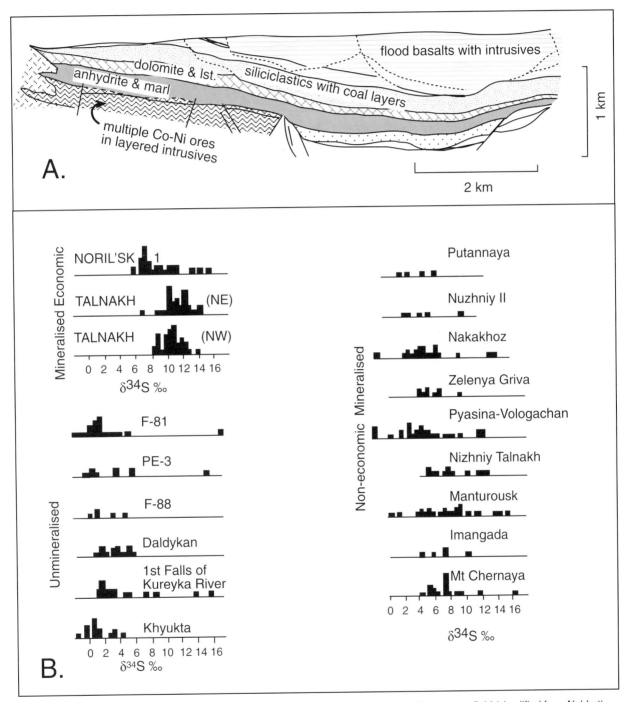

Figure 9.20. Noril'sk region. A) Geological section through the northwestern part of the Talnakh ore field (simplified from Naldrett, 1991). B) Isotopic composition of sulphur in unmineralised, mineralised economic and mineralised subeconomic intrusions of the Noril'sk - Talnakh region (after Grinenko, 1985; Naldrett, 1993).

evaporites. The sulphur in the calcium sulphate was reduced to sulphide, CaO entered the magma, and iron from the magma reacted with the reduced sulphur so that the end result was droplets of immiscible iron sulphide dispersed through the melt (Naldrett and Macdonald, 1980). These droplets acted as collectors for Ni, Cu and the platinum group elements that are now so enriched in the Noril'sk ores.

Naldrett (1991,1997) suggests that the prehnite + biotite + anhydrite + carbonate + zeolite + chlorite ± sulphide globules that occur in the chromite agglomerations in the picrite of the intrusions could represent remnants of partially assimilated sulphate-rich country rock. The assimilation of anhydrite-rich rocks, coupled with the reduction of sulphate to sulphide, would have introduced considerable oxygen into the silicate melt, which could have caused the precipitation of the chrome spinel. Inclusions of anhydrite-rich material floating in the magma could have served as loci for chromite crystallisations thus giving rise to the association between the agglomerations and the globules. Tarasov (1970) pointed out that Middle Carboniferous coal measures were also assimilated and may have assisted in the reduction of calcium sulphate. In summary, this world class Ni-PGE deposit at Noril'sk, with its anomalous Mesozoic age, perhaps reflects a fortuitous occurrence of thick sulphate evaporites atop a later active set of deep mantle-tapping rift grabens.

Hydrothermal salts

Anhydrite's retrograde solubility means the solubility of anhydrite decreases rapidly with increasing temperature (Figure 9.21a; Blount and Dickson, 1969). It explains why hydrothermal anhydrite precipitates within submarine volcanics and volcaniclastics during the heating of seawater caught up in submarine hydrothermal circulation cells (Figure 9.21b). Its retrograde solubility also explains why anhydrite is most obvious in the upper portions of young active vents in black and white "smokers". Its heating behaviour is the opposite of baryte, another common hydrothermal sulphate precipitate. Simple heating of seawater adjacent to seafloor vents, even without fluid mixing, will precipitate anhydrite, while simple cooling of hydrothermal waters will precipitate baryte.

Once buried, hydrothermal calcium sulphate, in the presence of organic matter or hydrocarbons and circulating hydrothermal brines, acts as a sulphur source to create H_2S, which then interacts with metal-carrying pore waters to precipitate metal sulphides. As discussed in Chapter 1, the hydrothermal form of $CaSO_4$ is not a true evaporite, but needs to be discussed in the metals context of this chapter.

Hydrothermal anhydrite, or indicators of its former presence, are commonplace within volcanogenic hosted massive sulphide (VHMS) deposits. VHMS deposits typically form in submarine depressions as circulating seawater becomes an ore-forming hydrothermal fluid through interaction with the heated upper crustal rocks. Submarine depressions, especially those created by submarine calderas and/or by large-scale tectonic activity (median rift valleys), are favourable sites. Fundamental processes leading to the formation of VHMS deposits include (Ohmoto, 1996):

• Intrusion of a heat source (typically $\approx 10^3$ km size pluton) into an oceanic crust or a submarine continental crust, which causes deep convective circulation of seawater around the pluton (Figure 9.2). The radius of a circulation cell is ≈ 5 km. The temperatures of fluids that discharge on the seafloor increase with time from the ambient temperature to a typical maximum of $\approx 350°$ C, and then decrease gradually to the ambient temperatures on a time scale of ≈ 100 - 10,000 years. The majority of subsurface sulphide and sulphate mineralization occurs during the waxing stage of hydrothermal activity,
• Reactions between warm country rocks and downward percolating seawater cause seawater SO_4^{2-} to precipitate as disseminated hydrothermal anhydrite in the country rocks in areas where temperatures are greater than 150°C.
• Reactions of the "modified" seawater with higher-temperature rocks during the waning stages of hydrothermal circulation transform this sulphate-depleted "seawater" into metal-rich or H_2S-rich ore-forming fluids. Metals are leached from the country rocks, while previously formed hydro thermal $CaSO_4$ is reduced by Fe^{2+}-bearing minerals and organic matter to provide H_2S. The combined mass of high temperature rocks that provide the metals and reduced sulphur in each VHMS marine deposit is typically $\approx 10^{11}$ tonnes (≈ 40 km^3 in volume). Except for SO_2, which produces acid-type alteration in some systems, the roles of magmatic fluids or gases are minor in most massive sulfide systems.
• Reactions between the ore-forming fluids and cooler rocks in the discharge zone cause alteration of the rocks and precipitation of ore minerals in the stockwork ores.
• Mixing of the ore-forming fluids with local seawater within unconsolidated sediments and/or on the seafloor causes precipitation of "primitive ores" with the black ore mineralogy (sphalerite + galena + pyrite + baryte + anhydrite).
• Reactions between the "primitive ores" with later and hotter hydrothermal fluids beneath a sulphate-rich thermal blanket causes transformation of "primitive ores" to "matured ores" that are enriched in chalcopyrite and pyrite.

In order to form VHMS deposits on the seafloor, throughflushing hydrothermal fluids must transport sufficient amounts of metals and reduced sulphur, each at concentration levels > 1ppm (Ohmoto, 1996). For a hydrothermal fluid with the salinity of normal seawater ($\approx$0.7m ΣCl) to be capable of transporting this amount of Cu and other base metals, it must be heated to temperatures > 300°C. Fluids with temperatures in excess of 300°C will boil at pressures >200 bars. Under such conditions the resulting vapour cannot carry sufficient quantities of metals to form a VHMS deposit. Boiling of a metalliferous hydrothermal brine outflow is prevented when the fluid vents into water that is deep enough to generate a sufficient confining pressure. At 350°C, a minimum seawater depth of 1550m is necessary to prevent boiling. If the fluid passes through a sedimentary package where it loses temperature and metals (Cu, Ba) before emanating, the water depth beneath which boiling is prevented is less ($\approx$1375m). Once vented, the turbulent mixing of hot hydrothermal waters with cooler seawater causes rapid precipitation of sulphides and calcium and barium sulphate, which produces the familiar black and white smokers (Blum and Puchelt, 1991).

VHMS deposit styles of mineralisation are always related to submarine volcanism and hydrothermally driven circulation with adjacent deepwater sediments, but can form within a variety of tectonic settings and sediment types:

• VHMS deposits form in subduction-related island-arc settings (Kuroko-type deposits);
• VHMS deposits form at mid-oceanic or back-arc spreading centres (Cyprus or TAG mound type deposits);
• VHMS deposits form at spreading centres, but, due to the proximity of one or more landmasses, the deposit is sediment-hosted (Besshi-type deposits).

For brevity and relevance, the emphasis in the following discussion of VHMS deposits is on the location and significance of the hydrothermal salts, rather than on a general discussion of mineralisation style. For excellent summaries of the relevant mineralisation literature, the interested reader is referred to Pirajno (1992) and Ohmoto (1996).

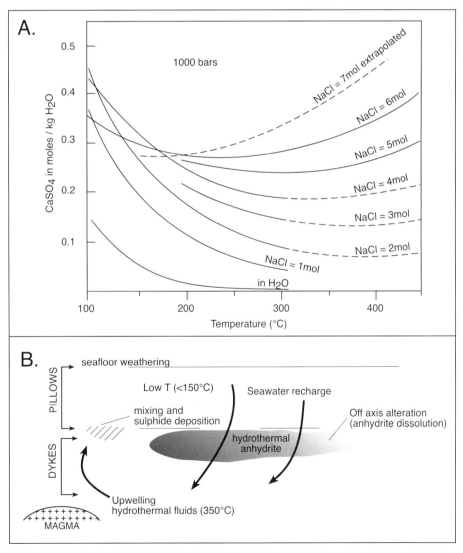

Figure 9.21. Hydrothermal anhydrite. A) Anhydrite solubility as a function of temperature at 1000 bars in H_2O and $NaCl-H_2O$ solutions of various concentrations. Dashed lines indicate extrapolated values (after Blount and Dickson, 1969). B) Formation of hydrothermal anhydrite from heated seawater in oceanic crust.

VHMS deposits in subduction-related island-arc settings: Kuroko-style deposits

Kuroko-style deposits are composed of copper- and zinc-bearing massive sulphide deposits and typically hosted in felsic to intermediate volcanics (Ohmoto and Skinner, 1983). Deposits tend to occur near felsic domes and toward the stratigraphic top of the volcanic or volcano-sedimentary sequences. Host strata can be marine rhyolites, dacites, subordinate basal and organic-rich mudstones and shale. Often the host strata are brecciated or made up of volcanic flows, tuffs, or, in some cases, are actually part of a preserved felsic dome. Pyritic siliceous rock (exhalite) may mark the horizon at which the deposit occurs, and proximity to a deposit is often indicated by sulphide clasts in a volcanic breccia. Some deposits appear to be gravity transported. Ancient examples range in age from Archaean to Cenozoic and the type example of this Kuroko-style of mineralisation is in the Miocene volcanics of northern Japan.

Hydrothermal $CaSO_4$ is widespread in the Miocene Hokuroko Basin in the Green Tuff Belt of northern Japan, where it forms an integral part of the Kuroko-type ores. Kuroko polymetallic massive sulphide deposits occur at the discharge site of submarine hydrothermal systems associated with caldera structures (Ohmoto, 1996). Calderas

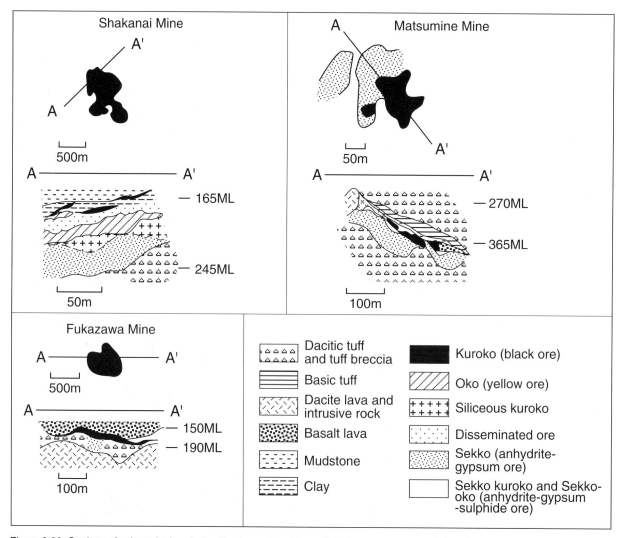

Figure 9.22. Geology of selected mines in the Kuroko region, Japan. Note the common association of ore zones with hydrothermal calcium sulphate (after Shikazano et al., 1983b).

Name	Mineral Phase
Kuroko ore	sphalerite + galena + barite
Oko ore	chalcopyrite + pyrite
Sekko	anhydrite + gypsum
	anhydrite + gypsum + sphalerite + galena + barite
	anhydrite + gypsum + chalcopyrite + pyrite
Keiko ore	quartz + pyrite + chalcopyrite

Table 9.5. Mineralogy of mineralised zones in classic Kuroko deposits.

Table 9.5) from the lowermost siliceous ore zone (keiko), through the overlying yellow ore zone (oko), to the upper black ore zone (kuroko). Baryte occurs mainly in the black ore zone and the overlying bedded baryte zone. Keiko ores indicate the highest fluid temperatures (>300°C) and were the feeder zone to the yellow and black ore systems. Most ores are compact, and massive but bedded, brecciated, or colloform textures dominate locally and graded sedimentary bedding and other sedimentary features can be seen in the black ore.

create depressions that became the loci for discharging hydrothermal solutions. In addition, the faulting associated with caldera formation is responsible for providing permeable channelways for the circulating hydrothermal fluids. Type localities are in the Miocene Green Tuff Belt of Honshu and Hokkaido in Japan (Figure 9.22; the classic Kuroko-type deposits). There the metal production is dominated by ore clusters centred around caldera structures. Similar base metal deposits, in terranes characterised by submarine felsic volcanics and secondary ores containing varying amounts of hydrothermal anhydrite, have now been identified around the entire Pacific Rim (Sawkins, 1990; Shikazono et al., 1983; Kusakabe and Chiba, 1983). The isotopic signature of anhydrite and baryte in most Kuroko-style deposits is intermediate between that of Miocene seawater and sulphate in hydrothermal fluids, suggesting that both types of sulphate precipitated from mixtures of hydrothermal fluids and local unaltered seawater (Ohmoto, 1996).

The Japanese Kuroko deposits are, in general, characterised by a common stratigraphic mineral zonation (Figure 9.23;

Each of the major mines is centred on a cluster of ore lenses that range in size from less than 0.1 million tonnes to ≈ 10 million tonnes. The stratabound lenses are typically elongate, with sharp upper boundaries and more diffuse lower boundaries that grade downward through lower grade stockworks into unmineralised footwall volcanics and tuffs. In many cases, hydrothermal anhydrite, with gypsum, occurs at the periphery of the black ore zone, either as near-monomineralogic aggregates or as units

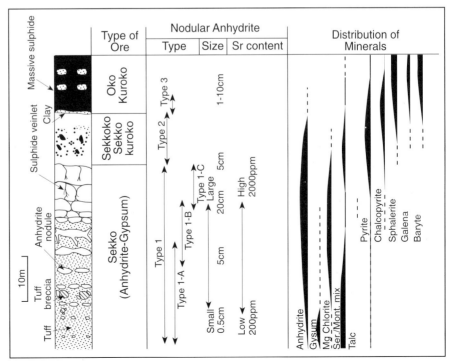

Figure 9.23. Vertical zonation of anhydrite in Kuroko ore, see text for discussion of the anhydrite types (after Shikazono et al., 1983).

365

containing sulphide minerals (sekko ore). Anhydrite is rare in the central part of the sericite-chlorite alteration zone where temperatures were >300°C, but is increasingly common in the more peripheral montmorillonite and zeolite alteration zones (T<300°C). Calcium sulphate underlies almost all mines in the Kuroko region and forms a mineable horizon of sekko ore; between 10^5 to 10^6 tonnes of sekko ore have been removed from individual deposits (Figure 9.22). Sekko ore can be further subdivided into sekko-kuroko and sekko-oko ores (Table 9.5).

Much of the gypsum that outcrops or subcrops in the region (Figure 9.22) is a secondary rehydration product from original hydrothermal anhydrite (Kuroda, 1983; Shikazono et al., 1983). The proportion of gypsum to anhydrite in the sekko ores probably reflects the rate of penetration of the modern meteoric groundwater system into the massive stratiform anhydrite units.

Hydrothermal anhydrite in the sekko ores occurs as four types: 1) nodular anhydrite, 2) anhydrite associated with sulphide minerals, 3) anhydrite in sulphide ore, and 4) vein anhydrite (Figure 9.23). Type 1, or nodular anhydrite, is the most abundant and typically constitutes more than 90% of the anhydrite in individual deposits. It occurs within dacitic tuff breccias and in tuff hosts, and changes up section into type 2 massive nodular anhydrite at the uppermost part of the sekko ore horizon. The diameters of the anhydrite nodules in sekko ore increase upward from a few mm to several metres. The thickness of the sekko ore varies within and between different deposits. The maximum thickness of sekko in the Matsumine and Shakanai mines is more than 100m, but is less than 50 m in the Fukazawa Mine (Figure 9.22). Thicknesses of sekko-kuroko or sekko-oko are generally less than 10 m.

Reactions between warm country rocks (T > 150° C) and percolating seawater precipitate seawater sulphate as dis-

seminated anhydrite. Kuroda (1983) in an inclusion-based study of the Furutobe and Matsuki deposits showed how convecting seawater was heated as it approached the high temperature zone nearer the ore deposits and precipitated anhydrite in the interstices of dacitic pyroclastic rocks and fine grained tuffs. This is the texture that is illustrated in the lower two-thirds of the type 1 texture in Figure 9.23.

The nature of the up section increase in anhydrite nodule size and purity, along with an associated upsection increase in Sr content. in many Kuroko deposits is not as well understood (Shikazono and Holland, 1983). Given the hot hydrothermal nature of the precipitating setting and the deep seafloor setting of adjacent sediments, the zones of coalesced nodules (upper part of type 1 and type 2) cannot reflect ghosting of a gypsum precursor. Changes in nodule size and purity are more likely to reflect rates of hydrothermal anhydrite crystallisation/replacement or the

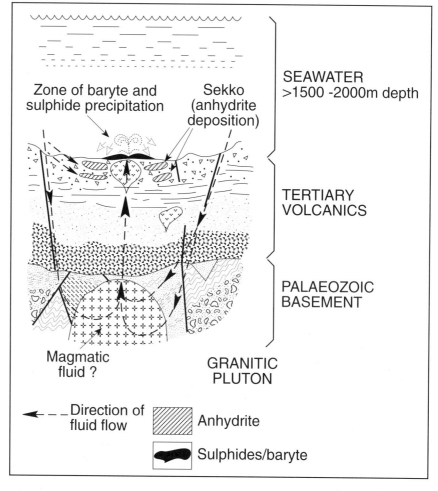

Figure 9.24. Hydrothermal circulation that characterises Kuroko deposits (after Shikazono et al., 1983).

stability of the anhydrite-precipitating environment within the thixotropic slurry, which is normal deepwater bottom sediment (larger nodules probably indicate slower, more stable, growth settings).

The ongoing preservation of widespread anhydrite, adjacent to, and beneath the massive sulphide bodies in the Hokuroko district requires more than the accumulation of anhydrite derived from chimney vent remains. It supports the notion of a separate zone of widespread subsurface anhydrite precipitation in most hydrothermal seafloor deposits (Figure 9.24). But anhydrite does quickly dissolve on the seafloor, or in hydrothermal mixtures below temperatures of 150°C. Once active circulation ceases, hydrothermal anhydrite is liable to flushing and leaching by cooler seawater. Its preservation in the Kuroko ores implies hydrologic isolation of the anhydrite from cold seawater. Its partial conversion to gypsum indicates subsequent meteoric flushing during uplift. Anhydrite's solubility in cool seawater pore brines (<150°C) also explains the lack of abundant $CaSO_4$ textures in massive sulphide bodies of older Kuroko-style deposits outside of Japan. There are, however, some intriguing breccias in these deposits.

Although hydrothermal anhydrite ultimately dissolves it can, while the mound is active, act as a concentrated subsurface sulphur source, which plays an fundamental role in the growth and maturation of sulphide mineralisation in the mound (Ohmoto, 1996). The question that have yet to be addressed in this and other styles of VHMS deposits are: "What are the indicators of the hydrothermal salt that was?" Are they similar to those of dissolved low temperature evaporites (Chapter 4) or are there a whole new and distinctive suite of indicators and breccias that, once they are fully documented, can be used to refine models of hydrothermal deposits?

VHMS deposits at spreading centres

The significance of hydrothermal anhydrite during the active mineralisation stages of a VHMS deposit forming at a mid oceanic spreading centre is clearly seen in the lithologies actively accumulating beneath the seafloor surface of TAG mound. TAG mound is an active hydrothermal mound, located 2.4 km east of the neovolcanic zone at 26° N, Mid-Atlantic Ridge. It is 200m in diameter, exhibits 50m of relief, and is covered entirely by hydrothermal precipitates.

Surface samples were recovered from the mound surface by the submersibles "Alvin" and "Mir" in 1986, 1990, and 1991. In September-November 1994, the Ocean Drilling

Program Leg 158 drilled a series of holes into the active mound to reveal unexpected high volumes of hydrothermal anhydrite within the sulphide mound. In fact, anhydrite occurs in breccias and as late-stage veins throughout most of the mound and its stockwork, down to at least 125 metres below sea floor (m.b.s.f.). As stated by Humphris et al., (1995): "These observations demonstrate the role of anhydrite in the growth of massive sulphide deposits, despite its absence in those preserved on land".

Present day activity on the mound surface includes highly focused outflow of a 363°C fluid from a chimney cluster on the top of the mound and deposition of a high S_2-O_2 mineral assemblage that reflects low concentrations of H_2S in the black smoker fluid (Tivey et al., 1995). Blocks of sulphide and white smoker chimneys, enriched in Zn, Au, Ag, Sb, Cd, and Pb, are accumulating atop the mound as venting black smoker fluid, which has been modified by mixing with entrained seawater, precipitates of sulphides and anhydrite. The amount of anhydrite precipitating from the venting fluids of TAG mound is 5-15 x 10^7 kg/yr (James and Elderfield, 1996). This is an order of magnitude greater than that of the Cu and Zn sulphides (10^6-10^7 kg/yr), clearly illustrating that much of the anhydrite forming on the mound surface is rapidly redissolved into the ambient seawater.

Away from the active smoker vents, the bathing of the mound surface in normal seawater dissolves the anhydrite and oxidizes the sulphides to leave behind a carapace of ochreous and iron oxide-rich material. Such outer oxidised layers, which formed during previous hydrothermal episodes, are now exposed on the steep outer walls of the mound. These flank deposits include a novel form of moss agate dominated by ferric iron oxyhydroxides (Hopkinson et al., 1998). The agate has an isopachous concentric laminar texture that formed during a series of rapid irreversible fluid changes as cooling fluid passed across a redox and pH front that separated oxidizing highly viscous siliceous gels from a mixed pyrite-iron oxide sediment. The laminar texture is a totally inorganic construct, and yet, based on its laminar accretionary texture, the agate could easily be confused with a biogenic structure. If such features are also present in ancient vent mounds they may easily be confused with stromatolitic features (e.g., see discussion of North Pole, Pilbara, in Chapter 2).

Beneath the mound surface lies a complex assemblage of sulphide-anhydrite-silica breccias (Figure 9.25; Humphris et al., 1995). Indeed, drilling has shown that sulphide seafloor mounds grow largely as an *in situ* accretionary breccia pile, driven by ongoing episodes of hydrothermal

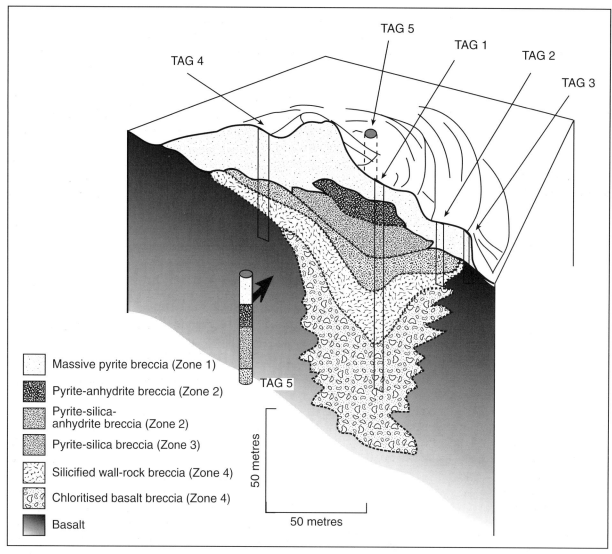

Figure 9.25. Internal mineral zonation of Tag mound. Note the importance of hydrothermal anhydrite in the internal structure of the mound. Note that well Tag 5 is out of the plane of section (after Humphris et al., 1995).

Legend:
- Massive pyrite breccia (Zone 1)
- Pyrite-anhydrite breccia (Zone 2)
- Pyrite-silica-anhydrite breccia (Zone 2)
- Pyrite-silica breccia (Zone 3)
- Silicified wall-rock breccia (Zone 4)
- Chloritised basalt breccia (Zone 4)
- Basalt

activity. Mounds do not grow simply by the accumulation of sulphide fallout onto the sea floor. During times of active hydrothermal circulation the internal mixing of black smoker fluids and heated seawater precipitates widespread sulphides and anhydrites. Inactive periods are marked by dissolution of anhydrite and collapse and brecciation of the mound interior. Subsequent episodes of hydrothermal activity form new hydrothermal precipitates, as well as cementing earlier breccias.

Breccia in the mound can be broken down into 4 zones (Figure 9.25; Humphris et al., 1995). Zone 1 is dominated by clast-supported massive pyrite breccias that dominate in the upper 10-20m of the mound. These are underlain by

the anhydritic zone 2 in the central part of the mound. Pyrite-anhydrite breccias are present down to 30 metres below sea floor (m.b.s.f.) and are underlain by pyrite-silica-anhydrite breccias that extend to the bottom of the anhydrite-rich zone 2 at about 45 m.b.s.f. Anhydrite veining is extremely well developed throughout zone 2, where composite anhydrite veins up to 45 cm thick are present. They comprise multistage fracture fillings and linings some of which include disseminated, fine-grained pyrite and chalcopyrite, as well as trace amounts of haematite. Zone 3 consists of intensely silicified and brecciated wall rock, which comprises the upper part of the upflow zone beneath the mound. Pyrite-silica breccias occur in the upper part of this zone and are underlain by silicified wallrock breccias. Below about 100 m.b.s.f. the silicified wall rock breccia grades into a chloritised basalt breccia (zone 4), where chloritised and weakly mineralised basalt fragments (1-5 cm diam) are cemented and veined by quartz ± pyrite.

Dating of the sulphides and anhydrites within the mound breccias clearly shows that sulphide accumulation and mound growth is not a continuous process (You and Bickle, 1998). The oldest material (11,000 to 37,000 years old) forms a layer across the centre of the mound stratigraphy, with younger material (2,300 to 7,800 years old) both above and below. This age distribution implies that TAG mound accreted via episodic activity of the hydrothermal system, with growth recurring at intervals of up to 2,000 years.

Extensive precipitation of anhydrite within the mound indicates extensive entrainment and heating of seawater. Teagle et al. (1998) found that the Sr^{87}/Sr^{86} ratios indicate that most of the anhydrite formed from ≈ 2:1 mixture of seawater and black smoker fluids (65% ± 15% seawater). The oxygen-isotopic compositions imply that anhydrite precipitated at temperatures between 147 and 270°C. This means the circulating seawater must have been conductively heated to between 100 and 180°C, before mixing and precipitation occurred. Precipitation of anhydrite within the TAG mound is also the major mechanism for removal of REE during mound circulation. Something like 0.15-0.35 gm of anhydrite is inferred to precipitate from every kg of fluid venting from the white smoker chimneys (Mills and Elderfield, 1995).

Once the hydrothermal circulations slows or stops, and the "in-mound" temperature falls below 150°C, the anhydrite in that region is likely to dissolve. During such inactive periods, the collapse of sulphide chimneys, the internal dissolution of mound anhydrite, and ongoing disruption by faulting will all cause an internal brecciation of the deposit. Through dissolution, former zones of hydrothermal anhydrite create intervals of enhanced porosity and cavities in the mound. Such intervals initiate fracture and collapse in the adjacent lithologies, which become permeable pathways during later renewed fluid circulation episodes. The alternating "coming and going" role of hydrothermal anhydrite within the mound hydrology is similar to that of sedimentary evaporites in the sedimentary mineralising systems. TAG mound, along with other active seafloor mounds (smokers), are considered to be modern analogues for Cyprus-style deposits.

Prior to the drilling of the TAG mound, the importance of widespread hydrothermal anhydrite in modern and ancient seafloor mounds, both as a sulphur source and as an integral mechanism for internal brecciation, was not well understood. Take away the anhydrite, and the internal structure of the TAG mound bears a striking resemblance to ancient volcanic-hosted massive sulphide exposed in ophiolite terranes (Cyprus-type deposits). Or, if we mentally replace the sea floor basalt with dacite, the zonation and distribution of the anhydrite and mineralisation is near identical to that in a Kuroko-style deposit. This underlies the significance of seawater-fed hydrothermal circulation cells, rather than the mineralogy of the host matrix, in controlling both sulphide mineralisation and the precipitation of anhydrite.

VHMS deposits at sediment-covered spreading centres: Guaymas Basin and Besshi-style deposits

The type Besshi deposit is in Japan, is metamorphosed, and has a Precambrian age. This and other ancient Besshi-style deposits typically form stratiform lenses and sheetlike accumulations of semi-massive to massive sulphides, which lack any preserved calcium sulphate. Given the typically Precambrian age of this deposit style and its seafloor genesis, this lack of preserved anhydrite is not surprising, even if precipitation of hydrothermal anhydrite was a fundamental part of its formation. Ore minerals are dominantly pyrite and/or pyrrhotite with variable amounts of chalcopyrite and sphalerite and minor to trace galena, arsenopyrite, and gold or electrum; baryte is generally absent (Slack, 1993). Many deposits have significant cobalt and high cobalt/nickel ratios. Typically associated with the deposits are a variety of wall rocks, including metachert, magnetite iron formation, sericite- and chlorite-rich schist, tourmalinite and albitite. According to Slack (1993), these wallrock lithologic units formed by premetamorphic hydrothermal alteration and/or by chemical sedimentation coeval with massive sulphide deposition.

Besshi-type deposits typically form when a hydrothermal ridge or axis is proximal to a source of terrigenous sediment, so that the axial ridge is always covered by beds of finer-grained deep sea sediment (Fox, 1984). In this situation the ascending magmas do not make it to the seafloor, but are intruded as sills into this sedimentary material. These sills then act as a heat source that sets up hydrothermal circulation cells within the sediments. The presence of ubiquitous host sediment distinguishes this type of deposit from the more open-marine sediment-free Cyprus-style of marine rift deposit exemplified by the modern TAG mound and from the sediment-starved style of rift deposit exemplified by the metalliferous deeps of the Red Sea.

A modern day counterpart to Besshi-style ores comes from the Guaymas Basin in the Gulf of California (Figure 9.26a). There muddy biogenic and turbiditic sediments up to 500m thick cover a series of spreading ridges and transform faults in the Guaymas Basin. The shallowest intrusions of magma occur some 400 m below the sediment surface in water depths of more than 1800m. Along a section extending over a little more than 120 km, over 120 hydrothermal vent sites with mineral deposits were detected by side scan sonar, while the deep tow vehicle passed through 20 thermal plumes (Lonsdale and Becker, 1985; Peter and Scott, 1991). Temperatures as high as 315°C have been measured at the vent exits. A dredge sample of one of these sites -the Seacliff hydrothermal site- contained sulphates (anhydrite and baryte), talc, calcite, pyrrhotite, sphalerite, wurtzite, marcasite, chalcopyrite and galena, all soaked in hydrothermal petroleum condensates.

The muddy sediments and hydrothermal fluids of the Guaymas Basin are enriched in organics derived from diatomaceous oozes. Buried diatom-rich sediments contain up to 3-4% oil-prone TOC, which is cracked by the circulating hydrothermal fluids to generate thermally derived hydrocarbons. Petroleum is carried as droplets entrained in the venting hydrothermal fluid or as a bulk phase (Simoneit et al., 1992). It also occurs in pore spaces and cavities in the chimneys and mounds and in hydrocarbon-bearing inclusions within amorphous silica and anhydrite.

Cooling hydrothermal outflow waters construct chimneys, mounds and spires with local clogging by worm tubes. The mineral chimneys and other constructs can rise up to 5 metres on mounds that in turn rise 5 to 25 metres above the surrounding seafloor. Chimneys are dominated by mixtures of anhydrite and pyrrhotite with lesser amounts of the other metal sulphides. The mounds can grow upwards, sideways and downward depending on how the mineral precipitates clog the outflow vents. The resulting vent shapes range from simple conical chimneys to multi-tiered pagoda-like spires with multiple horizontal eaves. Ultimately, the chimneys become unstable and collapse, causing the broken vent base to reopen and chimney growth to begin anew.

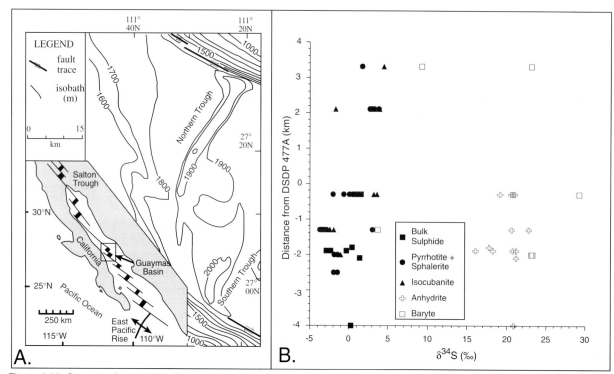

Figure 9.26. Guaymas Basin. A) Regional setting of drill hole 477A. B) Distribution of sulphur isotopes for samples from all Guaymas dive sites (replotted from data in Peter and Shanks, 1992).

Weblike mats of the filamentous sulphide-oxidising bacterium *Beggiatoa sp.* occur among the chimneys and worm tubes and on the surface of zones of any anoxic bottom sediment (Jannasch et al., 1989; Gundersen et al., 1992). This monospecific bacterial buildup constructs mats up to 30 cm thick in areas between the wormtubes, while thicknesses of up to 3 cm are common as bright yellow and orange mats on the seafloor adjacent to hydrothermal vents. Individual bacteria are gigantic, made up of cell filaments up to 120 μm in diameter. They are so large that individual filaments can be seen with the unaided human eye and resemble glass fibres in the lights of a submersible.

Because hydrothermal anhydrite dissolves under open seafloor conditions, it is usually absent in the material that covers inactive vents in the Guaymas Basin. It does, however, convert to gypsum which can be found in sediment cores from areas where the vents are no longer active (Sturz et al., 1996) $\delta^{34}S$ values for the sulphate minerals in the mounds range from 16.1‰ to 23.1‰ for the anhydrite, and from 23.1‰ to 29.3‰ for the baryte (Figure 9.26b; Peter and Shanks, 1992). Sulphur-isotopic composition indicates an unfractionated seawater sulphate source, although some samples from within the chimneys have more negative values, which suggest local sulphate reduction and sulphide oxidation related to bacterial processes. Differences in $\delta^{34}S$ between coexisting sulphide minerals from the mounds are generally small, suggesting sequential deposition of copper, zinc and iron sulphide minerals from a hydrothermal fluid of nearly constant $\delta^{34}S_{H2S}$ composition. During mineral replacement processes, which are common in the chimneys, secondary sulphides undoubtedly inherit some of their sulphur from precursor sulphide minerals (Peter and Scott, 1991).

Overall, the hydrothermal processes involving anhydrite in the vent region are similar to those acting in the TAG mound; but without drilling of some of these features in the Guaymas Basin, the volume and significance of high-temperature anhydrite preserved within the mineralised mound features cannot be quantified. It would not be surprising if subsurface hydrothermal anhydrite made up a large proportion of the hydrothermal precipitates in zones of actively circulating hydrothermal seawater, as it does in TAG mound and the Kuroko-style deposits. One possible difference between vent deposits of the Guaymas Basin compared with Cyprus- and Besshi-style deposits may well be the volume of entrained organics in sediment-covered hydrothermal cells.

Extraterrestial evaporites, "freeze-dried" salts, and the meaning of life

In order to complete this chapter, with its more speculative discussion of evaporite salts, I would be remiss if I did not include some mention of possible extraterrestrial salts and evaporites, as well as the influence of evaporites on extinction events here on Earth.

Analogues to the Earth's salt pans and saline lakes may have existed in crater-basins during the early Noachian epoch on the planet Mars (Forsythe and Zimbelman, 1995). Evidence includes images of terraced and channelised crater-basins, which imply ancient ponding of surface water, as well as possible prolonged and evolving erosional base levels. Williams and Zimbelman (1994) argue that White Rock, an enigmatic highly reflective crater interior deposit, measuring 12.5 x 15 km across, with an eroded thickness of 180 to 540 m, and a volume of 40 km³, may be the remnant of a saline lacustrine deposit. They infer the exposed sequence may include deposits similar to those that outline layered rocks and concentric desiccation rings formed during drying of continental salt lakes on earth. Edgett and Parker (1997), in a discussion of surface imagery of cratered terrain in the Sinus Meridiani region of Mars, recognised a combination of aeolian dunes and polygonally patterned ground. They argued, using a comparison with similar features that form today in saline seepage areas of the present-day Saudi Arabian desert, that Mars once had flowing surface waters and that preserved evaporite sediments are a distinct possibility.

Radar imagery of the surface of Venus has defined the geomorphic outline of venusian canali or outflow channels, as well as other associated volcanic deposits (Kargel et al., 1994). Most of these features resemble fluvial landforms much more than they resemble volcanic features on Earth and Mars. Yet surface temperatures on Venus are above 500°C, and are well above that of the boiling point of water. Some of the canali show meandering habits, crevasse splays and other geomorphic features that are very similar to fluvial channels and flood plains here on Earth. Given the high temperatures on the venusian surface these feature must be related to lava landforms, yet they more closely resemble the water-carved outflow channels of Mars and the channelled scablands here on Earth. Likewise, the collapsed terrain at the sources of some venusian channels resembles chaotic terrain at the meltwater sources of

371

martian outflow channels. There are also venusian lava deltas with shapes identical to bird's-foot deltas such as the Mississippi delta on Earth.

According to Kargel et al. (1994) these features suggest the involvement of an unusual lava and/or extraordinary eruption conditions, which they relate to a carbonatite lava rather than a silicate melt. Venusian nearsurface rocks may contain 4-19% calcite and anhydrite, this is in part a reflection of the likelihood that Ca-bearing hydrous silicates are unstable on the surface of Venus due to widespread sulphatization to anhydrite (Zolotov et al., 1997). Unlike silicate lavas, some carbonatites (including carbonate-sulphate-rich liquids) have a water-like rheology and would have a melting point that was only slightly greater than the venusian surface temperature. Such lavas could flow great distances while retaining high fluidity, significant mechanical erosiveness, and a substantial capacity to transport and deposit entrained sediment.

Such mixtures of crustal salts would melt at temperatures a few tens to a few hundred Kelvins above the venusian surface temperature; hence, melting may be induced by modest endogenetic or impact heating (Kargel et al., 1994). Salts on Venus may play many of the same geologic roles as water and ice do on Mars and Earth. A molten salt (carbonatite) "aquifer" may exist beneath a few hundred metres to several kilometres of solidified salt-rich "permafrost". In this way many geologic features on the venusian surface can be explained by carbonatite magmatism and impact melting. They include: 1) impact melting of crustal salts to form crater outflows, 2) small, sustained eruptions from molten salt aquifers to create sapping valleys, 3) large, sustained eruptions that form canali and their flood plans, and 4) catastrophic outbursts forming outflow channels and chaotic terrain.

Further out in the solar system on the surface of Europa, a moon of Jupiter, and a celestial body much more distant from the Sun than Mars, there is an even more spectacular and unusual development of salts. Using infrared spectral analysis on images of the cracked and icy surface of Europa, McCord et al. (1998) identified spectra that are best interpreted as hydrated salts dominated by trona and epsomite, rather than clays or other alternate minerals. The salts characterise distorted spectral bands that are only observed in the optically darker areas of Europa. This includes regions along many of the lineaments that crisscross the icy surface.

According to the authors the spectra may represent "evaporite" deposits formed by water, rich in dissolved salts, that reached the surface from a water-rich layer underlying the ice crust. Whether the sub-ice ocean still exists is a matter of speculation. Such a salt precipitating mechanism visualises freeze-drying the salts, that is separating the salts from the water as it escapes via breaches and lineaments to freeze onto the extremely cold Europan surface (-160°C). Such a mechanism means these salts are "freeze-dried" and not true evaporites as defined in Chapter 1. But, as noted by McCord et al., if the spectra are truly those of hydrated salts escaping from a saline liquid ocean beneath the icy cap, then it offers exciting evidence for the possibility of extraterrestrial life on Europa. It suggests that there may be, or once was, active hydrothermal activity on Europa and that water rock interaction has created an ionic solution beneath the icy surface similar to that of the early Archaean bicarbonate ocean (Chapter 2). It may well indicate hydrothermal vent conditions on a sub-ice ocean floor similar to those on earth when life first evolved in the early Archaean.

Evaporites do not just indicate the possibility of life on Europa, they may also have played a major role in the end-Cretaceous extinction here on earth (Brett, 1992). At that time a large asteroid collided with the earth to create the Chicxulub impact crater in Yucatan, Mexico. This region is underlain by widespread Jurassic anhydrites and their vaporization during the impact added a good deal to the ensuing climatic mayhem. Pope et al. (1994) calculated that $0.4 - 7.0 \times 10^{17}$ grams of sulphur were vaporized by the impact into the anhydrites. A large portion of sulphur was released as SO_2, which, due to the rate-limiting oxidation of SO_2, very slowly converted to aerosol. Their radiative transfer calculations of the effects of this aerosol, when combined with rates of acid production, coagulation and diffusion, show that solar transmission was reduced to 10-20% of normal for a period of 8-13 years. This reduction in solar transmission produced a climate forcing (cooling) of -300 Wm^2, so producing a decade of freezing and near-freezing temperatures.

The direct modelling of the results of shock vaporisation of anhydrite, when applied to the Chicxulub geology and using bolide diameters of 10 or 20 km across, yield volumes of 0.5×10^{17} grams and 2×10^{17} grams of degassed sulphur from the impact (Yang and Ahrens, 1998). These results add further veracity to the models of Pope et al. (1994). Passage of these volumes of sulphur into the atmosphere (as SO_2 or SO_3) are sufficient to cause global cooling of 10°C or more. A prolonged impact winter, with a major contribution from the degassing of evaporites, is a major contributing cause to the K/T extinctions. And who said evaporite studies were esoteric?

Summary

This chapter is all about salts that "went elsewhere" as they contributed to various mineralising systems. The salts disappeared in a number of ways and under a variety of temperature-pressure regimes. I have attempted to illustrate the diversity and complexity of the evaporite-metal association using some worldclass deposits and how, even in higher temperature terranes, evaporite bodies or their daughter products can play important roles in the mineralisation regimes of both base and precious metals.

A dissolving bedded or allochthonous salt body in a sedimentary diagenetic system can feed refluxing chloride-rich waters through shales or basalts and so form metalliferous basinal brines. At the same time, the dissolving evaporite mass can focus and channel the escape of any ascending metalliferous plumes into particular outflow zones and conduits. In the case of the allochthons on the seafloor this can form brine pools floored by laminites that are capable of accumulating metal sulphides. Much of early formed sulphide probably precipitates at the halocline by bacterially induced sulphate reduction. These newly formed crystals also carry pellicular organic material with them as they sink to the bottom.

In active incipient oceanic rifts, the stability of the deep seafloor brine pools, atop seafloor basalts located at the edge of allochthon sheets, can allow substantial volumes of metalliferous laminites (which can include Cu-sulphides) to accumulate over large brine-covered areas located well away from the metalliferous vents. Where allochthons have actively migrated through passive margin sediments so that they no longer cap seafloor basalts, the metal sulphides accumulating on the floor of the brine pools are dominated by Fe, Zn and Pb sulphides. Copper, with its much lower solubility in the temperature range of diagenetic brines, does not make it to the seafloor. Even as the metalliferous laminites are buried into warmer shallow diagenetic settings, the replacement process (pyrite framboids replaced by sphalerite and galena) can continue, fed by the same pore fluids that are still issuing onto the seafloor from nearby vents. Deeper in the subsurface, where there is little on no sulphate/sulphide in the pore waters, the feeder conduits of these ascending Fe-chloride rich basinal brines may be outlined by replacement haloes dominated by various ferroan carbonates. Such evaporite associated vent systems were probably active in the formation of worldclass base metal deposits in the Proterozoic basins of northern Australia.

When chloride-rich evaporite beds and allochthons dissolve at elevated temperatures and pressures, the resultant brines are capable of carrying gold and copper. Evaporite dissolution in various metasedimentary basin settings in the Proterozoic contributed to the formation of significant Cu-Au deposits, typically in association with the formation of ironstone hosts. In other deposits, metamorphosed lacustrine schists host the mineralisation. Stratiform albitites are common features of such systems, and indicate widespread sodic metasomatism driven by the disappearance of the evaporites or their sodic daughter products, such as marialitic scapolite.

Dissolving or altering evaporites in magmatic terrains can feed hot chloride-rich brines into hydrothermal circulation cells driven by the emplacement of magmatic bodies. This style of circulation precipitates a suite of characteristic sodic-rich metal accumulations hosted in Fe-oxide rich lithologies. Pyrite is a relatively uncommon precipitate in such systems, reflecting the paucity of sulphur in a brine formed by the alteration and dissolution of halite-dominated evaporites. In the much less common situation where substantial volumes of sedimentary anhydrite are assimilated into a magma chamber, a substantial Ni deposit may form.

Anhydrite is a common hydrothermal precipitate where seawater is heated beyond 150°C. Then it forms beds and irregular masses composed of nodules of various sizes from millimetres to metres in diameter as well as vein fills and carapaces atop sulphide mounds. Criteria to distinguish this nonevaporite form of calcium sulphate from sedimentary sulphate in ancient metasedimentary and igneous deposits have yet to be established.

If nothing else, the complexity of the evaporite-metal association outlined in this chapter means that to classify and target an ore deposit within an evaporite-influenced sequence clearly requires a model that is more than a simple listing of the properties and textures of the deposited sulphides.

References

Abraham, K., H. Mielke, and P. Povondra, 1972, On the enrichment of tourmaline in metamorphic sediments of the Arzberg Series, W. Germany (N.E. Bavaria): Neues Jahrbuch für Mineralogie Monatschefte, p. 209-219.

Achauer, C. W., 1982, Sabkha anhydrite; the supratidal facies of cyclic deposition in the Upper Minnelusa Formation (Permian), Rozet Fields area, Powder River basin, Wyoming, in C. R. Handford, Loucks, R. G., Davies, G. R., ed., Depositional and diagenetic spectra of evaporites; a core workshop., Tulsa, OK, SEPM Core Workshop No. 3, p. 193-209.

Ackermann, R. V., P. W. Schlische, and P. E. Olsen, 1995, Synsedimentary collapse of portions of the Lower Blomidon Formation (Late Triassic), Fundy Rift Basin, Nova Scotia: Canadian Journal of Earth Sciences, v. 32, p. 1965 - 1976.

Adams, J. E., and H. N. Frenzel, 1950, Capitan barrier reef, Texas and New Mexico: Journal of Geology, v. 58, p. 289-312.

Adams, J. F., and M. L. Rhodes, 1960, Dolomitisation by seepage refluxion: American Association of Petroleum Geologists Bulletin, v. 44, p. 1912-1920.

Addy, K. S., and E. W. Behrens, 1980, Time of accumulation of hypersaline anoxic brine in Orca basin (Gulf of Mexico): Marine Geology, v. 37, p. 241-252.

Aharon, P., 1988, A stable isotope study of magnesites from the Rum Jungle Uranium field. Australia: Implications for the origin of stratabound massive magnesites: Chemical Geology, v. 69, p. 127-145.

Aharon, P., H. H. Roberts, and R. Snelling, 1992, Submarine venting of brines in the deep Gulf of Mexico: observations and geochemistry: Geology, v. 20, p. 483-486.

Ahr, W. M., 1973, The carbonate ramp: an alternative to the shelf model: Transactions of Gulf Coast Association of Geological Societies, v. 23, p. 221-225.

Aitken, B. G., 1983, $T-X_{CO_2}$ stability relations and phase equilibria of a calcic carbonate scapolite: Geochimica et Cosmochimica Acta, v. 47, p. 351-362.

Akande, S. O., and M. Zentilli, 1984, Geologic, fluid inclusion, and stable isotope studies of the Gays River lead-zinc deposit, Nova Scotia, Canada: Economic Geology, v. 79, p. 1187-1211.

Alexander, L. L., and P. B. Flemings, 1995, Geologic evolution of a Pliocene-Pleistocene salt-withdrawal minibasin - Eugene Island Block 330, offshore Louisiana: American Association Petroleum Geologists Bulletin, v. 79, p. 1737-1756.

Allison, G. B., and C. J. Barnes, 1985, Estimation of evaporation from the normally 'dry' Lake Frome in South Australia: Journal of Hydrology, v. 78, p. 229-242

Alonso, H., and F. Risacher, 1996, Geochemistry of the Salar de Atacama. 1. Origin of components and salt balance [Spanish]: Revista Geologica de Chile, v. 23, p. 113-122.

Alonso, R. N., 1991, Evaporitas neógenas de los Andes Centrales: In, Pueyo, J. J. (edt.), Génesis de Formaciones Evaporíticas. Modelos Andinos e Ibéricos, Universitat de Barcelona. Estudi General, v. 2, p. 267-329.

Alonso, R. N., T. E. Jordan, K. T. Tabbutt, and D. S. Vandervoort, 1991, Giant evaporite belts of the Neogene central Andes: Geology, v. 19, p. 401-404.

Alsharhan, A. S., and C. G. S. C. Kendall, 1994, Depositional setting of the Upper Jurassic Hith Anhydrite of the Arabian Gulf; an analog to Holocene evaporites of the United Arab Emirates and Lake MacLeod of Western Australia: American Association of Petroleum Geologists Bulletin, v. 78, p. 1075-1096.

Alsop, G. I., D. J. Blundell, and I. Davison, 1996, Salt Tectonics: Special Publication, v. 100: London, Geological Society, 310 p.

Amery, G. B., 1969, Structure of Sigsbee Scarp, Gulf of Mexico: American Association of Petroleum Geologists, Bulletin, v. 53, p. 2480-2482.

Amery, G. B., 1978, Structure of continental slope, northern Gulf of Mexico, in A. H. Bouma, C. T. Moore, and J. M. Coleman, eds., Framework, facies, and oil-trapping characteristics of the upper continental margin, American Association Petroleum Geologists, Studies in Geology, v. 7, p. 141-153.

Amicux, P., 1980, Exemple d'un passage des 'black shales' aux evaporites dans le Ludien (Oligocene inferieur) du bassin de Mormoiron (Vaucluse, Sud- Est de la France). (Example of a transition of black shales to evaporites in the Ludian (Lower Oligocene) of the Mormoiron Basin, Vaucluse, south- eastern France): Bulletin, Centres de Recherches Exploration-Production Elf- Aquitaine, v. 4, p. 281-307.

Anderle, J. P., K. S. Crosby, and D. C. E. Waugh, 1979, Potash at Salt Springs, New Brunswick: Economic Geology, v. 74, p. 389-396.

REFERENCES

Anderson, G. M., and R. W. MacQueen, 1990, Mississippi valley-type lead-zinc deposits, in R. G. Roberts, and P. A. Sheahan, eds., Ore Deposit Models, St. John's, Geoscience Canada Reprint Series, p. 79-90.

Anderson, M. R., A. H. Rankin, and B. Spiro, 1992, Fluid mixing in the generation of mesothermal gold mineralisation in the Transvaal Sequence, Transvaal, South Africa: European Journal of Mineralogy, v. 4, p. 933-948. Anderson, N. L., R. J. Brown, and R. C. Hinds, 1988, Geophysical aspects of Wabamun salt distribution in southern Alberta: Canadian Journal of Exploration Geophysics, v. 24, p. 166-178.

Anderson, N. L., J. Hopkins, A. Martinez, R. W. Knapp, P. A. Macfarlane, W. L. Watney, and R. Black, 1994, Dissolution of bedded rock salt; a seismic profile across the active eastern margin of the Hutchinson Salt Member, central Kansas: Computers and Geosciences, v. 20, p. 889-903.

Anderson, N. L., and R. Knapp, 1993, An overview of some of the larger scale mechanisms of salt dissolution in Western Canada: Geophysics, v. 58, p. 1375-1387.

Anderson, N. L., R. J. Brown, and R. C. Hinds, 1988, Geophysical aspects of Wabamun salt distribution in southern Alberta: Canadian Journal of Exploration Geophysics, v. 24, p. 166-178.

Anderson, R. Y., W. E. J. Dean, D. W. Kirkland, and H. I. Snider, 1972, Permian Castile Varved Evaporite Sequence, West Texas and New Mexico: Geological Society of America Bulletin, v. 83, p. 59-85.

Andreason, M. W., 1992, Coastal siliciclastic sabkhas and related evaporative environments of the Permian Yates Formation, North Ward-Estes field, Ward County, Texas: American Association of Petroleum Geologists Bulletin, v. 76, p. 1735-1759.

Anschutz, P., and G. Blanc, 1995, Chemical mass balances in metalliferous deposits from the Atlantis II Deep, Red Sea: Geochimica et Cosmochimica Acta, v. 59, p. 4205-4218.

Anschutz, P., and G. Blanc, 1996, Heat and salt fluxes in the Atlantis II Deep (Red Sea): Earth and Planetary Science Letters, v. 142, p. 147-159.

Arakel, A. V., and T. Hong Jun, 1994, Seasonal evaporite sedimentation in desert playa lakes of the Karinga Creek drainage system, central Australia, in R. W. Renaut, and W. M. Last, eds., Sedimentology and geochemistry of modern and ancient saline lakes, Tulsa, OK, Society for Sedimentary Geology; Special Publication, p. 91-100.

Arbey, N., 1980, Silicification des évaporites: Bulletin, Centres de Recherches Exploration-Production Elf-Aquitaine, v. 4, p. 309-365.

Aref, M. A. M., 1998a, Holocene stromatolites and microbial laminites associated with lenticular gypsum in a marine-dominated environment, Ras el Shetan area, Gulf of Aqaba, Egypt: Sedimentology, v. 45, p. 245-262.

Aref, M. A. M., 1998b, Biogenic carbonates - are they a criterion for underlying hydrocarbon accumulations - an example from the Gulf of Suez region: American Association of Petroleum Geologists Bulletin, v. 82, p. 336-352.

Armstrong, A. K., 1995, Facies, diagenesis, and mineralogy of the Jurassic Todilto Limestone Member, Grants uranium district, New Mexico, v. 153, New Mexico Bureau of Mines Bulletin, 41 p.

Armstrong, J. P., F. J. Longstaffe, and F. J. Hein, 1993, Carbon and oxygen isotope geochemistry of calcite from the Jubilee Zn-Pb deposit, Breton Island: Canadian Mineralogist, v. 313, p. 755-766.

Arne, D., 1996, Thermal setting of the Cadjebut Zn-Pb deposit, Western Austraila: Journal of Geochemical Exploration, v. 57, p. 45-56.

Arne, D. C., L. W. Curtis, and S. A. Kissin, 1991, Internal zonation in a carbonate-hosted Zn-Pb-Ag deposit. Nanisivik, Baffin Island, Canada: Economic Geology, v. 86, p. 699-717.

Arp, G., 1995, Lacustrine bioherms, spring mounds, and marginal carbonates of the Ries-impact-crater (Miocene, southern Germany): Facies, v. 33, p. 35-89.

Arthurton, R. S., 1973, Experimentally produced halite compared with Triassic layered halite-rock from Cheshire, England: Sedimentology, v. 20, p. 145-160.

Assereto, R. I. A. M., and C. G. S. C. Kendall, 1977, Nature, origin, and classification of peritidal teepee structures and related breccias,: Sedimentology, v. 24, p. 153-210.

Augustithis, S. S., 1980, On the textures and treatment of the sylvinite ore from the Danakil Depression, Salt Plain (Piano del Sale), Tigre, Ethiopia: Chem. Erde., v. 39, p. 91-95.

REFERENCES

Aulstead, K. L., and R. J. Spencer, 1985, Diagenesis of Keg River Formation, northwestern Alberta; fluid inclusion evidence: Bulletin of Canadian Petroleum Geology, v. 33, p. 167-183.

Ayora, C., J. Garciaveigas, and J. Pueyo, 1994, The chemical and hydrological evolution of an ancient potash-forming evaporite basin as constrained by mineral sequence, fluid inclusion composition, and numerical simulation: Geochimica et Cosmochimica Acta, v. 58, p. 3379-3394.

Ayres, M. G., M. Bilal, R. W. Jones, L. W. Slenz, M. Tartir, and A. O. Wilson, 1982, Hydrocarbon Habitat in Main Producing Areas, Saudi Arabia: American Association of Petroleum Geologists Bulletin, v. 66, p. 1-6.

Baadsgaard, H., 1987, Rb-Sr and K-Ca isotope systematics in minerals from potassium horizons in the Prairie Evaporite Formation, Saskatchewan, Canada: Chemical Geology, v. 66, p. 1-15.

Babel, M., 1990, Crystallography and genesis of the giant intergrowths of gypsum from the middle Miocene evaporites of southern Poland: Arch. Min., v. 44, p. 103-35.

Babel, M., 1991, Dissolution of halite within the middle Miocene (Badenian) laminated gypsum of southern Poland: Acta Geologica Polonica, v. 41, p. 165-182.

Backer, H., and K. Lange, 1987, Recent hydrothermal metal accumulation, products and conditions of formation, in P. G. Telecki, M. R. Dobson, J. R. Moore, and U. von Stackelberg, eds., Marine Minerals, Dordrecht, Reidel, p. 317-337.

Bailey, R. K., 1949, Talc in the salines of the potash field near Carlsbad, Eddy County, New Mexico: Am. Mineralogist., p. 9-10.

Baker, D. M., R. J. Lillie, R. S. Yeats, G. D. Johnson, M. Yousuf, and A. S. H. Zamin, 1988, Development of the Himalayan frontal thrust zone: Salt Range, Pakistan: Geology, v. 16, p. 3-7.

Baker, P. A., and S. H. Bloomer, 1988, The origin of celestite in deep water carbonate sediments: Geochim. Cosmochim. Acta, v. 52, p. 335-339.

Barley, M. E., J. S. R. Dunlop, J. E. Glover, and D. I. Groves, 1979, Sedimentary evidence for an Archaean shallow-water volcanic-sedimentary facies, eastern Pilbara Block, Western Australia: Earth and Planetary Science Letters, v. 43, p. 74-84.

Barrett, T. J., and G. M. Anderson, 1982, The solubility of sphalerite and galena in NaCl brines: Economic Geology, v. 77, p. 1923-1933.

Barton, D. C., 1933, Mechanics of formation of salt domes with special reference to Gulf Coast salt domes of Texas and Louisiana: American Association Petroleum Geologists Bulletin, v. 17, p. 1025-1083.

Barton, M. D., and D. A. Johnson, 1996, Evaporitic-source model for igneous-related Fe-oxide—(REE-Cu-Au-Au) mineralisation: Geology, v. 24, p. 259-262.

Barton, M. D., R. P. Ilchik, and M. A. Marikos, 1991, Metasomatism, in D. M. Kerrick, ed., Contact Metamorphism, Mineralogical Society of America, Reviews in Mineralogy, 26, p. 321-350.

Bates, R. L., 1969, Potash Minerals: In Geology of the industrial rocks and minerals: New York, Dover Publ., pages 370-385 and 439-440.

Beaumont, C., and A. J. Tankard, 1987, Sedimentary basins and basin forming mechanisms: Canadian Society Petroleum Geologists Memoir, v. 12, p. 527 pp.

Bebout, D. G., and W. R. Maiklem, 1973, Ancient anhydrite facies and environments; their role in reconstructing geologic history of middle Devonian Elk Point Basin, Alberta: Amer. Assoc. Petrol. Geol., Bull., v. 57, p. 769-770.

Bechtel, A., M. Pervaz, and W. Puttmann, 1998, Role of organic matter and sulphate-reducing bacteria for metal sulphide precipitation in the Bahloul Formation at the Bou Grine Zn/Pb deposit (Tunisia): Chemical Geology, v. 144, p. 1-21.

Bechtel, A., and W. Puttmann, 1997, Palaeoceanography of the early Zechstein Sea during Kupferschiefer deposition in the Lower Rhone Basin (Germany) - A reappraisal from stable isotope and organic geochemical investigations: Palaeogeography Palaeoclimatology Palaeoecology, v. 136.

Bechtel, A., Y. N. Shieh, M. Pervaz, and W. Puttmann, 1996, Biodegradation of hydrocarbons and biogeochemical sulfur cycling in the salt dome environment - inferences from sulfur isotope and organic geochemical investigations of the Bahloul Formation at the Bou Grine Zn/Pb ore deposit, Tunisia: Geochimica et Cosmochimica Acta, v. 60, p. 2833-2855.

REFERENCES

Bechtel, A., W. Puttmann, and S. Hoernes, 1995, Reconstruction of the thermal history of the Kupferschiefer within the Zechstein Basin of Central Europe: a stable isotope and organic geochemical approach: Ore Geology Reviews, v. 9, p. 371-389.

Bechtel, A., and S. Hoernes, 1993, Stable isotopic variations of clay minerals: a key to the understanding of Kupferschiefer-type mineralization, Germany: Geochimica et Cosmochimica Acta, v. 57, p. 1799-1816.

Behr, H. J., H. Ahrendt, H. Martin, H. Porada, J. Rohrs, and K. Weber, 1983, Sedimentology and mineralogy of Upper Proterozoic playa-lake deposits in the Damara orogen, in H. Martin, and F. W. Eder, eds., Intracontinental Fold Belts, Berlin, Springer-Verlag, p. 577-610.

Behr, H. J., and H. N. Horn, 1982, Fluid inclusion systems in meta-playa deposits and their relationship to mineralization and tectonics: Chem. Geol., v. 37, p. 173-198.

Bein, A., and A. R. Dutton, 1993, Origin, distribution, and movement of brine in the Permian Basin (USA) - A model for displacemtn of connate brine: Geological Society of America Bulletin, v. 105, p. 695-707.

Bein, A., and L. S. Land, 1983, Carbonate sedimentation and diagenesis associated with Mg-Ca-chloride brines; the Permian San Andres Formation in the Texas Panhandle: Journal of Sedimentary Petrology, v. 53, p. 243-260.

Bein, A., and L. S. Land, 1982, San Andres carbonates in the Texas Panhandle: sedimentation and diagenesis associated with magnesium- calcium-chloride brines: Bureau of Economic Geology, University of Texas, Report of Investigation, v. 121, p. 48 p.

Bell, C. M., 1997, Saline lake turbidites in the La Coipa area, Northern Chile: Revista Geologica de Chile, v. 24, p. 259-267.

Bell, K., and J. B. Dawson, 1995, An assessment of the alleged role of evaporites and saline brines in the origin of natrocarbonatite, in K. Bell, and J. Keller, eds., Carbonatite volcanism: Oldoinyo Lengai and the petrogenesis of natrocarbonatites, Berlin, Springer-Verlag. p. 137-147.

Belous, I. R., S. I. Kirikilitsa, M. L. Levenshteyn, E. K. Rodina, and V. N. Florinskaya, 1984, Occurrences of mercury in northeastern Donbass salt dome: Internat. Geol. Rev., v. 26, p. 573-582.

Ben-Avraham, Z., T. M. Niemi, D. Neev, J. K. Hall, and Y. Levy, 1993, Distribution of Holocene sediments and neotectonics in the deep north basin of the Dead Sea: Marine Geology, v. 113, p. 219-231.

Benson, R. H., K. Rakic-El Bied, and L. W. McKenna, 1997, Eustatic implications of late Miocene depositional sequences in the Melilla Basin, northeastern Morocco: Sedimentary Geology, v. 107, p. 147-165.

Berry, R. F., R. B. Flint, and A. E. Grady, 1978, Deformation history of the Outalpa area and its application to the Olary Province, South Australia: Transactions Royal Society of South Australia, Adelaide, South Aust., Australia, v. 102, p. 43-53.

Bethke, C. M., 1989, Modeling subsurface flow in sedimentary basins: Geologische Rundschau, v. 78, p. 129-154.

Bierlein, F. P., P. M. Ashley, and I. R. Plimer, 1995, Sulphide mineralisation in the Olary Block, South Australia - Evidence for syn-tectonic to late-stage mobilisation: Mineralium Deposita, v. 30, p. 424-438.

Binns, R. A., J. McAndrew, and S. Sun, 1980, Origin of uranium mineralisation in Jabiluka, in J. Ferguson, and A. Goleby, eds., Uranium in the Pine Creek Geosyncline, Vienna, International Atomic Energy Agency, p. 543-562.

Birnbaum, S. J., and J. W. Wireman, 1985, Sulfate-reducing bacteria and silica solubility; a possible mechanism for evaporite diagenesis and silica precipitation in banded iron formations: Canadian Journal of Earth Sciences, v. 22, p. 1904-1909.

Blanc, G., and P. Anschutz, 1995, New stratification in the hydrothermal brine system of the Atlantis II Deep, Red Sea: Geology, v. 23, p. 543-546.

Blount, C. W., and F. W. Dickson, 1969, The solubility of anhydrite ($CaSO4$) in $NaCl-H_2O$ from 100 to 450 degrees C and 1 to 1000 bars: Geochim. Cosmochim. Acta., v. 33, p. 227-245.

Blum, N., and H. Puchelt, 1991, Sedimentary-hosted polymetallic massive sulphide deposits of the Kebrit and Shaban Deeps, Red Sea: Mineralium Deposita, v. 26, p. 217-227.

Bogoch, R., B. Buchbinder, and M. Magaritz, 1994, Sedimentology and geochemistry of lowstand peritidal lithofacies at the Cenomanian-Turonian boundary in the

REFERENCES

Cretaceous carbonate platform of Israel: Journal of Sedimentary Research Section A-Sedimentary Petrology and Processes, v. 64, p. 733-740.

Bone, Y., 1983, Interpretation of magnesites at Rum Jungle, N. T. using fluid inclusions: Geol. Soc. Australia, Journ., v. 30, p. 375-381.

Borchert, H., 1977, On the formation of Lower Cretaceous potassium salts and tachyhydrite in the Sergipe Basin (Brazil) with some remarks on similar occurrences in West Africa (Gabon, Angola etc.), in D. D. Klemm, and H. J. Schneider, eds., Time and strata bound ore deposits., Berlin, Germany, Springer-Verlag, p. 94-111.

Borchert, H., and R. O. Muir, 1964, Salt deposits—The origin, metamorphism and deformation of evaporites: London, D. Van Nostrand Co., Ltd., 338 p.

Borer, J. M., and P. M. Harris, 1991, Lithofacies and cyclicity of the Yates Formation, Permina Basin: Implications for reservoir heterogeneity: American Association Petroleum Geologists Bulletin, v. 75, p. 726-779.

Bouladon, J., 1984, Syngenesis and epigenesis at the Largentiere (Ardeche, France) Pb- Zn-Ag deposit, in A. Wauschkuh, ed., Syngenesis and Epigenesis in the Formation of Mineral Deposits, Springer-Verlag, p. 422-430.

Boulter, C. A., and J. E. Glover, 1986, Chert with relict hopper moulds from Rocklea Dome, Pilbara Craton, Western Australia; an Archean halite-bearing evaporite: Geology (Boulder), v. 14, p. 128-131.

Bouzenoune, A., and P. Lecolle, 1997, Petrographic and geochemical arguments for hydrothermal formation of the Ouenza siderite deposit (NE Algeria): Mineralium Deposita, v. 32, p. 189-196.

Boys, C., 1993, A geological approach to potash mining problems in Saskatchewan, Canada: Exploration and Mining Geology, v. 2, p. 129-138.

Boys, C., 1990, The geology of potash deposits at PCS Cory Mine, Saskatchewan: Master's thesis, University of Saskatchewan; Saskatoon, SK; Canada.

Bozkurt, U., 1989, A fluid inclusion study of selected boreholes, Salton Sea geothermal system, Imperial Valley, California: Master's thesis, University of California, Riverside, CA; United States.

Braithwaite, C. J. R., and V. Zedef, 1996, Hydromagnesite stromatolites and sediments in an alkaline lake, Salda Golu, Turkey: Journal of Sedimentary Research A: Sedimentary Petrology and Processes, v. 66, p. 991-1002.

Braithwaite, C. J. R., and V. Zedef, 1994, Living hydromagnesite stromatolites from Turkey: Sedimentary Geology, v. 92, p. 1-5.

Breck, D. W., 1974, Zeolite Molecular Sieves: New York, Wiley-Interscience, 771 p.

Bregant, D., G. Catalano, G. Civitarese, and A. Luchetta, 1990, Some chemical characteristics of the brines in Bannock and Tyro Basins: salinity, sulphur compounds, Ca^{2+}, F^-, pH, A_t, PO_4^{3-}, SiO_2, NH_3: Marine Chemistry, v. 31, p. 35-62.

Breit, G. N., M. B. Goldhaber, D. R. Shawe, and E. C. Simmons, 1990, Authigenic barite as an indicator of fluid movement through sandstones within the Colorado Plateau: Journal of Sedimentary Petrology, v. 60, p. 884-896.

Brett, R., 1992, The Cretaceous-Tertiary extinction; a lethal mechanism involving anhydrite target rocks: Geochimica et Cosmochimica Acta, v. 56, p. 3603-3606.

Broadbent, G. C., R. E. Myers, and J. V. Wright, 1996, Geology and origin of shale-hosted Zn-Pb-Ag mineralisation at the Century Mine, Northwest Queensland: MIC'96: New Developments in Metallogenic Research - The McArthur-Mt Isa - Cloncurry Minerals Province, p. 24-27.

Brodylo, L. A., and R. J. Spencer, 1987, Depositional environment of the Middle Devonian Telegraph Salts, Alberta, Canada: Bulletin Canadian Petroleum Geology, v. 35, p. 186-196.

Brown, A. C., 1993, Sediment-hosted Stratifrom Copper Deposits, in P. Sheahan, and M. E. Cherry, eds., Ore Deposit Models, volume II, St Johns, Newfoundland, Geological Association of Canada Reprint Series, p. 99-115.

Brown, C. E., and R. A. Ayuso, 1985, Significance of tourmaline-rich rocks in the Grenville Complex of St. Lawrence County, New York (USA): US Geological Survey Bulletin, v. 1626-C, p. 33.

Bryant, R. G., B. W. Sellwood, A. C. Millington, and N. A. Drake, 1994, Marine-like potash evaporite formation on a continental playa; case study from Chott el Djerid, southern Tunisia: Sedimentary Geology, v. 90, p. 269-291.

REFERENCES

Budai, J. M., K. C. Lohmann, and R. M. Owen, 1984, Burial dedolomite in the Mississippian Madison Limestone, Wyoming and Utah thrust belt: Journal of Sedimentary Petrology, v. 54, p. 276-288.

Budd, D. A., A. H. Saller, and P. M. Harris, 1995, Unconformities and porosity in carbonate strata, American Association of Petroleum Geologists Memoir v. 63: Tulsa, Okla.

Buick, R., 1992, The antiquity of oxygenic photosynthesis; evidence from stromatolites in sulphate-deficient Archaean lakes: Science, v. 255, p. 74-77.

Buick, R., and J. S. R. Dunlop, 1990, Evaporitic sediments of early Archaean age from the Warrawoona Group, North Pole, Western Australia: Sedimentology, v. 37, p. 247-277.

Burban, B., 1990, Kunwarara magnesite deposit, in F. E. Hughes, ed., Geology of the mineral deposits of Australia and Papua New Guinea; Volume 2, Australasian Institute of Mining and Metallurgy Monograph Series, p. 1675-1677.

Burchette, T. P., and V. P. Wright, 1992, Carbonate ramp depositional systems: Sedimentary Geology, v. 79, p. 3-57.

Burliga, S., 1996, Kinematics within the Klodawa salt diapir, central Poland, in G. I. Alsop, D. J. Blundell, and I. Davison, eds., Salt tectonics, London, United Kingdom, Geological Society of London Special Publication No.100, p. 11-21.

Burrus, J., and F. Audebert, 1990, Thermal and compaction processes in a young rifted basin containing evaporites; Gulf of Lions, France: Bulletin, v. 74, p. 1420-1440.

Busson, G., 1980, Evaporite deposits; illustration and interpretation of some environmental sequences: Paris, France, Ed. Technip, 266 p.

Butchins, C. S., and R. Mason, 1973, Metamorphic anhydrite in a kyanite-bearing schist from Churchill Falls, Labrador: Mineralogical Magazine, v. 39, p. 488-490.

Butler, R. W. H., W. H. Lickorish, M. Grasso, H. M. Pedley, and L. Ramberti, 1995, Tectonics and sequence stratigraphy in Messinian basins, Sicily; constraints on the initiation and termination of the Mediterranean salinity crisis: Geological Society of America Bulletin, v. 107, p. 425-439.

Byerly, G. R., and M. R. Palmer, 1991, Tourmaline mineralization in the Barberton greenstone belt, South Africa; early Archaean metasomatism by evaporite-derived boron: Contributions to Mineralogy and Petrology, v. 107, p. 387-402.

Cailteux, J., 1994, Lithostratigraphy of the Neoproterozoic Shaba-type (Zaire) Roan Supergroup and metallogenesis of associated stratiform mineralization: Journal of African Earth Sciences, v. 19, p. 279-301.

Cailteux, J., P. L. Binda, W. M. Katekesha, A. B. Kampunzu, M. M. Intiomale, D. Kapenda, C. Kaunda, K. Ngongo, T. Tshiauka, and M. Wendorff, 1994, Lithostratigraphical correlation of the Neoproterozoic Roan Supergroup from Shaba (Zaire) and Zambia, in the central African copper-cobalt metallogenic province: Journal of African Earth Sciences, v. 19, p. 265-278.

Camerlenghi, A., 1990, Anoxic Basins of the eastern Mediterranean: geological framework: Marine Chemistry, v. 31, p. 1-19.

Camerlenghi, A., and F. W. McCoy, 1990, Physiography and structure of Bacino Bannock (eastern Mediterranean): Geo-Marine Letters, v. 10, p. 23-30.

Cañaveras, J. C., S. Sanchezmoral, J. P. Calvo, M. Hoyos, and O. S., 1996, Dedolomites associated with karstification - an example of early dedolomitisation in lacustrine sequences from the Tertiary Madrid basin, central Spain: Carbonates and Evaporites, v. 11, p. 85-103.

Carlson, C. A., F. M. Phillips, D. Elmore, and H. W. Bentley, 1990, Chlorine-36 tracing of salinity sources in the Dry Valleys of Victoria Land, Antarctica: Geochimica et Cosmochimica Acta, v. 54, p. 311-318.

Carlson, E. H., 1992, Reactivated interstratal karst - example from the late Silurian rocks of western Lake Erie (USA): Sedimentary Geology, v. 76, p. 273-283.

Carlson, E. H., 1987, Celestite replacements of evaporites in the Salina Group: Sedimentary Geology, v. 54, p. 93-112.

Carlson, E. H., 1983, The occurrence of Mississippi Valley-type Mineralization in northwestern Ohio, in G. Kisvarsanyi, S. K. Grant, W. P. Pratt, and J. W. Koenig, eds., International Conference on Mississippi Valley Type lead-zinc deposits; Proceedings, University of Missouri Rolla, Mo, p. 424-435.

REFERENCES

Carpenter, A. B., M. L. Trout, and E. E. Pickett, 1974, Preliminary report on the origin and chemical evolution of oil field brines in central Mississippi: Economic Geology, v. 69, p. 1191 - 1206.

Casanova, J., 1986, East African Rift stromatolites, in L. E. Frostick, R. W. Renaut, I. Reid, and J. J. Tiercelin, eds., Sedimentation in the African Rifts, Journ. Geol. Soc. London Spec. Publ.

Casanova, J., and D. Nury, 1989, Biosédimentologie des stromatolites fluvio-lacustres du fossé oligocene de Marseille: Bull. Soc. Geol. Fr., v. 8, p. 1111-1128.

Casas, E., 1992, Modern carnallite mineralisation and Late Pleistocene to Holocene brine evolution in the nonmarine Qaidam Basin, China: Doctoral thesis, State University of New York at Binghampton.

Casas, E., T. K. Lowenstein, R. J. Spencer, and P. Zhang, 1992, Carnallite mineralization in the nonmarine Qaidam Basin, China; evidence for the early diagenetic origin of potash evaporites: Journal of Sedimentary Petrology, v. 62, p. 881-898.

Casas, E., and T. K. Lowenstein, 1989, Diagenesis of saline pan halite; comparison of petrographic features of modern, Quaternary and Permian halites: Journal of Sedimentary Petrology, v. 59, p. 724-739.

Castanier, S., J. P. Perthuisot, J. M. Rouchy, A. Maurin, and O. Guelorget, 1992, Halite ooids in Lake Asal, Djibouti; biocrystalline build-ups: Geobios, v. 25, p. 811-821.

Cathro, D. L., J. K. Warren, and G. E. Williams, 1992, Halite saltern in the Canning Basin, Western Australia; a sedimentological analysis of drill core from the Ordovician-Silurian Mallowa Salt: Sedimentology, v. 39, p. 983-1002.

Cendon, D. I., C. Ayora, and J. P. Pueyo, 1998, The origin of barren bodies in the Subiza potash deposit, Navarra, Spain - Implications for sylvite formation: Journal of Sedimentary Research Section A-Sedimentary Petrology and Processes, v. 68, p. 43-52.

Chabu, M., 1995, The geochemistry of phlogopite and chlorite from the Kipushi Zn-Pb-Cu deposit, Zaire: Canadian Mineralogist, v. 33, p. 547-558.

Chabu, M., and J. Boulegue, 1992, Barian feldspar and muscovite from the Kipushi Zn-Pb-Cu deposit, Shaba, Zaire: Canadian Mineralogist, v. 30, p. 1143-1152.

Chafetz, H. S., and J. L. Zhang, 1998, Authigenic euhedral megaquartz crystals in a Quaternary dolomite: Journal of Sedimentary Research Section A-Sedimentary Petrology and Processes, v. 68, p. 994-1000.

Charef, A., and S. M. F. Sheppard, 1991, The diapir related Bou Grine Pb-Zn deposit (Tunisia); evidence for role of hot sedimentary basin brines, in M. Pagel, and J. L. Leroy, eds., Source, transport and deposition of metals., Balkema, p. 269-272.

Charef, A., and S. M. F. Sheppard, 1987, Pb-Zn mineralization associated with diapirism: fluid inclusion and stable isotope (H, C, O) evidence for the origin and evolution of the fluids at Fedj-el-Adoum, Tunisia: Chemical Geology, v. 61, p. 113-134.

Chartrand, F. M., A. C. Brown, and R. V. Kirkham, 1989, Diagenesis, sulphides, and metal zoning in the Redstone copper deposit, Northwest Territories: Geological Association of Canada Special Paper, v. 36, p. 189-206.

Chi, G. X., and M. M. Savard, 1995, Fluid evolution and mixing in the Gays River Carbonate-hosted Zn-Pb deposit and its surrounding barren areas, Nova Scotia: Atlantic Geology, v. 31, p. 141-152.

Chipley, D. B. L., 1995, Fluid history of the Saskatchewan sub-basin of the western Canada sedimentary basin: Evidence from the geochemistry of evaporites: Doctoral thesis, University of Saskatchewan.

Chipley, D. B. L., and T. K. Kyser, 1989, Fluid inclusion evidence for the deposition and diagenesis of the Patience Lake Member of the Devonian Prairie Evaporite Formation, Saskatchewan, Canada: Sedimentary Geology, v. 64, p. 287-295.

Chovan, M., V. Hurai, H. K. Sachan, and J. Kantor, 1995, Origin of the fluids associated wih granodiorite-hosted, Sb-As-Au-W mineralisation at Dubrava (Nizkc Tatry Mts., Western Carpathians): Mineralium Deposita, v. 30, p. 48-54.

Chowns, T. M., and J. E. Elkins, 1974, The origin of quartz geodes and cauliflower cherts through the silicification of anhydrite nodules: J. Sediment. Petrol., v. 44, p. 885-903.

Cita, M. B., 1983, The Messinian salinity crisis in the Mediterranean: A review, in H. Berckhemer, and K. J. Hsu, eds., Alpine Mediterranean Geodynamics Review, American Geophysical Union; Geodynamics Series, p. 113-140.

REFERENCES

Clark, D. N., 1980, The sedimentology of the Zechstein 2 Carbonate Formation of eastern Drenthe, the Netherlands, in H. Fuchtbauer, and T. Peryt, eds., The Zechstein basin with emphasis on carbonate sequences, E. Schweizerbart'sche VbH: Contributions to Sedimentology 9, p. 131-165.

Clark, J. A., S. A. Stewart, and J. A. Cartwright, 1998, Evolution of the NW margin of the North Permian Basin, UK North Sea: Journal of the Geological Society, v. 155, p. 663-676.

Clarke, G. L., M. Guiraud, R. Powell, and J. P. Burg, 1987, Metamorphism in the Olary Block, South Australia; compression with cooling in a Proterozoic fold belt: Journal of Metamorphic Geology, v. 5, p. 291-306.

Clauzon, G., J. P. Suc, F. Gautier, A. Berger, and M. F. Loutre, 1996, Alternate interpretation of the Messinian salinity crisis: controversy resolved?: Geology, v. 24, p. 363-366.

Clayton, R. H., and L. Thorpe, 1982, Geology of the Nanisivik zinc- lead deposit: Geological Association of Canada, Special Paper, v. 25, p. 739-758.

Clement, G. P., and W. T. Holser, 1988, Geochemistry of Moroccan evaporites in the setting of the North Atlantic Rift: Journal of African Earth Sciences, v. 7, p. 375-382.

Clemmey, H., and N. Badham, 1982, Oxygen in the Precambrian atmosphere; an evaluation of the geological evidence: Geology, v. 10, p. 141-146.

Cloud, P. E., 1972, A working model of the primitive earth: Am. J. Sci., v. 272, p. 537-548.

Cluzel, D., 1986, Contribution a l'etude du metamorphisme des gisements cupro-cobaltiferes stratiformes du Sud-Shaba, Zaire. Le district minier de Lwishia. (Contribution to the study of the metamorphosed stratiform copper/cobalt deposits of southern Shaba, Zaire. The mining district of Lwishia): Journal of African Earth Sciences, v. 5, p. 557-574.

Cobbold, P. R., P. Szatmari, L. S. Demercian, D. Coelho, and E. A. Rossello, 1995, Seismic and experimental evidence for thin-skinned horizontal shortening by convergent radial gliding on evaporites, deepwater Santos Basin, Brazil, in M. P. A. Jackson, D. G. Roberts, and S. Snelson, eds., Salt tectonics: a global perspective, American Association Petroleum Geologists Memoir, p. 305-322.

Cocherie, A., J. Y. Calvez, and E. Oudin-Dunlop, 1994, Hydrothermal activity as recorded by Red Sea sediments; Sr-Nd isotopes and REE signatures: Marine Geology, v. 118, p. 291-302.

Cody, R. D., 1991, Organo-crystalline interactions in evaporite systems; the effects of crystallization inhibition: Journal of Sedimentary Petrology, v. 61, p. 851-859.

Cody, R. D., 1979, Lenticular gypsum; occurrences in nature, and experimental determinations of effects of soluble green plant material on its formation: J. Sediment. Petrol., v. 49, p. 1015-1028.

Cohen, A. S., 1989, Facies relationships and sedimentation in large rift lakes and implications for hydrocarbon exploration: Examples from Lake Turkana and Tanganyika: Palaeogeogr. Palaeoclimat. Palaeoecol., v. 70, p. 65-80.

Cohen, A. S., and C. Thouin, 1987, Nearshore carbonate deposits in Lake Tanganyika: Geology, v. 15, p. 414-418.

Comodi, P. M., M.; Zanazzi, P. F., 1990, Scapolites: variation of structure with pressure and possible role in the storage of fluids: European Journal of Mineralogy, v. 2, p. 195-202.

Cook, D. J., A. F. Randazzo, and C. L. Sprinkle, 1985, Authigenic fluorite in dolomitic rocks of the Floridan Aquifer: Geology, v. 13, p. 390-391.

Cook, N. D., 1993, Composition and source of fluids in metamorphosed non-marine evaporites, Olary Block, South Australia: Mid- to lower-crustal metamorphism and fluids conference, Mount Isa, Mount Isa, Queensl., Australia, July 26-Aug. 1, 1993, p. 42.

Cook, N. D., 1992, Geology of metamorphosed Proterozoic playa lake deposits, Olary Block, South Australia: PhD thesis, University of New England.

Cook, N. D. J., and P. M. Ashley, 1992, Meta-evaporite sequence, exhalative chemical sediments and associated rocks in the Proterozoic Willyama Supergroup, South Australia: implications for metallogenesis: Precambrian Research, v. 56, p. 211-226.

Cooper, A. M., 1991, Late Proterozoic hydrocarbon potential and its association with diapirism in Blinman #2, Central Flinders Ranges: Honours thesis, University of Adelaide - National Centre Petroleum Geology and Geophysics.

REFERENCES

Corselli, C., and F. S. Aghib, 1987, Brine formation and gypsum precipitation in the Bannock Basin (eastern Mediterranean): Marine Geology, v. 75, p. 185-199.

Coshell, L., and M. R. Rosen, 1994, Stratigraphy and Holocene history of Lake Hayward, Swan Coastal Plain, Western Australia, in R. Renaut, and W. Last, eds., Sedimentology and Geochemistry of Modern and Ancient Lakes, Tulsa OK, SEPM Special Publ. 50, p. 173-188.

Council, T. C., and P. C. Bennett, 1993, Geochemistry of ikaite formation at Mono Lake, California: implications for the origin of tufa mounds: Geology, v. 21, p. 971-974.

Coveney, R. M., 1989, A review of the origins of metal-rich Pennsylvanian black shales, central USA, with an inferred role for basinal brines: Applied Geochem., v. 4, p. 347 - 367.

Crick, I. H., and M. D. Muir, 1980, Evaporites and uranium mineralization in the Pine Creek Geosyncline: Symposium on the Pine Creek Geosyncline, jointly sponsored by the Bureau of Mineral Resources, Geology and Geophysics and the CSIRO Institute of Earth Resources, in cooperation with the International Atomic Energy Agency, I. A. E. A., Proceedings series; STI; PUB; 555, p. 531-542.

Crocker, I. T., 1979, Fluorite mineralisation in the dolomite of the Transvaal Supergroup, South Africa: Geological Society of South Africa, Special Publication, v. 6, p. 73-82.

Cunningham, K. J., R. H. Benson, K. Rakicelbied, and L. W. Mckenna, 1997, Eustatic implications of Late Miocene depositional sequences in the Melilla Basin, Northeastern Morocco: Sedimentary Geology, v. 107, p. 147-165.

Czapowski, G., 1987, Sedimentary facies in the oldest rock salt (Na 1) of the Leba elevation (northern Poland), in T. M. Peryt, ed., The Zechstein Facies in Europe, Berlin, Springer-Verlag Lecture Notes in Earth Sciences, p. 207-224.

D'Onfro, P., 1988, Mechanics of salt tongue formation with examples from Louisiana slope (abs): American Association of Petroleum Geology, Bulletin, v. 72, p. 175.

Dailly, G. C., 1976, A possible mechanism relating progradation, growth faulting, clay diapirism and overthrusting in a regressive sequence of sediments: Canadian Petroleum Geology Bulletin, v. 24, p. 92-116.

Dalgarno, C. R., and J. E. Johnson, 1968, Diapiric structures and late Precambrian-early Cambrian sedimentation in Flinders ranges, South Australia: American Association Petroleum Geologists, Memoir, v. 8, p. 301 -314.

Damore, F., D. Giusti, and A. Abdallah, 1998, Geothermics of the high-salinity geothermal field of Asal, Republic of Djibouti, Africa: Geothermics, v. 27, p. 197-210.

Dardeau, G., and P. C. Graciansky, 1990, Halokinesis and alpine rifting in the Alpes-Maritimes (France): Bull. Centres Rech. Explr.-Prod. Elf Aquitaine, v. 14, p. 443-464.

Davidson, G. J., 1994, Hostrocks to the stratabound iron-formation-hosted Starra gold-copper deposit, Australia: Mineralium Deposita, v. 29, p. 237-249.

Davidson, G. J., and R. R. Large, 1994, Gold metallogeny and the copper-gold association of the Australian Proterozoic: Mineralium Deposita, v. 29, p. 208 - 223.

Davidson, G. J., R. R. Large, G. L. Kary, and R. Osborne, 1989, The deformed iron formation-hosted Starra and Trough Tank Au-Cu mineralisation: a new association from the Proterozoic Eastern Succession of Mount Isa, Australia: Economic Geology Monograph, v. 6, p. 135 - 150.

Davies, G. R., and S. D. Ludlam, 1973, Origin of laminated and graded sediments, Middle Devonian of western Canada: Geological Soc. America, Bull., v. 84, p. 3527-3546.

Davis, D. M., and T. Engelder, 1987, Thin-skinned deformation over salt, in I. Lerche, and J. J. O'Brien, eds., Dynamical geology of salt and related structures, Orlando, Florida, Academic Press, p. 301-3budd37.

Davis, D. M., and T. Engelder, 1985, The role of salt in fold-and- thrust belts (Canada): Tectonophysics, v. 119, p. 67-88.

Davis, J. B., and D. W. Kirkland, 1979, Bioepigenetic sulfur deposits: Economic Geology, v. 74, p. 462-468.

Davis, J. R., I. Friedman, and J. D. Gleason, 1986, Origin of the lithium-rich brine, Clayton Valley, Nevada: US Geological Survey Bulletin, v. 1622, p. 131-138.

Davison, I. D., G. I. Alsop, and D. J. Blundell, 1996a, Salt tectonics:some aspects of deformation mechanics, in G. I. Alsop, D. J. Blundell, and I. Davison, eds., Salt tectonics, Geological Society, London; Special Publication, p. 1-10.

Davison, I., D. Bosence, G. I. Alsop, and M. H. Al-Aawah, 1996b, Deformation and sedimentation around active Miocene salt diapirs on the Tihama Plain, northwest Yemen,

REFERENCES

in G. I. Alsop, D. J. Blundell, and I. Davison, eds., Salt tectonics, Geological Society, London; Special Publication, p. 23-39.

De Brodtkorb, M. K., V. Ramos, M. Barbieri, and S. Ametrano, 1982, The evaporitic celestite-barite deposits of Neuquen, Argentina: Mineralium Deposita, v. 17, p. 423-436.

De Celles, P. G., and G. Mitra, 1995, History of the Sevier orogenic wedge in terms of critical taper models, Northeast Utah and Southwest Wyoming: Geological Society of America Bulletin, v. 107, p. 454-462.

De Deckker, P., and M. C. Geddes, 1980, Seasonal fauna of ephemeral saline lakes near the Coorong Lagoon, South Australia: Aust. J. Marine and Freshwater Research, v. 31, p. 677-699.

De Lange, G. J., J. J. Middleburg, C. H. van der Weijden, G. Catalano, G. W. Luther, III, D. J. Hydes, J. R. W. Woittiez, and G. P. Klinkhammer, 1990, Composition of anoxic hypersaline brines in the Tyro and Bannock Basins, eastern Mediterranean: Marine Chemistry, v. 31, p. 63-88.

De Magnée, I., and A. Francois, 1988, The origin of the Kipushi (Cu,Zn,Pb) Deposit in direct relation with a Proterozoic salt diapir; Copperbelt of Central Africa, Shaba, Republic of Zaire, in G. H. Friedrich, and P. M. Herzig, eds., Base metal sulfide deposits in sedimentary and volcanic environments, Berlin, Springer Verlag, p. 74-93.

De Putter, T., J. M. Rouchy, A. Herbosch, E. Keppens, C. Pierre, and E. Groessens, 1994, Sedimentology and palaeo-environment of the upper Visean anhydrite of the Franco-Belgian Carboniferous basin (Saint-Ghislain borehole, southern Belgium): Sedimentary Geology, v. 90, p. 77-93.

De Ronde, C. E. J., D. M. D. Channer, and S. E. T. C., 1996, Fluids of the Archean, in M. J. de Wit, and L. D. Ashwal, eds., Tectonic Evolution of Greenstone belts, Oxford, UK, Oxford University Press, Monographs on Geology and Geophysics.

De Ronde, C. E. J., M. J. Dewit, and E. T. C. Spooner, 1994, Early Archean (>3.2 Ga) Fe-oxide rich, hydrothermal discharge vents in the Barberton Greenstone belt, South Africa: Geological Society of America Bulletin, v. 106, p. 86-104.

De Ruiter, P. A. C., 1979, The Gabon and Congo basins salt deposits: Economic Geology, v. 74, p. 419-431.

Dean, W. E., 1978, Theoretical versus observed successions from evaporation of seawater, in W. E. Dean, and B. C. Schreiber, eds., Marine evaporites., Tulsa, OK, Soc. Econ. Paleontol. Mineral., Short Course Notes No. 4, p. 74-85.

Dean, W. E., and R. Y. Anderson, 1982, Continuous subaqueous deposition of the Permian Castile evaporites, Delaware Basin, Texas and New Mexico: in Handford, C. R., Loucks, R. G., Davies, G. R., (eds), Depositional and diagenetic spectra of evaporites; a core workshop. S. E. P. M. Core Workshop, v. 3, p. 324-353.

Dean, W. E., and T. D. Fouch, 1983, Lacustrine Environment, in P. A. Scholle, D. G. Bebout, and C. H. Moore, eds., Carbonate Depositional Environments, Tulsa, OK, American Assoc. Petroleum Geologists Memoir 33, p. 96-130.

Decima, A., and F. Wezel, 1973, Late Miocene evaporites of the central Sicilian Basin; Italy: Initial reports of the Deep Sea Drilling Project, v. 13, p. 1234-1240.

Degens, E. T., and D. A. Ross, 1969, Hot Brines and recent heavy metal deposits in the Red Sea: New York, N.Y., Springer Verlag, 600 p.

Dejonghe, L., 1990, The sedimentary structures of barite: examples from the Chaudfontaine ore deposit, Belgium: Sedimentology, v. 37, p. 303-323.

Demaison, G. J., and G. T. Moore, 1980, Anoxic environments and oil source bed genesis: American Association of Petroleum Geologists Bulletin, v. 64, p. 1179-1209.

Demercian, S., P. Szatmari, and P. R. Cobbold, 1993, Style and pattern of salt diapirs due to thin-skinned gravitational gliding, Campos and Santos basins, offshore Brazil: Tectonophysics, v. 228, p. 393-423.

Derrick, G. M., 1982, A Proterozoic rift zone at Mount Isa, Queensland, and implications for mineralisation: BMR Journal of Australian Geology and Geophysics, v. 7, p. 81-92.

Dilles, J. H., and J. M. Proffett, 1995, Metallogenesis of the Yerington Batholith, Nevada: Arizona Geological Society Digest, v. 20, p. 306-315.

Dimroth, E., and M. M. Kimberley, 1975, Precambrian atmospheric oxygen; evidence in the sedimentary distributions of carbon, sulfur, uranium, and iron: Can. J. Earth Sci., v. 13, p. 1161-1185.

REFERENCES

Draper, J. J., and A. R. Jensen, 1976, The geochemistry of Lake Frome, a playa lake in South Australia: Bureau of Mineral Resources Journal of Australian Geology Geophysics, v. 1, p. 83-104.

Druckman, Y., 1981, Sub-recent manganese-bearing stromatolites along shorelines of the Dead Sea, in C. Monty, ed., Phanerozoic stromatolites; case histories, Berlin, Federal Republic of Germany, Springer-Verlag, p. 197-208.

Dunham, R. J., 1972, Capitan Reef, New Mexico and Texas: Facts and questions to aid interpretation and group discussion: Midland, Texas, Permian Basin Section SEPM Publication 72-14.

Dupre, B., G. Blanc, J. Boulegue, and C. J. Allegre, 1988, Metal remobilization at a spreading centre studied using lead isotopes: Nature, v. 333, p. 165-167.

Dutkiewicz, A., and C. C. von der Borch, 1995, Lake Greenly, Eyre Peninsula, South Australia; sedimentology, palaeoclimatic and palaeohydrologic cycles: Palaeogeography, Palaeoclimatology, Palaeoecology, v. 113, p. 43-56.

Duval, B., C. Cramez, and M. P. A. Jackson, 1992, Raft tectonics in the Kwanza Basin, Angola: Marine and Petroleum Geology, v. 9, p. 389-404.

Dworkin, S. I., and L. S. Land, 1994, Petrographic and geochemical constraints on the formation and diagenesis of anhydrite cements, Smackover sandstones, Gulf of Mexico: Journal of Sedimentary Research, Section A: Sedimentary Petrology and Processes, v. 64, p. 339-348.

Eardley, A. J., 1966, Sediments of Great Salt Lake: Geol. Soc. Guidebook Geol. Utah, v. 20, p. 1305-1411.

Ece, O. I., and F. Coban, 1994, Geology, occurrence and genesis of Eskisesehir Sepiolites, Turkey: Clays and Clay Minerals, v. 42, p. 81-92.

Edgell, H. S., 1996, Salt tectonism in the Persian Gulf Basin, in G. I. Alsop, D. J. Blundell, and I. Davison, eds., Salt tectonics, London, United Kingdom, Geological Society of London Special Publication 100.

Edgell, H. S., 1991, Proterozoic salt basins of the Persian Gulf area and their role in hydrocarbon generation: Precambrian Research, v. 54, p. 1-14.

Edgett, K. S., and T. J. Parker, 1997, Water on early Mars - Possible subaqueous sedimentary deposits covering ancient cratered terrain in western Arabia and Sinus Meridiani: Geophysical Research Letters, v. 24, p. 2897-2900.

Eggink, J. W., D. E. Riegstra, and P. Suzanne, 1996, Using 3D seismic to understand the structural evolution of the UK Central North Sea: Petroleum Geoscience, v. 2, p. 83-96.

Ehrmann, F., 1922, De la situation du Trias et son rôle tectonique dans la Kabyliedes Babourgs: Société Géologique de France Bulletin, v. 22, p. 26-47.

Einaudi, M. T., and N. Oreskes, 1990, Progress toward an occurrence model for Proterozoic iron oxide deposits - a comparison between the ore provinces of South Australia and southeast Missouri: US Geological Survey Bulletin, v. 1932, p. 58-69.

El Anbaawy, M. I. H., M. A. H. Al Aawah, K. A. Al Thour, and M. E. Tucker, 1992, Miocene evaporites of the Red Sea Rift, Yemen Republic; sedimentology of the Salif halite: Sedimentary Geology, v. 81, p. 61-71.

El Tabakh, M., B. C. Schreiber, and J. K. Warren, 1998a, Origin of fibrous gypsum in the Newark rift basin, eastern North America: Journal of Sedimentary Research Section A-Sedimentary Petrology and Processes, v. 68, p. 88-99.

El Tabakh, M., B. C. Schreiber, C. Utha-aroon, L. Coshell, and J. K. Warren, 1998b, Diagenetic origin of Basal Anhydrite in the Cretaceous Maha Sarakham Salt - Khorat Plateau, NE Thailand: Sedimentology, v. 45, p. 579-594.

El Tabakh, M., C. Utha-aroon, L. Coshell, and J. K. Warren, 1995, Cretaceous saline deposits of the Maha Sarakham Formation in the Khorat Basin, Northeastern Thailand: International Conference on Geology, Geochronology and Mineral Resources of Indochina 22-25 November 1995, Khon Kaen, Thailand, v. Core Workshop Notes, p. 20 pp.

Eldridge, C. S., N. Williams, and J. L. Walshe, 1993, Sulfur isotope variability in sediment-hosted massive sulfide deposits as determined using the ion microprobe SHRIMP: II. A study of the H.Y.C. deposit at McArthur River, Northern Territory, Australia: Economic Geology, v. 88, p. 1-26.

Elliott, L. A., and J. K. Warren, 1989, Stratigraphy and depositional environment of lower San Andres Formation in subsurface and equivalent outcrops; Chaves, Lincoln, and Roosevelt counties, New Mexico: American

REFERENCES

Association Petroleum Geologists Bulletin, v. 73, p. 1307-1325.

Ellis, A. J., 1968, Natural hydrothermal systems and experimental hot-water/rock interaction; reactions with NaCl solutions and trace metal extraction: Geochim. Cosmochim. Acta, v. 32, p. 1356-1363.

Ellis, D. M., 1978, Stability and phase equilibra of chloride-bearing scapolites at 750°C and 4000 bar: Geochimica et Cosmochimica Acta, v. 42, p. 1271-1281.

Emery, D., and A. Robinson, 1993, Inorganic Chemistry: Applications to Petroleum Geology: Oxford, Blackwell Scientific Publications, 254 p.

Enami, M., J. G. Liou, and D. K. Bird, 1992, Cl-bearing amphibole in the Salton Sea geothermal system, California: Canadian Mineralogist, v. 30, p. 1077-1092.

Erba, E., 1991, Deep mid-water bacterial mats from anoxic basins of the eastern Mediterranean: Marine Geology, v. 100, p. 83-101.

Ericksen, G. E., 1993, Upper Tertiary and Quaternary saline deposits in the Central Andean region: Geological Association of Canada, Special Paper, v. 40, p. 89-102.

Ericksen, G. E., 1983, The Chilean nitrate deposits: American Scientist, v. 71, p. 366-374.

Ericksen, G. E., 1981, Geology and origin of the Chilean nitrate deposits: US Geological Survey Prof. Paper, v. 1188, p. 37 pp.

Esteban, M., 1996, An overview of Miocene reefs from Mediterranean areas: general trends and facies models, in E. K. Franseen, ed., Models for carbonate stratigraphy from Miocene reef complexes of Mediterranean regions, SEPM/Society for Sedimentary Geology; Concepts in Sedimentology and Paleontology, p. 3-53.

Esteban, M., and L. C. Pray, 1983, Pisoids and pisolite facies (Permian), Guadalupe Mountains, New Mexico and West Texas, in T. M. Peryt, ed., Coated Grains, Berlin, Springer-Verlag, p. 503-537.

Ettner, D. C., A. Bjorlykke, and T. Andersen, 1994, A fluid inclusion and stable isotope study of the Proterozoic Bidjovagge Au-Cu deposit, Finnmark, northern Norway: Mineralium Deposita, v. 29, p. 16-29.

Eugster, H. P., 1989, Geochemical environments of sediment-hosted Cu-Pb-Zn deposits, in R. W. Boyle, A. C. Brown, C. W. Jefferson, and E. C. Jowett, eds., Sediment

hosted stratiform copper deposits, Geological Association of Canada, Special Paper, p. 111-126.

Eugster, H. P., 1986, Lake Magadi, Kenya; a model for rift valley hydrochemistry and sedimentation?, in L. E. Frostick, R. W. Renaut, I. Reid, and J. J. Tiercelin, eds., Sedimentation in the African rifts, Geological Society Special Publications, p. 177-189.

Eugster, H. P., 1985, Oil shales, evaporites and ore deposits: Geochimica et Cosmochimica Acta, v. 49, p. 619-635.

Eugster, H. P., 1980, Geochemistry of evaporitic lacustrine deposits: Annual Review of Earth and Planetary Sciences, v. 8, p. 35-63.

Eugster, H. P., 1970, Chemistry and origin of the brines of Lake Magadi, Kenya: Special Publ. Geol. Soc., v. 3, p. 215-235.

Eugster, H. P., 1969, Inorganic bedded cherts from the Magadi area, Kenya: Contrib. Mineral. Petrology Beitr. Mineral. Petrologie., v. 22, p. 1-31.

Eugster, H. P., and L. A. Hardie, 1978, Saline Lakes, in A. Lerman, ed., Lakes; chemistry, geology, physics, New York, NY, Springer-Verlag, p. 237-293.

Evans, D. M., A. Cowden, and R. M. Barratt, 1989, Deformation and thermal erosion at the Foster nickel deposit, Kambalda-St. Ives, Western Australia, in M. S. Prendergast, and M. J. Jones, eds., Magmatic sulfides; the Zimbabwe volume; Fifth magmatic sulfide field conference, Zimbabwe, 1987, London, United Kingdom, Inst. Min. and Metall., p. 215-219.

Evans, H. T., Jr., 1979, The thermal expansion of anhydrite to 1000° C, Physics and Chemistry of Minerals, Berlin, Springer-Verlag, p. 77-82.

Evans, R., 1978, Origin and significance of evaporites in basins around Atlantic margin (abs): American Association Petroleum Geologists Bulletin, v. 62, p. 734.

Evans, R., 1970, Genesis of sylvite- and carnallite-bearing rocks from Wallace, Nova Scotia: Third Symposium on Salt, v. 1, p. 239-245.

Evans, R., 1967, The structure of the Mississippian evaporite deposit at Pugwash, Cumberland County, Nova Scotia: Economic Geology, v. 62, p. 262-273.

Ewing, M., and J. Antoine, 1966, New seismic data concerning sediments and diapiric structures in Sigsbee Deep and upper continental slope, Gulf of Mexico:

REFERENCES

American Association Petroleum Geologists, Bulletin, v. 50, p. 470-504.

Eyde, T. H., and D. T. Eyde, 1987, The Bowie chabazite deposit, in H. W. Peirce, ed., 21st Forum on the Geology of Industrial Minerals, Arizona Bur. Geol. and Min. Tech Spec. Paper., p. 133.

Fallara, F., M. M. Savard, and S. Paradis, 1998, A structural, petrographic, and geochemical study of the Jubilee Zn-Pb deposit, Nova Scotia, Canada and a new metallogenic model: Economic Geology, v. 93, p. 757-778.

Fegan, N. E., D. T. Long, W. B. Lyons, M. E. Hines, and P. G. Macumber, 1992, Metal partitioning in acid hypersaline sediments: Lake Tyrrell, Victoria, Australia: Chemical Geology, v. 96, p. 167-181.

Ferguson, J., H. Etminan, and F. Ghassemi, 1993, Geochemistry of deep formation waters in the Canning Basin, Western Australia, and their relationship to Pb-Zn mineralisation: Australian Journal of Earth Sciences, v. 40, p. 471-483.

Ferguson, J., H. Etminan, and F. Ghassemi, 1992, Salinity of deep formation water in the Canning Basin, Western Australia: BMR Journal of Australian Geology and Geophysics, v. 13, p. 93-105.

Ferguson, F., and G. W. Skyring, 1995, Redbed-associated sabkhas and tidal flats at Shark Bay, Western Australia: their significance for genetic models of stratiform Cu-(pb-Zn) deposits: Australian Journal of Earth Sciences, v. 42, p. 321-333.

Fischer, A. G., and L. T. Roberts, 1991, Cyclicity in the Green River Formation (lacustrine Eocene) of Wyoming: Journal of Sedimentary Petrology, v. 61, p. 1146-1154.

Fisher, W. L., and P. U. Rodda, 1969, Edwards Formation (lower Cretaceous) Texas: Dolomitization in a carbonate platform system: Amer. Assoc. Petroleum Geol. Bull., v. 52, p. 55-72.

Flanigen, E. M., 1984, Adsorption properties of molecular sieve zeolites, in W. G. Pond, and F. A. Mumpton, eds., Zeo-agriculture; use of natural zeolites in agriculture and aquaculture, Boulder, CO, United States, Westview Press., p. 55-68.

Fletcher, R. C., M. R. Hudec, and I. A. Watson, 1995, Salt glacier and composite sediment-salt glacier models for the emplacement and early burial of allochthonous salt sheets, in M. P. A. Jackson, D. G. Roberts, and S. Snelson, eds.,

Salt tectonics: a global perspective, American Association Petroleum Geologists Memoir, p. 77-108.

Flinch, J. F., A. W. Bally, and S. G. Wu, 1996, Emplacement of a passive-margin evaporitic allochthon in the Betic Cordillera of Spain: Geology, v. 24, p. 67-70.

Flint, S. S., 1989, Sediment-hosted stratabound copper deposits of the Central Andes: Geological Association of Canada Special Paper, v. 36, p. 371-398.

Folk, R. L., 1993, SEM imaging of bacteria and nannobacteria in carbonate sediments and rocks: Journal Sedimentary Petrology, v. 63, p. 990-999.

Folk, R. L., and A. Siedlecka, 1974, The ''schizohaline'' environment: its sedimentary and diagenetic fabrics as exemplified by late Paleozoic rocks of Bear Island, Svalbard: Sediment. Geol., v. 11, p. 1-15.

Folk, R. L., and J. S. Pittman, 1971, Length-slow chalcedony; a new testament for vanished evaporites: Journal Sedimentary Petrology, v. 41, p. 1045-1058.

Fontboté, L., and H. Gorzawski, 1990, Genesis of the Mississippi Valley-type Zn-Pb deposit of San Vicente, Central Peru: Geologic and isotopic (Sr, O, C, S, Pb) evidence: Economic Geology, v. 85, p. 1402-1437.

Forbes, B. G., 1991, Olary, South Australia; Sheet S1 54-2; Explanatory Notes: Geological Survey of South Australia.

Forrestal, P. J., 1990, Mount Isa and Hilton silver-lead-zinc deposits, in F. E. Hughes, ed., The Mineral Deposits of Australia and Papua New Guinea: Volume 1, Australasian Institute of Mining and Metallurgy: Monograph, p. 927 - 934.

Forsythe, R. D., and J. R. Zimbelman, 1995, A case for ancient evaporite basins on Mars: Journal of Geophysical Research-Planets, v. 100, p. 5553-5563.

Fox, J. S., 1984, Besshi-type volcanogenic sulphide deposits a review: Canadian Mining and Metallurgical Bulletin, v. 77, p. 57-68.

Frazier, W. J., 1975, Celestite in the Mississippian Pennington Formation, central Tennessee: Southeast. Geol., v. 16, p. 241-248.

Freyberger, S. G., C. J. Schenk, and L. F. Krystinik, 1988, Stokes surfaces and the effects of near-surface groundwater-table on aeolian deposition: Sedimentology, v. 35, p. 21-41.

References

Friedman, G. M., and V. Shukla, 1980, Significance of quartz euhedra after sulphates; example from the Lockport Formation (Middle Silurian) of New York: Journal of Sedimentary Petrology, v. 50, p. 1299-1304.

Frietsch, R., P. Tuisku, O. Martinsson, and J. Perdahl, 1997, Early Proterozoic Cu-(Au) and Fe ore deposits associated with regional NaCl metasomatism in northern Fennoscandinavia: Ore Geology Reviews, v. 12, p. 1-34.

Frimmel, H., 1988, Strontium isotopic evidence for the origin of siderite, ankerite and magnesite mineralisations in the Eastern Alps: Mineralium Deposita, v. 23, p. 268-275.

Frimmel, H. E., and W. Papesch, 1990, Sr, O and C isotope study of the Brixlegg barite deposit, Tyrol (Austria): Economic Geology, v. 85, p. 1162-1171.

Frumkin, A., 1994, Hydrology and denudation rates of halite karst: Journal of Hydrology, v. 162.

Fuzesy, A., 1982, Potash in Saskatchewan: Report Saskatchewan, Department of Mineral Resources.

Fuzesy, L. M., 1983, Petrology of potash ore in the Middle Devonian Prairie Evaporite, Saskatchewan: Potash '83; Potash technology; mining, processing, maintenance, transportation, occupational health and safety, environment., p. 47-57.

Gablina, I. F., 1981, New data on the formation conditions of the Dzhezkazgan copper deposit: International Geology Review, v. 23, p. 1313-1311.

Gao, G., and L. S. Land, 1991, Nodular chert from the Arbuckle Group, Slick Hills, SW Oklahoma: a combined field, petrographic and isotopic study: Sedimentology, v. 38, p. 857-870.

Gao, G., S. D. Hovorka, and H. H. Posey, 1990, Limpid dolomite in Permian San Andres halite rocks, Palo Duro Basin, Texas Panhandle; characteristics, possible origin, and implications for brine evolution: Journal of Sedimentary Petrology, v. 60, p. 118-124.

Garber, R. A., 1980, The sedimentology of the Dead Sea: PhD thesis, Rennsaler Polytechnic.

Garber, R. A., P. M. Harris, and J. M. Borer, 1990, Occurrence and significance of magnesite in Upper Permian (Guadalupian) Tansill and Yates formations, Delaware Basin, New Mexico: American Association Petroleum Geologists Bulletin, v. 74, p. 119-134.

Garcia-Veigas, J., F. Orti, L. Rosell, C. Ayora, R. J. M., and S. Lugli, 1995, The Messinian salt of the Mediterranean: geochemical study of the salt from the central Sicily Basin and comparison with the Lorca Basin (Spain): Bulletin de la Societe Geologique de France, v. 166, p. 699-710.

Garea, B. B., and C. J. R. Braithwaite, 1996, Geochemistry, isotopic composition and origin of the Beda dolomites, block NC74F, SW Sirt Basin, Libya: Journal of Petroleum Geology, v. 19, p. 289-304.

Garfunkel, Z., and G. Almagor, 1987, Active salt dome development in the Levant Basin, southeast Mediterranean, in I. Lerche, and J. J. O'Brien, eds., Dynamical geology of salt and related structures, New York, Academic Press, p. 263-300.

Garlick, W. G., 1981, Sabkhas, slumping and compaction at Mufubra, Zambia: Economic Geology, v. 76, p. 1817-1847.

Garrels, R. M., and C. L. Christ, 1965, Solutions, minerals, and equilibria: San Francisco, W. H. Freeman, 450 p.

Gaupp, R., A. Matter, J. Platt, K. Ramseyer, and J. Walzebuck, 1993, Diagenesis and fluid evolution of deeply buried Permian (Rotliegende) gas reservoirs, northwest Germany: American Association Petroleum Geologists Bulletin, v. 77, p. 1111-1128.

Ge, H. X., M. P. A. Jackson, and B. C. Vendeville, 1997, Kinematics and dynamics of salt tectonics driven by progradation: American Association Petroleum Geologists Bulletin, v. 81, p. 398-423.

Geeslin, J. H., and H. S. Chafetz, 1982, Ordovician Aleman ribbon cherts; an example of silicification prior to carbonate lithification: Journal of Sedimentary Petrology, v. 52, p. 1283-1293.

Gely, J. P., V. M. M. Blanc, D. F. Fache, M. Schuler, and M. Ansart, 1993, Characterization of organic-rich material in an evaporitic environment; the lower Oligocene of the Mulhouse Basin (Alsace, France): Geologische Rundschau, v. 82, p. 718-725.

Gendzwill, D. J., 1978, Winnipegosis mounds and Prairie evaporite formation of Saskatchewan; seismic study: In; Rehrig, D., The economic geology of the Williston Basin; Montana, North Dakota, South Dakota, Saskatchewan, Manitoba.

Gendzwill, D., and N. Martin, 1996, Flooding and loss of the Patience Lake potash mine: CIM Bulletin, v. 89, p. 62-73.

REFERENCES

Gerdes, G., K. Dunajtschik-Piewak, H. Riege, A. G. Taher, W. E. Krumbein, and H.-E. Reineck, 1994, Structural diversity of biogenic carbonate particles in microbial mats: Sedimentology, v. 41, p. 1273-1294.

Ghazban, F., H. P. Schwarcz, and D. C. Ford, 1992, Multistage dolomitization in the Society Cliffs Formation, northern Baffin Island, Northwest Territories, Canada: Canadian Journal of Earth Sciences, v. 29, p. 1459-1473.

Ghazban, F., H. P. Schwarcz, and D. C. Ford, 1990, Carbon and sulfur isotope evidence for in situ reduction of sulfate, Nanisivik lead-zinc deposits, Northwest Territories, Baffin Island, Canada: Economic Geology, v. 85.

Giuliani, G., A. Cheilletz, C. Arboleda, V. Carrillo, F. Rueda, and J. H. Baker, 1995, An evaporitic origin of the parent brines of Colombian emeralds; fluid inclusion and sulphur isotope evidence: European Journal of Mineralogy, v. 7, p. 151-165.

Giuliani, G., G. R. Olivo, O. J. Marini, and D. Michel, 1993, The Santa Rita gold deposit in the Proterozoic Paranoa Group, Goias, Brazil; an example of fluid mixing during ore deposition: Ore Geology Reviews, v. 8, p. 503-523.

Gize, A. P., H. L. Barnes, and J. S. Bell, 1991, A critical evaluation of organic processes in Mississippi valley-type genesis: Source, transport and deposition of metals, p. 527-530.

Glennie, K. W., 1989a, A summary of tropical desert sedimentary environments, present and past: Geological Association of Canada Special Paper, v. 36, p. 67-84.

Glennie, K. W., 1989b, Some effects of the Late Permian Zechstein transgression in northwestern Europe: Geological Association of Canada Special Paper, v. 36, p. 557-565.

Glennie, K. W., and A. T. Buller, 1983, The Permian Weissliegend of NW Europe: the partial deformation of eolian sands caused by the Zechstein transgression: Sedimentary Geology, v. 35, p. 43-81.

Godin, L., and I. Lager, 1986, Viscaria— a copper deposit in a Proterozoic playa-lake complex (abs): IAGOD Symposium 7th.

Godlevsky, M. N., and L. N. Grinenko, 1963, Some data on the isotopic composition of sulphur in the sulphides of the Noril'sk deposit: Geochemica, v. 1, p. 335-341.

Gómez-Pugnaire, M. T., G. Franz, and V. L. Sanchez-Vizcaino, 1994, Retrograde formation of NaCl-scapolite in high pressure metaevaporites from the Cordilleras-Beticas (Spain): Contributions to Mineralogy and Petrology, v. 116, p. 448-461.

Goodall, I. G., G. M. Harwood, A. C. Kendall, T. McKie, and M. E. Tucker, 1992, Discussion on sequence stratigraphy of carbonate-evaporite basins; models and application to the Upper Permian (Zechstein) of Northeast England and adjoining North Sea: Journal of the Geological Society of London, v. 149, p. 1050-1054.

Goodfellow, W. D., J. W. Lydon, and R. J. W. Turner, 1993, Geology and genesis of stratiform sediment-hosted (SEDEX) zinc-lead-silver sulphide deposits: Geological Association of Canada Special Paper, v. 40, p. 201-251.

Gordon, W. A., 1975, Distribution by latitude of Phanerozoic evaporite deposits: Journal of Geology, v. 83, p. 671-684.

Gornitz, V. M., and B. C. Schreiber, 1981, Displacive halite hoppers from the Dead Sea; some implications for ancient evaporite deposits: Journal of Sedimentary Petrology, v. 51, p. 787-794.

Gott, G. B., D. F. Wolcott, and C. G. Bowles, 1974, Stratigraphy of the Inyan Kara Group and localisation of Uranium deposits, Southern Black Hills, South Dakota and Wyoming: US Geol Surv. Prof. Paper, v. 763, p. 1-57.

Götzinger, M. A., and W. Grum, 1992, Die Pb-Zn-F Mineralisationen in der Umgebung von Evaporiten der Nördlichen Kalkalpen Österreich - Herkunft und Zusammensetzung der fluiden Phase: Mitt. Ges. Geol. Bergbaustud Österr (Vienna), v. 38, p. 47-56.

Gresens, R. L., 1978, Evaporites as precursors of massif anorthosite: Geology, v. 6, p. 46-50.

Grinenko, L. N., 1985, Sources of sulphur of the nickeliferous and barren gabbro-dolerite intrusions of the northwest Siberian platform: International Geology Review, p. 695-708.

Griswold, G. B., 1982, Geologic overview of the Carlsbad potash-mining district: Circular New Mexico Bureau of Mines and Mineral Resources, v. 182, p. 17-22.

Grossman, I. G., 1968, Origin of the sodium sulphate deposits of the northern Great Plains of Canada and the United States: U. S. Geol. Surv. Prof. Pap., v. 600-B, p. 104-109.

Grotzinger, J. P., 1995, Trends in Precambrian carbonate sediments and their implication for understanding

REFERENCES

evolution, in S. Bengtson, ed., Early Life on Earth: Nobel Symposium No. 84, New York, Columbia Univ. Press, p. 245-258.

Grotzinger, J. P., 1986, Shallowing upward cycles of the Wallace Formation, Belt Supergroup, northwestern Montana and northern Idaho: Montana Bureau of Mines Geol. Spec. Pub., v. 94, p. 143-160.

Grotzinger, J. P., and A. H. Knoll, 1995, Anomolous carbonate precipitates - Is the Precambrian the key to the Permian: Palaios, v. 10, p. 578-596.

Grotzinger, J. P., and J. F. Kasting, 1993, New constraints on Precambrian ocean composition: Journal of Geology, v. 101, p. 235-243.

Grotzinger, J. P., and J. F. Read, 1983, Evidence for primary aragonite precipitation, lower Proterozoic (1.9 Ga) Rocknest dolomite, Wopmay orogen, northwest Canada: Geology, v. 11, p. 710-713.

Gundersen, J. K., B. B. Jorgensen, E. Larsen, and H. W. Jannasch, 1992, Mats of giant sulphur bacteria on deep-sea sediments due to fluctuating hydrothermal flow: Nature, v. 360, p. 454-456.

Gundogdu, M. N., H. Yalcin, A. Temel, and N. Clauer, 1996, Geological, mineralogical and geochemical characteristics of zeolite deposits associated with borates in the Bigadic, Emet and Kirka Neogene lacustrine basins, western Turkey: Mineralium Deposita, v. 31, p. 492-513.

Gussow, W. C., 1968, Salt diapirism: importance of temperature, and energy source of emplacement, in J. Braunstein, and G. D. O'Brien, eds., Diapirism and diapirs, Tulsa OK, American Association of Petroleum Geologists Memoir 8, p. 16-52.

Gustavson, T. C., S. D. Hovorka, and A. R. Dutton, 1994, Origin of satin spar veins in evaporite basins: Journal of Sedimentary Research, Section A: Sedimentary Petrology and Processes, v. 64, p. 88-94.

Habermehl, M. A., 1988, Springs of the Great Artesian Baisn: Aspects of Hydrogeologic, Hydrochemical, and Isotopic Characteristics: SLEADS Conference 88, Salt Lakes in Arid Lands, p. 30 pp.

Habicht, J. K. A., 1979, Paleoclimate, Paleomagnetism and Continental drift: American Association of Petroleum Geologists, Studies in Geology, v. 9, p. 31 p.

Habicht, K. S., and D. E. Canfield, 1996, Sulphur isotope fractionation in modern microbial mats and the evolution of the sulphur cycle: Nature, v. 382, p. 342-343.

Hallager, W. S., M. R. Ulrich, J. R. Kyle, P. E. Price, and W. A. Gose, 1990, Evidence for episodic basin dewatering in salt-dome cap rocks: Geology, v. 18, p. 716-719.

Hallam, A., 1981, Facies interpretation and the stratigraphic record: San Francisco, W. H. Freeman and Co., 291 p.

Hammer, J., F. Junge, H. J. Rosler, S. Niese, B. Gleisberg, and G. Stiehl, 1990, Element and isotope geochemical investigations of the Kupferschiefer in the vicinity of 'Rote Faule', indicating copper mineralization (Sangerhausen Basin, G.D.R.): Chemical Geology, v. 85, p. 345-360.

Handford, C. R., 1991, Marginal marine halite; sabkhas and salinas, in J. L. Melvin, ed., Evaporites, petroleum and mineral resources, Elsevier Developments in Sedimentology No. 50, p. 1-66.

Handford, C. R., 1981, Coastal sabkha and salt pan deposition of the lower Clear Fork Formation (Permian), Texas: Journal of Sedimentary Petrology, v. 51, p. 761-778.

Handford, C. R., A. C. Kendall, D. R. Prezbindowski, J. B. Dunham, and B. W. Logan, 1984, Salina-margin tepees, pisoliths and aragonite cements, Lake MacLeod, Western Australia; Their significance in interpreting ancient analogs: Geology, v. 12, p. 523-527.

Hanor, J. S., 1994, Origin of saline fluids in sedimentary basins, in J. Parnell, ed., Geofluids; origin, migration and evolution of fluids in sedimentary basins: Geological Society Special Publications, London, United Kingdom, Geological Society of London, p. 151-174.

Hanson, R. E., T. J. Wilson, and H. Munyanyiwa, 1994, Geologic evolution of the Neoproterozoic Zambesi Orogenic belt in Zambia [Review]: Journal of African Earth Sciences and the Middle East., v. 18, p. 135-150.

Harben, P. W., and R. L. Bates, 1990, Industrial Minerals; Geology and world deposits: London, UK, Industrial Minerals Division, Metal Bulletin Plc, 312 p.

Harben, P. W., and M. Kuzvart, 1996, Industrial Minerals: A world Geology: London, Industrial Minerals Information Ltd., Metal Bulletin Plc, 462 p.

REFERENCES

Hardie, L. A., 1996, Secular variation in seawater chemistry: an explanation for the coupled secular variation in the mineralogies of marine limestones and potash evaporites over the past 600 m.y: Geology, v. 24, p. 279 - 283.

Hardie, L. A., 1991, On the significance of evaporites: Annual Review Earth and Planetary Science, v. 19, p. 131-168.

Hardie, L. A., 1990, The roles of rifting and hydrothermal $CaCl_2$ brines in the origin of potash evaporites: an hypothesis: American Journal of Science, v. 290, p. 43-106.

Hardie, L. A., 1984, Evaporites: Marine or non-marine?: American Journal of Science, v. 284, p. 193-240.

Hardie, L. A., J. P. Smoot, and H. P. Eugster, 1978, Saline lakes and their deposits; A sedimentological approach, in A. Matter, and M. A. Tucker, eds., Modern and ancient lake sediments, International Association Sedimentologists Special Publication 2, p. 7-41.

Hardie, L. A., and H. P. Eugster, 1971, The depositional environment of marine evaporites: A case for shallow, clastic accumulation: Sedimentology, v. 16, p. 187-220.

Hardie, L. A., and H. P. Eugster, 1970, The evolution of closed-basin brines: Spec Pub. Mineral. Soc. Am., v. 3, p. 273-290.

Harrison, J. C., 1995, Tectonics and kinematics of a foreland fold belt influenced by salt, Arctic Canada, in M. P. A. Jackson, D. G. Roberts, and S. Snelson, eds., Salt tectonics: a global perspective, American Association Petroleum Geologists Memoir, p. 379-412.

Hartley, A. J., and G. May, 1998, Miocene gypcretes from the Calama Basin, northern Chile: Sedimentology, v. 45, p. 351-364.

Hartley, A. J., S. Flint, and P. Turner, 1991, Analcime: a characteristic authigenic phase of Andean alluvium, northern Chile: Geological Journal, v. 26, p. 189-202.

Hartmann, M., J. C. Scholten, P. Stoffers, and K. F. Wehner, 1998, Hydrographic structure of the brine-filled deeps in the Red Sea - New results from the Shaban, Kebrit, Atlantis II, and Discovery deeps: Marine Geology, v. 144, p. 311-330.

Harvie, C. E., and J. H. Weare, 1980, The prediction of mineral solubilities in natural waters: the Na-K-Mg-Ca-SO_4-Cl-H_2O from zero to high concentration at 25°C: Geochim. Cosmochim. Acta, v. 44.

Harvie, C. E., J. H. Weare, L. A. Hardie, and H. P. Eugster, 1980, Evaporation of sea water; calculated mineral sequences: Science, v. 208, p. 498-500.

Harwood, G. M., 1980, Calcitized anhydrite and associated sulphides in the English Zechstein First Cycle Carbonate (EZ1 Ca), in H. Fuechtbauer, Peryt, T. M., ed., The Zechstein Basin, with emphasis on carbonate sequences., E. Schweizerbart'sche VbH: Contributions to Sedimentology, p. 61-72.

Hauer, K. L., 1995, Protoliths, diagenesis, and depositional history of the Upper Marble, Adirondack Lowlands, New York: Doctoral thesis, Miami University.

Hay, R. L., 1981, Geology of zeolites in sedimentary rocks, in F. A. Mumpton, ed., Mineralogy and geology of natural zeolites, Min. Soc. America Reviews in Mineralogy 4, p. 165-175.

Hay, W. W., M. J. Rosol, and J. L. Sloan II, 1988, Plate tectonic control of global patterns of detrital and carbonate sedimentation, in L. J. Doyle, and H. H. Roberts, eds., Carbonate - Clastic Transitions; Developments in Sedimentology 42, New York, Elsevier, p. 1-34.

Haynes, D. W., and M. S. Bloom, 1987, Stratiform copper deposits hosted by low-energy sediments: IV. Aspects of sulfide precipitation: Economic Geology, v. 82, p. 875-893.

Haynes, D. W., K. C. Cross, R. T. Bills, and M. H. Reed, 1995, Olympic Dam ore genesis; a fluid-mixing model: Economic Geology, v. 90, p. 281-307.

Hayward, A. B., and R. H. Graham, 1989, Some geometrical characteristics of inversion: Geological Society of London Special Publication, v. 44, p. 17-39.

Heaney, P. J., 1995, Moganite as an indicator for vanished evaporites: a testament reborn?: Journal of Sedimentary Research A: Sedimentary Petrology and Processes, v. A65, p. 633-638.

Heard, H. C., and W. W. Rubey, 1966, Tectonic implications of gypsum dehydration: Geol. Soc. America Bull, v. 77, p. 741-760.

Heaton, R. C., M. P. A. Jackson, M. Bahahmoud, and A. S. O. Nani, 1995, Superposed Neogene extension,

REFERENCES

contraction, and salt canopy emplacement in the Yemeni Red Sea, in M. P. A. Jackson, D. G. Roberts, and S. Snelson, eds., Salt Tectonics: a global perspective, Tulsa, American Association Petroleum Geologists Memoir 65, p. 333-351.

Hein, F. J., M. C. Graves., and A. Ruffman, 1991, The Jubilee zinc-lead deposit, Nova Scotia: the role of synsedimentary faults, in A. L. Sangster, ed., Mineral deposit studies in Nova Scotia, Geol. Surv. Canada Paper 90-8(2).

Helvaci, C., 1995, Stratigraphy, mineralogy, and genesis of the Bigadic Borate deposits, Western Turkey: Economic Geology, v. 90, p. 1237-1260.

Henneke, E., G. W. Luther, G. J. Delange, and J. Hoefs, 1997, Sulphur speciation in anoxic hypersaline sediments from the Eastern Mediterranean Sea: Geochimica et Cosmochimica Acta, v. 61, p. 307-321.

Henry, D. J., and C. V. Guidotti, 1985, Tourmaline as a petrogenetic indicator mineral: an example from the staurolite grade metapelites of NW Maine: American Mineralogist, v. 70, p. 1-15.

Heppenheimer, H., H. W. Hagemann, and W. Puttmann, 1995, A comparative study of the influence of organic matter on metal accumulation processes in the Kupferschiefer from the Hessian Depression and the north Sudetic Syncline: Ore Geology Reviews, v. 9, p. 391-409.

Heroux, Y., A. Chagnon, and M. Savard, 1996, Organic matter and clay anomalies associated with base-metal sulphide deposits: Ore Geology Reviews, v. 11, p. 157-173.

Heroux, Y., A. Chagnon, and M. M. Savard, 1994, Anomalies in properties of organic material as associated clay mineral assemblages - Pb-Zn of Gays River, Nova Scotia [French]: Exploration and Mining Geology, v. 3, p. 67-79.

Hesse, R., 1986, Early diagenetic pore water/sediment interaction: Modern offshore basins: Geoscience Canada, v. 13, p. 165-197.

Heydari, E., 1997, The role of burial diagenesis in hydrocarbon destruction and H_2S accumulation, Upper Jurassic Smackover Formation, Black Creek Field, Mississippi: American Association Petroleum Geologists Bulletin, v. 81, p. 26-45.

Heydari, E., and C. H. Moore, 1989, Burial diagenesis and thermochemical sulfate reduction, Smackover Formation, Southeast Mississippi salt basin: Geology, v. 17, p. 1080-1084.

Hietanen, A. H., 1967, Scapolite in the Belt Series in the St. Joe-Clearwater Region, Idaho: Geol. Soc. Amer. Special Paper, v. 86, p. 56 pp.

Hill, B. F., 1993, Magnesite and magnesia production by Queensland Magnesia (Operations) Pty Ltd at Kunwarara and Rockhampton, Qld, in J. T. Woodcock, and J. K. Hamilton, eds., Australasian mining and metallurgy; the Sir Maurice Mawby memorial volume, Melbourne, Victoria, Australia, Australasian Institute of Mining and Metallurgy,, p. 1388-1393.

Hill, C., 1995, Sulfur redox reactions: Hydrocarbons, native sulfur, Mississippi Valley-type deposits, and sulfuric acid karst in the Delaware Basin, New Mexico and Texas: Environmental Geology, v. 25, p. 16-23.

Hill, R. E. T., M. J. Gole, and S. J. Barnes, 1989, Olivine adcumulates in the Norseman-Wiluna greenstone belt, Western Australia; implications for volcanology of komatiites, in M. S. Prendergast, and M. J. Jones, eds., Magmatic sulfides; the Zimbabwe volume; Fifth magmatic sulfide field conference, Zimbabwe, 1987, London, United Kingdom, Inst. Min. and Metall., p. 189-206.

Hite, R. J., and D. E. Anders, 1991, Petroleum and evaporites, in J. L. Melvin, ed., Evaporites, petroleum and mineral resources, Amsterdam, Elsevier Developments in Sedimentology, p. 477-533.

Hite, R. J., and T. Japakasetr, 1979, Potash deposits of the Khorat Plateau, Thailand and Laos: Economic Geology, v. 74, p. 448-458.

Hitzman, M. W., N. Oreskes, and M. T. Einaudi, 1992, Geological characteristics and tectonic setting of Proterozoic iron oxide (Cu-U-Au-REE) deposits: Precambrian Research, v. 58, p. 241-287.

Hodge, B. L., 1986, Occurrence and exploitation of fluorite, in R. W. Nesbitt, and I. Nichol, eds., Geology in the real world - the Kingsley Dunham volume, London, Inst. Mining Metallurgy, p. 165.

Hoffman, P. F., I. R. Bell, R. S. Hildebrand, and L. Thorstad, 1977, Geology of the Athapuscow Aulocogen, East Arm of the Great Slave Lake, District of Mackenzie:

Current Research Part A: Geol. Surv. Canada Paper, v. 84-1A, p. 117-146.

Hogarth, D. D., 1979, Lapis lazuli from Edwards, New York; a possible metaevaporite: Geol. Assoc. Can. Mineral. Assoc. Can., Jt. Annu. Meet., Program Abstr.

Hogarth, D. D., and W. L. Griffin, 1978, Lapis lazuli from Baffin Island; a Precambrian meta-evaporite: Lithos, v. 11, p. 37-60.

Holland, H. D., 1984, The chemical evolution of the atmosphere and the oceans: Princeton, Princeton University Press, 582 p.

Holliday, D. W., 1970, The petrology of secondary gypsum rocks; a review: J. Sediment. Petrology., v. 40, p. 734-744.

Holwerda, J. G., and R. W. Hutchinson, 1968, Potash-bearing evaporites in the Danakil area, Ethiopia: Economic Geology, v. 63, p. 124-150.

Hooper, R. J., Leng Siang Goh, and F. Dewey, 1995, The inversion history of the northeastern margin of the Broad Fourteens Basin, in J. G. Buchanan, and P. G. Buchanan, eds., Basin inversion, Geological Society, London; Special Publication, 88, p. 307-317.

Hopkins, J. C., 1987, Contemporaneous subsidence and fluvial channel sedimentation: Upper Mannville C Pool, Berry Field, Lower Cretaceous of Alberta: Amer. Assoc. Petrol. Geol. Bulletin, v. 71, p. 334-345.

Hopkinson, L., S. Roberts, R. Herrington, and J. Wilkinson, 1998, Self-organisation of submarine hydrothermal siliceous deposits - Evidence from the TAG hydrothermal mound, 26°N Mid-Atlantic Ridge: Geology, v. 26, p. 347-350.

Hovorka, S. D., 1992, Halite pseudomorphs after gypsum in bedded anhydrite; clue to gypsum-anhydrite relationships: Journal of Sedimentary Petrology, v. 62, p. 1098-1111.

Hovorka, S., 1987, Depositional environments of marine-dominated bedded halite, Permian San Andres Formation, Texas: Sedimentology, v. 34, p. 1029-1054.

Hsu, K. J., W. B. F. Ryan, and M. B. Cita, 1973, Late Miocene Desiccation of the Mediterranean: Nature, v. 242, p. 240-244.

Huber, R., M. Kurr, H. W. Jannasch, and K. O. Stetter, 1989, A novel group of methanogenic archaebacteria (Methanopyrus) growing at 110°C: Nature, v. 342, p. 833-834.

Huichi, R., H. Renmin, and D. P. Cox, 1991, Copper deposition by fluid mixing in deformed strata adjacent to a salt diapir, Dongchuan area, Yunnan Province, China: Economic Geology, v. 86, p. 1539-1545.

Humphris, C. C., Jr., 1978, Salt movement on continental slope, northern Gulf of Mexico, in A. H. Bouma, C. T. Moore, and J. M. Coleman, eds., Framework, facies, and oil-trapping characteristics of the upper continental margin, American Association Petroleum Geologists, Studies in Geology, v. 7, p. 69-85.

Humphris, S. E., P. M. Herzig, D. J. Miller, J. C. Alt, K. Becker, D. Brown, G. Brugmann, H. Chiba, Y. Fouquet, J. B. Gemmell, G. G., M. D. Hannington, N. G. Holm, J. J. Honnorez, G. J. Iturrino, R. Knott, R. Ludwig, K. Nakamura, S. Petersen, A. L. Reysenbach, P. A. Rona, S. Smith, A. A. Sturz, M. K. Tivey, and X. Zhao, 1995, The internal structure of an active sea-floor massive sulphide deposit: Nature, v. 377, p. 713-716.

Hunt, D., and M. E. Tucker, 1992, Stranded parasequences and forced regressive wedge systems tract; deposition during base level fall: Sedimentary Geology, v. 81, p. 1-9.

Hurley, N. F., and R. Budros, 1990, Albion-Scipio and Stoney Point fields, U.S.A., Michigan Basin, in E. A. Beaumont, and N. H. Foster, eds., Stratigraphic traps; I, Tulsa, OK, American Association Petroleum Geologists, Treatise of Petroleum Geology, Atlas of Oil and Gas Fields, p. 1-37.

Hussain, M., and J. K. Warren, 1989, Nodular and enterolithic gypsum; the ''sabkha-tization'' of Salt Flat Playa, West Texas: Sedimentary Geology, v. 64, p. 13-24.

Huyck, H. L. O., and R. W. Chorey, 1991, Stratigraphic and petrographic comparison of the Creta and Kupferschiefer copper shale deposits: Mineralium Deposita, v. 26, p. 132-142.

Iijima, A., 1980, Geology of natural zeolites and zeolitic rocks, in L. V. Rees, ed., Fifth International Conference on Zeolites, Heyden Publishers, p. 103-117.

Ilavsky, J., J. Turan, and L. Turanova, 1991, Magnesite deposits and occurrences in the west Carpathians of Czechoslovakia: Ore Geology Reviews, v. 6, p. 537-561.

Intiomale, M. M., 1982, Le gisement Zn-Pb-Cu de Kipushi (Shaba, Zaire). Etudes geologique et metallogenique:

REFERENCES

Doctoral thesis, Univ. Catholique de Louvain, Louvain la-Neuve, Belgique.

Irwin, M. L., 1965, General theory of epeiric clear water sedimentation: American Association of Petroleum Geologists Bulletin, v. 49, p. 445-459.

Jackson, G. D., and T. R. Ianelli, 1981, Rift-related cyclic sedimentation in the Neohelikian Borden Basin, northern Baffin Island, in F. H. A. Campbell, ed., Proterozoic Basins of Canada, Geological Survey Canada Paper 81-10, p. 269-302.

Jackson, M. J., M. D. Muir, and K. A. Plumb, 1987, Geology of the southern McArthur Basin: BMR Bulletin, v. 220, Bureau Mineral Resources, Canberra, Australia, 173 p.

Jackson, M. P. A., 1995, Retospective salt tectonics, in M. P. A. Jackson, D. G. Roberts, and S. Snelson, eds., Salt tectonics: A global perspective, Tulsa OK, American Association Petroleum Geologists Memoir 65, p. 1-28.

Jackson, M. P. A., D. G. Roberts, and S. Snelson, 1995, Salt tectonics: American Association Petroleum Geologists Memoir, v. 65, 454 p.

Jackson, M. P. A., and C. J. Talbot, 1994, Advances in salt tectonics, in P. L. Hancock, ed., Continental deformation, Tarrytown, NY, Pergamon Press, p. 159-179.

Jackson, M. P. A., and B. C. Vendeville, 1994, Regional extension as a geologic trigger for diapirism: Geological Society of America Bulletin, v. 106, p. 57-73.

Jackson, M. P. A., B. C. Vendeville, and D. D. Shultz-Ela, 1994a, Structural dynamics of salt systems: Annual Review of Earth and Planetary Sciences, v. 22, p. 93-117.

Jackson, M. P. A., B. C. Vendeville, and D. D. Shultz-Ela, 1994b, Salt-related structures in the Gulf of Mexico: A field guide for geophysicists: The Leading Edge, v. August 1994, p. 837-842.

Jackson, M. P. A., and C. J. Talbot, 1991, A glossary of salt tectonics: Geological Circular 91-4, Bureau of Economic Geology, University of Texas at Austin, 44 p.

Jackson, M. P. A., R. R. Cornelius, C. H. Craig, A. Gansser, J. Stocklin, and C. J. Talbot, 1990, Salt diapirs of the Great Kavir, central Iran: Memoir - Geological Society of America, v. 177, p. 139 pp.

Jackson, M. P. A., and C. Cramez, 1989, Seismic recognition of salt welds in salt tectonic regimes: CGSSEPM Found.

10th A. Res. Conf. Prog. Extended Abs., Houston, Texas, p. 72-78.

Jackson, M. P. A., and R. B. Cornelius, 1987, Stepwise centrifuge modelling of the effects of differential sediment loading on the formation of salt structures, in I. Lerche, and J. J. O'Brien, eds., Dynamical geology of salt and related structures, Orlando, Florida, Academic Press, p. 163-259.

Jackson, M. P. A., and C. J. Talbot, 1986, External shapes, strain rates, and dynamics of salt structures: Geological Society of America Bulletin, v. 97, p. 305-323.

Jacobson, G., and J. Jankowski, 1989, Groundwater-discharge processes at a central Australian playa: Journal of Hydrology, v. 105, p. 275-295.

Jacobson, G., G. C. Lau, P. S. McDonald, and J. Jankowski, 1989, Hydrogeology and groundwater resources of the Lake Amadeus and Ayers Rock region, Northern Territory: Bulletin - Bureau of Mineral Resources, Geology and Geophysics, Australia, v. 230, 77 p.

James, N. P., and A. C. Kendall, 1992, Introduction to carbonate and evaporite facies models, in R. G. Walker, and N. P. James, eds., Facies Models: Responses to sea level change, Geological Association of Canada, p. 265-275.

James, R. H., and H. Elderfield, 1996, Chemistry of the ore-forming fluids and mineral formation rates in an active hydrothermal sulphide deposit on the Mid-Atlantic ridge: Geology, v. 24, p. 1147-1150.

Jankowski, J., and G. Jacobson, 1990, Hydrochemical processes in groundwater-discharge playas, central Australia: Hydrological Processes, v. 4, p. 59-70.

Jankowski, J., and G. Jacobson, 1989, Hydrochemical evolution of regional groundwaters to playa brines in central Australia: Journal of Hydrology, v. 108, p. 123-173.

Jannasch, H. W., D. C. Nelson, and C. O. Wirsen, 1989, Massive natural occurrence of unusually large bacteria (Beggiatoa sp.) at a hydrothermal deep-sea vent site: Nature, v. 342, p. 834-836.

Jenyon, M. K., 1986, Salt Tectonics: Barking, Essex, UK, Elsevier Applied Science Publishers.

Jiang, S. Y., M. R. Palmer, Q. M. Peng, and J. H. Yang, 1997, Chemical and stable isotopic compositions of Proterozoic metamorphosed evaporites and associated

References

tourmalines from the Houxianyu borate deposit, Eastern Liaoning, China: Chemical Geology, v. 135, p. 189-211.

Jiang, S. Y., M. R. Palmer, C. J. Xue, and Y. H. Li, 1994, Halogen-rich scapolite-biotite rocks from the Tongmugou Pb-Zn Deposit, Qinling, Northwestern China - Implications for the ore-forming processes: Mineralogical Magazine, v. 58, p. 543-552.

Johnson, J. P., and K. C. Cross, 1991, Geochronological and Sm-Nd isotopic constraints on the genesis of the Olympic Dam Cu-U-Au-Ag deposit, South Australia, in M. Pagel, and J. L. Leroy, eds., Source, transport and deposition of metals, Rotterdam, Netherlands, A. A. Balkema, p. 395-400.

Jones, B. F., and R. W. Renaut, 1994, Crystal fabrics and microbiota in large pisoliths from Laguna Pastos Grandes, Boliva: Sedimentology, v. 41, p. 1171-1202.

Jones, B. F., J. S. Hanor, and W. R. Evans, 1993, Sources of dissolved salts in the central Murray Basin, Australia: Chemical Geology, v. 111, p. 135-154.

Jordan, P., T. Noack, and T. Widmer, 1990, The evaporite shear zone of the Jura Boundary Thrust; new evidence from Wisen Well (Switzerland): Eclogae Geologicae Helvetiae, v. 83, p. 525-542.

Jordan, P. T., and R. Nuesch, 1989, Deformation structures in the Muschelkalk anhydrites of the Schafisheim Well (Jura Overthrust, northern Switzerland): Eclogae Geologicae Helvetiae, v. 82, p. 429-454.

Jørgensen, B. B., M. F. Isaksen, and H. W. Jannasch, 1992, Bacterial sulfate reduction above 100°C in deep-sea hydrothermal vent sediments: Science, v. 258, p. 1756-1757.

Jowett, E. C., 1992, Role of organics and methane in sulfide ore formation, exemplified by Kupferschiefer Cu-Ag deposits, Poland: Chemical Geology, v. 99, p. 51-63.

Jowett, E. C., 1989, Effects of continental rifting on the location and genesis of stratiform copper-silver deposits: Geological Association of Canada Special Paper, v. 36, p. 53-66.

Jowett, E. C., 1987, Formation of sulphide-calcite veinlets in the Kupferschiefer in Poland by natural hydrofracturing during basin subsidence: Journal of Geology, v. 95, p. 513-526.

Jowett, E. C., A. Rydzewski, and R. J. Jowett, 1987, The Kupferschiefer Cu-Ag ore deposits in Poland: a re-appraisal of the evidence of their origin and presentation of a new genetic model: Can Journ. Earth Sci., v. 24, p. 2016-2037.

Kaldi, J., and J. Gidman, 1982, Early diagenetic dolomite cements; examples from the Permian Lower Magnesian Limestone of England and the Pleistocene carbonates of the Bahamas: Journal of Sedimentary Petrology, v. 52, p. 1073-1085.

Kamali, M. R., N. M. Lemon, and S. N. Apak, 1995, Porosity generation and reservoir potential of Ouldburra Formation carbonates, Officer Basin, South Australia: APEA Journal, v. 35, p. 106-120.

Karakitsios, V., and F. Pomonipapaioannou, 1998, Sedimentological study of the Triassic solution collapse breccias of the Ionian Zone (NW Greece): Carbonates and Evaporites, v. 13, p. 207-218.

Kargel, J. S., J. F. Schreiber, and C. P. Sonett, 1996, Mudcracks and dedolomitisation in the Wittenoom Dolomite, Hamersley Group, Western Australia: Global and Planetary Change, v. 14, p. 73-96.

Kargel, J. S., R. L. Kirk, B. Fegley, and A. H. Treiman, 1994, Carbonate sulphate on Venus: Icarus, v. 112, p. 219-252.

Kashfi, M. S., 1985, The Pre-Zagros integrity of the Iranian Platform: Journal of Petroleum Geology, v. 8, p. 353-360.

Katz, B. J., K. K. Bissada, and J. W. Wood, 1987, Factors limiting potential of evaporites as hydrocarbon source rocks (abs): American Association Petroleum Geologists Bulletin, v. 71, p. 575.

Kaufman, J., 1994, Numerical models of fluid flow in carbonate platforms - Implications for dolomitization: Journal of Sedimentary Research Section A-Sedimentary Petrology and Processes, v. 64, p. 128-139.

Kazakov, A. V., and E. I. Sokolova, 1950, Conditions of formation of fluorite in sedimentary rocks: Akad. Nauk SSSR Geologicheskiy Institut Trudy, v. 114, p. 22-64.

Keays, R. R., 1995, The role of komatiitic and picritic magmatism and S-saturation in the formation of ore deposits: Lithos, v. 34, p. 1-18.

Keith, J., D., T. J. Thompson, and S. Ivers, 1996, The Uinta emerald and the emerald-bearing potential of the Red Pine Shale, Uinta Mountains, Utah: Abstracts with Programs - Geological Society of America, v. 28, p. 85.

REFERENCES

Kempe, S., and E. T. Degens, 1985, An early soda ocean?: Chemical Geology, v. 53, p. 95-108.

Kendall, A. C., 1992, Evaporites, in N. P. James, ed., Facies Models: Responses to sea level change, Geological Association of Canada, p. 375-409.

Kendall, A. C., 1989, Brine mixing in the Middle Devonian of Western Canada and its possible significance to regional dolomitization: Sedimentary Geology, v. 64, p. 271-285.

Kendall, A. C., 1988, Aspects of evaporite basin stratigraphy, in B. C. Schreiber, ed., Evaporites and hydrocarbons, New York, Columbia University Press, p. 11-65.

Kendall, A. C., and G. M. Harwood, 1989, Shallow water gypsum in the Castile Formation - significance and implications, in P. M. Harris, and G. A. Grover, eds., Subsurface and outcrop examination of the Capitans Shelf margin, northern Delaware Basin, SEPM Core Workshop, p. 451-457.

Kendall, C. G. S. C., and J. K. Warren, 1987, A review of the origin and setting of tepees and their associated fabrics: Sedimentology, v. 34, p. 1007-1027.

Kendall, C. G. S. C., and S. P. A. D. E. Skipwith, 1969, Holocene shallow-water carbonate and evaporite sediments of Khor al Bazam, Abu Dhabi, southwest Persian Gulf: American Association Petroleum Geologists Bulletin, v. 53, p. 841-869.

Kennedy, M., 1993, The Undoolya sequence - Late Proterozoic salt-influenced deposition, Amadeus Basin, central Australia: Australian Journal of Earth Sciences, v. 40, p. 217-228.

Kennicutt II, M. C., J. M. Brooks, R. R. Bidigare, R. R. Fay, T. L. Wade, and T. J. McDonald, 1985, Vent-type taxa in a hydrocarbon seep region on the Louisiana slope: Nature, v. 317, p. 351-353.

Kent, P. E., 1987, Island salt plugs in the Middle East and their tectonic implications, in Lerch, I. and O'Brien, J. J. (eds.), Dynamical geology of salt and related structures, in I. Lerche, and J. J. O'Brien, eds., Dynamical geology of salt and related structures, New York, Academic Press, p. 3-37.

Kent, P. E., 1979, The emergent Hormuz salt plugs of southern Iran: Journal of Petroleum Geology, v. 2, p. 117-144.

Kesler, S. E., H. D. Jones, F. C. Furman, R. Sassen, W. H. Anderson, and J. R. Kyle, 1994, Role of crude oil in the genesis of Mississippi Valley-Type deposits; evidence from the Cincinnati Arch: Geology, v. 22, p. 609-612.

Kesler, S. E., J. Ruiz, and L. M. Jones, 1988, Strontium isotope geochemistry of the Galeana Barite District, Nuevo Leon, Mexico: Economic Geology, v. 83, p. 1907-1917.

Kezao, C., and J. M. Bowler, 1985, Preliminary study on sedimentary characteristics and evolution of paleoclimate of Qarhan Salt Lake, Qaidam Basin: Scientia Sinica (series B), v. 28, p. 1218-1232.

Kharaka, Y. K., W. W. Carothers, L. M. Law, P. J. Lamonthe, and T. L. Fries, 1987, Geochemistry of metal-rich brines from the central Mississippi Salt Dome Basin, USA: Applied Geochem., v. 2, p. 543-562.

Kinsman, D. J., 1976, Evaporites; relative humidity control of primary mineral facies: Journal of Sedimentary Petrology, v. 46, p. 273-279.

Kinsman, D. J., 1975a, Salt floors to geosynclines: Nature, v. 255, p. 375-378.

Kinsman, D. J., 1975b, Rift valley basins and sedimentary history of trailing continental margins, in A. G. Fisher, and S. Judson, eds., Petroleum and plate tectonics, Princeton, NJ, Princeton University Press, p. 83-126.

Kirkham, R. V., 1989, Distribution, settings, and genesis of sediment-hosted stratiform copper deposits: Geological Association of Canada Special Paper, v. 36, p. 3-38.

Kirkland, D. W., R. E. Denison, and R. Evans, 1995, Middle Jurassic Todilto Formation of northern New Mexico, v. 147, New Mexico Bureau of Mines Bulletin, 39 p.

Kirkland, D. W., and R. Evans, 1981, Source-rock potential of evaporitic environment: American Association of Petroleum Geologists Bulletin, v. 65, p. 181-190.

Kisler, R. B., and W. C. Smith, 1983, Boron and borates, in S. J. Lefond, ed., Industrial Minerals and Rocks, New York, AIME, p. 533-560.

Ko, S. C., D. L. Olgaard, and T. F. Wong, 1997, Generation and maintenance of pore pressure excess in a dehydrating system. 1. Experimental and microstructural observations: Journal of Geophysical Research-Solid Earth, v. 102, p. 825-839.

REFERENCES

Ko, S. C., D. L. Olgaard, and B. Ueli, 1995, The transition from weakening to strengthening in dehydrating gypsum: Evolution of excess pore pressures: Geophys. Res. Lett., v. 22, p. 1009-1012.

Koehler, G., T. K. Kyser, R. Enkin, and E. Irving, 1997, Paleomagnetic and isotopic evidence for the diagenesis and alteration of evaporites in the Paleozoic Elk Point Basin, Saskatchewan, Canada: Canadian Journal of Earth Sciences, v. 34, p. 1619-1629.

Kontak, D. J., 1998, A study of fluid inclusions in sulphide and nonsulphide mineral phases from a carbonate-hosted Zn-Pb deposit, Gays River, Nova Scotia, Canada: Economic Geology, v. 93, p. 793-816.

Kontak, D. J., and S. Jackson, 1995, Laser-ablation ICP-MS microanalysis of calcite cement from a Mississippi-Valley type Zn-Pb deposit, Nova Scotia - Dramatic variability in REE content on macro- and micro-scales: Canadian Mineralogist, v. 33, p. 445-467.

Kontak, D. J., E. Farrar, and S. L. McBride, 1994, Ar40-Ar39 dating of fluid migration in a Mississsppi-Valley type deposit - The Gays River Zn-Pb deposit, Nova Scotia, Canada: Economic Geology, v. 89, p. 1501-1517.

Kontorovich, A. E., A. L. Pavlov, G. A. Tretyakov, and A. V. Khomenko, 1996, Physicochemical modelling of thermodynamic equilibria in a carbonate-evaporite sedimentary rock water-hydrocarbon system during contact metamorphism and catagenesis (Russian): Geokhimiya, v. 7, p. 598-610.

Kotwicki, V., 1986, The floods of Lake Eyre: Adelaide, Engineering and Water Supply Dept., 99 p.

Kovalevich, V. M., T. M. Peryt, and O. I. Petrichenko, 1998, Secular variation in seawater chemistry during the Phanerozoic as indicated by brine inclusions in halite: Journal of Geology, v. 106, p. 695-712.

Koyi, H., 1991, Gravity overturns, extension, and basement fault activation: Journal of Petroleum Geology, v. 12, p. 117-242.

Koyi, H., 1988, Experimental modeling of role of gravity and lateral shortening in Zagros mountain belt: American Association of Petroleum Geologists Bulletin, v. 72, p. 1381-1394.

Kribek, B., 1991, Metallogeny, structural, lithological and time controls of ore deposition in anoxic environments: Mineralium Deposita, v. 26, p. 122-131.

Kribek, B., J. Hladikova, K. Zak, J. Bendl, M. Pudilova, and Z. Uhlik, 1996, Barite-hyalophane sulfidic ores at Rozna, Bohemian Massif, Czech Republic metamorphosed black shale-hosted submarine exhalative mineralisation: Economic Geology, v. 91.

Krouse, H. R., C. A. Vian, L. S. Eliuk, A. Ueda, and S. Halas, 1988, Chemical and isotopic evidence fo thermochemical sulphate reduction by light hydrocarbon gases in deep carbonate reservoirs: Nature, v. 333, p. 415-419.

Krupp, R., T. Oberthür, and W. Hirdes, 1994, The early Precambrian atmosphere and hydrosphere: Thermodynamic constraints from mineral deposits: Economic Geology, v. 89, p. 1581-1598.

Kucha, H., 1990, Geochemistry of the Kupferschiefer, Poland: Geologische Rundschau, v. 79, p. 387-399.

Kucha, H., W. Prohaska, and E. F. Stumpfl, 1995, Deposition and transport of gold by thiosulphates, Veitsch, Austria: Mineralogical Magazine, v. 59, p. 253-258.

Kuehn, R., and K. J. Hsu, 1978, Chemistry of halite and potash salt cores, DSDP sites 374 and 376, Leg 42A, Mediterranean Sea;: Initial Reports of the Deep Sea Drilling Project, v. 42a, p. 613-619.

Kulke, H., 1978, Tektonik und Petrographie einer Salinarformation am Beispiel der Trias des Atlassystems (NW-Afrika), in W. Zeil, ed., Geotektonische Forschungen, p. 1-158.

Kunasz, I. A., 1983, Lithium raw materials, in S. J. Lefond, ed., Industrial minerals and rocks, New York, AIME, p. 869-880.

Kuroda, H., 1983, Geologic characteristics and formation environments of the Furutobe and Matsuki kuroko deposits, Akita Prefecture, Northeast Japan: Economic Geology Monographs, v. 5, p. 149-156.

Kusakabe, M., and H. Chiba, 1983, Oxygen and sulfur isotope composition of barite and anhydrite from the Fukazawa Deposit, Japan: Economic Geology Monographs, v. 5, p. 292-301.

Kushnir, J., 1982, The partitioning of seawater cations during the transformation of gypsum to anhydrite: Geochimica et Cosmochimica Acta, v. 46, p. 433-446.

Kushnir, S. V., 1986, The epigenetic celestite formation mechanism for rocks containing $CaSO_4$: Geochemistry International, v. 23, p. 1-9.

REFERENCES

Kwak, T. A. P., 1977, Scapolite compositional change in a metamorphic gradient and its bearing on the identification of meta-evaporite sequences: Geological Magazine, v. 114, p. 343-354.

Kyle, J. R., 1991, Evaporites, evaporitic processes and mineral resources., in J. L. Melvin, ed., Evaporites, petroleum and mineral resources: Developments in Sedimentology, Amsterdam, Elsevier, p. 477-533.

Kyle, J. R., and H. H. Posey, 1991, Halokinesis, Cap rock Development and salt dome mineral resources, in J. L. Melvin, ed., Evaporites, petroleum and mineral resources: Developments in Sedimentology, Amsterdam, Elsevier, p. 413-474.

Kyle, J. R., and P. E. Price, 1986, Metallic sulphide mineralization in salt-dome cap rocks, Gulf Coast, U.S.A: Institution of Mining and Metallurgy. Transactions, Section B: Applied Earth Sciences, v. 95, p. B6-B16.

Land, L. S., 1995a, Na-Ca-Cl saline formation waters, Frio Formation (Oligocene), south Texas, USA: Products of diagenesis: Geochimica et Cosmochimica Acta, v. 59, p. 2163-2174.

Land, L. S., 1995b, The role of saline formation water in crustal cycling: Aquatic Geochemistry, v. 1, p. 137-145.

Land, L. S., 1985, The origin of massive dolomite: Journal of Geological Education, v. 33, p. 112-125.

Land, L. S., R. A. Eustice, L. E. Mack, and J. Horita, 1995, Reactivity of evaporites during burial - An example from the Jurassic of Alabama: Geochimica et Cosmochimica Acta, v. 59, p. 3765-3778.

Land, L. S., and D. R. Prezbindowski, 1981, The origin and evolution of saline formation water, Lower Cretaceous carbonates, South-central Texas, U.S.A: J. Hydrol., v. 54, p. 51-74.

Landes, K. K., 1945, The Mackinac breccia: Michigan Geol. Surv. Div. Publ., v. 44, p. 121-154.

Lange, I. M., and R. C. Murray, 1977, Evaporite brine reflux as a mechanism for moving deep warm brines upward in the formation of Mississippi valley-type base metal deposits: Economic Geology, v. 72, p. 107-109.

Large, D. J., and A. P. Gize, 1996, Pristane/phytane ratios in the mineralized Kupferschiefer of the Fore-Sudetic Monocline, southwest Poland: Ore Geology Reviews, v. 11, p. 89-103.

Large, D. J., J. Macquaker, D. J. Vaughan, Z. Sawlowicz, and A. P. Gize, 1995, Evidence for low temperature alteration of sulfides in the Kupferschiefer copper deposits of southwestern Poland: Economic Geology, v. 90, p. 2143 - 2155.

Large, R. R., and P. J. McGoldrick, 1998, Lithogeochemical halos and geochemical vectors to stratiform sediment hosted Zn-Pb-Ag deposits,1 - Lady Loretta deposit, Queensland: Journal of Geochemical Exploration, v. 63, p. 37-56.

Large, R. R., D. L. Huston, P. J. McGoldrick, and P. A. Ruxton, 1989, Gold distribution and genesis in Australian volcanogenic massive sulphide deposits and their significance in gold transport models: Economic Geology Monographs, v. 6, p. 520-535.

Larsen, D., 1994, Origin and paleoenvironmental significance of calcite pseudomorphs after ikaite in the Oligocene Creede Formation, Colorado: Journal of Sedimentary Research A: Sedimentary Petrology and Processes, v. A64, p. 593-603.

Last, W. M., 1994, Deep-water evaporite mineral formation in lakes of Western Canada, in R. W. Renaut, and W. M. Last, eds., Sedimentology and geochemistry of modern and ancient saline lakes, SEPM/Society for Sedimentary Geology; Special Publication, p. 51-59.

Last, W. M., 1992, Petrology of carbonate hardgrounds from East Basin Lake, a saline maar lake, southern Australia: Sedimentary Geology, v. 81, p. 215-229.

Last, W. M., 1989, Sedimentology of a saline playa in the northern Great Plains, Canada: Sedimentology, v. 36, p. 109-123.

Laube, N., H. E. Frimmel, and S. Hoernes, 1995, Oxygen and carbon isotopic study on the genesis of the Steirischer Erzberg siderite deposit (Austria): Mineralium Deposita, v. 30, p. 285-293.

Lavoie, D., D. F. Sangster, M. M. Savard, and F. F., 1998, Breccias in the lower part of the Mississippian Windsor Group and their relation to Pb-Zn mineralisation: a summary: Economic Geology, v. 93, p. 734-735.

Laznicka, P., 1988, Breccias and coarse fragmentites: petrology, environments, associations, ores: Developments in Economic Geology, v. 25: New York, Elsevier, 832 p.

Leake, B. E., C. M. Farrow, and R. Townend, 1979, A pre-2,000 Myr old granulite facies metamorphosed evaporite

References

from Caraiba, Brazil?: Nature, v. 277, p. 49-50.

Leary, D. A., and J. N. Vogt, 1986, Diagenesis of the San Andres Formation (Guadalupian), Central Basin Platform, Permian Basin, in D. G. Bebout, Harris, P. M., ed., Hydrocarbon Reservoir Studies San Andres/Grayburg Formations, Permian Basin, SEPM Special Publ. 26, p. 67-68.

Lebedev, L. M., 1972, Minerals of contemporary hydrotherms of Cheleken: Geochem. Inter., v. 9, p. 485-504.

Lee, M. R., 1995, Calcite concretions in carbonate rocks of the late Permian Raisby Formation, north-east England: Proceedings - Yorkshire Geological Society, v. 50, p. 245-253.

Lee, M. R., 1994, Emplacement and diagenesis of gypsum and anhydrite in the late Permian Raisby Formation, north-east England: Proceedings - Yorkshire Geological Society, v. 50, p. 143-155.

Lee, M. R., and G. M. Harwood, 1989, Dolomite calcitization and cement zonation related to uplift of the Raisby Formation (Zechstein carbonate), Northeast England: Sedimentary Geology, v. 65, p. 285-305.

Lefond, S. J., 1983, Industrial Minerals and rocks: New York, AIME.

Lemon, N. M., 1988, Diapir recognition and modelling with examples from the Late Proterozoic Adelaide Geosyncline, Central Flinders Ranges, South Australia: Doctoral thesis, Univesity of Adelaide.

Lemon, N. M., 1985, Physical Modelling of Sedimentation Adjacent to Diapirs and Comparison with Late Precambrian Oratunga Breccia Body in Central Flinders Ranges, South Australia: American Association Petroleum Geologists Bulletin, v. 69, p. 1327 - 1328.

Lerche, I. and K. Petersen, 1995, Salt and Sediment Dynamics, CRC Press, New York, 322 pp.

Lerman, A., 1970, Chemical equilibria and evolution of chloride brines: Special Paper - Mineralogical Society of America, Fiftieth Anniversary Symposia, Mineralogy and Geochemistry of Non-Marine Evaporites, v. 3, p. 291-306.

Letouzey, J., B. Colletta, R. Vially, and J. C. Chermette, 1995, Evolution of salt-related structures in compressional settings, in M. P. A. Jackson, D. G. Roberts, and S. Snelson, eds., Salt tectonics: a global perspective, American Association Petroleum Geologists Memoir 65, p. 41-60.

Letouzey, J., P. Werner, and Y. Marty, 1990, Fault reactivation and structural inversion; backarc and intraplate compressive deformations; example of the eastern Sunda Shelf (Indonesia): Tectonophysics, v. 183, p. 341-362.

Lewis, S., and M. Holness, 1996, Equilibrium halite-H_2O dihedral angles: High rock salt permeability in the shallow crust: Geology, v. 24, p. 431-434.

Li, J. R., T. K. Lowenstein, and I. R. Blackburn, 1997, Responses of evaporite mineralogy to inflow water sources and climate during the past 100 Ky in Death Valley California: Geological Society of America Bulletin, v. 109, p. 1361-1371.

Light, M. P. R., and H. H. Posey, 1992, Diagenesis and its relation to mineralization and hydrocarbon reservoir development; Gulf Coast and North Sea basins, in Chilingar, K. H., G. V. Wolf, (eds), Diagenesis, III, Elsevier, Amsterdam-Oxford-New York, Netherlands, p. 511-541.

Lightfoot, P. C., R. R. Keays, G. G. Morrison, A. Bite, and K. P. Farrell, 1997, Geochemical relationships in the Sudbury igneous complex - Origin of the main mass and offset dykes: Economic Geology, v. 92, p. 289-307.

Lindblom, S., C. Broman, and O. Martinsson, 1996, Magmatic-hydrothermal fluids in the Pahtohavare Cu-Au deposit in greenstone at Kiruna, Sweden: Mineralium Deposita, v. 31, p. 307-318.

Lindsay, J. F., 1987, Upper Proterozoic evaporites in the Amadeus Basin, central Australia, and their role in basin tectonics: Geological Society of America Bulletin, v. 99, p. 852-865.

Linn, K. O., and S. S. Adams, 1966, Barren halite zones in potash deposits, Carlsbad, New Mexico: Second Symposium on Salt, p. 59-68.

Liro, L. M., 1992, Distribution of shallow salt structures, lower slope of the northern Gulf of Mexico, USA: Marine and Petroleum Geology, v. 9, p. 433-451.

Lock, D. E., 1986, The formation of modern epsomite deposits near Lake Eyre, and their significance for Early-Cainozoic weathering in central Australia: Sediments down under. 12th International Sedimentological Congress, sponsored by the International Association of Sedimentologists, Bureau of Mineral Resources, Geology and Geophysics, Geological Society of Australia and the Geological Society of New Zealand, Canberra, Australia,

REFERENCES

24-30 August, 1986. Abstracts. Canberra: International Sedimentological Congress., p. 189.

Logan, B. W., 1987, The MacLeod Evaporite Basin, Western Australia: Tulsa, OK, American Association of Petroleum Geologists Memoir 44, 140 p.

Logan, G. A., J. M. Hayes, G. B. Hieshima, and R. E. Summons, 1995, Terminal Proterozoic reorganisation of biogeochemical cycles: Nature, v. 376, p. 53-56.

Logan, R. G., W. J. Murray, and N. Williams, 1990, HYC silver-lead-zinc deposit, McArthur River, in F. E. Hughes, ed., Geology of the mineral deposits of Australia and Papua New Guinea, Monograph Series - Australasian Institute of Mining and Metallurgy no. 14, p. 907-911.

Lonsdale, P., and K. Becker, 1985, Hydrothermal plumes, hot springs and conductive heat flow in the southern trough of the Guaymas Basin: Earth Planet. Sci. Lett., v. 73, p. 211-225.

Loope, D. B., 1984, Discussion: Origin of extensive bedding planes in aeolian sandstones: a defence of Stoke's hypothesis: Sedimentology, v. 31, p. 123-125.

Lorenz, J. C., 1988, Synthesis of late Paleozoic and Triassic redbed sedimentation in Morocco, in V. H. Jacobshagen, ed., The Atlas system of Morocco; studies on its geodynamic evolution, Berlin, Springer-Verlag, Berlin-Heidelberg-New York, Lecture Notes in Earth Sciences, p. 139-168.

Loucks, R. G., and J. F. Sarg, 1993, Carbonate sequence stratigraphy : recent developments and applications, American Association of Petroleum Geologists Memoir 57: Tulsa, Oklahoma, U.S.A.

Loucks, R. G., and M. W. Longman, 1982, Lower Cretaceous Ferry Lake Anhydrite, Fairway Field, East Texas; product of shallow-subtidal deposition, in C. R. Handford, R. G. Loucks, and G. R. Davies, eds., Depositional and diagenetic spectra of evaporites; Core workshop no. 3., Tulsa, OK, SEPM, p. 130-173.

Lavoie, D., D. F. Sangster, M. M. Savard, and F. F., 1998, Breccias in the lower part of the Mississippian Windsor Group and their relation to Pb-Zn mineralisation: a summary: Economic Geology, v. 93, p. 734-735.

Lowe, D. R., 1983, Restricted shallow-water sedimentation of early Archean stromatolitic and evaporitic strata of the Strelley Pool Chert, Pilbara Block, Western Australia: Precambrian Research, v. 19, p. 239-283.

Lowenstein, T. K., 1987, Origin of depositional cycles in a Permian ''saline giant''; the Salado (McNutt Zone) evaporites of New Mexico and Texas: Geological Society of America Bulletin, v. 100, p. 592-608.

Lowenstein, T. K., and R. J. Spencer, 1990, Syndepositional origin of potash evaporites; petrographic and fluid inclusion evidence: American Journal of Science, v. 290, p. 43-106.

Lowenstein, T. K., R. J. Spencer, and P. Zhang, 1989, Origin of ancient potash evaporites; clues from the modern Qaidam Basin, western China: Science, v. 245, p. 1090-1092.

Lowenstein, T. K., and L. A. Hardie, 1985, Criteria for the recognition of salt-pan evaporites: Sedimentology, v. 32, p. 627-644.

Lu, F. H., and W. J. Meyers, 1998, Massive dolomitisation of a Late Miocene carbonate platform - A case of mixed evaporative brines with meteoric water, Nijar, Spain: Sedimentology, v. 45, p. 263-277.

Lucia, F. J., 1972, Recognition of evaporite-carbonate shoreline sedimentation, in J. K. Rigby, and W. K. Hamblin, eds., Recognition of ancient sedimentary environments, Soc. Econ. Paleontol. Mineral., Spec. Publ., p. 190-191.

Lucia, F. J., 1961, Dedolomitization in the Tansill (Permian) formation: Geol. Soc. America Bull., v. 72, p. 1107-1109.

Lydon, J. W., 1996, Sedimentary exhalative sulphides (sedex), in O. R. Eckstrand, W. D. Sinclair, and R. I. Thorpe, eds., Geology of Canadian Mineral Deposit Types, Geological Survey of Canada, Geology of Canada, p. 130-152.

Lydon, J. W., 1988, Volcanogenic Massive Sulphide Deposits Part 2: Genetic Models: Geoscience Canada, v. 15, p. 43-65.

Lynch, G., and J. V. A. Keller, 1998, Association between detachment faulting and salt diapirs in the Devonian-Carboniferous Maritimes Basin, Atlantic Canada: Bulletin of Canadian Petroleum Geology, v. 46, p. 189-209.

Lynch, G., J. V. A. Keller, and P. S. Giles, 1998, Influence of the Ainslie Detachment on the stratigraphy of the Maritimes Basin and mineralisation in the Windsor Group of Northern Nova Scotia, Canada: Economic Geology, v. 93, p. 703-718.

Lyons, W. B., S. W. Tyler, H. E. Gaudette, and D. T. Long, 1995, The use of strontium isotopes in determining

REFERENCES

groundwater mixing and brine fingering in a playa spring zone, Lake Tyrrell, Australia: Journal of Hydrology, v. 167, p. 225-239.

Lyons, W. B., S. Welch, D. T. Long, M. E. Hines, A. M. Giblin, A. E. Carey, P. G. Macumber, R. M. Lent, and A. L. Herczeg, 1992, The trace-metal geochemistry of the Lake Tyrrell system brines (Victoria, Australia): Chemical Geology, v. 96, p. 115-132.

MacDonald, I. R., 1992, Sea-floor brine pools affect behavior, mortality, and preservation of fishes in the Gulf of Mexico: lagerstatten in the making?: Palaios, v. 7, p. 383-387.

Machel, H. G., 1993, Anhydrite nodules formed during deep burial: Journal of Sedimentary Petrology, v. 63, p. 659-662.

Machel, H. G., H. R. Krouse, and R. Sassen, 1995, Products and distinguishing criteria of bacterial and thermochemical sulfate reduction: Applied Geochemistry, v. 10, p. 373-389.

Machel, H. G., and E. A. Burton, 1991, Burial-diagenetic sabkha-like gypsum and anhydrite nodules: Chemical Geology, v. 90, p. 211-231.

Macumber, P. G., 1991, Interaction between groundwater and surface systems in northern Victoria, Victoria Department of Conservation and Environment, Report, 345 p.

Magara, K., 1986, Geological models of petroleum entrapment: London, United Kingdom, Elsevier Appl. Sci. Publ.

Magee, J. W., J. M. Bowler, G. H. Miller, and D. L. G. Williams, 1995, Stratigraphy, sedimentology, chronology and palaeohydrology of Quaternary lacustrine deposits at Madigan Gulf, Lake Eyre, South Australia: Palaeogeography Palaeoclimatology Palaeoecology, v. 113, p. 3-42.

Maiklem, W. R., 1971, Evaporative drawdown - a mechanism for water level lowering and diagenesis in the Elk Point Basin: Bull. Canadian Assoc. Petrol. Geol., v. 19, p. 487-503.

Maisonneuve, J., 1982, The composition of the Precambrian ocean waters: Sedimentary Geology, v. 31, p. 1-11.

Maliva, R. G., 1987, Quartz geodes; early diagenetic silicified anhydrite nodules related to dolomitization: Journal of Sedimentary Petrology, v. 57, p. 1054-1059.

Mann, A. W., and R. L. Deutscher, 1978, Genesis principles for the precipitation of carnotite in calcrete drainages in Western Australia: Economic Geology, v. 73, p. 1724-1737.

Markl, G., and S. Piazolo, 1998, Halogen-bearing minerals in syenites and high-grade marbles of Dronning Maud Land, Antarctica - Monitors of fluid compositional changes during late-magmatic fluid-rock interaction processes: Contributions to Mineralogy and Petrology, v. 132, p. 246-248.

Martens, C. S., J. P. Chanton, and C. K. Paull, 1991, Biogenic methane from abyssal brine seeps at the base of the Florida Escarpment: Geology, v. 19, p. 851-854.

Martin, J. M., J. C. Braga, C. Betzler, and T. Brachert, 1996, Sedimentary model and high-frequency cyclicity in a Mediterranean, shallow-shelf, temperate-carbonate environment (uppermost Miocene, Agua Amarga Basin, southern Spain: Sedimentology, v. 43, p. 263-277.

Matsubara, S., A. Kato, and E. Hashimoto, 1992, Celestine from the Asaka gypsum mine, Koriyama City, Fukushima Prefecture, Japan: Min. J. (Japan), v. 16, p. 16-20.

Matsumoto, R., 1989, Isotopic composition of siderite derived from the decomposition of methane hydrate: Geology, v. 17, p. 707-710.

Mattes, B. W., and M. S. Conway, 1990, Carbonate/evaporite deposition in the Late Precambrian-Early Cambrian Ara Formation of southern Oman, in A. H. F. Robertson, M. P. Searle, and A. C. Ries, eds., The geology and tectonics of the Oman region, Geological Society Special Publication No. 49, p. 617-636.

Matthai, S. K., and R. W. Henley, 1996, Geochemistry and depositional environment of the gold-mineralised Proterozoic Koolpin Formation, Pine Creek Inlier, Northern Australia, Northern Australia - A comparison with modern shale sequences: Precambrian Research, v. 78, p. 211-235.

Mazzoli, S., 1994, Early deformation features in syn-orogenic Messinian sediments of the northern Marchean Apennines (Italy): Annales Tectonicae, v. 8, p. 134-147.

Mazzullo, S. J., and B. A. Birdwell, 1989, Syngenetic formation of grainstones and pisolites from fenestral carbonates in peritidal settings: Journal of Sedimentary Petrology, v. 59, p. 605-611.

McBride, E. F., L. S. Land, and L. E. Mack, 1987, Diagenesis of aeolian and fluvial feldspathic sandstones, Norphlet

REFERENCES

Formation (Upper Jurassic), Rankin County, Mississippi, and Mobile County, Alabama: American Association Petroleum Geologists Bulletin, v. 71, p. 1019-1034.

McCaffrey, M. A., B. Lazar, and H. D. Holland, 1987, The evaporation path of seawater and the coprecipitation of Br⁻ and K⁺ with halite: Journal of Sedimentary Petrology, v. 57, p. 928-937.

McClay, K. R., 1990, Deformation mechanics in analogue models of extensional fault systems, in R. J. R. Knipe, ed., Deformation mechanisms, rheology and tectonics, London, United Kingdom, Geological Society of London, p. 445-453.

McCord, T. B., G. B. Hansen, F. P. Fanale, R. Carlson, W. D. L. Matson, T. V. Johnson, W. D. Smythe, J. K. Crowley, P. D. Martin, A. Ocampo, C. A. Hibbitts, and J. C. Granahan, 1998, Salts on Europa's surface detected by Galileo's near infrared mapping Spectrometer: Science, v. 280, p. 1242-1245.

McGuinness, D. B., and J. R. Hossack, 1993, The development of allochthonous salt sheets as controlled by the rates of extension, sedimentation, and salt supply, in J. M. Armentrout, R. Bloch, H. C. Olson, and B. F. Perkins, eds., Rates of Geologic Processes, Gulf Coast Section SEPM Foundation, 14th Annual Research Conference, Houston, Texas, p. 127-139.

McIntosh, R. A., and N. C. Wardlaw, 1968, Barren halite bodies in the sylvinite mining zone at Esterhazy, Saskatchewan: Canadian Journal Earth Sciences, v. 5, p. 1221-1238.

McKibben, M. A., J. P. Andes Jr, and A. E. Williams, 1988a, Active ore formation at a brine interface in metamorphosed deltaic lacustrine sediments: the Salton Sea geothermal system, California: Economic Geology, v. 83, p. 511-523.

McKibben, M. A., A. E. Williams, and S. Okubo, 1988b, Metamorphosed Plio-Pleistocene evaporites and the origins of hypersaline brines in the Salton Sea geothermal system, California; fluid inclusion evidence: Geochimica et Cosmochimica Acta, v. 52, p. 1047-1056.

McManus, K. M., and J. S. Hanor, 1993, Diagenetic evidence for massive evaporite dissolution, fluid flow, and mass transfer in the Louisiana Gulf Coast: Geology, v. 21, p. 727-730.

McNaughton, K., and T. E. Smith, 1986, A fluid inclusion study of sphalerite and dolomite from the Nanisivik lead-

zinc deposit, Baffin Island, Northwest Territories, Canada: Economic Geology, v. 81, p. 713-720.

McPhie, J., 1993, The Tennant Creek Porphyry revisited - A synsedimentary sill with peperite margins, Early Proterozoic, Northern Territory: Australian Journal of Earth Sciences, v. 40, p. 545-558.

McWhae, J. R. H., 1953, The Carboniferous breccia of Bellefjordan: Geol. Mag., v. 90, p. 287-298.

Mees, F., 1998, The alteration of glauberite in lacustrine deposits of the Taoudenni-Agorgott basin, northern Mali: Sedimentary Geology, v. 117, p. 193-205.

Meister, F. M., and N. Aurich, 1972, Geologic outline and oil fields of Sergipe Basin, Brazil: Americal Assoc. Petrol. Geol., v. 56, p. 1034-1047.

Mello, U. T., G. D. Karner, and R. Anderson, 1995, Role of salt in restraining the maturation of sub-salt source rocks: Marine and Petroleum Geology, v. 12, p. 697-716.

Mello, U. T., R. N. Anderson, and G. D. Karner, 1994, Salt restrains maturation in subsalt play: Oil and Gas Journal, v. 92, p. 101-102,104-107.

Melvin, J. L. (ed), 1991, Evaporites, petroleum and mineral resources: Developments in sedimentology: Amsterdam, Elsevier.

Menduina, J., S. Ordonez, and M. A. Garcia del Cura, 1984, Geologia del yacimiento de glauberita de Cerezo del Rio Tiron (Provincia de Burgos): Boletin Geologico y Minero, Instituto Geologico y Minero de Espana, v. 95, p. 33-51.

Mernagh, T. P., C. A. Heinrich, J. F. Leckie, D. P. Carville, D. J. Gilbert, R. K. Valenta, and L. A. I. Wyborn, 1994, Chemistry of low-temperature hydrothermal gold, platinum, and palladium (+/- uranium) mineralization at Coronation Hill, Northern Territory, Australia: Economic Geology, v. 89, p. 1053 -1073.

Meyers, W. J., F. H. Lu, and J. K. Zachariah, 1997, Dolomitization by mixed evaporitic brines and freshwater, Upper Miocene carbonates, Nijar, Spain: Journal of Sedimentary Research Section A-Sedimentary Petrology and Processes, v. 67 A, p. 898-912.

Miall, A. D., 1997, The geology of stratigraphic sequences: Berlin, Springer-Verlag, 433 p.

Michalzik, D., 1996, Lithofacies, diagenetic spectra and sedimentary cycles of Messinian (Late Miocene) evaporites in SE Spain: Sedimentary Geology, v. 106, p. 203-222.

Middleton, G. V., 1961, Evaporite solution breccias from the Mississippian of southwest Montana: Jour. Sed. Petrology, v. 31, p. 189-195.

Middleton, H. A., and C. S. Nelson, 1996, Origin and timing of siderite and calcite concretions in Late Palaeocene non- to marginal-marine facies of the Tekuiti Group, New Zealand: Sedimentary Geology, v. 102, p. 93-115.

Middleton, K., M. Coniglio, R. Sherlock, and S. K. Frape, 1993, Dolomitization of Middle Ordovician carbonate reservoirs, southwestern Ontario: Bulletin of Canadian Petroleum Geology, v. 41, p. 150-163.

Milliken, K. L., 1979, The silicified evaporite syndrome; two aspects of silicification history of former evaporite nodules from southern Kentucky and northern Tennessee: J. Sediment. Petrol., v. 49, p. 245-256.

Mills, R., and H. Elderfield, 1995, Rare earth element geochemistry of hydrothermal deposits from the active TAG mound, 26 degrees N Mid-Atlantic Ridge: Geochimica et Cosmochimica Acta, v. 59, p. 3511-3524.

Milton, C., 1989, Oldoinyo Lengai natrocarbonatite lava: its history (abs.): 28th International Geological Congress, Washington 1989, p. 2-441.

Milton, C., 1968, The "Natro-carbonatite Lava" of Oldoinyo Lengai, Tanzania: Geol. Soc. Amer. Program with Abstracts, p. 202.

Moine, B., P. Sauvan, and J. Jarousse, 1981, Geochemistry of evaporite-bearing series; a tentative guide for the identification of meta-evaporites: Contributions to Mineralogy and Petrology, v. 76, p. 401-412.

Monnin, C., and C. Ramboz, 1996, The anhydrite saturation index of the ponded brines and sediment pore waters of the Red Sea deeps: Chemical Geology, v. 127, p. 141-159.

Montacer, M., J. R. Disnar, J. J. Orgeval, and J. Trichet, 1988, Relationship between Zn-Pb ore and oil accumulation processes; example of the Bou Grine Deposit (Tunisia): Organic Geochemistry, v. 13, p. 423-431.

Montgomery, S. L., and D. Moore, 1997, Subsalt play, Gulf of Mexico - A review: American Association Petroleum Geologists Bulletin, v. 81, p. 871-896.

Moore, C. H., A. Chowdhury, and L. Chan, 1988, Upper Jurassic platform dolomitization, northwestern Gulf of Mexico; a tale of two waters, in V. Skula, and P. A. Baker, eds., Sedimentology and geochemistry of dolostones, Tulsa, Okl., SEPM Special Publication No 43., p. 175-189.

Moore, C. M., 1989, Carbonate diagenesis and porosity: Developments in sedimentology, v. 46: Amsterdam ; New York, Elsevier Science Pub. Co.

Mora, C. I., and J. W. Valley, 1989, Halogen-rich scapolite and biotite: implications for metamorphic fluid-rock interaction: American Mineralogist, v. 74, p. 721-737.

Moritz, R., L. Fontbote, J. Spangenberg, S. Rosas, Z. Sharp, and D. Fontignie, 1996, Sr, C and O isotope systematics in the Pucara Basin, central Peru: Comparison between Mississippi Valley-type deposits and barren areas: Mineralium Deposita, v. 31, p. 147-162.

Morley, C. K., and G. Guerin, 1996, Comparison of gravity-driven deformation styles and behaviour associated with mobile shales and salt: Tectonics, v. 15, p. 1154-1170.

Morrison, S. J., and W. T. Parry, 1986, Formation of carbonate- sulfate veins associated with copper ore deposits from saline basin brines, Lisbon Valley, Utah: fluid inclusion and isotopic evidence (USA): Economic Geology, v. 81, p. 1853-1866.

Morrow, D. W., 1982, Descriptive field classification of sedimentary and diagenetic breccia fabrics in carbonate rocks: Bulletin of Canadian Petroleum Geology, v. 30, p. 227-229.

Morteani, G., P. Möller, and F. Schley, 1982, The rare earth element contents and the origin of the sparry magnesite mineralisations of Tux-Lanersbach, Entachen Alm, Spiessnägel, and Hochfilsen, Austria, and the lacustrine magnesite deposits of Aiani-Kozani, Greece, and Bela Stena, Yugoslavia: Economic Geology, v. 77, p. 617-631.

Morteani, G., and H. Neugebauer, 1990, Chemical and tectonic controls on the formation of sparry magnesite deposits - the deposits of the northern Greywacke Zone (Austria): Geologische Rundschau, v. 79, p. 337-344.

Mossman, D. J., and M. J. Brown, 1986, Stratiform barite in sabkha sediments, Walton-Cheverie, Nova Scotia: Economic Geology, v. 81, p. 2016-2021.

Mossman, D. J., R. N. Delabio, and A. D. Mackintosh, 1982, Mineralogy of clay marker seams in some Saskatchewan potash mines: Canadian Journal of Earth Sciences, v. 19, p. 2126-2140.

REFERENCES

Mostler, H., 1984, An jungpalaozoischen Karst gebundene Vererzungen mit einem Beitrag zur Genese de Siderite des Steinischen Erzberges: Geol. Palaont. Mitt. Innsbruck, v. 13, p. 97-111.

Mount, T., 1975, Diapirs and diapirism in the Adelaide 'Geosyncline'': Doctoral thesis, University of Adelaide.

Mouret, C., 1994, Geological history of NE Thailand since the Carboniferous: Relations with Indochina and the Carboniferous to early Cenozoic evolution model: Proc. Internat. Sympos. on Stratigraphic Correlation of Southeast Asia, Bangkok, p. 132-158.

Muffler, L. J., and D. E. White, 1969, Active metamorphism of upper Cenozoic sediments in the Salton Sea geothermal fields and the Salton Trough, southeastern California: Geol. Soc. Am. Bull., v. 80, p. 157-182.

Muir, M. D., 1987, Facies models for Australian Precambrian evaporites: Peryt, Tadeusz M. Evaporite basins. Inst. Geol., Warsaw, Poland. Lecture Notes in Earth Sciences, v. 13, p. 5-21.

Muir, M. D., 1983, Depositional environments of host rocks to northern Australian lead-zinc deposits, with special reference to McArthur River, in D. F. Sangster, ed., Short Course in Sediment-hosted Stratiform Lead-zinc Deposits, Toronto, Min Assoc. Can. Handbook 9, p. 141 - 174.

Muir, M. D., 1979, A sabkha model for the deposition of part of the Proterozoic McArthur Group of the Northern Territory, and its implications for mineralisation: BMR Journal of Australian Geology and Geophysics, v. 4, p. 149-162.

Munoz, M., A. J. Boyce, P. Courjaultrade, A. E. Fallick, and F. Tollon, 1994, Multi-stage fluid incursion in the Palaeozoic basement-hosted Saint-Salvy ore deposit (NW Montagne Noire, Southern France): Applied Geochemistry, v. 9, p. 609 ff.

Munyanyiwa, H., 1990, Mineral assemblages in calc-silicates and marbles in the Zambezi Mobile Belt: their implications on mineral-forming reactions during metamorphism: Journal of African Earth Sciences, v. 10, p. 693-700.

Murrell, S. A. F., and I. A. H. Ismail, 1976, The effect of decomposition of hydrous minerals on the mechanical properties of rocks at high pressures and temperatures: Tectonophysics, v. 31, p. 207-258.

Myers, D. M., and C. W. Bonython, 1958, The theory of recovering salt from sea water by solar evaporation: Journal of Applied Chemistry (Australia), v. 8, p. 207-219.

Naldrett, A. J., 1997, Key factors in the genesis of Noril'sk, Sudbury, Jinchuan, Voiseys Bay and other worldclass Ni-Cu-PGE deposits - Implications for exploration: Australian Journal of Earth Sciences, v. 44, p. 283-315.

Naldrett, A. J., 1993, Ni-Cu-PGE ores of the Noril'sk region Siberia: a model for giant magmatic sulfide deposits associated with flood basalts: Society of Economic Geologists Special Publication, v. 2, p. 81-123.

Naldrett, A. J., 1981, Nickel sulfide deposits: classification, composition, and genesis: Economic Geology, v. 75, p. 628-685.

Naldrett, A. K., and A. J. Macdonald, 1980, Tectonic settings of Ni-Cu sulphide ores: their importance in genesis and exploration., in D. W. Strangway, ed., The Continental Crust and its Mineral Deposits. Proc. symposium held to honour J. Tuzo Wilson, Toronto, May 1979, Geological Association of Canada: Special Paper 20, p. 633-657.

Nalpas, T., and J. P. Brun, 1993, Salt flow and diapirism related to extension at crustal scale: Tectonophysics., v. 228, p. 349-362.

Nash, J. T., and J. J. Connor, 1993, Iron and chlorine as guides to stratiform Cu-Co-Au deposits, Idaho Cobalt Belt, USA: Mineralium Deposita, v. 28, p. 99-106.

Needham, R. S., and G. J. De Ross, 1990, Pine Creek Inlier - regional geology and mineralization, in F. E. Hughes, ed., The Mineral Deposits of Australia and Papua New Guinea: Volume 1, Australasian Institute of Mining and Metallurgy: Monograph, p. 727 - 737.

Nelson, T. H., 1991, Salt tectonics and listric normal faulting, in A. Salvador, ed., The Gulf of Mexico Basin, Boulder, Geol. Soc. America, p. 73-89.

Nelson, T. H., and L. H. Fairchild, 1989, Emplacement and evolution of salt sills in northern Gulf of Mexico (abs): American Association of Petroleum Geologists Bulletin, v. 73, p. 395.

Nely, G. 1994, Evaporite Sequences in Petroleum Exploration: 2. Geophysical Methods. Éditions Technip, Paris. p. 141 - 252.

Neudert, M., 1986, A depositional model for the upper Mt Isa Group and implications for ore formation: PhD thesis,

REFERENCES

Research School of Earth Sciences, Australian National University, Canberra.

Neudert, M. K., and R. E. Russell, 1981, Shallow water and hypersaline features from the Middle Proterozoic Mt Isa Sequence: Nature, v. 293, p. 284-286.

Nickless, E. F. P., S. J. Booth, and P. N. Mosley, 1975, Celestite deposits of the Bristol area: Inst. Min. Metall., Trans., Sect. B., v. 84, p. 62-63.

Nicolaides, S., 1995, Origin and modification of Cambrian dolomites (Red Heart Dolomite and Arthur Creek Formation), Georgina Basin, central Australia: Sedimentology, v. 42, p. 249 - 266.

Nielsen, P., R. Swennen, J. A. D. Dickson, A. E. Fallick, and E. Keppens, 1997, Spheroidal dolomites in a Visean karst system - Bacterial induced origin: Sedimentology, v. 44, p. 177-195.

Nijman, W., K. H. Debruijne, and M. E. Valkering, 1998, Growth fault control of Early Archaean cherts, barite mounds and chert-barite veins, North Pole Dome, eastern Pilbara, Western Australia: Precambrian Research, v. 88, p. 25-52.

North, F. K., 1985, Petroleum Geology: Winchester, Mass, Unwin Hyman Ltd, 631 p.

Noth, S., 1997, High H_2S contents and other effects of thermochemical sulphate reduction in deeply buried carbonate reservoirs: Geologische Rundschau, v. 86, p. 275-287.

Ohmoto, H., 1996, Formation of volcanogenic massive sulphide deposits - The Kuroko perspective: Ore Geology Reviews, v. 10, p. 135-177.

Ohmoto, H., and B. J. Skinner, 1983, The Kuroko and related volcanogenic massive sulphide deposits: Economic Geology Monograph, v. 5, 605 p.

Olaussen, S., 1981, Formation of celestite in the Wenlock, Oslo region Norway - evidence for evaporitic depositional environments: Journal of Sedimentary Petrology, v. 51.

Olgaard, D. L., S. C. Ko, and T. F. Wong, 1995, Deformation and pore pressure in dehydrating gypsum under transiently drained conditions: Tectonophysics, v. 245, p. 237-248.

Oliver, J., 1986, Fluids expelled tectonically from orogenic belts; their role in hydrocarbon migration and other geologic phenomena: Geology, v. 14, p. 99-102.

Oliver, N. H. S., 1995, Hydrothermal history of the Mary Kathleen Fold belt, Mt Isa Block, Queensland: Australian Journal of Earth Sciences, v. 42, p. 267-279.

Oliver, N. H. S., T. J. Rawling, I. Cartwright, and P. J. Pearson, 1994, High temperature fluid-rock interaction and scapolitisation in an extension related hydrothermal system, Mary Kathleen, Australia: Journal of Petrology, v. 35, p. 1455-1491.

Oliver, N. H. S., V. J. Wall, and I. Cartwright, 1992, Internal control of fluid compositions in amphibolite-facies scapolitic calc-silicates, Mary Kathleen, Australia: Contributions to Mineralogy and Petrology, v. 111, p. 94-112.

Oliver, N. H. S., and V. J. Wall, 1987, Metamorphic plumbing system in Proterozoic calc-silicates, Queensland, Australia: Geology, v. 15, p. 793-796.

Olson, R. A., 1984, Genesis of paleokarst and strata-bound zinc-lead sulfide deposits in a Proterozoic dolostone, northern Baffin Island, Canada: Economic Geology, v. 79, p. 1056-1103.

Oreskes, N., and M. T. Einaudi, 1992, Origin of hydrothermal fluids at Olympic Dam: preliminary results from fluid inclusions and stable isotopes: Economic Geology, v. 87, p. 64-90.

Oreskes, N., and M. T. Einaudi, 1990, Origin of rare earth element-enriched hematite breccias at the Olympic Dam Cu-U-Au-Ag deposit, Roxby Downs, South Australia: Economic Geology, v. 85, p. 1-28.

Orgeval, J. J., 1994, Peridiapiric metal concentration: example of the Bou Grine deposit (Tunisian Atlas)., in Fontboté, and Boni, eds., Sediment-hosted Zn-Pb ores, SGA Special Publication No. 10, p. 354-388.

Orgeval, J. J., 1991, A new type of Zn-Pb concentration related to salt dome activity: Example of the Bou Grine deposit (Tunisian Atlas)., in M. Pagel, and J. L. Leroy, eds., Source, transport and deposition of metals., Balkema, p. 269-272.

Ortega-Gutierrez, F., 1984, Evidence of Precambrian evaporites in the Oaxacan granulite complex of southern Mexico: Precambrian Research, v. 23, p. 377-393.

Ortí, F., Helvací, C., Rosell, L. and Gündogan, I., 1998, Sulphate-borate relations in an evaporitic lacustrine environment: the Sultancayir Gypsum (Miocene, western Anotalia): Sedimentology, v. 45, p. 697-710.

References

Orti-Cabo, F., and D. J. Shearman, 1977, Estructuras y fabricas deposicionales en las evaporitas del mioceno superior (Messiniense) de San Miguel de Salinas (Alicante, Espana): Barc., Inst. Invest. Geol., Publ., v. 32, p. 5-54.

Orville, P. M., 1975, Stability of scapolite in the system Ab-An-NaCl-CaCO$_3$ at 4 kbar and 750°C: Geochimica et Cosmochimica Acta, v. 39, p. 1091-1105.

Osborn, W. L., 1989, Formation, diagenesis, and metamorphism of sulfate minerals in the Salton Sea geothermal system, California, U.S.A: Masters thesis, University of California, Riverside; Riverside, CA.

Oszczepalski, S., 1994, Oxidative alteration of the Kupferschiefer in Poland: oxide- sulphide parageneses and implications for ore-forming models: Kwartalnik Geologiczny, v. 38, p. 651-671.

Oszczepalski, S., 1989, Kupferschiefer in southwestern Poland: sedimentary environments, metal zoning, and ore controls: Geological Association of Canada Special Paper, v. 36, p. 571-600.

Palmer, M. R., 1991, Boron isotope systematics of hydrothermal fluids and tourmalines: a synthesis: Chemical Geology (Isotope Geoscience Section), v. 94, p. 111-121.

Palmer, M. R. and C. Helvaci, 1997, The boron isotope geochemistry of the Neogene borate deposits of western Turkey: Geochimica et Cosmochimica Acta, v. 61, p. 3161-3169.

Pan, Y. M., M. E. Fleet, and G. E. Ray, 1994, Scapolite in two Canadian gold deposits - Nickel Plate, British Columbia and Hemlo, Ontario: Canadian Mineralogist, v. 32, p. 825-837.

Papioanou, F. P., and Z. Carotsieris, 1993, Dolomitization patterns in Jurassic-Cretaceous dissolution-collapse breccias of Mainalon Mountain (Tripolis Unit, Central Peloponnesus-Greece): Carbonates and Evaporites, v. 8, p. 9-22.

Parafiniuk, J., 1989, Strontium and barium minerals in the sulfur deposits from the Tarnobrzeg region (SE Poland): Arch. Min., v. 43, p. 41-60.

Parák, T., 1991, Volcanic sedimentary rock-related metallogenesis in the Kiruna-Skellefte Belt of northern Sweden: Economic Geology Monograph, v. 8, p. 20-50.

Paraschiv, D., and G. Olteanu, 1970, Oil fields in the Miocene-Pliocene zone of eastern Carpathians (district of Ploiesti), in M. T. Halbouty, ed., Geology of giant petroleum fields, Tulsa, American Association Petroleum Geologists Memoir 14, p. 399-427.

Parnell, J., 1986, Devonian Magadi-type cherts in the Orcadian Basin, Scotland: Journal of Sedimentary Petrology, v. 56, p. 495-500.

Patterson, R. J., and D. J. J. Kinsman, 1981, Hydrologic framework of a sabkha along the Arabian Gulf: American Association of Petroleum Geologists Bulletin, v. 65, p. 1457-1475.

Paull, C. K., J. R. Chanton, A. C. Neumann, J. A. Coston, and C. S. Martens, 1992, Indicators of methane-derived carbonates and chemosynthetic organic carbon deposits: examples from the Florida Escarpment: Palaios, v. 7, p. 361-375.

Pavlov, D. I., 1970, Halite anatectite and some of its less highly metamorphosed analogs: Acad. Sci. Ussr, Dokl., Earth Sci. Sect., v. 195, p. 171-173.

Pawlowski, S., K. Pawlowska, and B. Kubica, 1979, Geology and genesis of the Polish sulfur deposits: Economic Geology, v. 74, p. 475-483.

Pedone, V. A., and R. L. Folk, 1996, Formation of aragonite cement by nannobacteria in the Great Salt Lake, Utah: Geology, v. 24, p. 763-765.

Peng, Q. M., and M. R. Palmer, 1995, The Palaeoproterozoic boron deposits in eastern Liaoning, China: a metamorphosed evaporite: Precambrian Research, v. 72, p. 185-197.

Perkins, W. G., 1990, Mount Isa copper orebodies., in F. E. Hughes, ed., The Mineral Deposits of Australia and Papua New Guinea: Volume 1, Australasian Institute of Mining and Metallurgy: Monograph, p. 935 - 941.

Perkins, W. G., 1984, Mount Isa silica dolomite and copper orebodies: the result of a syntectonic hydrothermal alteration system: Economic Geology, v. 79, p. 601 - 637.

Peryt, T. M., 1989, Basal Zechstein in southwestern Poland: sedimentation, diagenesis, and gas accumulations: Geological Association of Canada Special Paper, v. 36, p. 601-625.

Peryt, T. M., C. Pierre, and S. P. Gryniv, 1998, Origin of polyhalite deposits in the Zechstein (Upper Permian) Zdrada Platform (northern Poland): Sedimentology, v. 45, p. 565-578.

REFERENCES

Peter, J. M., and W. C. Shanks III, 1992, Sulfur, carbon, and oxygen isotope variations in submarine hydrothermal deposits of Guaymas Basin, Gulf of California, USA: Geochimica et Cosmochimica Acta, v. 56, p. 2025-2040.

Peter, J. M., and S. D. Scott, 1991, Hydrothermal mineralization in the Guaymas Basin, Gulf of California, in J. P. Dauphin, and B. R. T. Simoneit, eds., The Gulf and Peninsula Province of the Californias, American Association Petroleum Geologists Memoir 47, p. 721-741.

Philippe, Y., 1994, Transfer zone in the southern Jura thrust belt (eastern France): geometry, development and comparison with analogue modelling experiments, in A. Mascle, ed., Exploration and petroleum geology of France, Berlin, Springer Verlag, EAPG Memoir 4, p. 327-346.

Phillips, G. N., P. J. Williams, and G. De Jong, 1994, The nature of metamorphic fluids and significance for metal exploration, in J. Parnell, ed., Geofluids: Origin, Migration and Evolution of fluids in Sedimentary Basins, London, Geological Society Special Publication 78, p. 55-68.

Pierre, C., and J. M. Rouchy, 1988, Carbonate replacements after sulphate evaporites in the Middle Miocene of Egypt: Journ Sedt. Petrol., v. 58, p. 446-456.

Pietsch, B. A., D. J. Rawlings, P. M. Creaser, P. D. Kruse, M. Ahmad, P. A. Ferenczi, and T. L. R. Findhammer, 1991, Explanatory Notes-Bauhinia Downs SE53-3: Northern Territory Geological Survey.

Pirajno, F., 1992, Hydrothermal Mineral Deposits: Berlin, Springer Verlag, 709 p.

Pittman, J. G., 1985, Correlation of beds within the Ferry Lake Anhydrite of the Gulf Coastal Plain: Transactions Gulf Coast Association of Geological Societies, v. 35, p. 251-260.

Plimer, I. R., 1994, Strata-bound scheelite in meta-evaporites, Broken Hill, Australia: Economic Geology and the Bulletin of the Society of Economic Geologists, v. 89, p. 423-437.

Plimer, I. R., 1988, Tourmalinites associated with Australian Proterozoic submarine exhalative ores, in G. H. Friedrich, P. M. Herzig, eds., Base Metal Sulphide Deposits, Berlin, Springer Verlag, p. 255-283.

Plimer, I. R., 1984, The role of fluorine in submarine exhalative systems with special reference to Broken Hill, Australia: Mineralium Deposita, v. 19, p. 19-25.

Plimer, I. R., 1977, The origin of the albite-rich rocks enclosing the cobaltian pyrite deposit at Thackaringa, NSW, Australia: Mineralium Deposita, v. 12, p. 175 - 187.

Podladchikov, Y., C. Talbot, and A. N. B. Poliakov, 1993, Numerical models of complex diapirs: Tectonophysics, v. 228, p. 189-198.

Pohl, W., 1990, Genesis of magnesite deposits - models and trends: Geologische Rundschau, v. 79, p. 291-299.

Pohl, W., M. Amouri, O. Kolli, R. Scheffer, and D. Zachmann, 1986, A new genetic model for the North African metasomatic siderite deposits: Mineralium Deposita, v. 21, p. 228 - 233.

Pohl, W., and W. Siegl, 1986, Sediment-hosted magnesite deposits, in K. H. Wolh, ed., Handbook of statabound and stratiform ore deposits, Amsterdam, Elsevier Science Publishers, p. 233-310.

Poliakov, A. N. B., Y. Podladchikov, E. C. Dawson, and C. Talbot, 1996, Salt diapirism with simultaneous brittle faulting and viscous flow, in Alsop, G. I., D. J. Blundell, I. Davison, eds., Salt tectonics, Geological Society, London; Special Publication 100, p. 291-302.

Pope, K. O., K. H. Baines, A. C. Ocampo, and B. A. Ivanov, 1994, Impact winter and the Cretaceous/Tertiary extinctions: results of a Chicxulub asteroid impact model: Earth and Planetary Science Letters, v. 128, p. 719-725.

Posamentier, H. W., and D. P. James, 1993, An overview of sequence-stratigraphic concepts; uses and abuses, in H. W. Posamentier, C. P. Summerhayes, B. U. Haq, and G. P. Allen, eds., Sequence stratigraphy and facies associations, Oxford, UK, Special Publication of the International Association of Sedimentologists No. 18, p. 3-18.

Pottorf, R. J., and H. L. Barnes, 1983, Mineralogy, geochemistry, and ore genesis of hydrothermal sediments from the Atlantis II Deep, Red Sea: Economic Geology Monographs, v. 5, p. 198-223.

Preiss, W. V., 1988, The Adelaide Geosyncline: Bulletin, Geological Survey of South Australia, v. 53.

Price, P. E., J. R. Kyle, and G. R. Wessel, 1983, Salt Dome Related Zn-Pb Deposits, in G. Kisvarsanyi, ed., Proceedings, International Conference on Mississippi Valley Type Lead-Zinc Deposits, Univ. Missouri-Rolla, p. 558-571.

REFERENCES

Pueyo, J. J., Chong, G. and Vega, M., 1998, Mineralogy and parental brine evolution in the Pedro de Valdivia nitrate deposit, Antofagasta, Chile (Spanish): Revista Geologica de Chile, v. 25, p. 3-15.

Purser, B. H., 1973, Sedimentation around bathymetric highs in the southern Persian Gulf, in B. H. Purser, ed., The Persian Gulf: Holocene carbonate sedimentation and diagenesis in a shallow epicontinental sea, New York, Springer Verlag, p. 157-177.

Purvis, K., 1989, Zoned authigenic magnesites in the Rotliegend Lower Permian, southern North Sea: Sedimentary Geology, v. 65, p. 307-318.

Radke, B. M., and R. L. Mathis, 1980, On the formation and occurrence of saddle dolomite: Journal of Sedimentary Petrology, v. 50, p. 1149-1168.

Railsback, L. B., 1992, A geological numerical model for Paleozoic global evaporite deposition: Journal of Geology, v. 100, p. 261-277.

Raiswell, R., and H. J. Al-Biatty, 1992, Depositional and diagenetic C-S-Fe signatures and the potential of shales to generate metal-rich fluids, in M. Schidlowski, ed., Early organic evolution: Implications for mineral and energy resources, Berlin- Heidelberg, Springer-Verlag, p. 415-425.

Raiswell, R., F. Buckley, R. A. Berner, and T. F. Anderson, 1988, Degree of pyritization of iron as a palaeoenvironmental indicator of bottom water oxygenation: Journal of Sedimentary Petrology, v. 58, p. 812-819.

Ramage, J. R., 1987, Lithofacies, regional stratigraphy, and depositional systems of the Clear Fork Group (Permian) Palo Duro Basin, Texas Panhandle: Masters thesis, University of Texas at Austin.

Ramboz, C., and M. Danis, 1990, Superheating in the Red Sea? The heat-mass balance of the Atlantis II Deep revisited: Earth and Planetary Science Letters, v. 97, p. 190-210.

Ramsay, C. R., and J. Ridgway, 1977, Metamorphic patterns in Zambia and their bearing on problems of Zambia tectonic history: Precambrian Research, v. 4, p. 321-337.

Ramsay, C. R., and L. R. Davidson, 1970, The origin of scapolite in the regionally metamorphosed rocks of Mary Kathleen, Queensland: Contrib. Mineralogy and Petrology, v. 25, p. 41 - 51.

Rattenbury, M. S., 1994, A linked fold-thrust model for the deformation of the Tennant Creek goldfield, northern Australia: Mineralium Deposita, v. 29, p. 301-308.

Raup, O. B., 1970, Brine mixing - an additional mechanism for formation of basin evaporites: American Association Petroleum Geologists Bulletin, v. 54, p. 2246-2259.

Ravenhurst, C. E., P. H. Reynolds, M. Zentilli, and S. O. Akande, 1987, Isotopic constraints on the genesis of Zn-Pb mineralization at Gays River, Nova Scotia, Canada: Economic Geology, v. 82, p. 1294-1308.

Read, J. F., 1985, Carbonate platform facies models: Amer. Assoc. Petrol. Geol. Bull., v. 69, p. 1-21.

Read, J. F., and A. D. Horbury, 1993, Eustatic and tectonic controls on porosity evolution beneath sequence-bounding unconformities and parasequence disconformities on carbonate platforms, in A. D. Horbury, and A. G. Robinson, eds., Diagenesis and basin development, American Association Petroleum Geologists, Studies in Geology No. 36, p. 155-197.

Read, J. F., C. Kerans, L. J. Weber, J. F. Sarg, and F. M. Wright, 1995, Milankovitch sea-level changes, cycles, and reservoirs on carbonate platforms in greenhouse and ice-house worlds: SEPM/Society for Sedimentary Geology Short Course Notes, v. 35, 81 p.

Reeve, J. S., K. C. Cross, R. N. Smith, and N. Oreskes, 1990, Olympic Dam copper-uranium-gold-silver deposit, in F. E. Hughes, ed., The Mineral Deposits of Australia and Papua New Guinea: Volume 2, Australian Institute of Mining and Metallurgy, p. 1009 - 1035.

Reilly, J. F., I. R. MacDonald, E. K. Biegert, and J. M. Brooks, 1996, Geologic controls on the distribution of chemosynthetic communities in the Gulf of Mexico: American Association Petroleum Geologists Bulletin, v. 66, p. 39-62.

Reimold, W. U., 1993, Further debate on the origin of the Sudbury structure -: South African Journal of Science, v. 89, p. 546-551.

Reinhardt, J., 1992, The Corella Formation of the Rosebud Syncline (central Mount Isa Inlier): deposition, deformation, and metamorphism: Australian Geological Survey Organisation (AGSO), Bulletin, v. 243, p. 229-255.

Renaut, R. W., 1993, Morphology, distribution and preservation potential of microbial mats in the

REFERENCES

hydromagnesite-magnesite playas of the Cariboo Plateau, British Columbia, Canada: Hydrobiologia, v. 267, p. 75 - 98.

Renfro, A. R., 1974, Genesis of Evaporite-Associated Stratiform Metalliferous Deposits; A Sabkha Process: Economic Geology, v. 69, p. 33-45.

Rezak, R., 1985, Local carbonate production on a terrigenous shelf: Gulf Coast Association of Geological Societies Transactions, v. 35, p. 477-483.

Rezak, R., and T. J. Bright, 1981, Seafloor instability at the East Flower Garden Bank, northwest Gulf of Mexico: Geo-marine Letters, v. 1, p. 97-103.

Ribeiro de Almeida, T. I., and H. Bergamin Filho, 1993, Um estudo quimico-analitico de magnesitas do tipo veitsch da Serras das Eguas, Bahia: uma proposta metodologica para pesquisas geneticas (An analytic-chemical study of magnesite of the type found at the Serra das Eguas ridge, Bahia: a proposed methodology for research on ore genesis): Cadernos IG/UNICAMP, v. 3, p. 39-54.

Riccioni, R. M., P. W. G. Brock, and B. C. Schreiber, 1996, Evidence for early aragonite in paleo-lacustrine sediments: Journal of Sedimentary Research Section A-Sedimentary Petrology and Processes, v. 66, p. 1003-1010.

Richter-Bernburg, G., 1957, Isochrone Warven im Anhydrit des Zechstein 2: Germany, Geol. Landesanst., Geol. Jb., v. 74, p. 601-610.

Richter-Bernburg, G., 1986, Zechstein 1 and 2 anhydrites; facts and problems of sedimentation, in G. M. Harwood, and D. B. Smith, eds., The English Zechstein and related topics, London, UK, Geological Society Special Publication No. 22, p. 157-163.

Riciputi, L. R., D. R. Cole, and H. G. Machel, 1996, Sulfide formation in reservoir carbonates of the Devonian Nisku Formation, Alberta, Canada - An ion microprobe study: Geochimica et Cosmochimica Acta, v. 60, p. 325-336.

Riding, R., 1991, Classification of Microbial Carbonates, in R. Riding, ed., Calcareous Algae and Stromatolites, Berlin, Springer Verlag, p. 21-51.

Riding, R., 1979, Origin and diagenesis of lacustrine algal bioherms at the margin of the Ries Crater, Upper Miocene, southern Germany: Sedimentology, v. 26, p. 645-680.

Riding, R., J. C. Braga, J. M. Martin, and I. M. Sanchezalmazo, 1998, Mediterranean Messinian salinity crisis- Constraints from a coeval marginal basin, Sorbas, southeastern Spain: Marine Geology, v. 146, p. 1-20.

Riding, R., J. M. Martin, and J. C. Braga, 1991, Coral-stromatolite reef framework, Upper Miocene, Almeria, Spain: Sedimentology, v. 38, p. 799-819.

Risacher, F., and H. P. Eugster, 1979, Holocene pisoliths and encrustations associated with spring-fed surface pools, Pastos Grandes, Bolivia: Sedimentology, v. 26, p. 253-270.

Roberts, D. E., and G. R. T. Hudson, 1983, The Olympic Dam copper-uranium -gold deposit, Roxby Downs, South Australia: Economic Geology, v. 78, p. 799 - 822.

Roberts, H. H., P. Aharon, R. Carney, J. Larkin, and R. Sassen, 1990, Sea floor responses to hydrocarbon seeps, Louisiana continental slope: Geo-Marine Letters, v. 10, p. 232-243.

Roberts, W., and P. F. Williams, 1993, Evidence for early Mesozoic extensional faulting in Carboniferous rocks, southern New Brunswick, Canada: Canadian Journal of Earth Science, v. 30, p. 1324-1331.

Robertson, A. H. F., S. Eaton, E. J. Follows, and A. S. Payne, 1995, Depositional processes and basin analysis of Messinian evaporites in Cyprus: Terra Nova, v. 7, p. 233-253.

Roca, E., P. Anadon, R. Utrilla, and A. Vazquez, 1996, Rise, closure and reactivation of the Bicorb-Quesa evaporite diapir, eastern Prebetics, Spain: Journal of the Geological Society, v. 153, p. 311-321.

Rodriguezaranda, J. P., and J. Calvo, 1998, Trace fossils and rhizoliths as a tool for sedimentological and palaeoenvironmental analysis of ancient continental evaporite successions: Palaeogeography Palaeoclimatology Palaeoecology, v. 140, p. 383-399.

Roedder, E., 1984, The fluids in salt: American Mineralogist, v. 69, p. 413-439.

Ronov, A. B., V. E. Khain, A. N. Balukhovsky, and K. B. Seslavinsky, 1980, Quantitative analysis of Phanerozoic sedimentation: Sedimentary Geology, v. 25, p. 311-325.

Rose, A. W., A. T. Smith, R. L. Lustwerk, H. Ohmoto, and I. D. Hoy, 1986, Geochemical aspects of stratiform and red-bed copper deposits in the Catskill Formation (Pennsylvania, USA) and Redstone area (Canada). Sequence of mineralisation in sediment-hosted copper

References

deposits (part 3), in G. H. Friedrich, A. D. Genkin, A. J. Naldrett, J. D. Ridge, R. H. Sillitoe, and F. M. Vokes, eds., Geology and Metallogeny of Copper Deposits, Berlin, Springer-Verlag, p. 412-421.

Rosen, M. R., 1994, The importance of groundwater in playas: A review of playa classifications and the sedimentology and hydrology of playas, in M. R. Rosen, ed., Paleoclimate and the Basin Evolution of Playa Systems, Boulder, Co, Geological Society of America Special Paper 289, p. 1-18.

Rosen, M. R., J. V. Turner, L. Coshell, and V. Gailitis, 1995, The effects of water temperature, stratification, and biological activity on the stable isotopic composition and timing of carbonate precipitation in a hypersaline lake: Geochimica et Cosmochimica Acta, v. 59, p. 979-990.

Rosen, M. R., and J. K. Warren, 1990, The origin and significance of groundwater-seepage gypsum from Bristol Dry Lake, California, USA: Sedimentology, v. 37, p. 983-996.

Rotherham, J. F., 1997, A metasomatic origin for the iron-oxide Au-Cu Starra orebodies, Eastern Fold Belt, Mt Isa Inlier: Mineralium Deposita, v. 32, p. 205-218.

Rouchy, J. M., 1986, Sedimentologie des formations anhydritiques givetiennes et dinantiennes du segment varisque franco-belge: Bulletin de la Societe Belge de Geologie, v. 95, p. 111-127.

Rouchy, J. M., C. Pierre, and F. Sommer, 1995, Deep-water resedimentation of anhydrite and gypsum deposits in the Middle Miocene (Belayim Formation) of the Red Sea: Sedimentology, v. 42, p. 267-282.

Rouchy, J. M., M. C. Bernet-Rollande, and A. F. Maurin, 1994, Descriptive petrography of evaporites: Applications to the subsurface and laboratory. French Oil and Gas Industry Association Technical Committee, Evaporite Sequnces in Petroleum Exploration: 1. Geological Methods. Paris, France, Ed. Technip, p. 70-123.

Rouvier, H., V. Perthuisot, and A. Mansouri, 1985, Pb-Zn deposits and salt-bearing diapirs in southern Europe and North Africa: Economic Geology, v. 80, p. 666-687.

Rowan, M. G., 1995, Structural styles and evolution of allochthonous salt, central Louisiana outer shelf and upper slope, in M. P. A. Jackson, D. G. Roberts, and S. Snelson, eds., Salt tectonics; a global perspective, Tulsa, American Association Petroleum Geologists Memoir, v. 65, p. 199-228.

Rozen, O. M., 1979, Scapolite-plagioclase schists and the problem of Precambrian sulfate as illustrated by geochemical comparison of deposits of evaporite basins and metamorphic rocks of calcareous series: Doklady of the Academy of Sciences of the USSR, Earth Science Sections, v. 244, p. 138-141.

Ruckmick, J. C., B. H. Wimberly, and A. F. Edwards, 1979, Classification and genesis of biogenic sulfur deposits: Economic Geology, v. 74, p. 469-474.

Ruelle, J. C. L., 1982, Depositional environments and genesis of stratiform copper deposits of the Redstone copper belt, Mackenzie Mountains, N. W. T., in R. W. Hutchison, C. D. Spence, and J. M. Franklin, eds., Precambrian Sulphide Deposits, Geological Association of Canada Special Paper No. 25, p. 701-737.

Ruppel, S. C., and H. S. Cander, 1988, Dolomitization of shallow water carbonates by seawater and seawater-derived brines, San Andres Formation (Guadalupian), West Texas, in V. J. Shukla, and P. A. Baker, eds., Sedimentology and Geochemistry of Dolostones, Tulsa Okla, SEPM Special Publ. 43, p. 245-262.

Sage, L., and J. Letouzey, 1990, Convergence of the African and Eurasian plates in the eastern Mediterranean, in J. Letouzey, ed., Petroleum and tectonics in mobile belts, Paris, Technips, p. 49-68.

Salter, D. I., and I. M. West, 1966, Calciostrontianite in the basal Purbeck Beds of Durlston Head, Dorset: Min. Mag., v. 35, p. 146-150.

Salvan, H. M., 1972, Les niveaux salifers marocains, leurs caracteristques problemes, in G. Richter-bernburg, ed., Geology of Saline deposits, Paris, UNESCO, p. 147-168.

Salvany, J. M., and F. Orti, 1994, Miocene glauberite deposits of Alcanadre, Ebro Basin, Spain: sedimentary and diagenetic processes, in R. W. Renaut, and W. M. Last, eds., Sedimentology and geochemistry of modern and ancient saline lakes, SEPM/Society for Sedimentary Geology Special Publication 50, p. 203-215.

Sanchezmoral, S., S. Ordonez, M. A. G. Delcura, M. Hoyos, and J. C. Canaveras, 1998, Penecontemporaneous diagenesis in continental saline sediments - Bloeditization in Quero playa lake (La Mancha, Central Spain): Chemical Geology, v. 149, p. 189-204.

Sanford, W. E., and W. W. Wood, 1991, Brine evolution and mineral deposition in hydrologically open evaporite basins: American Journal of Science, v. 291, p. 687-710.

REFERENCES

Sangster, D. F., M. M. Savard, and D. J. Kontak, 1998, A genetic model for the mineralisation of Lower Windsor (Visean) carbonate rocks of Nova Scotia, Canada: Economic Geology, v. 93, p. 932-952.

Sans, M., A. L. Sanchez, and P. Santanach, 1996a, Internal structure of a detachment horizon in the most external part of the Pyrenean fold and thrust belt (northern Spain), in Alsop, G. I., D. J. Blundell, and I. Davison, (eds.), Salt tectonics, London, Geological Society Special Publication 100, p. 65-76.

Sans, M., J. A. Munoz, and J. Verges, 1996b, Triangle zone and thrust wedge geometries related to evaporitic horizons (southern Pyrenees): Bulletin of Geological Petroleum Geology, v. 44, p. 375-384.

Sarkar, A., J. A. Nunn, and J. S. Hanor, 1995, Free thermohaline convection beneath allochthonous salt sheets - An agent for salt dissolution and fluid flow in Gulf Coast sediments: Journal of Geophysical Research-Solid Earth, v. 100, p. 18085-18092.

Sass, E., and A. Bein, 1988, Dolomites and salinity; a comparative geochemical study: Shukla, Vijai, Baker, Paul A. Sedimentology and geochemistry of dolostones, based on a symposium. Special Publication Society of Economic Paleontologists and Mineralogists, v. 43, p. 223-233.

Sass, E., and Ben-Yaakov, 1977, The carbonate system in hypersaline solutions: The Dead Sea Brines: Marine Chem., v. 5, p. 83-109.

Savard, M. M., 1996, Pre-ore burial dolomitization adjacent to the carbonate-hosted Gays River Zn-Pb deposit, Nova Scotia: Canadian Journal of Earth Sciences, v. 33, p. 302-315.

Savard, M. M., B. Beauchamp, and J. Veizer, 1996, Significance of aragonite cements around Cretaceous marine methane seeps: Journal of Sedimentary Research A: Sedimentary Petrology and Processes, v. 66, p. 430-438.

Sawkins, F. J., 1990, Metal deposits in relation to plate tectonics: New York, NY, Springer-Verlag.

Sawlowicz, Z., 1992, Primary sulphide mineralization in Cu-Fe-S zones of Kupferschiefer, Fore-Sudetic monocline, Poland: Transactions - Institution of Mining and Metallurgy, Section B, v. 101, p. B1-B8.

Schaad, W., 1995, The origin of Rauhwackes (Cornieules) by the karstification of gypsum [German]: Eclogae Geologicae Helvetiae, v. 88, p. 59-90.

Schenk, P., P. Vonbitter, and R. Matsumoto, 1994, Deep-basin Deep-water carbonate-evaporite deposition of a saline giant - Loch Macumber (Visean), Atlantic Canada: Carbonates and Evaporites, v. 9, p. 187-210.

Schlager, W., and H. Bolz, 1977, Clastic accumulation of sulphate evaporites in deep water: J. Sediment. Petrol., v. 47, p. 600-609.

Schmid, I. H., 1987, Turkey's Salda Lake: a genetic model for Australia's newly discovered magnesite deposits: Ind. Min., v. 239, p. 19-31.

Schmid, R. M., 1988, Lake Torrens halite accumulation (South Australia): Zeitschrift der Deutschen Geologischen Gesellschaft, v. 139, p. 289-296.

Schmidt-Mumm, A., H. J. Behr, and E. E. Horn, 1987, Fluid systems in metaplaya sequences in the Damara Orogen (Namibia): evidence for sulfur-rich brines-general evolution and first results: Chemical Geology, v. 61, p. 135-145.

Scholle, P. A., L. Stemmerik, D. Ulmer-Scholle, G. Diliegro, and F. H. Henk, 1993, Palaeokarst-influenced depositional and diagenetic patterns in Upper Permian carbonates and evaporites, Karstryggen Area, Central East Greenland: Sedimentology, v. 40, p. 895-918.

Scholle, P. A., D. S. Ulmer, and L. A. Melin, 1992, Late-stage calcites in the Permian Capitan Formation and its equivalents, Delaware Basin margin, West Texas and New Mexico; evidence for replacement of precursor evaporites: Sedimentology, v. 39, p. 207-234.

Scholle, P. A., L. Stemmerik, and O. Harpoth, 1990, Origin of major karst-associated celestite mineralization in Karstryggen, central East Greenland: Journal of Sedimentary Petrology, v. 60, p. 397-410.

Schreiber, B. C. (ed), 1988a, Evaporities and hydrocarbons, Columbia University Press, 475 p.

Schreiber, B. C., 1988b, Subaqueous evaporite deposition. in B. C. Schreiber, ed., Evaporites and hydrocarbons, New York, Columbia University Press

Schreiber, B. C., and E. Schreiber, 1977, The salt that was: Geology, v. 5, p. 527-528.

REFERENCES

Schreyer, W., 1976, Whiteschists: their compositions and pressure-temperature regimes based on experimental field and petrographic evidence: Tectonophysics, v. 43, p. 127-144.

Schreyer, W., O. Medenbach, K. Abraham, W. Gebert, and W. F. Muller, 1982, Kulkeite, a new metamorphic phyllosilicate mineral: ordered 1:1 chlorite/talc mixed-layer: Contributions to Mineralogy and Petrology, v. 80, p. 103-109.

Schreyer, W., K. Abraham, and H. Kulke, 1980, Natural sodium phlogopite coexisting with potassium phlogopite and sodian aluminian talc in a metamorphic evaporite sequence from Derrag, Tell Atlas, Algeria: Contrib. Mineral. Petrol. Beitr. Mineral. Petrol., v. 74, p. 223-233.

Schreyer, W., and K. Abraham, 1976, Three-stage metamorphic history of a whiteschist from Sar e Sang, Afghanistan, as part of a former evaporite deposit: Contrib. Mineral. Petrol., v. 59, p. 111-130.

Schubel, K. A., and T. K. Lowenstein, 1997, Criteria for the recognition of shallow-perennial-saline-lake halites based on Recent sediments from the Qaidam Basin, western China: Journal of Sedimentary Research Section A-Sedimentary Petrology and Processes, v. 67, p. 74-87.

Schubel, K. A., and B. M. Simonson, 1990, Petrography and diagenesis of cherts of Lake Magadi, Kenya: Journal of Sedimentary Petrology, v. 60, p. 761-776.

Schultz-Ela, D. D., 1992, Restoration of cross-sections to constrain deformation processes of extensional terranes: Marine and Petroleum Geology, v. 9, p. 372-388.

Schultz-Ela, D. D., and M. P. A. Jackson, 1996, Relation of subsalt structures to suprasalt structures during extension: American Association of Petroleum Geologists Bulletin, v. 80, p. 1896-1924.

Schultz-Ela, D. D., M. P. A. Jackson, and B. C. Vendeville, 1993, Mechanics of active salt diapirism: Tectonophysics, v. 228, p. 275-312.

Schuster, D. C., 1995, Deformation of allochthonous salt and evolution of related salt-structural systems, Eastern Louisiana Gulf Coast, in M. P. A. Jackson, D. G. Roberts, and S. Snelson, eds., Salt tectonics: a global perspective, American Association Petroleum Geologists Memoir 65, p. 177-198.

Schwerdtner, W. M., 1964, Genesis of potash rocks in Middle Devonian Prairie Evaporite Formation of Saskatchewan: American Association Petroleum Geologists Bulletin, v. 48, p. 1108-1115.

Scrivener, R. C., and R. W. Sanderson, 1982, Talc and aragonite from the Triassic halite deposits of the Burton Row Borehole, Brent Knoll, Somerset: United Kingdom, Inst. Geological Sciences, Report, v82/1, v. 1, p. 58-60.

Searl, A., and S. Rankin, 1993, A preliminary petrographic study of the Chilean nitrates: Geological Magazine, v. 130, p. 319-333.

Seni, S. J., and M. P. A. Jackson, 1983a, Evolution of salt structures, East Texas diapir province; Part 1, Sedimentary record of halokinesis: American Association Petroleum Geologists Bulletin, v. 67, p. 1219-1244.

Seni, S. J., and M. P. A. Jackson, 1983b, Evolution of salt structures, East Texas diapir province; Part 2, Patterns and rates of halokinesis: American Association Petroleum Geologists Bulletin, v. 67, p. 1245-1274.

Seni, S. J., and M. P. A. Jackson, 1992, Segmentation of salt allochthons: Geology, v. 20, p. 169-172.

Serdyuchenko, D. P., 1975, Some Precambrian scapolite bearing rocks evolved from evaporites: Lithos, v. 8, p. 1-7.

Sessler, W., 1990, Influence of subrosion on three different types of salt deposits, in D. Heling, P. Rothe, U. Förstner, and P. Stoffers, eds., Sediments and Environmental Geochemistry, Berlin, Springer-Verlag, p. 179-196.

Shanks III, W. C., 1983, Economic and exploration significance of Red Sea metalliferous brine deposits, in W. C. Shank, ed., Cameron Volume on Unconventional Mineral Deposits, Am. Inst. Mining Eng. Monograph, p. 157-171.

Shanks III, W. C., and J. L. Bischoff, 1980, Geochemistry, sulfure isotope composition, and accumulation rates of Red Sea geothermal deposits: Economic Geology, v. 75, p. 445-459.

Sharma, R. S., 1981, Mineralogy of a scapolite-bearing rock from Rajasthan, northwest peninsular India: Lithos, v. 14, p. 165-172.

Shaw, A. B., 1977, A review of some aspects of evaporite deposition: Mountain Geologist, v. 14, p. 1-16.

REFERENCES

Shaw, A. B., 1964, Time in Stratigraphy: New York, McGraw-Hill, 365 p.

Shearman, D. J., 1978, Evaporites of coastal sabkhas, in W. E. Dean, and B. C. Schreiber, eds., Marine evaporites., Tulsa Ok., Soc. Econ. Paleontol. Mineral., Short Course Notes.

Shearman, D. J., 1970, Recent halite rock, Baja California, Mexico: Inst. Min. Metall., Trans., Sect. B., v. 79, p. B155-B162.

Shearman, D. J., 1966, Origin of marine evaporites by diagenesis: Inst. Mining and Metallurgy, Transactions, Series B, v. 75, p. B208-B215.

Shearman, D. J., A. McGugan, C. Stein, and A. J. Smith, 1989, Ikaite, $CaCO_3.6H_2O$, precursor of the thinolites in the Quaternary tufas and tufa mounds of the Lahontan and Mono Lake Basins, western United States: Geol. Soc. Am. Bull., v. 101, p. 913-7.

Shearman, D. J., and A. J. Smith, 1985, Ikaite, the parent mineral of jarrowite-type pseudomorphs: Proceedings of the Geologists' Association, v. 96, p. 305-314.

Shearman, D. J., G. Mossop, H. Dunsmore, and M. Martin, 1972, Origin of gypsum veins by hydraulic fracture: Inst. Min. Metall., Trans., Sect. B., v. 81, p. B149-B155.

Shearman, D. J., and J. G. Fuller, 1969, Anhydrite diagenesis, calcitization, and organic laminites, Winnipegosis Formation, Middle Devonian, Saskatchewan: Bull. Can. Pet. Geol., v. 17, p. 496-525.

Shearman, D. J., J. Khouri, and S. Taha, 1961, On the replacement of dolomite by calcite in some Mesozoic limestones from the French Jura: Proceedings Geologists' Assoc., London, v. 72, p. 1-12.

Sheppard, R. A., A. J. Gude, and G. M. Elson, 1978, Bowie zeolite deposit, Cochise and Graham Counties, Arizona, in L. B. Sand, and F. A. Mumpton, eds., Natural Zeolites — Occurrence, Properties, Use, Pergamon, p. 319-328.

Sheu, D. D., 1987, Sulfur and organic carbon contents in sediment cores from the Tyro and Orca basins: Marine Geology, v. 75, p. 157-164.

Sheu, D. D., and B. J. Presley, 1986a, Formation of hematite in the euxinic Orca basin, northern Gulf of Mexico: Marine Geology, v. 69, p. 309-321.

Sheu, D. D., and B. J. Presley, 1986b, Variations of calcium carbonate, organic carbon and iron sulfides in anoxic sediment from the Orca Basin, Gulf of Mexico: Marine Geology, v. 70, p. 103-118.

Shields, M. J., and P. V. Brady, 1995, Mass balance and fluid flow constraints on regional-scale dolomitization, Late Devonian, Western Canada Sedimentary Basin: Bulletin of Canadian Petroleum Geology, v. 43, p. 371-392.

Shikazono, N., and H. D. Holland, 1983, The partitioning of strontium between anhydrite and aqueous solutions from 150 to 250°C: Economic Geology Monographs, v. 5, p. 320-328.

Shikazono, N., H. D. Holland, and R. F. Quirk, 1983, Anhydrite in kuroko deposits; mode of occurrence and depositional mechanisms: Economic Geology Monographs, v. 5, p. 329-344.

Shinn, E. A., 1983, Tidal Flat Environment, in P. A. Scholle, D. G. Bebout, and C. H. Moore, eds., Carbonate Depositional Environments, Tulsa, OK, American Assoc. Petroleum Geologists Memoir 33, p. 172-210.

Sidle, W. C., D. Sayala, and T. Steinborn, 1985, Natural analogues in evaporites for the interpretation of mineralogic and geochemical variations in a nuclear waste repository: EOS, v. 66, p. 1153.

Siedlecka, A., 1972, Length-slow chalcedony and relicts of sulphates; evidences of evaporitic environments in the upper Carboniferous and Permian beds of Bear Island, Svalbard: J. Sediment. Petrol., v. 42, p. 812-816.

Sighinolfi, G. P., B. I. Kronberg, C. Gorgoni, and W. S. Fyfe, 1980, Geochemistry and genesis of sulphide-anhydrite-bearing Archean carbonate rocks from Bahia (Brazil): Chem. Geol., v. 29, p. 323-331.

Simo, J. A., C. M. Johnson, M. R. Vandrey, P. E. Brown, E. Castrogiovanni, P. E. Drzewiecki, J. W. Valley, and J. Boyer, 1994, Burial dolomitization of the Middle Ordovician Glenwood Formation by evaporitic brines, Michigan Basin, in B. Purser, M. Tucker, and D. Zenger, eds., Dolomites: a volume in honour of Dolomieu, Blackwell Scientific; IAS Special Publication, 21, p. 169-186.

Simoneit, B. R. T., R. N. Leif, A. A. Sturz, A. E. Sturdivant, and J. M. Gieskes, 1992, Geochemistry of shallow

REFERENCES

sediments in Guaymas Basin, Gulf of California: hydrothermal gas and oil migration and effects of mineralogy: Organic Geochemistry, v. 18, p. 765-784.

Simpson, E. L., and K. A. Eriksson, 1993, Thin eolianites interbedded within a fluvial and marine succession: early Proterozoic Whitworth Formation, Mount Isa Inlier, Australia: Sedimentary Geology, v. 87, p. 39-62.

Skall, H., 1975, The paleoenvironment of the Pine Point lead-zinc district: Economic Geology, v. 70, p. 22-47.

Skippen, G., and V. Trommsdorf, 1986, The influence of NaCl and KCl on phase relations in metamorphosed carbonate rocks: American Journal Science, v. 286, p. 81-104.

Slack, J. F., M. R. Palmer, B. P. J. Stevens, and R. G. Barnes, 1993, Origin and significance of tourmaline-rich rocks in the Broken-Hill District, Australia: Economic Geology, v. 88, p. 505-541.

Slack, J. F., M. R. Palmer, and B. P. J. Stevens, 1989, Boron isotope evidence for the involvement of non-marine evaporites in the origin of the Broken Hill ore deposits: Nature, v. 342, p. 913-916.

Smith, D. B., 1996, Deformation in the Late Permian Boulby Halite (EZ3Na) in Teesside, NE England, in G. I. Alsop, D. J. Blundell, and I. Davison, eds., Salt tectonics, London, United Kingdom, Geological Society of London special Publication, p. 77-88.

Smith, D. B., 1972, Foundered strata, collapse breccias and subsidence features of the English Zechstein, in G. Richter-Bernberg, ed., Geology of saline deposits, Paris, UNESCO, p. 255-269.

Smith, D. G., and J. R. Pullen, 1967, Hummingbird structure of southeast Saskatchewan: Bull. Can. Pet. Geol., v. 15, p. 468-482.

Smith, G. I., 1979, Subsurface stratigraphy and geochemistry of Late Quaternary evaporites, Searles Lake, California: US Geological Survey, Professional Paper, v. 1043, p. 130 pp.

Smoot, J. P., and T. K. Lowenstein, 1991, Depositional environments of non-marine evaporites, in J. D. Melvin, ed., Evaporites, petroleum and mineral resources, Elsevier Science, Developments in Sedimentology, p. 189-347.

Southgate, P. N., 1982, Cambrian skeletal halite crystals and experimental analogues: Sedimentology, v. 29, p. 391-407.

Southgate, P. N., I. B. Lambert, T. H. Donnelly, R. Henry, H. Etminan, and G. Weste, 1989, Depositional environments and diagenesis in Lake Parakeelya: a Cambrian alkaline playa from the Officer Basin, South Australia: Sedimentology, v. 36, p. 1091-1112.

Spangenberg, J. E., and S. A. Macko, 1998, Organic geochemistry of the San Vicente zinc-lead district, eastern Pucara Basin, Peru: Chemical Geology, v. 146, p. 1-23.

Spangenberg, J. E., L. Fontbote, S. Z. D., and H. J., 1996, Carbon and oxygen isotope study of hydrothermal carbonates in the zinc-lead deposits of the San Vicente district, Central Peru - Quantitative modeling of mixing processes and CO_2 degassing: Chemical Geology, v. 133, p. 289-315.

Spathopoulos, F., 1996, An insight on salt tectonics in the Angola Basin, South Atlantic, in G. I. Alsop, D. J. Blundell, and I. Davison, eds., Salt tectonics, Geological Society, London; Special Publication 100, p. 153-174.

Speczik, S., A. Bechtel, Y. Z. Sun, and W. Puttmann, 1995, A stable isotope and organic geochemical study of the relationship between Anthracosia shale and Kupferschiefer mineralization (SE Poland): Chemical Geology, v. 123, p. 133-151.

Speed, R. C., 1975, Carbonate breccia (rauhwacke) nappes of the Carson Sink region, Nevada: Geol. Soc. America Bull., v. 86, p. 473-486.

Speed, R. C., and R. N. Clayton, 1975, Origin of marble by replacement of gypsum in carbonate breccia nappes, Carson Sink region, Nevada: J. Geol., v. 83, p. 223-237.

Spencer, R. J., and L. A. Hardie, 1990, Contol of seawater composition by mixing of river waters and mid-ocean ridge hydrothermal brines, in R. J. Spencer, and I. M. Chou, eds., Fluid Mineral Interactions: A Tribute to H. P. Eugster, San Antonio, Geochem. Soc. Spec. Publ. 2, p. 409-419.

Spencer, R. J., and T. K. Lowenstein, 1990, Evaporites: Geoscience Canada. Reprint Series, v. 4, p. 141-163.

Spindler, W. M., 1977, Structure and stratigraphy of a small Plio-Pleistocene depocenter, Louisiana continental

REFERENCES

shelf: Gulf Coast Assoc. Geol. Soc. Trans, v. 27, p. 180-196.

Spötl, C., 1992, Clay minerals in Upper Permian evaporites from the northern Calcareous Alps (Alpine Haselgebirge Formation, Austria): European Journal of Mineralogy, v. 4, p. 1407-1419.

Spötl, C., 1989a, Complex zoning and resorption in breunnerite from Hall in Tyrol, Austria: Evidence from back-scattered electron microscopy: Mineralogy and Petrology, v. 40, p. 225-233.

Spötl, C., 1989b, The Alpine Haselgebirge Formation, Northern Calcareous Alps (Austria): Permo-Scythian evaporites in an alpine thrust system: Sedimentary Geology, v. 65, p. 113-125.

Spötl, C., 1988, Evaporitsche Fazies der Reichenhaller Formation (Skyth/ANis) im Haller Salzberg (Nördliche Kalkalpen, Tirol): Jahrb Geol Bundesanst, v. 131, p. 153-168.

Stanton, R. J., Jr, 1966, The solution brecciation process: Geol. Soc. America Bull., v. 77, p. 843-847.

Stengele, F., and W. Smykatz-Kloss, 1995, Mineralogical and geochemical study of Holocene sebkha sediments in southeastern Tunisia: Chemie der Erde-Geochemistry, v. 55, p. 241-256.

Stevens, B. P. J., R. G. Barnes, R. E. Brown, W. J. Stroud, and I. L. Willis, 1988, The Willyama Supergroup in the Broken Hill and Euriowie blocks, New South Wales: Precambrian Research, v. 40-41, p. 297 - 327.

Stewart, A. J., and D. H. Blake, 1992, Detailed studies of the Mount Isa Inlier: Australian Geological Survey Organisation, Bulletin, v. 243.

Stewart, A. J., R. Q. Oaks Jr, J. A. Deckelman, and R. D. Shaw, 1991, 'Mesothrust' versus 'megathrust' interpretations of the structure of the northeastern Amadeus Basin, central Australia: Bulletin - Bureau of Mineral Resources, Geology and Geophysics, Australia, v. 236, p. 361-383.

Stewart, J. I., 1994, The role of evaporitic-shale sediment packages in the localisation of copper-gold deposits: Copper Canyon area, Cloncurry: Australian mining looks north - the challenges and choices, p. 207-214.

Stewart, J. I., 1991, Proterozoic geology and gold geochemistry of the Marimo Basin area, Cloncurry, NW Queensland: Masters thesis, James Cook University of Queensland.

Stewart, S. A., and M. P. Coward, 1995, Synthesis of salt tectonics in the southern North Sea, UK: Marine and Petroleum Geology, v. 12, p. 457-475.

Stokes, W. L., 1968, Multiple parallel truncation bedding planes — a feature of wind-deposited sandstone formations: Journal of Sedimentary Petrology, v. 38, p. 510-515.

Strohmenger, C., E. Voigt, and J. Zimdars, 1996a, Sequence stratigraphy and cyclic development of Basal Zechstein carbonate-evaporite deposits with emphasis on Zechstein 2 off-platform carbonates (Upper Permian, Northeast Germany): Sedimentary Geology, v. 102, p. 33-54.

Strohmenger, C., M. Antonini, G. Jager, K. Rockenbauch, and C. Strauss, 1996b, Zechstein 2 Carbonate reservoir facies distribution in relation to Zechstein sequence stratigraphy (upper Permain, northwest Germany) - an intergrated approach: Bulletin des Centres de Recherches Exploration-Production Elf Aquitaine, v. 20, p. 1-35.

Sturz, A. A., A. E. Sturdivan, R. N. Leif, B. R. T. Simoneit, and J. Gieskes, 1996, Evidence for retrograde hydrothermal reactions in near surface sediments of the Guaymas Basin, Gulf of California: Applied Geochemistry, v. 11, p. 645 ff.

Sullivan, L. A., and A. J. Koppi, 1993, Barite pseudomorphs after lenticular gypsum in a buried soil from central Australia: Australian Journal of Soil Research, v. 31, p. 393-396.

Sullivan, M. D., R. S. Haszeldine, A. J. Boyce, G. Rogers, and A. E. Fallick, 1994, Late anhydrite cements mark basin inversion; isotopic and formation water evidence, Rotliegend Sandstone, North Sea: Marine and Petroleum Geology, v. 11, p. 46-54.

Sun, D. and Li, B., 1993, Origin of borates in saline lakes of China: Elsevier, Amsterdam, Seventh Symposium on Salt, v. 1, p 177-193.

Sun, S. Q., 1995, Dolomite reservoirs; porosity evolution and reservoir characteristics: American Association of Petroleum Geologists Bulletin, v. 79, p. 186-204.

Sun, S. Q., and M. Esteban, 1994, Paleoclimatic controls on sedimentation, diagenesis, and reservoir quality -

REFERENCES

Lessons from Miocene carbonates: American Association of Petroleum Geologists Bulletin, v. 78, p. 519-543.

Sun, Y., and W. Puettmann, 1997, Metal accumulation during and after deposition of the Kupferschiefer from the Sangerhausen Basin, Germany: Applied Geochemistry, v. 12, p. 577-592.

Sun, Y., W. Puttmann, and S. Speczik, 1995, Differences in the depositional environment of basal Zechstein in southwest Poland - Implications for base metal mineralization: Organic Geochemistry, v. 23, p. 819 - 835.

Supajanya, T., and M. C. Friederich, 1992, Salt tectonics of the Sakon Nakhon Basin, northeastern Thailand: Seventh regional conference on Geology, mineral and hydrocarbon resources of Southeast Asia (GEOSEA VII), Nov. 5-8, 1991.

Surdam, R. C., and R. A. Sheppard, 1978, Zeolites in alkaline lake deposits, in L. B. Sand, and F. A. Mumpton, eds., Natural Zeolites: Occurrence, properties, use, New York, Pergamon, p. 145-174.

Surdam, R. C., and C. A. Wolfbauer, 1974, Green River Formation, Wyoming: a playa lake complex: Geol. Soc. America Bulletin, v. 86, p. 335-345.

Surdam, R. C., H. P. Eugster, and R. H. Mariner, 1972, Magadi-type chert in Jurassic and Eocene to Pleistocene rocks Wyoming: Geol. Soc. Am. Bull., v. 83, p. 1739-1752.

Susura, B. B., V. O. Glybovsky, and A. V. Kislitsin, 1986, Red-colored terrigenous sediments; specific copper-forming systems, in G. H. Friedrich, A. D. Genkin, A. J. Naldrett, J. D. Ridge, R. H. Sillitoe, and F. M. Vokes, eds., Geology and metallogeny of copper deposits; proceedings of the Copper symposium; 27th international geological congress, Berlin, Springer-Verlag, p. 504-512.

Suwanich, P., 1986, Structural geology of potash and rock salt in Nachuak area, Khorat Plateau, Thailand: Fertilizer minerals in Asia and the Pacific, mineral concentrations and hydrocarbon accumulations in the Escap region, v. 1.

Svenningsen, O. M., 1995, Extensional deformation along the Late Precambrian - Cambrian Baltoscandian passive margin: the Saektjåkkå Nappe, Swedish Caledonides: Geol Rundsch, v. 84, p. 649-664.

Svenningsen, O. M., 1994, Tectonic significance of meta-evaporitic magnesite and scapolite deposits in the Seve-Nappes, Sarek Mountains, Swedish Caledonides: Tectonophysics, v. 231, p. 33-44.

Sverjensky, D. A., 1989, Chemical evolution of basinal brines that formed sediment-hosted Cu-Pb-Zn deposits, in R. W. Boyle, A. C. Brown, C. W. Jefferson, and E. C. Jowett, eds., Sediment hosted stratiform copper deposits, Geological Association of Canada, Special Paper 36, p. 127- 134.

Sverjensky, D. A., 1986, Genesis of Mississippi Valley-type lead-zinc deposits: Annual Review of Earth and Planetary Sciences, v. 14, p. 177-199.

Swager, C. P., 1985, Syndeformational carbonate-replacement model for the copper mineralization at Mount Isa, northwest Queensland: a microstructural study: Economic Geology, v. 80, p. 107- 125.

Sweeney, M. A., P. L. Binda, and D. J. Vaughan, 1991, Genesis of the ores of the Zambian Copperbelt: Ore Geology Reviews, v. 6, p. 51-76.

Swennen, R., W. Viaene, and C. Cornelissen, 1990, Petrography and geochemistry of the Belle Roche breccia (lower Visean, Belgium): evidence for brecciation by evaporite dissolution: Sedimentology, v. 37, p. 859-878

Swennen, R., W. Viaene, L. Jacobs, and O. J. Van, 1981, Occurrence of calcite pseudomorphs after gypsum in the Lower Carboniferous of the Vesder region (Belgium): Bulletin de la Societe Belge de Geologie, v. 90, p. 231-247..

Swett, K., and A. H. Knoll, 1989, Marine pisolites from Upper Proterozoic carbonates of East Greenland and Spitsbergen: Sedimentology, v. 36, p. 75-93.

Swihart, G. H., P. B. Moore, and E. L. Callis, 1986, Boron isotopic composition of marine and nonmarine evaporite borates: Geochimica et Cosmochimica Acta, v. 50, p. 1297-1301.

Talbot, C. J., 1993, Spreading of salt structures in the Gulf of Mexico: Tectonophysics, v. 228, p. 151-166.

Talbot, C. J., 1992a, Centrifuged models of Gulf of Mexico profiles: Marine and Petroleum Geology, v. 9, p. 412-432.

Talbot, C. J., 1992b, Quo vadis tectonophysics? With a pinch of salt!: Tectonophysics, v. 16, p. 1-20.

Talbot, C. J., 1979, Fold trains in a glacier of salt in southern Iran: Journal of Structural Geology, v. 1, p. 5-18.

REFERENCES

Talbot, C. J., 1978, Halokinesis and thermal convection: Nature, v. 273, p. 739-741.

Talbot, C. J., and M. Alavi, 1996, The past of a future syntaxis across the Zagros, in G. I. Alsop, D. J. Blundell, and I. Davison, eds., Salt tectonics, Geological Society, London; Special Publication 100, p. 89-110.

Talbot, C. J., and M. P. A. Jackson, 1987a, Internal kinematics of salt diapirs: American Association of Petroleum Geologists Bulletin, v. 71, p. 1068–1093.

Talbot, C. J., and M. P. A. Jackson, 1987b, Salt tectonics: Scientific American, v. 257, p. 70-79.

Talbot, C. J., and R. J. Jarvis, 1984, Age, budget and dynamics of an active salt extrusion in Iran: Journal of Structural Geology, v. 6, p. 521-533.

Talbot, M. R., K. Holm, and M. A. J. Williams, 1994, Sedimentation in low-gradient desert margin systems; a comparison of the Late Triassic of Northwest Somerset (England) and the late Quaternary of east-central Australia, in M. R. Rosen, ed., Paleoclimate and basin evolution of playa systems, Special Paper - Geological Society of America, p. 97-117.

Tarasov, A. V., 1970, Structural control of copper and nickel mineralization in Noril'sk I deposit: Int. Geol. Rev., v. 12, p. 933-941.

Taylor, J. C. M., 1990, Upper Permian-Zechstein, in K. W. Glennie, ed., Introduction to the Petroleum Geology of the North Sea (3rd Edition), Oxford, Blackwell, p. 153-190.

Taylor, K. G., and C. D. Curtis, 1995, Stability and facies association of early diagenetic mineral assemblages; an example from a Jurassic ironstone-mudstone succession, U.K: Journal of Sedimentary Research, Section A: Sedimentary Petrology and Processes, v. 65, p. 358-368.

Teagle, D. A. H., J. C. Alt, H. Chiba, S. E. Humphris, and A. Halliday, 1998, Strontium and oxygen isotopic constraints on fluid mixing, alteration and mineralisation in the TAG hydrothermal deposit: Chemical Geology, v. 149, p. 1-24.

Thibieroz, J., and M. Madre, 1976, Le gisement de siderite du Djebel Ouenza (Algérie) est controlée par un golfe de la mer aptienne: Bull. Soc. Hist. nat. Afr. Nord Alger 67, v. 314, p. 125-149.

Thiede, D. S., and E. N. Cameron, 1978, Concentration of heavy metals in the Elk Point evaporite sequence, Saskatchewan: Economic Geology, v. 73, p. 405-415.

Thrasher, J., A. J. Fleet, S. J. Hay, M. Hovland, and S. Düppenbecker, 1996, Understanding geology as the key to using seepage in exploration: the spectrum of seepage styles, in Schumacher, D., and M. A. Abrams, eds., Hydrocarbon migration and its near-surface expression, Tulsa, American Association Petroleum Geologists Memoir, p. 223-241.

Tivey, M. K., S. E. Humphris, G. Thompson, M. D. Hannington, and P. A. Rona, 1995, Deducing patterns of fluid flow and mixing within the TAG active hydrothermal mound using mineralogical and geochemical data: Journal of Geophysical Research-Solid Earth, v. 100, p. 12527-12555.

Tompkins, L. A., M. J. Rayner, D. I. Groves, and M. T. Roche, 1994, Evaporites; in situ sulfur source for rhythmically banded ore in the Cadjebut Mississippi Valley-type Zn-Pb deposit, Western Australia: Economic Geology, v. 89, p. 467-492.

Torres, R. R. L., 1978, Scapolite-bearing and related calc-silicate layers from the Alpujarride Series, (Betic Cordilleras of southern Spain); a discussion on their origin and some comments: Geol. Rundsch., v. 67, p. 342-355.

Trefry, J. H., B. J. Presley, W. L. Keeney-Kennicutt, and R. P. Trocine, 1984, Distribution and chemistry of manganese, iron, and suspended particulates in Orca Basin: Geomarine Lett., v. 4, p. 125-130.

Trommsdorff, V., G. Skippen, and P. Ulmer, 1985, Halite and sylvite as solid inclusions in high-grade metamorphic rocks: Contributions to Mineralogy and Petrology, v. 89, p. 24-29.

Truc, G., 1978, Lacustrine sediments in an evaporitic environment: the Ludian (Palaeogene) of the Mormoiron Basin, SE France, in A. Matter, and M. E. Tucker, eds., Modern and ancient lake sediments, Intern. Assoc. Sedt. Spec. Publ. 2, p. 187-202.

Trudinger, P. A., L. A. Chambers, and J. W. Smith, 1985, Low-temperature sulphate reduction; biological versus abiological: Canadian Journal of Earth Sciences = Journal Canadien des Sciences de la Terre, v. 22, p. 1910-1918.

Trusheim, F., 1960, Mechanism of salt migration in Northern Germany: American Association Petroleum Geologists Bulletin, v. 44, p. 1519-1540.

Trusheim, F., 1957, Über halokinese und ihre bedeutung für die strukturelle Entwicklung Norddeutschlands: Zeitschr. Deutsch. Geol. Ges., v. 109, p. 111-151.

417

REFERENCES

Tucker, M. E., 1993, Carbonate diagenesis and sequence stratigraphy: Sedimentology Review, v. 1, p. 51-72.

Tucker, M. E., 1991, Sequence stratigraphy of carbonate-evaporite basins; models and application to the Upper Permian (Zechstein) of Northeast England and adjoining North Sea: Journal of the Geological Society of London, v. 148, p. 1019-1036.

Tucker, M. E., 1976a, Quartz replaced anhydrite nodules ('Bristol Diamonds') from the Triassic of the Bristol District: Geol. Mag., v. 113, p. 569-574.

Tucker, M. E., 1976b, Replaced evaporites from the late Precambrian of Finnmark, Arctic Norway: Sediment. Geol., v. 16, p. 193-204.

Tuisku, P., 1985, The origin of scapolite in the central Lapland schist area, northern Finland: Preliminary results: Finland Geol. Surv. Bull., v. 331, p. 159-173.

Turner, R. W. J., 1992, Formation of Phanerozoic stratiform sediment-hosted zinc-lead deposits: Evidence for the critical role of oceanic anoxic events: Chemical Geology, v. 99, p. 165-188.

Tyler, N., 1986, The origin of gold mineralisation in the Pilgrim's Rest goldfield, eastern Transvaal: Information Circular, Economic Geology Research Unit, University of Witwatersrand, Johannesburg, v. 179, p. 34 pp.

Ullman, W. J., 1995, The fate and accumulation of bromide during playa salt deposition: an example from Lake Frome, South Australia: Geochimica et Cosmochimica Acta, v. 59, p. 2175-2186.

Ullman, W. J., and K. D. Collerson, 1994, The Sr-isotope record of late Quaternary hydrologic changes around Lake Frome, South Australia: Australian Journal of Earth Sciences, v. 41, p. 37-45.

Ulmer-Scholle, D. S., and P. A. Scholle, 1994, Replacement of evaporites within the Permian Park City Formation, Bighorn Basin, Wyoming, USA: Sedimentology, v. 41, p. 1203-1222.

Unrug, R., 1988, Mineralisation controls and source of metals in the Lufilian fold belt, Shaba (Zaire), Zambia and Angola: Econ. Geol., v. 83, p. 1247 - 1258.

Urai, J. L., C. J. Spiers, H. J. Zwart, and G. S. Lister, 1986, Weakening of rock salt by water during long-term creep: Nature, v. 324, p. 554-557.

Usiglio, M. J., 1849, Etudes sur la composition de l'eau de la Mediterranee et sur l'exploitation des sel quy'elle conteint: Ann. Chim. Phys., v. 27, p. 172-191.

Utha-aroon, C., 1992, Continental origin of the Maha-Sarakham evaporites, northeastern Thailand: Journal of Southeast Asian Earth Sciences, v. 7, p. 268.

Utha-aroon, C., L. Coshell, and J. K. Warren, 1995, Early and late dissolution in the Maha Sarakham Formation: Implications for basin stratigraphy: International Conference on Geology, Geochronology and Mineral Resources of Indochina, 22-25 November 1995, Khon Kaen, Thailand, p. 275-286.

Van Alstine, R. E., 1976, Continental rifts and lineaments associated with major fluorospar districts: Economic Geology, v. 71, p. 977-987.

Van der Sloot, H. A., D. Hoede, G. Hamburg, J. R. W. Woittiez, and C. H. Van der Weijden, 1990, Trace elements in suspended matter from the anoxic hypersaline Tyro and Bannock Basins (eastern Mediterranean): Marine Chemistry, v. 31, p. 187-203.

Van Houten, F. R., 1977, Triassic-Liassic deposits of Morocco and eastern North America; Comparison: American Assoc. Petroleum Geol. Bulletin, v. 61, p. 79-99.

Van Wagoner, J. C., R. M. Mitchum, K. M. Campion, and V. D. Rahmanian, 1990, Siliciclastic Sequence Stratigraphy in Well Logs, Cores and Outcrops, Amer. Assoc. Petrol. Geol. Methods in Exploration Series, No. 7, Tulsa.

Van Wagoner, J. C., H. W. Posamentier, R. M. Mitchum Jr., P. R. Vail, J. F. Sarg, T. S. Loutit, and J. Hardenbol, 1988, An overview of the fundamentals of sequence stratigraphy and key definitions, in C. K. Wilgus, B. S. Hastings, C. G. S. C. Kendall, H. W. Posamentier, C. A. Ross, and J. C. Van Wagoner, eds., Sea-Level Changes: An Integrated Approach, Tulsa, Soc. Econ. Paleontol. Mineral. Spec. Publ. 42, p. 39-45.

Vandervoort, D. S., 1997, Stratigraphic response to saline lake-level fluctuations and the origin of cyclic nonmarine evaporite deposits - The Pleistocene Blanca Lila Formation, Northwest Argentina: Geological Society America Bulletin, v. 109, p. 210-224.

Vanko, D. A., and F. C. Bishop, 1982, Occurrence and origin of marialitic scapolite in the Humboldt lopolith, N.W. Nevada: Contributions to Mineralogy and Petrology, v. 81, p. 277-289.

REFERENCES

Vaughan, D. J., M. Sweeney, G. Friedrich, R. Diedel, and C. Haranczyk, 1989, The Kupferschiefer: an overview with an appraisal of the different types of mineralization: Economic Geology, v. 84, p. 1003-1027.

Vearncombe, S., M. E. Barley, D. I. Groves, N. J. McNaughton, E. J. Mikucki, and J. R. Vearncombe, 1995, 3.26 Ga black smoker-type mineralization in the Strelley Belt, Pilbara Craton, Western Australia: Journal of the Geological Society of London, v. 152, p. 587-590.

Velasco, F., A. Pesquera, R. Arce, and F. Olmedo, 1987, A contribution to the ore genesis of the magnesite deposit of Eugui, Navarra (Spain): Mineralium Deposita, v. 22, p. 33-41.

Vendeville, B. C., and M. P. A. Jackson, 1993, Some dogmas in salt tectonics challenged by modeling: American Association Petroleum Geologists Hedberg Research Conference on Salt tectonics, p. 163-165.

Vendeville, B. C., and M. P. A. Jackson, 1992a, The rise of diapirs during thin-skinned extension: Marine and Petroleum Geology, v. 9, p. 331-353.

Vendeville, B. C., and M. P. A. Jackson, 1992b, The fall of diapirs during thin-skinned extension: Marine and Petroleum Geology, v. 9, p. 354-371.

Vengosh, A., A. R. Chivas, A. Starinsky, Y. Kolodny, B. Zhang, and P. Zhang, 1995, Chemical and boron isotope compositions of non-marine brines from the Qaidam Basin, Qinghai, China: Chemical Geology, v. 120, p. 135-154.

Vernon, R. H., and P. F. Williams, 1988, Distinction between intrusive and extrusive or sedimentary parentage of felsic gneisses; examples from the Broken Hill Block, NSW: Australian Journal of Earth Sciences, v. 35, p. 379-388.

Vially, R., J. Letouzey, F. Bénard, N. Haddadi, D. G., H. Askri, and A. Boudjema, 1994, Basin inversion along the North African margin, the Saharan Atlas (Algeria), in F. Roure, ed., Peritethyan Platforms, Paris, Technip, p. 79-118.

Vila, J. M., 1995, Premiere étude de surface d'un grand ≈ glacier de sel≈ sous-marin: l'est de la strcture Ouenza-Ladjebel-Méridef (confins algéro-tunisiens). Proposition d'un scénario de mise en placeet comparaisons: Bull. Soc. Geol. France, v. 166, p. 149-167.

Vila, J. M., M. Benyoussef, M. Chikhaoui, and M. Ghanmi, 1996a, A large submarine Middle Albian salt glacier in north-western Tunisia (250 sq. km) - the Triassic rocks of the Ben Gasseur diapir and of El Kef anticline [French]: Comptes Rendus de l Academie des Sciences Serie II Fascicule A-Sciences de la Terre et des Planetes, v. 322, p. 221 - 227.

Vila, J.-M., M. Ghanmi, and F. Kechid-Benkherouf, 1996b, Donnees nouvelles sur l'anticlinal d'El Ouasta-Sakiet (frontiere est-algerienne) et interpretation de son Trias comme un 'glacier de sel' sous-marin albien le long d'un block bascule, plisse au Tertiaire. [New data on the El Ouasta-Sakiet anticline (Algerian-Tunisian border) and interpretation of its Triassic rocks as a submarine Albian 'salt glacier' along a tilted block folded during the Tertiary]: Comptes Rendus - Academie des Sciences, Serie II: Sciences de la Terre et des Planetes, v. 323, p. 1035-1042.

von Bitter, P. H., S. D. Scott, and P. E. Schenk, 1990, Early Carboniferous low-temperature hydrothermal vent communities from Newfoundland: Nature, v. 344, p. 145-148.

von der Borch, C. C., 1965, Source of ions for Coorong dolomite formation: Am. J. Sci., p. 684-688.

von der Borch, C. C., B. Bolton, and J. K. Warren, 1977, Environmental setting and microstructure of subfossil lithified stromatolites associated with evaporites, Marion Lake, South Australia: Sedimentology, v. 24, p. 693-708.

von Engelhardt, W. H., H. Fuchtbauer, and G. Muller, 1977, Sedimentary petrology, Halsted Press, New York.

von Gehlen, K., H. Nielsen, and W. Ricke, 1962, S-Isotopen Verhaltnisse in Baryt und Sulfiden aus hydrothermalen Gangen im Schwarzwald und jüngeren Barytgangen in Suddeutschland und ihre genetische Bedeutung: Geochim Cosmochim. Acta, v. 26, p. 1189-1207.

Walker, R. N., R. G. Logan, and J. G. Binnekamp, 1978, Recent geological advances concerning the H.Y.C. and associated deposits, McArthur River, N.T: Journal of the Geological Society of Australia, v. 24, p. 365-380.

Walker, R. N., M. D. Muir, W. L. Diver, N. Williams, and N. Wilkins, 1977, Evidence of major sulphate evaporite deposits in the Proterozoic McArthur Group, Northern Territory, Australia: Nature, v. 265, p. 526-529.

Wall, V. J., and R. K. Valenta, 1990, Ironstone-related gold-copper mineralisation: Tennant Creek and elsewhere: Pacific Rim '90 Congress Proceedings, v. III, p. 855-864.

REFERENCES

Wallick, E. I., 1981, Chemical evolution of groundwater in a drainage basin of Holocene age, east central Alberta: Journal of Hydrology, v. 54, p. 245-283.

Walraven, F., and M. Chabu, 1994, Pb-isotope constraints on base-metal mineralisation at Kipushi (southeastern Zaire): Journal of African Earth Sciences and the Middle East, v. 18, p. 73-82.

Waltho, A. E., and S. J. Andrews, 1993, The Century zinc-lead deposit, northwest Queensland: Proc. Australasian Institute of Mining and Metallurgy conference, Adelaide, 1993, p. 41 - 61.

Wang, A. J., Q. M. Peng, and M. R. Palmer, 1998, Salt dome-controlled sulphide precipitation of Paleoproterozoic Fe-Cu sulphide deposits, eastern Liaoning, Northeastern China: Economic Geology, v. 93, p. 1-14.

Wardlaw, N. C., 1972, Unusual marine evaporites with salts of calcium and magnesium chloride in Cretaceous basins of Sergipe, Brazil: Economic Geology, v. 67, p. 156-168.

Wardlaw, N. C., 1968, Carnallite-sylvite relationships in the Middle Devonian Prairie evaporite formation, Saskatchewan: Geol. Soc. Amer. Bull., v. 79, p. 1273-1294.

Wardlaw, N. C., and G. D. Nicholls, 1972, Cretaceous evaporites of Brazil and West Africa and their bearing on the theory of continental separation: Internat. Geol. Congress, 24th, Section 6, p. 43-55.

Wardlaw, N. C., and D. W. Watson, 1966, Middle Devonian salt formations and their bromide content, Elk Point area, Alberta: Canadian Jour. Earth Sci., v. 3, p. 263-275.

Warrak, M., 1974, The petrography and origin of dedolomitized, veined or brecciated carbonate rock, the "cornieules" in the Frejus region, French Alps: Journ. Geol. Soc. Lond., v. 130, p. 229-247.

Warren, J. K., 1997, Evaporites, brines and base metals: brines, flow and "the evaporite that was": Australian Journal of Earth Sciences, v. 44, p. 149-183.

Warren, J. K., 1996, Evaporites, brines and base metals: What is an evaporite? Defining the rock matrix: Australian Journal of Earth Sciences, v. 43, p. 115 - 132.

Warren, J. K., 1991, Sulfate dominated sea-marginal and platform evaporative settings, in J. L. Melvin, ed., Evaporites, petroleum and mineral resources: Developments in Sedimentology 50, Amsterdam, Elsevier, p. 477-533.

Warren, J. K., 1990, Sedimentology and mineralogy of dolomitic Coorong lakes, South Australia: Journal of Sedimentary Petrology, v. 60, p. 843-858.

Warren, J. K., 1989, Evaporite sedimentology: Importance in hydrocarbon accumulation: Englewood Clifs, Prentice-Hall, 285 p.

Warren, J. K., 1988, Sedimentology of Coorong dolomite in the Salt Creek region, South Australia: Carbonates and Evaporites, v. 3, p. 175-199.

Warren, J. K., 1986, Source rock potential of shallow water evaporite settings: Journal Sedimentary Petrology, v. 56, p. 442-454.

Warren, J. K., 1985, On the significance of evaporite lamination: Schreiber, B. C., Harner, H. L., (eds) Sixth international symposium on salt, v. 6, p. 161-170.

Warren, J. K., 1983a, Tepees, modern (southern Australia) and ancient (Permian - Texas and New Mexico) - a comparison: Sedimentary Geology, v. 34, p. 1-19.

Warren, J. K., 1983b, Pedogenic calcrete as it occurs in Quaternary calcareous dunes in coastal South Australia: Journal of Sedimentary Petrology, v. 53, p. 787-796.

Warren, J. K., 1982a, The hydrological significance of Holocene tepees, stromatolites, and boxwork limestones in coastal salinas in South Australia: Journal of Sedimentary Petrology, v. 52, p. 1171-1201.

Warren, J. K., 1982b, Hydrologic setting, occurrence, and significance of gypsum in late Quaternary salt lakes, South Australia: Sedimentology, v. 29, p. 609-637.

Warren, J. K., and R. H. Kempton, 1997, Evaporite Sedimentology and the Origin of Evaporite-Associated Mississippi Valley-type Sulfides in the Cadjebut Mine Area, Lennard Shelf, Canning Basin, Western Australia., in I. P. Montanez, J. M. Gregg, and K. L. Shelton, eds., Basinwide diagenetic patterns: Integrated petrologic, geochemical, and hydrologic considerations, Tulsa OK, SEPM Special Publication 57, p. 183-205.

Warren, J. K., K. G. Havholm, M. R. Rosen, and M. J. Parsley, 1990, Evolution of gypsum karst in the Kirschberg Evaporite Member near Fredericksburg, Texas: Journal of Sedimentary Petrology, v. 60, p. 721-734.

REFERENCES

Warren, J. K., and C. G. S. C. Kendall, 1985, Comparison of sequences formed in marine sabkha (subaerial) and salina (subaqueous) settings; modern and ancient: American Association Petroleum Geologists Bulletin, v. 69, p. 1013-1023.

Weijermars, R., M. P. A. Jackson, and B. Vendeville, 1993, Rheological and tectonic modeling of salt provinces: Tectonophysics, v. 217, p. 143-174.

Wells, A. T., 1980, Evaporites in Australia: BMR Bulletin, v. 198: Canberra, Australia, Bureau of Mineral Resources, 104 p.

Wenkert, D., 1979, Flow of salt glaciers: Geophysical Research Letters, v. 6, p. 523-526.

Werner, M. L., M. D. Feldman, and L. P. Knauth, 1988, Petrography and geochemistry of water-rock interactions in Richton Dome cap rock (southeastern Mississippi, U.S.A.): Chemical Geology, v. 74, p. 113-135.

Werner, W., 1990, Examples of structural control of hydrothermal mineralization: fault zones in epicontinental sedimentary basins - a review: Geologische Rundschau, v. 79, p. 279-290.

West, I., 1973, Vanished evaporites; significance of strontium minerals: J. Sediment. Petrol., v. 43, p. 278-279.

West, I. , 1964, Evaporite diagenesis in the lower Purbeck beds of Dorset [with discussion]: Proc. Yorkshire Geol. Soc., v. 34, p. 315-330.

Whelan, J. F., R. O. Rye, W. de Lorraine, and H. Ohmoto, 1990, Isotopic geochemistry of the mid-Proterozoic evaporite basin; Balmat, New York: American Journal of Science, v. 290, p. 396-424.

Whittaker, S. G., and E. W. Mountjoy, 1996, Diagenesis of an Upper Devonian carbonate-evaporite sequence: Birdbear Formation, southern Interior Plains, Canada: Journal of Sedimentary Research A: Sedimentary Petrology and Processes, v. 66, p. 965-975.

Williams, D. F., and I. Lerche, 1987, Salt domes, organic-rich source beds and reservoirs in intraslope basins of the Gulf Coast region, in I. Lerche, and J. J. O'Brien, eds., Dynamical geology of salt and related structures, New York, Academic Press, p. 751-830.

Williams, P. J., 1994, Iron Mobility during synmetamorphic alteration in the Selwyn Range area, NW Queensland - Implications for the origin of ironstone-hosted Au-Cu deposits: Mineralium Deposita, v. 29, p. 250-260.

Williams, P., G. de Jong, and T. Verran, 1993, Evolution of Na-K-Fe-Si metasomatism and mineralization associated with the Cloncurry Fault, SE Mount Isa Inlier; a comparison with Kiruna-Olympic Dam type systems: Mid- to lower-crustal metamorphism and fluids conference, Mount Isa, Queensl., Australia, July 26-Aug. 1, 1993, p. 56-57.

Williams, S. H., and J. R. Zimbelman, 1994, White Rock - An eroded Martian lacustrine deposit: Geology, v. 22, p. 107-110.

Willis, I. L., R. E. Brown, W. J. Stroud, and B. P. J. Stevens, 1983, The Early Proterozoic Willyama supergroup: stratigraphic subdivision and interpretation of high to low-grade metamorphic rocks in the Broken Hill block, New South Wales: Journal, Geological Society of Australia, v. 30, p. 195 - 224.

Wilson, A. H., and J. A. Versfeld, 1994, The early Archaean Nondweni greenstone belt, southern Kaapvaal Craton, South Africa; Part I, Stratigraphy, sedimentology, mineralization and depositional environment: Precambrian Research, v. 67, p. 243-276.

Wilson, T. P., and D. T. Long, 1993, Geochemistry and isotope chemistry of Ca-Na-Cl brines in Silurian Strata, Michigan Basin, USA: Applied Geochemistry, v. 8, p. 507-524.

Wodzicki, A., and A. Piestrzynski, 1994, An ore genetic model for the Lubin-Sieroszowice mining district: Mineralium Deposita, v. 29, p. 30-43.

Wood, M. W., and H. F. Shaw, 1976, The geochemistry of celestites from the Yates area near Bristol (U.K.): Chem. Geol., v. 17, p. 179-193.

Wooden, J. L., G. K. Czamanske, R. M. Bouse, A. P. K. Likhachev, V. E., and V. Lyul'ko, 1992, Pb isotope data indicate a complex, mantle origin for the Noril'sk-Talnakh ores, Siberia: Economic Geology, v. 87, p. 1153-1165.

Woods, P. J. E., 1979, The geology of Boulby Mine: Economic Geology, v. 74, p. 409-418.

Worden, R. H., 1996, Controls on halogen concentrations in sedimentary formation waters: Mineralogical Magazine, v. 60, p. 259-274.

REFERENCES

Worden, R. H., P. C. Smalley, and A. E. Fallick, 1997, Sulfur cycle in buried evaporites: Geology, v. 25, p. 643-646.

Worden, R. H., and P. C. Smalley, 1996, H_2S-Producing reactions in deep carbonate gas reservoirs - Khuff Formation, Abu Dhabi: Chemical Geology, v. 133, p. 157-171.

Worden, R. H., P. C. Smalley, and N. H. Oxtoby, 1996, The effects of thermochemical sulphate reduction upon formation water salinity and oxygen isotopes in carbonate gas reservoirs: Geochimica et Cosmochimica Acta, v. 60, p. 3925-3931.

Worden, R. H., P. C. Smalley, and N. H. Oxtoby, 1995, Gas souring by thermochemical sulfate reduction at 140° C: American Association Petroleum Geologists Bulletin, v. 79, p. 854-863.

Woronick, R. E., and L. S. Land, 1985, Late burial diagenesis, Lower Cretaceous Pearsall and Lower Glen Rose Formations, south Texas (USA)., in N. Schneidermann, and P. M. Harris, eds., Carbonate cements, SEPM, Tulsa; Special Publication 36, p. 265-275.

Worrall, D. M., and S. Snelson, 1989, Evolution of the northern Gulf of Mexico, with emphasis on Cenozoic growth faulting and the role of salt, in A. W. Bally, and A. R. Palmer, eds., Geology of North America - An Overview, Boulder, CO, Geological Soc. America, p. 97-138.

Worsley, N., and A. Fuzesy, 1979, The potash-bearing members of the Devonian Prairie Evaporite of southeastern Saskatchewan, south of the mining area: Economic Geology, v. 74, p. 377-388.

Wu, S., 1993, Salt and slope tectonics offshore Louisiana: Doctoral thesis, Rice Univ. Houston.

Wu, S., P. R. Vail, and C. Cramez, 1990a, Allochthonous salt, structure and stratigraphy of the north-eastern Gulf of Mexico. Part I: stratigraphy: Marine and Petroleum Geology, v. 7, p. 318-333.

Wu, S., A. W. Bally, and C. Cramez, 1990b, Allochthonous salt, structure and stratigraphy of the north- eastern Gulf of Mexico. Part II: structure: Marine and Petroleum Geology, v. 7, p. 334-370.

Wu Moude, and L. Xiji, 1981, Two types of diapiric structure of the Kunyang Group, Yunnan. (Chinese): Acta Geologica Sinica, v. 55, p. 105-117.

Yang, W. B., and T. J. Ahrens, 1998, Shock vaporization of anhydrite and global effects of the K/T bolide: Earth and Planetary Science Letters, v. 156, p. 125-140.

Yang, W. B., R. J. Spencer, H. R. Krouse, T. K. Lowenstein, and E. Cases, 1995, Stable isotopes of lake and fluid inclusion brines, Dabusun Lake, Qaidam Basin, Western China - Hydrology and paleoclimatology in arid environments: Palaeogeography Palaeoclimatology Palaeoecology., v. 117, p. 279-290.

Yardley, B. W. D., and G. E. Lloyd, 1995, Why metasomatic fronts are really metasomatic sides: Geology, v. 23, p. 53-56.

Yeats, R. S., and R. J. Lillie, 1991, Contemporary tectonics of the Himalayan frontal fault system: folds, blind thrusts and the 1905 Kangra earthquake: Journal of Structural Geology, v. 13, p. 215-225.

You, C. F., and M. J. Bickle, 1998, Evolution of an active seafloor massive sulphide deposit: Nature, v. 394, p. 668-671.

Young, G. M., 1981, The Amundsen embayment, Northwest Territories: relevance to the upper Proterozoic evolution of North America, in F. H. A. Campbell, ed., Proterozoic Basins of Canada, Geol. Surv. Canada Paper 81-10, p. 203-218.

Zaikowski, A., B. J. Kosanke, and N. Hubbard, 1987, Noble gas composition of deep brines from the Palo Duro Basin, Texas: Geochimica et Cosmochimica Acta, v. 51, p. 73-84.

Zaw, K., D. L. Huston, R. R. Large, T. Mernagh, and C. F. Hoffmann, 1994, Microthermometry and geochemistry of fluid inclusions from the Tennant Creek Au-Cu deposits - Implications for ore deposition and exploration: Mineralium Deposita, v. 29, p. 288-300.

Zharkov, M. A., 1981, History of Paleozoic Salt Accumulation: Berlin, Springer Verlag, 308 pp p.

Zierenberg, R. A., and W. C. I. Shanks, 1983, Mineralogy and geochemistry of epigenetic features in metalliferous sediment, Atlantis II Deep, Red Sea: Economic Geology, v. 78, p. 57-72.

Zimmermann, J. L., and R. Moretto, 1996, Release of water and gases from halite crystals: European Journal of Mineralogy, v. 8, p. 413-422.

Zolotov, M. Y., B. Fegley, and K. Lodders, 1997, Hydrous silicates and water on Venus: Icarus, v. 130, p. 475-494.

Index

E